Advanced Engineering Mathematics

Advanced Engineering Mathematics

LADIS D. KOVACH • Naval Postgraduate School

 **ADDISON-WESLEY PUBLISHING COMPANY**

Reading, Massachusetts • Menlo Park, California
London • Amsterdam • Don Mills, Ontario • Sydney

Sponsoring Editor: *Stephen H. Quigley*
Production Editor: *Martha K. Morong*
Designer: *Adrianne Dudden*
Illustrator: *Kater Print*
Cover Design: *T. A. Philbrook*

Library of Congress Cataloging in Publication Data

Kovach, Ladis D.
  Advanced engineering mathematics.

  Includes index.
  1. Engineering mathematics.   I. Title.
TA330.K68   515′.02462   81-14936
ISBN 0-201-10340-0   AACR2

ISBN 0-201-10340-0
  BCDEFGHIJ-AL-898765432

preface

It is generally conceded that *design* is the primary function of an engineer. Engineering design ranges from the design of a versatile food processor to the design of a factory to produce sewing machines, radios, and electric motors, with the possibility that the plant might be expanded in two years to turn out a line of pumps and a line of internal combustion engines. Future design projects may be expected to include huge solar power plants and habitable satellites.

A prerequisite for design, however, is *analysis*. There are many facets to engineering analysis; included are the construction and study of models, modeling with computers, and the application of mathematical analysis. It has been said,* "Mathematics is a tool that forces precision on the imagination and focuses attention on central issues without the undue embellishment of extraneous ideas. It provides a sound basis for analysis wherever its usefulness is established." Our objective is to demonstrate in a number of ways how an engineer might strip a problem of worldly features that are unimportant complexities, approximate the problem by means of a mathematical representation, and analyze *this*. At the same time, the problem solver must be constantly aware of the simplifications that have been made and how these may affect the final result.

A number of mathematical techniques are used in engineering analysis and it is these techniques that are discussed in this text. Since the mathematics presented here is useful in solving problems that arise in different contexts, the book is not directed solely to the engineering student. Physicists and applied mathematicians will also find the material pertinent, since a variety of topics that normally follow a course in calculus are included.

The subjects discussed, however, are not presented in a haphazard manner so that the text is not a potpourri of topics in applied analysis. Rather, it is designed to *flow* from one subject to the next. A topic is not introduced until a need for it has been demonstrated. Many sections end with a challenging problem that cannot be solved until fur-

---

*Saaty, Thomas L., *Mathematical Models of Arms Control and Disarmament*. New York: Wiley, 1968.

ther techniques are developed. Among other things, this means that the "accustomed order" may not always be followed although all the usual topics can be found in the book. A serious attempt has been made to keep chapters—and even sections within chapters—independent of each other. In this way instructors have some flexibility in choosing the topics they wish to stress.

Most chapters start with an overview and many sections begin with a microreview. At the end of each section can be found a list of "key words and phrases." This serves as both a review of the material contained in the section and a handy index.

Exercises at the end of each section are divided into three parts. Those in the first group are related directly to the text in that they help to clarify the textual material or to supply computational steps that have been omitted. In the second group are exercises that provide the necessary practice to help the student become familiar with the methods and techniques presented. The last group includes exercises of a more theoretical or challenging nature, and may be considered as supplementary to the text. Since there are over 2000 exercises, it becomes a simple matter to choose those appropriate for a particular class.

Throughout the text there is emphasis on *problem solving* and on translating physical problems into mathematical language and conversely. The existence of the computer is acknowledged and separate sections on numerical methods are included. Answers to exercises and examples are given in the form in which they would be obtained with a hand-held calculator. For example, if a result is $\sqrt{3} \log 3$, then the approximation 1.9029 is also given. Exercises that require the use of a calculator are so marked.

A number of courses can be taught using this text. A moderately complete course in ordinary differential equations might include the material in Chapters 1, 2, and 3 and Sections 4.5, 7.1, 7.2, and 9.2. A brief course might consist of Chapters 1 and 2 and Section 9.2. Brief courses in linear algebra, vector analysis, and complex variables could use Chapters 4, 5, and 10, respectively. A fairly complete course in boundary-value problems might use Chapters 6, 7, 8, and 9 whereas for a shorter course Sections 6.1, 9.2, and 9.6 might be omitted. All the sections dealing with numerical methods may be considered optional. It is also relatively easy to stress applications since many of these are contained in separate sections.

Much of the material in this text has been used in various courses for a number of years. The revisions required to put the material in its present form were done using valuable suggestions of a number of teachers, including Peter A. Dashner, James Dowdy, and John Lucey. Some of the pedagogy can be traced back to my own teachers. I was particularly fortunate in having excellent teachers and I hereby acknowledge a debt I owe to R. P. Agnew, R. S. Burington, Michael

Golomb, Elizabeth Michalske, Max Morris, Sam Perlis, R. F. Rinehart, Arthur Rosenthal, Merrill Shanks, and C. C. Torrance for their inspiration. I am also grateful to two students, Bill Logan and Jim Simpkins, for their valuable assistance. Finally, I wish to thank Stephen H. Quigley, Roger E. Vaughn, and other members of the staff of Addison-Wesley for their cooperation and constant encouragement.

Monterey, California   L. D. K.
January 1982

contents

1   First-order Ordinary Differential Equations

1.1	Separating the variables	1
1.2	Exact differential equations	12
1.3	First-order linear equations	18
1.4	Miscellaneous techniques	24
1.5	Applications	32
1.6	Numerical methods	49
1.7	Some theoretical considerations	64

2   Higher-order Ordinary Differential Equations

2.1	Homogeneous second-order equations with constant coefficients	70
2.2	Homogeneous higher-order equations with constant coefficients	81
2.3	Nonhomogeneous equations	86
2.4	The Cauchy–Euler equation	98
2.5	Applications	104
2.6	Numerical methods	119

3   The Laplace Transformation

3.1	Basic properties and definitions	126
3.2	Transforms of derivatives and integrals	140
3.3	Inverse transforms	147
3.4	Convolution	157
3.5	Step, impulse, and periodic functions	164
3.6	Solving differential equations and systems	172

4   Linear Algebra

4.1	Matrices	184
4.2	Systems of linear algebraic equations	195
4.3	Linear transformations	211
4.4	Eigenvalues and eigenvectors	226

4.5   Applications to systems of ordinary differential equations   237
4.6   Numerical methods   251
4.7   Additional topics   261

5   Vector Calculus

5.1   Vector algebra   275
5.2   Vector differentiation   295
5.3   The del operator: Gradient, divergence, and curl   308
5.4   Orthogonal coordinate systems   318
5.5   Line, surface, and volume integrals   330
5.6   Integral theorems   348
5.7   Applications   363

6   Partial Differential Equations

6.1   First-order equations   378
6.2   Higher-order equations   387
6.3   Separation of variables   396
6.4   The vibrating-string equation   405

7   Fourier Series and Fourier Integrals

7.1   Fourier coefficients   414
7.2   Sine, cosine, and exponential series   430
7.3   Fourier integrals and transforms   439
7.4   Applications   448

8   Boundary-value Problems in Rectangular Coordinates

8.1   Laplace's equation   459
8.2   The wave equation   470
8.3   The diffusion equation   479
8.4   Transform methods   487
8.5   Sturm–Liouville problems   496

9   Boundary-value Problems in Other Coordinate Systems

9.1   Boundary-value problems in circular regions   506
9.2   Series solutions of ordinary differential equations   514
9.3   Bessel functions   529
9.4   Legendre polynomials   542
9.5   Applications   561
9.6   Numerical methods   571

10   Complex Variables

10.1   The algebra of complex numbers   578
10.2   Elementary functions   588
10.3   Differentiation   595

10.4 Mapping   604
10.5 The complex integral   613
10.6 Applications   626

General References   642

Tables

1   Laplace transforms   643
2   Exponential and hyperbolic functions   646

Answers and Hints to Selected Exercises   648

Index   687

1

first-order ordinary differential equations

1.1 SEPARATING THE VARIABLES

It has been determined experimentally that radioactive substances decompose at a rate proportional to the amount of the substance present. This simple fact is just one example of what we know about the world around us—a world that is sometimes referred to as the "real" world. The quotation marks indicate that we cannot discern all the complexities of our environment and that, at best, our concept of the world must be greatly simplified if we are to express its properties in quantitative terms.

Our opening statement does not need to be qualified as much as other similar statements do. For example, we often have to add phrases like "neglecting all friction" or "in a perfect vacuum" or "in the absence of any sources and sinks." In other words, we must *idealize* a number of factors before we can solve many of the problems that arise in engineering analysis. We will point out these simplifications and idealizations in the following chapters, since an awareness of them is essential to successful problem solving.

Returning to the statement about the behavior of radioactive substances, we note the phrase "at a rate." We have information here about the rate (a *derivative* with respect to time) at which a radioactive substance decomposes (loses mass). It is a common situation that we know something about a derivative of a function and, given this information, we seek the function itself.

The field of *differential equations* is the branch of mathematics in which we study methods for obtaining functions when we know the derivatives of these functions. It is the process of integration that takes us from the known derivative to the unknown function. Thus it seems natural to follow a calculus course with one that deals with differential equations. We will consider *ordinary* differential equations first, that is, those in which the dependent variable is a function of a *single* independent variable. In later chapters we will study *partial* differential equations, in particular those involving two and three independent variables.

We begin by posing and solving the problem discussed above.

EXAMPLE 1.1–1   If a radioactive substance decomposes at a rate proportional to the amount of the substance present, find the amount present at any time t.

Solution. The word "amount" is a clue that the dependent variable is mass, or m. The independent variable is, of course, the time t. Then m is actually $m(t)$, that is, a function of t. We will remember this as we write

$$\frac{dm}{dt} \propto m,$$

expressing the fact that the rate of change of m is proportional to m. This leads to the equation

$$\frac{dm}{dt} = -km,$$

where k (>0) is the constant of proportionality and the negative sign indicates that the mass is *decreasing* (the substance is decomposing).

The differential form of the equation to be solved (in contrast to the derivative form above) is

$$dm = -km\,dt.$$

We cannot integrate both members even though m is a function of t because we do not know how m and t are related. That is precisely what we are trying to find. Hence we resort to a technique called "separating the variables" and divide the last equation by m. Then

$$\frac{dm}{m} = -k\,dt$$

and integration produces*

$$\log |m| = \log m = -kt + \log C.$$

Foresight (which the student will later develop to some extent) prompted us to write the *arbitrary* constant of integration as $\log C$ ($C > 0$) in order to simplify the subsequent algebra. Note also that $m > 0$ so that the absolute value symbol may be omitted. Thus we have

$$\log m - \log C = \log \frac{m}{C} = -kt$$

or

$$\frac{m}{C} = e^{-kt} \quad \text{or} \quad m(t) = Ce^{-kt} \tag{1.1–1}$$

---

* We shall consistently write "log" for "$\log_e$" or "ln."

as the solution to the problem. Differential equations, such as the one in this example, which can be solved by separating the variables, are called **separable** differential equations. ∎

There are two *positive* constants in the solution that is shown in Eq. (1.1–1). The nature of the substance determines the value of k, which is called the radioactive (decay) constant. It should be mentioned that values of k are found by measurement of relatively large samples. In other words, this value is a statistic that does not hold on the atomic level.

The arbitrary constant C can be found if we know the value of m for some value of t, that is, if we are given an *initial condition*. For example, if $m = m_0$, a constant, when $t = 0$, then

$$m(t) = m_0 e^{-kt}. \tag{1.1–2}$$

This function represents the *solution* to the *initial-value problem*

$$\frac{dm}{dt} = -km, \qquad m(0) = m_0,$$

meaning that Eq. (1.1–2) satisfies the given differential equation and also the given initial condition.

The above initial-value problem can also be solved by evaluating *definite* integrals. We may write

$$\int_{m_0}^{m(t)} \frac{dm}{m} = -k \int_0^t dt,$$

expressing the relation that $m = m_0$ when $t = 0$ as shown by the lower limits of integration. Then integration yields

$$\log m(t) - \log m_0 = -kt$$

or

$$\log \frac{m(t)}{m_0} = -kt$$

or

$$m(t) = m_0 e^{-kt}$$

as before. When using this method we do not obtain the arbitrary constant of integration since we are dealing with definite integrals.

In connection with this problem, physicists are interested in the *half-life* of a radioactive substance. This is defined as the time it takes a substance to decompose to one-half its original mass. For example, the half-life of strontium 90 ($_{38}\text{Sr}^{90}$) is 28 years whereas that of radium 88 ($_{88}\text{Ra}^{226}$) is 1620 years. If we have a mass of m_0 at time $t = 0$, then the half-life for a given substance $T_{1/2}$ is obtained by solving the equation

$$\int_{m_0}^{m_0/2} \frac{dm}{m} = -k \int_0^{T_{1/2}} dt$$

for $T_{1/2}$. We have

$$\log \frac{1}{2} m_0 - \log m_0 = -kT_{1/2};$$

that is,

$$kT_{1/2} = -\log \frac{1}{2} = \log 2 \doteq 0.693*$$

or

$$T_{1/2} \doteq 0.693/k.$$

Instead of finding the half-life, we could find the time required for a substance to lose any given percentage of its mass. For example, to find the time required for a substance to lose 90 percent of its mass we would replace the upper limit of integration, $\frac{1}{2}m_0$, by $0.1m_0$ (see Exercise 3 of this section).

Figure 1.1–1 shows a graph of $m(t)$ that is characteristic of substances exhibiting exponential decay.

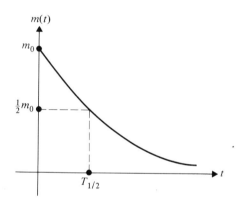

FIGURE 1.1–1 Radioactive decay.

EXAMPLE 1.1–2   Newton's law[†] of cooling states that, under certain ideal conditions, the rate at which the surface of a body cools is proportional to the difference between the surface temperature of the body and that of the surrounding medium (the so-called *ambient* temperature). Let u be the surface temperature (°C) of the body and u_0 the ambient temperature with $u > u_0$. Then Newton's law of cooling can be written

---

* The symbol $\doteq$ is read "is approximately equal to."
† After Sir Isaac Newton (1642–1727), an English mathematician and physicist.

as

$$\frac{du}{dt} = -k(u - u_0), \qquad k > 0,$$

the negative sign indicating that u is decreasing with time t. Separating the variables, we have

$$\frac{du}{u - u_0} = -k\,dt$$

and integrating gives us

$$\log (u - u_0) = -kt + \log C, \qquad C > 0,$$

or

$$u = u_0 + Ce^{-kt}.$$

It is apparent that as $t \to \infty$, $u \to u_0$, a fact that agrees with our everyday experience as well as with scientific experiment. ■

Additional information is needed to evaluate the constants k and C. Suppose it is known that the ambient temperature is $20°C$ and that $u = 45°C$ at $t = 0$. Then we find $C = 25°C$ and

$$u = 20° + 25°e^{-kt}.$$

If it is also known that $u = 40°C$ when $t = 10$ min, then $k \doteq 0.022$ and

$$u = 20° + 25°e^{-0.022t},$$

so that the problem is completely solved for the given set of conditions. When using this last equation to find the temperature at various times one must keep in mind that t is to be expressed in minutes since the value of k was found under this assumption. Figure 1.1–2 shows a graph of the above equation.

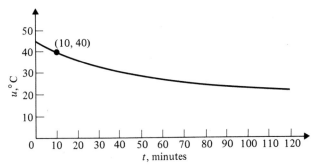

FIGURE 1.1–2 The graph of $u = 20° + 25°e^{-0.022t}$.

We continue with an example that shows that separating the variables may lead to certain difficulties.

EXAMPLE 1.1–3   Solve the differential equation

$$\frac{dy}{dx} = \frac{4xy}{x^2 + 1}.$$

Solution. In differential form we have

$$4xy\, dx = (x^2 + 1)\, dy.$$

Dividing by $y(x^2 + 1)$ separates the variables and we obtain

$$\frac{4x\, dx}{x^2 + 1} = \frac{dy}{y}.$$

Integrating each member produces

$$2 \log (x^2 + 1) = \log |y| + \log C, \qquad C > 0,$$

or

$$(x^2 + 1)^2 = Cy, \tag{1.1–3}$$

a family of curves shown in Fig. 1.1–3. Equation (1.1–3) appears to provide the correct result, for, on checking it by differentiation, we have

$$2(x^2 + 1)(2x) = C\frac{dy}{dx}$$

$$\frac{dy}{dx} = \frac{4x(x^2 + 1)}{C} = \frac{4x(x^2 + 1)y}{(x^2 + 1)^2} = \frac{4xy}{x^2 + 1}.$$

When we divided by $y(x^2 + 1)$ to separate the variables, however, we lost the result $y = 0$, which is clearly a solution. Thus we need to take into account that a part of the solution may be lost when we divide by a variable, a situation similar to that which exists in solving algebraic equations. ■

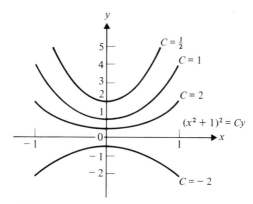

FIGURE 1.1–3 The solutions to Example 1.1–3.

When an arbitrary constant appears in the solution of a first-order differential equation such as the C in Eq. (1.1–1), the solution is called the **general solution**. This term indicates that *every* solution may be obtained from the general solution by assigning a specific value to the arbitrary constant. In Eq. (1.1–3), however, there is no finite value that can be assigned to C that will produce the solution $y = 0$, hence the latter is called a **singular solution**. A **particular** (or *unique*) **solution** is obtained from the general solution by assigning a value to the arbitrary constant. Thus solutions to initial-value problems are particular solutions.

General solutions to first-order equations are also called one-parameter *families* of curves. This is illustrated by the next example.

EXAMPLE 1.1–4 The general solution of the equation

$$\frac{dy}{dx} = \frac{-x}{y}$$

is $x^2 + y^2 = C^2$. This is a family of circles centered at the origin, the parameter being the radius C. By specifying a point through which a member of the family passes, we can determine a value of C and choose a particular circle from the entire family, as shown in Fig. 1.1–4. ∎

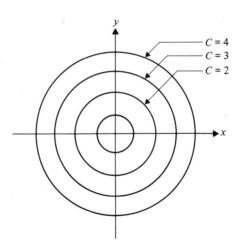

FIGURE 1.1–4 The one-parameter family $x^2 + y^2 = C$.

We have been discussing *first-order* differential equations. In classifying ordinary differential equations we specify the order first. Thus

$$\frac{d^2y}{dx^2} + \left(\frac{dy}{dx}\right)^2 + x^2y = 0 \qquad \text{(1.1–4)}$$

is a second-order equation, whereas

$$EI\frac{d^4y}{dx^4} = w(x)$$

is of fourth order, and so on. In short, the order of an equation is the same as the order of the highest derivative appearing in the equation.

Besides the order it is important to determine whether an equation is *linear* or *nonlinear*. The method of solution chosen may depend on this fact. The word "linear" refers to the *dependent* variable and its various derivatives. For instance, the equation in Example 1.1–3 is a linear equation whereas Eq. (1.1–4) is nonlinear because of the term $(dy/dx)^2$. The following are also examples of nonlinear equations:

$$y\frac{dy}{dx} - x^2y = 2,$$

$$\frac{d^2y}{dx^2} - x\frac{dy}{dx} + y = \sin y.$$

The first is nonlinear because of the product $y(dy/dx)$; no products of the dependent variable and its derivatives may appear in a linear equation. The second is nonlinear because of the term $\sin y$. We may think of $\sin y$ replaced by its Maclaurin expansion,

$$y - \frac{y^3}{3!} + \frac{y^5}{5!} - + \cdots,$$

and again we have exponents different from one applied to the dependent variable, which makes the equation nonlinear.

Some of the later sections dealing with ordinary differential equations will present methods of solution for *linear* equations. So that there will be no question about what is meant, we will *define* a first-order linear differential equation as one that can be written in the form

$$A(x)\frac{dy}{dx} + B(x)y + C(x) = 0,$$

where $A(x)$, $B(x)$, and $C(x)$ are functions of x and $A(x) \neq 0$.

Solutions of differential equations may appear in implicit or explicit form. The solutions we obtained in Examples 1.1–1 and 1.1–2 are in *explicit* form since the dependent variables are given explicitly in terms of the independent variables. We could easily have solved Eq. (1.1–3) for y and presented it in explicit form also. On the other hand, the solution obtained in Example 1.1–4 is in *implicit* form and it is best left that way. Solving it for y would introduce a radical and an ambiguous sign. The one-parameter family of circles can be more easily recognized from the implicit form given.

We conclude with an example that will lead us into the next topic.

EXAMPLE 1.1–5   Solve the differential equation

$$\frac{dy}{dx} = \frac{-(x^2 + y^2)}{2xy}.$$

Solution. This first-order equation cannot be solved by separating the variables. The equation is nonlinear (because of the term y^2 and the product $y\, dy/dx$) but this is not why the method of this section will not succeed. It is simply because of the *addition* of the x^2- and y^2-terms. Accordingly, some different technique is required in this case and we shall postpone this example until the next section. ■

We observe that the technique of separating the variables is useful for solving first-order differential equations that can be written in the form

$$\frac{dy}{dx} = M(x)N(y).$$

Singular solutions can be obtained from $N(y) = 0$.

KEY WORDS AND PHRASES

separable   singular solution $\sqrt{7}$
initial condition   particular solution
initial-value problem $\sqrt{7}$ order of a differential equation
half-life   linear and nonlinear
ambient   differential equations
general solution $\sqrt{7}$ explicit and implicit solutions

EXERCISES 1.1*

▶   1. Solve Example 1.1–1 with $m = m_0$ at $t = 0$ by using C instead of $\log C$ for the arbitrary constant of integration.

2. In Example 1.1–1 C is designated as an *arbitrary* constant. Explain why it can be written as $\log C$ rather than as C.

3. How long will it take strontium 90 to lose 2 percent of its original mass? 98 percent of its original mass?

4. Solve Example 1.1–2 with $u_0 = 20°C$, u = 45°C when $t = 0$, and with C instead of $\log C$ for the arbitrary constant of integration.

---

   * Exercises are divided into three sections. The exercises in the first group are related directly to the text and are intended to clarify the textual material and to supply any steps that have been omitted. In the second group are exercises that will provide the necessary practice so that the student will become familiar with the methods and techniques presented. The last group includes exercises of a more theoretical nature and those considered supplementary to the text.

5. a) Referring to Example 1.1–2, solve the following problem:

$$\frac{du}{dt} = -k(u - 25), \qquad u(0) = 60, \qquad u(20) = 50.$$

b) Graph the solution for $0 \le t \le 120$.

c) What is the value of u at the end of two hours?

6. Under what reasonable assumptions does Newton's law of cooling hold?

7. In Example 1.1–3 explain why the absolute value symbol could be left off the terms $x^2 + 1$ and y.

▶▶ **8.** Solve the following differential equations.

a) $t(y - 1) \, dt + y(t + 1) \, dy = 0$

b) $x^2 y \dfrac{dy}{dx} - e^y = 0$

c) $y \, dx - x \log x \, dy = 0, \quad x > 1$

d) $\sqrt{1 + y^2} \, dx - y(1 + x^2) \, dy = 0$

9. Solve the following initial-value problems.

a) $\sin \theta \, d\theta - 2y \, dy = 0, \quad y(\pi/2) = 0$

b) $\dfrac{dr}{ds} = \dfrac{(1 + r)^2}{1 - s}, \quad r(0) = 1$

c) $\dfrac{dy}{dx} = \dfrac{1 + y^2}{xy}, \quad y(2) = 3$

d) $v \, dv - g \, dt = 0, \quad v(t_0) = v_0$ and g is the (constant) acceleration due to gravity

10. Sketch a graph of the solution to Exercise 9(c).

11. a) Solve the equation

$$\tan t \frac{dy}{dt} - y \sec^2 t = 0.$$

b) Graph the solution for various values of C.

12. a) Solve the equation

$$\frac{dy}{dx} \log x - \frac{1}{x} y = 0, \qquad x > 0.$$

b) Obtain the particular solution for the case $y(e) = 2$.

c) Graph the solution in part (b).

13. Solve the differential equation

$$\frac{dy}{dx} = xy^2,$$

obtaining the general solution and any singular solutions that exist.

14. Solve the equation

$$x \frac{dy}{dx} = (1 - x^2) \cot y,$$

obtaining an implicit solution.

15. Obtain the solution to

$$\frac{du}{dt} = e^{2t+u}$$

in explicit form.

16. Solve the initial-value problem

$$\begin{cases} \dfrac{dy}{dt} = e^{-y} + 1, \\[2mm] y = 1 \quad \text{when } t = 0. \end{cases}$$

17. Obtain the solution to

$$\frac{dx}{dt} = 1 - x^2$$

in explicit form.

18. Solve the equation

$$\frac{du}{dt} = \frac{\csc u}{\sec t}.$$

19. Solve the equation

$$\frac{dy}{dx} = -\frac{x}{y},$$

and obtain the particular solution that passes through the point $(3, -4)$.

20. Solve the equation

$$(1 + y^2)\, dt + (1 + t^2)\, dy = 0,$$

subject to the condition $y = -1$ when $t = 0$.

21. Obtain an implicit solution to

$$(r + 1)\, ds - 2rs\, dr = 0.$$

What can you say about the signs of s and the arbitrary constant?

22. Solve the initial-value problem

$$\frac{du}{dt} = -2ut, \quad u(0) = u_0, \text{ a constant.}$$

23. Find an implicit solution to the differential equation

$$\cos^2 u \, \frac{du}{dv} = v(1 + v^2).$$

[*Hint*: $\cos^2 \theta = \frac{1}{2}(1 + \cos 2\theta)$.]

24. Identify the solutions of

$$x\, dx - \sqrt{16 - x^2}\, dy = 0.$$

25. Solve the differential equation

$$\frac{dy}{dx} + xy = xy^2.$$

(*Hint*: Recall the method of integration using partial fractions.)

26. A substance cools from 50°C to 35°C in an ambient temperature of 15°C in 45 min. How much longer will it take to reach 25°C?

▶▶▶ **27.** In the differential equation

$$\frac{dy}{dx} = \frac{xy - x \log y}{xy + y \log x},$$

separate the variables. Thus show that, although the variables *can* be separated, the method has some drawbacks.

28. What form must a first-order differential equation have if the method of separating the variables is to succeed?

29. A bedsheet drying on a line loses its moisture at a rate proportional to the amount of moisture present under certain ideal conditions (constant wind and humidity conditions, for example). If the sheet is 50 percent dry after the first hour, when will it be 98 percent dry?

30. Interest on invested money may be compounded annually, quarterly, daily, etc. If the money increases at a rate that is proportional to the amount present, the interest is said to be *compounded continuously*.

a) Obtain a formula that gives the amount of money $A(t)$ present at time t. Assume that $A(0) = p$, the original principal.

b) If the annual interest rate is 7 percent, how much will a principal of $2000 amount to in 8 years?

31. Of some historical interest are differential equations of the form

$$y = xy' + f(y'),$$

called **Clairaut equations** (after Alexis C. Clairaut (1713–1765), a French mathematician).

a) Verify that $y = cx + f(c)$ is the general solution of Clairaut's equation.

b) Find the general solution of

$$y = xy' + \cos y'.$$

c) Show that

$$y = x \arcsin x + \sqrt{1 - x^2}$$

is a singular solution of the equation in part (b).

d) Graph the general solution of the equation in part (b) for $c = 0, \pi/3$, and $\pi/2$, and also graph the singular solution in part (c), thus showing the geometrical relationship between the one-parameter family (general solution) and the singular solution.

1.2 EXACT DIFFERENTIAL EQUATIONS

We recall from calculus that if $z = f(x, y)$ is a function of the two independent variables x and y, then the total differential dz is given by

$$dz = \frac{\partial f}{\partial x} dx + \frac{\partial f}{\partial y} dy.$$

If $dz = 0$ at all points (x, y) for all choices of dx and dy, then $z = C$, a constant. Thus first-order differential equations of the form

$$M(x, y)\, dx + N(x, y)\, dy = 0 \qquad (1.2\text{--}1)$$

can be readily solved if $M(x, y)$ and $N(x, y)$ are the partial derivatives with respect to x and y, respectively, of some function f with the family of solutions expressed in the form $f(x, y) = C$. When this is the case, we say that the expression

$$M(x, y)\, dx + N(x, y)\, dy$$

is an **exact differential** and Eq. (1.2–1) is called an **exact differential equation**.

We now return to Example 1.1–5.

EXAMPLE 1.2–1 Solve the differential equation

$$\frac{dy}{dx} = \frac{-(x^2 + y^2)}{2xy}.$$

Solution. Rewriting the equation in differential form we have

$$(x^2 + y^2)\, dx + 2xy\, dy = 0. \qquad (1.2\text{--}2)$$

If this is an exact differential equation, then, for one thing,

$$\frac{\partial f}{\partial x} = x^2 + y^2$$

for some function $f(x, y)$. To solve this partial differential equation for f, we integrate with respect to x, holding y constant. Thus

$$f(x, y) = \frac{x^3}{3} + xy^2 + g(y),$$

where, instead of an arbitrary *constant* of integration, we have added an arbitrary *function* of y. On the other hand, we must also have

$$\frac{\partial f}{\partial y} = 2xy,$$

which leads to

$$f(x, y) = xy^2 + h(x)$$

with $h(x)$ an arbitrary function of x.

The only way that the above two (apparently different) expressions for $f(x, y)$ can be the same is if we take

$$g(y) = -C', \quad \text{a constant,}$$

and

$$h(x) = \frac{x^3}{3} - C'.$$

Then

$$f(x, y) = \frac{x^3}{3} + xy^2 - C'$$

and it is a simple matter to show that the original Eq. (1.2–2) is equivalent to the condition $df = 0$ so that the given differential equation *is* exact and its solution is*

$$\frac{x^3}{3} + xy^2 = C. \quad\blacksquare$$

Under what condition is a first-order differential equation exact? We have from Eq. (1.2–1)

$$M(x, y) = \frac{\partial f}{\partial x} \quad \text{and} \quad N(x, y) = \frac{\partial f}{\partial y}$$

as a condition for exactness. If, in addition, M and N are differentiable, it follows that

$$\frac{\partial M}{\partial y} = \frac{\partial^2 f}{\partial y \, \partial x} = \frac{\partial^2 f}{\partial x \, \partial y} = \frac{\partial N}{\partial x}$$

whenever the mixed partial derivatives of f exist and are continuous.[†] Hence, whenever $\partial M/\partial y$ and $\partial N/\partial x$ exist, are continuous, and are *equal*, then Eq. (1.2–1) is exact. Equality of the partial derivatives is not only a sufficient condition for exactness, it is also a necessary condition. We state the following theorem without proof.

Theorem 1.2–1 A necessary and sufficient condition that the first-order differential equation

$$M(x, y) \, dx + N(x, y) \, dy = 0$$

be exact is that

$$\frac{\partial M}{\partial y} = \frac{\partial N}{\partial x},$$

provided these partial derivatives are continuous.[‡]

A word of caution may be in order here. The differential equation being tested for exactness must be in precisely the form given in Theorem 1.2–1 before the criterion of that theorem can be applied.

---

* See Section 1.4 (Exercise 9) for another method of solving this equation.
  † See Robert C. James, *Advanced Calculus*, p. 298. Belmont, CA: Wadsworth, 1966.
  ‡ A proof of this theorem can be found in William E. Boyce and R. C. DiPrima, *Elementary Differential Equations and Boundary Value Problems*, 3d ed., p. 38. New York: Wiley, 1977.

In the next example we present another method for solving an exact differential equation.

EXAMPLE 1.2–2 Solve the differential equation

$$\frac{dy}{dx} = \frac{x+y}{y-x}.$$

Solution. Rewriting the equation as

$$(x + y)\,dx + (x - y)\,dy = 0,$$

and observing that

$$\frac{\partial}{\partial y}(x + y) = \frac{\partial}{\partial x}(x - y) = 1$$

shows that the given equation is exact. Thus, for some function $f(x, y)$, we must have

$$\frac{\partial f}{\partial x} = x + y$$

from which

$$f(x, y) = \frac{x^2}{2} + xy + g(y).$$

Differentiating this function partially with respect to y produces

$$\frac{\partial f}{\partial y} = x + \frac{dg}{dy} = x - y.$$

Hence

$$\frac{dg}{dy} = -y$$

so that

$$g = -\frac{y^2}{2} + C'$$

and

$$f(x, y) = \frac{x^2}{2} + xy - \frac{y^2}{2} + C'.$$

From this we conclude that the general solution of the given differential equation can be written as

$$x^2 + 2xy - y^2 = C. \quad \blacksquare$$

Exact differential equations do not occur too frequently and, when they do, they can sometimes be solved by other methods, as the exercises will show. However, the concept of an exact differential is useful in analysis. We will show in Chapter 5, for example, that finding the work done in moving against a force is very much simplified when we are dealing with exact differentials. In such cases the work done will be independent of the path between the two points and will depend only on the coordinates of the points themselves.

It should also be noted that there are specialized methods of solution in which the end result is an exact differential equation. In this connection see Example 1.4–3 as well as Exercises 6, 7, and 8 in Section 1.4.

Now we give an example to show that the concepts in this section and Section 1.1 are still inadequate.

EXAMPLE 1.2–3   Solve the first-order differential equation

$$\frac{dy}{dx} + \frac{y}{x} = x^2, \quad x > 0.$$

Solution. In differential form the given equation becomes

$$\left(\frac{y}{x} - x^2\right) dx + dy = 0$$

and it is apparent that the equation is not exact. Moreover, the variables cannot be separated, hence neither of the two methods studied so far can be applied. Yet the equation is *linear* and looks simple enough. Why, then, can we not solve it? The answer is that obtaining a solution depends on a knowledge of a general method that can be applied to *any* first-order linear differential equation. ■

It is unfortunate that the theory of ordinary differential equations has not been (perhaps, cannot be!) developed to the point where *one* method can be applied to solve *all* equations. Lacking such a general theory we must content ourselves with finding methods of solution that are applicable to certain *types* of equations. Hence we have a method for equations in which the variables can be separated and another method for exact equations. Next we will consider a method that can be applied to first-order linear equations.

The situation is not as hopeless as it may seem, however. A great many of the equations encountered in engineering analysis are of the type for which solutions will be developed in this and succeeding chapters. Other equations can be converted into solvable types by making certain assumptions and/or substitutions. This last procedure requires a careful consideration of the possible errors resulting from simplifying assumptions. *Numerical* methods, of course, will continue to be most useful when the standard techniques cannot be utilized.

KEY WORDS AND PHRASES

exact differential   **exact differential equation**

EXERCISES 1.2

▶ **1.** Solve the differential equation

$$\frac{dy}{dx} = -\frac{y}{x}$$

by separating the variables. Are any singular solutions introduced? Explain.

2. Obtain the solution $xy = C$ in Exercise 1 by using the method given in this section.

In Exercises 3–8, test the given equation for exactness. *Do not solve.*

✓ ③. $\dfrac{dy}{dx} = \dfrac{e^x(1 + e^y)}{e^y(1 + e^x)}$ **4.** $y(y - 1)\, dx - x(x + 1)\, dy = 0$

⑤. $(ye^{xy} + y^3)\, dx + (xe^{xy} - 3xy^2)\, dy = 0$ **6.** $(x - 2y \cos 2x)\, dx - \sin 2x\, dy = 0$

7. $(\cosh^2 x \sinh y)\, dx + (\sinh^2 x \cosh y)\, dy = 0$

✓ **8.** $\dfrac{df}{dw} = \dfrac{f(2f^2 + 3)}{w(3 - 2f^2)}$

▶▶ In Exercises 9–21, use either of the methods given in Sections 1.1 and 1.2 to find the general solution. Comment on singular solutions where appropriate.

9. $\dfrac{dy}{dx} = \dfrac{y^2 - 1}{1 - x^2}$ ✓**10.** $\dfrac{du}{dt} = \dfrac{t^2 - u}{t}$

⑪. $\dfrac{dy}{dt} = \dfrac{y}{y^2 - t}$ **12.** $(2xy^2 - x)\, dx + (x^2y + y)\, dy = 0$

13. $\dfrac{dy}{dx} = -\dfrac{x}{y}(1 + y)$ **14.** $(r - 3s)\, dr + 3(s - r)\, ds = 0$

15. $(2u - 5v)\, du + (2v - 5u)\, dv = 0$ **16.** $(1 + y^2)\, dt + y(t^2 + 1)\, dy = 0$

17. $\dfrac{dy}{dx} = \cot x \cot y - \dfrac{\cot x}{\sin x \sin y}$ **18.** $2rs\, dr + (r^2 + s^2)\, ds = 0$

✓**19.** $\dfrac{dr}{d\theta} = \dfrac{-r(\sin \theta + \cos \theta)}{r + \sin \theta - \cos \theta}$ **20.** $\dfrac{du}{dt} = \dfrac{t - u \cos t}{\sin t}$

21. $\dfrac{dx}{dy} = \dfrac{x}{x^3 - y}$

Solve the initial-value problems in Exercises 22–25.

22. $(2r\theta - \tan \theta)\, dr + (r^2 - r \sec^2 \theta)\, d\theta = 0, \quad r = 1$ when $\theta = \pi/4$

㉓. $\cos \theta\, dr - (r \sin \theta + 1)\, d\theta = 0, \quad r = 2$ when $\theta = 0$

24. $\dfrac{dx}{dy} = \dfrac{x^2 y}{2 - xy^2}$, $x = 2$ when $y = 1$

25. $\dfrac{du}{dv} = \dfrac{1 - u^2 v - v^3}{u(u^2 + v^2)}$, $u = 1$ when $v = 1$

26. a) Solve the equation

$$\frac{dy}{dx} = \frac{-(ye^x + e^y)}{e^x + xe^y}.$$

b) Is this differential equation linear or nonlinear? Explain.

27. Obtain the general solution of

$$(x^4 - 2xy^3)\, dx - 3x^2 y^2\, dy = 0.$$

▶▶▶ **28.** Graph the solution of the initial-value problem

$$\frac{dy}{dx} = \frac{x - 3}{9(1 - y)}, \quad y(7) = 0.$$

29. Show that a differential equation in which the variables can be separated is necessarily exact.

1.3 FIRST-ORDER LINEAR EQUATIONS

We look again at Example 1.2–3 since it involves a differential equation that cannot be solved by the methods presented thus far.

EXAMPLE 1.3–1 Solve the first-order linear differential equation

$$\frac{dy}{dx} + \frac{y}{x} = x^2, \qquad x > 0.$$

Solution. We have arbitrarily chosen the interval $x > 0$, although the equation is defined on every interval that does not include $x = 0$. It is natural to multiply by x in order to clear the equation of fractions. Then

$$x \frac{dy}{dx} + y = x^3$$

and we observe that the left member of the equation is the derivative of xy with respect to x. Hence we can write

$$\frac{d}{dx}(xy) = x^3$$

or

$$d(xy) = x^3\, dx;$$

integrating both members produces

$$xy = \frac{x^4}{4} + C.$$

This can now be solved for y and we have the explicit solution

$$y = \frac{x^3}{4} + \frac{C}{x}. \quad \blacksquare$$

It was fortuitous that multiplying by x enabled us to integrate the given equation. A function (such as x in the last example) that converts a portion of a differential equation into a derivative is called an **integrating factor**. We now seek an answer to the question, "Under what conditions can an integrating factor be found?"

A first-order differential equation that can be written in the form

$$A(x) \frac{dy}{dx} + B(x)y + C(x) = 0$$

is said to be *linear*. In an interval in which $A(x) \neq 0$ we can divide the equation by $A(x)$ to obtain the form

$$\frac{dy}{dx} + p(x)y = q(x), \tag{1.3-1}$$

where $p(x) = B(x)/A(x)$ and $q(x) = -C(x)/A(x)$. Guided by Example 1.3–1, we seek an integrating factor $g(x)$ such that

$$g(x) \frac{dy}{dx} + g(x)p(x)y = \frac{d}{dx}(yg(x)).$$

It is evident that *any* function g for which

$$p(x)g(x) = \frac{d}{dx} g(x)$$

will serve the purpose. A *positive* function g, however, will satisfy the separable equation

$$\frac{dg(x)}{g(x)} = p(x)\, dx.$$

Hence,

$$\int \frac{dg(x)}{g(x)} = \int p(x)\, dx$$

and, by virtue of the fundamental theorem of the integral calculus,

$$\int^x \frac{dg(t)}{g(t)} = \int^x p(t)\, dt$$

or

$$\log g(x) = \int^x p(t)\, dt.$$

From this we obtain $g(x)$ as

$$g(x) = \exp\left(\int^x p(t)\, dt\right), \tag{1.3-2}$$

where we have used the notation exp u for e^u as a matter of convenience. Attention is called to the fact that a lower limit of integration in Eq. (1.3–2) would merely provide a constant factor in $g(x)$ that could eventually be eliminated (see Exercise 5).

We now see that finding the integrating factor x in Example 1.3–1 was no accident, since in the equation

$$\frac{dy}{dx} + \frac{y}{x} = x^2$$

we had $p(x) = 1/x$. Hence the integrating factor was

$$g(x) = \exp\left(\int^x \frac{dt}{t}\right) = \exp\,(\log x) = x.$$

Another example may be helpful at this point.

EXAMPLE 1.3–2   Solve the equation

$$y' + (2 \cot x)y = 4 \cos x,$$

and find the particular solution for which $y(\pi/2) = 1/3$.

Solution. The integrating factor is given by

$$g(x) = \exp\left(2 \int^x \cot t\, dt\right) = \exp\,(2 \log\,(\sin x))$$

$$= \exp\,(\log\,(\sin^2 x)) = \sin^2 x.$$

Multiplying the differential equation by this factor, we have

$$y' \sin^2 x + (2 \sin x \cos x)y = 4 \sin^2 x \cos x$$

or

$$d(y \sin^2 x) = 4 \sin^2 x \cos x\, dx.$$

Then integration produces the equation

$$y \sin^2 x = \frac{4}{3} \sin^3 x + C,$$

which can be solved for y to obtain the general solution

$$y = \frac{4}{3} \sin x + C \csc^2 x.$$

Applying the given condition $y(\pi/2) = 1/3$, we find $C = -1$ and the particular solution

$$y = \frac{4}{3} \sin x - \csc^2 x.$$

A graph of this solution is shown in Fig. 1.3–1. ∎

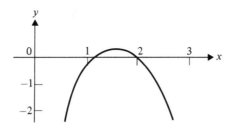

FIGURE 1.3–1 The graph of $y = \frac{4}{3}\sin x - \csc^2 x$.

Now that we have a formula for finding an integrating factor for a first-order linear differential equation, we can write the solution in the general case. Beginning with the equation

$$\frac{dy}{dx} + p(x)y = q(x), \tag{1.3–1}$$

we have the integrating factor

$$g(x) = \exp\left(\int^x p(t)\,dt\right). \tag{1.3–2}$$

Multiplying Eq. (1.3–1) by this function and integrating produces

$$y = \frac{1}{g(x)}\left(\int^x g(s)q(s)\,ds + C\right), \tag{1.3–3}$$

where we have changed the dummy variable of integration to s to prevent confusion with the independent variable x.

We do not recommend the use of Eq. (1.3–3), however. We prefer that the procedures used in Examples 1.3–1 and 1.3–2 be followed. These procedures are outlined in the following steps.

a) Write the differential equation in the form of Eq. (1.3–1).

b) Using the coefficient of y, find the integrating factor $g(x)$ as defined in Eq. (1.3–2).

c) Multiply Eq. (1.3–1) by $g(x)$; it is easy to forget that $q(x)$ must also be multiplied by $g(x)$.

d) Integrate the resulting equation, remembering that the left member becomes the product $yg(x)$.

e) Solve for y, being aware that the arbitrary constant C must be divided by $g(x)$ also.

We point out that some difficulties may arise in the course of solving a first-order linear equation. Consider the following example.

EXAMPLE 1.3–3   Find the general solution of the equation

$$\frac{dy}{dx} - \frac{1}{x^2} y = x, \qquad x > 0.$$

Solution. We readily obtain the integrating factor

$$g(x) = \exp\left(\frac{1}{x}\right).$$

Multiplying the differential equation by this function produces

$$\exp\left(\frac{1}{x}\right)\frac{dy}{dx} - \frac{1}{x^2}\exp\left(\frac{1}{x}\right)y = x\exp\left(\frac{1}{x}\right)$$

or

$$d\left(y\exp\left(\frac{1}{x}\right)\right) = x\exp\left(\frac{1}{x}\right)dx.$$

Hence,

$$y\exp\left(\frac{1}{x}\right) = \int x\exp\left(\frac{1}{x}\right)dx$$

and this is where we have difficulty. The integral on the right cannot be expressed in terms of elementary functions* nor does the integral appear in standard tables of integrals. Whenever this situation arises we will write

$$y = \exp\left(-\frac{1}{x}\right)\left(\int^x t\exp\left(\frac{1}{t}\right)dt + C\right)$$

as the general solution. ■

Another method that can occasionally be used is to change the roles of the dependent and the independent variables. In other words, an equation in the form

$$\frac{dx}{dy} + p(y)x = q(y)$$

is a first-order linear differential equation that can be solved explicitly for x by the methods given in this section. Other miscellaneous techniques and applications of first-order equations will be considered in Section 1.4.

---

* The class of **elementary functions** includes polynomials; the exponential and logarithmic functions; the trigonometric functions and their inverses; sums (and differences), products, and quotients of the foregoing; elementary functions of an elementary function. For example, $(\tan 3x)\exp(\sin x)$ is an elementary function. Non-elementary functions include the gamma function (Section 3.1, Exercise 25) and the error function (Section 7.4, Exercise 13).

KEY WORDS AND PHRASES

integrating factor   **first-order linear differential
                                            equation**

EXERCISES 1.3

▶ **1.** Find the general solution of

$$\frac{dy}{dx} + 2y = e^x.$$

✓ **2.** Solve the initial-value problem

$$\frac{dy}{dx} - xy = x, \quad y(0) = 0.$$

3. Find an integrating factor for

$$\frac{dy}{dx} + 2y = \exp(-x).$$

Then find the general solution of the given equation.

✓ **4. a)** Solve the initial-value problem

$$x\frac{dy}{dx} + 2y = x^2, \quad y(-1) = \frac{5}{4}.$$

b) Graph the solution.

5. Show that introducing a constant lower limit b in the expression for $g(x)$ in Example 1.3–1 does not affect the final result.

▶▶ **6.** Solve

$$x(1 + 2y)\,dx - dy = 0.$$

7. Solve the initial-value problem

$$dx + (x - \sin y)\,dy = 0, \qquad y = \pi \text{ when } x = 1.$$

8. Solve

$$\tan x\frac{dy}{dx} + y = 3x \sec x.$$

9. a) Solve the initial-value problem

$$x^2\frac{dy}{dx} + 2xy = \cos x, \qquad y(\pi) = 0.$$

b) Graph the solution.

In Exercises 10–16, find the general solutions by any of the methods discussed so far.

10. $x\dfrac{dy}{dx} + 2y = 8x^2$ ✓ **11.** $\dfrac{du}{dt} - u \cot t = \csc t$

12. $\dfrac{ds}{dt} = \cos t - s \sec t$

13. $(y + 1)\, dx + (2x - y)\, dy = 0$

14. $\dfrac{dx}{dy} = y(1 - x)$

15. $\dfrac{dy}{dx} = \exp(x - y)$

16. $(xy + 2)\, dx - x^2\, dy = 0$

In Exercises 17–20, solve the given initial-value problem.

17. $\dfrac{dy}{dx} - \dfrac{2y}{x} = x^3$, $\quad y(2) = 16$

18. $x\dfrac{dy}{dx} + 2y = \dfrac{\log x}{x^2}$, $\quad y(1) = 3$

19. $\sin\theta\,\dfrac{du}{d\theta} - u\cos\theta = \sin 2\theta - 2\theta$, $\quad u\!\left(\dfrac{\pi}{2}\right) = 1$

20. $\sin\theta\,\dfrac{dr}{d\theta} + r\cos\theta = 0$, $\quad r = \sqrt{2}$ when $\theta = \dfrac{\pi}{4}$

21. Solve

$$\dfrac{dy}{dx} - 3y\tan x = 2.$$

22. Solve

$$\dfrac{dx}{dy} + 4xy = y.$$

▶▶▶ **23.** Show that the general solution given for the differential equation in Example 1.3–3 satisfies the equation. Recall Leibniz's rule for differentiating an integral: If

$$F(x) = \int_{a(x)}^{b(x)} f(x, t)\, dt,$$

then

$$F'(x) = \int_{a(x)}^{b(x)} \dfrac{\partial}{\partial x} f(x, t)\, dt + f(b, t)\dfrac{d}{dx} b(x) - f(a, t)\dfrac{d}{dx} a(x).$$

(*Note*: The formula for the nth derivative of the product of two functions is also called Leibniz's rule.)

24. a) Solve the initial-value problem

$$x\dfrac{dy}{dx} - 2y = -x^2, \qquad y(1) = 0.$$

b) Graph the solution.

c) For what values of x is the solution valid?

1.4 MISCELLANEOUS TECHNIQUES

There are many types of first-order differential equations other than those we have discussed in the preceding sections. A few of these will be considered in this section and we will include these and the other types in the exercises. This is done not only to provide a review but also to give an opportunity for practice. It can easily happen that students may be able to solve differential equations as they are presented by individual types but have difficulty when confronted with an equation after the individual methods have been learned. Then they must mentally run through a checklist of the various types and, after *identifying* the type of the equation, *choose* the appropriate method of solution. This process requires familiarity with the methods, which, in turn, comes with practice.

We present an example of yet another type of first-order equation.

EXAMPLE 1.4–1 Solve the equation

$$(x + 3y)\, dx + (y - x)\, dy = 0, \quad x > 0.$$

Solution. In the given form it is easy to check that the equation is not exact and that the variables cannot be separated. Moreover, the equation cannot be put into the form we treated in Section 1.3. In this case we have

$$\frac{dy}{dx} = \frac{x + 3y}{x - y} = \frac{x\left(1 + 3\dfrac{y}{x}\right)}{x\left(1 - \dfrac{y}{x}\right)} = \frac{1 + 3\dfrac{y}{x}}{1 - \dfrac{y}{x}} = F\left(\frac{y}{x}\right).$$

The fact that dy/dx can be expressed as a function of y/x suggests the substitution

$$v = y/x \quad \text{or} \quad y = vx,$$

from which it follows that

$$\frac{dy}{dx} = v + x\frac{dv}{dx}.$$

Hence the given differential equation can be written as

$$v + x\frac{dv}{dx} = \frac{1 + 3v}{1 - v}$$

or

$$x\frac{dv}{dx} = \frac{1 + 3v}{1 - v} - v = \frac{(1 + v)^2}{1 - v},$$

which implies that

$$\frac{(1 - v)\, dv}{(v + 1)^2} = \frac{dx}{x}$$

and the variables have been separated. Thus integrating (see Exercise 1) gives us

$$\frac{-2}{v + 1} = \log C|x(v + 1)|$$

or

$$\frac{-2x}{x+y} = \log C|x+y|,$$

which is the general solution in implicit form. Note that $y = -x$ is a singular solution since we divided by $(v+1)$. ■

EXAMPLE 1.4–2   Solve

$$\frac{dy}{dx} = \frac{y}{x + \sqrt{xy}}, \qquad x > 0.$$

Solution. It may be easier sometimes to interchange the dependent and independent variables as we have suggested before. We use the present example to illustrate the procedure. Consider

$$\frac{dx}{dy} = \frac{x + \sqrt{xy}}{y} = \frac{x}{y} + \sqrt{\frac{x}{y}}$$

and make the substitutions

$$u = \frac{x}{y}$$

$$x = uy$$

$$\frac{dx}{dy} = u + y\frac{du}{dy}.$$

Then we have

$$y\frac{du}{dy} = u^{1/2}$$

and after separating the variables, we obtain the general solution in implicit form:

$$2\sqrt{\frac{x}{y}} = \log C|y|.$$

This time $y = 0$ is a singular solution. ■

A natural question is: Under what conditions can the methods of the two previous examples be applied? There is a simple answer to this question. If $f(x, y)$ has the property that

$$f(tx, ty) = t^n f(x, y) \tag{1.4–1}$$

for n a nonnegative integer, then $f(x, y)$ is called a "homogeneous function of x and y of degree n." If the differential equation

$$M(x, y)\, dx + N(x, y)\, dy = 0 \tag{1.4–2}$$

is such that *both* $M(x, y)$ and $N(x, y)$ have the property shown in Eq. (1.4–1), then we can write

$$\frac{dy}{dx} = F\left(\frac{y}{x}\right) \quad \text{or} \quad G\left(\frac{x}{y}\right)$$

and the foregoing technique can be applied.

It may happen that there is a change of variable that will transform an equation into a form in which the last method can be applied. For example, if

$$(x - 2y - 8)\, dx + (2x - 3y - 13)\, dy = 0$$

is the differential equation to be solved, then the change of variables

$$x = X + 2, \qquad y = Y - 3,$$

will transform the equation into

$$\frac{dY}{dX} = -\frac{X - 2Y}{2X - 3Y},$$

which can be readily solved (Exercise 3). The conditions under which transformations of this kind can be made will be treated next.

Suppose that the differential equation has the form

$$\frac{dy}{dx} = \frac{ax + by + c}{dx + ey + f},$$

where $a, b, c, d, e,$ and f are constants. We introduce a change of variable,

$$x = X + h, \qquad y = Y + k,$$

and the differential equation becomes

$$\frac{dY}{dX} = \frac{aX + bY + ah + bk + c}{dX + eY + dh + ek + f}.$$

Now h and k can be found so that

$$ah + bk + c = 0$$

and

$$dh + ek + f = 0,$$

provided that $ae - bd \neq 0$. The result is then

$$\frac{dY}{dX} = \frac{aX + bY}{dX + eY}$$

and the substitution,

$$Y = vX, \qquad \frac{dY}{dx} = v + X\frac{dv}{dX},$$

transforms the equation into

$$\frac{dX}{X} = \frac{dv}{g(v)},$$ **(1.4–3)**

where

$$g(v) = \frac{a + (b - d)v - ev^2}{d + ev}.$$

The final result is a separable differential equation. Details are left for the exercises (see Exercises 4 and 5).

In Section 1.3 we introduced an integrating factor for the first-order linear differential equation. It is possible to find integrating factors in other cases also, as shown in the next example.

EXAMPLE 1.4–3

a) The equation $(x^2 - y^2)\,dx + 2xy\,dy = 0$ can be solved by the method given in Example 1.4–1. We observe, however, that multiplying the equation by $(x^2 + y^2)^{-2}$ makes the equation exact (Exercise 6). Unfortunately, whenever more than one method can be used to solve an equation, it is not always possible to know a priori which method will entail the least amount of computation.

b) The equation $y\,dx + (2x - ye^y)\,dy = 0$ presents a challenge until we realize that multiplying by y converts the equation into an exact one.

c) Multiplying the equation $(x^2 + 2y)\,dx - x\,dy = 0$ by x^{-3} makes it exact. ■

It can be proved that if the equation

$$M(x, y)\,dx + N(x, y)\,dy = 0$$

has a solution, then an integrating factor $\mu(x, y)$ exists that makes the equation exact and, moreover, μ satisfies the linear partial differential equation*

$$N\frac{\partial \mu}{\partial x} - M\frac{\partial \mu}{\partial y} + \mu\left(\frac{\partial N}{\partial x} - \frac{\partial M}{\partial y}\right) = 0.$$

The difficulty, of course, is that it may require more labor to find the integrating factor than to solve the ordinary differential equation in the first place. Hence the use of integrating factors cannot be considered a general method for solving differential equations. It can be

---

* See Michael Golomb and Merrill Shanks, *Elements of Ordinary Differential Equations*, 2d ed., p. 53. New York: McGraw-Hill, 1965.

said, however, that if the expression $x\,dy + y\,dx$ appears in an equation, then xy or a function of xy may be a candidate for an integrating factor. If $x\,dx + y\,dy$ appears, then $x^2 + y^2$ or a function of $x^2 + y^2$ may work.

KEY WORDS AND PHRASES

homogeneous function of degree n integrating factor

EXERCISES 1.4

▶ **1.** In Example 1.4–1 solve the differential equation

$$\frac{(1 - v)\,dv}{(v + 1)^2} = \frac{dx}{x}.$$

(*Hint*: The left member can be written as

$$-\frac{1}{2}\frac{2(v + 1)}{(v + 1)^2} + \frac{2}{(v + 1)^2}.)$$

✓ **2.** Write dy/dx in Example 1.4–1 as

$$\frac{dx}{dy} = G\!\left(\frac{x}{y}\right)$$

and solve.

3. Solve the equation

$$\frac{dY}{dX} = -\frac{X - 2Y}{2X - 3Y}.$$

4. Solve the system

$$ah + bk + c = 0,$$
$$dh + ek + f = 0,$$

for h and k, assuming that $ae - bd \neq 0$.

5. Obtain the separable equation (1.4–3)

$$\frac{dX}{X} = \frac{dv}{g(v)}.$$

✓ **6. a)** Make the substitution $y = vx$ to solve the equation

$$(x^2 - y^2)\,dx + 2xy\,dy = 0.$$

b) Multiply the equation in part (a) by $(x^2 + y^2)^{-2}$ and show that the result is an exact differential equation.

c) Solve the equation in part (b).

7. a) Show that y is an integrating factor of the equation

$$y \, dx + (2x - ye^y) \, dy = 0.$$

(Compare with Example 1.4–3.)

b) Solve the equation in part (a).

8. a) Use the integrating factor x^{-3} to make the equation

$$(x^2 + 2y) \, dx - x \, dy = 0$$

exact. (Compare with Example 1.4–3.)

b) Solve the differential equation in part (a) by the method given in Section 1.2.

c) Solve the differential equation in part (a) by the method given in Section 1.3.

▶▶ **9. a)** Show that

$$\frac{x^2 + y^2}{2xy}$$

is a homogeneous function of x and y of degree 2.

b) Solve the differential equation

$$\frac{dy}{dx} = \frac{-(x^2 + y^2)}{2xy}.$$

(Compare with Example 1.2–1.)

10. a) Show that

$$\frac{y + x^2 + y^2}{x}$$

is not a homogeneous function of x and y.

b) By making the substitution $u = y/x$, solve the differential equation

$$x \, dy = (y + x^2 + y^2) \, dx.$$

11. Solve

$$x \frac{dy}{dx} - y = x \sin \frac{y}{x}.$$

12. Solve each of the following.

a) $\dfrac{dy}{dx} = \dfrac{2x^2 + y^2}{-2xy + 3y^2}$

b) $x \, dy - y \, dx = \sqrt{xy} \, dx$

c) $(x \exp(y/x) + y) \, dx - x \, dy = 0$

d) $(y^2 - 2xy) \, dx + (2xy - x^2) \, dy = 0$

13. In each of the following, find an integrating factor that will make the given equation exact. Then solve the equation.

a) $(x^2 + y^2 + x) \, dx + y \, dy = 0$

b) $(x \, dy + y \, dx) + x^2 y^3 \, dy = 0$

c) $2y \, dx - (x^2 + y^2 + 2x) \, dy = 0$

Hint:

$$d\left(\arctan\left(\frac{x}{y}\right)\right) = \frac{y\,dx - x\,dy}{x^2 + y^2}.$$

14. Solve each of the following.

a) $\dfrac{dy}{dx} = \dfrac{x - y}{x + 3y}$

b) $\dfrac{dx}{dy} = \dfrac{5x - y}{2x + 2y}$

c) $\dfrac{dy}{dx} = \dfrac{x + y - 1}{x + 4y + 2}$

d) $\dfrac{dy}{dx} = \dfrac{x - 3y + 3}{2x - 6y + 1}$

▶▶▶ **15.** Make an obvious substitution to solve

$$\frac{dy}{dx} = (x + y)^2.$$

16. Prove that $\cos x \cos y$ is an integrating factor for the equation

$$(2x \tan y \sec x + y^2 \sec y)\,dx + (2y \tan x \sec y + x^2 \sec x)\,dy = 0.$$

17. Prove that $(x^2 + y^2)^{-1}$ is an integrating factor for the equation

$$(y + xf)\,dx - (x - yf)\,dy = 0,$$

where $f = f(x^2 + y^2)$.

18. Solve the equation

$$\frac{dy}{dx} = \frac{y + \sqrt{y^2 - x^2}}{x}.$$

19. In solving

$$\frac{dy}{dx} = \frac{ax + by + c}{dx + ey + f}$$

we assumed that $ae - bd \neq 0$. Given that $ae = bd$, show that

$$ax + by = \frac{a}{d}(dx + ey)$$

and that the substitution $u = ax + by$ transforms the differential equation into

$$\frac{du}{dx} = \frac{a((b + d)u + af + bc)}{du + af}.$$

20. Show that

$$\left(y - xf\left(\frac{y}{x}\right)\right)^{-1}$$

is an integrating factor of the equation

$$f\left(\frac{y}{x}\right)dx - dy = 0.$$

21. a) Graph some of the solutions of Exercise 14(a).

b) Show that the solutions of Exercise 14(a) are a family of hyperbolas.

1.5 APPLICATIONS

We have considered the first-order differential equation

$$\frac{dy}{dx} = f(x, y),$$

and discussed solution techniques for various forms of the function $f(x, y)$. It should be apparent by now, however, that this function can take so many forms that we cannot possibly consider more than a few of them. This we have done and now we concentrate on *applications* of first-order differential equations.

Radioactive decay

A radioactive substance decomposes at a rate proportional to the amount present at time t:

$$\frac{dm}{dt} = -km, \qquad k > 0, \qquad m = m(t).$$

The statement applies statistically in the large, not at the atomic level. The radioactive decay constant, k, covers a wide range in the thousands of known nuclides. No two have exactly the same decay constant, which ranges from 1.6×10^{-25} sec^{-1} for lead-204 ($_{82}Pb^{204}$) to 2.3×10^6 sec^{-1} for polonium-212 ($_{84}Po^{212}$).

Radiocarbon dating is an interesting application of radioactive decay and, at the same time, gives an example of the (sometimes controversial) assumptions that need to be made before a problem can be solved. In 1949 Willard F. Libby announced his unique method for dating archaeological finds for which he received the Nobel prize in 1960. The method depends on determination of the proportion of radioactive carbon in an object whose age is to be found.

A rare isotope of ordinary carbon, C-12, is radiocarbon, C-14, which occurs in the atmosphere in the ratio of one atom of C-14 for every 10^{12} atoms of C-12. The C-14 atoms combine with oxygen to form carbon dioxide just as the C-12 atoms do. The carbon dioxide is absorbed by plants in the process of photosynthesis, plants are eaten by animals, and so C-14 becomes a link in a complicated food chain. Thus all carbon ultimately comes from the carbon dioxide in the atmosphere and consequently contains a percentage of C-14. When a tree, for example, is cut down and used to build a house, the chain is broken and the process of radioactive decay of the C-14 begins. Since the half-life of C-14 is 5580 ± 45 years,* a determination many centuries later of the ratio of C-14 to C-12 can give an indication of the age of the house.

---

* See W. F. Libby, *Radiocarbon Dating*, 2d ed., p. 36. Chicago: University Press, 1955.

There are a number of factors that must be taken into account if archaeologists and anthropologists are to give us a picture of the past.* First, there is a statistical aspect to which we have already referred. The half-life of C-14 as given by Libby depends on an experimental determination of the radioactive decay constant discussed in Section 1.1. To say that the half-life is 5580 ± 45 years means that it lies in the given range with probability of 0.5. Those familiar with statistics refer to this range as the 50-percent *confidence interval*. Naturally, any measurement of the percentage of C-14 in some artifact is subject to a similar deviation, which may be considerably larger if a small sample is being measured.

A second factor that must be considered in radiocarbon dating comes from the underlying hypothesis of the method itself. It must be assumed that the total number of C-14 atoms is invariant (unchanged) at any time and at any location, in other words, that the rate of formation of the atoms by bombardment of the earth's atmosphere by cosmic radiation is exactly equal to their decay rate.

Finally, it must be assumed that samples that are dated by the radiocarbon method have not been contaminated and that there has not been an unusual delay between the time the living thing died and the time it became incorporated in the environment in which it was found. For example, a particular house was built from a recently cut tree and not from wood used in building a house many years before. Hence there can be a discrepancy whenever historical, geological, and archaelogical measurements of time are compared with radiocarbon measurements. In fact, more recent work has shown that dates previously determined to be 3000 B.C. may be closer to 2300 B.C.†

Bacterial growth

Bacteria increase at a rate proportional to the number present at time t:

$$\frac{dN}{dt} = kN, \qquad k > 0, \qquad N = N(t).$$

The statement applies statistically in the large in a closed ecosystem. If we keep this in mind, N can be considered a continuous function of t although it is, in fact, an integer-valued function. This is not an unreasonable assumption since N is usually "large" and it is apparent that the larger the value of N, the smaller the percentage difference between successive integers. The assumption that each bacterium has

*See Colin Renfrew, *Before Civilization: The Radiocarbon Revolution and Prehistoric Europe*. New York: Alfred A. Knopf, 1974.

† See, especially, H. E. Suess, "Bristlecone-pine calibrations of the radiocarbon time-scale 5200 B.C. to the present," in I. U. Olsson (ed.), *Radiocarbon Variations and Absolute Chronology*, pp. 3003–3012. New York: Wiley, 1970.

the same chance of reproducing or dying is also present here so that individual variation is ignored.

The basic differential equation of growth,

$$\frac{dN}{dt} = kN,$$

used here and in many other problems, was studied by Thomas R. Malthus (1766–1834), an English economist and demographer. Figure 1.5–1 shows the form of the solution of the above differential equation.

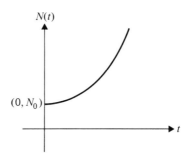

FIGURE 1.5–1 Exponential growth.

Compartmental analysis

In studying the effects of drugs in a biological system it is convenient to divide that system into *compartments*. For example, the organs and the blood in an animal can be thought of as compartments. Thus the model of the system might consist of several compartments with provisions for transport of various substances from one compartment to several others via various routes. Figure 1.5–2 shows a simplified model of iron metabolism in humans.

One of the most important simplifying assumptions that is usually made is that the rate of transport, da/dt, of a drug or a tracer A out of a given compartment is proportional to its concentration a within that compartment. The resulting differential equation,

$$\frac{da}{dt} = ka,$$

is thus the same as those in the preceding two examples. The assumption is valid for tracer kinetics, provided the drug being traced is at a constant concentration throughout the particular compartment of

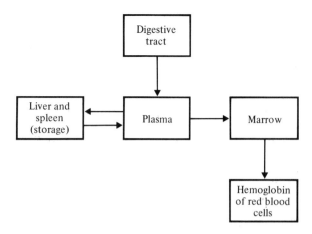

FIGURE 1.5–2 Compartmental analysis in biokinetics.

the animal. Another way of saying this is that the animal is in a steady state. The simplified model is also valid for pharmacokinetics, provided the concentration of the drugs is relatively small.

Law of cooling (Newton)

The rate at which the surface of a substance cools is proportional to the difference between the surface temperature and the ambient temperature. This is written mathematically as

$$\frac{du}{dt} = -k(u - u_0), \qquad k > 0, \qquad u(t) > u_0,$$

where u_0 is the ambient temperature and $u(t)$ is the temperature at time t. The statement applies to an isotropic substance immersed in an environment of constant temperature. When dealing with good conductors of heat such as silver and copper, we can assume that the temperature of the surface is practically the same as that of the interior since conduction takes place in such a short time. Further, it is almost self-evident that the law applies to an object in air only if the air is circulating so that its constant temperature is maintained. Although the law is called a law of *cooling*, it applies equally well to heating. In the latter case we would have $u(t) < u_0$ and $k < 0$ so that the differential equation would remain precisely the same. It is customary to refer to "cooling," however, since this is in agreement with the second law of thermodynamics, which states that heat "flows" in a direction from a point of higher temperature to one of lower temperature. See Fig. 1.5–3.

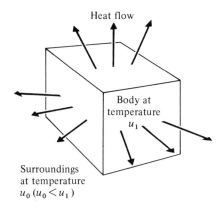

FIGURE 1.5–3 The law of cooling.

The capstan

If a rope is slung around a drum rotating at constant angular velocity, it provides a mechanical advantage that can be used to raise and lower heavy loads (Fig. 1.5–4). Consider the drum of radius r rotating in a counterclockwise direction (Fig. 1.5–5a). Suppose the rope is kept taut by tensions T_0 and T_1, and makes contact with the drum at P_0 and P_1, respectively. If the coefficient of sliding friction between the rope and the drum is μ, then we must have $T_0 < T_1$.

We can obtain the relation between the two tensions by considering an arbitrary point P on the drum between P_0 and P_1 shown in Fig. 1.5–5(b). A small portion of the rope making central angle $\Delta\theta$ has tension T and $T + \Delta T$ at two adjacent points. Let N be the normal force per unit length at P. Then the total normal force at P is $Nr\,\Delta\theta$. The tangential or frictional force at P is $\mu Nr\,\Delta\theta$. We can resolve the forces at P in two perpendicular directions and it is most convenient to choose the normal and tangential directions. In the normal direction

FIGURE 1.5–4 The capstan.

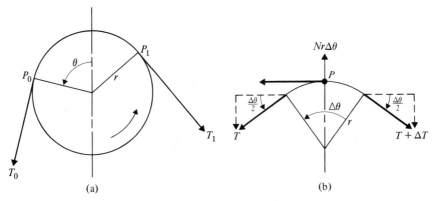

FIGURE 1.5–5 The capstan shown diagrammatically.

we must have

$$T \sin \frac{\Delta\theta}{2} + (T + \Delta T) \sin \frac{\Delta\theta}{2} = Nr\,\Delta\theta$$

for equilibrium, whereas in the tangential direction we must have

$$T \cos \frac{\Delta\theta}{2} + \mu NR\,\Delta\theta = (T + \Delta T) \cos \frac{\Delta\theta}{2}.$$

Since $\Delta\theta$ is small, we can make the approximations

$$\sin \frac{\Delta\theta}{2} \doteq \frac{\Delta\theta}{2} \qquad \text{and} \qquad \cos \frac{\Delta\theta}{2} \doteq 1$$

and eliminate the term $\Delta T \sin \Delta\theta/2$ so that the two equations become

$$T = Nr,$$
$$\mu Nr\,\Delta\theta = \Delta T.$$

These can now be combined to form the single first-order differential equation

$$\mu T\,d\theta = dT,$$

which can be solved by using the initial condition $T = T_0$ when $\theta = 0$. We have tacitly assumed that the frictional force is the product of the normal force and the coefficient of sliding friction. This is a useful empirical relation but is not a physical law as is Newton's second law. We have also neglected the fact that μ varies with the relative velocity between the two surfaces and is only approximately independent of the contact area involved.

Orthogonal trajectories

The one-parameter family of curves that is the solution of

$$\frac{dy}{dx} = f(x, y)$$

has a family of orthogonal trajectories that is the solution of

$$\frac{dy}{dx} = -\frac{1}{f(x, y)}.$$

A curve that intersects every member of a family at right angles is called an *orthogonal trajectory* of the given family. Thus the family

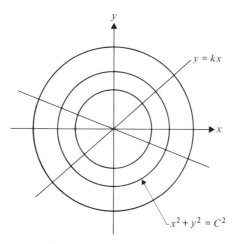

FIGURE 1.5–6 Orthogonal trajectories.

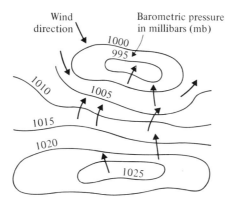

FIGURE 1.5–7 Orthogonal trajectories in meteorology.

of concentric circles centered at the origin,

$$x^2 + y^2 = C^2,$$

and the lines through the origin,

$$y = kx,$$

shown in Fig. 1.5–6 are orthogonal trajectories of each other. In potential theory the equipotential lines and streamlines are orthogonal to each other. In meteorology the orthogonal trajectories of the isobars (curves of equal barometric pressure) denote the wind direction from high- to low-pressure areas (Fig. 1.5–7).

Chemical reactions of the first order

If in a chemical reaction the rate of change of concentration of a reactant is proportional to the concentration, the reaction is defined to be one of the first order (Fig. 1.5–8). The recognition of this fact (by Wilhelmy* in 1850) was the historical beginning of chemical kinetics. Many chemical decompositions of molecules containing more than three atoms are first-order reactions, provided the concentration is not low. An example is the decomposition of nitrogen pentoxide, which is transformed into oxygen and nitrogen tetroxide according

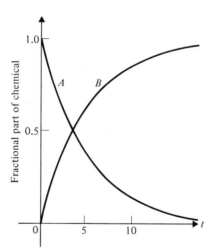

FIGURE 1.5–8 The chemical reaction

$A \xrightarrow{k_1} B.$

* Ludwig F. Wilhelmy (1812–1864), a German physical chemist.

to the chemical equation

$$2N_2O_5 \xrightarrow{k_1} O_2 + 2N_2O_4.$$

If a is the concentration at time $t = 0$ and x is the concentration at time t, then a first-order reaction is characterized by the differential equation

$$-\frac{d}{dt}(a - x) = \frac{dx}{dt} = k_1(a - x).$$

Here $k_1 > 0$ represents the fractional number of molecules decomposing per unit of time. This constant can also be thought of as the average probability per unit of time that one molecule taken at random will decompose. In the decomposition of nitrogen pentoxide the average value of k_1 is 0.0909 with dimension $(\text{min})^{-1}$.

Chemical reactions of the second order

When molecules of types A and B react to form molecules of type P, we have a chemical reaction of the second order given by

$$A + B \xrightarrow{k_2} P.$$

If the original concentrations of A and B are given by a and b, respectively, and if x represents the concentration of P at time t, then a second-order reaction is characterized by the differential equation

$$\frac{dx}{dt} = k_2(a - x)(b - x).$$

Most chemical reactions are of the second order (Fig. 1.5–9).

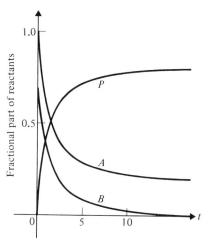

FIGURE 1.5–9 The chemical reaction $A + B \xrightarrow{k_2} P$.

Hydraulics (Torricelli's* law)

When a vessel filled with water is allowed to discharge through an orifice under the action of gravity, the mean velocity of the water flowing through the orifice is given by

$$v = \sqrt{2gh},$$

where g is the acceleration due to gravity and h is the instantaneous height or *head* of the liquid above the orifice. The given equation represents an ideal condition that cannot be attained. It has been shown experimentally that because of friction at the orifice and contraction of the discharging flow (the so-called *vena contracta* of hydrodynamics), a more realistic equation is

$$v \doteq 0.61\sqrt{2gh}.$$

Note that v and h are both functions of t. See Fig. 1.5–10.

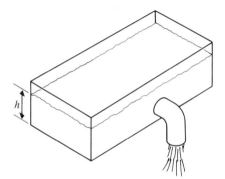

FIGURE 1.5–10 Flow discharge.

Absorption of light (Lambert's law)

When light enters an absorbing medium, equal fractions of the light are absorbed along equal paths. This law has been attributed to both Bouguer (Pierre Bouguer (1698–1758), Royal Professor of Hydrography at LeHavre) and Lambert (Johann H. Lambert (1728–1777), a German physicist, astronomer, and mathematician who worked primarily in the field of heat radiation). It is assumed that the absorbing material is isotropic. Moreover, true absorption would imply that *all* the energy of the light is converted into *heat motion* of the molecules of the absorbing medium. This does not occur because, in fact, some of the light is *scattered* rather than absorbed.

---

* Evangelista Torricelli (1608–1647), an Italian physicist.

In the ideal case, however, light of intensity I is reduced to $I - \Delta I$ along a path of length Δx in a medium, as shown in Fig. 1.5–11. Lambert's law states that $\Delta I/I$ is the same for all paths of length Δx; that is, this ratio is proportional to Δx and we can write

$$\frac{\Delta I}{I} \propto \Delta x \quad \text{or} \quad \frac{dI}{I} = -k\,dx,$$

where the (positive) constant k is called the *absorption coefficient*.

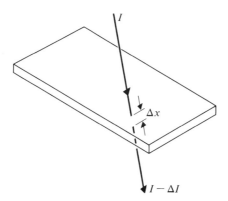

FIGURE 1.5–11 Absorption of light.

Electric series circuit (Kirchhoff's* second law)

Consider the series circuit shown in Fig. 1.5–12, which consists of a resistance R, an inductance L, and a constant impressed voltage E. According to Kirchhoff's second law, the impressed voltage on a closed loop must equal the sum of the voltage drops in the loop. Thus when the switch S is closed, the current i satisfies the differential equation

$$L\frac{di}{dt} + Ri = E.$$

This equation assumes that there is no capacitance in the circuit, that is, the resistance and inductance are capacitance-free, a condition not attainable with ordinary components. In the above equation, Ri is the voltage drop across the resistance and $L(di/dt)$ is the voltage drop across the inductance.

---

* Gustave R. Kirchhoff (1824–1887), a German physicist.

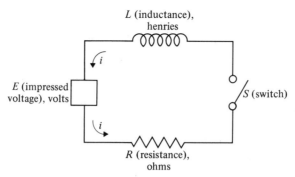

FIGURE 1.5–12 A series circuit.

Population growth (the logistic equation)

Consider an animal population given by $N(t)$ at time t. Assume that the population consists of a single species and exists in a closed system, that is, there is no migration in or out. Assume further that each animal in the population has the same probability of reproducing or dying. This assumption eliminates the need to deal with individual variation and, in fact, also eliminates both age and sex so that we can consider a homogeneous group.

Let b and d represent the number of births and deaths, respectively, per animal per unit of time. Then

$$N(t + \Delta t) = N(t) + bN(t)\, \Delta t - dN(t)\, \Delta t$$

or

$$\frac{N(t + \Delta t) - N(t)}{\Delta t} = rN(t),$$

where $r = b - d$. Assume that r is a function of N alone. Assume also that as N increases, b decreases and d increases or, what is the same thing, that $dr/dN < 0$. This last assumption can be justified on the basis that in a limited environment crowding and malnutrition will be present. Since the simplest function is a linear one, we assume that

$$r = r_0\left(1 - \frac{N(t)}{K}\right),$$

where r_0 and K are positive parameters (Fig. 1.5–13). Biologists call K the "carrying capacity of the environment" and r_0 the growth rate.

With all the above assumptions we have the "logistics equation,"

$$\frac{d}{dt}N(t) = r_0\left(1 - \frac{N(t)}{K}\right)N(t), \quad t \geq 0, \quad N(0) = N_0.$$

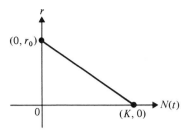

FIGURE 1.5–13 A linear function for population growth.

This equation is also known as the Verhulst equation (after Pierre-François Verhulst (1804–1849), a Belgian sociologist and mathematician who used its solution in his study of population dynamics).

Mathematical economics

Suppose that the total dollar value of an enterprise at time t is given by $u(t)$ and assume that the rate at which $u(t)$ increases is proportional to $u(t)$. Then

$$\frac{d}{dt}u(t) = au(t), \qquad u(0) = c,$$

where a is a positive constant and c represents the initial capitalization of the enterprise. The solution of the above initial-value problem shows that $u(t)$ is growing exponentially. On the face of it, this would appear to be a desirable situation but it is unrealistic. The principal objective in the management of an enterprise is profit and this must somehow be brought into the picture.

Assume that it is decided to reduce the dollar value of the enterprise at a rate proportional to the total value. If $p(t)$ denotes the total profit over the time interval $[0, t]$, then

$$\frac{d}{dt}p(t) = ku(t), \qquad p(0) = 0,$$

where $0 \leq k \leq 1$. Hence the governing differential equation becomes

$$\frac{d}{dt}u(t) = a(1 - k)u(t), \qquad u(0) = c.$$

The solution of this initial-value problem is

$$u(t) = c \exp [a(1 - k)t],$$

and from this we obtain

$$p(t) = \frac{kc}{a(1 - k)} \{\exp [a(1 - k)t] - 1\}.$$

Now it is apparent that if $k = 0$, nothing is taken out of the business, hence the profit is zero. On the other hand, if $k = 1$, then everything is being taken out of the business and it cannot increase in size. Thus there must be an optimum value of k that can be found by fixing t to some value T and maximizing the resulting $p(T)$.

The above examples are meant to convey the idea that there is a great variety of areas in which first-order differential equations play a significant role.

KEY WORDS AND PHRASES

radioactive decay   orthogonal trajectory
radiocarbon dating

EXERCISES 1.5

▶ 1. Find the half-life of lead-204; of polonium-212.

2. An object is found to contain 12.3 percent of the original amount of C-14. How old is the object, assuming that the half-life of C-14 is 5600 years?

3. **a)** Given that the number of bacteria double in six hours, find the proportionality constant k.

   b) How many hours will it take the number of bacteria to triple?

4. A thermometer reading 20°C indoors is placed outdoors. In 15 minutes it reads 15°C and in another 15 minutes it reads 12°C. What is the outside temperature?

5. If the thermometer in Exercise 4 is now brought indoors (still at 20°C), how long will it be before the thermometer registers 15°C?

6. A thermometer reading 18°C indoors is placed outdoors. In 10 minutes it reads 23°C and in another 10 minutes it reads 26°C. What is the outside temperature?

7. Solve the differential equation of the capstan and compute the ratio T/T_0 for $\theta = \pi/2$, $3\pi/2$, 2π, 4π, and 5π, using a coefficient of friction $\mu = 0.4$. How heavy a load can a man weighing 80 kg control, if the rope is wound two and one-half times around the revolving drum?

8. Find the family of orthogonal trajectories of the family $xy = c$.

9. Solve the differential equation of the first-order chemical reaction explicitly for k_1. Also solve the equation for x.

10. Obtain the time $T_{1/2}$ corresponding to half completion ($x = \frac{1}{2}a$) of a first-order chemical reaction. Also compute $T_{1/2}$ for the decomposition of nitrogen peroxide.

11. Solve the differential equation of the second-order chemical reaction explicitly for k_2. Also solve the equation for x.

12. When concentrations of the reactants in a second-order reaction are greatly different (for example, when $a \gg b$),* the term $a - x$ remains essentially constant and is very close in value to a during the reaction. Solve the differential equation under this assumption.

13. Solve the differential equation of the second-order chemical reaction in the special case when the concentrations of the two reactants are equal $(a = b)$.

14. Solve the differential equation expressing Lambert's law if the light intensity is I_0 when the light enters an absorbing medium $(x = 0)$.

15. A point source of light is situated in water having an absorption coefficient $k = 0.07 \text{ m}^{-1}$. As the light spreads out from this source it is found that the intensity is 78 percent of the original at a distance x from the source. Find x.

16. Solve the differential equation of the series circuit containing a resistance R, an inductance L, and a constant impressed voltage E.

17. In Exercise 16 what is the limiting value of the current? What is $T_{1/2}$, the time necessary to reach one-half this limiting value?

18. Solve Exercise 16 if the constant impressed voltage E is replaced by $E_0 \cos \omega t$, an alternating voltage of amplitude E_0.

19. An inductance of 2 henries and a resistance of 10 ohms are connected in series with an emf of 100 volts (Fig. 1.5–14). Given that current is zero initially (when $t = 0$), compute the current when $t = 0.1$ sec.

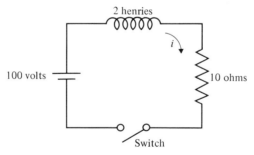

FIGURE 1.5–14

20. Solve Exercise 19 if the emf is changed to $100 \sin 60t$ volts, everything else remaining the same.

21. Show that the solution of the logistics equation can be written

$$\frac{N(t)}{K} = \frac{\exp{(r_0 t)}}{(K/N_0) + \exp{(r_0 t)} - 1}.$$

22. Referring to Exercise 21, plot the dimensionless variables $N(t)/K$ against $r_0 t$ for initial values $N_0/K = 0.25, 0.5, 0.75, 1.0,$ and 1.5.

---

* The symbol $\gg$ means "is very much greater than"; this is a comparative, not a mathematical, term.

23. Solve the initial-value problem

$$\frac{d}{dt}u(t) = a(1-k)u(t), \qquad u(0) = c, \qquad a > 0, \qquad 0 \le k \le 1$$

of the mathematical economics application.

24. For fixed T the profit of an enterprise is given by

$$p(T) = \frac{kc}{a(1-k)} \{\exp [a(1-k)T] - 1\},$$

where a, c, and T are positive constants and $0 < k < 1$. Find the value of k that maximizes p.

▶▶ **25.** Find the orthogonal trajectories of the family of curves (Fig. 1.5–15)

$$(x - c)^2 + y^2 = c^2.$$

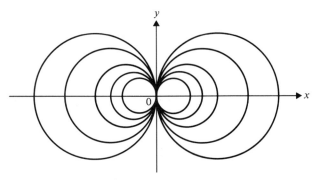

FIGURE 1.5–15

26. An inductance of 3 henries and a resistance of 6 ohms are placed in series with a voltage source $E(t) = 3 \sin t$. Given that the current in the circuit is initially 10 amps, set up and solve the appropriate differential equation to find the current at any time t.

27. An object at 70°F is placed outdoors where the temperature is 40°F. It is observed that in three minutes the temperature of the object has dropped to 60°F.

a) When will the temperature of the object reach 50°F?

b) What is the temperature of the object five minutes after it has been placed outdoors?

▶▶▶ **28.** The amount $A(t)$ of a principal P at compound interest is given by

$$A(t) = P\left(1 + \frac{r}{n}\right)^{nt},$$

where r is the annual interest rate (as a decimal, not a percent), n is the number of computing periods in a year ($n = 4$ for interest compounded quarterly, etc.), and t is the time in years. Find the equivalent annual simple interest rate for $n = 2, 4, 8, 16, 32, 64, 128, 256,$ and 512. Deduce from this the equivalent annual simple interest rate for a principal compounded *continuously.*

Continuously compounded interest   Money grows at a rate proportional to the amount present at any time t when interest is compounded continuously. Thus

$$\frac{dA}{dt} = rA,$$

where r is the annual rate of interest (as a decimal) and A is the amount at time t.

29. Given that a principal of \$1000 doubles in 9 years when interest is compounded continuously, find the interest rate.

30. What annual simple interest rate is equivalent to $7\frac{1}{2}$ percent compounded continuously?

31. What interest rate compounded continuously is equivalent to an annual simple interest rate of $6\frac{1}{4}$ percent?

32. Given that an object of mass m is dropped from rest in a medium that offers a resistance proportional to $|v|$, the instantaneous magnitude of the velocity of the object, find the velocity v at any time t. Also find the limiting velocity.

33. Yeast cells grow in a culture at a rate proportional to the number $N(t)$ present at time t. If there are N_0 cells initially, show that

$$N(t) = N_0(2)^{t/K},$$

where $K > 0$.

34. Referring to Exercise 33, show that the time required for the total number of cells to increase from N_0 to N_1 is given by

$$t = K \log_2 \left(\frac{N_1}{N_0}\right).$$

35. Referring to Exercise 33, find the time required for the total number of cells to double.

36. A capacitance of 5×10^{-6} farad and a resistance of 2000 ohms are connected in series with an emf of 100 volts (Fig. 1.5–16). If $I(0) = 0.01$ amp,

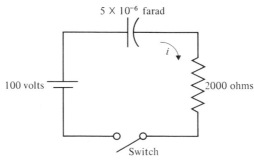

FIGURE 1.5–16

find $I(0.01)$. (*Hint*: The voltage drop across the capacitance is Q/C, where $I = dQ/dt$, C is in farads, and Q is in coulombs.)

37. Solve Exercise 36 if the emf is $100 \sin(120\pi t)$ volts, everything else remaining the same.

38. By offering two free books, a firm manages to enroll N_0 members in a book club offering specially bound classics. It is estimated that the membership would grow at a rate proportional to the membership except for one thing. A market survey showed that there are at most N_1 people who are interested in joining such a book club. Hence a better estimate is that the growth rate is proportional to the number of members and the number of remaining people who are interested. State and solve the appropriate differential equation.

39. In deriving the logistic equation assume that $dr/dN > 0$. Hence show that this has the effect of replacing r_0 by $-m$, where $m > 0$, resulting in the solution

$$N(t) = \frac{K}{1 - \left(\dfrac{N_0 - K}{N_0}\right)e^{mt}}.$$

Thus show that if $N_0 > K$, then as t increases $N(t)$ will increase to the point where

$$1 - \left(\frac{N_0 - K}{N_0}\right)e^{mt} = 0.$$

This is a model of *population explosion* as well as of a *nuclear fission process*.

40. A skydiver falls from rest toward the earth. The total weight of the skydiver and equipment is 192 lb. Before the parachute opens, the air resistance (in pounds) is numerically equal to $\frac{1}{2}v$, where v is the velocity (in feet per second). The parachute opens 5 sec after the fall begins.

a) Use Newton's second law, $F = ma$, to formulate the differential equation satisfied by v. Note that $m = w/g$ and use $g = 32$ ft/sec^2.

b) Find the velocity of the skydiver at any time t for $0 \le t \le 5$.

c) Graph the solution.

41. a) Read the amusing application of Newton's law of cooling to the field of criminology, "Elementary (My Dear Watson) Differential Equation" by Brian J. Winkel, *Math. Mag.* **52**, no. 5, November 1979, p. 315.

b) Solve the problem given there using 98.6°F as the normal body temperature.

c) Comment on the admissibility in court of evidence obtained in this manner.

1.6 NUMERICAL METHODS

In earlier sections we have presented a number of techniques for solving first-order differential equations. In Section 1.5 we discussed several applications of first-order equations for which our methods were adequate. Unfortunately, however, we may find an equation for which this is not the case. See, for example, R. Bellman's *Introduction to the Mathematical Theory of Control Processes*, vol. 1 (New York: Academic Press, 1968), where it is pointed out that in multidimensional control processes the use of dynamic programming leads to a **Riccati*** **equation**:

$$\frac{dy}{dx} = a_1(x) + a_2(x)y + a_3(x)y^2, \quad y(0) = y_0.$$

(In this connection see also Section 2.6 for a special case of the above equation.) In addition, J. J. Stoker's *Nonlinear Vibrations in Mechanical and Electrical Systems* (New York: Interscience Publishers, 1950) contains numerous examples of nonlinear differential equations.

We have emphasized that, by making certain simplifying assumptions, it is possible to transform a nonlinear equation into a linear one or into one that falls into one of the standard categories. This procedure will, in fact, continue to be an important phase of our study of mathematical analysis. At the same time, however, there will be occasions when it will not be possible to simplify or linearize the differential equation that must be solved. You will find that as you are required to solve more sophisticated problems they will seldom fall into the categories that we have been discussing.

We consider next a simple example that cannot be solved by the methods given in the preceding sections.

EXAMPLE 1.6–1   Solve the initial-value problem

$$\frac{dy}{dx} = x^2 - y^2, \quad y(0) = 1.$$

Solution. You may quickly verify that the given (nonlinear) equation does not fall into any of the standard categories we have presented. Since we do have a starting value—the point $(0, 1)$—we can use the given equation to find the slope at the initial point; call it y_0'. Thus $y_0' = -1$ and, if we don't go too far from the point $x_0 = 0$, we can assume that the slope will be approximately constant over this small interval. In other words, if $x_1 = x_0 + h$, then

$$\begin{aligned}
y_1 &= y_0 - (x_1 - x_0)\\
&= y_0 - (x_0 + h - x_0)\\
&= y_0 - h,
\end{aligned}$$

and we have a second point (x_1, y_1). We can now obtain a new value, y_1', from

$$y_1' = x_1^2 - y_1^2$$

---

* Jacopo Riccati (1676–1754), an Italian mathematician.

and proceed to a third point (x_2, y_2) as before. If we add equal increments, h, to the value of x at each step, then we have after n steps

$$\frac{y_{n+1} - y_n}{x_{n+1} - x_n} = x_n^2 - y_n^2$$

or

$$y_{n+1} = y_n + h(x_n^2 - y_n^2). \tag{1.6-1}$$

In Table 1.6–1 we have listed the values of x_n and y_n from the initial point $(0, 1)$ to the point where $x = 1$, using steps of $h = 0.1$. ∎

TABLE 1.6–1 Solution to Example 1.6–1, $h = 0.1$, using Euler's method.

n	x_n	y_n'	y_n
0	0	−1.000	1.000
1	0.100	−0.800	0.900
2	0.200	−0.632	0.820
3	0.300	−0.483	0.757
4	0.400	−0.342	0.708
5	0.500	−0.205	0.674
6	0.600	−0.067	0.654
7	0.700	0.071	0.647
8	0.800	0.212	0.654
9	0.900	0.354	0.675
10	1.000	0.495	0.711

The numerical method described above can be readily generalized to the initial-value problem

$$\frac{dy}{dx} = f(x, y), \quad y(x_0) = y_0. \tag{1.6-2}$$

Using a constant step of size h for the x variable, we have

$$y_{n+1} = y_n + hf(x_n, y_n), \quad n = 0, 1, 2, \ldots . \tag{1.6-3}$$

This iterative procedure was originally obtained by Euler and is known as **Euler's* method**. Its advantages are that it is simple and easy to use with either a hand-held electronic calculator or a large-scale computer. Computation can proceed from one step to the next without even having to use a memory if the function $f(x, y)$ is simple enough. On the other hand, it is this very simplicity that gives the Euler method a serious disadvantage. Small errors at each step (which are present

---

* Leonhard Euler (1707–1783), a Swiss mathematician.

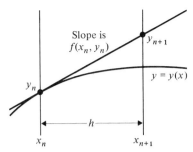

FIGURE 1.6–1 Euler's method.

to a greater or lesser degree) are magnified as the computation progresses so that a considerable error may eventually result (Fig. 1.6–1).

The main source of error in the Euler method is due to the basic assumption that the solution can be approximated by straight-line segments that are obtained by using the information at a single point. It can be shown that the error is proportional to h^2.* Thus it would not be of any use to carry the computations to more than three decimal places, as we have done in Table 1.6–1. On the other hand, cutting down the step size should reduce the error and we will see the effect of this in Table 1.6–2. By building the solution along straight-line segments we are also assuming that the absolute value of the local curvature of the solution is not too large. Since the curvature at a point is proportional to the value of the second derivative at the point, the error due to this factor can usually be estimated only at best.

In Table 1.6–2 we show the results of applying Euler's method to the problem in Example 1.6–1 using a step size of $h = 0.05$. The value of y for $x = 0.6$ is about 2.4 percent higher in Table 1.6–2 than the corresponding value in Table 1.6–1. At present we don't know which figure is more accurate but we shall return to this question as we study other numerical methods.

A modification of Euler's method has been developed to Heun.[†] It should be noted that **Heun's method** is also known as the **improved Euler method**. It seems reasonable that, instead of using the straight line between two points, we use an *average* value of the slopes at the two points. If we wish to solve the initial-value problem

$$\frac{dy}{dx} = f(x, y), \qquad y(x_0) = y_0,$$

* See, for example, William R. Derrick and Stanley I. Grossman, *Elementary Differential Equations with Applications*, 2d ed., p. 288ff. (Reading, Mass.: Addison-Wesley, 1981).

[†] Karl L. W. Heun (1859–1929), a German mathematician.

TABLE 1.6–2 Solution to
Example 1.6–1, $h = 0.05$, using
Euler's method.

n	x_n	y'_n	y_n
0	0	− 1.0000	1.000
1	0.0500	− 0.9000	0.9500
2	0.1000	− 0.8090	0.9050
3	0.1500	− 0.7249	0.8646
4	0.2000	− 0.6461	0.8283
5	0.2500	− 0.5711	0.7960
6	0.3000	− 0.4990	0.7674
7	0.3500	− 0.4288	0.7425
8	0.4000	− 0.3599	0.7211
9	0.4500	− 0.2918	0.7031
10	0.5000	− 0.2240	0.6885
11	0.5500	− 0.1562	0.6773
12	0.6000	− 0.0882	0.6695

then the Heun method uses the recursive relation

$$y_{n+1} = y_n + \frac{h}{2}(y'_n + f(x_n + h, y_n + hy'_n))$$

or

$$y_{n+1} = y_n + \frac{h}{2}\{f[x_n, y_n] + f[x_n + h, y_n + hf(x_n, y_n)]\}. \qquad \textbf{(1.6–4)}$$

See Fig. 1.6–2 for a comparison of the Euler and Heun methods. Using the problem of Example 1.6–1 we present the computations for the latter method in Table 1.6–3.

Numerical methods for solving first-order differential equations are often categorized as to "order." Thus Euler's method is classified as

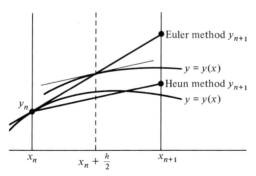

FIGURE 1.6–2 A comparison of Euler's and Heun's methods.

TABLE 1.6–3 Solution to Example 1.6–1, $h = 0.05$, using Heun's method.

n	x_n	y_n	y'_n	$f(x_n + h, y_n + hy'_n)$
0	0	1.0000	-1.0000	-0.9000
1	0.0500	0.9525	-0.9048	-0.8131
2	0.1000	0.9096	-0.8173	-0.7321
3	0.1500	0.8708	-0.7358	-0.6556
4	0.2000	0.8360	-0.6589	-0.5824
5	0.2500	0.8050	-0.5855	-0.5117
6	0.3000	0.7776	-0.5146	-0.4426
7	0.3500	0.7536	-0.4455	-0.3749
8	0.4000	0.7331	-0.3775	-0.3077
9	0.4500	0.7160	-0.3101	-0.2407
10	0.5000	0.7022	-0.2431	-0.1737
11	0.5500	0.6918	-0.1761	-0.1065
12	0.6000	0.6847	-0.1089	-0.0389
13	0.6500	0.6810	-0.0413	0.0290
14	0.7000	0.6807	0.0266	0.0973
15	0.7500	0.6838	0.0949	0.1659
16	0.8000	0.6904		

of **first order** since it gives exact results if the solution is a curve of *first* degree (a straight line); Heun's method is of **second order** since it gives exact results if the solution is a curve of *second* degree (a parabola) or less, etc. In general, a numerical method is said to be of **order** n if it gives exact results for problems whose solutions are polynomials of degree n or less. We have already mentioned that the local error (error at a point) in Euler's method is proportional to h^2. In Heun's method the local error is proportional to h^3.

We should caution the reader about assuming that one can obtain any desired degree of accuracy by simply making h small enough. This is not the case, since decreasing the step size not only results in longer computational time but may also create a point at which the roundoff error of the individual calculations offsets any gain in accuracy due to reducing the size of h. There are a number of other problems of stability and convergence that arise in the use of numerical methods. The references cited in this section contain valuable information for the nonexpert in the field.

More sophisticated methods of solving the initial-value problem (Eq. 1.6–2) include a method of Adams* called a **predictor-corrector method**. Starting with points *beyond* the initial value, this method uses one formula to **predict** the next point and another to **correct** the

---

* John C. Adams (1819–1892), an English astronomer.

TABLE 1.6-4 Solution to Example 1.6-1, $h = 0.1$, using the Adams–Bashforth and Adams–Moulton formulas.

n	x_n	y_n, predicted	y_n, corrected
4	0.4	0.73312	0.73299
5	0.5	0.70210	0.70197
6	0.6	0.68451	0.68439
7	0.7	0.68038	0.68030
8	0.8	0.68992	0.68984
9	0.9	0.71317	0.71311
10	1.0	0.75012	0.75007
11	1.1	0.80049	0.80044
12	1.2	0.86371	0.86367
13	1.3	0.93891	0.93889
14	1.4	1.02491	

prediction. We will illustrate this method by solving the problem of Example 1.6–1.

For the predictor we use a formula called the *Adams–Bashforth* predictor*, given by

$$y_{n+1} = y_n + \frac{h}{24} (55y'_n - 59y'_{n-1} + 37y'_{n-2} - 9y'_{n-3}). \qquad \textbf{(1.6-5)}$$

This formula requires a knowledge of the derivative at *three* previous points. For these we will use values computed by Heun's method, which appear in Table 1.6–3, except that we will carry the calculations to five decimals. For the corrections we will use the *Adams–Moulton[†] corrector*, given by

$$y_{n+1} = y_n + \frac{h}{24} (9y'_{n+1} + 19y'_n - 5y'_{n-1} + y'_{n-2}). \qquad \textbf{(1.6-6)}$$

For details of the derivations of the last two formulas the reader is referred to Curtis F. Gerald's *Applied Numerical Analysis*, 2d ed., pp. A.14ff (Reading, Mass.: Addison-Wesley, 1978).

Using a step size of $h = 0.1$ we list the values obtained by using the Adams formulas in Table 1.6–4. We can see that the values in this table differ but little from those shown in Table 1.6–3. Note also the close agreement between the predicted and the corrected values in Table 1.6–4. A graph of the results in Table 1.6–4 are shown in Fig. 1.6–3 together with the results from Table 1.6–1 for comparison.

The use of predictor-corrector methods requires some "starting" values. These can be obtained by using the Heun method, as we have

---

* Francis Bashforth (1819–1912), a British mathematician.
† Forest R. Moulton (1872–1952), an American astronomer.

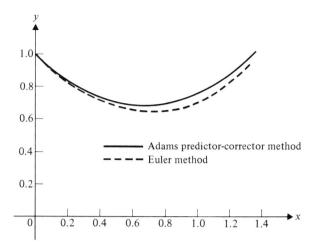

FIGURE 1.6-3 Solutions to Example 1.6–1.

done in Table 1.6–3, since the accuracy of this method is quite good in the neighborhood of the initial point. There are other methods for obtaining starting values and we will look at one of these next.

If we expand $y = F(x)$, the solution of the differential equation

$$\frac{dy}{dx} = f(x, y),$$

into a Taylor's series about $x = x_0$, we have

$$F(x) = F(x_0) + F'(x_0)(x - x_0) + \frac{F''(x_0)}{2!}(x - x_0)^2 + \cdots.$$

Truncating this after three terms gives us an iterative formula for successive values of y, namely,

$$y_{n+1} = y_n + hy'_n + \frac{h^2}{2} y''_n, \qquad \textbf{(1.6–7)}$$

where we have put $h = x - x_0$. Returning to Example 1.6–1, we have

$$y'_n = x_n^2 - y_n^2,$$
$$y''_n = 2x_n - 2y_n y'_n$$
$$= 2x_n - 2x_n^2 y_n + 2y_n^3.$$

Hence, using $h = 0.05$, we have

$$y_{n+1} = y_n + 0.05(x_n^2 - y_n^2) + 0.0025(x_n - x_n^2 y_n + y_n^3).$$

This three-term Taylor's series was used to obtain the values given in Table 1.6–5. A comparison of values in Table 1.6–5 with those in Table 1.6–4 shows excellent agreement.

TABLE 1.6–5 Solution to
Example 1.6–1, $h = 0.05$,
using the three-term
Taylor's series.

n	x_n	y_n
0	0	1.00000
1	0.05	0.95250
2	0.10	0.90954
3	0.15	0.87079
4	0.20	0.83598
5	0.25	0.80492
6	0.30	0.77745
7	0.35	0.75348
8	0.40	0.73293
9	0.45	0.71576
10	0.50	0.70195
11	0.55	0.69149
12	0.60	0.68439
13	0.65	0.68066
14	0.70	0.68031
15	0.75	0.68337
16	0.80	0.68986

If more accuracy per unit of computational effort is required, then
a group of methods known as **Runge–Kutta methods*** is recommended. Of these the most popular is the *fourth-order Runge–Kutta
method*. In contrast to the predictor-corrector methods this is a self-
starting method. It uses a weighted average of the slopes computed at
the points where $x = x_n$, $x = x_n + \frac{1}{2}h$, and $x = x_n + h$.[†] The recursion
formula is given by

$$y_{n+1} = y_n + \frac{h}{6}(y'_{n1} + 2y'_{n2} + 2y'_{n3} + y'_{n4}), \qquad \textbf{(1.6–8)}$$

where

$$y'_{n1} = f(x_n, y_n),$$
$$y'_{n2} = f(x_n + \tfrac{1}{2}h, y_n + \tfrac{1}{2}hy'_{n1}),$$
$$y'_{n3} = f(x_n + \tfrac{1}{2}h, y_n + \tfrac{1}{2}hy'_{n2}),$$
$$y'_{n4} = f(x_n + h, y_n + hy'_{n3}).$$

We illustrate the use of the method in the following example.

---

* After the German mathematicians Carl Runge (1856–1927), who began the work
in 1895, and Martin W. Kutta (1867–1944), who continued it in 1901.

[†] A derivation of the method can be found in Anthony Ralston and Philip Rabinowitz, *A First Course in Numerical Analysis*, 2d ed., pp. 208ff. (New York: McGraw-Hill,
1978).

EXAMPLE 1.6–2 Solve the initial-value problem

$$\frac{dy}{dx} = y^2 - x, \qquad y(1) = 0,$$

by the fourth-order Runge–Kutta method. Use $h = 0.1$ and carry the solution to $x = 2$.

Solution. The solution is shown in Table 1.6–6 and in Fig. 1.6–4. ■

It might appear that a Taylor's series expansion about a point (x_0, y_0) would be useful numerically since any desired number of terms

TABLE 1.6–6 Solution to Example 1.6–2, $h = 0.1$, using the fourth-order Runge–Kutta method.

n	x_n	y_n	y'_{n1}	y'_{n2}	y'_{n3}	y'_{n4}
0	1.0	0	−1.00000	−1.04750	−1.04726	−1.08903
1	1.1	−0.10464	−1.08905	−1.12469	−1.12412	−1.15289
2	1.2	−0.21697	−1.15293	−1.17459	−1.17399	−1.18820
3	1.3	−0.33427	−1.18826	−1.19501	−1.19475	−1.19411
4	1.4	−0.45364	−1.19421	−1.18647	−1.18687	−1.17244
5	1.5	−0.57219	−1.17259	−1.15206	−1.15336	−1.12730
6	1.6	−0.68737	−1.12752	−1.09684	−1.09912	−1.06434
7	1.7	−0.79710	−1.06463	−1.02693	−1.03014	−0.98979
8	1.8	−0.89991	−0.99016	−0.94860	−0.95255	−0.90965
9	1.9	−0.99495	−0.91008	−0.86746	−0.87189	−0.82898
10	2.0	−1.08191				

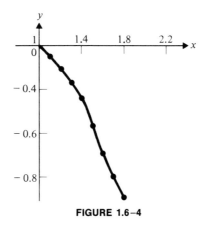

FIGURE 1.6–4

may be generated. We illustrate this method by what might be called a "horrible example."

EXAMPLE 1.6–3 Given the initial-value problem

$$\frac{dy}{dx} = x^2 - y^2, \qquad y(0) = 1,$$

use a Taylor's series expansion about $(0, 1)$ to solve the problem numerically in order to obtain the value of y when $x = 1$.

Solution. From $y' = x^2 - y^2$ we compute successive derivatives:

$$y'' = 2x - 2yy',$$
$$y''' = 2 - 2yy'' - 2(y')^2,$$
$$y^{(iv)} = -2yy''' - 6y'y'', \quad \text{etc.}$$

Thus we have

$$y_0' = -1, \qquad y_0'' = 2, \qquad y_0''' = -4, \qquad y_0^{(iv)} = 20, \quad \text{etc.}$$

Substituting these values into the Taylor's series

$$y = y_0 + (x - x_0)y_0' + (x - x_0)^2 \frac{y_0''}{2!} + (x - x_0)^3 \frac{y_0'''}{3!} + (x - x_0)^4 \frac{y_0^{(iv)}}{4!} + \cdots$$

produces

$$y_n = 1 - x_n + x_n^2 - \tfrac{2}{3}x_n^3 + \tfrac{5}{6}x_n^4 + \cdots. \qquad (1.6-9)$$

The results of using 3, 4, and 5 terms of this series are shown in Table 1.6–7. Figure 1.6–5 shows a comparison of the Euler method and the five-term Taylor's series method. ∎

TABLE 1.6–7 Solution to Example 1.6–3, using the Taylor's series.

n	x_n	y_n (3 terms)	y_n (4 terms)	y_n (5 terms)
0	0	1.00000	1.00000	1.00000
1	0.1	0.91000	0.90933	0.90941
2	0.2	0.84000	0.83467	0.83600
3	0.3	0.79000	0.77200	0.77875
4	0.4	0.76000	0.71733	0.73866
5	0.5	0.75000	0.66667	0.71875
6	0.6	0.76000	0.61600	0.72400
7	0.7	0.79000	0.56133	0.76141
8	0.8	0.84000	0.49867	0.84000
9	0.9	0.91000	0.42400	0.97075
10	1.0	1.00000	0.33333	1.16667

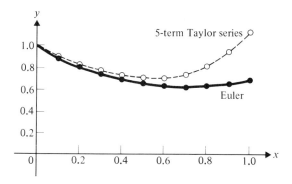

FIGURE 1.6–5 A comparison of the Euler and Taylor series methods.

It should be clear from a comparison of the values in Table 1.6–7 with those in Table 1.6–4 that the Taylor's series method is of questionable value. This is due to the fact that we have gone a considerable distance from the initial point and also that the interval of convergence of the series being used is unknown. Thus there are many factors that must be taken into account when using numerical methods. The numerical analyst is cognizant of the pitfalls and errors involved in the various methods. The errors, in particular, are discussed in detail in P. Henrici's *Discrete Variable Methods in Ordinary Differential Equations* (New York: Wiley, 1962) and *Error Propagation for Difference Methods* (New York: Wiley, 1963).

There is a graphical method that is sometimes helpful for finding the shape of the solution curve of a differential equation. Starting with

$$\frac{dy}{dx} = f(x, y),$$

we can graph the curves

$$f(x, y) = m,$$

where m is a constant. These curves are called **isoclines** (that is, having the same inclination) and everywhere on such a curve the slope dy/dx is the same. Such a family of isoclines is shown in Fig. 1.6–6 for the problem in Example 1.6–1, the short lines having the constant slope indicated by the value of m shown. Thus, starting at the given initial-value point $(0, 1)$ with a slope of -1, we are led (with increasing x) to a slope of $-\frac{1}{2}$, 0, $\frac{1}{2}$, and 1 as shown. The solution curve is shown as a solid curve to distinguish it from the isoclines. In using this method the spacing of the isoclines is important to success.

We have emphasized numerical methods at this point because the solutions of first-order differential equations will have great importance as we consider higher-order equations in Chapter 2. We will see,

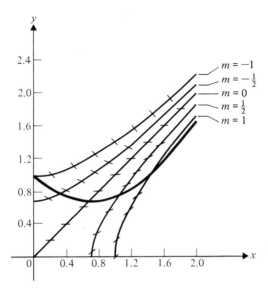

FIGURE 1.6–6 Isoclines for Example 1.6–1.

for example, that second-order equations are reducible to a system of two first-order equations for which the methods of the present section will be applicable.

KEY WORDS AND PHRASES

Riccati equation
Euler method
Heun method

predictor-corrector methods
Runge–Kutta methods
isoclines

EXERCISES 1.6

▶ **1.** Solve the initial-value problem

$$\frac{dy}{dx} = y^2 - x, \qquad y(0) = 0.$$

Use Euler's method with $h = 0.1$ and continue the solution to $x = 4$.

2. Solve the initial-value problem

$$\frac{dy}{dx} = y^2 - x, \qquad y(1) = 0.$$

Use Euler's method with $h = 0.1$ and continue the solution to $x = 4$.

3. Solve the initial-value problem

$$\frac{dy}{dx} = y^2 - x, \qquad y(2) = 0.$$

Use Euler's method with $h = 0.1$ and continue the solution to $x = 4$.

4. In Exercise 3 change the step size to $h = 0.05$ and obtain the solution.

5. Graph the solutions to Exercises 1, 2, and 3 on the same graph and discuss the appearance of the graphs.

6. In the initial-value problem

$$\frac{dy}{dx} = y^2 - x, \qquad y(0) = 0,$$

the term y^2 is small for $0 \le y \le 0.4$. Use this fact to obtain values of y' for $x = 0.05, 0.10, 0.15, 0.20$, and 0.25. Could these results be used as starting values for the Adams method?

7. Obtain a three-term Taylor's series for the initial-value problem of Exercise 2.

8. Use the results of Exercise 7 to solve Exercise 2 by the Adams method. Compare the solution with that of Exercise 2.

9. Solve Exercise 3 by using a three-term Taylor's series.

10. Use a three-term Taylor's series to solve the initial-value problem

$$\frac{dy}{dx} = x + y^2, \qquad y(0) = 1.$$

Use $h = 0.1$ and carry the solution to the point where $x = 2$.

11. Solve the initial-value problem

$$\frac{dy}{dx} = e^y + xy, \qquad y(0) = -1,$$

using $h = 0.1$ to obtain $y(2)$ (a) by Euler's method; (b) by Heun's method.

12. Solve by any method:

$$\frac{dy}{dx} = y - \frac{2x}{y}, \qquad y(0) = 1.$$

Carry the solution to $x = 2$.

13. Find $y(0.5)$ by Euler's method using $h = 0.1$:

$$\frac{dy}{dx} = x + y + xy, \qquad y(0) = 1.$$

14. With the same step size as that in Exercise 13 find $y(0.5)$ using Heun's method.

15. The first-order linear differential equation

$$\frac{dy}{dx} = x + y$$

can be solved analytically. Solve this equation for $y(0.5)$ with $y(0) = 1$ by

a) Euler's method with $h = 0.1$,

b) Euler's method with $h = 0.05$,

c) Heun's method with $h = 0.1$,

d) Heun's method with $h = 0.05$;

and compare your results with the analytical solution $y(0.5) = 1.79744$.

16. Use starting values from Exercise 15 to obtain $y(0.5)$ in that exercise by

a) the Adams method with $h = 0.1$;

b) the Runge–Kutta method with $h = 0.1$.

17. Obtain five terms of the Taylor's series that represents the solution to

$$\frac{dy}{dx} = x + y, \qquad y(0) = 1.$$

18. Use the series obtained in Exercise 17 to compute $y(0.5)$. Comment on the possible error in this computation.

19. Use the method of isoclines to obtain a qualitative sketch of the general solution to

$$\frac{dy}{dx} = -0.1xy.$$

20. Use the method of isoclines to investigate the nature of the solutions of the logistic equation (Section 1.5)

$$\frac{d}{dt} N(t) = (1 - N(t))N(t)$$

for $N(0) = 0.25, 0.75, 1$, and 1.5.

21. Sketch some of the isoclines for

$$\frac{dy}{dt} + y^3 = 1.$$

22. Find $y(0.6)$ if

$$\frac{dy}{dt} + y^3 = 1, \qquad y(0) = 0,$$

a) by Euler's method with $h = 0.1$;

b) by Heun's method with $h = 0.05$.

▶▶ 23. Use the method of isoclines to show that solutions of

$$\frac{dy}{dx} = 0.1xy$$

exhibit symmetry about both axes.

24. By setting y' and y'' equal to zero and then using the method of isoclines, graph some of the solutions of the nonlinear differential equation

$$y' + y^4 = 16.$$

25. Use the method of isoclines to obtain solutions of

$$\frac{dy}{dx} = x^2 + y^2$$

that pass through (a) (1, 2) and (b) (0, 0).

1.7 SOME THEORETICAL CONSIDERATIONS

In the preceding sections we have stressed problem solving in our study of first-order ordinary differential equations and have touched only lightly on the theoretical aspects. This is in accordance with our objective to provide the mathematical tools necessary to solve problems in various areas. Now we will present a brief discussion of three important mathematical topics, namely, **existence**, **uniqueness**, and **continuity**.

Given a differential equation, it is not entirely a trivial question to ask: Does the equation *have* a solution? Some problems do not have solutions, at least not *real* solutions. For example, it is not possible to find real solutions to

$$\left(\frac{dy}{dx}\right)^2 + 1 = 0$$

or

$$\frac{dy}{dx} = \sqrt{1 - y^2}, \qquad y\left(\frac{\pi}{2}\right) = 3.$$

Of course we know that solutions can be found (at least up to a point) to first-order *linear* differential equations, as discussed in Section 1.3. For nonlinear equations, however, it would appear that we must examine the subject of existence of solutions. Hand in hand with existence is the subject of *uniqueness*. In other words, once we have obtained a solution by some method are we assured that it is the *only* solution? Might not a different method produce a different solution? Might there not be an infinite number of solutions? If the latter were the case, then finding a solution would not have much significance. Uniqueness is all-important when numerical methods are used, since otherwise it would be entirely possible to jump from one solution to another as an iterative procedure is used.

There is a fundamental theorem that answers questions about existence and uniqueness. It can be stated as follows.

Theorem 1.7–1 Given a first-order differential equation

$$\frac{dy}{dx} = f(x, y)$$

with initial condition $y(x_0) = y_0$, a sufficient condition that a unique solution $y = F(x)$ exist is that $f(x, y)$ and $\partial f/\partial y$ be real, finite, single-valued, and continuous in some rectangular region of the xy-plane containing the point (x_0, y_0).

We will not give a proof of this theorem since one can be found in most texts dealing with differential equations. See, for example, Martin Braun's *Differential Equations and Their Applications*, 2d ed. (New York: Springer-Verlag, 1978).

Note that the theorem does not say *how* $F(x)$ may be found so that the methods of solution previously discussed will continue to be of practical value. Moreover, the conditions given are sufficient but not necessary. Actually the continuity of $\partial f / \partial y$ may be replaced by a somewhat weaker condition and the theorem will still hold. Finally, the existence of a solution requires only the continuity of $f(x, y)$ but it seems desirable to combine existence and uniqueness in the theorem as stated.

Theorem 1.7–1 can be rephrased in more concise mathematical language as follows.

Theorem 1.7–1(a) Given that the functions $f(x, y)$ and $\partial f / \partial y$ are continuous in some rectangle, $x_1 < x < x_2, y_1 < y < y_2$, which contains the point (x_0, y_0). Then, in some interval (see Fig. 1.7–1) $x_0 - \delta < x < x_0 + \delta$ contained in $x_1 < x < x_2$, there exists a unique solution $y = F(x)$ of the initial-value problem

$$\frac{dy}{dx} = f(x, y), \qquad y(x_0) = y_0.$$

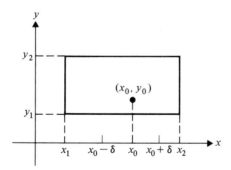

FIGURE 1.7–1 A unique solution interval.

EXAMPLE 1.7–1 Find the solution of

$$\frac{dy}{dx} = y^{2/3}, \qquad y(0) = 0.$$

Solution. Since the variables can be separated, we can obtain the solution $y = x^3/27$. It is apparent, however, that $y = 0$ is also a solution so that neither solution is unique. In this case $f(x, y) = y^{2/3}$ so that $\partial f/\partial y$ is not defined at $y = 0$; hence the conditions of Theorem 1.7–1 cannot be satisfied in any region that contains some portion of the x-axis. ∎

EXAMPLE 1.7–2 Find the solution of

$$\frac{dy}{dx} = y^2, \qquad y(0) = \frac{1}{2}.$$

Solution. This time the solution is $y = 1/(2 - x)$, which is not continuous at $x = 2$. Hence the quantity δ of Theorem 1.7–1(a) is equal to 2 and the given initial-value problem has a *guaranteed* unique solution if $-2 < x < 2$. Actually a solution exists for all x satisfying $-\infty < x < 2$ (see Fig. 1.7–2). ∎

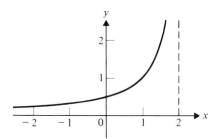

FIGURE 1.7–2 The solution to Example 1.7–2.

It is comforting to know when a solution to an initial-value problem exists even though we may have to labor considerably to determine its nature. It is also important to know that the solution, when found, is unique. There is one more property, however, that we would like the solution to have, namely, *continuity*. In particular, the solution should be a continuous function of the initial condition. Loosely speaking, this means that if the initial value is changed by a "small" amount, then the solution will also change only by a "small" amount. The two "small" amounts need not be equal but are related in the same way that the δ and ε are related in the definition of a continuous function. To expand on this idea we need the following definition.

Definition 1.7–1 Let $f(x, y)$ be continuous in the rectangle $R: x_1 \leq x \leq x_2, y_1 \leq y \leq y_2$. Then $f(x, y)$ is said to satisfy a **Lipschitz condition** in R if

$$|f(\xi, \eta_1) - f(\xi, \eta_2)| \leq k|\eta_1 - \eta_2|$$

for (ξ, η_1) and (ξ, η_2) in R and for some constant k, called a **Lipschitz constant**.

* After Rudolf O. S. Lipschitz (1832–1903), a German mathematician.

Now we can state a theorem that describes the conditions under which the solution to an initial-value problem is continuous.

Theorem 1.7–2 Let $f(x, y)$ satisfy a Lipschitz condition in R. Then $y = F(x)$, the solution of the initial-value problem

$$\frac{dy}{dx} = f(x, y), \qquad y(x_0) = y_0,$$

over the rectangle R, is a continuous function of y_0.

We omit the proof and refer the reader for it and other topics in this section to William R. Derrick and Stanley I. Grossman's *Elementary Differential Equations with Applications*, 2d ed., pp. 420–431 (Reading, Mass.: Addison-Wesley, 1981).

KEY WORDS AND PHRASES

existence
uniqueness
continuity

Lipschitz condition
Lipschitz constant

EXERCISES 1.7

▶ 1. For what values of x and y does each of the following first-order differential equations possess unique solutions?

a) $\dfrac{dy}{dx} = \sqrt{xy}$

b) $\dfrac{dy}{dx} = \log(4 - y^2)$

c) $\dfrac{dy}{dx} = \log|4 - y^2|$

d) $\dfrac{dy}{dx} = \dfrac{1}{x^2 - y^2}$

2. Solve the initial-value problem

$$\frac{dy}{dx} = \frac{1}{2}y, \qquad y(1) = 0.$$

3. Solve each of the following initial-value problems and describe the rectangle in the tu-plane in which a unique solution exists.

a) $t^2 \dfrac{du}{dt} + \dfrac{ut}{\sqrt{(1 - t^2)}} = 1 - \sqrt{1 - t^2}, \quad u(1/2) = 0$

b) $t \dfrac{du}{dt} + u = u \log(ut), \quad u(1) = e$

c) $(1 - t^2) \dfrac{du}{dt} - 2ut = 0, \quad u(0) = 1$

4. Solve

$$\frac{dy}{dx} = xy^{1/2}, \qquad y(0) = 0,$$

thus showing that a solution exists. Show, however, that the solution is not unique. Explain.

5. Show that

$$\frac{dy}{dx} = 2y^{1/2}, \qquad y \geq 0, \qquad y(0) = 0,$$

has two solutions. Graph the solutions. Does this problem violate the uniqueness theorem? Why?

6. Analyze the solution of the equation

$$x\frac{dy}{dx} - y = 0$$

on any interval containing the origin.

▶▶ **7.** A Lipschitz condition is weaker than the condition of continuity; that is, a discontinuous function can satisfy a Lipschitz condition. The following function is discontinuous at each point of the y-axis except at the origin:

$$f(x, y) = \begin{cases} y \sin (1/x), & \text{if } x \neq 0, \\ 0, & \text{if } x = 0. \end{cases}$$

Show that $f(x, y)$ satisfies a Lipschitz condition with Lipschitz constant $k = 1$ on every rectangle.

8. Show that $f(x, y) = xy^2$ satisfies a Lipschitz condition on the unit square centered at the origin. What is the Lipschitz constant?

9. Repeat Exercise 8 for $f(x, y) = y^{2/3}$ over the rectangle $-1 \leq x \leq 1, 1 \leq y \leq 2$.

10. Show that $f(x, y) = xy^2$ does not satisfy a Lipschitz condition if $-1 \leq x \leq 1$ and y is unrestricted.

11. Explain why the initial-value problem

$$\frac{dy}{dt} = ty^2$$

has a unique solution for each of the given initial conditions.

a) $y(0) = 1$ **b)** $y(1) = 1$

c) $y(0) = 0$ **d)** $y(1) = 0$

12. Find the solutions to each of the initial-value problems in Exercise 11.

13. Picard's method* gives an approximation,

$$y_1(x) = y_0 + \int_{x_0}^{x} f(s, y_0(s)) \, ds,$$

to the initial-value problem

$$\frac{dy}{dx} = f(x, y), \qquad y(x_0) = y_0.$$

* After Charles Émile Picard (1856–1941), a French mathematician, who developed it in 1890.

By guessing an initial approximation $y(x) = y_0(x)$ the above method can be iterated and it can be shown that under certain assumptions the process will converge in the neighborhood of x_0. Although the method is not considered a practical numerical method, it is one of the methods used to prove existence in Theorem 1.7–1(a). Use Picard's method for the problem

$$\frac{dy}{dx} = x^2 + xy, \qquad y(0) = 0, \qquad y_0(x) = x^3/3$$

to obtain

$$y_1(x) = \frac{x^2}{3}\left(x + \frac{2x}{5}\right).$$

Then compute $y_2(x)$ and $y_3(x)$. Generalize to obtain $y(x)$ and use the ratio test to show that the resulting series converges for all x.

14. Use Picard's method to solve

$$\frac{dy}{dx} = y - 2xy, \qquad y(1) = 1, \qquad x > 0,$$

and thus show that $y(x) = \exp(x - x^2)$, $x \ge 0$, is a unique solution.

15. Use Picard's method to solve

$$\frac{dy}{dx} = 2xy, \qquad y(0) = 1,$$

and obtain $y(x) = \exp(x^2)$ valid for all x.

16. Consider the problem of Example 1.6–1:

$$\frac{dy}{dx} = x^2 - y^2, \qquad y(0) = 1.$$

Apply Picard's method to this problem and compare the result with those obtained in Section 1.6.

17. Given the initial-value problem

$$\frac{dy}{dx} = f(x, y), \qquad y(x_0) = y_0,$$

with solution $y = F(x)$, determine what is meant by the statement, "The solution $F(x, x_0, y_0)$ is a continuous function of the ordered triple (x, x_0, y_0)."

2
higher-order ordinary differential equations

2.1 HOMOGENEOUS SECOND-ORDER EQUATIONS WITH CONSTANT COEFFICIENTS

In Chapter 1 we discussed the solution of first-order ordinary differential equations:

$$\frac{dy}{dx} = f(x, y).$$

We were able to develop solution techniques for various special forms of the function f. Fortunately, most of these special forms are frequently encountered in such diverse fields as physics, chemistry, mechanics, biology, hydrodynamics, and economics. When necessary we could resort to a number of numerical methods in order to obtain solutions.

When it comes to higher-order differential equations, the complexities become much greater. For example, the most general second-order ordinary differential equation has the form

$$y'' = \phi(x, y, y'),$$

and it should be obvious that only the simplest forms of the function ϕ can be considered. We shall dispose of these forms first.

Since it requires *two* integrations to obtain y from d^2y/dx^2 and since each integration introduces an arbitrary constant, *two* constants of integration appear in the general solution of a second-order equation, as we would expect. Thus in an initial-value problem, initial values of both y and y' must be specified. This is illustrated in the following example.

EXAMPLE 2.1–1 A brick is dropped from the top of a building 80 m high. Find the time it takes the brick to reach the ground and determine its terminal velocity.

Solution. It is convenient to take a coordinate system with y representing the (positive) height of the brick above the ground at any time

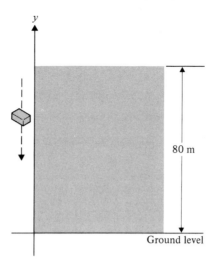

FIGURE 2.1–1 A falling object.

t, as shown in Fig. 2.1–1. Since the brick is "dropped" we can assume that the initial velocity is zero, that is,

$$\dot{y}(0) = 0,$$

where we have indicated differentiation with respect to t by a dot. We also have

$$y(0) = 80$$

so that there are enough conditions given to find the two constants of integration. If we neglect air resistance, the only force on the brick is the (downward) gravitational force, so that by Newton's second law,

$$\ddot{y} = -g = -9.81 \text{ m/sec}^2.$$

Integrating with respect to t gives us

$$\dot{y} = -9.81t + C_1$$

and using the condition that the initial velocity is zero, we find $C_1 = 0$ and $\dot{y} = -9.81t$. Integrating again, we have

$$y = -4.91t^2 + C_2$$

and the condition $y(0) = 80$ leads to $C_2 = 80$ and

$$y(t) = -4.91t^2 + 80.$$

This last expression gives y, the height of the brick above the ground at any time t. Equating y to zero, we find $t = 4$ sec for the time it takes

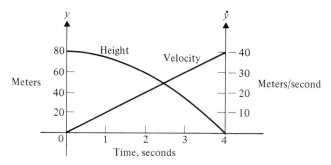

FIGURE 2.1-2 Height and velocity vs. time.

the brick to reach the ground. The velocity at $t = 4$ sec is 39.2 m/sec and this is the terminal velocity. Graphs of y and $\dot{y}$ are shown in Fig. 2.1-2. ■

The most common second-order differential equations are the linear ones with constant coefficients, that is, those of the form

$$y''(x) + ay'(x) + by(x) = f(x),$$

where a and b are real constants. In this section we will consider solutions of the so-called *homogeneous* equation, that is, those for which $f(x) \equiv 0$.

It was probably Euler who first discovered that

$$y = \exp(rx)$$

was a solution of the homogeneous linear second-order differential equation with constant coefficients,

$$y'' + ay' + by = 0, \qquad\qquad (2.1\text{--}1)$$

if the constant r is properly chosen. Substituting this value of y and its first and second derivatives into Eq. (2.1-1) produces

$$r^2 \exp(rx) + ar \exp(rx) + b \exp(rx) = 0$$

or

$$(r^2 + ar + b) \exp(rx) = 0.$$

Since $\exp(rx) \neq 0$, we must have

$$f(r) = r^2 + ar + b = 0. \qquad\qquad (2.1\text{--}2)$$

This quadratic equation is called the **characteristic equation** of Eq. (2.1-1) since it characterizes the type of solution that the differential equation has. We consider three cases according to the three forms of the discriminant, $a^2 - 4b$, of Eq. (2.1-2).

Case I $a^2 - 4b > 0$ (Fig. 2.1–3).

The roots of Eq. (2.1–2) are *real* and *unequal* in this case. If these roots are designated by r_1 and r_2, then $y_1 = \exp(r_1 x)$ and $y_2 = \exp(r_2 x)$ are two solutions of Eq. (2.1–1). Since, for any solution $y(x)$ of the differential equation, $cy(x)$ is also a solution for an arbitrary constant c (Exercise 1), the two solutions y_1 and y_2 above are "different" only if one is *not* a constant times the other. Functions that are "different" in this sense are said to be **linearly independent**. In mathematical terms, y_1 and y_2 are linearly independent if

$$c_1 y_1 + c_2 y_2 = 0$$

can be satisfied only by taking the real or complex constants c_1 and c_2 *both* to be zero.

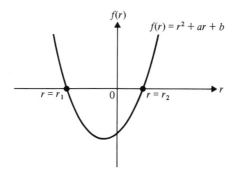

FIGURE 2.1–3 Case I: Real and unequal roots.

We call $c_1 y_1 + c_2 y_2$ a **linear combination** of y_1 and y_2. More precisely, we say that $y_1(x)$ and $y_2(x)$ are linearly independent on an interval $x_0 \leq x \leq x_1$ if, for every x in this interval,

$$c_1 y_1(x) + c_2 y_2(x) = 0$$

holds only if $c_1 = c_2 = 0$. This definition can be immediately extended to any (finite) number of functions. Thus we say the n functions,

$$f_1(x), f_2(x), \ldots, f_n(x),$$

are **linearly dependent** on an interval $x_0 \leq x \leq x_1$ if, for every x in this interval, the equation

$$c_1 f_1 + c_2 f_2 + \cdots + c_n f_n = 0$$

holds with at least *one* of the c_i ($i = 1, 2, \ldots, n$) different from zero. It follows that a set of functions is linearly dependent on an interval if at least *one* of the functions can be expressed as a linear combination of the others on the interval. We will return to this concept in Chapter 4.

In the present case of the functions $\exp(r_1 x)$ and $\exp(r_2 x)$, we have

$$c_1 \exp(r_1 x) + c_2 \exp(r_2 x) = 0$$

and consequently

$$c_1 r_1 \exp(r_1 x) + c_2 r_2 \exp(r_2 x) = 0,$$

the second equation being obtained from the first by differentiation. Recalling Cramer's rule* from algebra, we can solve the above system of linear algebraic equations for unique values of c_1 and c_2 if and only if the determinant of coefficients

$$\begin{vmatrix} \exp(r_1 x) & \exp(r_2 x) \\ r_1 \exp(r_1 x) & r_2 \exp(r_2 x) \end{vmatrix}$$

is *different from zero*. Since the determinant has the value

$$(r_2 - r_1) \exp(r_1 + r_2)x,$$

which *is* nonzero for all x (by virtue of the fact that r_1 and r_2 are unequal), we obtain $c_1 = c_2 = 0$, showing that y_1 and y_2 are linearly independent on every interval. The determinant above is called the **Wronskian**[†] of the functions $\exp(r_1 x)$ and $\exp(r_2 x)$.[‡]

It is now a simple matter to show that if y_1 and y_2 are solutions of Eq. (2.1–1), then so is any linear combination of these functions (see Exercise 2). Hence the *general* solution is given by

$$y = c_1 y_1(x) + c_2 y_2(x)$$

or, in this case,

$$y = c_1 \exp(r_1 x) + c_2 \exp(r_2 x),$$

since this last has two arbitrary constants.

EXAMPLE 2.1–2 Solve the initial-value problem

$$y'' - y' - 6y = 0, \qquad y(0) = 3, \qquad y'(0) = -4.$$

Solution. The characteristic equation is

$$r^2 - r - 6 = 0$$

and its roots are $r_1 = -2$ and $r_2 = 3$. Hence the general solution is

$$y = c_1 \exp(-2x) + c_2 \exp(3x).$$

Differentiating this gives us

$$y' = -2c_1 \exp(-2x) + 3c_2 \exp(3x)$$

---

* See Theorem 4.2–1.
† After Hoene Wronski (1778–1853), a Polish mathematician and philosopher.
‡ The Wronskian of $f(x)$ and $g(x)$ is also written $W(f(x), g(x))$, a simplified notation for the 2×2 determinant.

and substituting the initial conditions into the last two equations produces the system

$$c_1 + c_2 = 3$$
$$-2c_1 + 3c_2 = -4.$$

This system yields $c_1 = 13/5$ and $c_2 = 2/5$ so that the solution of the given problem is

$$y = \frac{1}{5}(13 \exp(-2x) + 2 \exp(3x)). \quad \blacksquare$$

Case II $a^2 - 4b = 0$ (Fig. 2.1–4).

In this case the roots of the characteristic equation are *equal*, that is, $r_1 = r_2 = r$, say. Thus this method yields only *one* solution, namely,

$$y_1 = \exp(rx),$$

where $r = -a/2$. We use the method of **variation of parameters** originated by Joseph Louis Lagrange (1736–1813), a French mathematician, to find a second solution, $y_2(x)$. Let

$$y_2(x) = u(x)y_1(x) = u(x) \exp(rx).$$

$$y = e^{rx}(c_1 + c_2 x)$$

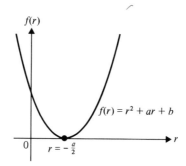

$f(r)$

$f(r) = r^2 + ar + b$

0

$r = -\frac{a}{2}$

r

FIGURE 2.1–4 Case II: Real and equal roots.

Then omitting the subscript and the argument x, we have

$$y' = ur \exp(rx) + u' \exp(rx)$$

and

$$y'' = ur^2 \exp(rx) + 2u'r \exp(rx) + u'' \exp(rx).$$

Substituting into Eq. (2.1–1) and collecting terms produces

$$u'' + u'(2r + a) + u(r^2 + ar + b) = 0.$$

But both terms in parentheses are zero, the first because of the relation $r = -a/2$ and the second because it is the left member of the characteristic equation of the differential equation. Hence $u'' = 0$, which has solution $u = c_2 x$ (the second constant of integration may be omitted) (see Exercise 3) so that the second solution in this case is

$$y_2(x) = c_2 x \exp\left(-\frac{a}{2}x\right)$$

and the general solution can now be written

$$y(x) = c_1 \exp\left(-\frac{a}{2}x\right) + c_2 x \exp\left(-\frac{a}{2}x\right)$$

or

$$y(x) = \exp\left(-\frac{a}{2}x\right)(c_1 + c_2 x). \tag{2.1–3}$$

In Exercise 4 the student is asked to show that $\exp(rx)$ and $x \exp(rx)$ are linearly independent for all x.

We will examine the method of variation of parameters more closely in Section 2.3 in connection with finding solutions of non-homogeneous equations. For the present, we remark that the method is also known as the method of *reduction of order* since a solution to a *second*-order equation was obtained by solving a *first*-order equation, namely, $u' = c_2 x$.

EXAMPLE 2.1–3   Solve the initial-value problem

$$y'' - 8y' + 16y = 0, \qquad y(0) = 1, \qquad y'(0) = 6.$$

Solution. We have the characteristic equation

$$r^2 - 8r + 16 = 0,$$

which has the double root $r = 4$. Hence the general solution, by Eq. (2.1–3), is

$$y(x) = (c_1 + c_2 x) \exp(4x).$$

We observe from Eq. (2.1–3) and its derivative that

$$c_1 = y(0)$$

and

$$c_2 = y'(0) - \frac{a}{2}y(0)$$

so that the solution to the initial-value problem is

$$y(x) = (1 + 2x) \exp(4x). \quad \blacksquare$$

Case III $a^2 - 4b < 0$ (Fig. 2.1–5).

This is the case when the characteristic equation has complex roots that we know occur in complex conjugate pairs. Accordingly, let $r_1 = \alpha + \beta i$ and $r_2 = \alpha - \beta i$ be the roots of $r^2 + ar + b = 0$. Then the two solutions of Eq. (2.1–1) are given by

$$y_1(x) = \exp (\alpha + \beta i)x$$

[handwritten: $r = \dfrac{-a \pm \sqrt{a^2 - 4b}}{2}$]

and

$$y_2(x) = \exp (\alpha - \beta i)x$$

so that the general solution is

[handwritten: not this one] $\quad y(x) = (C_1 \exp (\beta ix) + C_2 \exp (-\beta ix)) \exp (\alpha x).$ **(2.1–4)**

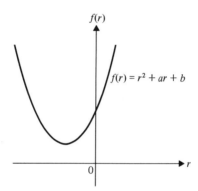

FIGURE 2.1–5 Case III: Complex roots.

This can be put into a more tractable form. To do this we use *Euler's formula*,

$$\exp (i\theta) = \cos \theta + i \sin \theta,$$

and replace C_1 and C_2 by $\frac{1}{2}(c_1 - ic_2)$ and $\frac{1}{2}(c_1 + ic_2)$, respectively. The justification for this last is that the two constants in the general solution are *arbitrary* and by choosing C_1 and C_2 in this manner we obtain *real* solutions to Eq. (2.1–1). Then Eq. (2.1–4) becomes

$$y(x) = \left(\left(\frac{c_1 - ic_2}{2} \right)(\cos \beta x + i \sin \beta x) \right.$$

$$\left. + \left(\frac{c_1 + ic_2}{2} \right)(\cos \beta x - i \sin \beta x) \right) \exp (\alpha x)$$

$$= (c_1 \cos \beta x + c_2 \sin \beta x) \exp (\alpha x), \qquad \textbf{(2.1–5)}$$

which is the most useful form. See Exercise 6, which shows that $\exp(\alpha x)$ $\cos \beta x$ and $\exp(\alpha x) \sin \beta x$ are linearly independent for all x. See also Exercise 20.

EXAMPLE 2.1–4 Solve the initial-value problem

$$y'' - 4y' + 13y = 0, \qquad y(0) = 1, \qquad y'(0) = 8.$$

Solution. The characteristic equation

$$r^2 - 4r + 13 = 0$$

has roots $2 \pm 3i$. Thus the general solution, by Eq. (2.1–5), is

$$y = \exp(2x)(c_1 \cos 3x + c_2 \sin 3x).$$

We also have

$$y' = \exp(2x)(-3c_1 \sin 3x + 3c_2 \cos 3x)$$
$$+ 2 \exp(2x)(c_1 \cos 3x + c_2 \sin 3x),$$

hence the initial conditions produce the simultaneous equations

$$c_1 = 1$$
$$2c_1 + 3c_2 = 8,$$

and the required solution is

$$y(x) = \exp(2x)(\cos 3x + 2 \sin 3x). \quad \blacksquare$$

In this section we considered second-order linear differential equations and made two simplifying assumptions: first, the coefficients were constant, and second, the equations were homogeneous. In Section 2.3 we will consider nonhomogeneous equations with constant coefficients. In Section 2.2 we extend the methods of this section to higher-order equations.

KEY WORDS AND PHRASES

characteristic equation   **linearly dependent**
linear combination   **Wronskian**
linearly independent   **variation of parameters**

EXERCISES 2.1

▶ **1.** Show that if $y_1(x)$ satisfies

$$y'' + ay' + by = 0,$$

where a and b are constants, then $cy_1(x)$ where c is a constant also satisfies the differential equation.

2. Show that if $y_1(x)$ and $y_2(x)$ are *any* two solutions of Eq. (2.1–1), then so is a linear combination of these functions.

3. In Case II write the solution to $u'' = 0$ as $u = c_2 x + c_3$. Use this function to find the general solution and compare with Eq. (2.1–3).

4. Show that the functions $\exp(rx)$ and $x \exp(rx)$ are linearly independent for all x by computing their Wronskian.

5. Show that if the characteristic equation has a double root, then the constants in Eq. (2.1–3) are given by

$$c_1 = y(0),$$

$$c_2 = y'(0) - \frac{a}{2} y(0).$$

6. Show that the functions $\exp(\alpha x) \cos \beta x$ and $\exp(\alpha x) \sin \beta x$ are linearly independent for all x by computing their Wronskian.

7. Carry out the computational details required to obtain Eq. (2.1–5).

8. Find the general solution of each of the following equations.
   a) $y'' - 3y' + 2y = 0$
   b) $y'' - 6y' + 9y = 0$
   c) $y'' - 6y' + 25y = 0$

9. Find the general solution of

$$y'' - 3y' = 0.$$

10. Solve each of the following completely.
    a) $y'' + 2y' + 2y = 0$
    b) $y'' + y' + 2y = 0$
    c) $8y'' + 4y' + y = 0, \quad y(0) = 0, \quad y'(0) = 1$
    d) $x'' + 4x = 0, \quad x(\pi/4) = 1, \quad x'(\pi/4) = 3$

In Exercises 11–13, solve the given initial-value problem.

11. $y'' + 3y' + 2y = 0, \quad y(0) = 0, \quad y'(0) = 2$

12. $y'' + 9y = 0, \quad y(0) = 2, \quad y'(0) = 9$

13. $y'' - 4y' + 4y = 0, \quad y(0) = 3, \quad y'(0) = -6$

14. Find the general solutions in each case.
    a) $\ddot{y} - 4y = 0$, the dot representing d/dt
    b) $u'' + 6u' + 9u = 0$, with $u = u(v)$
    c) $y'' + 5y' = 0$, with $y = y(r)$
    d) $y'' - 2y' - 35y = 0$, with $y = y(x)$

15. Solve each of the following initial-value problems.
    a) $x'' + x' - 3x = 0, \quad x(0) = 0, \quad x'(0) = 1$
    b) $u'' + 5u' + 6u = 0, \quad u(0) = 1, \quad u'(0) = 2$
    c) $\ddot{\theta} + 2\pi\dot{\theta} + \pi^2\theta = 0, \quad \theta(1) = 1, \quad \dot{\theta}(1) = 1/\pi$
    d) $4y'' + 20y' + 25y = 0, \quad y(0) = 1, \quad y'(0) = 2$

16. Solve Example 2.1–1 given that the initial velocity of the brick is 5 m/sec.

17. Show that the general solution of

$$y'' - y = 0$$

can be written

$$y(x) = c_1 \cosh x + c_2 \sinh x.$$

Hint:

$$\cosh x = \frac{e^x + e^{-x}}{2},$$

$$\sinh x = \frac{e^x - e^{-x}}{2}.$$

See Fig. 2.1–6.

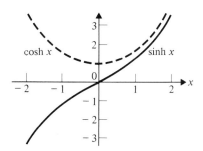

FIGURE 2.1–6 Hyperbolic functions.

▶▶▶ **18.** Obtain the general solution to Example 2.1–1 for the initial conditions $y(0) = s_0$, $y'(0) = v_0$, where s_0 and v_0 are constants.

19. Find the solution of

$$y'' - y' - 6y = 0, \qquad y(0) = -1, \qquad y(1) = 1.$$

(*Note:* In contrast with an *initial-value* problem, the above problem is called a two-point *boundary-value* problem.)

20. a) In Case III show that the only solutions of the form $y = \exp(rx)$ are

$$y_1(x) = (\cos \beta x + i \sin \beta x) \exp(\alpha x) = u(x) + iv(x)$$

and

$$y_2(x) = (\cos \beta x - i \sin \beta x) \exp(\alpha x) = u(x) - iv(x).$$

Here $u(x)$ is the *real part* of $y_1(x)$ and $v(x)$ is the *imaginary part* of $y_1(x)$. Note that both the real and imaginary parts are *real* functions.

b) Writing $u(x) = \operatorname{Re}(y_1)$ and $v(x) = \operatorname{Im}(y_1)$, show that $u(x) = \operatorname{Re}(y_1) = \operatorname{Re}(y_2)$ satisfies the differential equation (2.1–1).

c) Similarly, show that $v(x) = \operatorname{Im}(y_1) = -\operatorname{Im}(y_2)$ satisfies Eq. (2.1–1).

d) Show that $(\cos \beta x) \exp(\alpha x)$ and $(\sin \beta x) \exp(\alpha x)$ have a nonvanishing Wronskian on every interval.

e) Write the general solution of Eq. (2.1–1).

2.2 HOMOGENEOUS HIGHER-ORDER EQUATIONS WITH CONSTANT COEFFICIENTS

We mentioned earlier that there is no unified theory that can be applied to solve ordinary differential equations. Instead, the *type* of an equation determines the *method* of solution. In Section 2.1 we considered the homogeneous second-order equations with constant coefficients. We will see in Section 2.3 that the methods presented for solving homogeneous equations are essential for solving nonhomogeneous equations as well. In this section we will extend the methods of Section 2.1 to higher-order homogeneous equations.

Consider the nth-order equation with (real) constant coefficients:

$$y^{(n)}(x) + a_1 y^{(n-1)}(x) + \cdots + a_{n-1} y'(x) + a_n y(x) = 0. \qquad \textbf{(2.2–1)}$$

There is no loss of generality in taking the coefficient of $y^{(n)}(x)$ to be unity since if it were some other nonzero constant, we could divide the equation by it. Substituting $y = \exp(rx)$ into the above gives us the *characteristic equation*:

$$r^n + a_1 r^{n-1} + \cdots + a_{n-1} r + a_n = 0. \qquad \textbf{(2.2–2)}$$

The roots of this polynomial equation may be real and unequal; some may be equal and some may be complex, occurring in complex conjugate pairs. All these possible cases were discussed in Section 2.1 and the results obtained there are applicable now. We keep in mind that the general solution of Eq. (2.2–1) must contain n arbitrary constants. Stated another way, if we can find a set of n linearly independent solutions of Eq. (2.2–1), then the general solution may be expressed as a linear combination of these. Some examples will clarify the extension of the methods of Section 2.1.

EXAMPLE 2.2–1   Find the general solution of

$$y^{(v)} - 2y^{(iv)} + 2y''' = 0.$$

Solution. Here the characteristic equation is

$$r^5 - 2r^4 + 2r^3 = 0$$

or

$$r^3(r^2 - 2r + 2) = 0.$$

Thus $r = 0$ is a **three-fold root** (or a root of multiplicity three) and the other roots are $r = 1 \pm i$. With this information we can write the general solution,

$$y = c_1 + c_2 x + c_3 x^2 + (c_4 \cos x + c_5 \sin x) \exp x.$$

We have here a special case ($r = 0$) of the fact that if r is a three-fold root of the characteristic equation, then the functions

$$y_1 = \exp(rx), \qquad y_2 = x \exp(rx), \qquad y_3 = x^2 \exp(rx)$$

are linearly independent solutions. See John H. Staib's *An Introduction to Matrices and Linear Transformations*, p. 229 (Reading, Mass.: Addison-Wesley, 1969) for a proof that a nonzero Wronskian implies the linear independence of a set of functions. In this case the Wronskian is

$$\begin{vmatrix} y_1 & y_2 & y_3 \\ y_1' & y_2' & y_3' \\ y_1'' & y_2'' & y_3'' \end{vmatrix}$$

and we will show later that this determinant has the value $2 \exp(3rx)$ (see Exercise 34 in Section 4.2). ∎

It should be apparent that the last example presented no difficulty since we were able to find the roots of the fifth-degree characteristic equation by using the quadratic formula. Niels Henrik Abel (1802–1829) proved in 1824 that, *in general*, it is not possible to find the roots of polynomial equations of degree greater than four by means of a finite number of algebraic operations involving the coefficients. In other words, there is no algorithm such as the quadratic formula for equations of degree higher than four. Even for polynomial equations of degree four some ingenuity may be required to find the roots, as illustrated in the following example.

EXAMPLE 2.2–2 Find the general solution of the equation

$$y^{(iv)} - 4y''' + 8y'' - 8y' + 4y = 0.$$

Solution. We have the fourth-degree characteristic equation

$$r^4 - 4r^3 + 8r^2 - 8r + 4 = 0.$$

If we can find one quadratic factor of this, then we can obtain the roots. With this in mind, we try (guess) the product:

$$(r^2 + ar + b)(r^2 + cr + d).$$

Expanding this and collecting terms results in

$$r^4 + (a + c)r^3 + (ac + b + d)r^2 + (ad + bc)r + bd$$

and matching the coefficients in the characteristic equation, we find

$$a + c = -4$$
$$ac + b + d = 8$$
$$ad + bc = -8$$
$$bd = 4.$$

This last is a system of *nonlinear* algebraic equations and an easy solution seems hopeless. As in factoring quadratic equations, however, we take a cue from $bd = 4$. We can try $b = 1$ and $d = 4$ or $b = 4$ and

$d = 1$ or the negatives of these. We can also try $b = d = \pm 2$ and this last seems an easier approach. Then the system becomes

$$a + c = -4$$
$$ac + 2b = 8$$
$$ab + bc = -8$$
$$b^2 = 4,$$

and from the first and third equations we have $b = 2$. Thus the system is reducible to

$$a + c = -4$$
$$ac = 4,$$

and this can be solved to obtain $c = -2$ and $a = -2$. Thus the factorization of the characteristic equation is

$$(r^2 - 2r + 2)^2 = 0.$$

We now have repeated complex roots $1 \pm i$, which yield the general solution

$$y = (c_1 \cos x + c_2 \sin x) \exp x + (c_3 \cos x + c_4 \sin x)x \exp x.$$

We emphasize that this example is not a typical one but serves merely to illustrate *one* method that may sometimes be of use. ■

An algorithm from algebra known as **synthetic division** (Fig. 2.2–1) is also very helpful in this phase of the work where we need to find roots of polynomial equations. Recall that if r_1 is a **root** of the polynomial equation (2.2–2), then $r - r_1$ is a **factor** and each time a root is found, division by this factor reduces the degree of the polynomial. The discussion following Example 2.2–2 is, of course, relevant if the roots of the characteristic equation are *integers*. In applications this is seldom the case and it is generally necessary to use a computer. Software packages usually contain a subroutine for obtaining roots of polynomial equations.

We conclude with an example of an initial-value problem.

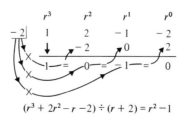

$$(r^3 + 2r^2 - r - 2) \div (r + 2) = r^2 - 1$$

FIGURE 2.2–1 An example of synthetic division.

EXAMPLE 2.2–3 Solve the initial-value problem

$$y''' + 2y'' - y' - 2y = 0, \qquad y(0) = 3, \qquad y'(0) = 4, \qquad y''(0) = 6.$$

Solution. The characteristic equation

$$r^3 + 2r^2 - r - 2 = 0$$

has one root, $r = -2$, which can be found by trial. (See also Exercise 3.) Hence one factor is $r + 2$ and we find by division that the other factor is $r^2 - 1$ so that the roots are $r = -2$ and ± 1 and the general solution is

$$y = c_1 \exp(-2x) + c_2 \exp(-x) + c_3 \exp x.$$

Differentiating twice gives us

$$y' = -2c_1 \exp(-2x) - c_2 \exp(-x) + c_3 \exp x,$$
$$y'' = 4c_1 \exp(-2x) + c_2 \exp(-x) + c_3 \exp x,$$

and substituting the initial values yields the following system of equations:

$$c_1 + c_2 + c_3 = 3$$
$$-2c_1 - c_2 + c_3 = 4$$
$$4c_1 + c_2 + c_3 = 6.$$

In Chapter 4 we will present a method (Gaussian elimination) for solving systems of linear algebraic equations. For the present, however, we will rely on Cramer's rule studied in algebra or elimination of one or more variables. In the above system, for example, we can eliminate c_2 by adding the first and second equations and also adding the second and third. The resulting two equations can then be solved for c_1 and c_3. Proceeding in this way, we obtain $c_1 = 1$, $c_2 = -2$, and $c_3 = 4$ and the solution to the initial-value problem is

$$y = \exp(-2x) - 2\exp(-x) + 4\exp x. \quad \blacksquare$$

KEY WORDS AND PHRASES

three-fold root   **factor**
synthetic division

EXERCISES 2.2

▶   1. Show that the functions 1, x, and x^2 obtained in Example 2.2–1 are linearly independent by showing that they have a nonzero Wronskian.

   2. Verify the solution given in Example 2.2–1.

3. If the coefficients of a polynomial equation $p_n(x) = 0$ add to zero, then $x = 1$ is a root. If the coefficients of $p_n(-x) = 0$ add to zero, then $x = -1$ is a root. Use these facts to find all the roots of each of the following characteristic equations.

a) $r^3 + 2r^2 - r - 2 = 0$
b) $r^3 + r^2 - 5r + 3 = 0$
c) $r^4 + 3r^3 + 3r^2 + 3r + 2 = 0$
d) $2r^4 - 7r^3 - 2r^2 + 13r + 6 = 0$

4. Show that the functions 1, $\cos bx$, and $\sin bx$ are linearly independent if b is a nonzero constant.

5. Solve the initial-value problem

$$y''' + 6y'' + 9y' = 0, \qquad y(0) = 2, \qquad y'(0) = 6, \qquad y''(0) = 9.$$

6. Find the general solution of the equation

$$y''' - 2y'' - y' + 2y = 0.$$

In Exercises 7–15, find the general solution in each case.

7. $y^{(iv)} - 8y'' + 16y = 0$
8. $y''' - 2y'' - 4y' + 8y = 0$
9. $y''' + 9y' = 0$
10. $y^{(iv)} + 4y'' = 0$
11. $y''' - y'' - 4y' + 4y = 0$
12. $y^{(iv)} + 5y'' - 36y = 0$
13. $y^{(iv)} + 4y''' = 0$
14. $y''' + 3y'' - y' - 3y = 0$
15. $y''' - 2y'' - y' + 2y = 0$

16. Solve the initial-value problem

$$y''' - 3y'' + 9y' + 13y = 0, \qquad y(0) = 0, \qquad y'(0) = 12, \qquad y''(0) = -6.$$

17. Find the general solution of each of the given equations in which the independent variable is t.

a) $y''' - y'' - y' + y = 0$
b) $u''' + 5u'' - u' - 5u = 0$
c) $x''' - 6x'' + 3x' + 10x = 0$
d) $y^{(v)} - 5y^{(iv)} + 10y''' - 10y'' + 5y' - y = 0$

[*Hint*: Recall Pascal's triangle from algebra (Fig. 2.2–2).]

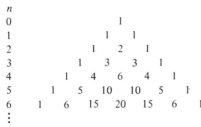

n							
0				1			
1			1		1		
2		1		2		1	
3		1	3		3	1	
4	1		4	6	4		1
5	1	5	10		10	5	1
6	1	6	15	20	15	6	1
⋮							

FIGURE 2.2–2 Pascal's triangle. Each row gives the coefficients in the expansion of $(x + y)^n$. For $(x - y)^n$ the signs alternate: $+ - + - \cdots$.

18. Solve the following initial-value problems in which the independent variable is t.

 a) $y''' - 3y'' + 3y' - y = 0$, $y(0) = 1$, $y'(0) = 2$, $y''(0) = 3$

 b) $y''' - 9y' = 0$, $y(0) = 1$, $y'(0) = 0$, $y''(0) = 2$

 c) $y''' - 3y'' + 4y' - 2y = 0$, $y(0) = 1$, $y'(0) = 2$, $y''(0) = 3$

 d) $x''' - 3x' - 2x = 0$, $x''(0) = 0$, $x'(0) = 9$, $x(0) = 0$

19. Solve the initial-value problem

$$y^{(iv)} + y''' - 4y'' - 4y' = 0,$$

$$y(0) = 5, \qquad y'(0) = 0, \qquad y''(0) = 10, \qquad y'''(0) = 18.$$

20. Find the general solution of each equation.

 a) $y^{(iv)} - 4y''' + 4y'' = 0$ \qquad\qquad b) $y^{(iv)} + 6y'' + 9y = 0$

 c) $y^{(iv)} - 16y = 0$ \qquad\qquad d) $y^{(iv)} - 4y''' + 6y'' - 4y' + y = 0$

▶▶▶ 21. If it is known that $y_1(x) = \cos x \exp(x)$ is one solution of a homogeneous linear differential equation with constant coefficients, then this information can be used to find the general solution. Use this method to find the general solution of

$$y^{(iv)} - y''' - 6y'' + 14y' - 12y = 0.$$

22. Find the general solution of

$$\frac{d^4y}{dx^4} - 5\frac{d^2y}{dx^2} - 10\frac{dy}{dx} - 6y = 0.$$

23. Obtain the general solution of

$$y^{(iv)} + y = 0.$$

24. Solve completely:

$$3y''' - 16y'' + 23y' - 6y = 0.$$

25. Given that ω and ω^2 are both two-fold complex roots of

$$r^5 - r^4 - 2r^3 + 2r^2 + r - 1 = 0,$$

 find ω and ω^2.

2.3 NONHOMOGENEOUS EQUATIONS

In the preceding sections we dealt with solutions of *homogeneous* differential equations with constant coefficients. For example, we found solutions of

$$y''(x) + a_1 y'(x) + a_2 y(x) = 0,$$

where a_1 and a_2 were (real) constants. The general solution of this equation contains two arbitrary constants and we will refer to it as the **complementary solution** (also called the complementary *function*), denoted by $y_c(x)$. It will play an important role as we shall soon see.

In this section we will seek solutions to nonhomogeneous equations, beginning with second-order equations of the form

$$y''(x) + a_1 y'(x) + a_2 y(x) = f(x), \qquad (2.3-1)$$

where a_1 and a_2 are constants. Assume that $y_p(x)$ is *any* solution of this equation. We call $y_p(x)$ a **particular solution** (also called a particular *integral*). It is easy to prove (see Exercise 1) that the *sum* $y_c(x) + y_p(x)$ is a solution of Eq. (2.3–1). Since this sum contains two arbitrary constants it is, by definition, the *general* solution of Eq. (2.3–1), the nonhomogeneous equation. We already know from Section 2.2 how to find $y_c(x)$; consequently, this section will be devoted mainly to finding a particular solution, $y_p(x)$, to Eq. (2.3–1).

We face the same problem here that we have met a number of times before, namely, how to handle the many different forms of a function $f(x)$. This function may be as simple as a constant or it may be piecewise continuous like a sawtooth. We will proceed cautiously and first consider a class of functions that consists of polynomials, exponentials, sines, and cosines. Although this may seem unnecessarily restrictive, functions belonging to the above class occur frequently enough in applications to merit some special attention. This class of functions has the property that it is closed under differentiation, meaning that successive differentiation of these functions produces functions of the same type. It is this property that makes the following **method of undetermined coefficients** a useful one. We illustrate the method with a number of examples.

EXAMPLE 2.3–1 Find a particular solution of the equation

$$y'' + 4y' - y = x^2 - 3x + 4.$$

Solution. Here $f(x)$ is a polynomial—a quadratic. We *assume* a particular solution of the form

$$y_p(x) = Ax^2 + Bx + C$$

and seek to find the constants A, B, and C. Differentiating gives us

$$y_p'(x) = 2Ax + B,$$
$$y_p''(x) = 2A,$$

and substituting into the given differential equation produces

$$2A + 8Ax + 4B - Ax^2 - Bx - C = x^2 - 3x + 4.$$

Next we equate coefficients of like powers of x. The justification for this is that the functions 1, x, and x^2 are linearly independent (see Exercise 1 in Section 2.2); hence the equation

$$c_1 + c_2 x + c_3 x^2 = 0$$

can be satisfied only by taking $c_1 = c_2 = c_3 = 0$.* In the present case we have

$$(2A + 4B - C - 4) + (8A - B + 3)x + (-A - 1)x^2 = 0,$$

which leads to the values $A = -1, B = -5, C = -26$. Thus a particular solution is

$$y_p(x) = -x^2 - 5x - 26. \quad \blacksquare$$

EXAMPLE 2.3–2   Find a particular solution of the equation

$$y'' - 2y' - 3y = 2 \exp x - 3 \exp (2x).$$

Solution. Before solving this example we mention the important **principle of superposition**, which applies to *linear* differential equations of any order. This principle can be stated as follows. If $y_1(x)$ is a solution of

$$y^{(n)}(x) + a_1(x)y^{(n-1)}(x) + \cdots + a_n(x)y(x) = f_1(x)$$

and $y_2(x)$ is a solution of

$$y^{(n)}(x) + a_1(x)y^{(n-1)}(x) + \cdots + a_n(x)y(x) = f_2(x),$$

then $y_3(x) = y_1(x) + y_2(x)$ is a solution of

$$y^{(n)}(x) + a_1(x)y^{(n-1)}(x) + \cdots + a_n(x)y(x) = f_1(x) + f_2(x).$$

You are asked to prove this last statement in Exercise 2 for the case $n = 2$. By induction it can be shown that the principle of super-position applies to the sum of any finite number of functions, $f_1(x), f_2(x), \ldots, f_m(x)$.

Hence to solve the problem of Example 2.3–2 we assume a particular solution

$$y_p(x) = A \exp x + B \exp (2x),$$

and substitute into the differential equation to obtain

$$A \exp x + 4B \exp (2x) - 2A \exp x - 4B \exp (2x) - 3A \exp x - 3B \exp (2x)$$
$$= 2 \exp x - 3 \exp (2x).$$

Comparing coefficients of $\exp x$ and $\exp (2x)$ yields $A = -\frac{1}{2}, B = 1$, and a particular solution,

$$y_p(x) = -\tfrac{1}{2} \exp x + \exp (2x). \quad \blacksquare$$

EXAMPLE 2.3–3   Find a particular solution of the equation

$$y'' - 2y' + y = 3 \sin 2x.$$

---

* See also the discussion on Case I in Section 2.1.

Solution. Since differentiating $\sin 2x$ produces $2 \cos 2x$ we assume a particular solution that includes both $\sin 2x$ and $\cos 2x$, that is, one of the form

$$y_p(x) = A \sin 2x + B \cos 2x.$$

Then differentiating and substituting into the differential equation produces

$$-4A \sin 2x - 4B \cos 2x - 4A \cos 2x + 4B \sin 2x + A \sin 2x + B \cos 2x$$
$$= 3 \sin 2x$$

and this leads to

$$(-3A + 4B) \sin 2x + (-4A - 3B) \cos 2x = 3 \sin 2x.$$

Thus we have the system of equations

$$-3A + 4B = 3$$
$$-4A - 3B = 0,$$

whose solution is $A = -9/25$ and $B = 12/25$ so that a particular solution is

$$y_p(x) = -\tfrac{3}{25}(3 \sin 2x - 4 \cos 2x). \quad \blacksquare$$

In the three previous examples we have been careful not to use functions $f(x)$ in Eq. (2.3–1) that included terms present in the complementary solution (see Exercise 5). If such terms *do* appear in $f(x)$, the above procedure must be modified as shown in the following example.

EXAMPLE 2.3–4   Find the general solution of the equation

$$y'' + 3y' - 4y = 3 \exp x.$$

Solution. Here the complementary solution is

$$y_c(x) = c_1 \exp x + c_2 \exp (-4x)$$

so that we cannot choose for a particular solution a constant times $\exp x$ since we already have such a term present in $y_c(x)$. Instead we assume, as in Case II in Section 2.1, that

$$y_p(x) = Ax \exp x,$$

which leads to

$$(Ax + 2A + 3Ax + 3A - 4Ax) \exp x = 3 \exp x$$

and $A = \tfrac{3}{5}$. With this value the general solution becomes

$$y(x) = c_1 \exp x + c_2 \exp (-4x) + \tfrac{3}{5}x \exp x. \quad \blacksquare$$

We remark that, if both $\exp x$ and $x \exp x$ had appeared in the complementary solution of the last example, we would have assumed that

$$y_p(x) = Ax^2 \exp x$$

for a particular solution (see Exercise 6). This procedure is to be followed in the case of polynomials and sine and cosine functions as well. In other words, we multiply the functions assumed for the particular solution by the *lowest* power of x so that there will be *no duplication* of the functions in the complementary solution.

The method of undetermined coefficients given here is not restricted to second-order equations but can be used with higher-order equations also. It is only necessary that $f(x)$ in Eq. (2.3–1) have the prescribed form. Moreover, *products* of functions can also be treated as shown in the next example.

EXAMPLE 2.3–5   Determine the general solution of the equation

$$y''' + y'' = (\cos x) \exp x.$$

Solution. The complementary solution is easily found to be

$$y_c(x) = c_1 + c_2 x + c_3 \exp (-x).$$

We assume a particular solution,

$$y_p(x) = (A \cos x + B \sin x) \exp x,$$

differentiate, and substitute into the given equation. Then

$$(-4A \sin x - 2A \cos x + 2B \cos x - 2B \sin x) \exp x = (\cos x) \exp x,$$

which leads to

$$2A + B = 0$$
$$-2A + 4B = 1,$$

and $A = -\frac{1}{10}$ and $B = \frac{1}{5}$, so that a particular solution is

$$y_p(x) = -\frac{1}{10}(\cos x - 2 \sin x) \exp x$$

and the general solution is

$$y(x) = c_1 + c_2 x + c_3 \exp (-x) - \frac{1}{10}(\cos x - 2 \sin x) \exp x. \quad \blacksquare$$

A word of caution is in order here concerning initial-value problems. The arbitrary constants can *not* be found from the complementary solution; the general solution must be used.

When $f(x)$ is a product of a polynomial and an exponential, sine, or cosine function, then care must be taken in assuming the proper degree for the polynomial. The polynomial must be multiplied by the lowest possible power of x to ensure that none of the product terms duplicates terms that are present in the complementary solution. We illustrate this in the next example.

EXAMPLE 2.3–6   Find a particular solution of

$$y'' + 4y' + 4y = (2 + x) \exp(-2x).$$

Solution. Since the complementary solution is

$$y_c(x) = c_1 \exp(-2x) + c_2 x \exp(-2x),$$

we must not include these terms in a particular solution. Ordinarily, looking only at the right-hand side of the differential equation, we would assume a particular solution of the form

$$y_p(x) = (Ax + B) \exp(-2x),$$

but here we must multiply by x^2 to obtain

$$y_p(x) = (Ax^3 + Bx^2) \exp(-2x)$$

whose terms do not duplicate any terms in the complementary solution. Then

$$
\begin{aligned}
y_p'(x) &= -2(Ax^3 + Bx^2) \exp(-2x) + (3Ax^2 + 2Bx) \exp(-2x) \\
&= (-2Ax^3 + (-2B + 3A)x^2 + 2Bx) \exp(-2x)
\end{aligned}
$$

and

$$y_p''(x) = (4Ax^3 + (-12A + 4B)x^2 + (-8B + 6A)x + 2B) \exp(-2x).$$

Substituting these values into the given differential equation gives us

$$(6Ax + 2B) \exp(-2x) = (2 + x) \exp(-2x),$$

hence $A = \frac{1}{6}$ and $B = 1$. Thus a particular solution is

$$y_p(x) = (\tfrac{1}{6}x^3 + x^2) \exp(-2x). \qquad \blacksquare$$

We emphasize that we have been dealing with a *special* class of functions: polynomials, exponentials, sines, and cosines. These functions have the remarkable property that derivatives of their sums *and products* are again sums and products. It can be shown that the method illustrated in the above examples can be applied in general if $f(x)$ in Eq. (2.3–1) has the proper form.[*]

The method of undetermined coefficients is convenient to use but it fails in the case when $f(x) = \sec x \tan x$ in Eq. (2.3–1). Consequently,

---

[*] Proof of this, "The undetermined coefficients theorem," can be found in Martinus Esser, *Differential Equations*, p. 51ff. (Philadelphia: W. B. Saunders, 1968).

it is desirable to have a more general method for cases like this. Such a method is the one called **variation of parameters**, which was introduced in Section 2.1. We illustrate with an example before presenting the general method.

EXAMPLE 2.3–7   Find the general solution of the equation

$$y'' + y = \sec x \tan x.$$

Solution. The complementary solution is

$$y_c(x) = c_1 \cos x + c_2 \sin x$$

and we assume a particular solution of the form

$$y_p(x) = u(x) \cos x + v(x) \sin x.$$

Note that this last is just the complementary solution with the arbitrary constants (parameters) c_1 and c_2 replaced by functions $u(x)$ and $v(x)$, respectively. This accounts for the name of the method: variation of parameters. Then, differentiating and simplifying the notation by omitting the independent variable from u and v, we have

$$y_p' = u' \cos x - u \sin x + v' \sin x + v \cos x.$$

At this point it is apparent that some simplification is desirable since differentiating again will introduce terms involving u'' and v''. Hence we *arbitrarily* impose the condition

$$u' \cos x + v' \sin x = 0. \qquad \textbf{(2.3–2)}$$

This will not only simplify y_p' considerably but it will also provide a second equation in the system to be solved for u' and v'. One condition, of course, is that y_p satisfy the given differential equation. Continuing, we have

$$y_p' = -u \sin x + v \cos x$$

and

$$y_p'' = -u' \sin x - u \cos x + v' \cos x - v \sin x.$$

Substituting in the given equation and collecting terms produces

$$-u' \sin x + v' \cos x = \sec x \tan x. \qquad \textbf{(2.3–3)}$$

Now Eqs. (2.3–2) and (2.3–3) can be solved simultaneously to obtain

$$u' = -\tan^2 x \quad \text{and} \quad v' = \tan x.$$

Integration then yields the results

$$u = x - \tan x \quad \text{and} \quad v = -\log |\cos x|$$

and the particular solution

$$y_p(x) = x \cos x - \sin x - \sin x \log |\cos x|.$$

Hence the general solution is

$$y(x) = c_1 \cos x + c_2 \sin x + x \cos x - \sin x - \sin x \log |\cos x|$$

or

$$y(x) = c_1 \cos x + c_2' \sin x + x \cos x - \sin x \log |\cos x|$$

when the terms in $\sin x$ are combined. ∎

The last example can be generalized in the following way. Suppose that $y_1(x)$ and $y_2(x)$ are linearly independent solutions of the homogeneous linear equation

$$y''(x) + a_1(x)y'(x) + a_2(x)y(x) = 0. \qquad (2.3\text{--}4)$$

Then the complementary solution of

$$y''(x) + a_1(x)y'(x) + a_2(x)y(x) = f(x) \qquad (2.3\text{--}5)$$

can be written

$$y_c(x) = c_1 y_1(x) + c_2 y_2(x).$$

Assume a particular solution of the form

hod of Variation

$$y_p(x) = u(x)y_1(x) + v(x)y_2(x).$$

Then

Parameters $y_p'(x) = u(x)y_1'(x) + v(x)y_2'(x) + u'(x)y_1(x) + v'(x)y_2(x)$

and setting

$$u'(x)y_1(x) + v'(x)y_2(x) = 0 \qquad (2.3\text{--}6)$$

gives us

$$y_p'(x) = u(x)y_1'(x) + v(x)y_2'(x),$$
$$y_p''(x) = u'(x)y_1'(x) + u(x)y_1''(x) + v'(x)y_2'(x) + v(x)y_2''(x).$$

Substituting the above values of $y_p(x)$, $y_p'(x)$, and $y_p''(x)$ into Eq. (2.3–5) yields

$$u(y_1'' + a_1 y_1' + a_2 y_1) + v(y_2'' + a_1 y_2' + a_2 y_2) + u'y_1' + v'y_2' = f(x).$$

But since y_1 and y_2 are solutions of the homogeneous equation (2.3–4), the terms in parentheses are zero and the last equation reduces to

$$u'(x)y_1'(x) + v'(x)y_2'(x) = f(x).$$

This, together with Eq. (2.3–6), yields the system of equations

$$u'(x)y_1(x) + v'(x)y_2(x) = 0$$
$$u'(x)y_1'(x) + v'(x)y_2'(x) = f(x), \qquad (2.3\text{--}7)$$

which can be solved for $u'(x)$ and $v'(x)$. We are assured that the system (2.3–7) has a unique solution because the determinant of coefficients,

$$W(y_1(x), y_2(x)) = \begin{vmatrix} y_1(x) & y_2(x) \\ y_1'(x) & y_2'(x) \end{vmatrix},$$

is different from zero since $y_1(x)$ and $y_2(x)$ are linearly independent. This determinant, in fact, is the Wronskian of $y_1(x)$ and $y_2(x)$ (see Section 2.1). Thus

$$u'(x) = \frac{\begin{vmatrix} 0 & y_2(x) \\ f(x) & y_2'(x) \end{vmatrix}}{W(y_1(x), y_2(x))} = \frac{-y_2(x)f(x)}{W(y_1(x), y_2(x))}$$

and

$$v'(x) = \frac{\begin{vmatrix} y_1(x) & 0 \\ y_1'(x) & f(x) \end{vmatrix}}{W(y_1(x), y_2(x))} = \frac{y_1(x)f(x)}{W(y_1(x), y_2(x))}.$$

From these we can obtain $u(x)$ and $v(x)$ by integration:

$$u(x) = - \int^x \frac{y_2(s)f(s)\, ds}{W(y_1(s), y_2(s))} \tag{2.3-8}$$

and

$$v(x) = \int^x \frac{y_1(s)f(s)\, ds}{W(y_1(s), y_2(s))}. \tag{2.3-9}$$

Finally, we can substitute the values from Eqs. (2.3–8) and (2.3–9) into $y_p(x)$ and add the result to $y_c(x)$ to obtain the general solution of Eq. (2.3–5). It should be noted that if we add arbitrary constants when solving for $u(x)$ and $v(x)$ from Eqs. (2.3–8) and (2.3–9), then the resulting $y_p(x)$ will be the general rather than a particular solution of Eq. (2.3–5).

The above elegant method called *variation of parameters*, due to Lagrange, is quite general. It can be used irrespective of the coefficients, $a_1(x)$ and $a_2(x)$ in Eq. (2.3–5), since these terms do not enter into the calculation of $y_p(x)$. Further, the method can be extended to linear nonhomogeneous differential equations of higher order in a natural way. When solving a third-order differential equation, for example, we would have another equation in addition to Eqs. (2.3–8) and (2.3–9) and the Wronskian would be a 3×3 determinant. Hence the computational difficulties increase considerably as the order of the equation to be solved increases.

There is another drawback to the method, however, which is present even when solving a second-order equation. The integrands in Eqs. (2.3–8) and (2.3–9) can become so involved, even for the most simple functions $f(x)$, that $u(x)$ and $v(x)$ cannot be found in closed form.

When using the method of variation of parameters it is not recommended that Eqs. (2.3–8) and (2.3–9) be employed. It is much better to proceed as we have done in Example 2.3–7 and thus obtain the two equations involving $u'(x)$ and $v'(x)$. If Eqs. (2.3–8) and (2.3–9) *are* used, then the differential equation to be solved must be in the proper form, as shown in Eq. (2.3–5); that is, the coefficient of $y''(x)$ must be unity.

KEY WORDS AND PHRASES

complementary solution
particular solution
method of undetermined
   coefficients

principle of superposition
method of variation
   of parameters

EXERCISES 2.3

▶ 1. Prove that if $y_p(x)$ is any solution of Eq. (2.3–1) and $y_c(x)$ is the general solution of

$$y''(x) + a_1 y'(x) + a_2 y(x) = 0,$$

then $y(x) = y_c(x) + y_p(x)$ is the general solution of Eq. (2.3–1).

2. Prove that if $y_1(x)$ is a solution of

$$y'' + a_1(x)y' + a_2(x)y = f_1(x)$$

and $y_2(x)$ is a solution of

$$y'' + a_1(x)y' + a_2(x)y = f_2(x),$$

then $y_1(x) + y_2(x)$ is a solution of

$$y'' + a_1(x)y'(x) + a_2(x)y = f_1(x) + f_2(x).$$

3. Find a particular solution of the equation

$$y'' - 2y' - 3y = 2 \exp x - 3 \exp (2x).$$

(Compare Example 2.3–2.)

4. Obtain a particular solution of the equation

$$y'' - 2y' + y = 3 \sin 2x.$$

(Compare Example 2.3–3.)

5. Obtain the complementary solutions for the equations of each of the following examples.

  a) Example 2.3–1

  b) Example 2.3–2

  c) Example 2.3–3

6. Find the general solution of

$$y'' - 2y' + y = 3 \exp x.$$

(Compare Example 2.3–4.)

7. Solve completely:

$$y''' + y'' = e^x \cos x.$$

(Compare Example 2.3–5.)

8. Solve completely:

$$y'' + 4y' + 4y = (2 + x)e^{-2x}.$$

(Compare Example 2.3–6.)

9. Solve Eqs. (2.3–2) and (2.3–3) simultaneously.

10. Show that adding arbitrary constants to Eqs. (2.3–8) and (2.3–9) will produce the general solution of Eq. (2.3–5).

▶▶ 11. Find a particular solution of

$$y''' - 3y'' + 3y' - 3y = 4 \exp x.$$

In Exercises 12–16, indicate the correct form that a particular solution must have in order to solve the problem by the method of undetermined coefficients. Do not find the particular solution.

12. $y'' + y' - 6y = (x^2 - 4) \sin 2x$

13. $y'' - y = 3 \sinh x$ (*Note:* $\sinh x = \frac{1}{2}(e^x - e^{-x})$.)

14. $y'' + 4y = (x^2 - 4) \sin 2x$

15. $y'' - y = (3 + x) \sinh x$

16. $y^{(iv)} - y = \sinh x \cos x$

In Exercises 17–24, find the general solutions of the given differential equations.

17. $y'' - 4y = 4 \exp (2x)$

18. $\dfrac{d^2N}{dt^2} - 4\dfrac{dN}{dt} + 4N = 4 \exp (2t)$

19. $y''' - 2y'' - 4y' + 8y = x \exp (2x)$

20. $y''' - y'' - 4y' + 4y = \exp x + \exp (2x)$

21. $y''' + 2y'' - y' - 2y = 5 \sin 2x$

22. $y^{(iv)} + 2y'' + y = x^2 - 2x$

23. $y'' + 4y' + 4y = 4x^2 - 8x$

24. $4y'' - 4y' - 3y = \cos 2x$

25. Illustrate the principle of superposition by finding the general solution of

$$y'' + 2y' + y = \sin x + \cos 2x.$$

26. Find a particular solution of

$$y'' + 3y' + 2y = 2 \exp (3x).$$

27. Find a particular solution of

$$y'' + 4y = 3 \sin 2x.$$

28. Solve Example 2.3–5 by the method of variation of parameters. Comment on the work required to obtain a solution by the two methods given in this section.

29. Find the general solution of

$$y'' - 4y' + 5y = x \cosh 2x.$$

[*Note:* $\cosh 2x = \frac{1}{2}(\exp (2x) + \exp (-2x))$.]

30. Solve the initial-value problem

$$y'' - 2y' + y = xe^x, \qquad y(0) = 3, \qquad y'(0) = 5.$$

31. Find a particular solution for each of the following equations by the method of variation of parameters.

a) $y'' + y = \tan x$ **b)** $y'' + y = \sec x$ **c)** $y'' - 2y' + y = e^x/x^2$

32. Find the general solution of each of the following equations.

a) $y'' + 2y' + y = (e^x - 1)^{-2}$ **b)** $y'' - 3y' + 2y = e^x/(e^x + 1)$

c) $y'' + 4y = \cot 2x$ **d)** $y'' + 9y = \csc 3x$

33. Solve the initial-value problem

$$y'' - 2y' + y = e^x/(1 - x)^2, \qquad y(0) = 2, \qquad y'(0) = 6.$$

▶▶▶ **34.** Attempt to solve Example 2.3–7 by the method of undetermined coefficients. What are some of the difficulties encountered?

35. What step in the method of variation of parameters explains why this method has also been called "reduction of order"?

36. Verify that $y_1(x) = x$ and $y_2(x) = 1/x$ are solutions of

$$x^3 y'' + x^2 y' - xy = 0.$$

Then use this information to find the general solution of

$$x^3 y'' + x^2 y' - xy = x/(1 + x).$$

37. a) Solve the equation

$$y'' - y = x \sin x$$

by the method of undetermined coefficients. (*Hint:* If the method appears to fail, try the following scheme. Using Euler's formula,

$$\exp(ix) = \cos x + i \sin x,$$

we have

$$\exp(-ix) = \cos x - i \sin x$$

and, subtracting,

$$\sin x = (\exp(ix) - \exp(-ix))/2i.$$

Then use this representation of $\sin x$ and recall that two complex numbers are equal if and only if their real parts are equal and their imaginary parts are equal.)

b) Solve by using the method of variation of parameters.

38. Prove the following Principle of Superposition Theorem: If $F_i(x)$ are solutions of

$$y'' + a(x)y' + b(x)y = f_i(x)$$

for $i = 1, 2, \ldots, m$, then

$$y = F_1(x) + F_2(x) + \cdots + F_m(x)$$

is a solution of

$$y'' + a(x)y' + b(x)y = f_1(x) + f_2(x) + \cdots + f_m(x).$$

2.4 THE CAUCHY–EULER EQUATION

One type of linear differential equation with variable coefficients that can be reduced to a form already considered is the Cauchy–Euler equation, one form of which is

$$x^2 y'' + axy' + by = f(x), \qquad x > 0, \qquad\qquad \textbf{(2.4–1)}$$

where a and b are constants. This equation is also called a *Cauchy** equation, an *Euler* equation, and an *equidimensional* equation. The last term comes from the fact that the physical dimension of x in the left member of Eq. (2.4–1) is immaterial since replacing x by cx where c is a nonzero constant leaves the dimension of the left member unchanged. We will meet a form of this equation later (Chapter 9) in our study of boundary-value problems having spherical symmetry. We will assume throughout this section that $x \neq 0$. For the most part we will assume that $x > 0$ although we shall also deal with the case $x < 0$ later. We begin with an example to illustrate the method of solution.

EXAMPLE 2.4–1   Find the complementary solution of the equation

$$x^2 y'' + 2xy' - 2y = x^2 e^{-x}, \qquad x > 0.$$

Solution. This is a Cauchy–Euler equation and we make the substitutions

$$y_c(x) = x^m, \qquad y_c' = mx^{m-1}, \qquad y_c'' = m(m-1)x^{m-2},$$

so the homogeneous equation becomes

$$(m(m-1) + 2m - 2)x^m = 0.$$

Because of the restriction $x \neq 0$, we must have

$$m^2 + m - 2 = 0,$$

which has roots $m_1 = -2$ and $m_2 = 1$. Thus

$$y_c(x) = c_1 x^{-2} + c_2 x. \quad \blacksquare$$

We remark that the substitution, $y_c(x) = x^m$, did not come from thin air. It was dictated by the *form* of the left member of the differential equation which, in turn, ensured that each term of the equation would contain the common factor x^m.

It should be pointed out also that, if we were interested in obtaining the *general* solution of the equation in Example 2.4–1, we could use the complementary solution above and the method of variation of parameters given in Section 2.3.

---

* Augustin-Louis Cauchy (1789–1857), a French mathematician.

EXAMPLE 2.4–2   Obtain the complementary solution of

$$x^2 y'' + 3xy' + y = x^3, \qquad x > 0.$$

Solution. This time the substitution $y_c = x^m$ leads to

$$m^2 + 2m + 1 = 0,$$

which has a double root $m = -1$. Hence we have *one* solution of the homogeneous equation, namely,

$$y_1(x) = x^{-1}.$$

One might guess that a second linearly independent solution could be obtained by multiplying $y_1(x)$ by x. This procedure, however, is limited to the case of equations with *constant* coefficients and is thus not applicable here. It is easy to check that $y = 1$ is *not* a solution. But we can use the method of reduction of order to find a second solution.

We set $y_2(x) = u(x)/x$ and compute two derivatives. Thus

$$y_2'(x) = \frac{u'x - u}{x^2},$$

$$y_2''(x) = \frac{x^2(xu'' - 2u') + 2ux}{x^4},$$

and substitution into the homogeneous equation results in

$$xu'' + u' = 0,$$

which can be solved* by setting $u' = v$ and separating the variables. Then

$$v = \frac{du}{dx} = \frac{c_2}{x}$$

and

$$u = c_2 \log x.$$

Hence

$$y_2(x) = \frac{c_2}{x} \log x$$

and the complementary solution is

$$y_c(x) = \frac{c_1}{x} + \frac{c_2}{x} \log x.$$

We shall see later that the function $\log x$ is a characteristic one in the case of repeated roots. ∎

EXAMPLE 2.4–3   Find the complementary solution of

$$x^2 y'' + xy' + y = \cos x, \qquad x > 0.$$

---

* An alternative method is to note that $xu'' + u' = d(xu') = 0$, leading to $xu' = c_2$.

Solution. In this example we have, after substituting $y_c(x) = x^m$,

$$m^2 + 1 = 0$$

so that the solutions are x^i and x^{-i}. Hence the complementary solution is

$$y_c(x) = C_1 x^i + C_2 x^{-i}. \tag{2.4-2}$$

A more useful form can be obtained, however, by replacing C_1 and C_2 by $\frac{1}{2}(c_1 - ic_2)$ and $\frac{1}{2}(c_1 + ic_2)$, respectively, and noting that

$$x^i = \exp(i \log x) = \cos(\log x) + i \sin(\log x).$$

Note that the above is similar to the development in Section 2.1. With these changes the complementary solution (2.4–2) can be written

$$y_c(x) = c_1 \cos(\log x) + c_2 \sin(\log x). \quad \blacksquare$$

We shall summarize the various cases that occur when solving the homogeneous Cauchy–Euler equation:

$$x^2 y'' + axy' + by = 0. \tag{2.4-3}$$

Substitution of $y_c(x) = x^m$ and its derivatives into Eq. (2.4–3) leads to the equation

$$m(m-1) + am + b = 0$$

or

$$m^2 + (a-1)m + b = 0. \tag{2.4-4}$$

This is called the **auxiliary equation** of the homogeneous Cauchy–Euler equation (2.4–3).

Case I $(a-1)^2 - 4b > 0$.

The roots of Eq. (2.4–4) are real and unequal, say, m_1 and m_2. Then

$$y_c(x) = c_1 x^{m_1} + c_2 x^{m_2} \tag{2.4-5}$$

since the Wronskian

$$\begin{vmatrix} x^{m_1} & x^{m_2} \\ m_1 x^{m_1 - 1} & m_2 x^{m_2 - 1} \end{vmatrix} = (m_2 - m_1)x^{m_1 + m_2 - 1} \neq 0,$$

showing that x^{m_1} and x^{m_2} are linearly independent.*

Case II $(a-1)^2 - 4b = 0$.

The roots of Eq. (2.4–4) are real and equal, say, $m_1 = m_2 = m$. Then

$$y_1(x) = x^m$$

is one solution of Eq. (2.4–3). A second solution can be found by the method of reduction of order. Let

$$y_2(x) = x^m u(x)$$

* Note that the assumption $x > 0$ is essential here.

be a second solution, differentiate twice, and substitute into Eq. (2.4–3). Then

$$u(m(m-1) + am + b)x^m + u'(2m + a)x^{m+1} + u''x^{m+2} = 0.$$

But the coefficient of u vanishes because x^m is a solution of Eq. (2.4–3) and $2m + a = 1$ from Eq. (2.4–4). Thus

$$xu'' + u' = 0,$$

which is satisfied by $u = \log x$ and

$$y_2(x) = x^m \log x.$$

In this case the complementary solution is

$$y_c(x) = x^m(c_1 + c_2 \log x). \tag{2.4–6}$$

Case III $(a-1)^2 - 4b < 0.$

The roots of Eq. (2.4–4) are complex conjugates, say, $m_1 = \alpha + \beta i$ and $m_2 = \alpha - \beta i$. Then two linearly independent solutions of the homogeneous equation are

$$y_1(x) = x^{\alpha + \beta i} = x^\alpha x^{\beta i} = x^\alpha \exp (i\beta \log x)$$

and

$$y_2(x) = x^{\alpha - \beta i} = x^\alpha x^{-\beta i} = x^\alpha \exp (-i\beta \log x).$$

Using Euler's formula to transform the exponential gives us

$$y_1(x) = x^\alpha(\cos (\beta \log x) + i \sin (\beta \log x))$$

and

$$y_2(x) = x^\alpha(\cos (\beta \log x) - i \sin (\beta \log x)).$$

Hence the complementary solution becomes

$$y_c(x) = x^\alpha(c_1 \cos (\beta \log x) + c_2 \sin (\beta \log x)). \tag{2.4–7}$$

If the general solution to Eq. (2.4–1) is required, it is necessary to add a particular solution to the appropriate complementary solution. A particular solution can be found by the method of variation of parameters, although difficulties may be encountered as previously mentioned in Section 2.3. Note that the method of undetermined coefficients is not applicable here since the Cauchy–Euler differential equation does not have constant coefficients.

There is an alternative method for solving Eq. (2.4–1). Since $x > 0$, we can make the substitution

$$u = \log x.$$

This leads to $x = \exp (u)$ and, using the chain rule, to

$$\frac{dy}{dx} = \frac{dy}{du}\frac{du}{dx} = \frac{1}{x}\frac{dy}{du}.$$

We also have

$$\frac{d^2y}{dx^2} = \frac{d}{dx}\left(\frac{dy}{dx}\right) = \frac{d}{dx}\left(\frac{1}{x}\frac{dy}{du}\right)$$

$$= \frac{1}{x}\frac{d}{du}\left(\frac{dy}{du}\right)\frac{du}{dx} - \frac{1}{x^2}\frac{dy}{du}$$

$$= \frac{1}{x^2}\left(\frac{d^2y}{du^2} - \frac{dy}{du}\right).$$

This method has the advantage that Eq. (2.4–1) is transformed into

$$\frac{d^2y}{du^2} + (a-1)\frac{dy}{du} + by = f(\exp u).$$

Thus the differential equation has constant coefficients and the methods of Section 2.1 are available for finding the complementary solution of the Cauchy–Euler equation. In fact, if $f(\exp u)$ has the proper form, then the method of undetermined coefficients may lead to the general solution of the nonhomogeneous equation quite easily.

The methods of this section can be extended in a natural way to Cauchy–Euler equations of order higher than two. In such equations the degree of the auxiliary equations is increased but the discussion in Section 2.2 is still applicable.

We have considered exclusively the case where $x > 0$. If solutions are desired for values of x satisfying $x < 0$, then x may be replaced by $-x$ in the differential equation and in the solution.

KEY WORDS AND PHRASES

Cauchy–Euler equation　　　　　　　　auxiliary equation

EXERCISES 2.4

▶ In the following exercises, assume that the independent variable is positive unless otherwise stated.

1. Find a particular solution of the equation of Example 2.4–1.
2. Find the general solution of each of the equations given in:
   a) Example 2.4–1;
   b) Example 2.4–2;
   c) Example 2.4–3.

▶▶ 3. Solve the initial-value problem

$$x^2y'' - 2y = 0, \qquad y(1) = 6, \qquad y'(1) = 3.$$

4. Find the general solution of
$$x^2 y'' + 5xy' - 5y = 0.$$

5. Find the general solution of
$$t^2 y'' + 5ty' + 5y = 0.$$

6. Find the complementary solution of
$$x^3 y''' - 2x^2 y'' - 2xy' + 8y = x^2 + 2.$$

7. Find the general solution of
$$r^2 u'' + 3ru' + u = 0.$$

8. Solve the equation
$$t^3 u''' + 3t^2 u'' + tu' = 0.$$

9. Obtain the general solution for
$$x^2 y'' + xy' - 9y = x^2 - 2x.$$

10. Find the general solution of
$$x^2 y'' + xy' + 4y = \log x.$$

11. Solve the initial-value problem
$$x^2 y'' - xy' + 2y = 5 - 4x, \qquad y(1) = 0, \qquad y'(1) = 0.$$

12. Find the general solution of
$$x^3 y''' + 4x^2 y'' - 8xy' + 8y = 0.$$

13. Solve Exercise 12 given the following conditions:
$$y(2) = 6, \qquad y'(2) = -6, \qquad y''(2) = 1.$$

14. Solve each of the following initial-value problems.
   a) $x^2 y'' + xy' = 0, \quad y(1) = 1, \quad y'(1) = 2$
   b) $x^2 y'' - 2y = 0, \quad y(1) = 0, \quad y'(1) = 1$

▶▶▶ 15. Show that the substitution $x = \exp u$ transforms $d^3 y/dx^3$ into
$$\frac{1}{x^3}\left(\frac{d^3 y}{du^3} - 3\frac{d^2 y}{du^2} + 2\frac{dy}{du}\right).$$

16. Use the result of Exercise 15 to solve
$$x^3 y''' + x^2 y'' - 4xy' = 3x^2.$$

17. Solve the initial-value problem
$$ty''' + y'' = 3/t - 5/t^2, \qquad y(1) = 3, \qquad y'(1) = -3, \qquad y''(1) = 9.$$

18. Obtain the general solution for the equation of Exercise 6.

19. Use the method of reduction of order to find a second solution for each differential equation given one solution as shown.
   a) $x^2 y'' - xy' + y = 0, \quad x > 0, \quad y_1(x) = x$
   b) $xy'' + 3y' = 0, \quad x > 0, \quad y_1(x) = 2$

    c) $x^2 y'' + xy' - 4y = 0,\quad x > 0,\quad y_1(x) = x^2$

    d) $x^2 y'' - xy' + y = 0,\quad x > 0,\quad y_1(x) = x \log x^2$

20. Show that the products xy' and $x^2 y''$ remain unchanged if x is replaced by cx, where c is a nonzero constant.

2.5 APPLICATIONS

In Section 1.5 we considered the applications of first-order differential equations to a variety of topics. We stressed not only the methods of solution but also the simplifying assumptions that had to be made in order to obtain solutions. This was done in keeping with our stated objective to provide the tools of mathematical analysis that will allow the modeling of real-life problems. We will follow this procedure in this section as we look at some applications of higher-order equations.

The simple pendulum

Consider a particle of mass m suspended by a "weightless" string of length L. This is a reasonable model of a pendulum if the mass of the particle is very much greater than the mass of the supporting string, which is usually the case. We suppose that the mass is constrained to move only in a vertical plane and that its position is determined by the angle θ measured in radians from a central position, as shown in Fig. 2.5–1. Also shown are the two forces that act on the particle: its weight, mg, and the restoring force, F. The distance traveled by the particle

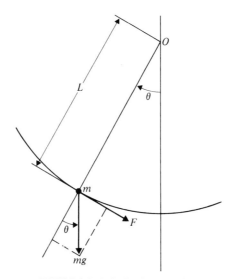

FIGURE 2.5–1 A simple pendulum.

along the arc is $L\theta$, hence the restoring force, F, by Newton's second law is

$$-F = ma = m \frac{d^2}{dt^2} (L\theta) = mL\ddot{\theta},$$

where differentiation with respect to t is denoted by a dot. We see by Fig. 2.5–1 that the directions of F and θ are opposite. Hence one tangential force is the negative of the other, that is, the sum of the *tangential* forces is zero:

$$mL\ddot{\theta} + mg \sin \theta = 0$$

or

$$\ddot{\theta} + \frac{g}{L} \sin \theta = 0. \qquad (2.5-1)$$

In getting to this point we have neglected not only the mass of the supporting string but also the friction of the pivot at O and the air resistance against the mass and string as the mass swings in its arc.

Now we must make another simplifying assumption since Eq. (2.5–1) is a nonlinear differential equation. Recall that the Maclaurin's series expansion of $\sin \theta$ is given by

$$\sin \theta = \theta - \frac{\theta^3}{3!} + \frac{\theta^5}{5!} - + \cdots .$$

This is an alternating series that converges for $-\infty < \theta < \infty$. It is known that the remainder after the nth term does not exceed the absolute value of the $(n + 1)$st term.* Hence in the present case, if we approximate $\sin \theta$ by θ, the first term of the series, the resulting error will be less than $\theta^3/3!$. For an angle of 0.1 radian (about 5.7°) this amounts to one part in 6000 or about 0.017 percent. Accordingly, for reasonably "small" values of θ, we may write Eq. (2.5–1) as

$$\ddot{\theta} + (g/L)\theta = 0$$

and obtain the general solution

$$\theta = c_1 \cos \sqrt{\frac{g}{L}} t + c_2 \sin \sqrt{\frac{g}{L}} t.$$

The two constants can be found by considering the actual operation of a pendulum (in clocks, for example). At $t = 0$ the mass is pulled over to a position $\theta = \theta_0$ and released. Thus the initial conditions are

$$\theta(0) = \theta_0 \quad \text{and} \quad \dot{\theta}(0) = 0,$$

and the solution to the initial-value problem is

$$\theta(t) = \theta_0 \cos \sqrt{\frac{g}{L}} t. \qquad (2.5-2)$$

---

* See the Leibniz theorem for real alternating series in David V. Widder, *Advanced Calculus*, 2d ed., p. 293ff. (Englewood Cliffs, N.J.: Prentice-Hall, 1961).

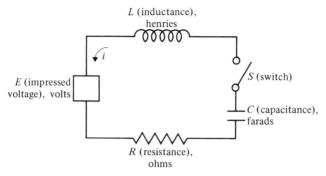

FIGURE 2.5–2 An *RLC* series circuit.

Electric circuits

The circuit in Fig. 1.5–6 contains an inductance and a resistance in series with a voltage source. If we now add a capacitance, C, in series with these, the voltage drop across the capacitance is given by

$$\frac{1}{C}\int i\,dt.$$

The series circuit is shown in Fig. 2.5–2.

Kirchhoff's second law, which states that the sum of the voltage drops around a closed circuit is zero, results in the equation

$$L\frac{di}{dt} + Ri + \frac{1}{C}\int i\,dt = E.$$

This is an *integrodifferential equation*, which can be put into a more recognizable form by differentiating it with respect to t. Thus

$$L\frac{d^2i}{dt^2} + R\frac{di}{dt} + \frac{1}{C}i = 0, \tag{2.5–3}$$

which is a second-order homogeneous equation with constant coefficients when E is a constant (dc) voltage source.

Newton's laws

Consider a mass m suspended from a spring which, in turn, is attached to a fixed support. We assume throughout that **Hooke's law*** holds, which means that the spring force F_s is directly proportional to the displacement y. Another way of saying this is that the displacement

---

* After Robert Hooke (1635–1703), an English physicist, who first published it in 1676 in the form of an anagram.

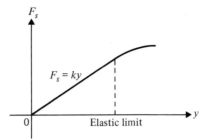

FIGURE 2.5–3 Spring force and elastic limit.

does not take on values that cause the spring to be elongated beyond its elastic limit. Let k denote the *spring constant*; that is, the spring force F_s is given by

$$F_s = ky,$$

where y is the distance the spring is stretched (or compressed) and k is determined experimentally. (See Fig. 2.5–3.)

We assume that the mass is concentrated at a point and we can consider its motion as the motion of a particle. We also ignore the mass of the spring, the air resistance, and the fact that none of the kinetic energy in the spring is lost in heat. The above are usually secondary effects and can thus be neglected in most situations, although it should be realized that the *relative* magnitudes of the quantities involved must be taken into account when making such simplifying assumptions.

In Fig. 2.5–4 we show the situation at time $t = 0$. The mass has been moved from its equilibrium position so that the initial displacement of the mass is given by y_0, that is, $y(0) = y_0$. We seek the displacement at any time t, namely, $y(t)$. If the velocity of the mass is denoted

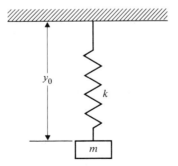

FIGURE 2.5–4 Spring–mass system.

by $v(t)$, then Newton's second law states that the force F acting on a moving particle is equal to the rate of change of its momentum, mv. Hence

$$F = \frac{d(mv)}{dt} = m\frac{dv}{dt} = m\ddot{y},$$

since m is constant. The spring force F_s is given by

$$F_s(t) = -ky(t),$$

the negative sign indicating that the force acts to restore the mass to its equilibrium position and is consequently opposite in sign to the downward (positive) displacement.

When we equate the two forces above, we have

$$m\ddot{y} = -ky$$

and from this we can phrase the initial-value problem,

$$m\ddot{y} + ky = 0, \qquad y(0) = y_0, \qquad \dot{y}(0) = 0. \tag{2.5–4}$$

The solution to the problem above gives the displacement $y(t)$ of an idealized spring–mass system undergoing *free vibrations* (so-called because the equation is homogeneous). Motion such as described by the differential equation in (2.5–4)—that is, where the acceleration is proportional to the negative of the displacement—is called **simple harmonic motion** (Fig. 2.5–5).

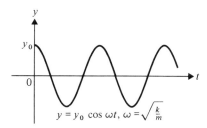

FIGURE 2.5–5 Simple harmonic motion.

If we now add a *viscous damper* (dashpot) to the system, then this adds a damping force given by

$$F_d(t) = -c\dot{y}.$$

Here again the force acts to restore the mass to its equilibrium position but this time the force is proportional to the velocity $\dot{y}$, the constant c being determined experimentally. Figure 2.5–6 shows the system.

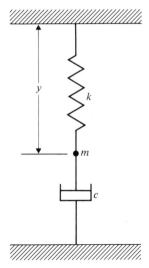

FIGURE 2.5–6 Spring–mass–
damper system.

With prescribed constant initial displacement and initial velocity, the initial-value problem becomes

$$m\ddot{y} + c\dot{y} + ky = 0, \qquad y(0) = y_0, \qquad \dot{y}(0) = v_0. \qquad \textbf{(2.5–5)}$$

If we add an external force $f(t)$ (also called a *forcing function*) acting on the mass in the above systems, then we have examples of *forced vibrations*. In these cases the differential equations are nonhomogeneous and Eqs. (2.5–4) and (2.5–5) become

$$m\ddot{y} + ky = f(t), \qquad y(0) = y_0, \qquad \dot{y}(0) = 0, \qquad \textbf{(2.5–4a)}$$

and

$$m\ddot{y} + c\dot{y} + ky = f(t), \qquad y(0) = y_0, \qquad \dot{y}(0) = v_0, \qquad \textbf{(2.5–5a)}$$

respectively. In many applications the force $f(t)$ is sinusoidal so that the required particular solution can be found by the method of undetermined coefficients. The exercises provide illustrations of the various types of problems encountered in practice. Attention is called to the analogy between a spring–mass–damper system, Eq. (2.5–5), and a capacitance–inductance–resistance system, Eq. (2.5–3). This analogy extends to the simplifying assumptions made in the two cases.

Bending of a beam

In the study of *elasticity* we find some good examples of the simplifying assumptions that must be made in order to put a differential equation into a tractable form. Consider a uniform, cantilever beam, that is,

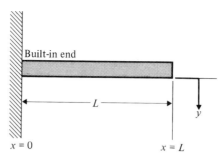

FIGURE 2.5-7 A cantilever beam.

one that has one end embedded in concrete and has constant cross section throughout its length. If there were no load on the beam (including the weight of the beam itself), then an imaginary line connecting the centroids of all cross sections would be perfectly straight. Because of the loading, however, the result is a curve called the *deflection curve* (or elastic curve), having equation $y = f(x)$. For convenience we take $x = 0$ at the built-in end and $x = L$ at the free end and measure the deflection y positive downward. See Fig. 2.5–7.

The Bernoulli–Euler law[*] states that the curvature K of the beam is proportional to the bending moment M, that is, $K \propto M$ or

$$K = \frac{M}{EI},$$

where E is Young's modulus (a function of the material of which the beam is made) and I is the moment of inertia of the cross section with respect to the y-axis. The quantity EI is a measure of the flexural rigidity of the beam.

We point out that there is an underlying assumption in our discussion. On the one hand, we assume that the beam is one-dimensional and we analyze its deflection curve. On the other hand, we must consider moments of inertia of cross sections that are two-dimensional. This type of simplification is common in the theory of structures.[†]

Recall from calculus that the curvature is given by

$$K = \frac{y''}{(1 + (y')^2)^{3/2}}.$$

---

[*] After Jakob (James) Bernoulli (1654–1705), a Swiss mathematician. See I. S. Sokolnikoff, *Mathematical Theory of Elasticity*, 2d ed. (New York: McGraw-Hill, 1956) for details.

[†] See Theodore van Kármán and M. A. Biot, *Mathematical Methods in Engineering*, Ch. 7 (New York: McGraw-Hill, 1940).

Now, if the deflection y is small, then the slope y' is also small and the curvature is approximately equal to y''. Hence $y'' = M/EI$. But the shear force S is given by

$$S = \frac{dM}{dx},$$

and the load per unit length of the beam, $w(x)$, is related to the shear force by

$$w(x) = \frac{dS}{dx}.$$

If there is no other load on the beam, that is, there are no concentrated forces along the beam, then

$$w(x) = \frac{dS}{dx} = \frac{d^2M}{dx^2} = \frac{d^2}{dx^2}\left(EI\frac{d^2y}{dx^2}\right).$$

Since the beam is uniform, E and I are constant and we obtain the differential equation

$$EI\frac{d^4y}{dx^4} = w(x). \tag{2.5-6}$$

This fourth-order equation has four arbitrary constants in its general solution; hence four conditions must be given. For the problem stated here the following conditions would be appropriate:

i) $y(0) = 0$ since there is no displacement at the built-in end;

ii) $y'(0) = 0$ since the elastic curve is tangent to its undistorted line at the built-in end;

iii) $y''(L) = 0$ since the curvature of the elastic curve is zero at the free end;

iv) $y'''(L) = 0$ since the shearing force vanishes at the free end.

This problem is an example of a *two-point boundary-value* problem in contrast to the *initial-value* problems with which we have been dealing. The latter require that the function and its derivatives be specified at a *single* point. This distinction between the two types of problems is important since the existence and uniqueness theorems for linear differential equations apply to initial-value problems.

Satellite falling to earth

Consider a satellite of mass m falling toward the earth. By Newton's law of gravitation the earth's attractive force, F, on the satellite is proportional to mM/r^2, where M is the mass of the earth and r is the distance between the two masses, as shown in Fig. 2.5–8.

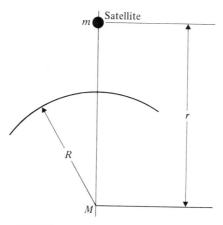

FIGURE 2.5–8 A satellite falling to earth.

We have tacitly assumed that the law of gravitation can be extended so that we can consider the mass of the earth to be concentrated at its center. This assumption is generally made although, strictly speaking, it is true only if the density of the earth were a function of its radius alone. We will also ignore the retarding effect of the earth's atmosphere since we will assume that the satellite is falling from a considerable height. Finally we will assume that g, the acceleration due to gravity, remains constant.

We know that at the earth's surface (when $r = R$) the force of attraction is mg; hence the exact form of F must be given by

$$F = \frac{gmR^2}{r^2}.$$

By Newton's second law,

$$m\frac{d^2r}{dt^2} = -\frac{gmR^2}{r^2}, \tag{2.5–7}$$

where the negative sign is necessary because $g > 0$ and the gravitational force acts toward the center of the earth—in the direction of decreasing r. Thus we have the second-order differential equation

$$\frac{d^2r}{dt^2} = \frac{-gR^2}{r^2}.$$

We choose the initial conditions

$$r = r_0 \quad \text{and} \quad \frac{dr}{dt} = 0 \quad \text{when } t = 0.$$

Although the given differential equation is nonlinear, we can solve it by making a substitution. Let $v = dr/dt$ so that

$$\frac{dv}{dr} = \frac{d}{dr}\left(\frac{dr}{dt}\right) = \frac{\dfrac{d}{dt}\left(\dfrac{dr}{dt}\right)}{\dfrac{dr}{dt}};$$

hence

$$\frac{d^2r}{dt^2} = v\frac{dv}{dr}.$$

Thus the equation becomes

$$v\frac{dv}{dr} = \frac{-gR^2}{r^2},$$

which, although still nonlinear, can be solved by separating the variables. Hence

$$v^2 = \frac{2gR^2}{r} + C$$

and the condition $r = r_0$ when $v = 0$ leads to

$$v^2 = 2gR^2\left(\frac{1}{r} - \frac{1}{r_0}\right).$$

The velocity at impact, v_e, can be found from

$$v_e^2 = 2gR^2\left(\frac{1}{R} - \frac{1}{r_0}\right)$$

and if r_0 is very great, we can write

$$v_e \doteq -\sqrt{2gR}.$$

Using $g = 0.0061$ mi/sec^2 and $R = 3960$ mi, we obtain $v_e \doteq -6.95$ mi/sec or approximately -7 mi/sec. Note that the negative of this figure can also be considered the *escape velocity*, that is, the minimum velocity with which a rocket, say, must leave the earth if it is to overcome the earth's gravitational field.

Free fall

What originated as a tactical maneuver—the delayed parachute jump—has become a sport as well, known as "free fall" or "sky diving." When determining the limiting velocity in a free fall it is necessary to take into account the resistance due to the earth's atmosphere. If we assume that this resistive force is proportional to the square of the velocity, then Eq. (2.5–7) in the preceding example becomes

$$m\frac{d^2r}{dt^2} = \frac{-gmR^2}{r^2} + kv^2.$$

Since we will now be dealing with values of r that differ from R by at most 0.25 percent, we can make the approximation $R = r$ and write

$$m\frac{dv}{dt} = -gm + kv^2. \tag{2.5–8}$$

This is a form of a Riccati equation (see Section 1.6), which can be solved by separating the variables. We obtain

$$-\frac{m}{k}\sqrt{\frac{k}{gm}}\tanh^{-1}\left(v\sqrt{\frac{k}{gm}}\right) = t + C$$

and the initial condition $v = 0$ when $t = 0$ produces $C = 0$. Then putting $a = \sqrt{gm/k}$ and solving for v, we have

$$v = a\tanh\left(\frac{-kta}{m}\right) = a\frac{\exp(-kta/m) - \exp(kta/m)}{\exp(-kta/m) + \exp(kta/m)}.$$

The limiting velocity V can now be obtained by letting $t \to \infty$ and we find $V \to -a$ (Exercise 29), that is,

$$V = -\sqrt{gm/k},$$

the negative sign again indicating motion toward the earth.

Variable mass

Newton's second law has the form

$$F = ma = m\frac{dv}{dt},$$

where m is constant. We now consider the case where m is variable, specifically, $m(t)$. This is the case of a rocket with fuelless mass m_r and fuel capacity m_f. If the fuel is burned at a rate of b g/sec, then the mass at any time t is given by

$$m(t) = m_r + m_f - bt.$$

This expression for $m(t)$ is valid so long as fuel is present, that is, for $0 \le t \le m_f/b$.

By Newton's first law the force propelling the rocket in the forward direction must be balanced by a force in the opposite direction (Fig. 2.5–9). The latter is due to the expelled gases at the rear. If the velocity of these gases is v_g cm/sec, then the retrothrust is bv_g and we can write

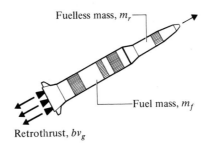

Fuelless mass, m_r

Fuel mass, m_f

Retrothrust, bv_g

FIGURE 2.5–9 A rocket with variable mass.

the differential equation:

$$\frac{dv}{dt}(m_r + m_f - bt) - bv_g = 0. \qquad (2.5\text{--}9)$$

This equation applies to a simplified model and assumes that m_r, m_f, v_g, and b are constants and that there are no external forces acting except the two shown.

KEY WORDS AND PHRASES

integrodifferential equation
Hooke's law
free vibrations
simple harmonic motion

forced vibrations
deflection curve
escape velocity

EXERCISES 2.5

▶ 1. Solve the simple-pendulum problem given that the mass is 1.2 kg and the length of the string is 0.8 m. Assume that $\theta(0) = 0.1$ radian and $g = 980$ cm/sec^2.

2. The frequency f of a simple pendulum is given by

$$f = \frac{1}{2\pi}\sqrt{\frac{g}{2L}}.$$

Find the frequency of the pendulum in Exercise 1.

3. For what length L will the pendulum of Exercise 1 have a period T ($T = 1/f$) of 1 sec?

4. Solve Eq. (2.5–3). For what values of the circuit parameters will the solution be periodic?

5. In the RLC-series circuit (Fig. 2.5–10), assume that the impressed voltage is $E_0 \cos \omega t$ and solve the resulting differential equation.

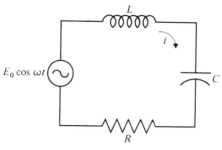

FIGURE 2.5–10

6. A 1.3-kg mass is suspended from a spring which, in turn, is attached to a solid support (Fig. 2.5–11). The mass increases the length of the spring by 2 cm (from this information the spring constant can be found). Given that the spring is pulled down another 3 cm and released (this defines the initial conditions), find the displacement of the mass at any time t.

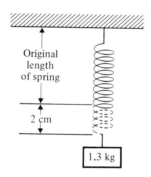

Original
length
of spring

2 cm

1.3 kg

FIGURE 2.5–11

7. Obtain the period (see Exercise 3) of the motion in Exercise 6.

8. Solve the problem in Exercise 6 if a dashpot is added to the system, the damping coefficient being equal to 1.5 kg/cm/sec.

9. A 75-gram mass stretches a spring 3 cm initially. With what frequency will the mass oscillate when set in motion?

10. Find the maximum velocity if the system in Exercise 9 oscillates with an amplitude of 6 cm.

11. A cantilever beam of length L m weighs 2.3 kg/m. If the beam carries no other load, then find the equation of the elastic curve. What is the maximum deflection of the beam?

12. Obtain the equation of the deflection curve of a cantilever beam of length 10 m if $w(x) = 16EI$, a constant.

13. In the example of the falling satellite solve for $v = dr/dt$ (using the negative square root). Then separate the variables and solve for t in terms of r.

14. Use the fact that $r \to r_0$ as $t \to 0$ to find the constant of integration in Exercise 13.

15. Referring to Exercise 14, find the time required for the satellite to reach the earth's surface. What is this time if the satellite is 200 km above the surface of the earth? (Use $R = 6376$ km.)

16. In a free-fall study, k has been found experimentally to have the value 0.1191 lb per ft. A jumper and his equipment weighed 261 lb. What was the limiting velocity?

17. Suppose that in the case of free fall in a gravitational field it is assumed that the air resistance is proportional to the velocity, v. Find the limiting velocity in this case.

18. Solve Eq. (2.5–9) under the condition that $v = 0$ when $t = 0$. Where does the restriction $t \le m_f/b$ enter into the computation?

19. a) Using the result of Exercise 18, put $v = ds/dt$ and find s if $s = 0$ when $t = 0$. *Hint*: Let

$$u = \frac{m_r + m_f - bt}{m_r + m_f}.$$

   b) Find the distance traveled at burnout, that is, when $t = m_f/b$.

▶▶▶ 20. Show that the general solution of the simple-pendulum problem can be written as

$$\theta = c_3 \cos\left(\sqrt{\frac{g}{L}}\, t + \delta\right)$$

for appropriate constants c_3 and δ. How are these constants related to c_1 and c_2 in the text?

21. An inductance of 2 henries, a resistance of 10 ohms, and a capacitance of 0.02 farad are connected in series with an impressed voltage of 6 sin 3t volts. The current i and di/dt are both zero initially. Find the magnitudes of the **transient current** (given by the complementary function) and the **steady-state current** (given by the particular solution) one second after the switch has been closed.

22. Consider the system in Eq. (2.5–3) with $L = 10$ henries, $R = 100$ ohms, $C = 10^{-3}$ farad, $i = 0$, and $di/dt = 5$ when $t = 0$, and find i as a function of t. What is the maximum value of i?

23. Solve Eq. (2.5–4) given that the initial displacement is 3 cm and the initial velocity is 5 cm/sec.

24. Solve Eq. (2.5–5) given that a forcing function of 3 sin 2t is present.

25. Solve Eq. (2.5–4) given that a forcing function $f(t) = a \cos \omega t$ is present. Note that two cases must be considered depending on whether the term $\cos \omega t$ appears in the complementary solution or not.

26. A 1-kg mass is suspended from a spring that is originally 1 m long and, when set in vibration, has a period of 1 sec. How long is the spring when the mass hangs in equilibrium?

27. Provide the details in the solution of Eq. (2.5–8).

28. Obtain the general solution of the **Riccati equation**,

$$\frac{dy}{dx} + y^2 + a(x)y + b(x) = 0,$$

for the case where $a(x)$ and $b(x)$ each have a constant value of 2.

29. In the free-fall example, show that $V \to -a$ when $t \to \infty$.

30. Referring to Eq. (2.5–5), show that when $c^2 > 4mk$, the resulting motion is **overdamped** as shown in Fig. 2.5–12. Explain the appearance of the curves in the figure on the basis of the initial conditions shown.

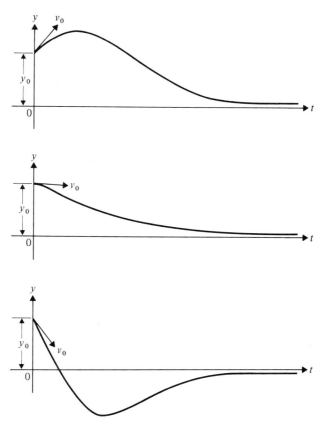

FIGURE 2.5–12 Overdamped motion.

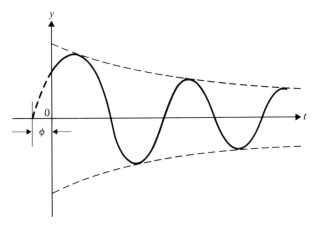

FIGURE 2.5–13 Underdamped motion.

31. Referring to Eq. (2.5–5), show that when $c^2 < 4mk$, the resulting motion is **underdamped** as shown in Fig. 2.5–13.

32. Sketch the solution of Eq. (2.5–5) when $c^2 = 4mk$, that is, the case of **critical damping**. Explain the effect that y_0 and v_0 have on the resulting motion.

2.6 NUMERICAL METHODS

We pointed out in Section 1.6 that the numerical solution of higher-order differential equations can be reduced to the problem of solving a *system* of first-order equations. For example, if we wish to solve the initial-value problem

$$y'' = f(x, y, y'), \qquad y(0) = y_0, \qquad y_0'(0) = y_0', \qquad \textbf{(2.6–1)}$$

we make the substitutions $u = y'$ and $u' = y''$ so that Eq. (2.6–1) becomes

$$\begin{aligned} u' &= f(x, y, u), & u(0) &= y_0', \\ y' &= u, & y(0) &= y_0. \end{aligned} \qquad \textbf{(2.6–2)}$$

This system of two equations can be solved in *parallel* fashion by starting with the initial value of the four-tuple (x, y, u, u'), incrementing x by h and using the relations in Eqs. (2.6–2) to obtain a second four-tuple, and so on. In other words, we begin with $(x, y, u, u') = (0, y_0, y_0', f(0, y_0, y_0'))$ and compute (h, y_1, u_1, u_1'), the next point, where y_1 and u_1 are found from their derivatives as given by Eqs. (2.6–2). For this last we can use any of the methods for solving first-order equations discussed in Section 1.6.

We take this opportunity to introduce another method for solving first-order differential equations. This method was not included among those presented in Section 1.6. A fourth-order predictor-corrector method known as **Milne's method*** uses the *predictor*,

$$y_{n+1} = y_{n-3} + \frac{4h}{3}(2y'_{n-2} - y'_{n-1} + 2y'_n), \qquad \text{(2.6–3)}$$

and the *corrector*,

$$y_{n+1} = y_{n-1} + \frac{h}{3}(y'_{n-1} + 4y'_n + y'_{n+1}). \qquad \text{(2.6–4)}$$

The latter is based on Simpson's rule, familiar to us from calculus. Note that Eq. (2.6–3) requires *four* starting values, which may be obtained by using Euler's method, shown in Eq. (1.6–3); Heun's (improved Euler) method, shown in Eq. (1.6–4); or a three-term series method, shown in Eq. (1.6–7). Attention is called to the fact that the y'_{n+1}-term in Eq. (2.6–4) is computed using the y_{n+1}-term in Eq. (2.6–3) and the differential equation being solved.

When solving the pair of first-order equations (2.6–2) by Milne's method, we would have a predictor and a corrector for each. Thus, in addition to Eqs. (2.6–3) and (2.6–4) we also have the following equations.

$$\text{Predictor:}\ u_{n+1} = u_{n-3} + \frac{4h}{3}(2u'_{n-2} - u'_{n-1} + 2u'_n) \qquad \text{(2.6–5)}$$

$$\text{Corrector:}\ u_{n+1} = u_{n-1} + \frac{h}{3}(u'_{n-1} + 4u'_n + u'_{n+1}) \qquad \text{(2.6–6)}$$

As an example of the use of Milne's method for solving a second-order initial-value problem, consider the following.

EXAMPLE 2.6–1 Solve the initial-value problem

$$y'' + 2x^2y' - y^2 = \log y, \qquad y(0) = 1, \qquad y'(0) = 0,$$

carrying the solution to $x = 0.5$ using a step of $h = 0.05$.

Solution. Note that the given second-order differential equation does not fit into any of the "types" presented in this chapter mainly because it is nonlinear. We make the substitutions

$$u' = y^2 - 2x^2u + \log y,$$
$$y' = u,$$

with initial conditions

$$x_0 = 0, \qquad y_0 = 1, \qquad u_0 = 0, \qquad u'_0 = 1.$$

---

* After William E. Milne (1890–1971), an American mathematician.

In order to obtain the necessary four starting values we will use Eq. (1.6–7). Thus

$$u'_n = y_n^2 - 2x_n^2 u_n + \log y_n,$$

$$u''_n = \left(2y_n + \frac{1}{y_n}\right) u_n - 2x_n(2u_n + x_n u'_n),$$

and we substitute these values into

$$u_{n+1} = u_n + hu'_n + \frac{h^2}{2} u''_n$$

to obtain successive values of u. The values of y are found from

$$y_{n+1} = y_n + hu_n + \frac{h^2}{2} u'_n.$$

Table 2.6–1 shows the results obtained in this way.

TABLE 2.6–1 Starting values for Example 2.6–1 using Eq. (1.6–7).

n	x_n	y_n	u_n	u'_n
0	0	1.0000	0	1.0000
1	0.05	1.0013	0.0500	1.0035
2	0.10	1.0051	0.1004	1.0131
3	0.15	1.0114	0.1511	1.0276

With the four starting values we can now use Eqs. (2.6–3) through (2.6–6) to continue the solution to the point where $x = 0.5$ as required. These values are shown in Table 2.6–2. Note the close agreement between the predicted and corrected values. Figure 2.6–1 shows the solutions to Example 2.6–1 by both the Milne and Euler methods. ■

TABLE 2.6–2 Milne's method for Example 2.6–1.

n	x_n	y_n	u_n	Predicted u_{n+1}	Predicted y_{n+1}	Corrected u_{n+1}	Corrected y_{n+1}
3	0.15	1.0114	0.1511	0.2033	1.0201	0.2032	1.0202
4	0.20	1.0202	0.2032	0.2559	1.0317	0.2556	1.0317
5	0.25	1.0317	0.2556	0.3096	1.0458	0.3096	1.0458
6	0.30	1.0458	0.3096	0.3638	1.0627	0.3641	1.0627
7	0.35	1.0627	0.3641	0.4196	1.0822	0.4208	1.0822
8	0.40	1.0822	0.4208	0.4753	1.1048	0.4770	1.1048
9	0.45	1.1048	0.4770	0.5323	1.1299	0.5338	1.1299
10	0.50	1.1299	0.5338				

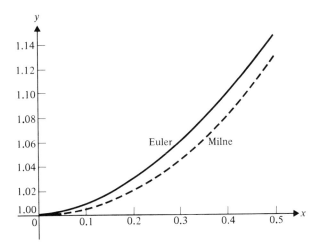

FIGURE 2.6–1 Solutions to Example 2.6–1.

We add a word of caution regarding numerical methods at this point. A numerical method for solving differential equations should be stable and convergent. A method is said to be **stable** if the propagated error is bounded. A method is said to be **convergent** if the numerical solution approaches the true solution as the step size h approaches zero. Excellent discussions of convergence and stability can be found in Anthony Ralston and P. Rabinowitz's *A First Course in Numerical Analysis*, 2d ed. (New York: McGraw-Hill, 1978). The authors point out, for example, that Milne's method is unstable and exhibit a solution to an equation that shows that near the initial point the method is excellent but the accuracy deteriorates rapidly after a large number of points (146 in the case shown) have been computed. This problem can be corrected by using Hamming's method* given in Ralston and Rabinowitz's work.

It is not uncommon in applied mathematics that in $y'' = \phi(x, y, y')$ the function ϕ is free of y'. For this case Fox† has given the following pair of fifth-order equations.

$$\text{Predictor: } y_{n+1} = y_n + y_{n-2} - y_{n-3} + \frac{h^2}{4} (5y''_n + 2y''_{n-1} + 5y''_{n-2}) \quad \textbf{(2.6–7)}$$

$$\text{Corrector: } y_{n+1} = 2y_n - y_{n-1} + \frac{h^2}{12} (y''_{n+1} + 10y''_n + y''_{n-1}) \quad \textbf{(2.6–8)}$$

---

* After Richard W. Hamming (1915–   ).
† Augustus H. Fox, *Fundamentals of Numerical Analysis*, p. 76 (New York: Ronald Press, 1963).

It should be noted that using a computer to solve differential equations provides an insight into the nature of the solutions that is difficult to obtain in any other way. The effect of changing initial conditions and forcing functions can be easily observed. In this connection see Richard Bronson and A. Jones's "Batch processing differential equations on a microcomputer," *Amer. Math. Monthly* **85**, no. 4 (April 1978), pp. 272–275.

KEY WORDS AND PHRASES

Milne's method   **convergent method**
stable method

EXERCISES 2.6

1. Convert the following initial-value problem involving a second-order equation to one involving two first-order equations.

$$\ddot{y} - (t + y)\dot{y} = t(2 - y^2), \qquad y(0) = -1, \qquad \dot{y}(0) = 1$$

2. In the theory of structures it is shown that the radius of curvature at any point of a beam is proportional to the bending moment,

$$EI \frac{y''}{(1 + (y')^2)^{3/2}} = M(x),$$

where y is the deflection of the elastic curve and x is the distance along the beam. In Section 2.5 we neglected the term $(y')^2$ since it was small compared to unity. Without neglecting this term for the cantilever beam for which $y(0) = y'(0) = 0$, express the differential equation above as a pair of simultaneous first-order equations.

3. Convert the following initial-value problem into a system of first-order initial-value problems.

$$y''' - 3y'' - 6xy' - 12x^2y = \cos x, \qquad y(0) = 1, \qquad y'(0) = 2, \qquad y''(0) = 3$$

4. Find starting values to $x = 0.5$ using a three-term Taylor's series and a step size of $h = 0.1$ for the initial-value problem

$$y'' + x^2y = 0, \qquad y(0) = 1, \qquad y'(0) = 1.$$

5. Using starting values from Exercise 4, extend the solution by means of Milne's method to $x = 1$ with $h = 0.1$.

6. Solve the initial-value problem

$$\frac{d^2y}{dx^2} + x^2 \frac{dy}{dx} + 3y = x, \qquad y(0) = 1, \qquad y'(0) = 2.$$

Use $h = 0.1$ and Euler's method to determine $y(0.5)$.

7. Find $y(0.5)$ to three decimals, given

$$y'' - 8x^2y' - 3e^xy = -xe^{-x}, \qquad y(0) = 1, \qquad y'(0) = -1.$$

Use a series method with $h = 0.1$.

8. Repeat Exercise 7 with $h = 0.05$.

9. Use the improved Euler method to obtain y and $\dot{y}$ for $t = 0.6$ using $h = 0.2$:

$$\ddot{y} - y\dot{y} - ty = t - 1, \qquad y(0) = 1, \qquad \dot{y}(0) = 0.$$

10. Continue the solution of Exercise 9 to $t = 1.0$ using the Adams–Moulton method and $h = 0.2$.

11. Use a Taylor's series method to solve

$$y'' = xy, \qquad y(0) = 1, \qquad y'(0) = 1$$

for $y(0.2)$, $y(0.4)$, and $y(0.6)$.

12. Solve the initial-value problem

$$xy'' - y' + 4x^3y = 0, \qquad y(1) = 1, \qquad y'(1) = 2.$$

Use $h = 0.1$ and find $y(1.5)$ to four decimals by a series method.

13. Use Euler's method to obtain four starting values for y, the solution of the initial-value problem

$$y'' - x^2y = 0, \qquad y(0) = 1, \qquad y'(0) = 1.$$

Use a step size $h = 0.1$.

14. Use the data obtained in Exercise 13 to compute $y(1)$ to four decimals from Eqs. (2.6–4) and (2.6–5).

15. Find $y(0.1)$ to four decimals, given

$$y'' - y^2 = x^2, \qquad y(0) = 1, \qquad y'(0) = 0.$$

16. Referring to Exercise 1, use a Taylor's series method to find $y(0.1)$ and $\dot{y}(0.1)$ to four decimals.

17. Repeat Exercise 16 using the improved Euler method.

18. Find $y(0.6)$ for

$$y'' = yy', \qquad y(0) = 1, \qquad y'(0) = -1.$$

Begin the solution using the Taylor's series method to obtain $y(0.1)$, $y(0.2)$, and $y(0.3)$. Then continue with the Adams–Moulton method using the same $h = 0.1$.

19. The analytical solution of

$$\ddot{y} + 64y = 16 \cos 8t, \qquad y(0) = \dot{y}(0) = 0,$$

is known to be $y = t \sin 8t$. Determine $y(0.2)$, $y(0.4)$, and $y(0.6)$ by one of the numerical methods presented and compare your results with the correct solution.

REFERENCES

Conte, S. D., and C. de Boor, *Numerical Analysis: An Algorithmic Approach*, 3d ed. New York: McGraw-Hill, 1980.
An excellent coverage of numerical methods including the solution of differential equations.

Creese, T. M., and R. M. Haralick, *Differential Equations for Engineers*. New York: McGraw-Hill, 1978.
Contains a wide variety of applications to science and engineering.

Finizio, N., and G. Ladas, *Ordinary Differential Equations with Modern Applications*. Belmont, Calif.: Wadsworth, 1978.
The modern applications include medicine, pharmacology, biokinetics, ecology, biology, psychology, and economics.

Haberman, R., *Mathematical Models, Mechanical Vibrations, Population Dynamics, and Traffic Control* (An Introduction to Applied Mathematics). Englewood Cliffs, N.J.: Prentice-Hall, 1977.

Jordan, D. W., and P. Smith, *Nonlinear Ordinary Differential Equations*. New York: Oxford University Press, 1977.

Kaplan, W., *Elements of Differential Equations*. Reading, Mass.: Addison-Wesley, 1964.

Luenberger, D. C., *Introduction to Dynamic Systems: Theory, Models, and Applications*. New York: Wiley, 1979.
Contains a variety of applications and blends the classical approach with the state-space approach.

Ortega, J. M., and W. G. Poole, Jr., *An Introduction to Numerical Methods for Differential Equations*. Marshfield, Mass.: Pitman, 1981.
Includes the standard methods as well as some recent ones for solving both ordinary and partial differential equations.

Plaat, O., *Ordinary Differential Equations*. San Francisco: Holden-Day, 1971.

Rabenstein, A. L., *Introduction to Ordinary Differential Equations*, 2d ed. New York: Academic Press, 1972.

Saaty, T. L., and J. Bram, *Nonlinear Mathematics*. New York: McGraw-Hill, 1964.

Wiley, C. R., *Differential Equations*. New York: McGraw-Hill, 1979.
An exceptionally complete coverage of the subject.

3

the laplace transformation

3.1 BASIC PROPERTIES AND DEFINITIONS

In earlier chapters we presented methods for solving ordinary differential equations. For the most part the equations were linear such as the second-order equation

$$\ddot{y}(t) + a_1\dot{y}(t) + a_2 y(t) = f(t), \qquad \textbf{(3.1--1)}$$

with constant coefficients a_1 and a_2. By finding the roots of the characteristic equation it was possible to write down the complementary solution $y_c(t)$, which contained two arbitrary constants. Adding a particular solution to this produced the general solution of Eq. (3.1–1).

When a_1 and a_2 were constants and $f(t)$ was of a special form, the method of undetermined coefficients could be used to obtain a particular solution. If a_1 and a_2 were functions of t, the method of variation of parameters was appropriate. Although both methods had some shortcomings, they could usually produce results when $f(t)$ was continuous. It happens, however, that equations of the form (3.1–1) are encountered in connection with mechanical and electrical (and other) systems in which the forcing function $f(t)$ is *not* continuous. For example, $f(t)$ may be an impulse function or it may have one value for a certain range of t-values and another value outside this range. In short, we need to have a method that can be used when $f(t)$ is a *piecewise continuous* function defined as follows.

Definition 3.1–1 The function f is **piecewise continuous*** on (a, b) if this interval can be partitioned into a finite number of subintervals such that

a) f is continuous on each open subinterval, and

b) f has finite left- and right-hand limits at each point of subdivision.

---

* Also called "sectionally continuous."

EXAMPLE 3.1–1   The function f defined as

$$f(t) = \begin{cases} 2t, & \text{for } 0 < t < 1, \\ 1, & \text{for } 1 \le t < 2, \\ -1, & \text{for } 2 \le t < 3, \end{cases}$$

is piecewise continuous on $(0, 3)$. Figure 3.1–1 shows the graph of this function. At $t = 1$ the left-hand limit is found to be

$$\lim_{t \to 1^-} f(t) = 2,$$

whereas the right-hand limit is given by

$$\lim_{t \to 1^+} f(t) = 1. \quad \blacksquare$$

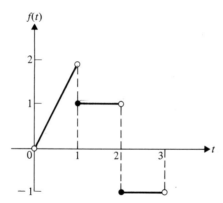

FIGURE 3.1–1 The function of Example 3.1–1.

We remark that continuous functions belong to the class of piecewise continuous functions and that points of continuity are characterized by the fact that the left- and right-hand limits are *equal*. The type of discontinuity at $t = 1$ and $t = 2$ exhibited by the function of Example 3.1–1 is called a (finite) *jump discontinuity*.

Another example of a piecewise continuous function is shown in Fig. 3.1–2. Note that this function satisfies Definition 3.1–1, although $g(1)$ does not exist and $g(2)$ differs from the left- and right-hand limits of $g(t)$ at $t = 2$. (See Exercise 1.)

It would be very desirable to have some means of dealing with the complexities due to the presence of discontinuities in a differential

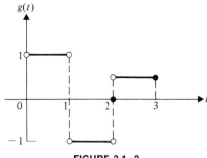

FIGURE 3.1–2

equation. What is needed is some way to transform a piecewise continuous function into a continuous function. This is not as far-fetched an idea as it seems because *transformations* play a very important role in much of mathematics. Following are some examples of familiar transformations.

1. Transforming positive real numbers into their *logarithms* converts the operations of multiplication and division into the simpler operations of addition and subtraction, respectively.
2. The coordinate transformation (*translation of axes*)

$$x = X + 2 \quad \text{and} \quad y = Y + 3$$

transforms the equation $x^2 - 4x - 4y + 16 = 0$ into $X^2 = 4Y$, which can be graphed more easily (Fig. 3.1–3).

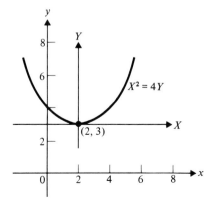

FIGURE 3.1–3 Translation of axes.

3. The coordinate transformation (*rotation of axes* through an angle of $\pi/4$ in this case)

$$x = (X - Y)/\sqrt{2} \quad \text{and} \quad y = (X + Y)/\sqrt{2}$$

transforms the equation $5x^2 - 6xy + 5y^2 = 8$ into the more easily recognizable form $X^2 + 4Y^2 = 4$ (Fig. 3.1-4).

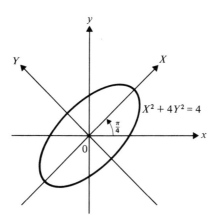

FIGURE 3.1-4 Rotation of axes.

4. The transformation of variable, $x = 3 \sin \theta$ (called *trigonometric substitution*), converts the integral

$$\int \frac{dx}{(9 - x^2)^{3/2}}$$

into the easily manageable

$$\frac{1}{9} \int \sec^2 \theta \, d\theta.$$

For solving differential equations we introduce a transformation that belongs to a class of **integral transforms** that have the form

$$F(s) = \int_a^b K(s, t) f(t) \, dt. \tag{3.1-2}$$

The function $K(s, t)$ is called the **kernel** (that is, the nucleus) of the transformation. One or both limits of integration may be infinite, in which case the transformation is accomplished by means of an *improper integral*. Equation (3.1-2) indicates that $f(t)$ has been transformed into $F(s)$ and we sometimes describe this change by saying that a function in the t-domain $(f(t))$ has been transformed into a function in the s-domain $(F(s))$.

The integral transform that is the topic of this chapter was conceived by Laplace in 1779 and developed into its present useful form in 1890 by Oliver Heaviside, (1850–1925), an English electrical engineer. The Laplace transformation, commonly called the **Laplace transform**, is defined as

$$F(s) = \mathscr{L}(f(t)) = \int_0^\infty \exp{(-st)}f(t)\, dt, \qquad s > 0. \qquad \textbf{(3.1–3)}$$

Following custom we will write

$$\int_0^\infty \exp{(-st)}f(t)\, dt$$

for the more cumbersome, but mathematically correct,

$$\lim_{b \to \infty} \int_0^b \exp{(-st)}f(t)\, dt.$$

The kernel of the Laplace transform is $\exp{(-st)}$ with $s > 0$. This last inequality indicates that we are considering s to be a real number although, in general, it may be complex (see Section 10.6).

Before we discuss the conditions under which the transformation shown in Eq. (3.1–3) can be carried out, we compute the Laplace transforms of some common functions.

EXAMPLE 3.1–2 Find the Laplace transform of $f(t) = 3$.

Solution.

$$\int_0^\infty 3 \exp{(-st)}\, dt = 3 \int_0^\infty \exp{(-st)}\, dt$$

$$= 3\, \frac{\exp{(-st)}}{-s}\bigg|_0^\infty = \frac{3}{s}. \quad \blacksquare$$

Observe that in the last example

$$\lim_{t \to \infty} \left(-\frac{1}{s} \exp{(-st)} \right) = 0$$

because $s > 0$. Note also that the constant 3 was unaffected by the transformation and we can say, in general, that if a is a constant, then we have the correspondence (Fig. 3.1–5)

$$f(t) = a \leftrightarrow F(s) = \frac{a}{s}. \qquad \textbf{(3.1–4)}$$

In Eq. (3.1–4) we have exhibited a **Laplace transform pair**. Our goal is to obtain enough of these pairs so that we will have a short

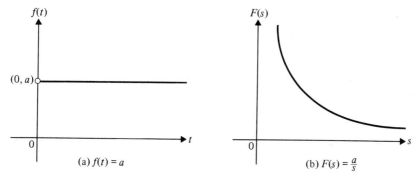

FIGURE 3.1-5 A Laplace transform pair.

table of Laplace transforms to which we can refer. Later we will use Laplace transform pairs in the reverse (or inverse) direction also. For example, if we know that $F(s) = a/s$, Eq. (3.1–4) tells us that $f(t) = a$. The question of uniqueness of the inverse transform will be discussed in Section 3.3. Uniqueness of the Laplace transform itself is the subject of Exercise 29.

EXAMPLE 3.1–3   Find the Laplace transform of $f(t) = t$.

Solution. We have

$$F(s) = \int_0^\infty t \exp(-st)\, dt$$

$$= \left. \frac{t \exp(-st)}{-s} \right|_0^\infty + \frac{1}{s} \int_0^\infty \exp(-st)\, dt$$

$$= \frac{1}{s^2}. \quad \blacksquare$$

Here we have used integration by parts, noted that $t \exp(-st) = 0$ at both limits by L'Hôpital's rule, and used Eq. (3.1–4) to evaluate the last integral. Thus we have another transform pair:

$$f(t) = t \leftrightarrow F(s) = \frac{1}{s^2}. \tag{3.1–5}$$

Using integration by parts twice and following the method of Example 3.1–3, we can obtain (Exercise 3a)

$$f(t) = t^2 \leftrightarrow F(s) = \frac{2}{s^3}. \tag{3.1–6}$$

If n is a positive integer, then we can use repeated integration by parts to obtain (Exercise 20)

$$f(t) = t^n \leftrightarrow F(s) = \frac{n!}{s^{n+1}}, \quad n \text{ a positive integer.} \qquad (3.1-7)$$

EXAMPLE 3.1–4 If $f(t) = \sin at$, find $F(s)$.

Solution. Integrating by parts gives us

$$F(s) = \int_0^\infty \exp(-st) \sin at \, dt$$

$$= \left. \frac{\exp(-st) \sin at}{-s} \right|_0^\infty + \frac{a}{s} \int_0^\infty \exp(-st) \cos at \, dt$$

$$= \frac{a}{s} \int_0^\infty \exp(-st) \cos at \, dt.$$

Here we have used the fact that because $\sin at$ is bounded, that is, $|\sin at| \leq 1$,

$$\lim_{t \to \infty} \exp(-st) \sin at = 0.$$

A second integration by parts results in

$$F(s) = \frac{a}{s} \left(\left. \frac{\exp(-st) \cos at}{-s} \right|_0^\infty - \frac{a}{s} \int_0^\infty \exp(-st) \sin at \, dt \right)$$

$$= \frac{a}{s} \left(\frac{1}{s} - \frac{a}{s} F(s) \right)$$

and, solving for $F(s)$, we have

$$F(s) = \frac{a}{s^2 + a^2}.$$

This gives us the pair (Fig. 3.1–6)

$$f(t) = \sin at \leftrightarrow F(s) = \frac{a}{s^2 + a^2} \qquad (3.1-8)$$

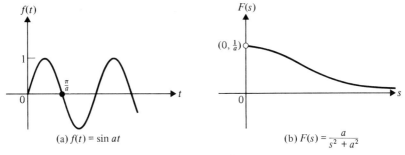

(a) $f(t) = \sin at$

(b) $F(s) = \dfrac{a}{s^2 + a^2}$

FIGURE 3.1–6

and the pair

$$f(t) = \cos at \leftrightarrow F(s) = \frac{s}{s^2 + a^2} \qquad (3.1\text{--}9)$$

is obtained in a similar fashion (Exercise 3b). ∎

The last example suggests that boundedness of a function is, in some way, connected with the existence of the Laplace transform of the function. We look into this matter further in the following example.

EXAMPLE 3.1–5 Find the Laplace transform of $\exp(at)$.

Solution. We have

$$\int_0^\infty \exp(at) \exp(-st)\, dt = \int_0^\infty \exp((a-s)t)\, dt$$

$$= \frac{\exp((a-s)t)}{a-s} \bigg|_0^\infty = \frac{1}{s-a},$$

provided that $s > a$. This restriction is required in order that

$$\lim_{t \to \infty} \frac{\exp((a-s)t)}{a-s} = 0.$$

Hence we have the pair

$$f(t) = \exp(at) \leftrightarrow F(s) = \frac{1}{s-a}, \qquad s > a. \quad \blacksquare \qquad (3.1\text{--}10)$$

We next give a definition that will be useful in stating the existence theorem for Laplace transforms.

Definition 3.1–2 A function $f(t)$ is of *exponential order* $\exp(at)$ if non-negative constants M and T exist such that

$$|f(t)| \le M \exp(at)$$

for all $t \ge T$.

This definition implies that if a function is of exponential order $\exp(at)$, then it will not increase faster than $M \exp(at)$ as $t \to \infty$. (*Note:* This is also expressed by saying that the function is $O(\exp(at))$. Thus $\sin t$ and $\cos t$ are of exponential order 1 ($a = 0$), whereas $\exp(bt)$ is of exponential order $\exp(at)$, provided that $b \le a$.

Theorem 3.1–1 (*Existence of the Laplace transform*). If a function $f(t)$ is

a) piecewise continuous on $t \ge 0$, and

w/t S vdu

b) of exponential order exp (at) with $a > 0$,

then its Laplace transform $F(s)$ exists for all $s > a$. Moreover, $\mathscr{L}(|f(t)|)$ exists for $s > a$.*

We remark that hypothesis (b) alone is not sufficient to ensure that a function have a Laplace transform since functions like $1/t$ and $1/t^2$ do *not* have Laplace transforms (see Exercise 5). On the other hand, the function $f(t) = t^{-1/2}$ does not satisfy condition (a) of Theorem 3.1–1 (why?), yet it possesses a Laplace transform (see Exercise 25). Thus the existence theorem gives *sufficient* conditions only. The question of uniqueness of the Laplace transform is related to the convergence of the defining improper integral (Exercise 29).

In order to obtain Laplace transform pairs more easily (without using the definition) we state two theorems that will be helpful.

Theorem 3.1–2 (*Scale change*). If $\mathscr{L}(f(t)) = F(s)$ when $s > a$, then for $b > 0$,

$$\mathscr{L}(f(bt)) = \frac{1}{b} F\left(\frac{s}{b}\right), \qquad \frac{s}{b} > a. \qquad (3.1\text{–}11)$$

PROOF. By definition,

$$F(s) = \int_0^\infty \exp(-st)f(t)\, dt,$$

hence

$$F(s/b) = \int_0^\infty \exp(-st/b)f(t)\, dt.$$

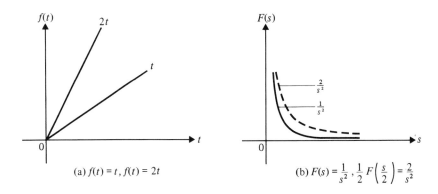

(a) $f(t) = t,\ f(t) = 2t$

(b) $F(s) = \dfrac{1}{s^2},\ \dfrac{1}{2} F\left(\dfrac{s}{2}\right) = \dfrac{2}{s^2}$

FIGURE 3.1–7 Scale change.

---

* We omit the proof of the existence theorem and refer the interested reader to C. Ray Wiley's *Differential Equations*, pp. 186ff. (New York: McGraw-Hill, 1979).

Now we replace t/b by u in the integral to obtain

$$F(s/b) = \int_0^\infty \exp(-su)f(bu)b\,du = b\mathscr{L}(f(bt))$$

as required (see Fig. 3.1–7). $\square$

Theorem 3.1–3 (*Shifting*). If $\mathscr{L}(f(t)) = F(s)$ when $s > a$, then

$$\mathscr{L}(\exp(bt)f(t)) = F(s - b) \qquad\qquad \textbf{(3.1–12)}$$

for $s \geq a + b$ where b is any constant.

PROOF. From the definition,

$$F(s - b) = \int_0^\infty \exp(-(s - b)t)f(t)\,dt$$

$$= \int_0^\infty \exp(-st)(\exp(bt)f(t))\,dt$$

$$= \mathscr{L}(\exp(bt)f(t)). \quad \square$$

We now give some examples showing how Theorem 3.1–3 can be used.

EXAMPLE 3.1–6 Obtain the Laplace transform of $\exp(bt)\cos at$.

Solution. From Eq. (3.1–9) we know that the transform of $\cos at$ is $s/(s^2 + a^2)$; hence the transform of

$$f(t) = \exp(bt)\cos at$$

is

$$\frac{s - b}{(s - b)^2 + a^2}$$

by Theorem 3.1–3. Thus we have the transform pair (Fig. 3.1–8)

$$f(t) = \exp(bt)\cos at \leftrightarrow F(s) = \frac{s - b}{(s - b)^2 + a^2}. \quad \blacksquare \qquad \textbf{(3.1–13)}$$

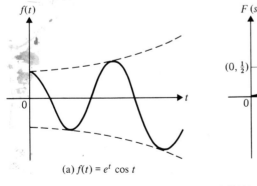

(a) $f(t) = e^t \cos t$

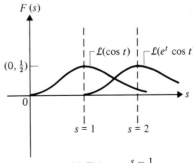

(b) $F(s) = \dfrac{s - 1}{(s - 1)^2 + 1}$

FIGURE 3.1–8 Shifting.

In a similar fashion we can obtain (Exercise 6)

$$f(t) = \exp{(bt)} \sin{at} \leftrightarrow F(s) = \frac{a}{(s-b)^2 + a^2}. \qquad \textbf{(3.1–14)}$$

EXAMPLE 3.1–7 Obtain the Laplace transform of $t^n \exp{(bt)}$, n a positive integer.

Solution. From Eq. (3.1–7) and Theorem 3.1–3 we have the pair (Fig. 3.1–9)

$$f(t) = t^n \exp{(bt)} \leftrightarrow F(s) = n!/(s-b)^{n+1}. \; \blacksquare \qquad \textbf{(3.1–15)}$$

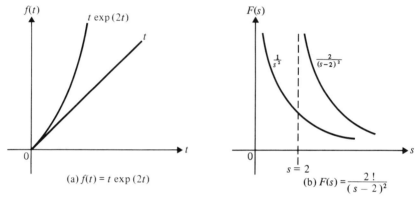

(a) $f(t) = t \exp{(2t)}$

(b) $F(s) = \dfrac{2\,!}{(s-2)^2}$

FIGURE 3.1–9

We close this section by stating a basic property of the Laplace transformation, namely, that it is a *linear transformation*. We can see from the definition that if the functions $f(t)$ and $g(t)$ have Laplace transforms $F(s)$ and $G(s)$, respectively, then

a) $\displaystyle\int_0^\infty \exp{(-st)}(kf(t))\,dt = kF(s)$, and

b) $\displaystyle\int_0^\infty \exp{(-st)}(f(t) + g(t))\,dt = F(s) + G(s)$

$$\textbf{(3.1–16)}$$

for any constant k. Properties (a) and (b) characterize a linear transformation or, as it is sometimes called, a *linear operator*. Differentiation and integration are familiar examples of linear operations.

Table 3.1–1 is a brief list of Laplace transforms to which we can refer when necessary.

TABLE 3.1–1 Laplace transforms.

$f(t)$	$\mathcal{L}(f(t)) = F(s)$	Reference
a	a/s	Eq. (3.1–4)
t	$1/s^2$	Eq. (3.1–5)
t^n	$n!/s^{n+1}$, n a positive integer	Eq. (3.1–7)
t^α	$\Gamma(\alpha+1)/s^{\alpha+1}$, $\alpha > -1$	Exercise 25
$\sin at$	$\dfrac{a}{s^2+a^2}$	Eq. (3.1–8)
$\cos at$	$\dfrac{s}{s^2+a^2}$	Eq. (3.1–9)
$\exp(at)$	$\dfrac{1}{s-a}$, $s > a$	Eq. (3.1–10)
$f(bt)$	$\dfrac{1}{b}F\left(\dfrac{s}{b}\right)$, $\dfrac{s}{b} > a$	Eq. (3.1–11)
$\exp(bt)f(t)$	$F(s-b)$, $s \geq a+b$	Eq. (3.1–12)
$\exp(bt)\cos at$	$\dfrac{s-b}{(s-b)^2+a^2}$	Eq. (3.1–13)
$\exp(bt)\sin at$	$\dfrac{a}{(s-b)^2+a^2}$	Eq. (3.1–14)
$t^n\exp(bt)$	$\dfrac{n!}{(s-b)^{n+1}}$, n a positive integer	Eq. (3.1–15)
$\cosh at$	$\dfrac{s}{s^2-a^2}$	Exercise 4
$\sinh at$	$\dfrac{a}{s^2-a^2}$	Exercise 4

KEY WORDS AND PHRASES

piecewise continuous
jump discontinuity
transformation
integral transform

kernel
Laplace transform
exponential order exp (at)
linear operator

EXERCISES 3.1

▶ 1. Compute each of the following limits for the piecewise continuous function $g(t)$ of this section.

a) $\lim\limits_{t\to 0^-} g(t)$ b) $\lim\limits_{t\to 0^+} g(t)$ c) $\lim\limits_{t\to 1^-} g(t)$

d) $\lim\limits_{t\to 1^+} g(t)$ e) $\lim\limits_{t\to 2^-} g(t)$ f) $\lim\limits_{t\to 2^+} g(t)$

g) $\lim\limits_{t\to 3^-} g(t)$ h) $\lim\limits_{t\to 3^+} g(t)$

2. Carry out the details in Example 3.1–3.

✓**3.** Obtain the Laplace transforms of each of the following functions. State any restrictions that apply.

a) t^2 　　　　　　　　　　　　　　**b)** $\cos at$

✓**4.** Obtain the Laplace transforms of each of the following functions. Note that you are using the fact that the Laplace transform is a linear transformation.

a) $\cosh at$ 　　　　　　　　　　　**b)** $\sinh at$

5. Prove that $1/t$ and $1/t^2$ do not have Laplace transforms. (*Hint*: Show that $\int_0^1 \exp(-st)f(t)\,dt$ *diverges* for these functions.)

✓**6.** Find the Laplace transform of $\exp(bt)\sin at$.

►► **7.** Graph each of the given functions.

a) $f(t) = \begin{cases} 1, & \text{for } 0 \leq t < 1, \\ -1, & \text{for } 1 \leq t < 2. \end{cases}$

✓**b)** $g(t) = \begin{cases} t, & \text{for } 0 < t \leq 2, \\ 4 - t, & \text{for } 2 \leq t < 4. \end{cases}$

c) $h(t) = \pi - t$ for $0 < t < \pi$.

8. Referring to Exercise 7, evaluate each of the following limits.

a) $\lim_{t \to 0^+} f(t);\ \lim_{t \to 1^+} f(t);\ \lim_{t \to 1^-} f(t);\ \lim_{t \to 2^-} f(t)$.

✓**b)** $\lim_{t \to 0^+} g(t);\ \lim_{t \to 2^-} g(t);\ \lim_{t \to 2^+} g(t);\ \lim_{t \to 4^-} g(t)$.

c) $\lim_{t \to 0^+} h(t);\ \lim_{t \to \pi^-} h(t);\ \lim_{t \to \pi^+} h(t);\ \lim_{t \to 2\pi^-} h(t)$.

9. Describe the discontinuities of the functions $f(t)$, $g(t)$, and $h(t)$ of Exercise 8.

✓**10.** Find the Laplace transforms of each of the following functions.

a) $3t^2 - 2t + 5$

b) $\sin 2t \cos t - \cos 2t \sin t$ (*Hint*: Use trigonometric identities to simplify first.)

c) $t \exp(at)$

✓**11.** Show that the two properties (a) and (b) in Eq. (3.1–16) are equivalent to

$$\int_0^\infty \exp(-st)(c_1 f(t) + c_2 g(t))\,dt = c_1 F(s) + c_2 G(s)$$

for arbitrary constants c_1 and c_2.

12. Find the Laplace transforms of each of the following functions.

a) $f(t) = \begin{cases} 0, & \text{for } 0 \leq t < 1, \\ t, & \text{for } t \geq 1. \end{cases}$

✓**b)** $f(t) = \begin{cases} \sin t, & \text{for } 0 \leq t < \pi, \\ 3, & \text{for } t \geq \pi. \end{cases}$

✓**13.** Find the function $f(t)$ corresponding to each of the following transforms. Assume $s > 0$.

a) $F(s) = 4/s^3$ 　　　　　　　　　**b)** $F(s) = 3/(s^2 + 4)$

c) $F(s) = 2s/(s^2 + 4)$ 　　　　　　**d)** $F(s) = \dfrac{s^2 + s + 1}{s(s^2 + 1)}$

✓ 14. For what values of s are the Laplace transforms of each of the following functions defined? Assume that a and b are constants.

    a) $3t - 2$ ✓**b)** $\sin(at + b)$

    c) $\cosh(at + b)$ ✓**d)** $t^2 + at + b$

15. Find the Laplace transform pair for each of the following functions.

    a) $\dfrac{a_1}{s} + \dfrac{a_2}{s^2} + \cdots + \dfrac{a_n}{s^n}$, $s > 0$ **b)** $\dfrac{s - 2}{s^2 - 2}$, $s > \sqrt{2}$

    c) $\dfrac{s - 2}{s^2 + 3}$, $s > 0$

16. Find the Laplace transform of the function

$$f(t) = \begin{cases} 1, & \text{for } 0 \le t < 1, \\ 0, & \text{for } t \ge 1. \end{cases}$$

✓ 17. Graph the function $f(t)$ of Exercise 16 and then graph its Laplace transform.

18. By using the Euler formula,

$$\exp(iat) = \cos at + i \sin at,$$

and equating real and imaginary parts, find $\mathcal{L}(\exp(iat))$ and from this obtain $\mathcal{L}(\cos at)$ and $\mathcal{L}(\sin at)$.

19. Use Theorem 3.1–3 to obtain the Laplace transforms of each of the following.

    a) $\cosh at \cos at$ **b)** $\sinh at \sin at$

    c) $\sinh at \cos at$ **d)** $\cosh at \sin at$

▶▶▶ 20. Show that if n is a positive integer, then the Laplace transform of t^n is $n!/s^{n+1}$.

21. Prove that the integral transform shown in Eq. (3.1–2) is a linear transformation.

22. Prove that if

$$\lim_{t \to \infty} \exp(-at)f(t) = 0,$$

then $f(t)$ is of exponential order $\exp(at)$.

23. Prove that $\exp(t^2)$ is not of exponential order and does not have a Laplace transform.

24. Although $\exp(t^2)$ is not of exponential order (Exercise 23), show that the function

$$f(t) = t \exp(t^2) \sin(\exp(t^2))$$

has a Laplace transform showing that the conditions of Theorem 3.1–1 are not necessary.

25. Show that the Laplace transform of t^α $(\alpha > -1)$ is given by

$$\frac{\Gamma(\alpha + 1)}{s^{\alpha+1}},$$

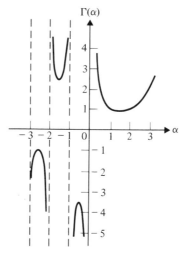

FIGURE 3.1–10 The gamma function.

where

$$\Gamma(\alpha + 1) \equiv \int_0^\infty t^\alpha \exp(-t)\, dt$$

is called the **gamma function** of $\alpha + 1$ (Fig. 3.1–10).

26. Referring to Exercise 25, show that

$$\Gamma(\alpha + 1) = \alpha\Gamma(\alpha).$$

(*Hint*: Apply integration by parts to $\Gamma(\alpha + 1)$.)

27. Referring to Exercise 25, show that

$$\Gamma(n + 1) = n!, \qquad n = 1, 2, 3, \ldots .$$

For this reason the gamma function is also called the **generalized factorial function**.

28. Prove that if $f(t)$ is piecewise continuous on (a, b), then $f(t)$ is integrable on (a, b).

29. Prove that the Laplace transformation is unique.

3.2 TRANSFORMS OF DERIVATIVES AND INTEGRALS

If the Laplace transform is to provide a useful method for solving differential equations such as that shown in Eq. (3.1–1), then we must be able to transform the left member of this equation as well as the right. This means that we must obtain the transform pairs for $f'(t)$ and $f''(t)$. To this end we state two theorems.

Theorem 3.2–1 If $f(t)$ is piecewise continuous on $t \geq 0$ and of exponential order $\exp(at)$ with $a > 0$ and if $f'(t)$ is piecewise continuous on $t \geq 0$, then*

$$\mathcal{L}(f'(t)) = s\mathcal{L}(f(t)) - f(0^+) - \sum_{i=1}^{n} \exp(-sT_i)(f(T_i^+) - f(T_i^-)), \quad \textbf{(3.2--1)}$$

where $s > a$ and the discontinuities of $f(t)$ are at $t = T_1, T_2, \ldots, T_n$.

We omit the proof of this theorem but remark that when $f(t)$ is *continuous*, then Eq. (3.2–1) is replaced by (Exercise 16)

$$\mathcal{L}(f'(t)) = s\mathcal{L}(f(t)) - f(0), \qquad s > a. \qquad \textbf{(3.2--2)}$$

In most of our work this will be the appropriate form for computing the Laplace transform of $f'(t)$. Equation (3.2–2) is also of value in obtaining transform pairs, as shown in the following examples.

EXAMPLE 3.2–1 Use Theorem 3.2–1 to show that

$$\mathcal{L}(t \exp(bt)) = \frac{1}{(s - b)^2}.$$

Solution. Note that this result can also be obtained by using Theorem 3.1–3 (see Eq. 3.1–15). If, however, we consider

$$f(t) = t \exp(bt),$$

then

$$f'(t) = bt \exp(bt) + \exp(bt)$$

and

$$\mathcal{L}(f'(t)) = \mathcal{L}(bt \exp(bt) + \exp(bt))$$
$$-f(0) + s\mathcal{L}(t \exp(bt)) = \mathcal{L}(bt \exp(bt)) + \mathcal{L}(\exp(bt))$$
$$= b\mathcal{L}(t \exp(bt)) + \mathcal{L}(\exp(bt)).$$

Hence

$$(s - b)\mathcal{L}(t \exp(bt)) = f(0) + \mathcal{L}(\exp(bt))$$

$$= \frac{1}{s - b}$$

so that

$$\mathcal{L}(t \exp(bt)) = \frac{1}{(s - b)^2}.$$

We remark that $f(t)$ and $f'(t)$ satisfy the hypotheses of Theorem 3.2–1 (Exercise 1). ∎

EXAMPLE 3.2–2 Find the Laplace transform of

$$f(t) = \cos^2 at.$$

---

* We use $f(T_i^+)$ for $\lim_{t \to T_i^+} f(t)$ and $f(T_i^-)$ for $\lim_{t \to T_i^-} f(t)$.

Solution. We have $f(0) = 1$ and

$$f'(t) = -2a \sin at \cos at = -a \sin 2at.$$

Thus, using Eq. (3.2–2), we get

$$\mathcal{L}(f'(t)) = s\mathcal{L}(f(t)) - 1$$

so that

$$\mathcal{L}(f'(t)) = s\mathcal{L}(f'(t)) + 1 / s$$

$$\mathcal{L}(\cos^2 at) = \frac{-2a^2}{s^2 + 4a^2} \cdot \frac{1}{s} + \frac{1}{s}$$

$$= \frac{1}{s}\left(\frac{s^2 + 2a^2}{s^2 + 4a^2}\right).$$

Again we note that the hypotheses of Theorem 3.2–1 are satisfied (Exercise 2). ∎

Theorem 3.2–2 If $f(t)$ and $f'(t)$ are of exponential order exp (at) with $a > 0$, and if $f'(t)$ is continuous and $f''(t)$ is piecewise continuous on $t \geq 0$, then $\mathcal{L}(f'(t)) = sF(s) - f(0)$

$$\mathcal{L}(f''(t)) = s^2 F(s) - sf(0) - f'(0), \qquad s > a. \qquad \textbf{(3.2–3)}$$

The proof of this theorem and the following one are left as exercises (3 and 4 of this section, respectively).

Theorem 3.2–3 If $f(t)$, $f'(t)$, ..., $f^{(n-1)}(t)$ are of exponential order exp (at) with $a > 0$, and $f^{(n-1)}(t)$ is continuous and $f^{(n)}(t)$ is piecewise continuous on $t \geq 0$, then

$$\mathcal{L}(f^{(n)}(t)) = s^n F(s) - s^{n-1} f(0) - s^{n-2} f'(0)$$
$$- \cdots - f^{(n-1)}(0), \qquad s > a. \qquad \textbf{(3.2–4)}$$

EXAMPLE 3.2–3   Transform the following initial-value problem to the s-domain.

$$4y''(t) - 4y'(t) + y(t) = 3 \sin 2t, \qquad y(0) = 1, \qquad y'(0) = -1$$

Solution. Denoting $\mathcal{L}(y(t))$ by $Y(s)$, we have

$$\mathcal{L}(4y''(t)) = 4(s^2 Y(s) - sy(0) - y'(0))$$
$$= 4s^2 Y(s) - 4s + 4;$$
$$\mathcal{L}(4y'(t)) = 4(sY(s) - y(0))$$
$$= 4s Y(s) - 4;$$
$$\mathcal{L}(3 \sin 2t) = \frac{6}{s^2 + 4};$$

using Eqs. (3.2–3) and (3.2–4) and Table 3.1–1. Thus the initial-value problem becomes

$$(2s - 1)^2 Y(s) = 4(s - 2) + \frac{6}{s^2 + 4}.$$

In Section 3.3 we will discuss the *inverse* transformation of $Y(s)$, as obtained from the last equation, to $y(t)$, the solution of the given initial-value problem. ■

In the proof of the existence theorem of the Laplace transform (Theorem 3.1–1) it is shown that the transform of $|f(t)|$ converges independently of the value of s provided that $s > a$. In other words, $\mathscr{L}(f(t)) = F(s)$ converges *uniformly* and *absolutely* in every interval $s \geq \alpha$ where $\alpha > a$. The uniform convergence of the improper integral permits us to interchange certain operations as shown below:

$$\lim_{s \to 0^+} \int_0^\infty \exp{(-st)}f(t)\,dt = \int_0^\infty \lim_{s \to 0^+} \exp{(-st)}f(t)\,dt, \quad \text{(3.2–5)}$$

$$\lim_{s \to \infty} \int_0^\infty \exp{(-st)}f(t)\,dt = \int_0^\infty \lim_{s \to \infty} \exp{(-st)}f(t)\,dt, \quad \text{(3.2–6)}$$

$$\frac{d}{ds}\int_0^\infty \exp{(-st)}f(t)\,dt = \int_0^\infty \frac{d}{ds}(\exp{(-st)}f(t))\,dt, \quad \text{(3.2–7)}$$

$$\int_s^b \left(\int_0^\infty \exp{(-st)}f(t)\,dt \right) ds = \int_0^\infty \left(\int_s^b \exp{(-st)}f(t)\,ds \right) dt, \quad b \geq s.$$
$$\text{(3.2–8)}$$

For a more complete discussion and proofs of the above see Walter Rudin's *Principles of Mathematical Analysis*, 3d ed. (New York: McGraw-Hill, 1976). Suffice it to say that these relations hold whenever $f(t)$ is piecewise continuous on $t \geq 0$ and of exponential order $\exp{(at)}$ with $a > 0$.

We can now prove the following theorem.

Theorem 3.2–4 (*Differentiation of $F(s)$*). If $f(t)$ is piecewise continuous on $t \geq 0$ and of exponential order $\exp{(at)}$ with $a > 0$, then for $n = 1$, $2, \ldots$,

$$F^{(n)}(s) = \frac{d^n F(s)}{ds^n} = \mathscr{L}((-t)^n f(t)), \quad s > a. \quad \text{(3.2–9)}$$

PROOF. If $n = 1$, we have, using Eq. (3.2–7),

$$\frac{dF(s)}{ds} = \frac{d}{ds}\int_0^\infty \exp{(-st)}f(t)\,dt$$

$$= \int_0^\infty \frac{d}{ds}(\exp{(-st)}f(t)\,dt$$

$$= \int_0^\infty -t\exp{(-st)}f(t)\,dt$$

$$= \mathscr{L}(-tf(t)).$$

But the functions $t^n f(t)$ are piecewise continuous and of exponential order $\exp{(at)}$ with $a > 0$ (Exercise 1); hence Eq. (3.2–7) may be applied repeatedly to the last result to complete the proof (Exercise 5). □

Theorem 3.2–4 shows that differentiating the transform of a function with respect to s is equivalent to multiplying the function by $-t$. This provides us with another method for finding Laplace transform pairs as shown in the next example.

EXAMPLE 3.2–4   Find the Laplace transform of

$$f(t) = t^2 \cosh at.$$

Solution. From Table 3.1–1 we have

$$\mathcal{L}(\cosh at) = \frac{s}{s^2 - a^2}.$$

Differentiating this twice with respect to s yields

$$\frac{2s(s^2 + 3a^2)}{(s^2 - a^2)^3};$$

hence

$$\mathcal{L}(t^2 \cosh at) = \frac{2s(s^2 + 3a^2)}{(s^2 - a^2)^3}, \qquad s > a. \quad \blacksquare$$

Contrast the above with the amount of computation that would be required if we proceeded from first principles using the definition of the Laplace transform (3.1–3).

Theorem 3.2–4 and the following theorem will also be useful in obtaining inverse Laplace transforms, as we will see in Section 3.3.

Theorem 3.2–5 If $f(t)$ is piecewise continuous on $t \geq 0$ and of exponential order $\exp(at)$ with $a > 0$, then

$$\int_s^\infty F(u)\, du = \mathcal{L}(f(t)/t) \qquad\qquad \textbf{(3.2–10)}$$

for $s > a$ provided that $\lim_{t \to 0^+} f(t)/t$ exists.

PROOF. From Eq. (3.2–8) we have

$$\int_s^b F(u)\, du = \int_s^b \left(\int_0^\infty \exp(-ut) f(t)\, dt \right) du$$

$$= \int_0^\infty \left(\int_s^b \exp(-ut) f(t)\, du \right) dt$$

$$= \int_0^\infty \frac{f(t)}{t} (\exp(-st) - \exp(-bt))\, dt$$

$$= \int_0^\infty \frac{f(t)}{t} \exp(-st)\, dt - \int_0^\infty \frac{f(t)}{t} \exp(-bt)\, dt$$

$$= G(s) - G(b),$$

where $G(s) = \mathcal{L}(f(t)/t)$. But $G(b) \to 0$ as $b \to \infty$ (Exercise 17); hence the desired result follows. Note that the above proof also depends on the fact that $f(t)/t$ is piecewise continuous on $t \geq 0$ and of exponential order $\exp(at)$ with $a > 0$ (Exercise 6). ☐

The last theorem indicates that integrating the transform of a function is equivalent to dividing the function by t. We can use this property to obtain transforms, as illustrated by the following example.

EXAMPLE 3.2–5   Find the Laplace transform of $(\sin t)/t$.

Solution. Consider $f(t) = \sin t$, which meets all the conditions of Theorem 3.2–5. From Table 3.1–1 we have

$$F(s) = \mathcal{L}(\sin t) = \frac{1}{s^2 + 1}.$$

Thus

$$\int_s^\infty \frac{du}{u^2 + 1} = \arctan u \Big|_s^\infty = \frac{\pi}{2} - \arctan s$$

and

$$\mathcal{L}\left(\frac{\sin t}{t}\right) = \frac{\pi}{2} - \arctan s$$

$$= \arctan \frac{1}{s}, \quad s > a. \quad \blacksquare$$

Theorem 3.2–5 will allow us to expand our brief table of Laplace transforms (Table 3.1–1) and will also be of use in finding inverse transforms in the following section.

EXERCISES 3.2

▶   **1.** Show that if $f(t)$ is piecewise continuous on $t \geq 0$ and of exponential order $\exp(at)$ with $a > 0$, then $t^n f(t)$ also has these properties.

2. Show that the function

$$f(t) = \cos^2 at$$

satisfies the hypotheses of Theorem 3.2–1.

3. Prove Theorem 3.2–2. (*Hint*: Use the definition of $\mathcal{L}(f''(t))$, integrate by parts, and apply Eq. 3.2–2.)

4. Prove Theorem 3.2–3.

5. Complete the proof of Theorem 3.2–4.

6. Prove that if $f(t)$ is piecewise continuous on $t \geq 0$ and of exponential order $\exp(at)$ with $a > 0$, and if

$$\lim_{t \to 0^+} f(t)/t$$

exists, then $f(t)/t$ is piecewise continuous on $t \geq 0$ and of exponential order exp (at) with $a > 0$.

▶▶ 7. Use Eq. (3.2–1) to find the Laplace transforms of the derivatives of each of the following discontinuous functions.

a) $f(t) = \begin{cases} 0, & \text{for } 0 < t < 1, \\ 1, & \text{for } 1 < t < 2, \\ 0, & \text{for } t > 2. \end{cases}$

(See Fig. 3.2–1.)

b) $f(t) = \begin{cases} t, & \text{for } 0 < t < 1, \\ 0, & \text{for } t > 1. \end{cases}$

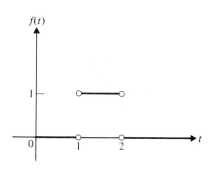

FIGURE 3.2–1

c) $f(t) = \begin{cases} 2 - t, & \text{for } 1 < t < 2, \\ 0, & \text{otherwise.} \end{cases}$

d) $f(t) = \begin{cases} 1, & \text{for } 0 < t < 1, \\ t, & \text{for } 1 < t < 2, \\ 0, & \text{for } t > 2. \end{cases}$

(See Fig. 3.2–2.)

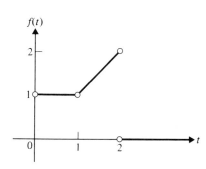

FIGURE 3.2–2

8. Obtain the Laplace transform of each of the following functions.

a) $\sin^2 at$

b) $t^2 \exp (bt)$

c) $t \sin at$

d) $t \cos at$

e) $t \sinh at$

f) $t \cosh at$

✓ **9.** Obtain the Laplace transform of each of the following functions.

a) $\dfrac{1}{t}(1 - \cos at)$ **b)** $\dfrac{1}{t}(1 - \exp(-t))$

✓ **c)** $\dfrac{1}{t}(\exp t - \cos t)$ ✓ **d)** $\dfrac{2}{t}(1 - \cosh at)$

e) $\dfrac{1}{t}(\exp(-at) - \exp(-bt))$

✓ **10.** Use Theorem 3.2–4 to obtain the Laplace transforms of each of the following functions.

a) $t^2 \cos at$ ✓ **b)** $t^2 \sin at$

c) $t \exp(t) \sin t$ ✓ **d)** $t \exp(t) \cos 2t$

11. Show that

$$\int_0^\infty \frac{\sin ax}{x}\,dx = \frac{\pi}{2}, \qquad a > 0.$$

(*Hint:* See Example 3.2–5 and use Eq. 3.2–5.)

12. Show that

$$\int_0^\infty \frac{e^{-ax}\sin x}{x}\,dx = \operatorname{arccot} a, \qquad a > 0.$$

13. Show that

$$\int_0^\infty \frac{\exp(-ax) - \exp(-bx)}{x}\,dx = \log\frac{b}{a}.$$

14. Show that

$$\int_0^\infty \frac{\arctan ax - \arctan bx}{x}\,dx = \frac{\pi}{2}\log\frac{a}{b}.$$

15. Show that

$$\int_0^\infty \frac{\cos ax - \cos bx}{x} = \log\frac{b}{a}.$$

▶▶▶ **16.** Prove that if $f(t)$ is continuous for $t \ge 0$ and of exponential order $\exp(at)$ with $a \ge 0$, and if $f'(t)$ is piecewise continuous for $t \ge 0$, then Eq. (3.2–2) holds. (*Hint:* Use the definition of $\mathcal{L}(f'(t))$ and integration by parts.)

17. Prove that if $f(t)$ is piecewise continuous for $t \ge 0$ and of exponential order $\exp(at)$ with $a \ge 0$, then

$$\lim_{s \to \infty} F(s) = 0,$$

where $F(s) = \mathcal{L}(f(t))$. (*Hint:* Use Eq. 3.2–6.)

3.3 INVERSE TRANSFORMS

In Sections 3.1 and 3.2 we obtained the Laplace transforms of a number of piecewise continuous functions. We also showed (see Eq. 3.2–4) that the transforms of derivatives of functions of interest can be expressed in terms of the transforms of the functions themselves. All of this leads to the fact that given an initial-value problem involving $y(t)$ and its derivatives, we can transform the problem to an algebraic expression involving $Y(s)$, the Laplace transform of $y(t)$. Solving this expression for $Y(s)$ is a necessary step to obtaining $y(t)$, the solution to the given problem. We have carried the computation to this point for an initial-value problem in Example 3.2–3.

The next natural question is, "Having found $Y(s)$, how do we find $y(t)$?" It would seem that Table 3.1–1 would be helpful in this regard. Although entries in the table were made *from* an $f(t)$ *to* a corresponding $F(s)$, we may consider finding an entry such as

$$F(s) = \frac{a}{s^2 + a^2}$$

and conclude that

$$f(t) = \sin at.$$

When we do this (use the table of transforms in reverse), we are saying that

$$\mathscr{L}^{-1}\left(\frac{a}{s^2 + a^2}\right) = \sin at,$$

that is, the *inverse Laplace transform* of $a/(s^2 + a^2)$ is $\sin at$.

Our experience with other tables such as integral tables and tables of logarithms indicates that some caution is called for. We ask the question, "Is the inverse Laplace transform unique, that is, does $F(s) = G(s)$ imply that $f(t) = g(t)$?" A partial answer to this question is contained in the following example.

EXAMPLE 3.3–1   Consider the two functions defined as follows:

$$f(t) = \begin{cases} 2, & \text{for } 0 \le t < 2, \\ 1, & \text{for } 2 \le t, \end{cases} \quad \text{and} \quad g(t) = \begin{cases} 2, & \text{for } 0 \le t \le 2, \\ 1, & \text{for } 2 < t. \end{cases}$$

Graphs of these functions are shown in Fig. 3.3–1. Clearly $f(t) \ne g(t)$

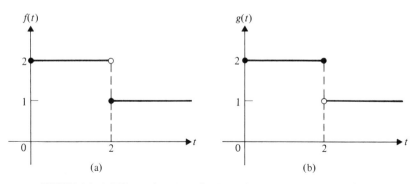

(a)   (b)

FIGURE 3.3–1 Different functions that have the same Laplace transform.

for $t > 0$ since $f(2) = 1$ and $g(2) = 2$. Both functions, however, have Laplace transforms, which we compute in parallel fashion below.

$$F(s) = 2 \int_0^{2^-} \exp(-st)\, dt \qquad G(s) = 2 \int_0^{2} \exp(-st)\, dt$$

$$+ \int_2^{\infty} \exp(-st)\, dt \qquad + \int_{2^+}^{\infty} \exp(-st)\, dt$$

$$= \frac{2\exp(-st)}{-s}\Big|_0^{2^-} \qquad = \frac{2\exp(-st)}{-s}\Big|_0^{2}$$

$$+ \frac{\exp(-st)}{-s}\Big|_2^{\infty} \qquad + \frac{\exp(-st)}{-s}\Big|_{2^+}^{\infty}$$

$$= \lim_{t \to 2^-} \frac{2\exp(-st)}{-s} \qquad = \frac{-2\exp(-2s)}{s} + \frac{2}{s}$$

$$+ \frac{2}{s} + \frac{\exp(-2s)}{s} \qquad + \lim_{t \to 2^+} \frac{\exp(-st)}{s}$$

$$= \frac{-2\exp(-2s)}{s} + \frac{2}{s} \qquad = \frac{-2\exp(-2s)}{s} + \frac{2}{s}$$

$$+ \frac{\exp(-2s)}{s} \qquad + \frac{\exp(-2s)}{s}$$

$$= \frac{2}{s} - \frac{\exp(-2s)}{s}; \qquad = \frac{2}{s} - \frac{\exp(-2s)}{s}.$$

Thus $F(s) = G(s)$ although $f(t) \neq g(t)$. From this single example we can conclude that the inverse transform is *not* unique. ■

There is a theorem, which we state without proof, that covers the uniqueness of the inverse transform. It was first proved* by M. Lerch (1860–1922) and named after him. A proof can be found in Donald L. Kreider, et al., *An Introduction to Linear Analysis*, p. 678 (Reading, Mass.: Addison-Wesley, 1966).

Theorem 3.3–1 (*Lerch's theorem*). If $\mathscr{L}(f(t)) = \mathscr{L}(g(t))$, then $f(t) = g(t) + n(t)$, where $n(t)$ satisfies

$$\int_0^t n(u)\, du = 0 \quad \text{for every } t > 0.$$

The function $n(t)$ in Theorem 3.3–1 is called a *null function*. In Example 3.3–1 $n(t)$ was defined at the single point $t = 2$. A null function may be nonzero at a finite number of points or even at an infinite

---

* Rozpravy ceské Akadamie, 2nd class, vol. 1, no. 33 (1892) and vol. 2, no. 9 (1893). Later published as "Sur un point de la Théorie des Fonctions Génératrices d'Abel," *Acta Mathematica*, vol. 27 (1903), pp. 339–351.

number of points such as $t = 1, 2, 3, \ldots$, but it cannot differ from zero over any interval of positive length on the t-axis.

Fortunately, functions that differ only by a null function do not cause any difficulties in problems of practical interest. Hence we will use the following simplified version of Lerch's theorem.

Theorem 3.3–2 If $f(t)$ and $g(t)$ are continuous on $t \geq 0$ and $\mathcal{L}(f(t)) = \mathcal{L}(g(t))$, then $f(t) \equiv g(t)$ for all $t \geq 0$.

Theorem 3.3–2 will allow us to use a table of transforms in all the situations that we will encounter. Recalling Example 3.2–3, we had

$$(2s - 1)^2 Y(s) = 4(s - 2) + \frac{6}{s^2 + 4}.$$

Solving for $Y(s)$ gave us

$$Y(s) = \frac{4s^3 - 8s^2 + 16s - 26}{4s^4 - 4s^3 + 17s^2 - 16s + 4}.$$

Now we face another difficulty. We cannot realistically hope to find the above function $Y(s)$ in a table of transform pairs, as complete as that table may be. Yet rational polynomial functions such as this occur frequently and we need to have a systematic method of dealing with them. First of all, we have "committed" too much algebra in solving for $Y(s)$. It is more desirable to leave the result in factored form and to have several rational functions rather than just a single one. Accordingly, we try

$$Y(s) = \frac{4(s - 2)}{(2s - 1)^2} + \frac{6}{(s^2 + 4)(2s - 1)^2}.$$

Since the above terms still do not appear in Table 3.1–1, we make use of a technique from algebra called *decomposition into partial fractions*. Although it is one of the techniques used when methods of integration are studied in calculus, we review the method now by means of a number of examples.

EXAMPLE 3.3–2   Find $f(t)$ if

$$F(s) = \frac{s - 2}{s^2 - s - 6}.$$

Solution. We write

$$\frac{s - 2}{s^2 - s - 6} = \frac{s - 2}{(s + 2)(s - 3)} = \frac{A}{s + 2} + \frac{B}{s - 3},$$

where A and B are to be determined. Thus

$$\frac{s - 2}{s^2 - s - 6} = \frac{A(s - 3) + B(s + 2)}{(s + 2)(s - 3)},$$

which leads to
$$A(s - 3) + B(s + 2) \equiv s - 2.$$

We have indicated that this last is actually an *identity*, meaning that it must hold for *all* values of s. Hence if $s = 3$, we obtain $B = \frac{1}{5}$, and if $s = -2$, then $A = \frac{4}{5}$. Now we can write*
$$F(s) = \frac{4}{5}\frac{1}{s + 2} + \frac{1}{5}\frac{1}{s - 3}$$

so that
$$f(t) = \frac{4}{5}e^{-2t} + \frac{1}{5}e^{3t}. \quad \blacksquare$$

EXAMPLE 3.3–3 Find $f(t)$ if
$$F(s) = \frac{s^2 - 4s + 7}{s^4 - 4s^3 + 6s^2 - 4s + 1}.$$

Solution. Here
$$F(s) = \frac{s^2 - 4s + 7}{(s - 1)^4} = \frac{A}{s - 1} + \frac{B}{(s - 1)^2} + \frac{C}{(s - 1)^3} + \frac{D}{(s - 1)^4}$$
$$= \frac{A(s - 1)^3 + B(s - 1)^2 + C(s - 1) + D}{(s - 1)^4}.$$

Hence equating *numerators* and putting $s = 1$ yields $D = 4$. The remaining constants can be obtained by equating like powers of s. Thus $A = 0,^\dagger$ $B = 1$, $-2B + C = -4$, and $C = -2$, and we have
$$F(s) = \frac{1}{(s - 1)^2} - \frac{2}{(s - 1)^3} + \frac{2}{3}\frac{6}{(s - 1)^4}$$

so that
$$f(t) = te^t - t^2e^t + \frac{2}{3}t^3e^t. \quad \blacksquare$$

The last example illustrates the fact that if the denominator of $F(s)$ contains a *repeated* linear factor, then all the various powers of that factor that may arise must be included in the partial fraction decomposition.

EXAMPLE 3.3–4 Find $f(t)$ if
$$F(s) = \frac{1}{s^3 + s}.$$

---

* We are tacitly assuming that the operator $\mathscr{L}^{-1}$ is also linear (see Exercise 15).
† Note that the updated problem to be solved now is
$$s^2 - 4s + 7 \equiv B(s - 1)^2 + C(s - 1) + 4.$$

Solution. Here

$$F(s) = \frac{1}{s(s^2 + 1)} = \frac{A}{s} + \frac{Bs + C}{s^2 + 1}$$

$$= \frac{A(s^2 + 1) + s(Bs + C)}{s(s^2 + 1)}.$$

Putting $s = 0$ after equating the numerators yields $A = 1$, whereas putting $s = i$ $(i = \sqrt{-1})$ produces $-B + Ci = 1$ from which $B = -1$ and $C = 0$. Thus

$$F(s) = \frac{1}{s} - \frac{s}{s^2 + 1}$$

so that

$$f(t) = 1 - \cos t. \quad \blacksquare$$

In the above example we have illustrated the fact that if there is an irreducible (that is, prime) *quadratic* factor in the denominator of $F(s)$, then the partial fraction decomposition of this term must have a *linear* numerator. We have also shown that complex values may be assigned to *s*. In some cases combinations of the foregoing techniques may be required.

Unfortunately, the method of decomposition into partial fractions has some drawbacks. Suppose, for example, that we wish to find the inverse transform of

$$F(s) = \frac{s}{(s^2 + 1)^2}.$$

We leave it as an exercise (Exercise 13) to show that

$$\frac{As + B}{s^2 + 1} + \frac{Cs + D}{(s^2 + 1)^2} = \frac{s}{(s^2 + 1)^2}$$

cannot be satisfied except trivially.* Thus some other method must be found for functions of this type that are products. We will return to this example in Section 3.4.

Going back to Example 3.2–3, we have

$$Y(s) = \frac{4(s - 2)}{(2s - 1)^2} + \frac{6}{(s^2 + 4)(2s - 1)^2}.$$

By the method illustrated in the previous examples we can write

$$Y(s) = \frac{265}{289} \frac{1}{s - 1/2} - \frac{39}{34} \frac{1}{(s - 1/2)^2} + \frac{24}{289} \frac{s}{s^2 + 4} - \frac{45}{289} \frac{2}{s^2 + 4}.$$

---

* See also Exercise 4(g) in Section 10.6.

Now with the aid of Table 3.1–1 we have

$$y(t) = \frac{265}{289} \exp\left(\frac{t}{2}\right) - \frac{39}{34} t \exp\left(\frac{t}{2}\right) + \frac{24}{289} \cos 2t - \frac{45}{289} \sin 2t.$$

The above result could also have been obtained by the methods of Chapter 2. We point out, however, that the Laplace transform method provides a *systematic* procedure involving the use of a *table*. Moreover, the initial conditions are incorporated into the transform.

A second-order initial-value problem,

$$A\ddot{y} + B\dot{y} + Cy = f(t), \tag{3.3–1}$$

with $y(0)$ and $\dot{y}(0)$ specified and where A, B, and C are constants, produces

$$Y(s) = \frac{1}{As^2 + Bs + C}(F(s) + (As + B)y(0) + A\dot{y}(0)). \tag{3.3–2}$$

If Eq. (3.3–1) describes the behavior of a mechanical, electrical, hydraulic, or biological system, then we refer to $Y(s)$ as the **response transform** (Fig. 3.3–2). Equation (3.3–2) emphasizes the fact that the response transform is a product of two factors. One, the term

$$\frac{1}{As^2 + Bs + C},$$

is called the **system function** since it characterizes the system. The second,

$$(F(s) + (As + B)y(0) + A\dot{y}(0)),$$

is the **excitation function**, which involves the initial conditions and the transform of the forcing function $f(t)$. Thus one can write Eq. (3.3–2) from Eq. (3.3–1) (usually with the aid of a table) to obtain $Y(s)$, use partial fraction decomposition, and then again refer to a table to find $y(t)$. (See Exercise 13.)

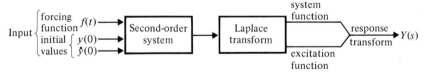

FIGURE 3.3–2 Solving a second-order initial-value problem by the Laplace transform method.

One of the advantages of the Laplace transform method over the methods presented in Chapter 2 can now be seen more clearly. In applied mathematics we often study the behavior of a system under various initial conditions and forcing functions. This study is easily done by changing only the excitation function in Eq. (3.3–2).

We give another example, this time extending the previous ideas to a third-order initial-value problem (see Exercise 1).

EXAMPLE 3.3–5 Solve the initial-value problem

$$\dddot{y} + \dot{y} = t + 1, \qquad y(0) = \dot{y}(0) = \ddot{y}(0) = 3.$$

Solution. Here the system function is

$$\frac{1}{s^3 + s} = \frac{1}{s(s^2 + 1)}$$

and the excitation function is

$$\frac{1}{s^2} + \frac{1}{s} + (s^2 + 1)(3) + s(3) + 3$$

or

$$\frac{3s^4 + 3s^3 + 6s^2 + s + 1}{s^2}.$$

Hence

$$Y(s) = \frac{3s^4 + 3s^3 + 6s^2 + s + 1}{s^3(s^2 + 1)}$$

and, using partial fraction decomposition, we have

$$Y(s) = \frac{5}{s} + \frac{1}{s^2} + \frac{1}{s^3} - 2\,\frac{s - 1}{s^2 + 1}.$$

Thus

$$y(t) = 5 + t + \tfrac{1}{2}t^2 - 2\cos t + 2\sin t. \quad\blacksquare$$

This last example illustrates the somewhat mechanical character of the Laplace transform method. This method is preferred by those who must *routinely* investigate the behavior of various systems under different forcing functions and initial conditions. Our objective in this chapter, however, is not to reduce problem solving to a routine but to provide mathematical tools for a wide spectrum of problems. To this end we still need to discuss forcing functions that are step, impulse, or periodic functions as well as methods of solving linear differential equations with nonconstant coefficients. These matters will be taken up in the remainder of this chapter. A more general method of going from $F(s)$ to $f(t)$ depends on the properties of functions of a complex variable and discussion will be postponed until Chapter 10.

KEY WORDS AND PHRASES

inverse transform

null function

partial fraction decomposition

response transform

system function

excitation function

EXERCISES 3.3

▶ **1.** Show that for the initial-value problem

$$A\dddot{y} + B\ddot{y} + C\dot{y} + Dy = f(t)$$

with $y(0)$, $\dot{y}(0)$, and $\ddot{y}(0)$ specified and A, B, C, and D constants, the transform of $y(t)$ is given by

$$Y(s) = \frac{1}{As^3 + Bs^2 + Cs + D}$$

$$\times (F(s) + (As^2 + Bs + C)y(0) + (As + B)\dot{y}(0) + A\ddot{y}(0)).$$

2. Carry out the details in Example 3.3–5.

3. Solve Example 3.3–5 without making use of the Laplace transform.

▶▶ ✓ **4.** For each of the following functions $F(s)$, find $\mathcal{L}^{-1}(F(s))$.

a) $\dfrac{s}{s^2 + 5}$ ✓b) $\dfrac{1}{(s + 3)^2}$ ✓c) $\dfrac{-2}{s^2 - 2}$

d) $\dfrac{1}{s^2 + 3}$ e) $\dfrac{2s^2 - 3s + 5}{s^2(s^2 + 1)}$

✓ **5.** Find the inverse transforms of each of the following functions.

a) $\dfrac{1}{(s^2 + 4)(s - 3)}$ ✓b) $\dfrac{s^2 + 8}{s(s^2 + 16)}$

✓c) $\dfrac{1}{(s + 3)(s - 2)}$ d) $\dfrac{s + 3}{(s + 2)(s - 3)}$

6. Obtain the Laplace transform pair for each of the following, where a, b, c, and d are constants.

a) $\dfrac{1}{s(s + a)(s + b)}$ b) $\dfrac{s + d}{s(s + a)(s + b)}$

c) $\dfrac{1}{(s + a)(s + b)(s + c)}$ d) $\dfrac{s + d}{(s + a)(s + b)(s + c)}$

7. Show that

$$F(s) = \frac{2s^2 + 5}{s^2 - 3}$$

cannot be the Laplace transform of any ordinary function.*

8. Use partial fraction decomposition to write each of the following as a sum of fractions. Here a, b, c, and d are constants.

a) $\dfrac{s + d}{(s + a)(s^2 + b^2)}$ b) $\dfrac{s + d}{s(s + a)(s^2 + b^2)}$

c) $\dfrac{s}{(s^2 + a^2)(s^2 + b^2)}$ d) $\dfrac{s + d}{(s^2 + a^2)(s^2 + b^2)}$

* We are not assuming any knowledge of a function having the properties listed in Eq. (3.5–9).

9. Find the inverse transform of each function given in Exercise 8.

10. Show that

$$\mathcal{L}^{-1}\left(\frac{s+2}{(s-2)^2+9}\right) = \frac{5}{3}\exp(2t)\sin(3t+\phi),$$

where $\phi = \arctan(-1/2)$. (Compare Eq. 3.1–13.)

11. Consider

$$F(s) = \frac{3s+9}{s^2+2s+10}.$$

This can be written

$$\frac{3(s+1)}{(s+1)^2+9} + \frac{6}{(s+1)^2+9}$$

by completing the square. Use Eqs. (3.1–13) and (3.1–14) to obtain $f(t)$.

12. Use the method suggested in Exercise 11 to find the inverse of each of the following.

   a) $F(s) = \dfrac{2s+3}{s^2-2s+5}$

   b) $F(s) = \dfrac{2s+15}{s^2+6s+25}$

   c) $F(s) = \dfrac{s+3}{s^2+2s+2}$

13. Solve each of the following initial-value problems by using the Laplace transform.

   a) $\ddot{y} - 4y = 0,\quad y(0) = 1,\quad \dot{y}(0) = 2$

   b) $\ddot{y} + \omega^2 y = 0,\quad y(0) = y_0,\quad \dot{y}(0) = v_0$

   c) $\ddot{y} - 5\dot{y} + 4y = \exp(2t),\quad y(0) = 1,\quad \dot{y}(0) = 0$

   d) $\ddot{y} - 4\dot{y} + 4y = t^2,\quad y(0) = 0,\quad \dot{y}(0) = 1$

   e) $\ddot{y} + 2\dot{y} + 2y = t,\quad y(0) = \dot{y}(0) = 1$

14. Solve each of the following initial-value problems involving linear differential equations. (*Hint*: Use Theorem 3.2–4.)

   a) $t\ddot{y} - t\dot{y} - y = 0,\quad y(0) = 0,\quad \dot{y}(0) = 1$

   b) $\ddot{y} + 2t\dot{y} - 4y = 1,\quad y(0) = \dot{y}(0) = 0$

▶▶▶ 15. Show that the method of partial fraction decomposition fails for the function

$$F(s) = \frac{s}{(s^2+1)^2}.$$

16. Use Theorem (3.2–4) to solve the problem in Exercise 15 (compare Exercise 1c of Section 3.2).

17. Let $f(t) = \mathcal{L}^{-1}(F(s))$ and $g(t) = \mathcal{L}^{-1}(G(s))$. Prove that $af(t) + bg(t) = \mathcal{L}^{-1}(aF(s) + bG(s))$ for any constants a and b. (*Hint*: $f(t) = \mathcal{L}^{-1}(F(s))$ can also be written as $\mathcal{L}(f(t)) = F(s)$. Note that this shows that the inverse transform is also a *linear* operation.)

18. The use of derivatives and integrals as outlined in Section 3.2 can be helpful in finding inverse transforms. Do so for each of the following functions, where a and b are constants.

a) $\frac{1}{s} \arctan \frac{1}{s}$ (Compare Example 3.2–5.)

b) $\log \dfrac{s-a}{s-b}$ **c)** $\log\left(1 + \dfrac{a^2}{s^2}\right)$

19. Show that if $\mathscr{L}(f(t)) = F(s)$ and $a > 0$, then

$$\mathscr{L}^{-1}(F(as + b)) = \frac{1}{a} \exp\left(-\frac{bt}{a}\right) f\left(\frac{t}{a}\right),$$

where b is constant.

20. Solve Example 3.2–3 without using the Laplace transform.

21. Carry out the details given in this section to solve Example 3.2–3 by using the Laplace transform.

22. Discuss the conditions that must exist before problems like those in Exercise 14 can be solved by the Laplace transform method.

23. Carry the solution of

$$\ddot{y} + t^2 y = 0, \qquad y(0) = 0, \qquad \dot{y}(0) = 1$$

as far as possible.

3.4 CONVOLUTION

We have seen that the Laplace transformation is a *linear* transformation. This means that the transform of a sum of functions is the sum of the transforms of the individual functions. The same is true, of course, of the inverse transformation, that is, the inverse transform of a sum of functions is the sum of the inverse transforms of the separate functions. We have used this linearity property of the Laplace transform a number of times in the preceding sections of this chapter.

Other properties of the Laplace transformation have enabled us to transform *products*. In Example 3.1–6 we obtained the transform of $\cos(at) \exp(bt)$; in Example 3.1–7 the transform of $t^n \exp(bt)$; in Example 3.2–2 the transform of $\cos^2(at)$; and in Example 3.2–5 the transform of $(\sin t)/t$. In Section 3.3 we saw that obtaining the inverse transform of products was far from a simple matter. One useful technique was to decompose the product into a sum and use a table to find the inverse of each term in the sum. Clearly, what is needed is a general method that can be applied to a product such as $F(s)G(s)$ in order to yield the appropriate function of t. Such a method exists and is called *convolution*. We introduce the concept in the following way.

Consider the two power series

$$f(t) = a_0 + a_1 t + a_2 t^2 + \cdots + a_n t^n + \cdots$$

and

$$g(t) = b_0 + b_1 t + b_2 t^2 + \cdots + b_n t^n + \cdots.$$

If we were to multiply these, we would have

$$f(t)g(t) = a_0b_0 + (a_0b_1 + a_1b_0)t + (a_0b_2 + a_1b_1 + a_2b_0)t^2 + \cdots.$$

Observe the coefficient of t^2. The subscripts in each product add to two and the a_i have been combined with the b_j in a systematic manner called **convolution**. The word comes from the folding together, one part upon another, as petals in a flower bud.

Using summation notation, we can express the above briefly as

$$f(t) = \sum_{n=0}^{\infty} a_n t^n,$$

$$g(t) = \sum_{n=0}^{\infty} b_n t^n,$$

$$f(t)g(t) = \left(\sum_{n=0}^{\infty} a_n t^n\right)\left(\sum_{n=0}^{\infty} b_n t^n\right)$$

$$= \sum_{n=0}^{\infty} \left(\sum_{m=0}^{n} a_m b_{n-m}\right) t^n.$$

Thus the coefficients in the product series $f(t)g(t)$ are given by

$$c_n = \sum_{m=0}^{n} a_m b_{n-m}, \qquad n = 0, 1, 2, \ldots .$$

If we replace the dummy index of summation m in this last by the dummy variable of integration u and, at the same time, "replace" the sequences $\{a_n\}$ and $\{b_n\}$ by $f(t)$ and $g(t)$ and the summation by an integral, we have the following definition.

Definition 3.4–1 The **convolution** of $f(t)$ and $g(t)$, denoted by $f * g(t)$, is given by

$$f * g(t) = \int_0^t f(u)g(t - u)\, du. \qquad (3.4-1)$$

It should be emphasized that the brief discussion preceding Definition 3.4–1 was given to provide motivation for the definition. Convolution is a useful concept and can be found in various places in applied mathematics since it plays an important role in heat conduction, wave motion, plastic flow and creep, and time series analysis.

Convolution can be considered as a generalized product of functions and, as such, has many of the properties of a product. We have, for example, for functions f, g, and h, which are piecewise continuous on $t \geq 0$ and of exponential order:

a) $f * g = g * f$, showing that convolution is *commutative* (Exercise 1a);

b) $f * (g * h) = (f * g) * h$, showing that convolution is *associative* (Exercise 1b);

c) $f * (g + h) = f * g + f * h$, showing that convolution is *distributive* over addition (Exercise 1c);

d) $f * 0 = 0 * f = 0$ (Exercise 1d);

e) $f * (cg) = (cf) * g = c(f * g)$ (Exercise 1e).

It is *not* true, however, that $f * 1 = f$ (Exercise 2).

Thus convolution has many of the properties of an ordinary product. We remark that the symbol $*$ has been used since 1936 and so antedates the use of the same symbol in programming languages (e.g., FORTRAN). Note that we will use the common functional notation denoting the convolution of two functions f and g by $f * g$, the value of this convolution at the point t by $(f * g)(t)$ or $f * g(t)$.

For our present purpose the importance of convolution is brought out in the following theorem.

Theorem 3.4–1 (*Convolution theorem**). Let $f(t)$ and $g(t)$ be piecewise continuous on $t \geq 0$ and of exponential order exp (at) with $a > 0$. Then the transform of the convolution $f * g(t)$ exists when $s > a$ and is equal to $F(s)G(s)$. In other words,

$$\mathcal{L}^{-1}(F(s)G(s)) = f * g(t). \tag{3.4–2}$$

PROOF. We begin by using different variables in the integrals in order to prevent confusion:

$$F(s)G(s) = \left(\int_0^\infty \exp(-su)f(u) \, du \right) \left(\int_0^\infty \exp(-sv)g(v) \, dv \right)$$
$$= \int_0^\infty \int_0^\infty \exp(-s(u+v))f(u)g(v) \, dv \, du.$$

Next, we make a change of variables:

$$u + v = t \qquad u = u$$
$$dv = dt, \qquad du = du,$$

so that

$$F(s)G(s) = \int_0^\infty \int_u^\infty \exp(-st)f(u)g(t-u) \, dt \, du.$$

The last double integral in the tu-plane is over the shaded region of Fig. 3.4–1. Interchanging the two integrals, but integrating over the

---

* Also called Borel's theorem, after Émile Borel (1871–1956), a French mathematician.

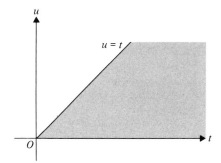

FIGURE 3.4–1

same region produces

$$F(s)G(s) = \int_0^\infty \int_0^t \exp{(-st)}f(u)g(t-u)\, du\, dt$$

$$= \int_0^\infty \exp{(-st)}\left(\int_0^t g(t-u)f(u)\, du\right) dt$$

$$= \int_0^\infty \exp{(-st)}g * f(t)\, dt$$

$$= \int_0^\infty \exp{(-st)}f * g(t)\, dt$$

$$= \mathscr{L}(f * g),$$

completing the proof. $\square$

We next give some examples in which Theorem 3.4–1 is used.

EXAMPLE 3.4–1 Find the inverse transform of

$$\frac{1}{s^2(s+1)^2}.$$

Solution. Let $F(s) = 1/s^2$ and $G(s) = 1/(s+1)^2$, then by Table 3.1–1,

$$f(t) = t \quad \text{and} \quad g(t) = t \exp{(-t)}.$$

Hence, by Theorem 3.4–1,

$$\mathscr{L}^{-1}\left(\frac{1}{s^2(s+1)^2}\right) = \int_0^t (t-u)u \exp{(-u)}\, du$$

$$= t\int_0^t u \exp{(-u)}\, du - \int_0^t u^2 \exp{(-u)}\, du$$

$$= t - 2 + (t+2) \exp{(-t)}. \quad \blacksquare$$

Note that we actually computed $g * f(t)$ rather than $f * g(t)$ for convenience. This example could also have been solved by using partial fractions (Exercise 3).

EXAMPLE 3.4–2 Find the inverse transform of

$$F(s) = \frac{s}{(s^2 + 1)^2}.$$

Solution. From Table 3.1–1 we know that

$$\mathscr{L}(\cos t) = \frac{s}{s^2 + 1} \quad \text{and} \quad \mathscr{L}(\sin t) = \frac{1}{s^2 + 1}.$$

Thus

$$
\begin{aligned}
\mathscr{L}^{-1}\left(\frac{s}{(s^2 + 1)^2}\right) &= \mathscr{L}^{-1}\left(\frac{1}{s^2 + 1} \cdot \frac{s}{s^2 + 1}\right) \\
&= \sin t * \cos t = \int_0^t \sin (t - u) \cos u \, du \\
&= \int_0^t (\sin t \cos u - \cos t \sin u) \cos u \, du \\
&= \sin t \int_0^t \cos^2 u \, du - \cos t \int_0^t \sin u \cos u \, du \\
&= \sin t \left(\frac{\sin u \cos u + u}{2}\right) - \cos t \left. \frac{\sin^2 u}{2} \right|_{u=0}^{u=t} \\
&= \frac{t \sin t}{2}. \quad \blacksquare
\end{aligned}
$$

Note that this example could not be solved by the method of partial fractions (Exercise 13 in Section 3.3).

We will use the convolution theorem (Theorem 3.4–1) in Section 3.6 to solve differential equations. This theorem is also of use, however, in solving integral and integrodifferential equations. An **integral equation** is one in which the unknown function appears in the integrand of a definite integral. An **integrodifferential equation** has derivatives of the unknown function as well as having the function in the integrand of a definite integral. One type of integral equation was used by Vito Volterra (1860–1940), an Italian mathematician, in 1931 in modeling population growth. Since then it has also been used to model phenomena in electrical systems theory, nuclear reactor theory, thermodynamics, and chemotherapy. The following example illustrates a method of solving it.

EXAMPLE 3.4–3 Solve the **Volterra integral equation**

$$f(t) = \exp(-t) - 2 \int_0^t \cos (t - u) f(u) \, du.$$

Solution. On taking the transforms of both members of the equation we have

$$F(s) = \frac{1}{s+1} - 2\frac{s}{s^2+1} F(s);$$

hence

$$F(s) = \frac{s^2+1}{(s+1)^3} = \frac{1}{s+1} - \frac{2}{(s+1)^2} + \frac{2}{(s+1)^3}$$

from which we find

$$f(t) = (1-t)^2 \exp(-t). \quad\blacksquare$$

We conclude this section with the remark that some authors use the terms "resultant" and "Faltung" ("folding" in German) for convolution. In earlier works* $f * g$ was also called "the composition of f and g."

KEY WORDS AND PHRASES

convolution   Volterra integral equation
integral equation   Faltung
integrodifferential equation

EXERCISES 3.4

▶ **1.** If $f, g,$ and h are piecewise continuous functions on $t \geq 0$ and of exponential order $\exp(at)$ with $a > 0$, and if c is a constant, then prove the following properties of convolution.

   a) $f * g = g * f$ (*Hint:* Make a change of variable $v = t - u$.)

   b) $f * (g * h) = (f * g) * h$
       (*Hint:* Write $f * (g * h) = \int_0^t f(u) \int_0^{t-u} g(t - u - v)h(v)\, dv\, du;$ then reverse the order of integration.)

   c) $f * (g + h) = f * g + f * h$

   d) $f * 0 = 0 * f = 0$

   e) $f * (cg) = (cf) * g = c(f * g)$

2. Show that

$$\cos t * 1 = \sin t;$$

hence it is *not* true in general that

$$f * 1 = f.$$

3. Solve the problem of Example 3.4–1 by using partial fractions instead of convolution.

▶▶ **4.** Obtain each of the following convolutions.

   a) $(\sin t) * (\cos t)$ **b)** $t * \exp(at)$

---

* Harold, T. Davis, *The Theory of Linear Operators*, p. 160 (Bloomington, Ind.: The Principia Press, 1936).

○. **c)** $\sin bt * \exp (at)$ ○ **d)** $\sin bt * \sin bt$

✓**e)** $t * \sin t$

✓ **5.** Use the results of Exercise 4, where possible, to find the inverse transform of each of the given functions.

✓ **a)** $\dfrac{1}{s^3 - as^2}$ ✓**b)** $\dfrac{1}{(s^2 + a^2)^2}$

c) $\dfrac{2}{(s - 3)(s^2 + 4)}$ **d)** $\dfrac{a}{s^2(s^2 + a^2)}$

e) $\dfrac{1}{s^3(s^2 + 1)}$

f) $\dfrac{1}{((s + b)^2 + a)^2}$ [*Hint*: Use the result in part (b).]

6. Find $F(s)$ in each of the following:

    a) $f(t) = \int_0^t (t - u)^3 \sin u \, du$

    ✓**b)** $f(t) = \int_0^t u^5 \exp (3(t - u)) \, du$

    c) $f(u) = \int_0^u (u - t)^3 t^5 \, dt$

✓ **7.** Solve each of the following Volterra integral equations.

    a) $f(t) = t^2 + \int_0^t \sin (t - u) f(u) \, du$

    b) $f(x) = 1 - \int_0^x (x - y) f(y) \, dy$

    ✓**c)** $x(t) = \exp (-t) - 2 \int_0^t \cos (t - u) x(u) \, du$

8. Solve the following initial-value problem by using the convolution theorem. (Primes indicate d/dt.)

$$y^{iv}(t) + 2y''(t) + y(t) = \sin t, \qquad y(0) = y'(0) = y''(0) = y'''(0) = 0$$

9. Find

$$\mathscr{L}^{-1}\left(\frac{a^2}{s(s^2 + a^2)}\right)$$

    a) by partial fraction decomposition;

    b) by using Theorem 3.4–1.

▶▶▶ **10.** Compute $\cos t * \cos t$ and thus show that $f * f$ is not necessarily non-negative.

11. a) Show that

$$\mathscr{L}^{-1}\left(\frac{F(s)}{s}\right) = \int_0^t f(u) \, du.$$

    b) Explain why convolution cannot be used to obtain $\mathscr{L}^{-1}(sF(s))$.

12. Given $f(t) = t^n$ and $g(t) = t^m$, show that

$$f * g = t^{m+n+1} \int_0^t u^n (1 - u)^m \, du.$$

What, if any, restrictions must be placed on m and n?

13. Referring to Exercise 12, use Theorem 3.4–1 to show that

$$\int_0^1 u^n (1 - u)^m \, du = \frac{m! \, n!}{(m + n + 1)!}.$$

14. Consider the integrodifferential equation

$$\frac{dy(t)}{dt} + 2y + \int_0^t y(u) \, du = 0.$$

Find $y(t)$ given that $y(0^+) = 1$.

15. Show that

$$t^m * t^n = \frac{\Gamma(m + 1)\Gamma(n + 1)}{\Gamma(m + n + 2)} t^{m + n + 1},$$

where $m, n > -1$ (compare Exercise 25 in Section 3.1).

16. Use the convolution theorem to show that

$$\frac{1}{s} F(s) = \mathscr{L}\left(\int_0^t f(u) \, du\right).$$

3.5 STEP, IMPULSE, AND PERIODIC FUNCTIONS

An important application of the Laplace transformation is found in solving ordinary differential equations in which the forcing function is discontinuous and/or periodic. In this section we will consider some special functions and obtain their Laplace transforms.

Step functions

We begin by defining a **unit step function**,* $u_c(t)$, as follows:

$$u_c(t) = \begin{cases} 0, & \text{for } t < c, \\ 1, & \text{for } t > c, \end{cases} \quad c \geq 0. \qquad (3.5\text{--}1)$$

The graph of this function is shown in Fig. 3.5–1. Clearly this function is piecewise continuous on $t \geq 0$ and is of exponential order $\exp{(at)}$. Hence $u_c(t)$ has a Laplace transform, which we can compute as

$$\mathscr{L}(u_c(t)) = \int_0^\infty u_c(t) \exp{(-st)} \, dt$$

$$= \int_c^\infty \exp{(-st)} \, dt$$

$$= \frac{\exp{(-sc)}}{s}, \quad s > 0.$$

---

* Also known as the **Heaviside function**. Some authors use the notation $U(t - c)$.

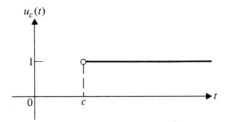

FIGURE 3.5–1 A unit step function.

This gives us the Laplace transform pair

$$u_c(t) \leftrightarrow \frac{\exp{(-sc)}}{s}, \qquad s > 0. \tag{3.5–2}$$

In Theorem 3.1–3 we proved the result

$$\mathscr{L}(\exp{(bt)}f(t)) = F(s - b), \qquad s \geq a + b,$$

where $\mathscr{L}(f(t)) = F(s)$ when $s > a$. We now have a result analogous to the above shifting theorem.

Theorem 3.5–1 If $F(s) = \mathscr{L}(f(t))$ for $s > a \geq 0$, then

$$\mathscr{L}(u_c(t)f(t - c)) = F(s) \exp{(-cs)}, \qquad s > a. \tag{3.5–3}$$

PROOF.

$$\mathscr{L}(u_c(t)f(t - c)) = \int_0^\infty \exp{(-st)}u_c(t)f(t - c)\, dt$$

$$= \int_c^\infty \exp{(-st)}f(t - c)\, dt.$$

Now we make a change of variable: $v = t - c$. Then

$$\mathscr{L}(u_c(t)f(t - c)) = \int_0^\infty \exp{(-s(v + c))}f(v)\, dv$$

$$= \exp{(-cs)} \int_0^\infty \exp{(-sv)}f(v)\, dv$$

$$= \exp{(-cs)}F(s), \qquad s > a. \quad \square$$

We remark that the most useful form* of Theorem 3.5–1 is

$$\mathscr{L}^{-1}(F(s) \exp{(-cs)}) = u_c(t)f(t - c). \tag{3.5–4}$$

We illustrate with some examples.

* See also Exercise 1.

EXAMPLE 3.5–1   Find the inverse transform of

$$\frac{1}{s^2}(1 - \exp(-s))^2.$$

Express the result in terms of unit step functions and graph.

Solution. Let

$$G(s) = \frac{1}{s^2}(1 - \exp(-s))^2$$

$$G(s) = \frac{1}{s^2} - \frac{2}{s^2}\exp(-s) + \frac{1}{s^2}\exp(-2s).$$

Then, using the linearity property of the inverse transform and Theorem 3.5–1, we have

$$g(t) = t - 2(t - 1)u_1(t) + (t - 2)u_2(t).$$

A graph of this function is given in Fig. 3.5–2 by the dotted lines. Note that the function shown can also be defined as

$$g(t) = \begin{cases} t, & \text{when } 0 \le t < 1, \\ 2 - t, & \text{when } 1 \le t < 2, \\ 0, & \text{when } 2 \le t. \end{cases} \quad \blacksquare$$

EXAMPLE 3.5–2   Find the Laplace transform of the function

$$u_1(t)(t^3 - 3t^2 + 4t + 4).$$

Solution. In order to take advantage of Theorem 3.5–1 we first express the polynomial in powers of $t - 1$. Thus

$$t^3 - 3t^2 + 4t + 4 = A(t - 1)^3 + B(t - 1)^2 + C(t - 1) + D$$

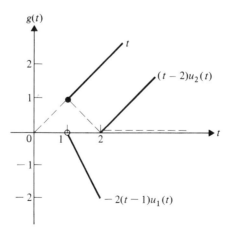

FIGURE 3.5–2

and, on equating coefficients of like powers of t, we find $A = 1$, $B = 0$, $C = 1$, and $D = 6$. Hence,

$$f(t) = u_1(t)(t^3 - 3t^2 + 4t + 4) = u_1(t)((t-1)^3 + (t-1) + 6)$$

and, using linearity and Theorem 3.5–1, we have

$$F(s) = \exp(-s)\left(\frac{6}{s^4} + \frac{1}{s^2} + \frac{6}{s}\right). \quad \blacksquare$$

Impulse functions

In many applications it is necessary to find the response of a system to an impulsive forcing function. By an *impulsive force* we mean a force that has a very large magnitude but acts over a very short period of time. Such forces are encountered when a circuit is subjected to a high voltage for a short period as in a lightning strike, when an object receives a hammer blow, or, in general, when two bodies collide.

We will arrive at a definition of an impulse function by starting with the following definition of a "pseudoimpulse function," $\delta_h(t)$:

$$\delta_h(t) = \frac{1}{h}(u_0(t) - u_h(t)), \qquad h > 0, \tag{3.5–5}$$

where $u_0(t)$ and $u_h(t)$ are unit step functions defined in Eq. (3.5–1). Thus $\delta_h(t)$ can be expressed as

$$\delta_h(t) = \begin{cases} 1/h, & \text{when } 0 \leq t \leq h, \\ 0, & \text{when } t > h, \end{cases} \qquad h > 0. \tag{3.5–6}$$

A graph of this function is shown in Fig. 3.5–3. We can see that if h is small, then the magnitude of the function, $1/h$, is large and its duration is small. The *strength* of the function $\delta_h(t)$ is given by

$$\int_0^h \delta_h(t)\,dt = \int_0^\infty \delta_h(t)\,dt = 1. \tag{3.5–7}$$

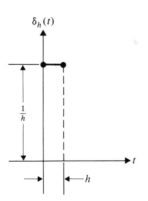

FIGURE 3.5–3

We can compute the transform of $\delta_h(t)$ to obtain

$$\mathscr{L}(\delta_h(t)) = \frac{1}{h} \int_0^h \exp(-st)\, dt = \frac{1}{hs}(1 - \exp(-sh)). \quad \textbf{(3.5–8)}$$

This result can also be found from Eq. (3.5–5) by using Eq. (3.5–2) (Exercise 2).

From $\delta_h(t)$ we can proceed to the **Dirac delta function**,* $\delta(t)$, by a limit process. We *define* this function $\delta(t)$ in the following way:

$$\delta(t) = \lim_{h \to 0} \delta_h(t).$$

This results in a useful entity although $\delta(t)$ is not a function in the mathematical sense; it is a "pseudofunction." It does, however, belong to a class of generalized functions and its use as a function may be studied by "distribution theory."[†]

The Dirac delta function has the following *unusual* properties:

a) $\delta(t) = 0$ for $t \neq 0$,

b) $\lim_{t \to 0} \delta(t) = \infty$,

c) $\int_0^\infty \delta(t)\, dt = 1$,

d) $\mathscr{L}(\delta(t)) = 1$.

$$\text{(3.5–9)}$$

Property (3.5–9d), in particular, which is obtained from Eq. (3.5–8) by applying l'Hôpital's rule (Exercise 3), shows that $\delta(t)$ is not an ordinary function. Recall that for all functions $f(t)$ discussed so far

$$\lim_{s \to \infty} \mathscr{L}(f(t)) = 0.$$

Accordingly, we must verify all results that are obtained by using $\delta(t)$.

One of the most useful properties of $\delta(t)$ is its *sifting property*, that is, its ability to pick (or sift) out the value of a function at a particular point. If $f(t)$ is continuous on $t \geq 0$, then

$$\int_0^\infty f(t)\, \delta_h(t)\, dt = \frac{1}{h} \int_0^h f(t)\, dt$$

$$= f(\hat{t}) \qquad (0 < \hat{t} < h),$$

by the mean value theorem for integrals. Hence

$$\int_0^\infty f(t)\, \delta(t)\, dt = \lim_{h \to 0} f(\hat{t}) = f(0)$$

---

* After Paul A. M. Dirac (1902–   ), a British theoretical physicist known for his work in quantum mechanics.

[†] See Arthur E. Danese, *Advanced Calculus: An Introduction to Applied Mathematics*, vol. 2, ch. 24 (Boston: Allyn and Bacon, 1965).

and

$$\int_0^\infty f(t)\, \delta(t - t_0)\, dt = f(t_0). \qquad (3.5\text{--}10)$$

Periodic functions

We have already considered some *periodic functions* as forcing functions in differential equations. Functions like sin (at) and cos (at) were discussed in Section 3.1, where we obtained their Laplace transforms. Now we look at periodic functions such as the square wave $g(t)$ shown in Fig. 3.5–4. This periodic function having period $2b$ can be defined as

$$g(t) = \begin{cases} a, & \text{when } 0 < t < b, \\ 0, & \text{when } t = 0 \text{ and } t = b, \\ -a, & \text{when } b < t < 2b, \end{cases}$$

$$g(t + 2b) = g(t).$$

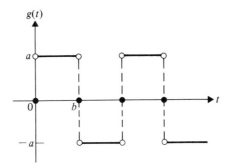

FIGURE 3.5–4 A square wave.

In order to compute the Laplace transform of a periodic function such as $g(t)$ above we need the following result.

Theorem 3.5–2 Let $f(t)$ have a Laplace transform $F(s)$ and let $f(t) = f(t + T)$. Then

$$F(s) = \frac{1}{1 - \exp(-sT)} \int_0^T \exp(-st)f(t)\, dt, \qquad s > 0. \quad (3.5\text{--}11)$$

PROOF. By definition,

$$F(s) = \int_0^\infty \exp(-st)f(t)\, dt$$

$$= \int_0^T \exp(-st)f(t)\, dt + \int_T^\infty \exp(-st)f(t)\, dt.$$

Now we make a change of variable in the second integral by putting

$t = u + T$. Then

$$\int_T^\infty \exp(-st)f(t)\,dt = \int_0^\infty \exp(-s(u + T))f(u + T)\,du$$

$$= \exp(-sT)F(s),$$

since $f(u + T) = f(u)$, and

$$F(s) = \int_0^T \exp(-st)f(t)\,dt + \exp(-st)F(s),$$

from which the desired result follows on solving for $F(s)$. □

EXAMPLE 3.5–3 Obtain the Laplace transform $G(s)$ of the periodic square wave $g(t)$ shown in Fig. 3.5–4.

Solution. We have

$$\int_0^{2b} \exp(-st)g(t)\,dt = a\int_0^b \exp(-st)\,dt - a\int_b^{2b} \exp(-st)\,dt$$

$$= \frac{a}{s}(1 - \exp(-sb))^2;$$

hence

$$F(s) = \frac{a}{s}\left(\frac{(1 - \exp(-sb))^2}{1 - \exp(-2sb)}\right)$$

$$= \frac{a}{s}\left(\frac{1 - \exp(-sb)}{1 + \exp(-sb)}\right). \quad \blacksquare$$

KEY WORDS AND PHRASES

unit step function	**impulse function**
Heaviside function	**Dirac delta function**
impulsive force	**periodic function**

EXERCISES 3.5

▶ **1.** Show that another form of Theorem 3.5–1 can be obtained by writing

$$\mathscr{L}(u_c(t)f(t)) = \exp(-cs)\mathscr{L}(f(t + c)).$$

2. Obtain the Laplace transform of $\delta_h(t)$ by starting with Eq. (3.5–5).

3. Use l'Hôpital's rule and Eq. (3.5–8) to show that $\mathscr{L}(\delta(t)) = 1$.

4. Show that the solution to Example 3.5–3 can be written

$$F(s) = \frac{a}{s}\tanh\frac{bs}{2}.$$

▶▶ 5. Graph each of the following functions.

    a) $u_0(t)$ **b)** $tu_2(t)$

    c) $u_c(t - t_0)$ **d)** $u_2(t) - u_1(t)$

6. Compute the Laplace transform of each of the functions in Exercise 5.

7. Find the inverse transform of each of the following, where a, b, and c are nonnegative constants.

    a) $\dfrac{1}{s}(1 - \exp(-cs))$ **b)** $\dfrac{1}{s}(\exp(-as) - \exp(-bs))$

    c) $\dfrac{1}{s^2}(\exp(-as) - \exp(-bs))^2$ **d)** $\dfrac{1}{s^2}(\exp(-as) - \exp(-bs))$

8. Graph each of the inverse transforms obtained in Exercise 7.

9. **a)** Find the Laplace transform of the function

$$f(t) = (\sin t)u_\pi(t).$$

    b) Graph $f(t)$.

10. Find the transform of each of the following.

    a) $u_2(t)(t^3 - 3t^2 + 4t + 4)$ **b)** $u_1(t)(t^4 + 2t^2 - 3t + 4)$

    c) $u_2(t) \sin(t - 2)$ **d)** $u_3(t) \exp(t - 3)$

11. Sketch the graph of the following function on the interval $0 \le t \le 10$:

$$f(t - 3)u_3(t), \quad \text{where } f(t) = \sin t.$$

12. Use Theorem 3.5–1 to obtain the transform of each of the following.

    a) $u_b(t) \sinh a(t - b)$ **b)** $u_b(t) \cosh a(t - b)$

    c) $u_b(t) \cos a(t - b)$

13. Show that

$$u_c(t) = \int_0^t \delta(v - c)\, dv.$$

14. Prove the following property of the Dirac delta function:

$$\int_0^\infty \delta(t - c)\, dt = 1, \quad c > 0.$$

15. Find the inverse transform of each of the following functions.

    a) $\dfrac{s - a}{s + a}$ **b)** $\dfrac{s^2}{s^2 + 1}$ **c)** $\dfrac{s^3}{s^3 + 1}$

16. Find the Laplace transform of each of the following periodic functions.

    a) $f(t) = \begin{cases} 1, & \text{when } 0 \le t \le a, \\ 0, & \text{when } a < t < 2a, \end{cases}$ **b)** $f(t) = \begin{cases} 0, & \text{when } 0 < t < a, \\ 1, & \text{when } a \le t \le 2a, \end{cases}$

    $f(t + 2a) = f(t)$ $f(t + 2a) = f(t)$

    c) $g(t) = \begin{cases} \dfrac{t}{a}, & \text{when } 0 < t < a, \\ 2 - \dfrac{t}{a}, & \text{when } a \le t \le 2a, \end{cases}$ **d)** $h(t) = |\sin at|$

    $g(t + 2a) = g(t)$

17. Find the inverse transform of each of the following.

a) $\dfrac{\exp(-as) + as - 1}{as^2(1 - \exp(-as))}$ **b)** $\dfrac{1 - \exp(-as)}{s(1 + \exp(as))}$ **c)** $\dfrac{as + 1 - \exp(as)}{as^2(1 - \exp(as))}$

▶▶▶ **18.** Prove that if $f(t)$ is piecewise continuous on $t \geq 0$ and of exponential order, then

$$\delta(t) * f(t) = f(t).$$

Hence $\delta(t)$ can be considered to be the *multiplicative unit* in convolution (see Section 3.4).

19. Represent the periodic function in Fig. 3.5–5 by unit step functions; then obtain its transform.

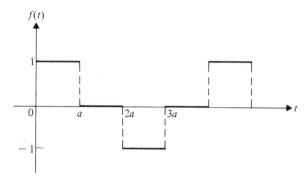

FIGURE 3.5–5

3.6 SOLVING DIFFERENTIAL EQUATIONS AND SYSTEMS

So far in this chapter we have presented various aspects of the Laplace transformation. We have transformed functions, derivatives, and integrals from the *t*-domain to the *s*-domain. We have also been concerned with the more difficult problem of transforming from the *s*-domain back to the *t*-domain. All of this has been directed toward the goal of solving differential equations. Our aim is to be able to solve differential equations with forcing functions that are piecewise continuous, periodic, or impulse functions.

This section will consequently use material presented in previous sections and will thus serve as a review of the entire chapter. In addition, we will apply the Laplace transform method to *systems* of ordinary differential equations. This is a subject that has been touched on only briefly (in Section 2.6) but has wide practical application.

Differential equations

We begin with some examples from Chapter 2 and apply the transform method to obtain their solutions. We use primes to indicate differentiation with respect to x.

EXAMPLE 3.6–1 (Compare Example 2.1–4.) Solve the initial-value problem

$$y'' - 4y' + 13y = 0, \qquad y(0) = 1, \qquad y'(0) = 8.$$

Solution. Using Eqs. (3.2–2) and (3.2–3) to transform the equation, we have

$$s^2 Y(s) - s(1) - 8 - 4(s Y(s) - 1) + 13 Y(s) = 0.$$

Solving for $Y(s)$ gives us

$$Y(s) = \frac{s+4}{s^2 - 4s + 13} = \frac{s+4}{(s-2)^2 + 9},$$

where we have used a technique from algebra called "completing the square." Hence,

$$Y(s) = \frac{s-2}{(s-2)^2 + 3^2} + 2\frac{3}{(s-2)^2 + 3^2}$$

and, using Eqs. (3.1–13) and (3.1–14), we have

$$y(x) = \cos 3x \exp (2x) + 2 \sin 3x \exp (2x). \quad \blacksquare$$

EXAMPLE 3.6–2 (Compare Example 2.3–4.) Find the general solution of the equation

$$y'' + 3y' - 4y = 3 \exp x.$$

Solution. Transforming the equation results in

$$s^2 Y(s) - sy(0) - y'(0) + 3(s Y(s) - y(0)) - 4 Y(s) = \frac{3}{s-1};$$

hence

$$Y(s) = \frac{s+3}{(s+4)(s-1)} y(0) + \frac{1}{(s+4)(s-1)} y'(0) + \frac{3}{(s+4)(s-1)^2},$$

where $y(0)$ and $y'(0)$ are arbitrary constants. The inverse transform can now be found by decomposition into partial fractions or (as we have done) referring to Table 1 at the back of the book. Thus

$$y(x) = \frac{\exp(-4x)}{5}\left(y(0) - y'(0) - \frac{12}{5}\right) + \frac{\exp x}{5}\left(4y(0) + y'(0) + \frac{12}{5}\right) + \frac{3x}{5} \exp x$$

$$= c_1 \exp x + c_2 \exp(-4x) + \frac{3}{5} x \exp x.$$

Note that the lack of specified numerical initial values in this problem complicates the algebra somewhat. ∎

EXAMPLE 3.6–3 Solve the initial-value problem

$$xy'' + 2y' + (2 - x)y = 2 \exp x, \qquad y(0) = 0.$$

Solution. Here we have a linear differential equation with *nonconstant* coefficients. Making use of Eq. (3.2–9) when transforming, we have (see Exercise 2),

$$-\frac{d}{ds}(s^2 Y(s)) + 2s Y(s) + 2 Y(s) + \frac{dY(s)}{ds} = \frac{2}{s-1}.$$

Hence,

$$(1 - s^2)\frac{dY}{ds} + 2Y = \frac{2}{s-1} \tag{3.6–1}$$

and this *first-order* linear differential equation can be solved by the method discussed in Section 1.3. Then,

$$Y(s) = \frac{1}{s^2 - 1} + C,$$

where C is an arbitrary constant. This constant must be zero, however, since otherwise $Y(s)$ would not satisfy Eq. (3.6–1). Moreover, $C = 0$ is the only way we can satisfy

$$\lim_{s \to \infty} Y(s) = 0.$$

Thus, inverting $Y(s)$, we find

$$y(x) = \sinh x$$

as a solution to the problem. In Exercise 2 you are asked to investigate why $y'(0)$ was not specified. ∎

Example 3.6–3 gives an indication that the Laplace transform method may *not* be applicable when the differential equation has nonconstant coefficients. The method will work *if* the coefficients are polynomials in the independent variable *and* the differential equation in $Y(s)$ can be readily solved *and* the inverse of $Y(s)$ can be found. In short, we must have a fortuitous set of conditions. For this reason we will confine our attention to differential equations with constant coefficients in the remainder of this section.

EXAMPLE 3.6–4 A forcing function in the form of an impulse is applied to the spring–mass system as shown in Fig. 3.6–1. Find the displacement at any time t.

Solution. We phrase this problem as an initial-value problem, using a dot to represent differentiation with respect to t:

$$m\ddot{y} + ky = \delta(t), \qquad y(0) = \dot{y}(0) = 0.$$

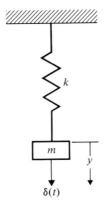

FIGURE 3.6–1

Taking transforms of both members of the differential equation gives us

$$ms^2 Y(s) + kY(s) = 1,$$

where we have used Eq. (3.5–9d). Hence,

$$Y(s) = \frac{1}{ms^2 + k} = \frac{1}{\sqrt{km}} \frac{\sqrt{k/m}}{s^2 + k/m}$$

and

$$y(t) = \frac{1}{\sqrt{km}} \sin\left(\sqrt{\frac{k}{m}}\, t\right).$$

If k has units of g/cm and m has units of g-sec^2/cm (dynes), then it would appear that the last result is not dimensionally correct, since $y(t)$ should have units of cm. Recalling the definition of $\delta(t)$, however, from Section 3.5, we see that the quantity 1 in the numerator of the coefficient of the sine has units of g-sec since it is an impulse. ■

EXAMPLE 3.6–5 Solve the following initial-value problem in which the forcing function is nonperiodic (Fig. 3.6–2).

$$\ddot{y} + 4y = \sin t + u_\pi(t) \sin(t - \pi), \qquad y(0) = \dot{y}(0) = 0.$$

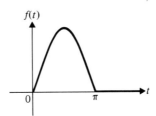

FIGURE 3.6–2 The function
$f(t) = \sin t + u_\pi(t) \sin(t - \pi)$.

Solution. Transforming the differential equation and using the given initial values, we have

$$s^2 Y(s) + 4Y(s) = \frac{1}{s^2 + 1} + \frac{\exp(-\pi s)}{s^2 + 1}.$$

Hence,

$$Y(s) = \frac{1 - \exp(-\pi s)}{(s^2 + 1)(s^2 + 4)}$$

$$= \frac{1/3}{s^2 + 1} - \frac{1/3}{s^2 + 4} + \frac{1/3 \exp(-\pi s)}{s^2 + 1} - \frac{1/3 \exp(-\pi s)}{s^2 + 4}$$

from which

$$y(t) = \frac{1}{3} \sin t - \frac{1}{6} \sin 2t + \frac{1}{3} u_\pi(t) \sin(t - \pi) - \frac{1}{6} u_\pi(t) \sin(2t - \pi)$$

$$= \frac{1}{6}(1 - u_\pi(t))(2 \sin t - \sin 2t). \quad\blacksquare$$

We can generalize the two previous examples in the following way. Consider the initial-value problem

$$\ddot{y} + y = g(t), \qquad y(0) = \dot{y}(0) = 0.$$

The transform of this problem is

$$s^2 Y(s) + Y(s) = G(s),$$

where $G(s) = \mathcal{L}(g(t))$. Hence

$$Y(s) = \frac{G(s)}{s^2 + 1}$$

and the inverse transform can be obtained by using convolution (compare Eq. 3.4–2). Thus

$$y(t) = \int_0^t \sin(t - u)\, g(u)\, du$$

$$= \sin t \int_0^t \cos u\, g(u)\, du - \cos t \int_0^t \sin u\, g(u)\, du. \qquad (3.6\text{–}2)$$

If instead of the homogeneous initial conditions we have

$$y(0) = y_0 \quad \text{and} \quad \dot{y}(0) = v_0,$$

then Eq. (3.6–2) becomes

$$y(t) = \sin t \int_0^t \cos u\, g(u)\, du - \cos t \int_0^t \sin u\, g(u)\, du + y_0 \cos t + v_0 \sin t.$$

$$(3.6\text{–}3)$$

It is apparent from Eq. (3.6–3) that the ease with which even a simple differential equation can be solved depends on how readily the integrals can be evaluated. This, in turn, depends on the form of the forcing function g. If $g(u)$ can be expressed in terms of sines and cosines, then the integrands will be simple enough. The integrations are similar to what we do when we represent an arbitrary function by a series of sines and cosines called *Fourier series*, a topic that we will discuss in Chapter 7.

Systems of differential equations

In the remainder of this section we will consider **systems of ordinary differential equations**. For simplicity we will deal only with first- and second-order equations having constant coefficients. The method of Laplace transforms can be applied to such systems, as shown by the following examples.

EXAMPLE 3.6–6   Solve the system

$$\dot{x} + 3x - 4y = \cos t,$$
$$\dot{y} - 3y + 2x = t,$$
$$x(0) = 0, \qquad y(0) = 1.$$

Solution. On transforming this system we have

$$sX(s) + 3X(s) - 4Y(s) = \frac{s}{s^2 + 1},$$

$$sY(s) - 1 - 3Y(s) + 2X(s) = \frac{1}{s^2}.$$

This is a pair of algebraic equations in $X(s)$ and $Y(s)$, which may be simplified to

$$(s + 3)X - 4Y = \frac{s}{s^2 + 1},$$

$$2X + (s - 3)Y = \frac{1 + s^2}{s^2}.$$

(3.6–4)

Next we eliminate one of the variables. We can eliminate Y, for example, by multiplying the first equation by $(s - 3)$ and the second by 4 and adding. Then

$$X(s) = \frac{s(s - 3)}{(s^2 + 1)(s^2 - 1)} + \frac{4(s^2 + 1)}{s^2(s^2 - 1)}$$

$$= \frac{7/2}{s - 1} - \frac{5}{s + 1} + \frac{1/2(3s + 1)}{s^2 + 1} - \frac{4}{s^2}$$

and, using Table 3.1–1 or Table 1 at the back of the book, we have

$$x(t) = \frac{7}{2} e^t - 5e^{-t} + \frac{3}{2} \cos t + \frac{1}{2} \sin t - 4t.$$

We can now follow one of three courses to obtain $y(t)$: (1) We can eliminate X from the system (3.6–4), solve for Y, and find its inverse; (2) we can substitute the $x(t)$ as found into the second of the given equations and solve for $y(t)$; or (3) we can substitute the $x(t)$ and $\dot{x}(t)$ into the first equation and solve for $y(t)$. In all cases we have

$$y(t) = \frac{7}{2}e^t - \frac{5}{2}e^{-t} + \cos t - 3t - 1$$

(see Exercise 7). We observe that this problem can also be solved without using the Laplace transform. The second differential equation can be solved for x and this value, as well as the $\dot{x}$ obtained by differentiation, can be substituted into the first differential equation. The result is *

$$\ddot{y} - y = 3t + 1 - 2\cos t, \qquad y(0) = 1, \qquad \dot{y}(0) = 3,$$

which can be solved by the methods given in Section 2.3. The value of $x(t)$ can be obtained in similar fashion. ∎

EXAMPLE 3.6–7   Solve the system

$$\ddot{x} - \ddot{y} + x - 4y = 0,$$
$$\dot{x} + \dot{y} = \cos t + 2\cos 2t,$$
$$x(0) = 0, \qquad \dot{x}(0) = 1, \qquad y(0) = 0, \qquad \dot{y}(0) = 2.$$

Solution. Upon transforming the equations and using the given initial conditions, we have

$$(s^2 + 1)X - (s^2 + 4)Y = -1,$$

$$sX \qquad + sY = \frac{s}{s^2 + 1} + \frac{2s}{s^2 + 4}.$$

We can eliminate Y by multiplying the first equation by s and the second by $(s^2 + 4)$ and adding. Then

$$X(s) = \frac{1}{s^2 + 1}$$

so that $x(t) = \sin t$, and from the second differential equation it follows that $y = \sin 2t$. It is uncanny how simple the algebra became in this problem! A more realistic situation will be found in Exercise 8. ∎

It should be self-evident that systems of differential equations become more difficult to solve as the number of equations and the number of dependent variables increases. The scheme of eliminating all but one of the variables involves algebraic complexities when we

---

* The initial condition $\dot{y}(0) = 3$ is found from the second differential equation of the system by putting $t = 0$.

consider three, six, or ten equations that make the methods discussed in this section impractical. What is needed for solving larger systems is a means of dealing with large amounts of data easily. Such a means is provided by *matrix algebra*. Accordingly, we shall turn to this topic and its applications in the next chapter.

EXERCISES 3.6

▶ 1. Solve the problem of Example 3.6–2 by using decomposition into partial fractions.

2. **a)** Explain why $y'(0)$ was not specified for the problem in Example 3.6–3.
   [*Hint*: $\mathcal{L}(xy'') = -(d/ds)(s^2 Y - sy(0) - y'(0))$.]

   b) Obtain Eq. (3.6–1) in Example 3.6–3.

3. Show that Eq. (3.6–3) is the correct result for the problem given in the text.

4. In the problem of Example 3.6–6 obtain $y(t)$ by each of the following methods.

   a) Eliminate X from the system (3.6–4).

   b) Substitute the value $x(t)$ obtained in the text into the second of the given differential equations and solve the resulting equation.

   c) Substitute the value $x(t)$ obtained in the text and $\dot{x}(t)$ into the first of the given differential equations and solve the resulting equation.

5. Verify that the solutions obtained for the system of Example 3.6–6 satisfy the given system.

6. Solve the problem in Example 3.6–6 without using the Laplace transform.

7. Solve the system of Example 3.6–7 by eliminating X, solving for $Y(s)$, and then finding $y(t)$. Then find $x(t)$.

8. Obtain the solution of the system
$$\dot{x} - x + 2y = -t^2,$$
$$\dot{y} - 5y - 4x = \exp t,$$
$$x(0) = 0, \qquad y(0) = 1.$$
(*Hint*: A calculator would be very useful here.)

▶▶ 9. Solve each of the following initial-value problems.

   a) $\ddot{y} + 4y = \sin t + u_\pi(t) \sin (t - \pi), \quad y(0) = 1, \quad \dot{y}(0) = 0$

   b) $\ddot{y} + 4y = \sin t + u_\pi(t) \sin (t - \pi), \quad y(0) = 0, \quad \dot{y}(0) = 1$

   c) $\ddot{y} + 4y = \sin t + u_\pi(t) \sin (t - \pi), \quad y(0) = 1, \quad \dot{y}(0) = 1$

10. Solve each of the following initial-value problems.

   a) $y'' + 3y' - 4y = 3 \exp x, \quad y(0) = \frac{1}{5}, \quad y'(0) = -\frac{1}{5}$

   b) $y'' + 3y' - 4y = 2 \exp (-4x), \quad y(0) = 2, \quad y'(0) = 1$

   c) $y'' + 3y' - 4y = \sin x - \cos x, \quad y(0) = 1, \quad y'(0) = 0$

11. Obtain the solution to each of the following.

   a) $t\ddot{y} - t\dot{y} - y = 0, \quad y(0) = 0, \quad \dot{y}(0) = 1$

b) $t\ddot{y} + (3t - 1)\dot{y} - (4t + 9)y = 0, \quad y(0) = \dot{y}(0) = 0$

c) $\ddot{y} - t\dot{y} + 2y = 0, \quad y(0) = -1, \quad \dot{y}(0) = 0$

12. **a)** Obtain the general solution of the initial-value problem

$$\dot{y}(t) + ay(t) = f(t), \quad y(0) = y_0.$$

Here a and y_0 are constants.

b) Under what conditions is the solution obtained in part (a) valid?

13. Solve the following:

$$\ddot{y} + at\dot{y} - 2ay = 1, \quad a > 0,$$
$$y(0) = \dot{y}(0) = 0.$$

14. Solve the system of equations

$$\dot{x} - y + z = 0,$$
$$\dot{y} - x - z = -\exp t$$
$$\dot{z} - x - y = \exp t,$$
$$x(0) = 1, \quad y(0) = 2, \quad z(0) = 3.$$

15. Solve the following problem:

$$t\ddot{y} + \dot{y} + ty = 0, \quad y(0) = 1.$$

(*Hint*: $\mathcal{L}(J_0(t)) = (s^2 + 1)^{-1/2}$, where $J_0(t)$ is the Bessel function of the first kind of order zero. We will meet this function again in Section 9.3.)

16. Solve the following initial-value problem of a spring–mass system (compare Example 3.6–4):

$$m\ddot{y} + ky = 0, \quad y(0) = y_0, \quad \dot{y}(0) = v_0,$$

where y_0 and v_0 are given constants.

17. If a viscous damping force proportional to the velocity also acts on the mass given in Exercise 16, then the equation of motion becomes

$$m\ddot{y} + c\dot{y} + ky = 0.$$

Find $y(t)$ for this case if

$$y(0) = 0 \quad \text{and} \quad \dot{y}(0) = v_0.$$

18. Consider the system

$$\dot{x} + \dot{y} \quad\quad = 2 \sinh t,$$
$$\dot{y} + \dot{z} = \exp t,$$
$$\dot{x} \quad\quad + \dot{z} = 2 \exp t + \exp(-t),$$
$$x(0) = y(0) = 1, \quad z(0) = 0.$$

a) Solve the system by the Laplace transform method.

b) Solve the system by back substitution, that is, solving the third equation for $\dot{z}$, substituting this value into the second equation, then solving the latter for $\dot{y}$ and substituting into the first. Note that this is the same as subtracting the third equation from the second (eliminating $\dot{z}$) and then subtracting the result from the first (eliminating $\dot{y}$).

▶▶▶ **19. a)** Solve the problem

$$\frac{d^2i}{dt^2} + 4\frac{di}{dt} + 5i = 5u_0(t), \qquad i(0) = 1, \qquad \frac{di(0)}{dt} = 2.$$

b) If $i(t)$ represents current in a network containing a resistance, an inductance, and a capacitance in series, then interpret the problem of part (a) physically.

20. a) A resistance of 1 ohm and an inductance of 1 henry are connected in series with a voltage source that applies a unit pulse of width a at time $t = 0$. Using the condition $i(0) = 0$, write the appropriate initial-value problem and solve it by using the Laplace transform. (*Hint*: The forcing function is $u_0(t) - u_a(t)$.)

b) Explain how the solution to the problem of part (a) could have been obtained by using the principle of superposition.

c) If the forcing function in part (a) is changed to $\delta(t)$, show that the current becomes $i(t) = e^{-t}$.

d) Show that, in general, a forcing function $\delta(t)$ in series with a resistance R and an inductance L will result in a current $i(t) = (1/L) \exp(-Rt/L)$. (*Note*: This problem has far-reaching consequences in network analysis for it shows that the response to an impulse is a function of the constants in the network. Accordingly, the impulse response can be used as a means of identifying the network.)

21. Show that the Cauchy–Euler equation

$$t^2\ddot{y} + t\dot{y} + y = f(t)$$

is transformed by the Laplace transform into a Cauchy–Euler equation.

22. Solve the integrodifferential equation

$$\dot{y} + a^2 \int_0^t y(u)\,du = a, \qquad y(0) = 0, \qquad a \text{ is a constant.}$$

23. A radioactive substance R_1 having decay rate k_1 disintegrates into a second radioactive substance R_2 having decay rate k_2. Substance R_2 disintegrates into R_3, which is stable. If $m_i(t)$, $i = 1, 2, 3$, represents the mass of substance R_i at time t, the applicable equations are

$$\frac{dm_1}{dt} = -k_1 m_1,$$

$$\frac{dm_2}{dt} = -k_2 m_2 + k_1 m_1,$$

$$\frac{dm_3}{dt} = k_2 m_2.$$

Solve the above system under the conditions

$$m_1(0) = m_0, \qquad m_2(0) = 0, \qquad m_3(0) = 0.$$

(See the discussions of radioactive decay in Sections 1.1 and 1.5.)

24. Consider the initial-value problem

$$\dot{y} + 3y = 0, \qquad y(3) = 1.$$

Solve this problem by using the Laplace transform after putting $\tau = t - 3$ in order to obtain an initial condition at $\tau = 0$.

25. In the transformer-coupled circuit of Fig. 3.6–3 the mutual inductance is M. If we assume zero initial currents, the resulting differential equations are

$$2\frac{di_1}{dt} + i_1 - \frac{di_2}{dt} = 10 \cos t,$$

$$2\frac{di_2}{dt} + i_2 - \frac{di_1}{dt} = 0.$$

Find i_1 and i_2.

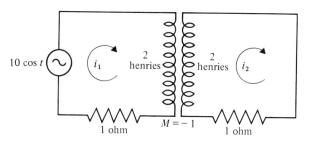

FIGURE 3.6–3

26. In the parallel network of Fig. 3.6–4 all resistances are 100 ohms and both inductances are 1 henry. With zero currents in the circuits, the switch is closed, which results in the equations

$$\frac{di_1}{dt} + 20i_1 - 10i_2 = 100,$$

$$\frac{di_2}{dt} + 20i_2 - 10i_1 = 0.$$

Solve the system for i_1 and i_2.

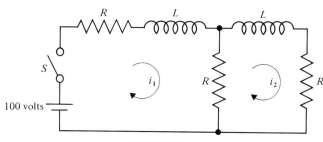

FIGURE 3.6–4

REFERENCES

Churchill, R. V., *Operational Mathematics*, 3d ed. New York: McGraw-Hill, 1972.
An excellent intermediate-level text with emphasis on applications. Contains a good table of Laplace transforms.

Danese, A. E., *Advanced Calculus: An Introduction to Applied Mathematics*, vol. 2. Boston: Allyn and Bacon, 1965.
Chapter 23 contains theoretical aspects of the subject. Rich in examples and problems.

Davies, B., *Integral Transforms and Their Applications*. New York: Springer-Verlag, 1978.
Part I contains a somewhat advanced treatment of the theoretical aspects.

Doetsch, G., *Guide to the Applications of Laplace Transforms*. New York: D. Van Nostrand, 1961.
This book contains an excellent table of transforms with graphs of many of the functions f(t).

Doetsch, G., *Introduction to the Theory and Application of the Laplace Transformation*. New York: Springer-Verlag, 1974.

Holl, D. L., C. G. Maple, and B. Vinograde, *Introduction to the Laplace Transform*. New York: Appleton-Century-Crofts, 1959.
An elementary monograph on the subject.

Kuhfittig, P. K. F., *Introduction to the Laplace Transform*, vol. 8 of Mathematical Concepts and Methods in Science and Engineering. New York: Plenum Press, 1978.
A readable text that contains many examples and exercises aimed at the engineering student.

Myers, G. A., *Analytical Methods in Conduction Heat Transfer*. New York: McGraw-Hill, 1971.
Chapter 6 contains a brief development and applications especially to partial differential equations.

Oberhettinger, F., and L. Badii, *Tables of Laplace Transforms*. New York: Springer-Verlag, 1973.
An unusually complete list of Laplace transforms and inverse Laplace transforms.

Sneddon, I. N., *The Use of Integral Transforms*. New York: McGraw-Hill, 1972.
Chapter 3 contains a mathematically rigorous treatment of the Laplace transform. Chapter 9 deals with generalized functions, including the Dirac delta function.

Thomson, W. T., *Laplace Transformation*, 2d ed. Englewood Cliffs, N.J.: Prentice-Hall, 1960.
Rich in dynamical, structural, and electrical applications.

Widder, D. V., *Advanced Calculus*, 2d ed. Englewood Cliffs, N.J.: Prentice-Hall, 1961.
Chapters 13 and 14 contain theoretical aspects of the subject.

4 *linear algebra*

4.1 MATRICES

After the development of calculus there resulted a surge in applied mathematics in the latter part of the seventeenth century. One of the topics that received considerable attention was the solution of differential equations. Various schemes were devised for solving different kinds of equations. One of the most popular was (and still is) the Laplace transformation, whereby certain differential equations (the linear ones with constant coefficients) were transformed into algebraic equations that could be more easily manipulated. Systems of differential equations, however, led to *systems* of linear algebraic equations. As the need developed to solve ever larger systems, it became necessary to find a more efficient method of dealing with large amounts of data. Such a method is provided by *matrix algebra*, which is an outgrowth of work done by Arthur Cayley ((1821–1895), a British mathematician) in 1858 in the theory of linear transformations.

We define a **matrix** (plural, *matrices*) as a rectangular array* of entities called **elements**. A matrix will first be distinguished by its size, the nature of its elements, and their position in the array. Thus the matrix

$$\begin{pmatrix} 2 & -3 & 5 \\ 0 & 1 & 4 \end{pmatrix}$$

is 2×3 and its elements are integers. By convention we designate the size of a matrix by giving *first* the number of its *rows* and *second* the number of its *columns*. Although all matrices are rectangular, certain ones have special names. For example, an $n \times n$ matrix is called a **square matrix**, an $n \times 1$ matrix is called a **column vector**, and a $1 \times n$ matrix is called a **row vector**. In this chapter we deal only with matrices that have *real numbers* as elements. We will also call real numbers **scalars**. Matrices having complex numbers as elements have applications in mathematical physics and quantum mechanics. Gen-

---

* This is actually a circular definition since "matrix" and "array" are synonyms in this case.

erally, the elements may be functions or even matrices, as we shall see later.

We use capital letters to designate matrices and double subscripts to distinguish their elements. Thus a *general* 2×3 matrix can be designated by A and written

$$\begin{pmatrix} a_{11} & a_{12} & a_{13} \\ a_{21} & a_{22} & a_{23} \end{pmatrix}.$$

Note that the first subscript names the row that the element occupies and the second names the column. It is sometimes convenient to shorten this notation to (a_{ij}) provided the range of values of i and j are given or understood. In a 2×3 matrix $i = 1, 2$, and $j = 1, 2, 3$.

Next we give some definitions and theorems pertaining to the algebra of matrices.

Definition 4.1–1 (*Equality*). Two matrices are equal if and only if they have the same size *and* corresponding elements are equal.

In other words, if $A = (a_{ij})$ and $B = (b_{ij})$, then $A = B$ implies that $a_{ij} = b_{ij}$ for all i and j involved. Hence,

$$\begin{pmatrix} 1 & 2 & 3 \\ 4 & 5 & 6 \end{pmatrix} \neq \begin{pmatrix} 1 & 2 & 3 \\ 4 & 6 & 5 \end{pmatrix},$$

whereas

$$\begin{pmatrix} a_{11} & a_{12} \\ a_{21} & a_{22} \end{pmatrix} = \begin{pmatrix} -5 & 3 \\ 0 & 2 \end{pmatrix}$$

implies that $a_{11} = -5$, $a_{12} = 3$, $a_{21} = 0$, and $a_{22} = 2$.

Definition 4.1–2 (*Addition*). If $A = (a_{ij})$ and $B = (b_{ij})$, then $A + B = (a_{ij} + b_{ij})$.

Inherent in this definition is the fact that addition is only possible when two matrices have the same size. The definition then states that the sum is obtained by adding corresponding elements. Thus

$$\begin{pmatrix} 1 & -2 & 3 \\ 0 & 4 & 5 \end{pmatrix} + \begin{pmatrix} 2 & 2 & -2 \\ -3 & 1 & -3 \end{pmatrix} = \begin{pmatrix} 3 & 0 & 1 \\ -3 & 5 & 2 \end{pmatrix}.$$

As a consequence of the definition of matrix addition we can prove a number of properties relating to this operation. Since these results follow directly from well-known properties of the real numbers, the proofs are left for the exercises.

Theorem 4.1–1 If A, B, and C are $m \times n$ matrices, then

a) $A + B$ is an $m \times n$ matrix (closure property);

b) $A + B = B + A$ (commutative property);

c) $(A + B) + C = A + (B + C)$ (associative property).

Definition 4.1–3 (*Multiplication by a scalar*). If c is any real number, then $c(a_{ij}) = (ca_{ij})$. We say that we have multiplied the matrix (a_{ij}) by the *scalar* (real number) c.*

The above definition states that, when multiplying a matrix by a scalar, every element of the matrix must be multiplied by that scalar. Hence we have for an arbitrary matrix $A = (a_{ij})$,

a) $1(a_{ij}) = (1a_{ij}) = (a_{ij})$,

b) $-1(a_{ij}) = (-1a_{ij}) = (-a_{ij}) = -(a_{ij}) = -A$, $\qquad$ **(4.1–1)**

c) $0(a_{ij}) = (0a_{ij}) = O$.

We call O, a matrix all of whose elements are zeros, a **zero matrix**. Hence it follows that

$$A + O = A, \qquad (4.1\text{–}2)$$

which shows that a zero matrix behaves like an *additive identity* in matrix algebra. Moreover, for every matrix A there exists a unique matrix $-A$ (defined in (4.1–1b)) such that

$$A + (-A) = O, \qquad (4.1\text{–}3)$$

which shows the existence of an *additive inverse* for every matrix.

In the foregoing we have emphasized the fact that, with the definitions given, matrices have the same properties under addition that real numbers do. The situation is much different, however, when it comes to matrix multiplication. Before defining multiplication we give a simple example of a *linear transformation*. (A definition will be given in Definition 4.3–7.) Consider the two equations

$$\begin{aligned} y_1 &= a_{11}x_1 + a_{12}x_2 + a_{13}x_3, \\ y_2 &= a_{21}x_1 + a_{22}x_2 + a_{23}x_3. \end{aligned} \qquad (4.1\text{–}4)$$

We can think of the variables y_1 and y_2 as being expressed in terms of the variables x_1, x_2, and x_3. Since both equations are linear, we call (4.1–4) a linear transformation that is characterized by the 2×3 matrix

$$A = \begin{pmatrix} a_{11} & a_{12} & a_{13} \\ a_{21} & a_{22} & a_{23} \end{pmatrix}.$$

If now we also have

$$\begin{aligned} x_1 &= b_{11}z_1 + b_{12}z_2, \\ x_2 &= b_{21}z_1 + b_{22}z_2, \\ x_3 &= b_{31}z_1 + b_{32}z_2, \end{aligned} \qquad (4.1\text{–}5)$$

---

* Later we will consider multiplying a matrix by a matrix.

a second linear transformation relating x_1, x_2, and x_3 to z_1 and z_2, then we can compute the relation between the y_i and z_i, $i = 1, 2$. The latter is given by (Exercise 5)

$$y_1 = (a_{11}b_{11} + a_{12}b_{21} + a_{13}b_{31})z_1 + (a_{11}b_{12} + a_{12}b_{22} + a_{13}b_{32})z_2,$$
$$y_2 = (a_{21}b_{11} + a_{22}b_{21} + a_{23}b_{31})z_1 + (a_{21}b_{12} + a_{22}b_{22} + a_{23}b_{32})z_2. \qquad \textbf{(4.1–6)}$$

Using matrix notation, we let A be as given above and

$$X = \begin{pmatrix} x_1 \\ x_2 \\ x_3 \end{pmatrix}, \qquad Y = \begin{pmatrix} y_1 \\ y_2 \end{pmatrix}, \qquad Z = \begin{pmatrix} z_1 \\ z_2 \end{pmatrix}, \qquad B = \begin{pmatrix} b_{11} & b_{12} \\ b_{21} & b_{22} \\ b_{31} & b_{32} \end{pmatrix}.$$

Then Eqs. (4.1–4) and (4.1–5) become

$$Y = AX \quad \text{and} \quad X = BZ,$$

respectively. Hence Eq. (4.1–6) can be written

$$Y = (AB)Z.$$

In using products of matrices above we have conformed to the following definition.

Definition 4.1–4 (*Matrix multiplication*). If $A = (a_{ij})$ is $m \times n$ and $B = (b_{jk})$ is $n \times r$, then the *product* $AB = C$ is $m \times r$ and is given by

$$C = (c_{ik}) = \left(\sum_{q=1}^{n} a_{iq} b_{qk} \right), \qquad \textbf{(4.1–7)}$$

meaning that the sum represents the element in the ith row and kth column in the product. (Recall that representing B by (b_{jk}) means that the element in the jth row and kth column of B is b_{jk}.) In Eq. (4.1–7) we have indicated that AB is defined only if the number of columns of A is the same as the number of rows of B.* The definition of multiplication also shows that multiplying matrices is a *row-by-column* process. We multiply elements of a row by corresponding (first by first, second by second, etc.) elements of a column and add these products. Thus the element in the ith row and kth column of the product AB is obtained by multiplying the ith row of A by the kth column of B in the above fashion. We illustrate with some examples.

EXAMPLE 4.1–1 Given the matrices

$$X = \begin{pmatrix} 2 \\ -3 \\ 1 \end{pmatrix}, \qquad Y = \begin{pmatrix} 3 \\ -2 \end{pmatrix}, \qquad A = \begin{pmatrix} 1 & -2 & 3 \\ 0 & 4 & 5 \end{pmatrix}, \qquad B = \begin{pmatrix} 2 & 2 \\ -3 & 1 \\ 4 & 5 \end{pmatrix},$$

find AX, BY, and AB.

---

* Some authors call A and B *conformable* in this case.

Solution. We have

$$AX = \begin{pmatrix} 1 & -2 & 3 \\ 0 & 4 & 5 \end{pmatrix} \begin{pmatrix} 2 \\ -3 \\ 1 \end{pmatrix} = \begin{pmatrix} 1(2) + (-2)(-3) + 3(1) \\ 0(2) + 4(-3) + 5(1) \end{pmatrix} = \begin{pmatrix} 11 \\ -7 \end{pmatrix},$$

$$BY = \begin{pmatrix} 2 & 2 \\ -3 & 1 \\ 4 & 5 \end{pmatrix} \begin{pmatrix} 3 \\ -2 \end{pmatrix} = \begin{pmatrix} 2(3) + 2(-2) \\ -3(3) + 1(-2) \\ 4(3) + 5(-2) \end{pmatrix} = \begin{pmatrix} 2 \\ -11 \\ 2 \end{pmatrix},$$

$$AB = \begin{pmatrix} 1 & -2 & 3 \\ 0 & 4 & 5 \end{pmatrix} \begin{pmatrix} 2 & 2 \\ -3 & 1 \\ 4 & 5 \end{pmatrix}$$

$$= \begin{pmatrix} 1(2) + (-2)(-3) + 3(4) & 1(2) + (-2)(1) + 3(5) \\ 0(2) + 4(-3) + 5(4) & 0(2) + 4(1) + 5(5) \end{pmatrix}$$

$$= \begin{pmatrix} 20 & 15 \\ 8 & 29 \end{pmatrix}. \quad ■$$

Note that BA in Example 4.1–1 is not defined, and, in general, *matrix multiplication is not commutative* (see Exercise 1b). We do, however, have the properties expressed in the following theorem. It is assumed that all indicated products and sums are defined.

Theorem 4.1–2 If A, B, and C are matrices, then

a) $(AB)C = A(BC)$ (associative property);

b) $AO = O$ (annihilator property);

c) $A(B + C) = AB + AC$ (left distributive property);

d) $(A + B)C = AC + BC$ (right distributive property).

Observe that the commutative property is missing from the above. Since addition of matrices *is* commutative, we use the statement "A and B commute" to mean $AB = BA$. There is another property related to multiplication that is different for matrices than for the algebra with which we are familiar. In the algebra of real and complex numbers if $ab = 0$, then it follows that either $a = 0$ and/or $b = 0$. However, if A and B are matrices with AB equal to a zero matrix, then it does *not* follow that $B = O$ if $A \neq O$. This is shown by the following example using 2×2 matrices:

$$\begin{pmatrix} 1 & 1 \\ 0 & 0 \end{pmatrix} \begin{pmatrix} 2 & 3 \\ -2 & -3 \end{pmatrix} = \begin{pmatrix} 0 & 0 \\ 0 & 0 \end{pmatrix}.$$

Hence the product of nonzero matrices may be equal to a zero matrix.

We will have occasion to interchange the rows and columns of a matrix, thus (usually) forming a new matrix. This process is called *transposition*, defined as follows.

Definition 4.1–5 (*Transposition*). If $A = (a_{ij})$ is $m \times n$, then the **transpose of** A, denoted by A^T, is $n \times m$ and is given by

$$A^T = (a_{ji}).$$

Using the matrices of Example 4.1–1, we have

$$X^T = (2 \quad -3 \quad 1), \qquad B^T = \begin{pmatrix} 2 & -3 & 4 \\ 2 & 1 & 5 \end{pmatrix}.$$

Properties of transposition are listed in the following theorem.

Theorem 4.1–3 For matrices A and B,

a) $(A + B)^T = A^T + B^T$;

b) $(AB)^T = B^T A^T$;

c) $(A^T)^T = A$.

The proofs are left for the exercises. Note that the transpose of a product is the product of the transposes in the *reverse order*.

When a matrix is square—that is, $n \times n$—we have some additional descriptive terms. We say that a matrix A is **symmetric** if $A = A^T$. In this case $(a_{ij}) = (a_{ji})$ for all i and j. That portion of a matrix that consists of the elements a_{ii} is referred to as the **principal diagonal** of the matrix. In transposition these diagonal elements do not change.

If a matrix is square and its elements off the diagonal are all zero, then the matrix is called a **diagonal matrix**. For example,

$$D = \begin{pmatrix} a_1 & 0 & 0 & 0 \\ 0 & a_2 & 0 & 0 \\ 0 & 0 & a_3 & 0 \\ 0 & 0 & 0 & a_4 \end{pmatrix}$$

is a diagonal matrix, which we can write in abbreviated form as

$$D = \operatorname{diag}(a_1 \quad a_2 \quad a_3 \quad a_4).$$

A diagonal matrix whose n diagonal elements are all equal, say c,

$$E = \operatorname{diag}(c \quad c \quad \ldots \quad c),$$

is called a **scalar matrix**. If A is $n \times n$, then $AE = EA = cA$, which agrees with the definition of multiplication by a scalar given in Definition 4.1–3. In the special case where $c = 1$, we have

$$I_n = \operatorname{diag}(1 \quad 1 \quad \ldots \quad 1),$$

which is called *the* $n \times n$ **identity matrix**. It has the property

$$AI = IA = A$$

for every $n \times n$ matrix A and thus acts as the unit in multiplication. When no confusion can result, we omit the subscript and denote the identity matrix by I as above.

The presence of identity and zero matrices can sometimes simplify matrix multiplication if we use a technique called **partitioning**. Any matrix can be divided by vertical and horizontal lines into *submatrices*. For example,

$$A_1 = \begin{pmatrix} a_{11} & a_{12} \\ a_{21} & a_{22} \end{pmatrix}$$

is a submatrix of the matrix

$$A = \left(\begin{array}{cc|c} a_{11} & a_{12} & a_{13} \\ a_{21} & a_{22} & a_{23} \\ \hline a_{31} & a_{32} & a_{33} \end{array} \right),$$

obtained by partitioning A as shown by the dashed lines. In this way a matrix can be considered as having elements that are themselves matrices. In the above example, we can write

$$A = \begin{pmatrix} A_1 & A_2 \\ A_3 & A_4 \end{pmatrix}$$

where, in addition to A_1 defined above, we have

$$A_2 = \begin{pmatrix} a_{13} \\ a_{23} \end{pmatrix}, \qquad A_3 = (a_{31} \quad a_{32}), \qquad A_4 = (a_{33}).$$

If A were one of the factors of a product, we could perform the multiplication by treating the A_i as *elements* of A. Thus if A_1 were an identity matrix and A_3 were a zero matrix, then the computation in obtaining the product of A and another matrix would be considerably simplified. This type of multiplication is called *block multiplication*. We illustrate the procedure in the following example.

EXAMPLE 4.1–2 Find AB given that

$$A = \left(\begin{array}{cc|cc|c} 1 & 0 & -2 & 3 & 1 \\ 0 & 1 & 5 & 2 & 4 \\ \hline 0 & 0 & 1 & 0 & 0 \\ 0 & 0 & 0 & 1 & 0 \end{array} \right)$$

and

$$B = \left(\begin{array}{ccc|ccc} 3 & 4 & -3 & 0 & 0 & 0 \\ -2 & 1 & 2 & 0 & 0 & 0 \\ \hline 0 & 0 & 0 & 1 & 2 & -1 \\ 0 & 0 & 0 & 3 & -2 & 4 \\ \hline 0 & 0 & 0 & 1 & 0 & 2 \end{array} \right).$$

Solution. We partition the two matrices as shown. The important aspect of this operation is that the *columns of A* must be partitioned in exactly the same way as the *rows of B*; otherwise the products of the resulting submatrices may not be defined. The *horizontal* partitioning of A and the *vertical* partitioning of B are immaterial. We have done this above, not arbitrarily, but in such a way as to take full advantage of zero and identity matrices. With the partitioning shown, we can now use block multiplication and consider

$$\begin{pmatrix} A_1 & A_2 & A_3 \\ A_4 & A_5 & A_6 \end{pmatrix}\begin{pmatrix} B_1 & B_2 \\ B_3 & B_4 \\ B_5 & B_6 \end{pmatrix} = \begin{pmatrix} I_2 & A_2 & A_3 \\ O & I_2 & O \end{pmatrix}\begin{pmatrix} B_1 & O \\ O & B_4 \\ O & B_6 \end{pmatrix}.$$

Hence the product can be written

$$\begin{pmatrix} B_1 & A_2 B_4 + A_3 B_6 \\ O & B_4 \end{pmatrix} = \begin{pmatrix} 3 & 4 & -3 & 8 & -10 & 16 \\ -2 & 1 & 2 & 15 & 6 & 11 \\ 0 & 0 & 0 & 1 & 2 & -1 \\ 0 & 0 & 0 & 3 & -2 & 4 \end{pmatrix}. \quad \blacksquare$$

In applied mathematics it often happens that large matrices (those having a large number of rows and columns) are also *sparse*; that is, they contain a large percentage of zero elements. Partitioning and block multiplication are especially useful when dealing with such sparse matrices. In addition to blocks of zeros, diagonal matrices (besides the identity matrices) should be considered when partitioning.

We will also have occasion to speak about square matrices that are triangular. If all the elements of a square matrix above (below) the principal diagonal are zero, then the matrix is called a **lower (upper) triangular matrix** (see Fig. 4.1–1). It follows that if a matrix is both upper and lower triangular, then it must be a diagonal matrix.

$$\begin{pmatrix} a_{11} & a_{12} & a_{13} & a_{14} \\ 0 & a_{22} & a_{23} & a_{24} \\ 0 & 0 & a_{33} & a_{34} \\ 0 & 0 & 0 & a_{44} \end{pmatrix} \qquad \begin{pmatrix} a_{11} & 0 & 0 & 0 \\ a_{21} & a_{22} & 0 & 0 \\ a_{31} & a_{32} & a_{33} & 0 \\ a_{41} & a_{42} & a_{43} & a_{44} \end{pmatrix}$$
$$\text{(a)} \qquad\qquad\qquad \text{(b)}$$

FIGURE 4.1–1 Triangular matrices: (a) upper triangular; (b) lower triangular.

Because matrix multiplication is not commutative we must specify whether we are multiplying a matrix on the right or on the left. In this connection the terms *premultiply* and *postmultiply* are also used. In the product AB, for example, B has been *pre*multiplied by A, or A has been *post*multiplied by B.

KEY WORDS AND PHRASES

matrix (pl., matrices)	**principal diagonal**
element	**diagonal matrix**
square matrix	**scalar matrix**
column vector	**identity**
row vector	**partitioning**
scalar	**block multiplication**
zero matrix	**triangular matrix**
transpose	**premultiply**
symmetric matrix	**postmultiply**

EXERCISES 4.1

▶ 1. If A, B, and C are arbitrary $m \times n$ matrices and O is the $m \times n$ zero matrix, then prove each of the following.

   a) $A + B = B + A$ **b)** $(A + B) + C = A + (B + C)$

   c) $A + O = A$ **d)** $A + (-A) = O$

2. **a)** Obtain Eq. (4.1–6) algebraically from Eqs. (4.1–4) and (4.1–5).

   b) Obtain Eq. (4.1–6) by multiplying the appropriate matrices in Eqs. (4.1–4) and (4.1–5).

3. **a)** Prove the associative property of matrix multiplication: part (a) of Theorem 4.1–2. (*Hint*: Use summation notation as in Eq. 4.1–7.)

   b) Prove the left distributive property: part (c) of Theorem 4.1–2.

4. Prove Theorem 4.1–3. (*Hint*: In part (b) use the summation notation of Eq. 4.1–7.)

▶▶ 5. Given the matrices

$$A = \begin{pmatrix} 3 & 0 \\ -1 & 4 \end{pmatrix}, \qquad B = \begin{pmatrix} 1 & 2 \\ 3 & 4 \end{pmatrix}, \qquad C = (2 \quad 3), \qquad D = \begin{pmatrix} 1 \\ -1 \end{pmatrix},$$

   compute each of the following.

   a) $A + B$, $A - B$, $B - A$ **b)** AB, BA, CD, DC, BD, CA

   c) $2A + 3B$, $4CB$, $(A + B)AB$, A^2

6. Given the matrices

$$A = \begin{pmatrix} 1 & 2 & 3 \\ 0 & -1 & 1 \\ 2 & 3 & 0 \end{pmatrix}, \qquad B = \begin{pmatrix} 3 & 1 & 0 \\ 1 & -1 & 2 \\ 0 & 2 & 1 \end{pmatrix},$$

   compute each of the following.

   a) AB, $(AB)^T$, $(A + B)^T$

   b) BA, $(BA - AB)^T$, $(B + A)A^T$

   c) A^3, $A^3 - 10A$ (*Note*: $A^3 = AAA$.)

   d) B^3, B^2, $B^3 - 3B^2 - 6B + 16I$

7. Using the matrices of Example 4.1–1 compute
$$B^T X, \quad A^T Y, \quad \text{and} \quad B^T A^T.$$

8. If

$$A = \begin{pmatrix} 0 & 0 & 0 & 1 \\ 0 & 0 & 2 & 0 \\ 1 & 0 & 0 & 0 \\ 0 & 1 & 0 & 0 \end{pmatrix} \quad \text{and} \quad B = \begin{pmatrix} 1 & 2 & 0 & 0 \\ 0 & 1 & 0 & 0 \\ 0 & 0 & 0 & 1 \\ 0 & 0 & 2 & 2 \end{pmatrix},$$

then compute AB and BA by partitioning A and B appropriately.

9. Given the matrix

$$A = \begin{pmatrix} 1 & 1 & -1 \\ -1 & 0 & 1 \\ 1 & -1 & -1 \end{pmatrix}:$$

a) find A^2, A^3, A^5, and A^{10};

b) find $A^3 + 2A$;

c) show that although $A(A^2 + 2I) = O$, it is not true that either $A = O$ or $A^2 = -2I$.

10. Show that if

$$B = \begin{pmatrix} 2 & -2 & -1 \\ 2 & 3 & 4 \\ 3 & 5 & 9 \end{pmatrix} \quad \text{and} \quad A = \begin{pmatrix} 1 & 1 & 1 \\ 1 & 1 & 1 \\ -1 & -1 & -1 \end{pmatrix},$$

then $BA = A$, yet $B \neq I$. What additional property must a matrix have before it can be called *the* identity matrix?

11. Show that the matrix

$$B = \begin{pmatrix} 5 & 4 \\ 1 & 2 \end{pmatrix}$$

satisfies the equation $x^2 - 7x + 6 = 0$. (*Hint*: Every term in the equation must be a 2×2 matrix.)

12. Find all matrices A satisfying the equation

$$\begin{pmatrix} 0 & 1 \\ 0 & 2 \end{pmatrix} A = \begin{pmatrix} 0 & 0 \\ 0 & 0 \end{pmatrix}.$$

13. Find all matrices B that commute with

$$A = \begin{pmatrix} 0 & 1 \\ 0 & 2 \end{pmatrix}.$$

14. Find all matrices A satisfying the equation

$$\begin{pmatrix} 0 & 1 \\ 0 & 2 \end{pmatrix} A = \begin{pmatrix} 0 & 0 & 1 \\ 0 & 0 & 2 \end{pmatrix}.$$

15. Prove that sums and products of triangular matrices are triangular.

16. a) Use either the left or right distributive property to expand

$$(A + B)(A - B).$$

b) Under what conditions does $(A + B)(A - B) = A^2 - B^2$?

17. If A commutes with B, show that A^T commutes with B^T.

18. Prove that if A and B are diagonal matrices, then they commute.

▶▶▶ **19.** Prove that $AA^T = O$ implies that $A = O$.

20. A *Markov** (or *stochastic*) matrix plays a role in probability theory. An $n \times n$ Markov matrix has the following two properties:

i) $0 \le a_{ij} \le 1$;

ii) $\displaystyle\sum_{j=i}^{n} a_{ij} = 1,$ for $i = 1, 2, \ldots, n$.

a) Show that

$$A = \begin{pmatrix} \frac{1}{2} & \frac{1}{4} & \frac{1}{4} \\ \frac{1}{3} & \frac{1}{3} & \frac{1}{3} \\ \frac{1}{6} & \frac{1}{6} & \frac{2}{3} \end{pmatrix} \quad \text{and} \quad B = \begin{pmatrix} \frac{1}{5} & \frac{2}{5} & \frac{2}{5} \\ \frac{1}{6} & \frac{1}{2} & \frac{1}{3} \\ \frac{1}{8} & \frac{2}{8} & \frac{5}{8} \end{pmatrix}$$

are Markov matrices.

b) Show that AB and BA are also Markov matrices.

c) Show that every 2×2 Markov matrix has the form

$$\begin{pmatrix} p & 1-p \\ 1-q & q \end{pmatrix}, \qquad 0 \le p \le 1, \qquad 0 \le q \le 1.$$

d) Given that

$$C = \begin{pmatrix} 0.1 & 0.9 \\ 0.9 & 0.1 \end{pmatrix},$$

compute C^2, C^3, C^4; then guess C^n.

21. If A is $p \times q$ and B is $r \times s$, show that AB exists if $q = r$ and is $p \times s$.

22. A square matrix A is said to be *skew-symmetric* if $A^T = -A$. Show that any square matrix can be written as the sum of a symmetric matrix and a skew-symmetric matrix. (*Hint:* Begin by showing that $\frac{1}{2}(A + A^T)$ is symmetric.)

23. Prove that the additive inverse of a matrix is unique. (*Hint:* If $A + (-A) = O$ and $A + B = O$, draw a conclusion involving $(-A)$ and B.)

24. If c and d are real numbers and A and B are $m \times n$ matrices, prove each of the following.

a) $c(A + B) = cA + cB$ **b)** $(c + d)A = cA + dA$ **c)** $c(dA) = (cd)A$

---

* After A. A. Markov (1856–1922), a Russian probabilist who first used the matrix in 1907.

25. A square matrix B such that $B^n = O$ for some positive integer n but $B^{n-1} \neq O$ is said to be *nilpotent* of index n. Show that

$$\begin{pmatrix} 0 & 0 & 0 & 0 \\ 3 & 0 & 0 & 0 \\ 4 & 1 & 0 & 0 \\ 1 & 2 & 6 & 0 \end{pmatrix}$$

is nilpotent of index 4.

26. If A and B are symmetric, prove that AB is symmetric if and only if A and B commute.

27. If A is any square matrix, show each of the following.

a) AA^T and A^TA are both symmetric.

b) $A + A^T$ is symmetric.

c) $A - A^T$ is skew-symmetric.

4.2 SYSTEMS OF LINEAR ALGEBRAIC EQUATIONS

We encounter systems of linear algebraic equations in a wide variety of situations. These range from finding the point of intersection of two lines in the plane by solving a pair of simultaneous equations to solving a partial differential equation by numerical methods. In the latter case we may have to deal with a large number of equations (say, one thousand) and many variables. Thus we have a need to manipulate vast amounts of data efficiently and we will see in this section that matrix methods are very helpful for this purpose.

We consider first systems of *two* equations in *two* variables. Such a system has the general form

$$\begin{aligned} a_{11}x_1 + a_{12}x_2 &= b_1 \\ a_{21}x_1 + a_{22}x_2 &= b_2, \end{aligned} \tag{4.2–1}$$

where b_1, b_2, and the a_{ij} are real numbers. The variables are denoted by x_1 and x_2 rather than x and y so that we can extend the methods we will develop to larger systems. By a *solution* of (4.2–1) we mean all the ordered pairs (x_1, x_2) that satisfy both equations. We refer to this totality of ordered pairs as **the solution set** of the system. In other words, the solution set of (4.2–1) consists of all ordered pairs (x_1, x_2) that satisfy the system.

Three cases may now be distinguished, as follows.

Case I: The solution set contains precisely *one* ordered pair.

Case II: The solution set contains *no* ordered pairs.

Case III: The solution set contains an *infinite number* of ordered pairs.

For any given system one, and only one, of the above cases can hold, that is, the three cases are *mutually exclusive*. Moreover, it is impossible for any other case to exist. For example, we cannot have a system (4.2–1) that has precisely *two* distinct ordered pairs in its solution set. Thus the above three cases are also *collectively exhaustive*. The terms "mutually exclusive" and "collectively exhaustive" play a role in probability theory as well as here. We further distinguish Case II by saying that this result implies that the system is **inconsistent**. In contrast, systems resulting in solution sets containing one or an infinite number of ordered pairs are said to be **consistent**. The implications of these terms will become more apparent later.

For two equations in two variables the three cases above have a simple geometric interpretation, which is shown in Fig. 4.2–1. The case of three equations and three variables is considered geometrically in Exercise 35.

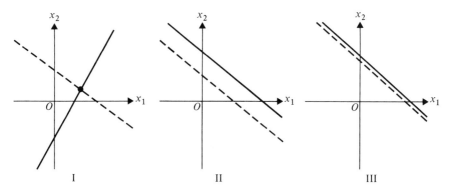

FIGURE 4.2–1 Two equations in two variables. Case I: one solution (intersecting lines). Case II: no solutions (parallel lines). Case III: infinite number of solutions (coincident lines).

EXAMPLE 4.2–1   Find the solution set of each of the following systems.

a) $3x_1 + 2x_2 = 0$
    $-2x_1 - x_2 = 1$

b) $2x_1 + 3x_2 = 2$
    $4x_1 + 6x_2 = 5$

c) $x_1 - 2x_2 = 3$
    $2x_1 - 4x_2 = 6$

Solution. Using our knowledge of algebra we can solve system (a) by *elimination* (eliminating either x_1 or x_2 and solving for the other) or by *substitution* (solving one equation for either x_1 or x_2 and substituting this value into the other equation). For example, multiplying the second equation by 2 and adding the result to the first equation yields $x_1 = -2$. Multiplying the first equation by 2 and the second by 3 and adding the results gives $x_2 = 3$. Thus the solution set of (a) is

$$\{(-2, 3)\},$$

a set consisting of a *single* ordered pair. We call this solution *unique* in the sense that no other (different) ordered pair can satisfy the two equations of the system. A graph of the equations in (a) will verify the algebraic result (Exercise 1).

When we attempt to solve system (b) by elimination, we are unsuccessful and obtain the inconsistent statement "$0 = 1$." Thus the system has no solution; we call it an inconsistent system and say that its solution set is the **empty set**, denoted by $\varnothing$, the symbol for a set containing no members whatever. A graph of the two lines constituting the system shows that the lines are parallel and thus have no points in common.

An examination of system (c) shows that the second equation is simply twice the first, hence provides no new information. Thus, in effect, we have only one equation and *every* ordered pair that satisfies this equation belongs to the solution set of the system. In this way the solution set contains an infinite number of ordered pairs, each pair corresponding to a point on the line that is the graph of the equation. The solution set can be written as

$$\{(3 + 2x_2, x_2)\,|\,x_2 \text{ a real number}\}$$

or as

$$\{(3 + 2k, k)\,|\,k \text{ a real number}\}. \quad \blacksquare$$

When the number of equations in a system is m and the number of variables is n, we speak of an $m \times n$ system and the elements of the solution set are *n-tuples* $(x_1, x_2, \ldots, x_n)$. The three cases discussed above still apply, however. In other words, the solution set of an $m \times n$ system contains zero, one, or an infinite number of n-tuples. Keeping this fact in mind will simplify the study of this and other sections in this chapter.

Cramer's rule

The unique solution of an $n \times n$ system can be found by a procedure known as Cramer's rule,* which has been in use since 1750. Applying

---

* After Gabriel Cramer (1704–1752), a Swiss mathematician.

this rule to system (a) in Example 4.2–1, we have

$$x_1 = \frac{\begin{vmatrix} 0 & 2 \\ 1 & -1 \end{vmatrix}}{\begin{vmatrix} 3 & 2 \\ -2 & -1 \end{vmatrix}} = \frac{0(-1) - (1)(2)}{3(-1) - (2)(-2)} = \frac{-2}{1} = -2,$$

$$x_2 = \frac{\begin{vmatrix} 3 & 0 \\ -2 & 1 \end{vmatrix}}{\begin{vmatrix} 3 & 2 \\ -2 & -1 \end{vmatrix}} = \frac{3(1) - (-2)(0)}{1} = 3.$$

Here we have expressed x_1 and x_2 as quotients of two **determinants**. Associated with every $n \times n$ matrix A there is a real number called the determinant of A, denoted by $|A|$ (the notation $\det(A)$ is also used). If

$$A = \begin{pmatrix} a_{11} & a_{12} \\ a_{21} & a_{22} \end{pmatrix},$$

then

$$|A| = \begin{vmatrix} a_{11} & a_{12} \\ a_{21} & a_{22} \end{vmatrix} = a_{11}a_{22} - a_{21}a_{12}, \tag{4.2–2}$$

which can be taken as the definition of the determinant of a 2×2 matrix.

The determinant of a 3×3 matrix can now be defined as a linear combination of the determinants of 2×2 matrices. Thus if

$$A = \begin{pmatrix} a_{11} & a_{12} & a_{13} \\ a_{21} & a_{22} & a_{23} \\ a_{31} & a_{32} & a_{33} \end{pmatrix},$$

then

$$|A| = a_{11} \begin{vmatrix} a_{22} & a_{23} \\ a_{32} & a_{33} \end{vmatrix} - a_{12} \begin{vmatrix} a_{21} & a_{23} \\ a_{31} & a_{33} \end{vmatrix} + a_{13} \begin{vmatrix} a_{21} & a_{22} \\ a_{31} & a_{32} \end{vmatrix}. \tag{4.2–3}$$

Note that the determinants in this definition are determinants of *submatrices* of A and are related to the coefficients. For example, a_{11} is multiplied by the determinant of the submatrix obtained by deleting the first row and first column of A; a_{12} is multiplied by the determinant of the submatrix obtained by deleting the first row and second column of A, etc. In general, a_{ij} multiplies the determinant of a submatrix obtained by deleting the ith row and jth column of A. Moreover, the coefficients in Eq. (4.2–3) are simply the elements of the first row of A with one change in sign. We say that $|A|$ has been expanded by its *first row* in this case. The sign of each coefficient a_{ij} is found from $(-1)^{i+j}$. If we had expanded $|A|$ by its second row, we would have obtained

$$|A| = -a_{21} \begin{vmatrix} a_{12} & a_{13} \\ a_{32} & a_{33} \end{vmatrix} + a_{22} \begin{vmatrix} a_{11} & a_{13} \\ a_{31} & a_{33} \end{vmatrix} - a_{23} \begin{vmatrix} a_{11} & a_{12} \\ a_{31} & a_{32} \end{vmatrix}.$$

From Eqs. (4.2–2) and (4.2–3) it is apparent that the determinant of the 3×3 matrix A is the sum of six terms—three of one sign and three of the opposite sign. Carrying these ideas one step further, we can define the determinant of a 4×4 matrix as the sum of four determinants of the appropriate 3×3 submatrices. Hence the determinant of a 4×4 matrix is the sum of 24 or 4! terms and, in general, the determinant of an $n \times n$ matrix is the sum of $n!$ terms. The determinants involved in these sums belong to submatrices, which, in turn, are obtained by deleting a row and a column from the original matrix. The determinant of a matrix can be computed by using *any* row or column of the matrix for the expansion (see Exercise 2).

We can now state a theorem pertaining to systems that have unique solutions.

Theorem 4.2–1 (*Cramer's rule*). Given the $n \times n$ system $AX = B$, where

$$A = \begin{pmatrix} a_{11} & a_{12} & \cdots & a_{1n} \\ a_{21} & a_{22} & \cdots & a_{2n} \\ \vdots & & & \\ a_{n1} & a_{n2} & \cdots & a_{nn} \end{pmatrix}, \qquad X = \begin{pmatrix} x_1 \\ x_2 \\ \vdots \\ x_n \end{pmatrix}, \qquad B = \begin{pmatrix} b_1 \\ b_2 \\ \vdots \\ b_n \end{pmatrix}.$$

The unique solution of this system is given by

$$x_k = \frac{|A_k|}{|A|}, \qquad k = 1, 2, \ldots, n,$$

provided $|A| \neq 0$, where $|A_k|$ is the determinant of the matrix obtained from A by replacing its kth column by the elements of B.

We postpone the proof of this theorem in order not to be deterred from our objective of presenting methods for solving systems of linear algebraic equations. A proof can be found in Section 4.7 (see Theorem 4.7–3).

Gaussian elimination

Clearly, the computation involved in solving large systems by Cramer's rule is excessive and other methods must be used. One of these, called **Gaussian elimination,*** will be considered next. First we need some

---

* After Karl F. Gauss (1777–1855), a German mathematician.

definitions. Suppose we seek solutions to the $m \times n$ system

$$AX = B, \tag{4.2-4}$$

where

$$A = \begin{pmatrix} a_{11} & a_{12} & \cdots & a_{1n} \\ a_{21} & a_{22} & \cdots & a_{2n} \\ \vdots & & & \\ a_{m1} & a_{m2} & \cdots & a_{mn} \end{pmatrix}, \qquad X = \begin{pmatrix} x_1 \\ x_2 \\ \vdots \\ x_n \end{pmatrix}, \qquad B = \begin{pmatrix} b_1 \\ b_2 \\ \vdots \\ b_m \end{pmatrix}.$$

It will be convenient to work with the $m \times (n + 1)$ matrix,

$$A|B = \begin{pmatrix} a_{11} & a_{12} & \cdots & a_{1n} & | & b_1 \\ a_{21} & a_{22} & \cdots & a_{2n} & | & b_2 \\ \vdots & & & & | & \\ a_{m1} & a_{m2} & \cdots & a_{mn} & | & b_m \end{pmatrix},$$

which is called the **augmented matrix** of the system (4.2–4). Note that all the essential information pertaining to the system is contained in its augmented matrix. Each row of this matrix represents one of the equations of the system and we can easily relate rows and equations. For example, the second row is a representation of the equation

$$a_{21}x_1 + a_{22}x_2 + \cdots + a_{2n}x_n = b_2.$$

We now remark that the solution set of the system (4.2–4) is unaffected by certain changes made to the augmented matrix. We call these changes **elementary row operations**, which are defined as follows.

E1. Any two rows may be interchanged (this simply changes the order in which the equations of the system are written).

E2. Any row may be multiplied by a *nonzero* constant (this is equivalent to multiplying both members of an equation by a nonzero constant).

E3. Any row may be multiplied by a constant and the result added to any other row (this can be expressed by the familiar statement, "equal quantities added to equal quantities result in equal quantities").

When the above elementary row operations are performed on an augmented matrix a *different* matrix results (most of the time). The important fact, however, is that this different matrix represents a system that has the *same* solution set as the original matrix. Thus, in a sense, the two matrices are equivalent. We call this type of equivalence **row equivalence** and denote it by $\sim$ (read "is row equivalent to"). We write $A \sim B$, that is, A is row equivalent to B, meaning that A can be transformed into B (or B into A) by means of one or more elementary row operations. We present an example that will show the advantages of using elementary row operations.

EXAMPLE 4.2–2 Find the solution set of the system

$$2x_1 - 4x_2 + x_3 = 0$$
$$x_1 + x_2 + 4x_3 = 5$$
$$3x_1 + x_2 - 3x_3 = -1.$$

Solution. We begin with the augmented matrix of the system,

$$\begin{pmatrix} 2 & -4 & 1 & 0 \\ 1 & 1 & 4 & 5 \\ 3 & 1 & -3 & -1 \end{pmatrix},$$

and list the elementary row operations that lead to the solution.

a) Interchange rows one and two.

$$\begin{pmatrix} 1 & 1 & 4 & 5 \\ 2 & -4 & 1 & 0 \\ 3 & 1 & -3 & -1 \end{pmatrix}$$

b) Multiply row one by -2 and add the result to row two.

$$\begin{pmatrix} 1 & 1 & 4 & 5 \\ 0 & -6 & -7 & -10 \\ 3 & 1 & -3 & -1 \end{pmatrix}$$

c) Multiply row one by -3 and add the result to row three.

$$\begin{pmatrix} 1 & 1 & 4 & 5 \\ 0 & -6 & -7 & -10 \\ 0 & -2 & -15 & -16 \end{pmatrix}$$

d) Multiply row two by $-\frac{1}{6}$.

$$\begin{pmatrix} 1 & 1 & 4 & 5 \\ 0 & 1 & \frac{7}{6} & \frac{10}{6} \\ 0 & -2 & -15 & -16 \end{pmatrix}$$

e) Multiply row two by 2 and add the result to row three.

$$\begin{pmatrix} 1 & 1 & 4 & 5 \\ 0 & 1 & \frac{7}{6} & \frac{10}{6} \\ 0 & 0 & -\frac{38}{3} & -\frac{38}{3} \end{pmatrix}$$

f) Multiply row three by $-\frac{3}{38}$.

$$\begin{pmatrix} 1 & 1 & 4 & 5 \\ 0 & 1 & \frac{7}{6} & \frac{10}{6} \\ 0 & 0 & 1 & 1 \end{pmatrix}$$

The third row of this last matrix now provides the information that $x_3 = 1$. The second row becomes

$$x_2 + \tfrac{7}{6}x_3 = \tfrac{10}{6} = x_2 + \tfrac{7}{6},$$

from which we have that $x_2 = \frac{1}{2}$. The first row states

$$x_1 + x_2 + 4x_3 = 5 = x_1 + \tfrac{1}{2} + 4,$$

so that $x_1 = \frac{1}{2}$ and the solution set is

$$\{(\tfrac{1}{2}, \tfrac{1}{2}, 1)\}. \quad \blacksquare$$

By definition each of the above augmented matrices is row equivalent to the next, since each was obtained from the previous one by an elementary row operation. In Example 4.2–2 we have illustrated the method of solving linear systems by **Gaussian elimination** (also called **Gaussian reduction**). This method provided the value of x_3, which was then used to find x_2. Finally the values of x_3 and x_2 were used to obtain x_1. This process is called **back substitution**. We call attention to certain features of the Gaussian elimination method.

1. The elements of the augmented matrix need not be integers.
2. The method is readily programmable and adapted to hand-held calculators.*
3. The method is easily extended to larger systems than the 3×3 of the example.
4. The method can be used with $m \times n$ systems.

Caution: Elementary row operations must be done one at a time. For example, it is incorrect to add row one to row two and row two to row one *simultaneously* since this can lead to a row of zeros where there should not be such a row.

Our goal when applying Gaussian elimination is to obtain a matrix such as the one in step (f) of Example 4.2–2. This matrix is said to be in **row echelon form** according to the following definition.

Definition 4.2–1 An $m \times n$ matrix is in *row echelon form* if it has the following properties.

a) Rows consisting entirely of zeros are all together in the bottom part of the matrix.

b) Any row that does not consist entirely of zeros has 1 as its first nonzero element. Further, all elements below this leading 1 are zeros.

c) Each successive leading 1 (reading from left to right) is to the right of and below the previous one.

It is the last property that gives a matrix its echelon character. We further clarify the definition by exhibiting the matrices in Fig. 4.2–2, all of which are in row echelon form. The matrices in Fig. 4.2–3, however, are *not* in row echelon form. In (a) the leading one in the second row does not have a zero below it; in (b) the row of zeros is not at the bottom; and in (c) the leading ones are not successively to the right of and below the ones above them.

---

* A computer program would not *begin* the solution of Example 4.2–2 as we have done (interchanging rows one and two) but would simply multiply row one of the original augmented matrix by $\frac{1}{2}$ in order to create a 1 in the a_{11} position. The remaining steps, however, would be the same.

$$
\begin{pmatrix} 1 & 2 & 3 \\ 0 & 1 & 5 \\ 0 & 0 & 1 \end{pmatrix}
\qquad
\begin{pmatrix} 1 & 2 & 0 & 3 \\ 0 & 1 & 0 & 2 \\ 0 & 0 & 0 & 1 \\ 0 & 0 & 0 & 0 \end{pmatrix}
\qquad
\begin{pmatrix} 0 & 1 & 4 & 3 & 0 \\ 0 & 0 & 1 & 5 & 0 \\ 0 & 0 & 0 & 0 & 1 \\ 0 & 0 & 0 & 0 & 0 \end{pmatrix}
$$

(a)   (b)   (c)

FIGURE 4.2–2 Matrices in row echelon form.

$$
\begin{pmatrix} 1 & 2 & 3 \\ 0 & 1 & 5 \\ 0 & 1 & 0 \end{pmatrix}
\qquad
\begin{pmatrix} 1 & 2 & 0 & 3 \\ 0 & 1 & 0 & 2 \\ 0 & 0 & 0 & 0 \\ 0 & 0 & 0 & 1 \end{pmatrix}
\qquad
\begin{pmatrix} 0 & 0 & 1 & 5 & 0 \\ 0 & 1 & 4 & 3 & 0 \\ 0 & 0 & 0 & 0 & 1 \\ 0 & 0 & 0 & 0 & 0 \end{pmatrix}
$$

(a)   (b)   (c)

FIGURE 4.2–3 Matrices not in row echelon form.

Gauss–Jordan reduction

The system of Example 4.2–2 can also be solved by another method that eliminates the need for back substitution. Beginning with step (f), we continue as follows:

g) Multiply row two by -1 and add the result to row one.
$$
\begin{pmatrix} 1 & 0 & \frac{17}{6} & \frac{20}{6} \\ 0 & 1 & \frac{7}{6} & \frac{10}{6} \\ 0 & 0 & 1 & 1 \end{pmatrix}
$$

h) Multiply row three by $-\frac{17}{6}$ and add the result to row one.
$$
\begin{pmatrix} 1 & 0 & 0 & \frac{3}{6} \\ 0 & 1 & \frac{7}{6} & \frac{10}{6} \\ 0 & 0 & 1 & 1 \end{pmatrix}
$$

i) Multiply row three by $-\frac{7}{6}$ and add the result to row two.
$$
\begin{pmatrix} 1 & 0 & 0 & \frac{1}{2} \\ 0 & 1 & 0 & \frac{1}{2} \\ 0 & 0 & 1 & 1 \end{pmatrix}.
$$

We can now obtain the solution set $\{(\frac{1}{2}, \frac{1}{2}, 1)\}$ from step (i). This last method is called **Gauss–Jordan* reduction** and the matrix in step (i) is said to be in *reduced row echelon form*. It appears that the amount of computation necessary to obtain the solution by using Gaussian elimination and Gauss–Jordan reduction is almost the same, but we will see in Section 4.6 that this is not always necessarily the case.

Inverse of a square matrix

We present one other method for solving systems of linear algebraic equations. *If the system $AX = B$ is $n \times n$,* meaning that A is an $n \times n$

---

* Camille Jordan (1838–1922), a French mathematician.

matrix, and *if* there exists a matrix A^{-1}, called the **inverse** of A, such that

$$AA^{-1} = A^{-1}A = I, \qquad (4.2\text{--}5)$$

then we can solve the system in the following manner. Premultiplying $AX = B$ by A^{-1} produces $A^{-1}(AX) = A^{-1}B$ or $(A^{-1}A)X = A^{-1}B$ because matrix multiplication is associative. But $A^{-1}A = I$ and $IX = X$; hence $X = A^{-1}B$ is the required solution. If A^{-1} exists so that Eq. (4.2–5) holds, then A is said to be **nonsingular** (a matrix that has no inverse is called **singular**). We state the following theorem relating to inverses and leave the proofs to the exercises. (See Exercises 5–8.)

Theorem 4.2–2 Let A and B be nonsingular matrices. Then

a) A^{-1} is unique;
b) $(A^{-1})^{-1} = A$;
c) $(AB)^{-1} = B^{-1}A^{-1}$;
d) $(A^T)^{-1} = (A^{-1})^T$.

We can compute the inverse of a matrix by means of Gauss–Jordan reduction. Define an **elementary matrix** as one obtained by performing one of the elementary row operations on I. For example,

$$\begin{pmatrix} 0 & 1 & 0 \\ 1 & 0 & 0 \\ 0 & 0 & 1 \end{pmatrix}, \quad \begin{pmatrix} 1 & 0 & 0 \\ 0 & 1 & 0 \\ 0 & 0 & c \end{pmatrix}, \quad \text{and} \quad \begin{pmatrix} 1 & 0 & 0 \\ 0 & 1 & 0 \\ 0 & c & 1 \end{pmatrix}$$

are elementary matrices obtained from I_3—the first by interchanging rows one and two, the second by multiplying row three by c ($\neq 0$), and the third by multiplying row two by c and adding the result to row three. If a matrix A is premultiplied by an elementary matrix E, the product EA will be the matrix A on which the elementary row operation symbolized by E has been performed (Exercise 9). Thus if A is nonsingular, a succession of elementary row operations can transform A to I, that is, $A \sim I$. Hence, after r elementary row operations we have

$$(E_1 E_2 \cdots E_r)A = I, \qquad (4.2\text{--}6)$$

where $E_1, E_2, \ldots, E_r$ are elementary matrices and $A^{-1} = E_1 E_2 \cdots E_r$ (Exercise 10). From a computational viewpoint it is easier to compute A^{-1} by computing the product $E_1 E_2 \cdots E_r$ rather than multiplying the individual E_i. The following example shows the procedure.

EXAMPLE 4.2–3 Find A^{-1} given that

$$A = \begin{pmatrix} 1 & -1 & 1 \\ 2 & 1 & 2 \\ 3 & 2 & -1 \end{pmatrix}.$$

Solution. We augment the matrix A with a 3×3 identity and perform elementary row operations until we have transformed A into *reduced row echelon form*:

$$
\begin{pmatrix}
1 & -1 & 1 & 1 & 0 & 0 \\
2 & 1 & 2 & 0 & 1 & 0 \\
3 & 2 & -1 & 0 & 0 & 1
\end{pmatrix}
\sim
\begin{pmatrix}
1 & -1 & 1 & 1 & 0 & 0 \\
0 & 3 & 0 & -2 & 1 & 0 \\
0 & 5 & -4 & -3 & 0 & 1
\end{pmatrix}
$$

$$
\sim
\begin{pmatrix}
1 & 0 & 1 & \frac{1}{3} & \frac{1}{3} & 0 \\
0 & 1 & 0 & -\frac{2}{3} & \frac{1}{3} & 0 \\
0 & 0 & -4 & \frac{1}{3} & -\frac{5}{3} & 1
\end{pmatrix}
\sim
\begin{pmatrix}
1 & 0 & 1 & \frac{1}{3} & \frac{1}{3} & 0 \\
0 & 1 & 0 & -\frac{2}{3} & \frac{1}{3} & 0 \\
0 & 0 & 1 & -\frac{1}{12} & \frac{5}{12} & -\frac{1}{4}
\end{pmatrix}
$$

$$
\sim
\begin{pmatrix}
1 & 0 & 0 & \frac{5}{12} & -\frac{1}{12} & \frac{1}{4} \\
0 & 1 & 0 & -\frac{2}{3} & \frac{1}{3} & 0 \\
0 & 0 & 1 & -\frac{1}{12} & \frac{5}{12} & -\frac{1}{4}
\end{pmatrix}.
$$

Hence,

$$
A^{-1} =
\begin{pmatrix}
\frac{5}{12} & -\frac{1}{12} & \frac{1}{4} \\
-\frac{2}{3} & \frac{1}{3} & 0 \\
\frac{1}{12} & \frac{5}{12} & -\frac{1}{4}
\end{pmatrix}
= \frac{1}{12}
\begin{pmatrix}
5 & -1 & 3 \\
-8 & 4 & 0 \\
1 & 5 & -3
\end{pmatrix}
$$

and it can be shown that $AA^{-1} = A^{-1}A = I$ (Exercise 11). ■

The inverse of a nonsingular matrix is unique (Exercise 5). The above method produces the inverse because it keeps track of the elementary row operations that are being performed. This is accomplished by performing these same operations on the identity matrix by which we augment A. Thus, after A has been row reduced to I, we have the situation shown in Eq. (4.2-6) with the identity being row reduced to A^{-1}. In brief,

$$
(A \mid I) \sim (I \mid A^{-1}).
$$

Whether a system $AX = B$ that has a unique solution is solved by Gaussian elimination or by computing A^{-1} and then $A^{-1}B$ often depends on other factors. If the system is to be solved once only, then Gaussian elimination may be the preferred method. In applications, however, we are often faced with the problem of solving the same system for different inputs, that is, different matrices B. If this is the case, then computing A^{-1} is the most efficient method.

Homogeneous systems

If $B = O$, then the system $AX = O$ is called a **homogeneous system**. Such a system is *always* consistent. It has the unique solution $X = O$ (called the **trivial solution**) if A is nonsingular (Exercise 12) and it has an infinite number of solutions if A is singular. In the latter case, row reducing A will result in a number, say r, of zero rows. Hence $n - r$ of the variables can be obtained in terms of r variables, which are arbitrary. Homogeneous systems that are not square may be solved as shown in the following example.

EXAMPLE 4.2–4 Obtain the solution set of the system

$$\begin{aligned}
x_1 - 2x_2 + 2x_3 &= 0 \\
2x_1 + x_2 - 2x_3 &= 0 \\
3x_1 + 4x_2 - 6x_3 &= 0 \\
3x_1 - 11x_2 + 12x_3 &= 0.
\end{aligned}$$

Solution. We row reduce the matrix of coefficients to obtain the row echelon form:

$$\begin{pmatrix} 1 & -2 & 2 \\ 2 & 1 & -2 \\ 3 & 4 & -6 \\ 3 & -11 & 12 \end{pmatrix} \sim \begin{pmatrix} 1 & -2 & 2 \\ 0 & 5 & -6 \\ 0 & 10 & -12 \\ 0 & -5 & 6 \end{pmatrix} \sim \begin{pmatrix} 1 & -2 & 2 \\ 0 & 5 & -6 \\ 0 & 0 & 0 \\ 0 & 0 & 0 \end{pmatrix} \sim \begin{pmatrix} 1 & -2 & 2 \\ 0 & 1 & -\frac{6}{5} \\ 0 & 0 & 0 \\ 0 & 0 & 0 \end{pmatrix}.$$

The zero row three of the third matrix was obtained by multiplying the second row of the previous matrix by -2 and adding the result to the third row, and the zero row four was obtained by multiplying the second row by 1 and adding the result to the fourth row. From the row echelon form we now obtain

$$\begin{aligned}
x_2 &= \tfrac{6}{5}x_3, \\
x_1 &= 2x_2 - 2x_3 = \tfrac{2}{5}x_3.
\end{aligned}$$

Hence the solution set can be written

$$\{(\tfrac{2}{5}k, \tfrac{6}{5}k, k) \mid k \text{ a real number}\}.$$

Thus the solution set contains an infinite number of ordered triples. In other words, every scalar multiple of (2, 6, 5) belongs to the solution set. ■

KEY WORDS AND PHRASES

solution set	row equivalence
consistent	row echelon form
inconsistent	back substitution
empty set	Gauss–Jordan reduction
Cramer's rule	reduced row echelon form
determinant	singular
Gaussian elimination	elementary matrix
augmented matrix	homogeneous system
elementary row operation	trivial solution

EXERCISES 4.2

▶ **1.** Graph the systems shown in Example 4.2–1.

2. Given the matrix

$$A = \begin{pmatrix} 5 & -6 & -1 & 1 \\ 0 & 2 & -3 & 2 \\ 1 & 2 & -1 & 4 \\ -1 & 0 & 2 & 1 \end{pmatrix},$$

find $|A|$ by each of the following methods:

a) expansion using the first column;

b) expansion using the fourth row;

c) expansion using the third column.

3. List the elementary row operations that will transform each of the matrices (a), (b), and (c) of Fig. 4.2–2 into *reduced* row echelon form.

4. List the elementary row operations that will transform each of the matrices (a), (b), and (c) of Fig. 4.2–3 into (a) row echelon form; (b) reduced row echelon form.

5. Prove part (a) of Theorem 4.2–2. (*Hint*: Let $AB = I$ and $AC = I$; then show that $B = C$.)

6. Prove part (b) of Theorem 4.2–2. (*Hint*: What is $(A^{-1})(A^{-1})^{-1}$?)

7. Prove part (c) of Theorem 4.2–2. (*Hint*: What is $(B^{-1}A^{-1})(AB)$?)

8. Prove part (d) of Theorem 4.2–2. (*Hint*: Write the transposes of $AA^{-1} = I$ and of $A^{-1}A = I$.)

9. For an *arbitrary* 3×3 matrix A, compute each of the following.

a) $\begin{pmatrix} 0 & 1 & 0 \\ 1 & 0 & 0 \\ 0 & 0 & 1 \end{pmatrix} A$

b) $\begin{pmatrix} 1 & 0 & 0 \\ 0 & 1 & 0 \\ 0 & 0 & c \end{pmatrix} A$

c) $\begin{pmatrix} 1 & 0 & 0 \\ 0 & 1 & 0 \\ 0 & c & 1 \end{pmatrix} A$

d) $\begin{pmatrix} 0 & 1 & 0 \\ 1 & 0 & 0 \\ 0 & 0 & 1 \end{pmatrix}\begin{pmatrix} 1 & 0 & 0 \\ 0 & 1 & 0 \\ 0 & c & 1 \end{pmatrix} A$

10. Show that if A is $n \times n$, then $A^{-1}A = I$ implies $AA^{-1} = I$.

11. Show that $AA^{-1} = A^{-1}A = I$ for the matrix A of Example 4.2–3.

12. Show that the homogeneous system $AX = O$, where A is $n \times n$, has only the trivial solution if A is nonsingular, by:

a) using the inverse of A; **b)** using Cramer's rule.

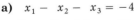

▶▶ **13.** Find the solution set of each of the following systems.

a) $\begin{aligned} x_1 - x_2 - x_3 &= -4 \\ 2x_1 + 3x_2 + 5x_3 &= 23 \\ x_1 - 2x_2 + 3x_3 &= 6 \end{aligned}$

b) $\begin{aligned} x_1 + 2x_2 - x_3 &= 0 \\ 2x_1 + x_2 + 2x_3 &= 3 \\ x_1 - x_2 + x_3 &= 3 \end{aligned}$

c) $\begin{aligned} 2x_1 + x_2 + x_3 &= 0 \\ x_1 + x_2 + 3x_3 &= 0 \end{aligned}$

14. Find the solution set of each of the following systems.

   a) $x_1 + 5x_2 - 5x_3 = 1$
   $x_1 - x_2 + x_3 = 2$
   $x_1 + x_2 - x_3 = 1$

   b) $x_1 - x_2 + x_3 = 0$
   $x_1 + x_2 + x_3 = 0$
   $x_1 - x_2 - x_3 = 0$

15. Use Cramer's rule to find the solution set of each of the following systems.

   a) $2x_1 + x_2 = -1$
   $3x_1 + 2x_2 = 0$

   b) $2x_1 - 3x_2 + x_3 = 0$
   $3x_1 - 2x_2 + 2x_3 = 0$
   $x_1 + 3x_2 - 2x_3 = 0$

   c) $5x_1 + 7x_2 = 3$
   $2x_1 + 3x_2 = 1$

   d) $3x_1 - 2x_2 + 2x_3 = 5$
   $2x_1 + x_2 - 3x_3 = 5$
   $5x_1 - 3x_2 - x_3 = 16$

16. Show that the determinant of a triangular matrix is the product of the elements on the principal diagonal.

17. Evaluate each of the following determinants.

   a) $\begin{vmatrix} 2 & 4 & 6 \\ 0 & 2 & 3 \\ 1 & 4 & 9 \end{vmatrix}$

   b) $\begin{vmatrix} 1 & 1 & 1 \\ 0 & 1 & 1 \\ 1 & 2 & 3 \end{vmatrix}$

   c) $\begin{vmatrix} 1 & 2 & 3 \\ 0 & -4 & 2 \\ -1 & 5 & 4 \end{vmatrix}$

   d) $\begin{vmatrix} 1 & 0 & -1 & 1 \\ 0 & 4 & 0 & 1 \\ 3 & 2 & -2 & 1 \\ 0 & -1 & 1 & 4 \end{vmatrix}$

   e) $\begin{vmatrix} 1 & \frac{1}{2} \\ \frac{1}{2} & \frac{1}{3} \end{vmatrix}$

   f) $\begin{vmatrix} 1 & \frac{1}{2} & \frac{1}{3} \\ \frac{1}{2} & \frac{1}{3} & \frac{1}{4} \\ \frac{1}{3} & \frac{1}{4} & \frac{1}{5} \end{vmatrix}$

   g) $\begin{vmatrix} 0 & 1 & -1 \\ 3 & 1 & -4 \\ 2 & 1 & 1 \end{vmatrix}$

   h) $\begin{vmatrix} 1 & 4 & -2 & 1 \\ -1 & 2 & -1 & 1 \\ 3 & 3 & 0 & 4 \\ 4 & -4 & 2 & 3 \end{vmatrix}$

18. In each part obtain the elementary 4×4 matrix E that performs the specified elementary row operation on A if EA is computed.

   a) Interchange rows 2 and 4.

   b) Multiply row 3 by a nonzero constant c.

   c) Multiply row 3 by a constant k and add the result to row 1.

   d) Multiply row 2 by -2 and add the result to row 3 *and* multiply row 2 by 3 and add the result to row 4.

19. Referring to Exercise 18, obtain the inverses of each of the elementary matrices E. (*Hint:* What elementary row operation will "undo" what E has done to A?)

20. Explain why the elementary row operation E2 of the text contains the word "nonzero" but E3 does not.

21. Show that the system $AX = B$, where

$$A = \begin{pmatrix} 3 & 2 & -1 & -4 \\ 1 & -1 & 3 & -1 \\ 2 & 1 & -3 & 0 \\ 0 & -1 & 8 & -5 \end{pmatrix} \quad \text{and} \quad B = \begin{pmatrix} 10 \\ -4 \\ 16 \\ 3 \end{pmatrix},$$

has no solution, but if $B = (2 \ 3 \ 1 \ 3)^T$, then the system has an infinite number of solutions.

22. Use Gaussian elimination to solve the system

$$\begin{pmatrix} 5 & -1 & 1 \\ 2 & 4 & 0 \\ 1 & 1 & 5 \end{pmatrix} X = \begin{pmatrix} 10 \\ 12 \\ -1 \end{pmatrix}.$$

23. Which of the following matrices are singular?

a) $\begin{pmatrix} 3 & 2 & -1 \\ 0 & -1 & 4 \\ 6 & 3 & 2 \end{pmatrix}$
  b) $\begin{pmatrix} 1 & 0 & -2 & 3 \\ 3 & 1 & 1 & 4 \\ -1 & 0 & 2 & -1 \\ 4 & 2 & 6 & 0 \end{pmatrix}$

c) $\begin{pmatrix} 1 & 0 & -2 & 3 \\ 3 & 1 & 1 & 4 \\ -1 & 0 & 2 & -1 \\ 4 & 3 & 6 & 0 \end{pmatrix}$

24. Show that the equation of the line determined by the points (a, b) and (c, d) can be written in determinant form as

$$\begin{vmatrix} x_1 & x_2 & 1 \\ a & b & 1 \\ c & d & 1 \end{vmatrix} = 0.$$

25. Find the solution set of each of the following systems.

*a) $\begin{aligned} 4x_1 - 7x_2 + x_3 - 6x_4 &= 0 \\ x_1 + 2x_2 - 5x_3 + 4x_4 &= 0 \\ 2x_1 - 3x_2 + 2x_3 + 3x_4 &= 0 \end{aligned}$
  b) $\begin{aligned} 2x_1 + 3x_2 &= 3 \\ 3x_1 + 2x_2 &= 7 \\ x_1 - 2x_2 &= 5 \end{aligned}$

c) $\begin{aligned} x_1 + 2x_2 - x_3 + 3x_4 &= 3 \\ 3x_1 + 6x_2 - x_3 + 8x_4 &= 10 \\ 2x_1 + 4x_2 + 4x_3 + 3x_4 &= 9 \end{aligned}$

*26. Find the solution set of each of the following systems, $AX = B$, with A and B as shown.

a) $A = \begin{pmatrix} 3.000 & -4.031 & -3.112 \\ -0.002 & 4.000 & 4.000 \\ -2.000 & 2.906 & -5.387 \end{pmatrix}, \quad B = \begin{pmatrix} -4.413 \\ 7.998 \\ -4.481 \end{pmatrix}$

* Calculator problem.

b) $A = \begin{pmatrix} 4.23 & -1.06 & 2.11 \\ -2.53 & 6.77 & 0.98 \\ 1.85 & -2.11 & -2.32 \end{pmatrix}$, $B = \begin{pmatrix} 5.28 \\ 5.22 \\ -2.58 \end{pmatrix}$

c) $A = \begin{pmatrix} 2.51 & 1.48 & 4.53 \\ 1.48 & 0.93 & -1.30 \\ 2.68 & 3.04 & -1.48 \end{pmatrix}$, $B = \begin{pmatrix} 0.05 \\ 1.03 \\ -0.53 \end{pmatrix}$

27. Consider the matrix

$$A = \begin{pmatrix} 4 & 1 & 0 \\ 2 & -1 & 2 \\ a & b & -1 \end{pmatrix}.$$

  a) Find values of a and b that make A singular.

  b) Find values of a and b that make A nonsingular.

28. Solve each of the following systems.

  a) $\begin{aligned} x_1 - 2x_2 + x_3 &= 1 \\ x_1 + x_2 - x_3 &= 0 \\ 2x_1 - 3x_2 + x_3 &= 0 \\ -3x_1 + 2x_2 + 3x_3 &= 1 \end{aligned}$
  b) $\begin{aligned} x_1 \quad\quad + x_3 &= 3 \\ x_2 + 3x_3 &= 2 \\ 2x_1 + x_2 + x_3 &= 4 \\ 2x_1 + x_2 + 2x_3 &= 5 \end{aligned}$

  c) $\begin{aligned} 5x_1 + 3x_2 - 2x_3 &= 0 \\ -2x_1 - 2x_2 + x_3 &= 4 \\ -8x_1 - 8x_2 + 3x_3 &= -4 \end{aligned}$
  d) $\begin{aligned} x_1 + x_2 + 2x_3 &= 2 \\ x_2 + 2x_3 &= 1 \\ 3x_1 + x_2 + 2x_3 &= 4 \end{aligned}$

▶▶▶ **29.** Prove that if A is a nonsingular symmetric matrix, then A^{-1} is symmetric.

30. Prove that if A commutes with B and A is nonsingular, then A^{-1} commutes with B.

31. Prove that if A is symmetric and nonsingular, then A^{-1} is symmetric.

32. Prove that if A is nonsingular, then A is symmetric if and only if A^{-1} is symmetric. (*Hint*: Two proofs are required: one to show that a nonsingular symmetric A implies a symmetric A^{-1}; the other to show that a nonsingular A and a symmetric A^{-1} implies a symmetric A.)

33. Prove that the relation "$\sim$" (is row equivalent to) is an equivalence relation, that is, it has the following properties:

  i) $A \sim A$ (reflexive property);

  ii) $A \sim B \to B \sim A$ (symmetric property);

  iii) $A \sim B$ and $B \sim C \to A \sim C$ (transitive property).

34. If $y_1 = \exp(rx)$, $y_2 = x \exp(rx)$, and $y_3 = x^2 \exp(rx)$, then show that $W(y_1, y_2, y_3) = 2 \exp(3\,rx)$. (Compare Example 2.2–1.)

35. A linear equation in three variables can be represented by a plane in (x_1, x_2, x_3)-space. Discuss the possible geometric interpretations of three equations in three variables, that is, the cases of a unique solution, no solution, and an infinite number of solutions.

4.3 LINEAR TRANSFORMATIONS

In Section 4.1 we showed how the definition of matrix multiplication followed naturally from successive linear transformations. Now we will consider transformations in greater detail, particularly with respect to the "spaces" involved. For example, the following 2×3 matrix

$$\begin{pmatrix} 1 & 2 & -3 \\ -2 & 4 & 1 \end{pmatrix} \begin{pmatrix} x_1 \\ x_2 \\ x_3 \end{pmatrix} = \begin{pmatrix} x_1 + 2x_2 - 3x_3 \\ -2x_1 + 4x_2 + x_3 \end{pmatrix} \qquad \textbf{(4.3–1)}$$

represents a transformation of the ordered *triples* $(x_1, x_2, x_3)^T$ into the ordered *pair* $(x_1 + 2x_2 - 3x_3, -2x_1 + 4x_2 + x_3)^T$. In particular, this transformation transforms the point $(14, 5, 8)$ to $(0, 0)$. Stated another way, a vector ($n \times 1$ and $1 \times m$ matrices are called *vectors*) having three components (elements) has been transformed into a vector having two components. We next define the space consisting of all vectors having n real components.

Definition 4.3–1 A *vector* is a $1 \times n$ matrix, denoted by

$$\mathbf{x} = (x_1, x_2, \ldots, x_n).$$

The totality of all such vectors $\mathbf{x}$ constitutes a real n-**dimensional vector space** $\mathbb{R}^n$ that has the following properties.

i) If $\mathbf{u}$ and $\mathbf{v}$ are any two vectors in $\mathbb{R}^n$, then $\mathbf{u} + \mathbf{v}$ is in $\mathbb{R}^n$. In other words, $\mathbb{R}^n$ is *closed under vector addition*. Further, vector addition has the following properties:
1. $\mathbf{u} + \mathbf{v} = \mathbf{v} + \mathbf{u}$ for all $\mathbf{u}$, $\mathbf{v}$ in $\mathbb{R}^n$;
2. $\mathbf{u} + (\mathbf{v} + \mathbf{w}) = (\mathbf{u} + \mathbf{v}) + \mathbf{w}$ for all $\mathbf{u}$, $\mathbf{v}$, $\mathbf{w}$ in $\mathbb{R}^n$;
3. There exists a unique vector $\mathbf{0}$ in $\mathbb{R}^n$ such that $\mathbf{u} + \mathbf{0} = \mathbf{0} + \mathbf{u} = \mathbf{u}$ for all $\mathbf{u}$ in $\mathbb{R}^n$;
4. For each $\mathbf{u}$ in $\mathbb{R}^n$ there exists a unique vector $-\mathbf{u}$ in $\mathbb{R}^n$ such that $\mathbf{u} + (-\mathbf{u}) = (-\mathbf{u}) + \mathbf{u} = \mathbf{0}$.

ii) If $\mathbf{u}$ is any vector in $\mathbb{R}^n$ and c is any real number, then $c\mathbf{u}$ is in $\mathbb{R}^n$. In other words, $\mathbb{R}^n$ is *closed under scalar multiplication*. Further, scalar multiplication has the following properties:
1. $c\mathbf{u} = \mathbf{u}c$ for any $\mathbf{u}$ in $\mathbb{R}^n$ and any real number c;
2. $c(\mathbf{u} + \mathbf{v}) = c\mathbf{u} + c\mathbf{v}$ for any $\mathbf{u}$, $\mathbf{v}$ in $\mathbb{R}^n$ and any real number c;
3. $(c + d)\mathbf{u} = c\mathbf{u} + d\mathbf{u}$ for any $\mathbf{u}$ in $\mathbb{R}^n$ and any real numbers c, d;
4. $(cd)\mathbf{u} = c(d\mathbf{u})$ for any $\mathbf{u}$ in $\mathbb{R}^n$ and any real numbers c, d;
5. $1\mathbf{u} = \mathbf{u}$ and $0\mathbf{u} = \mathbf{0}$ for any $\mathbf{u}$ in $\mathbb{R}^n$.

We remark that all of the above properties apply to matrices and have been used in previous sections of this chapter. Now we are restricting ourselves to $1 \times n$ matrices, calling these *vectors* and denoting them by boldface lower-case letters. We could also have used $n \times 1$ matrices (or column vectors) since all the properties in Definition 4.3–1 apply equally well to these. One reason for this change of notation is that it will be convenient in what follows to express equations such as Eq. (4.3–1) by writing $A\mathbf{x} = \mathbf{y}$, clearly distinguishing between the matrix of the transformation A and the vectors $\mathbf{x}$ and $\mathbf{y}$ involved. In the transformation (4.3–1) the vector being transformed is a vector of $\mathbb{R}^3$, whereas the transformed vector is in $\mathbb{R}^2$.

Definition 4.3–1 states that $\mathbb{R}^n$, a real n-dimensional vector space, consists of *all* vectors of the form

$$\mathbf{x} = (x_1, x_2, \ldots, x_n).$$

We will also be interested, however, in a *part* of $\mathbb{R}^n$; for example, all vectors of the form

$$\mathbf{u} = (u_1, u_2, \ldots, u_{n-1}, 0)$$

that constitute a *subset* of $\mathbb{R}^n$. Thus we need the following definition of a *subspace*.

Definition 4.3–2 If V is a vector space and W is a subset of V, then W is a subspace of V if it is a vector space.

We stress that not *all* subsets of vector spaces are subspaces. For example, the subset of $\mathbb{R}^3$ consisting of all vectors of the form $(x_1, x_2, 1)$ is not a subspace because it is not a vector space according to Definition 4.3–1 (Exercise 1). Every vector space has two subspaces: the space itself and the space consisting of the zero vector $\mathbf{0}$ (Exercise 2). Other subspaces are easily found by means of the following theorem, whose proof we omit.

Theorem 4.3–1 Let V be a vector space and W a subset of V. Then W is a subspace of V if and only if W is closed under vector addition and scalar multiplication.

Thus we do not need to check *all* the properties of a vector space listed in Definition 4.3–1, but only parts (i) and (ii). Theorem 4.3–1 assures us that if these properties hold, then all the other properties follow. The proofs are tedious but not difficult.

Our discussion of vector spaces continues since at this point we know what a vector space is (Definition 4.3–1) but we do not know how to form one. More definitions are required.

Definition 4.3–3 Let $S = \{\mathbf{u}_1, \mathbf{u}_2, \ldots, \mathbf{u}_j\}$ be a set of vectors in a vector space V. The vector

$$\mathbf{v} = c_1\mathbf{u}_1 + c_2\mathbf{u}_2 + \cdots + c_j\mathbf{u}_j,$$

where the c_i are real numbers, is called a **linear combination** of the vectors in S.

We remark that $\mathbf{v}$ is necessarily in the vector space V. For example, if

$$\mathbf{u}_1 = (2, 3, 4), \qquad \mathbf{u}_2 = (-1, 0, 5), \qquad \mathbf{u}_3 = (1, -4, 7),$$

then

$$\mathbf{v} = c_1(2, 3, 4) + c_2(-1, 0, 5) + c_3(1, -4, 7),$$

where c_1, c_2, and c_3 are real numbers, is a linear combination of $\mathbf{u}_1$, $\mathbf{u}_2$, and $\mathbf{u}_3$. In this example $\mathbf{u}_1$, $\mathbf{u}_2$, and $\mathbf{u}_3$ are vectors in $\mathbb{R}^3$ and it is natural to ask whether *every* vector of $\mathbb{R}^3$ can be expressed as a linear combination of these. If so, then it is clear that a description of $\mathbb{R}^3$ can be given using only the vectors $\mathbf{u}_1$, $\mathbf{u}_2$, and $\mathbf{u}_3$. We need some more definitions before we can make the above simplification.

Definition 4.3–4 Let $S = \{\mathbf{u}_1, \mathbf{u}_2, \ldots, \mathbf{u}_j\}$ be a set of vectors in a vector space V. The set S **spans** V if *every* vector in V is a linear combination of the vectors in S.

We say that V is *spanned by* S and call S a *spanning set* of V whenever S spans V.

EXAMPLE 4.3–1 Show that the set

$$S = \{(1, 0, 0), (1, 1, 0), (1, 1, 1)\}$$

spans $\mathbb{R}^3$.

Solution. We take an *arbitrary* vector in $\mathbb{R}^3$, say (x_1, x_2, x_3), and show that *it* can be expressed as a linear combination of the vectors in S. Thus

$$c_1(1, 0, 0) + c_2(1, 1, 0) + c_3(1, 1, 1) = (x_1, x_2, x_3)$$

and this leads to the system of linear equations

$$c_1 + c_2 + c_3 = x_1$$
$$c_2 + c_3 = x_2$$
$$c_3 = x_3.$$

The unique solution of this system is (Exercise 3)

$$c_1 = x_1 - x_2, \qquad c_2 = x_2 - x_3, \qquad c_3 = x_3,$$

and we can use these values to show, for example, that

$$(-2, 5, 7) = -7(1, 0, 0) - 2(1, 1, 0) + 7(1, 1, 1). \quad \blacksquare$$

The set S of Example 4.3–1 must have some special property that other sets do not have. For example, the set

$$T = \{(1, 0, 2), (1, 2, 0), (1, 1, 1)\}$$

does *not* span $\mathbb{R}^3$ (Exercise 4). Only those vectors (x_1, x_2, x_3) of $\mathbb{R}^3$ for which $2x_1 - x_2 - x_3 = 0$* can be expressed as linear combinations of the vectors in the set T. The special property referred to is called **linear independence** and is defined next.

Definition 4.3–5 Let $S = \{\mathbf{u}_1, \mathbf{u}_2, \ldots, \mathbf{u}_j\}$ be a set of *distinct* vectors in a vector space V. Then the set S is *linearly independent* over the real numbers if the equation

$$c_1\mathbf{u}_1 + c_2\mathbf{u}_2 + \cdots + c_j\mathbf{u}_j = 0 \qquad (4.3–2)$$

can be satisfied *only* by taking

$$c_1 = c_2 = \cdots = c_j = 0,$$

where the c_i are real numbers.

The phrase "over the real numbers" is often omitted since, if it is understood, no confusion can result. If Eq. (4.3–2) can be satisfied by nontrivial values of the c_i, then the set S is *linearly dependent*. In this case, any one of the $\mathbf{u}_i$ can be expressed as a linear combination of the others. In a linearly independent set *none* of the vectors can be expressed as a linear combination of the others. Following are some immediate consequences of Definition 4.3–5 with the proofs being left for the exercises (Exercise 5).

Theorem 4.3–2

a) Any nonzero vector is linearly independent.

b) The zero vector is linearly dependent.

c) Any set of vectors that contains the zero vector is linearly dependent.

A *basis* for a vector space V can now be defined by using the concepts of linear independence and a spanning set.

Definition 4.3–6 Let S be a set of vectors in the vector space V. Then S is a **basis** for V if

a) S is linearly independent, and

b) S spans V.

---

* Geometrically, this is a plane through the origin. The set T spans this plane.

As an example, the set

$$S = \{(1, 0, 0), (1, 1, 0), (1, 1, 1)\}$$

spans $\mathbb{R}^3$ (Example 4.3–1) and is linearly independent (Exercise 6), hence is a basis for $\mathbb{R}^3$. Adding *any other* vector to S, however, would produce a set that is *linearly dependent*, although it would still span $\mathbb{R}^3$ (Exercise 7). On the other hand, deleting a vector from the set S would produce a set that is still linearly independent but no longer spans $\mathbb{R}^3$ (Exercise 8). This suggests that the *number* of vectors in a basis for a vector space is invariant. We have, in fact, the following theorem.

Theorem 4.3–3 If a vector space V has a basis consisting of n vectors, then any other basis will also contain n vectors.

A proof of Theorem 4.3–3 can be found in John W. Dettman's *Introduction to Linear Algebra and Differential Equations*, p. 114 (New York: McGraw-Hill, 1974).

We can now give a more precise meaning to the phrase "*n*-dimensional vector space."

Definition 4.3–7 If a vector space V has a basis, then the number of vectors in this basis is called the **dimension** of V, written "dim (V)." The vector space consisting of the zero vector has dimension zero.

Thus $\mathbb{R}^n$ has dimension n, meaning that every basis for $\mathbb{R}^n$ contains n vectors. Of all the bases for $\mathbb{R}^n$ there is one that is of particular interest because of its simplicity. This is the one consisting of the set

$$\{\mathbf{e}_1, \mathbf{e}_2, \ldots, \mathbf{e}_n\},$$

where

$$\mathbf{e}_1 = (1, 0, 0, \ldots, 0),$$
$$\mathbf{e}_2 = (0, 1, 0, \ldots, 0),$$
$$\vdots$$
$$\mathbf{e}_n = (0, 0, 0, \ldots, 1);$$

that is, the $\mathbf{e}_i$ are rows of the identity matrix I_n. We call such a basis *the natural basis* for $\mathbb{R}^n$ (see Fig. 4.3–1).

We next call attention to the relationship between *functions* and transformations. When we write $y = f(x)$, the symbol f represents operations that have been performed involving x. Moreover, f is *single-valued*, meaning that the function assigns one and only one value to each value of x. We also speak of the domain and range of f, meaning the sets to which x and y, respectively, belong. For example, in $y = x \log x$ the domain is the set $\{x: x > 0\}$, whereas the range is the set

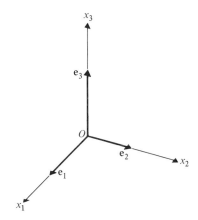

FIGURE 4.3–1 The natural basis for $\mathbb{R}^3$.

$\{y: -\infty < y < \infty\}$. We also call a function a *mapping* if it is convenient, for instance, to think of $y = x \log x$ as a mapping. This function maps the positive real half-line $(x > 0)$ into the real line $(-\infty < y < \infty)$.

With the above in mind we state the following definition of a linear transformation.

Definition 4.3–8 Let $\mathbb{R}^n$ and $\mathbb{R}^m$ be n-dimensional and m-dimensional vector spaces, respectively. A single-valued mapping $L: \mathbb{R}^n \to \mathbb{R}^m$ is called a **linear transformation*** of $\mathbb{R}^n$ into $\mathbb{R}^m$ if

a) $L(\mathbf{u} + \mathbf{v}) = L(\mathbf{u}) + L(\mathbf{v})$ for every $\mathbf{u}$, $\mathbf{v}$ in $\mathbb{R}^n$, and

b) $L(c\mathbf{u}) = cL(\mathbf{u})$ for every $\mathbf{u}$ in $\mathbb{R}^n$ and every real number c.

EXAMPLE 4.3–2 Examine each of the following transformations to determine whether or not they are linear.

a) $L: \mathbb{R}^3 \to \mathbb{R}^2$ defined by $L(u_1, u_2, u_3) = (u_1, u_2)$

b) $L: \mathbb{R}^3 \to \mathbb{R}^3$ defined by $L(u_1, u_2, u_3) = (u_1, u_2, 0)$

c) $L: \mathbb{R}^2 \to \mathbb{R}^3$ defined by $L(u_1, u_2) = (u_2, u_1, 1)$

d) $L: \mathbb{R}^4 \to \mathbb{R}^2$ defined by $L(u_1, u_2, u_3, u_4) = (u_1 u_2, u_3 - u_4)$

Solution. Let $\mathbf{u} = (u_1, u_2, u_3)$ and $\mathbf{v} = (v_1, v_2, v_3)$. Then, from the definition of the mapping in part (a),

$$L(\mathbf{u} + \mathbf{v}) = L(u_1 + v_1, u_2 + v_2, u_3 + v_3) = (u_1 + v_1, u_2 + v_2),$$

and

$$L(\mathbf{u}) + L(\mathbf{v}) = (u_1, u_2) + (v_1, v_2) = (u_1 + v_1, u_2 + v_2).$$

---

* Note the relationship to "linear operator" defined in Section 3.1.

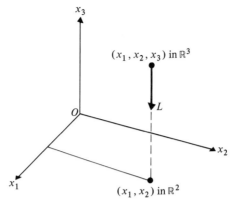

FIGURE 4.3–2 The transformation
$L(x_1, x_2, x_3) = (x_1, x_2)$.

Also,

$$L(c\mathbf{u}) = L(cu_1, cu_2, cu_3) = (cu_1, cu_2) = cL(\mathbf{u}).$$

Hence the mapping (a) is a linear transformation. It is called a *projection* because a line drawn from the point (x_1, x_2, x_3) perpendicular to the x_1x_2-plane will intersect the latter in the point (x_1, x_2). See Fig. 4.3–2.

We leave part (b) for the exercises (Exercise 9) and remark that this mapping, too, is a linear transformation called a projection. Although the points (x_1, x_2) and $(x_1, x_2, 0)$ appear to be the same geometrically, there is an important difference between the linear transformations in (a) and (b). The first can be accomplished by the matrix multiplication

$$(u_1, u_2, u_3)\begin{pmatrix} 1 & 0 \\ 0 & 1 \\ 0 & 0 \end{pmatrix} = (u_1, u_2),$$

whereas the second is given by

$$(u_1, u_2, u_3)\begin{pmatrix} 1 & 0 & 0 \\ 0 & 1 & 0 \\ 0 & 0 & 0 \end{pmatrix} = (u_1, u_2, 0),$$

showing that two *different* matrices are involved.

In part (c) we have for $\mathbf{u} = (u_1, u_2)$ and $\mathbf{v} = (v_1, v_2)$,

$$L(\mathbf{u} + \mathbf{v}) = L(u_1 + v_1, u_2 + v_2) = (u_2 + v_2, u_1 + v_1, 1)$$

and

$$L(\mathbf{u}) + L(\mathbf{v}) = L(u_1, u_2) + L(v_1, v_2) = (u_2, u_1, 1) + (v_2, v_1, 1)$$
$$= (u_2 + v_2, u_1 + v_1, 2).$$

Since these are not the same, the mapping is not a linear transformation. In part (d) we have

$$L(cu_1, cu_2, cu_3, cu_4) = (c^2 u_1 u_2, c(u_3 - u_4)),$$

and

$$cL(u_1, u_2, u_3, u_4) = c(u_1 u_2, u_3 - u_4) = (cu_1 u_2, c(u_3 - u_4)),$$

and since these differ, this is not a linear transformation. ∎

The linear transformation in Example 4.3–2(b) can be defined in matrix form by writing

$$(u_1, u_2, u_3) \begin{pmatrix} 1 & 0 \\ 0 & 1 \\ 0 & 0 \end{pmatrix} = (u_1, u_2),$$

as we have done above. Since we are calling $1 \times n$ and $n \times 1$ matrices vectors without distinguishing particularly between row vectors and column vectors, we can also express this transformation as

$$\begin{pmatrix} 1 & 0 & 0 \\ 0 & 1 & 0 \end{pmatrix} \begin{pmatrix} u_1 \\ u_2 \\ u_3 \end{pmatrix} = \begin{pmatrix} u_1 \\ u_2 \end{pmatrix}.$$

Each of these notations has advantages and disadvantages and we will use whichever one seems most appropriate.

Linear transformations have certain characteristics, which we will now examine in some detail. First, we consider linear transformations that are *one-to-one* as defined in the following.

Definition 4.3–9 A linear transformation $L: V \to W$ of a vector space V into a vector space W is **one-to-one** if $L(\mathbf{u}) = L(\mathbf{v})$ implies that $\mathbf{u} = \mathbf{v}$.

Note that the transformation of Example 4.3–2(a) is *not one-to-one* since

$$L(u_1, u_2, 2) = L(u_1, u_2, 3) = (u_1, u_2).$$

Definition 4.3–9 does not provide a practical method of determining whether a transformation is *one-to-one*. A better method is to determine which vectors are transformed into the zero vector. Note that a linear transformation always transforms the zero vector into the zero vector (Exercise 38). First we have a definition.

Definition 4.3–10 Let $L: V \to W$ be a linear transformation of a vector space V into a vector space W. The **kernel** of L, written ker L, is the subset of all those elements $\mathbf{v}$ of V such that $L(\mathbf{v}) = \mathbf{0}_W$.

We will use the notation $\mathbf{0}_W$ and $\mathbf{0}_V$ to denote the zero vectors of W and V, respectively. The following theorems can be stated.

Theorem 4.3–4 If $L: V \to W$ is a linear transformation of a vector space V into a vector space W, then ker L is a subspace of V.

The proof of Theorem 4.3–4 is left to the exercises (Exercise 10).

Theorem 4.3–5 If $L: V \to W$ is a linear transformation of a vector space V into a vector space W, then L is one-to-one if and only if dim (ker L) = 0.

PROOF. If dim (ker L) = 0, then ker $L = \{\mathbf{0}_v\}$. To show that L is one-to-one, suppose that $L(\mathbf{v}_1) = L(\mathbf{v}_2)$ for $\mathbf{v}_1$ and $\mathbf{v}_2$ in V. Then $L(\mathbf{v}_1) - L(\mathbf{v}_2) = L(\mathbf{v}_1 - \mathbf{v}_2) = \mathbf{0}_W$, which shows that $\mathbf{v}_1 - \mathbf{v}_2 = \mathbf{0}_V$ or $\mathbf{v}_1 = \mathbf{v}_2$, that is, L is one-to-one.

Conversely, if L is one-to-one, then, for $\mathbf{v}$ in ker L, $L(\mathbf{v}) = \mathbf{0}_W$. But since $L(\mathbf{0}_V) = \mathbf{0}_W$ (Exercise 38), $\mathbf{v} = \mathbf{0}_V$; that is, ker $L = \{\mathbf{0}_V\}$, which has dimension zero. $\square$

EXAMPLE 4.3–3 Find a basis for ker L if L is the linear transformation defined by

$$L: \mathbb{R}^5 \to \mathbb{R}^4$$

such that

$$L\begin{pmatrix} x_1 \\ x_2 \\ x_3 \\ x_4 \\ x_5 \end{pmatrix} = \begin{pmatrix} x_1 - x_3 + 3x_4 - x_5 \\ x_1 \quad\quad + 2x_4 - x_5 \\ 2x_1 - x_3 + 5x_4 - x_5 \\ -x_3 + x_4 \end{pmatrix}.$$

Solution. Since ker L consists of all those vectors in $\mathbb{R}^5$ that are transformed into the zero vector in $\mathbb{R}^4$, we have the linear homogeneous system

$$\begin{aligned} x_1 - x_3 + 3x_4 - x_5 &= 0 \\ x_1 \quad\quad + 2x_4 - x_5 &= 0 \\ 2x_1 - x_3 + 5x_4 - x_5 &= 0 \\ -x_3 + x_4 \quad\quad &= 0. \end{aligned}$$

The reduced row echelon form of the coefficient matrix is

$$\begin{pmatrix} 1 & 0 & 0 & 2 & 0 \\ 0 & 0 & 1 & -1 & 0 \\ 0 & 0 & 0 & 0 & 1 \\ 0 & 0 & 0 & 0 & 0 \end{pmatrix},$$

which shows that $x_5 = 0$, $x_3 = x_4$, and $x_1 = -2x_4$ with x_2 and x_4 arbitrary. Thus the solution set of the homogeneous system has the form

$$\{(-2t, s, t, t, 0): s \text{ and } t \text{ real}\}$$

and a basis of ker L is

$$\left\{ \begin{pmatrix} 0 \\ 1 \\ 0 \\ 0 \\ 0 \end{pmatrix}, \begin{pmatrix} -2 \\ 0 \\ 1 \\ 1 \\ 0 \end{pmatrix} \right\}.$$

We are assured of having linearly independent vectors in the basis set by first putting $t = 0$, $s = 1$ and then putting $t = 1$, $s = 0$. (Why?) Hence dim (ker L) = 2. The space spanned by the basis vectors above is also called the **null space** of L and the dimension of this space is called the **nullity** of L. ∎

We have studied linear transformations from a vector space V *into* a vector space W. The space V is called the **domain** of the transformation, whereas W is called its **codomain**. A certain subset of the codomain defined next is of particular importance.

Definition 4.3–11 Let L be a linear transformation from a vector space V into a vector space W, $L: V \to W$. The subset of W consisting of the totality of vectors $L(\mathbf{v})$ with $\mathbf{v}$ in V is called the **range** of L. In case range $L = W$, then the transformation is said to be *onto*.

Theorem 4.3–6 If $L: V \to W$ is a linear transformation of a vector space V into a vector space W, then range L is a subspace of W.

PROOF. We need to prove that range L is closed under vector addition and scalar multiplication (see Theorem 4.3–1). Let $\mathbf{u}$ and $\mathbf{v}$ be two vectors in range L. Then $L(\mathbf{u}_1) = \mathbf{u}$ and $L(\mathbf{v}_1) = \mathbf{v}$ for some vectors $\mathbf{u}_1$ and $\mathbf{v}_1$ in V. Consider $\mathbf{u} + \mathbf{v} = L(\mathbf{u}_1) + L(\mathbf{v}_1) = L(\mathbf{u}_1 + \mathbf{v}_1)$ which implies that $\mathbf{u} + \mathbf{v}$ is in range L. Also $c\mathbf{u} = cL(\mathbf{u}_1) = L(c\mathbf{u}_1)$ showing that $c\mathbf{u}$ is in range L. Thus range L is a subspace of W. □

EXAMPLE 4.3–4 Find a basis for range L if L is a linear transformation defined by

$$L: \mathbb{R}^5 \to \mathbb{R}^4$$

such that

$$L \begin{pmatrix} x_1 \\ x_2 \\ x_3 \\ x_4 \\ x_5 \end{pmatrix} = \begin{pmatrix} x_1 - x_3 + 3x_4 - x_5 \\ x_1 \qquad + 2x_4 - x_5 \\ 2x_1 - x_3 + 5x_4 - x_5 \\ -x_3 + x_4 \end{pmatrix}.$$

Solution. Since the domain of the transformation is $\mathbb{R}^5$, it is sufficient to examine what the transformation does to the vectors that form a natural basis for $\mathbb{R}^5$. (Why?) Hence we have

$$(1, 0, 0, 0, 0)L = (1, 1, 2, 0),$$
$$(0, 1, 0, 0, 0)L = (0, 0, 0, 0),$$
$$(0, 0, 1, 0, 0)L = (-1, 0, -1, -1),$$
$$(0, 0, 0, 1, 0)L = (3, 2, 5, 1),$$
$$(0, 0, 0, 0, 1)L = (-1, -1, -1, 0).$$

The vectors on the right span range L so that by sifting out the linearly dependent ones we can find a basis for range L. This sifting process can be accomplished by applying row reduction to a matrix whose rows are the vectors in question. Thus, omitting intermediate steps (see Exercise 11), we have

$$\begin{pmatrix} 1 & 1 & 2 & 0 \\ 0 & 0 & 0 & 0 \\ -1 & 0 & -1 & -1 \\ 3 & 2 & 5 & 1 \\ -1 & -1 & -1 & 0 \end{pmatrix} \sim \begin{pmatrix} 1 & 0 & 1 & 1 \\ 0 & 1 & 1 & -1 \\ 0 & 0 & 1 & 0 \\ 0 & 0 & 0 & 0 \\ 0 & 0 & 0 & 0 \end{pmatrix}.$$

Hence, a basis for range L is given by

$$\left\{ \begin{pmatrix} 1 \\ 0 \\ 1 \\ 1 \end{pmatrix}, \begin{pmatrix} 0 \\ 1 \\ 1 \\ -1 \end{pmatrix}, \begin{pmatrix} 0 \\ 0 \\ 1 \\ 0 \end{pmatrix} \right\},$$

showing that dim (range L) = 3. ∎

We conclude this section by stating three theorems that apply.

Theorem 4.3–7 (*Sylvester's law of nullity**). If $L: V \to W$ is a linear transformation from a vector space V into a vector space W, then

$$\dim (\text{range } L) + \dim (\ker L) = \dim (V).$$

Theorem 4.3–8 Let $L: V \to W$ be a linear transformation from a vector space V into a vector space W and let dim (V) = dim (W). Then L is onto if and only if it is one-to-one.

Theorem 4.3–9 Let $L: V \to W$ be a linear transformation from a vector space V into a vector space W. Then L is one-to-one if and only if dim (range L) = dim (V).

---

* After James J. Sylvester (1814–1897), an English mathematician.

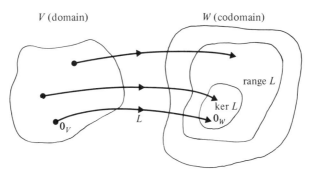

FIGURE 4.3–3 "Picture" of a linear transformation $L: V \rightarrow W$.

In order to "visualize" the above theorems, a diagram such as the one shown in Fig. 4.3–3 may be helpful. Note that the figure shows that $L(\mathbf{0}_v) = \mathbf{0}_w$ and that (possibly) other elements of V may be transformed by L into elements of ker L. Those elements of V *not* transformed into ker L are necessarily transformed into range L. Finally, range L is a subset of the codomain W.

KEY WORDS AND PHRASES

vector space	linear transformation
subspace	kernel
linear combination	null space
spanning set	nullity
linear independence	domain
basis	codomain
dimension	range

EXERCISES 4.3

▶ **1.** Show that the subset of $\mathbb{R}^3$ consisting of all vectors of the form $(x_1, x_2, 1)$ is not a subspace of $\mathbb{R}^3$.

2. Show that the space consisting of the single vector $\mathbf{0}$ is a vector space.

3. Solve the system of equations in Example 4.3–1.

4. Show that the set
$$T = \{(1, 0, 2), (1, 2, 0), (1, 1, 1)\}$$

does not span $\mathbb{R}^3$ by showing that only those vectors (x_1, x_2, x_3) of $\mathbb{R}^3$ for which $2x_1 - x_2 - x_3 = 0$ are spanned by the vectors in the set T.

5. Prove Theorem 4.3–2.

6. Show that the set
$$S = \{(1, 0, 0), (1, 1, 0), (1, 1, 1)\}$$
   is linearly independent.

7. Show that the set
$$S = \{(1, 0, 0), (1, 1, 0), (1, 1, 1), (-1, 2, 3)\}$$
   (a) spans $\mathbb{R}^3$ and (b) is linearly dependent.

8. Show that the set
$$S = \{(1, 0, 0), (1, 1, 1)\}$$
   (a) is linearly independent and (b) does not span $\mathbb{R}^3$.

9. Show that the mapping of Example 4.3–2(b) is a linear transformation.

10. Prove Theorem 4.3–4.

11. Show that
$$\begin{pmatrix} 1 & 1 & 2 & 0 \\ 0 & 0 & 0 & 0 \\ -1 & 0 & -1 & -1 \\ 3 & 2 & 5 & 1 \\ -1 & -1 & -1 & 0 \end{pmatrix} \sim \begin{pmatrix} 1 & 0 & 1 & 1 \\ 0 & 1 & 1 & -1 \\ 0 & 0 & 1 & 0 \\ 0 & 0 & 0 & 0 \\ 0 & 0 & 0 & 0 \end{pmatrix}.$$

▶▶ 12. Consider the mapping $L: \mathbb{R}^2 \to \mathbb{R}^3$ defined by
$$L(u_1, u_2) = (u_2, u_1, 1).$$
   a) Show that this mapping is single-valued.
   b) Is this mapping a linear transformation? Why?
   c) Describe the mapping geometrically.

13. Determine which of the following sets spans $\mathbb{R}^3$.
   a) $\{(2, 2, 0), (0, 2, 2), (0, 0, 2)\}$
   b) $\{(-1, 2, 3), (3, 2, 1), (0, 1, 2)\}$
   c) $\{(1, 1, 0), (0, 0, 1), (3, 3, 1)\}$

14. **a)** Prove that the vectors
$$\mathbf{v}_1 = (1, 1, 1), \qquad \mathbf{v}_2 = (1, 2, 3), \qquad \mathbf{v}_3 = (2, 2, 0)$$
   form a basis of $\mathbb{R}^3$.
   b) Express the vectors in the natural basis for $\mathbb{R}^3$ in terms of the $\mathbf{v}_i$ in part (a).

15. Consider the vectors
$$\mathbf{u}_1 = (1, 0, 2), \qquad \mathbf{u}_2 = (1, 2, 3), \qquad \mathbf{u}_3 = (-4, 3, 5), \qquad \mathbf{u}_4 = (11, -6, -16).$$
   a) Show that $\mathbf{u}_1$, $\mathbf{u}_3$, and $\mathbf{u}_4$ are linearly dependent and express u_3 as a linear combination of $\mathbf{u}_1$ and $\mathbf{u}_4$.
   b) Show that $\mathbf{u}_1$, $\mathbf{u}_2$, and $\mathbf{u}_3$ are linearly independent.

16. Show that the subset of $\mathbb{R}^n$ consisting of all vectors having the form $(x_1, x_2, \ldots, x_{n-1}, 0)$ is a subspace of $\mathbb{R}^n$.

17. Show that Definition 4.3–8 is equivalent to the following: A single-valued mapping $L: \mathbb{R}^n \to \mathbb{R}^m$ is called a *linear transformation* of $\mathbb{R}^n$ into $\mathbb{R}^m$ if

$$L(a\mathbf{u} + b\mathbf{v}) = aL(\mathbf{u}) + bL(\mathbf{v})$$

for every $\mathbf{u}$, $\mathbf{v}$ in $\mathbb{R}^n$ and any real numbers a and b.

18. Show that the set
$$\{(1, -1, 0), (2, 1, 0)\}$$

spans a subspace of $\mathbb{R}^3$. What is the dimension of this subspace?

19. Let (x_1, x_2, x_3) be an arbitrary vector of $\mathbb{R}^3$. Which of the following are subspaces?

   a) All vectors such that $x_1 = x_2 = x_3$.

   b) All vectors such that $x_2 = 1$.

   c) All vectors such that $x_1 = 0$.

   d) All vectors such that x_1, x_2, and x_3 are rational numbers.

20. Show that all the vectors (x_1, x_2, x_3, x_4) of $\mathbb{R}^4$ for which $x_1 + x_4 = x_2 + x_3$ form a subspace V of $\mathbb{R}^4$. Then show that the set

$$\{(1, 0, 0, -1), (0, 1, 0, 1), (0, 0, 1, 1)\}$$

spans V.

21. **a)** Show that

$$L\begin{pmatrix} x \\ y \end{pmatrix} = \begin{pmatrix} -x \\ y \end{pmatrix}$$

   is a linear transformation.

   b) Explain the geometric significance of L.

22. Find a basis for the kernel and range of each of the following transformations.

   a) $L\begin{pmatrix} 1 \\ 0 \\ 0 \end{pmatrix} = \begin{pmatrix} 0 \\ 0 \\ 1 \end{pmatrix}$, $L\begin{pmatrix} 0 \\ 1 \\ 0 \end{pmatrix} = \begin{pmatrix} 0 \\ 1 \\ 0 \end{pmatrix}$, $L\begin{pmatrix} 0 \\ 0 \\ 1 \end{pmatrix} = \begin{pmatrix} 1 \\ 0 \\ 0 \end{pmatrix}$

   b) $L\begin{pmatrix} 1 \\ 0 \\ 0 \\ 0 \end{pmatrix} = \begin{pmatrix} 1 \\ 1 \\ 1 \end{pmatrix}$, $L\begin{pmatrix} 0 \\ 1 \\ 0 \\ 0 \end{pmatrix} = \begin{pmatrix} -1 \\ 0 \\ 1 \end{pmatrix}$, $L\begin{pmatrix} 0 \\ 0 \\ 1 \\ 0 \end{pmatrix} = \begin{pmatrix} 1 \\ 2 \\ 3 \end{pmatrix}$, $L\begin{pmatrix} 0 \\ 0 \\ 0 \\ 1 \end{pmatrix} = \begin{pmatrix} 1 \\ -1 \\ -3 \end{pmatrix}$

   c) $\begin{pmatrix} 1 & 0 & 1 \\ 2 & 1 & 1 \\ -1 & 1 & -2 \end{pmatrix} \begin{pmatrix} x_1 \\ x_2 \\ x_3 \end{pmatrix}$

   d) $\begin{pmatrix} 1 & 2 & 0 & 1 \\ 2 & -1 & 2 & -1 \\ 1 & -3 & 2 & -2 \end{pmatrix} \begin{pmatrix} x_1 \\ x_2 \\ x_3 \\ x_4 \end{pmatrix}$

   e) $(1, 0, 0)L = (0, 1, 0, 2)$,
   $(0, 1, 0)L = (0, 1, 1, 0)$,
   $(0, 0, 1)L = (0, 1, -1, 4)$

23. Consider the subspace of $\mathbb{R}^3$ spanned by the vectors $(1, -3, 2)$ and $(0, 4, 1)$. Determine which of the following vectors belong to this subspace.

a) $(2, 9, 4)$ 　　　　**b)** $(\frac{1}{2}, 1, 1)$ 　　　　**c)** $(2, 9, 5)$

d) $(4, 7, 6)$ 　　　　**e)** $(0, \frac{1}{3}, -\frac{2}{3})$ 　　　　**f)** $(-\frac{1}{2}, -\frac{1}{2}, -\frac{3}{2})$

24. Describe the subspace spanned by each of the following sets of vectors.

a) $\{(2, 1, 3), (-1, 2, 1)\}$ 　　　　**b)** $\{(1, 0, 2), (2, 1, -2)\}$

25. Show that the following sets of vectors in $\mathbb{R}^4$ are linearly independent.

a) $\{(1, 1, 0, 0), (1, 0, 1, 0), (0, 1, 1, 0), (0, 1, 0, 1)\}$

b) $\{(1, 1, 1, 1), (0, 1, 1, 1), (0, 0, 1, 1), (0, 0, 0, 1)\}$

26. Express the vector $(2, 0, -1, 3)$ as a linear combination of the vectors in parts (a) and (b) of Exercise 25.

27. Determine which of the following sets of vectors in $\mathbb{R}^3$ are linearly dependent.

a) $\{(-1, 3, 2), (3, 4, 0), (1, 4, 4)\}$

b) $\{(2, -2, 1), (1, -3, 4), (-3, 1, 2)\}$

c) $\{(-2, 1, 3), (3, -2, -1), (-1, 0, 5)\}$

d) $\{(1, 0, -5), (4, 2, 2), (1, 1, 6)\}$

28. Prove that the vectors $(2, 2, 1)$, $(2, 1, 0)$, and $(2, 1, 1)$ form a basis for $\mathbb{R}^3$.

29. What is the dimension of the subspace of $\mathbb{R}^4$ spanned by the following vectors?

a) $(1, 0, 2, -1)$, $(3, -1, -2, 0)$, $(1, -1, -6, 2)$, $(0, 1, 8, -3)$

b) $(1, 1, 10, -4)$, $(\frac{1}{2}, 0, 1, -\frac{1}{2})$, $(-\frac{1}{2}, \frac{1}{2}, 3, -1)$

30. Which of the following sets of vectors are linearly independent in $\mathbb{R}^3$?

a) $\mathbf{u} = 2\mathbf{e}_1 - \mathbf{e}_2 + 2\mathbf{e}_3$,
$\mathbf{v} = -\mathbf{e}_1 + \mathbf{e}_2 - 3\mathbf{e}_3$,
$\mathbf{w} = -2\mathbf{e}_1 + 2\mathbf{e}_2 - \mathbf{e}_3$

b) $\mathbf{u} = -3\mathbf{e}_1 + 5\mathbf{e}_2 + 2\mathbf{e}_3$,
$\mathbf{v} = 4\mathbf{e}_1 - 3\mathbf{e}_2 - 3\mathbf{e}_3$,
$\mathbf{w} = 2\mathbf{e}_1 + 7\mathbf{e}_2 + \mathbf{e}_3$

c) $\mathbf{u} = -2\mathbf{e}_1 + 5\mathbf{e}_2 - 6\mathbf{e}_3$,
$\mathbf{v} = 5\mathbf{e}_1 + \mathbf{e}_2 - 3\mathbf{e}_3$,
$\mathbf{w} = -4\mathbf{e}_1 - \mathbf{e}_2 + 6\mathbf{e}_3$

d) $\mathbf{u} = 4\mathbf{e}_1 - 2\mathbf{e}_2 + 3\mathbf{e}_3$,
$\mathbf{v} = 2\mathbf{e}_1 - 8\mathbf{e}_2 + 7\mathbf{e}_3$,
$\mathbf{w} = \mathbf{e}_1 + 3\mathbf{e}_2 - 2\mathbf{e}_3$

31. Prove that each of the following is a linear transformation from $\mathbb{R}^2$ into $\mathbb{R}^2$ and describe each one geometrically. (*Hint*: A diagram might help.)

a) $(x_1, x_2)L = -(x_2, x_1)$ 　　　　**b)** $(x_1, x_2)L = 2(x_1, x_2)$

c) $(x_1, x_2)L = -(x_1, x_2)$ 　　　　**d)** $(x_1, x_2)L = (x_2, x_1)$

e) $(x_1, x_2)L = (x_1 + x_2, 0)$

▶▶▶ **32.** Consider all vectors (x_1, x_2, x_3, x_4) in $\mathbb{R}^4$ such that $x_4 = ax_1 + bx_2 + cx_3$, where a, b, and c are *fixed* real numbers. Show that these vectors form a subspace V. Then find a spanning set for V.

33. Find a basis for the subspace V of all vectors (x_1, x_2, x_3, x_4) in $\mathbb{R}^4$ such that $x_1 + x_2 + x_3 + x_4 = 0$.

34. Show that the following is a linear transformation:

$$L(x_1, x_2) = (x_1, x_2)\begin{pmatrix} \cos\theta & -\sin\theta \\ \sin\theta & \cos\theta \end{pmatrix}.$$

(*Note*: This transformation is a *rotation* in the $x_1 x_2$-plane through an angle θ measured counterclockwise.)

35. Prove that if a set of vectors is linearly independent, it does not include the zero vector. (*Hint*: Use contradiction.)

36. Given the two vectors $(2, 0, 1, 1)$ and $(1, 1, 0, 3)$ in $\mathbb{R}^4$, find two more vectors so that the four will form a basis for $\mathbb{R}^4$. Is your answer unique? Why?

37. Find LM and ML for each of the following linear transformations from $\mathbb{R}^3$ into $\mathbb{R}^3$. Which of these pairs commute?

   a) $(x_1, x_2, x_3)L = (x_2, x_1, x_1 + x_2 + x_3)$,

   $(x_1, x_2, x_3)M = (x_2, x_1, -x_1 - x_2 - x_3)$

   b) $(x_1, x_2, x_3)L = (x_2, x_1 + x_3, x_1 + x_2 + x_3)$,

   $(x_1, x_2, x_3)M = (x_1 + x_2 + x_3, x_2, -x_1 - x_2 - x_3)$

38. Prove that a linear transformation transforms a zero vector into a zero vector.

39. In Example 4.3–4 continue the row reduction until a reduced row echelon form is obtained and thus find another basis for range L.

4.4 EIGENVALUES AND EIGENVECTORS

If **x** is a column vector, then the product $A\mathbf{x}$, where A is a real $m \times n$ matrix, is another vector, say **y**. We have seen in the last section that

$$A\mathbf{x} = \mathbf{y}$$

defines a linear transformation $L: \mathbb{R}^n \to \mathbb{R}^m$ and we considered the characteristics of such transformations there.

In this section we will confine our attention to $n \times n$ matrices and to linear transformations having the form

$$A\mathbf{x} = \lambda\mathbf{x}, \tag{4.4-1}$$

where λ is a scalar. An important reason for studying Eq. (4.4–1) stems from the fact that a physical system may be represented in some sense by a matrix A and for such a system a vector **x** satisfying Eq. (4.4–1) has special significance. (We will elaborate on this in Section 4.5.) Thus, given a square matrix A, we seek one or more vectors **x** that satisfy Eq. (4.4–1). Such vectors are called **eigenvectors** or *characteristic vectors* or *proper vectors* or *latent vectors*. We will use the word "eigenvector," which is a slight anglicization of the German "Eigenvektor."

Associated with each eigenvector satisfying Eq. (4.4–1) there is a value of the scalar λ that is called an **eigenvalue** or *characteristic value* or *proper value* or *latent root* of A. The German word is "Eigenwert" but we will use the common hybridization "eigenvalue" since it parallels "eigenvector."

Pioneering work in this area was done by two English mathematicians, Arthur Cayley (1821–1895) and William Rowan Hamilton (1805–1865). Cayley studied linear transformations from the standpoint of invariance, that is, he was interested in those vectors **x** that satisfied $A\mathbf{x} = \lambda\mathbf{x}$ and he called these vectors *invariant vectors*. It was Sylvester (see Theorem 4.3–7) who first used the term "latent root." German mathematicians used the prefix "eigen-" which was later translated as "characteristic."

Equation (4.4–1) may be written

$$A\mathbf{x} - \lambda\mathbf{x} = \mathbf{0}$$

and, on factoring **x** out on the right (that is, using the right distributive property of matrix multiplication), we have

$$(A - \lambda I)\mathbf{x} = \mathbf{0}. \tag{4.4–2}$$

Note that the presence of the identify I is essential in this last equation. (Why?) Now Eq. (4.4–2) is actually a system of homogeneous linear equations:

$$(a_{11} - \lambda)x_1 + a_{12}x_2 + \cdots + a_{1n}x_n = 0$$
$$a_{21}x_1 + (a_{22} - \lambda)x_2 + \cdots + a_{2n}x_n = 0$$
$$\vdots$$
$$a_{n1}x_1 + a_{n2}x_2 + \cdots + (a_{nn} - \lambda)x_n = 0.$$

Such a system *always* has a solution, namely, $x_1 = x_2 = \cdots = x_n = 0$, called the *trivial solution*. In other words, the zero vector satisfies Eq. (4.4–2). We will be interested, however, only in *nonzero eigenvectors*. Accordingly, the matrix of coefficients in the system (4.4–2) must be singular (compare Section 4.2), that is, the determinant of this matrix must be zero. Hence,

$$|A - \lambda I| = 0. \tag{4.4–3}$$

This last is actually an nth-degree polynomial equation in λ,

$$(-1)^n\lambda^n + \cdots + |A| = 0, \tag{4.4–4}$$

which is called the **characteristic equation** of A. The roots $\lambda_1, \lambda_2, \ldots, \lambda_n$ are the eigenvalues of A and knowing these will enable us to obtain the eigenvectors of A. We illustrate with an example.

EXAMPLE 4.4–1 Find the eigenvalues and corresponding eigenvectors of the matrix

$$A = \begin{pmatrix} 0 & -1 & -3 \\ 2 & 3 & 3 \\ -2 & 1 & 1 \end{pmatrix}.$$

Solution. We first obtain the characteristic equation by expanding the determinant of the matrix $A - \lambda I$ using elements of the first row. Thus

$$|A - \lambda I| = \begin{vmatrix} -\lambda & -1 & -3 \\ 2 & 3-\lambda & 3 \\ -2 & 1 & 1-\lambda \end{vmatrix}$$

$$= -\lambda[(3-\lambda)(1-\lambda) - 3] + [2(1-\lambda) + 6] - 3[2 + 2(3-\lambda)]$$

$$= -\lambda(\lambda^2 - 4\lambda) - 2(\lambda - 4) + 6(\lambda - 4)$$

$$= (\lambda - 4)(-\lambda^2 + 4) = 0.$$

Hence the eigenvalues are $\lambda_1 = -2$, $\lambda_2 = 2$, and $\lambda_3 = 4$. Note carefully that we did not need to express the characteristic equation in the form $\lambda^3 - 4\lambda^2 - 4\lambda + 16 = 0$ since we were interested in the roots of the equation rather than the equation itself and the procedure shown above made our task simpler. Next we find the eigenvector corresponding to each eigenvalue. For $\lambda_1 = -2$, $Ax = \lambda x$ becomes

$$\begin{pmatrix} 0 & -1 & -3 \\ 2 & 3 & 3 \\ -2 & 1 & 1 \end{pmatrix} \begin{pmatrix} x_1 \\ x_2 \\ x_3 \end{pmatrix} = -2 \begin{pmatrix} x_1 \\ x_2 \\ x_3 \end{pmatrix}$$

or

$$2x_1 - x_2 - 3x_3 = 0,$$

$$2x_1 + 5x_2 + 3x_3 = 0.$$

Adding these provides the result $x_1 = -x_2$ and subtracting gives $x_2 = -x_3$, hence $(1, -1, 1)^T$ is an* eigenvector belonging to the eigenvalue $\lambda_1 = -2$. Similarly from $Ax = 2x$ we obtain $(1, 1, -1)^T$, an eigenvector belonging to $\lambda_2 = 2$. Finally, $(-1, 1, 1)^T$ is an eigenvector belonging to $\lambda_3 = 4$ (see Exercise 1). ∎

We remark that the eigenvalues need not necessarily be real. For example, the characteristic equation $\lambda^3 - 3\lambda^2 + 4\lambda - 2 = 0$ has one real and two complex roots (Exercise 2). In this section, however, we will be dealing with matrices whose characteristic equations have only real roots. In Example 4.4–1 the eigenvalues were *distinct* (different). The next two examples will show what can happen when an eigenvalue is repeated.

EXAMPLE 4.4–2 Find the eigenvalues and corresponding eigenvectors of the matrix

$$A = \begin{pmatrix} -4 & 5 & 5 \\ -5 & 6 & 5 \\ -5 & 5 & 6 \end{pmatrix}.$$

---

* Note that any scalar multiple of this eigenvector also qualifies as an eigenvector belonging to the eigenvalue -2.

Solution. We have

$$|A - \lambda I| = (-4 - \lambda)[(6 - \lambda)(6 - \lambda) - 25]$$
$$- 5[-5(6 - \lambda) + 25] + 5[-25 + 5(6 - \lambda)]$$
$$= (-4 - \lambda)(\lambda - 1)(\lambda - 11) - 25(\lambda - 1) - 25(\lambda - 1)$$
$$= (\lambda - 1)(-\lambda^2 + 7\lambda - 6) = -(\lambda - 1)^2(\lambda - 6) = 0.$$

Hence the eigenvalues of A are $\lambda_1 = \lambda_2 = 1$ and $\lambda_3 = 6$, that is, $\lambda = 1$ is an eigenvalue **of multiplicity two**. Substituting this value into $A\mathbf{x} = \lambda\mathbf{x}$ produces the *single* equation

$$-5x_1 + 5x_2 + 5x_3 = 0$$

or

$$x_1 = x_2 + x_3.$$

Hence x_2 and x_3 are arbitrary and we can find *two* linearly independent eigenvectors by first putting $x_2 = 0$ and $x_3 = 1$ and then putting $x_2 = 1$ and $x_3 = 0$. Thus eigenvectors belonging to the eigenvalue 1 are $(1, 0, 1)^T$ and $(1, 1, 0)^T$. For $\lambda_3 = 6$ we have an eigenvector $(1, 1, 1)^T$ (Exercise 3).

The linearly independent eigenvectors $(1, 0, 1)^T$ and $(1, 1, 0)^T$ span a space called the **eigenspace** of $\lambda = 1$. It is a subspace of $\mathbb{R}^3$. Similarly, the eigenvector $(1, 1, 1)^T$ spans the eigenspace of $\lambda = 6$, also a subspace of $\mathbb{R}^3$ (see Exercise 4). ■

EXAMPLE 4.4–3 Find the eigenvalues and corresponding eigenvectors of the matrix

$$B = \begin{pmatrix} 4 & 6 & 6 \\ 1 & 3 & 2 \\ -1 & -5 & -2 \end{pmatrix}.$$

Solution. The eigenvalues are (Exercise 5) $\lambda_1 = 1$ and $\lambda_2 = \lambda_3 = 2$. For $\lambda_1 = 1$ we have

$$x_1 + 2x_2 + 2x_3 = 0$$
$$-x_1 - 5x_2 - 3x_3 = 0,$$

so that $x_3 = -3x_2$, $x_1 = 4x_2$, and an eigenvector is $(4, 1, -3)^T$. For $\lambda_2 = \lambda_3 = 2$ we find

$$x_1 + 3x_2 + 3x_3 = 0$$
$$x_1 + x_2 + 2x_3 = 0$$
$$-x_1 - 5x_2 - 4x_3 = 0,$$

and, on using Gaussian reduction, we obtain $x_3 = -2x_2$ and $x_1 = 3x_2$. Hence a corresponding eigenvector is $(3, 1, -2)^T$. ■

The last two examples show that a repeated eigenvalue may have one or two eigenvectors corresponding to it. We are sure that there will be at least *one* nonzero eigenvector because for a given λ, the system of homogeneous equations $(A - \lambda I)\mathbf{x} = \mathbf{0}$ has a nontrivial solution. (Why?) Moreover, the eigenvectors found for each matrix formed linearly independent sets. That this is not merely due to coincidence will be investigated next.

Similarity

Definition 4.4–1 If A and B are $n \times n$ matrices, then A is *similar to B* if there is a nonsingular matrix P such that $P^{-1}AP = B$.

Similarity of matrices is a fruitful concept since significant ideas in geometry and dynamics are based on it. In Section 4.5 we will look at an example in which similarity simplifies the task of obtaining a solution to a system of differential equations. For the present we note that if A is similar to a *diagonal* matrix D, then we can write

$$P^{-1}AP = D = \text{diag}\,(a_1, a_2, \ldots, a_n).$$

Hence it follows that the characteristic polynomial of D is the same as that of A (see Exercise 23 in Section 4.7 for a proof of the fact that $|AB| = |A||B|$). Thus

$$\begin{aligned}
|D - \lambda I| &= |P^{-1}AP - \lambda P^{-1}P| \\
&= |P^{-1}(A - \lambda I)P| = |P^{-1}||A - \lambda I||P| \\
&= |P^{-1}P||A - \lambda I| = |A - \lambda I|.
\end{aligned}$$

But this implies that the diagonal elements $a_1, a_2, \ldots, a_n$ are the eigenvalues $\lambda_1, \lambda_2, \ldots, \lambda_n$ of A in some order. We have proved the following theorem.

Theorem 4.4–1 If a matrix A is similar to a diagonal matrix D, then the eigenvalues of A are the diagonal elements of D.

It now comes as no surprise that the matrix P above can be obtained from the eigenvectors of A. In fact, the matrices A of Example 4.4–2 and B of Example 4.4–3 differ in this respect. For matrix A we can find a nonsingular matrix P such that $P^{-1}AP = D$ and we say A is *diagonalizable*, whereas for B no such matrix P can be found and we say B is *not* diagonalizable. The reason for this difference is contained in the next theorem.

Theorem 4.4–2 An $n \times n$ matrix A is similar to a diagonal matrix D if and only if the set of eigenvectors of A includes a basis for R^n.

PROOF. Let $\mathbf{x}_1, \mathbf{x}_2, \ldots, \mathbf{x}_n$ be eigenvectors of A and suppose that they form a basis for R^n. Then

$$A\mathbf{x}_i = \lambda_i\mathbf{x}_i = \mathbf{x}_i\lambda_i, \qquad i = 1, 2, \ldots, n. \tag{4.4–5}$$

If P is the matrix whose ith column contains the elements of $\mathbf{x}_i$ and if

$$D = \text{diag}(\lambda_1, \lambda_2, \ldots, \lambda_n),$$

then Eq. (4.4–5) is equivalent to

$$AP = PD.$$

Since the eigenvectors of A are linearly independent, it follows that P is nonsingular* and

$$P^{-1}AP = D,$$

that is, A is similar to a diagonal matrix D. The converse is left for the exercises (see Exercise 6). $\square$

EXAMPLE 4.4–4 Show that the matrix A of Example 4.4–2 is diagonalizable.

Solution. Using the eigenvectors of A as found in Example 4.4–2, we have

$$P = \begin{pmatrix} 1 & 1 & 1 \\ 0 & 1 & 1 \\ 1 & 0 & 1 \end{pmatrix} \quad \text{and} \quad P^{-1} = \begin{pmatrix} 1 & -1 & 0 \\ 1 & 0 & -1 \\ -1 & 1 & 1 \end{pmatrix}$$

so that $P^{-1}AP = \text{diag}(1, 1, 6)$. The details are left for Exercise 7. ∎

Quadratic forms

An important application of diagonalization is found in connection with quadratic forms, which we shall discuss briefly.

Definition 4.4–2 An expression of the form

$$ax^2 + bxy + cy^2,$$

where a, b, and c are real numbers, at least one of which is nonzero, is called a **quadratic form** in x and y.

Quadratic forms occur in the study of conic sections in analytic geometry. A quadratic form in x, y, and z may occur in the study of quadratic surfaces (see Exercise 18). In addition to geometry, quadratic forms play an important role in dynamics, statistics, and problems of maxima and minima.

Every quadratic form can be expressed in matrix form using a *symmetric* matrix. For example, the quadratic form in Definition 4.4–2 can be written

$$(x, y) \begin{pmatrix} a & \frac{1}{2}b \\ \frac{1}{2}b & c \end{pmatrix} \begin{pmatrix} x \\ y \end{pmatrix}.$$

---

* Recall the process used in Section 4.2 to find the inverse of a matrix.

This fact, together with Theorem 4.4–3, gives us a means of identifying conic sections easily. But first we define what is meant by vectors being orthogonal.

If

$$\mathbf{x} = (x_1, x_2, \ldots, x_n) \quad \text{and} \quad \mathbf{y} = (y_1, y_2, \ldots, y_n),$$

then the product $\mathbf{x}\mathbf{y}^T(=\mathbf{y}\mathbf{x}^T)$ is a *scalar* (real number) called the **scalar product*** of $\mathbf{x}$ and $\mathbf{y}$. Of particular importance is the case when the scalar product is zero. We say that $\mathbf{x}$ and $\mathbf{y}$ are **orthogonal** when their scalar product is zero. It will be shown in Chapter 5 that orthogonal vectors may be interpreted geometrically in terms of perpendicular lines.

Theorem 4.4–3 Eigenvectors belonging to distinct eigenvalues of a (real) symmetric matrix are orthogonal.

PROOF. Let A be a symmetric matrix, with distinct eigenvalues λ_i and λ_j belonging to eigenvectors $\mathbf{x}_i$ and $\mathbf{x}_j$, respectively. Then

$$A\mathbf{x}_i = \lambda_i \mathbf{x}_i \quad \text{and} \quad A\mathbf{x}_j = \lambda_j \mathbf{x}_j.$$

Multiplying the first of these on the left by $\mathbf{x}_j^T$ produces

$$\mathbf{x}_j^T A\mathbf{x}_i = \lambda_i \mathbf{x}_j^T \mathbf{x}_i, \tag{4.4–6}$$

and multiplying the second on the left by $\mathbf{x}_i^T$ yields

$$\mathbf{x}_i^T A\mathbf{x}_j = \lambda_j \mathbf{x}_i^T \mathbf{x}_j.$$

Transposing this last equation produces

$$\mathbf{x}_j^T A^T \mathbf{x}_i = \mathbf{x}_j^T A\mathbf{x}_i = \lambda_j \mathbf{x}_j^T \mathbf{x}_i.$$

Subtracting Eq. (4.4–6) from this last yields

$$(\lambda_j - \lambda_i)\mathbf{x}_j^T \mathbf{x}_i = 0,$$

which proves the theorem. □

We use Theorem 4.4–3 to simplify the conic,

$$5x^2 - 6xy + 5y^2 = 8.$$

Consider the quadratic form

$$(x, y)\begin{pmatrix} 5 & -3 \\ -3 & 5 \end{pmatrix}\begin{pmatrix} x \\ y \end{pmatrix} = \mathbf{x}A\mathbf{x}^T,$$

which has a symmetric matrix A. The eigenvalues of A are 2 and 8 with eigenvectors $(1, 1)^T$ and $(-1, 1)^T$, respectively.

Recalling that eigenvectors can be found only to within a scalar multiple, take the eigenvectors to be $(1/\sqrt{2})(1, 1)^T$ and $(1/\sqrt{2})(-1, 1)^T$

---

* Also called "dot product" and "inner product."

and form the matrix

$$P = \frac{1}{\sqrt{2}} \begin{pmatrix} 1 & -1 \\ 1 & 1 \end{pmatrix}.$$

The inverse of this matrix is

$$P^{-1} = \frac{1}{\sqrt{2}} \begin{pmatrix} 1 & 1 \\ -1 & 1 \end{pmatrix} = P^T,$$

that is, the inverse of P is the same as its transpose. A matrix having this property is called an **orthogonal matrix**. The name comes from the fact that the columns of P, considered as vectors, are orthogonal. Moreover, each column vector **u** also has the property that $\mathbf{u}^T\mathbf{u} = 1$.

Hence, putting

$$\mathbf{X}^T = (X, Y) \quad \text{and} \quad \mathbf{x}^T = P\mathbf{X},$$

we have

$$\mathbf{x}A\mathbf{x}^T = \mathbf{X}^T P^T A P \mathbf{X} = (X, Y)\begin{pmatrix} 2 & 0 \\ 0 & 8 \end{pmatrix}\begin{pmatrix} X \\ Y \end{pmatrix}$$

$$= 2X^2 + 8Y^2,$$

and the equation $5x^2 - 6xy + 5y^2 = 8$ becomes $X^2 + 4Y^2 = 4$. This last equation is seen to represent an ellipse in standard form in a co-ordinate system in which the orthogonal **principal axes** are X and Y passing through $(1, 1)$ and $(-1, 1)$ in the xy-plane. The coordinate transformation is the same as a rotation of the x- and y-axes through an angle of $\pi/4$ in a counterclockwise direction.

Diagonalization of a quadratic form eliminates all except the squared terms. The method can be readily extended to quadric surfaces, as shown in the following example.

EXAMPLE 4.4–5 Transform the quadratic form in x, y, and z

$$f(x, y, z) = 4xy + 4xz + 4yz$$

to principal axes. Then identify the quadric surface $f(x, y, z) = 1$.

Solution. We can express $f(x, y, z)$ as follows:

$$(x, y, z) = \begin{pmatrix} 0 & 2 & 2 \\ 2 & 0 & 2 \\ 2 & 2 & 0 \end{pmatrix}\begin{pmatrix} x \\ y \\ z \end{pmatrix}.$$

The eigenvalues of the symmetric matrix are $\lambda_1 = \lambda_2 = -2$ and $\lambda_3 = 4$ with eigenvectors $\mathbf{x}_1 = (1, -1, 0)^T$, $\mathbf{x}_2 = (1, 0, -1)^T$, and $\mathbf{x}_3 = (1, 1, 1)^T$, respectively (see Exercise 11c). Let

$$P = \begin{pmatrix} 1 & 1 & 1 \\ -1 & 0 & 1 \\ 0 & -1 & 1 \end{pmatrix}$$

so that

$$P^{-1} = \frac{1}{3}\begin{pmatrix} 1 & -2 & 1 \\ 1 & 1 & -2 \\ 1 & 1 & 1 \end{pmatrix}$$

and

$$\frac{1}{3}\begin{pmatrix} 1 & -2 & 1 \\ 1 & 1 & -2 \\ 1 & 1 & 1 \end{pmatrix}\begin{pmatrix} 0 & 2 & 2 \\ 2 & 0 & 2 \\ 2 & 2 & 0 \end{pmatrix}\begin{pmatrix} 1 & 1 & 1 \\ -1 & 0 & 1 \\ 0 & -1 & 1 \end{pmatrix} = \begin{pmatrix} -2 & 0 & 0 \\ 0 & -2 & 0 \\ 0 & 0 & 4 \end{pmatrix}.$$

Thus $f(x, y, z) = 1$ becomes $-2X^2 - 2Y^2 + 4Z^2 = 1$, which is a hyperboloid of two sheets (see Exercise 8). ∎

It is not obvious how an *orthogonal* matrix can be used to transform the quadratic form of this last example to principal axes. This topic will be further developed in Section 4.7 in connection with the construction of orthogonal bases.

KEY WORDS AND PHRASES

eigenvector
eigenvalue
characteristic equation
eigenspace
diagonalizable matrix
similarity

quadratic form
scalar product
orthogonal vectors
principal axes
orthogonal matrix

EXERCISES 4.4

▶

1. Fill in the details involved in finding the eigenvectors belonging to the eigenvalues $\lambda_2 = 2$ and $\lambda_3 = 4$ in Example 4.4–1.

2. Find the eigenvalues given that the characteristic equation is

$$\lambda^3 - 3\lambda^2 + 4\lambda - 2 = 0.$$

(*Note*: Eigenvalues may be complex.)

3. Carry out the necessary details for finding the eigenvectors in Example 4.4–2.

4. Referring to Example 4.4–2, show that the eigenspaces of $\lambda = 1$ and $\lambda = 6$ are subspaces of $\mathbb{R}^3$. (*Hint*: Recall the definition of subspace as stated in Theorem 4.3–1.)

5. Find the eigenvalues of the matrix B of Example 4.4–3. Then show that the corresponding eigenvectors are linearly independent.

6. Complete the proof of Theorem 4.4–2, that is, prove that if A is similar to a diagonal matrix D so that $P^{-1}AP = D$, then the elements of the columns of P are the same as those of the eigenvectors of A, which are linearly independent.

7. Referring to Example 4.4–4, find P^{-1}. Also show that if

$$Q = \begin{pmatrix} 1 & 1 & 1 \\ 1 & 0 & 1 \\ 1 & 1 & 1 \end{pmatrix},$$

then $Q^{-1}AQ = \text{diag} (6, 1, 1)$.

8. In Example 4.4–5 carry out the work required to compute P^{-1} and verify that the given quadratic form can be diagonalized as shown.

▶▶ 9. Find the eigenvalues and corresponding eigenvectors of each of the following matrices.

a) $\begin{pmatrix} 1 & 2 \\ 2 & 1 \end{pmatrix}$ b) $\begin{pmatrix} 1 & 1 \\ 0 & 1 \end{pmatrix}$ c) $\begin{pmatrix} 1 & 2 \\ 1 & 2 \end{pmatrix}$

10. Find the eigenvalues and corresponding eigenvectors of each of the following matrices.

a) $\begin{pmatrix} 4 & 6 & 6 \\ 1 & 3 & 2 \\ -1 & -4 & -3 \end{pmatrix}$ b) $\begin{pmatrix} -3 & -7 & -5 \\ 2 & 4 & -3 \\ 1 & 2 & 2 \end{pmatrix}$ c) $\begin{pmatrix} 11 & -4 & -7 \\ 7 & -2 & -5 \\ 10 & -4 & -6 \end{pmatrix}$

11. Find the eigenvalues and eigenvectors of each of the following symmetric matrices.

a) $\begin{pmatrix} 3 & 0 & 0 \\ 0 & 3 & 4 \\ 0 & 4 & 3 \end{pmatrix}$ b) $\begin{pmatrix} 2 & 4 & 6 \\ 4 & 2 & -6 \\ -6 & -6 & -15 \end{pmatrix}$ c) $\begin{pmatrix} 0 & 2 & 2 \\ 2 & 0 & 2 \\ 2 & 2 & 0 \end{pmatrix}$

12. Find the eigenvalues and eigenvectors of each of the following matrices.

a) $\begin{pmatrix} 1 & 1 & 3 \\ 1 & 0 & 1 \\ -1 & 1 & 0 \end{pmatrix}$ b) $\begin{pmatrix} 4 & -2 & -1 \\ 6 & -3 & -2 \\ -3 & 2 & 2 \end{pmatrix}$ c) $\begin{pmatrix} 1 & 1 & 1 \\ 1 & 1 & 1 \\ 1 & 1 & 1 \end{pmatrix}$

13. Find the eigenvalues and corresponding eigenvectors of each of the following matrices, which have repeated eigenvalues.

a) $\begin{pmatrix} 7 & -2 & -4 \\ 3 & 0 & -2 \\ 6 & -2 & -3 \end{pmatrix}$ b) $\begin{pmatrix} -2 & 2 & -3 \\ 2 & 1 & -6 \\ -1 & -2 & 0 \end{pmatrix}$ c) $\begin{pmatrix} 4 & 2 & 3 \\ 2 & 1 & 0 \\ -1 & -2 & 0 \end{pmatrix}$

d) $\begin{pmatrix} 2 & 4 & -2 \\ 4 & 2 & -2 \\ -2 & -2 & -1 \end{pmatrix}$ e) $\begin{pmatrix} -7 & 0 & 6 \\ -6 & -1 & 6 \\ -8 & 0 & 7 \end{pmatrix}$ f) $\begin{pmatrix} 1 & 2 & 3 \\ 0 & 1 & 2 \\ 0 & 0 & 1 \end{pmatrix}$

g) $\begin{pmatrix} -3 & -7 & -5 \\ 2 & 4 & 3 \\ 1 & 2 & 2 \end{pmatrix}$

14. Obtain the eigenvalues and corresponding eigenvectors of each of the following matrices.

a) $\begin{pmatrix} 1 & 0 & 1 \\ 0 & 1 & 0 \\ 1 & 0 & 1 \end{pmatrix}$ **b)** $\begin{pmatrix} 1 & 0 & 1 \\ 0 & 2 & 0 \\ 1 & 0 & -1 \end{pmatrix}$

c) $\begin{pmatrix} 5 & -6 & -6 \\ -1 & 4 & 2 \\ 3 & -6 & -4 \end{pmatrix}$ **d)** $\begin{pmatrix} 1 & 2 & 1 \\ 1 & 2 & 1 \\ 0 & 1 & 2 \end{pmatrix}$

e) Which of the above are diagonalizable?

15. Obtain the eigenvalues and corresponding eigenvectors of each of the following matrices.

a) $\begin{pmatrix} 1 & 0 & -1 & 0 \\ 0 & 1 & 1 & 0 \\ -1 & 1 & 2 & 1 \\ 0 & 0 & 1 & -1 \end{pmatrix}$ **b)** $\begin{pmatrix} -1 & 2 & 1 & 3 \\ 0 & 0 & -2 & 1 \\ 0 & 0 & 2 & -3 \\ 0 & 0 & 0 & 4 \end{pmatrix}$

16. Reduce each of the following quadratic forms in x and y to principal axes.

a) $3x^2 + 3y^2 + 2xy$ **b)** $2x^2 - y^2 - 4xy$

c) $2x^2 + 5y^2 - 12xy$ **d)** $x^2 + y^2 - 2xy$

17. Reduce each of the following quadratic forms in x, y, and z to principal axes.

a) $2xz$

b) $7x^2 + 7y^2 + 4z^2 + 4xy + 2xz - 2yz$

c) $x^2 + 5y^2 + 2z^2 + 4xy + 2xz + 6yz$

d) $x^2 + 2y^2 + 4z^2 - 4xz$

*18. Find the eigenvalues and corresponding eigenvectors of the matrix

$$A = \begin{pmatrix} 7 & -2 & 0 \\ -2 & 6 & 2 \\ 0 & 2 & 5 \end{pmatrix}.$$

▶▶▶ 19. Prove that if λ is an eigenvalue of A^{-1}, then $1/\lambda$ is an eigenvalue of A.

20. Explain the statement, "The nonzero eigenvectors of A span the kernel of $A - \lambda I$."

21. Prove that zero is an eigenvalue of A if and only if A is singular.

22. Prove that if λ is an eigenvalue of the matrix A with eigenvector $\mathbf{x}$, then λ^n is an eigenvalue of A^n with eigenvector $\mathbf{x}$ for every positive integer n. (*Hint:* Use mathematical induction.)

23. A matrix A having the property $A^2 = A$ is said to be *idempotent*. Symmetric idempotent matrices play a role in mathematical statistics. Prove the following properties of an idempotent matrix A.

a) $A^n = A$ for every positive integer n.

b) The only nonsingular idempotent matrix is the identity.

c) The eigenvalues of A are either 0 or 1. (*Hint:* See Exercise 22.)

* Calculator problem.

d) The number of eigenvalues of A that are equal to 1 is the same as the number of linearly independent rows of A.

e) Illustrate the above properties using the matrix

$$A = \begin{pmatrix} 2 & 4 & 6 \\ 4 & 8 & 12 \\ -3 & -6 & -9 \end{pmatrix}.$$

24. Explain why the term free of λ in the characteristic equation (4.4–4) is $|A|$.

4.5 APPLICATION TO SYSTEMS OF ORDINARY DIFFERENTIAL EQUATIONS

In Chapters 1 and 2 we gave some indication of the wide application of ordinary differential equations. We find, however, that *systems* of ordinary differential equations play a most important role in applied mathematics. For one thing, as shown in Section 2.6, higher-order differential equations can be reduced to a system of first-order equations. Moreover, whenever more than one dependent variable is present in a problem involving differential equations, we need to solve a system of such equations.

We will use the material in this chapter for notational and computational convenience in obtaining solutions to systems of differential equations. For a first example consider the system of first-order differential equations with constant coefficients

$$\begin{aligned} \dot{x}_1 &= x_1 + x_2 + x_3 \\ \dot{x}_2 &= 2x_1 + x_2 - x_3 \\ \dot{x}_3 &= -8x_1 - 5x_2 - 3x_3, \end{aligned}$$

where the dot denotes differentiation with respect to t. In matrix notation this system can be written simply as

$$\dot{\mathbf{x}} = A\mathbf{x}, \tag{4.5–1}$$

where $\mathbf{x} = (x_1, x_2, x_3)^T$ and A is the matrix of coefficients.

If we can find a nonsingular matrix P such that $\mathbf{x} = P\mathbf{v}$, then $\dot{\mathbf{x}} = P\dot{\mathbf{v}}$ and this linear coordinate transformation can be used to convert Eq. (4.5–1) into $P\dot{\mathbf{v}} = AP\mathbf{v}$ from which $\dot{\mathbf{v}} = P^{-1}AP\mathbf{v}$. If A is diagonalizable, then such a nonsingular matrix P does exist and we have

$$\dot{\mathbf{v}} = \text{diag}\,(\lambda_1, \lambda_2, \lambda_3)\mathbf{v}, \tag{4.5–2}$$

where the λ_i are eigenvalues of A.

We can now see the advantage of diagonalization. Equation (4.5–2) represents a system of three ordinary first-order homogeneous differential equations that can be easily solved by separating the variables (see Section 1.1). In the present example we have $\lambda_1 = -2$, $\lambda_2 = -1$, and $\lambda_3 = 2$ with

$$P = \begin{pmatrix} -4 & -3 & 0 \\ 5 & 4 & 1 \\ 7 & 2 & 1 \end{pmatrix}$$

(see Exercise 1) and we are assured by Theorem 4.4–2 that P is non-singular (Exercise 2). Thus the first equation of (4.5–2) is $\dot{v}_1 = -2v_1$, whose solution is $v_1 = c_1 \exp(-2t)$, where c_1 is an arbitrary constant. In this fashion we obtain

$$\mathbf{v} = (c_1 \exp(-2t),\ c_2 \exp(-t),\ c_3 \exp(2t))^T$$

from Eq. (4.5–2) by integration. Then

$$\mathbf{x} = P\mathbf{v} = c_1 \begin{pmatrix} -4 \\ 5 \\ 7 \end{pmatrix} e^{-2t} + c_2 \begin{pmatrix} -3 \\ 4 \\ 2 \end{pmatrix} e^{-t} + c_3 \begin{pmatrix} 0 \\ 1 \\ -1 \end{pmatrix} e^{2t} \qquad \textbf{(4.5–3)}$$

is the solution of Eq. (4.5–1). From this *general* solution we can obtain particular ones. For example, if the additional information $\mathbf{x}(0) = (2, -12, 24)^T$ were given, we would have an initial-value problem with solution

$$\mathbf{x} = 4 \begin{pmatrix} -4 \\ 5 \\ 7 \end{pmatrix} e^{-2t} - 6 \begin{pmatrix} -3 \\ 4 \\ 2 \end{pmatrix} e^{-t} - 8 \begin{pmatrix} 0 \\ 1 \\ -1 \end{pmatrix} e^{2t}.$$

The details are left for the exercises (see Exercise 3).

Another way of looking at the last example is the following. Starting with $\dot{\mathbf{x}} = A\mathbf{x}$ we *assume* that there is a solution of the form $\mathbf{x} = \mathbf{c} \exp(\lambda t)$, where λ is a scalar and $\mathbf{c}$ is a constant vector. Then $\dot{\mathbf{x}} = \mathbf{c}\lambda \exp(\lambda t)$, hence we have $\mathbf{c}\lambda \exp(\lambda t) = A\mathbf{c} \exp(\lambda t)$ or $A\mathbf{c} = \lambda \mathbf{c}$. But this implies that a nontrivial solution may be found provided that $\mathbf{c}$ is an eigenvector of A belonging to the eigenvalue λ. Thus the methods of Section 4.4 are applicable whenever the matrix A is a diagonalizable matrix. The details are shown in the following example.

EXAMPLE 4.5–1 Solve the system of differential equations

$$\dot{\mathbf{x}} = \begin{pmatrix} -4 & 5 & 5 \\ -5 & 6 & 5 \\ -5 & 5 & 6 \end{pmatrix} \mathbf{x}.$$

Solution. We readily find the eigenvectors of the coefficient matrix to be $\lambda_1 = \lambda_2 = 1$ and $\lambda_3 = 6$. (Compare Example 4.4–2.) The respective eigenvectors are $(1, 0, 1)^T$, $(1, 1, 0)^T$, and $(1, 1, 1)^T$, which shows that $(1, 0, 1)^T \exp(t)$, $(1, 1, 0)^T \exp(t)$, and $(1, 1, 1)^T \exp(6t)$ are solutions. If we take a linear combination of these we find

$$\mathbf{x} = c_1 \begin{pmatrix} 1 \\ 0 \\ 1 \end{pmatrix} \exp(t) + c_2 \begin{pmatrix} 1 \\ 1 \\ 0 \end{pmatrix} \exp(t) + c_3 \begin{pmatrix} 1 \\ 1 \\ 1 \end{pmatrix} \exp(6t). \quad \blacksquare$$

This last expression is the *general solution* of the given system because it is a linear combination of three linearly independent solutions. This is true because

$$\begin{vmatrix} \exp(t) & \exp(t) & \exp(6t) \\ 0 & \exp(t) & \exp(6t) \\ \exp(t) & 0 & \exp(6t) \end{vmatrix} = \exp(8t),$$

which is never zero. The determinant is called the **Wronskian** of the three solutions of the system. That the solution found *is* the general solution follows from the nonvanishing of the Wronskian. This topic will be explored further in Section 4.7. For the present we state a theorem.

Theorem 4.5–1 Given the system of homogeneous equations

$$\dot{\mathbf{x}}(t) = A(t)\mathbf{x}(t), \qquad \mathbf{x}(t_0) = \mathbf{x}_0, \tag{4.5–4}$$

where the elements of the $n \times n$ matrix $A(t)$ are continuous on $a < t < b$, $\mathbf{x}_0$ is a prescribed vector and $a < t_0 < b$. Then there exists a unique solution of the initial-value problem (4.5–4). Moreover, if

$$\mathbf{x}_1(t), \mathbf{x}_2(t), \ldots, \mathbf{x}_n(t)$$

are linearly independent solutions of

$$\dot{\mathbf{x}}(t) = A(t)\mathbf{x}(t) \tag{4.5–5}$$

on $a < t < b$, then the unique general solution of Eq. (4.5–4) is given by

$$\mathbf{x}(t) = c_1\mathbf{x}_1(t) + c_2\mathbf{x}_2(t) + \cdots + c_n\mathbf{x}_n(t). \tag{4.5–6}$$

The set of linearly independent solutions on $a < t < b$ is said to form a **fundamental set of solutions** for that interval. It follows that a set of fundamental solutions on an interval has a nonvanishing Wronskian on that interval. In Example 4.5–1 the set

$$\{(1, 0, 1)^T \exp(t), (1, 1, 0)^T \exp(t), (1, 1, 1)^T \exp(6t)\}$$

is a fundamental set of solutions. If a system such as (4.5–5) has n linearly independent eigenvectors, then the general solution may be obtained in a straightforward manner. When there are fewer than n linearly independent eigenvectors (which can happen only if one or more eigenvalues are repeated), then a different approach is required, as shown in the next example.

EXAMPLE 4.5–2 Find the general solution of the system

$$\dot{\mathbf{x}} = \begin{pmatrix} 4 & 6 & 6 \\ 1 & 3 & 2 \\ -1 & -5 & -2 \end{pmatrix} \mathbf{x}.$$

(Compare Example 4.4–3.)

Solution. Here the eigenvalues are $\lambda_1 = 1$ and $\lambda_2 = \lambda_3 = 2$ with corresponding eigenvectors $(4, 1, -3)^T$ and $(3, 1, -2)^T$. Since there are only two linearly independent eigenvectors, we cannot immediately write down the general solution. We can, however, recall the procedure from Section 2.1 and make an educated guess that $\mathbf{c}t \exp(2t)$ is a solution for some constant vector $\mathbf{c}$. If so, then it remains only to find $\mathbf{c}$, hence put

$$\mathbf{x} = \mathbf{c}t \exp(2t), \qquad \dot{\mathbf{x}} = \mathbf{c}[\exp(2t)](2t+1)$$

and substitute into the equation $\dot{\mathbf{x}} = A\mathbf{x}$, where A is the matrix of coefficients. Thus

$$\mathbf{c}(2t+1)\exp(2t) = A\mathbf{c}t \exp(2t)$$

or

$$2t\mathbf{c}\exp(2t) + \mathbf{c}\exp(2t) - A\mathbf{c}t\exp(2t) = \mathbf{0}. \qquad (4.5\text{–}7)$$

But $\exp(2t)$ and $t\exp(2t)$ are linearly independent for all t (Exercise 4) so that the only way that the last equation can be satisfied is to take the coefficients of $\exp(2t)$ and $t\exp(2t)$ to be zero. Hence $\mathbf{c} = \mathbf{0}$ and we need to revise our educated guess since $\mathbf{x} = \mathbf{0}$ is *not* an acceptable solution. (Why?)

A reexamination of Eq. (4.5–7) indicates that a better guess might be

$$\mathbf{x} = \mathbf{b}\exp(2t) + \mathbf{c}t\exp(2t)$$

for some constant vectors $\mathbf{b}$ and $\mathbf{c}$. Then

$$\dot{\mathbf{x}} = 2\mathbf{b}\exp(2t) + \mathbf{c}\exp(2t) + 2\mathbf{c}t\exp(2t)$$

and substitution into $\dot{\mathbf{x}} = A\mathbf{x}$ produces

$$2\mathbf{c}t\exp(2t) + (2\mathbf{b} + \mathbf{c})\exp(2t) = A(\mathbf{b}\exp(2t) + \mathbf{c}t\exp(2t)).$$

Now if we equate coefficients of $\exp(2t)$ and $t\exp(2t)$, we obtain the two equations

$$(A - 2I)\mathbf{c} = \mathbf{0},$$
$$(A - 2I)\mathbf{b} = \mathbf{c}.$$

We already know that $\mathbf{c} = (3, 1, -2)^T$ from the first of these. (Why?) The second has a vanishing determinant, $|A - 2I| = 0$, hence its solution is not unique. We find, however, that $\mathbf{b} = (3, 1, -\frac{3}{2})^T$ is a solution (Exercise 5). Moreover, the vectors $(4, 1, -3)^T$, $(3, 1, -2)^T$, and $(3, 1, -\frac{3}{2})^T$ are linearly independent (Exercise 6) and can be used to form a fundamental set of solutions. Hence we can write the general solution as

$$\mathbf{x} = c_1 \begin{pmatrix} 4 \\ 1 \\ -3 \end{pmatrix} e^t + c_2 \left(\begin{pmatrix} 3 \\ 1 \\ -\frac{3}{2} \end{pmatrix} e^{2t} + \begin{pmatrix} 3 \\ 1 \\ -2 \end{pmatrix} te^{2t} \right). \quad \blacksquare$$

It is entirely possible that the eigenvalues of a matrix may be complex. When this happens, the eigenvectors and the fundamental set of solutions will have complex entries. We may, however, use Euler's formula to convert these to sines and cosines in a manner analogous to that used in Section 2.1. We illustrate with an example.

EXAMPLE 4.5–3   Find the general solution of the system

$$\dot{\mathbf{x}} = \begin{pmatrix} 2 & -1 \\ 2 & 4 \end{pmatrix} \mathbf{x}.$$

Solution. From the characteristic equation,

$$\begin{vmatrix} 2 - \lambda & -1 \\ 2 & 4 - \lambda \end{vmatrix} = (2 - \lambda)(4 - \lambda) + 2 = \lambda^2 - 6\lambda + 10 = 0,$$

we obtain the eigenvalues $\lambda_1 = 3 + i$ and $\lambda_2 = 3 - i$. For $\lambda_1 = 3 + i$ we have

$$\begin{pmatrix} 2 & -1 \\ 2 & 4 \end{pmatrix} \begin{pmatrix} x_1 \\ x_2 \end{pmatrix} = (3 + i) \begin{pmatrix} x_1 \\ x_2 \end{pmatrix}$$

or $2x_1 - x_2 = (3 + i)x_1$. Hence

$$x_1 = \frac{-x_2}{i + 1} = \frac{(i - 1)}{2} x_2$$

and an eigenvector belonging to λ_1 can be written $(i - 1, 2)^T$. Similarly an eigenvector belonging to $\lambda_2 = 3 - i$ can be written $(i + 1, -2)^T$. Thus the general solution can be written

$$\mathbf{x} = \begin{pmatrix} i - 1 & i + 1 \\ 2 & -2 \end{pmatrix} \begin{pmatrix} k_1 \exp (3 + i)t \\ k_2 \exp (3 - i)t \end{pmatrix}$$

$$= \begin{pmatrix} (k_2 - k_1)e^{3t}(\cos t + \sin t) + i(k_1 + k_2)e^{3t}(\cos t - \sin t) \\ -2(k_2 - k_1)e^{3t} \cos t + 2i(k_1 + k_2)e^{3t} \sin t \end{pmatrix}$$

$$= \begin{pmatrix} c_1 e^{3t}(\cos t + \sin t) + c_2 e^{3t}(\cos t - \sin t) \\ -2c_1 e^{3t} \cos t + 2c_2 e^{3t} \sin t \end{pmatrix}$$

$$= c_1 \left(\begin{pmatrix} 1 \\ -2 \end{pmatrix} \cos t + \begin{pmatrix} 1 \\ 0 \end{pmatrix} \sin t \right) e^{3t} + c_2 \left(\begin{pmatrix} 1 \\ 0 \end{pmatrix} \cos t + \begin{pmatrix} -1 \\ 2 \end{pmatrix} \sin t \right) e^{3t},$$

$$(4.5–8)$$

where we have replaced $k_2 - k_1$ by c_1 and $i(k_1 + k_2)$ by c_2. (See also Exercise 29, where it is shown that c_1 and c_2 are *real*.)

A word of caution is in order at this point. Since the eigenvectors are known to within only a scalar multiple, it is possible that a general solution can be obtained that doesn't resemble Eq. (4.5–8) at first glance. Theorem 4.5–1 assures us, however, that the solution is unique and that solutions that appear to be "different" are really equivalent. For example, it can be shown (Exercise 8) that the general solution to Example 4.5–3 can also be written

$$\mathbf{x} = b_1 \left(\begin{pmatrix} -1 \\ 1 \end{pmatrix} \cos t + \begin{pmatrix} 0 \\ -1 \end{pmatrix} \sin t \right) e^{3t}$$

$$+ b_2 \left(\begin{pmatrix} 0 \\ -1 \end{pmatrix} \cos t - \begin{pmatrix} -1 \\ 1 \end{pmatrix} \sin t \right) e^{3t}. \quad \blacksquare \qquad \textbf{(4.5–9)}$$

We remark that if $\mathbf{x}_1(t)$ and $\mathbf{x}_2(t)$ are linearly independent solutions of $\dot{\mathbf{x}} = A\mathbf{x}$, then the general solution can be expressed as

$$\mathbf{x}(t) = c_1 \mathbf{x}_1(t) + c_2 \mathbf{x}_2(t),$$

where c_1 and c_2 are *real* constants.

So far we have considered only homogeneous systems of equations. Now we give some examples of nonhomogeneous systems. The procedure is the same as that followed in Section 2.3; that is, we obtain a particular solution and add this to the solution of the associated homogeneous systems (the complementary solution). It will be found that the method of variation of parameters discussed in Section 2.3 will be helpful in obtaining a particular solution. Under certain conditions, however, the method of undetermined coefficients will simplify the computation. We give examples of both methods.

EXAMPLE 4.5–4 Find a particular solution of the system

$$\dot{\mathbf{x}} = \begin{pmatrix} 4 & -9 & 5 \\ 1 & -10 & 7 \\ 1 & -17 & 12 \end{pmatrix} \mathbf{x} + \begin{pmatrix} 1 + 13t \\ 3 + 15t \\ 2 + 26t \end{pmatrix}.$$

Solution. Using the method of undetermined coefficients we assume that

$$\mathbf{x}_p = \mathbf{a} + \mathbf{b}t, \qquad \dot{\mathbf{x}}_p = \mathbf{b},$$

where $\mathbf{a}$ and $\mathbf{b}$ are constant vectors to be determined, and substitute into the given system. Then

$$\mathbf{b} = \begin{pmatrix} 4 & -9 & 5 \\ 1 & -10 & 7 \\ 1 & -17 & 12 \end{pmatrix} (\mathbf{a} + \mathbf{b}t) + \begin{pmatrix} 1 + 13t \\ 3 + 15t \\ 2 + 26t \end{pmatrix},$$

which is actually a system of six linear equations in six unknowns—the components of **a** and **b**. It is simpler to solve first for **b** by equating the coefficients of t on both sides of the equation. We find $\mathbf{b} = (1, 3, 2)^T$ and then compute $\mathbf{a} = (0, 0, 0)^T$ (Exercise 10). Thus a particular solution is given by

$$\mathbf{x}_p = (t, 3t, 2t)^T. \quad \blacksquare$$

We next consider the method of variation of parameters as applied to

$$\dot{\mathbf{x}} = A\mathbf{x} + \mathbf{f}(t) \tag{4.5-10}$$

with the restriction that the $n \times n$ matrix A is diagonalizable. Although this restriction may appear to be severe, a great many problems of this kind occur in applied mathematics. Recall that every symmetric matrix is diagonalizable and it is a fact that symmetric matrices abound when modeling physical problems.

If

$$\mathbf{x} = c_1\mathbf{x}_1 + c_2\mathbf{x}_2 + \cdots + c_n\mathbf{x}_n$$

is the general solution of a *homogeneous* system $\dot{\mathbf{x}} = A\mathbf{x}$, then we assume that

$$\mathbf{x}_p = u_1\mathbf{x}_1 + u_2\mathbf{x}_2 + \cdots + u_n\mathbf{x}_n, \tag{4.5-11}$$

where the u_i are functions of t, is a particular solution of the non-homogeneous system (4.5–10). From

$$\mathbf{x}_p = \sum_{i=1}^{n} u_i\mathbf{x}_i$$

we obtain

$$\dot{\mathbf{x}}_p = \sum_{i=1}^{n} (u_i\dot{\mathbf{x}}_i + \dot{u}_i\mathbf{x}_i),$$

so that substituting these into Eq. (4.5–10) and rearranging gives us

$$\sum_{i=1}^{n} u_i(\dot{\mathbf{x}}_i - A\mathbf{x}_i) + \sum_{i=1}^{n} \dot{u}_i\mathbf{x}_i = \mathbf{f}(t).$$

But the first sum is zero because the $\mathbf{x}_i$ are solutions of the homogeneous system. Hence

$$\sum_{i=1}^{n} \dot{u}_i\mathbf{x}_i = \mathbf{f}(t), \tag{4.5-12}$$

which can be solved for the $\dot{u}_i$, and from these we can obtain the u_i by integration. These values can then be substituted into Eq. (4.5–11) to obtain a particular solution. Note that Eq. (4.5–12) is a system of *linear* equations in the u_i and can be solved by Gaussian elimination or some other method previously discussed. We are assured of a *unique* solution since the matrix of coefficients of the system is nonsingular; it is, in fact, the fundamental solution set of the system $\dot{\mathbf{x}} = A\mathbf{x}$. We illustrate the preceding ideas with an example.

EXAMPLE 4.5–5 The general solution of the homogeneous system

$$\dot{x} = \begin{pmatrix} -1 & 1 \\ -5 & 3 \end{pmatrix} \mathbf{x}$$

is given by

$$\mathbf{x} = c_1 \begin{pmatrix} \cos t \\ 2\cos t - \sin t \end{pmatrix} e^t + c_2 \begin{pmatrix} \sin t \\ 2\sin t + \cos t \end{pmatrix} e^t.$$

Find a particular solution to the nonhomogeneous system

$$\dot{\mathbf{x}} = \begin{pmatrix} -1 & 1 \\ -5 & 3 \end{pmatrix} \mathbf{x} + \begin{pmatrix} \cos t \\ 0 \end{pmatrix}.$$

Solution. We assume a particular solution

$$\mathbf{x}_p = u_1 \begin{pmatrix} \cos t \\ 2\cos t - \sin t \end{pmatrix} e^t + u_2 \begin{pmatrix} \sin t \\ 2\sin t + \cos t \end{pmatrix} e^t, \quad \textbf{(4.5–13)}$$

differentiate, and substitute into the given system. Then from Eq. (4.5–12), we have

$$\dot{u}_1 e^t \cos t + \dot{u}_2 e^t \sin t = \cos t,$$
$$\dot{u}_1 (2\cos t - \sin t) + \dot{u}_2 (2\sin t + \cos t) = 0,$$

and solving (by using Cramer's rule or elimination) for $\dot{u}_1$ and $\dot{u}_2$ we obtain (with the aid of some double-angle trigonometric identities),

$$\dot{u}_1 = \frac{e^{-t}}{2}(1 + \cos 2t + 2\sin 2t),$$

$$\dot{u}_2 = \frac{-e^{-t}}{2}(2 + 2\cos 2t - \sin 2t).$$

Integrating, we find

$$u_1 = -\tfrac{1}{2}e^{-t}(1 + \cos 2t),$$
$$u_2 = \tfrac{1}{2}e^{-t}(2 - \sin 2t).$$

Finally, substituting these values into Eq. (4.5–13) gives us

$$\mathbf{x}_p = \begin{pmatrix} 1 \\ 2 \end{pmatrix} \sin t - \begin{pmatrix} 1 \\ 1 \end{pmatrix} \cos t. \quad \blacksquare$$

We present one more example to show how some of the topics of this section fit together.

EXAMPLE 4.5–6 Find the general solution of the system

$$\dot{\mathbf{x}} = \begin{pmatrix} 4 & -2 \\ 1 & 1 \end{pmatrix} \mathbf{x} + \begin{pmatrix} -2e^t \\ e^{-2t} \end{pmatrix}.$$

Solution. Let

$$A = \begin{pmatrix} 4 & -2 \\ 1 & 1 \end{pmatrix}, \qquad \mathbf{f}(t) = \begin{pmatrix} -2e^t \\ e^{-2t} \end{pmatrix}.$$

Now

$$|A - \lambda I| = \begin{vmatrix} 4 - \lambda & -2 \\ 1 & 1 - \lambda \end{vmatrix} = \lambda^2 - 5\lambda + 6 = 0,$$

which shows that the eigenvalues of the homogeneous system are $\lambda_1 = 2$ and $\lambda_2 = 3$. For $\lambda_1 = 2$ we have an eigenvector $(1, 1)^T$, whereas for $\lambda_2 = 3$ an eigenvector is $(2, 1)^T$. Hence the complementary solution (the solution of the homogeneous system) is

$$\mathbf{x}_c = c_1 \begin{pmatrix} 1 \\ 1 \end{pmatrix} e^{2t} + c_2 \begin{pmatrix} 2 \\ 1 \end{pmatrix} e^{3t},$$

where c_1 and c_2 are arbitrary constants. Next, we define

$$P = \begin{pmatrix} 1 & 2 \\ 1 & 1 \end{pmatrix},$$

a nonsingular matrix in which the elements of the columns are the same as those in the eigenvectors of A. Then

$$P^{-1} = \begin{pmatrix} -1 & 2 \\ 1 & -1 \end{pmatrix}$$

and $P^{-1}AP = \text{diag}(2, 3)$. Hence, if we put $\mathbf{x} = P\mathbf{v}$ and $\dot{\mathbf{x}} = P\dot{\mathbf{v}}$, then the nonhomogeneous system becomes

$$P\dot{\mathbf{v}} = AP\mathbf{v} + \mathbf{f}$$

or

$$\dot{\mathbf{v}} = P^{-1}AP\mathbf{v} + P^{-1}\mathbf{f} = \begin{pmatrix} 2 & 0 \\ 0 & 3 \end{pmatrix} \mathbf{v} + P^{-1}\mathbf{f}$$

$$= \begin{pmatrix} 2 & 0 \\ 0 & 3 \end{pmatrix} \mathbf{v} + \begin{pmatrix} 2 \\ -2 \end{pmatrix} e^t + \begin{pmatrix} 2 \\ -1 \end{pmatrix} e^{-2t}.$$

The solution of *this* system is the sum of the complementary solution,

$$\mathbf{v}_c = c_1 \begin{pmatrix} 1 \\ 0 \end{pmatrix} e^{2t} + c_2 \begin{pmatrix} 0 \\ 1 \end{pmatrix} e^{3t},$$

and a particular solution. To find the latter, we assume that

$$\mathbf{v}_p = u_1 \begin{pmatrix} 1 \\ 0 \end{pmatrix} e^{2t} + u_2 \begin{pmatrix} 0 \\ 1 \end{pmatrix} e^{3t};$$

then, by Eq. (4.5–12),

$$\dot{u}_1 \binom{1}{0} e^{2t} + \dot{u}_2 \binom{0}{1} e^{3t} = \binom{2}{-2} e^t + \binom{2}{-1} e^{-2t}.$$

We readily find the solutions to these to be

$$\dot{u}_1 = 2e^{-t} + 2e^{-4t},$$
$$\dot{u}_2 = -2e^{-2t} - e^{-5t}.$$

Integration produces

$$u_1 = -2e^{-t} - \tfrac{1}{2}e^{-4t},$$
$$u_2 = e^{-2t} + \tfrac{1}{5}e^{-5t},$$

so that the particular solution is

$$\mathbf{v}_p = \binom{-2}{1} e^t + \frac{1}{10} \binom{-5}{2} e^{-2t}.$$

Thus

$$\mathbf{v} = c_1 \binom{1}{0} e^{2t} + c_2 \binom{0}{1} e^{3t} + \binom{-2}{1} e^t + \frac{1}{10} \binom{-5}{2} e^{-2t}$$

and

$$\mathbf{x} = P\mathbf{v} = c_1 \binom{1}{1} e^{2t} + c_2 \binom{2}{1} e^{3t} + \binom{0}{-1} e^t + \frac{1}{10} \binom{-1}{-3} e^{-2t}$$

is the general solution of the given system. ■

It should be pointed out that the Laplace transform methods discussed in Section 3.6 are also applicable to the types of systems discussed in this section.

If the coefficient matrix A in $\dot{\mathbf{x}} = A\mathbf{x}$ or $\dot{\mathbf{x}} = A\mathbf{x} + \mathbf{f}(t)$ is not constant but a function of t, say $A(t)$, then the work required to obtain a solution may be very much greater. In these cases the numerical methods presented in the next section may have more appeal. In this connection see also the material in Section 2.6.

Systems of differential equations may originate in problems having more than one dependent variable. An example of such a problem is the mass–spring system, shown in Fig. 4.5–1. Here m_1 and m_2 are masses suspended through springs having spring constants k_1 and k_2. The displacements of the two masses from points of equilibrium are given by y_1 and y_2 with positive displacements being measured downward. We assume that the masses can be considered as being concentrated at two points, the springs are weightless, there is no damping present, and the motions are such that the elastic limits of the springs are not exceeded. The system can be put into oscillation by specifying initial displacements and/or initial velocities of the two masses or by applying a forcing function to one or both of the masses.

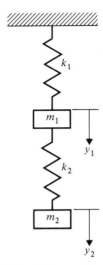

FIGURE 4.5–1 Mass–spring system.

Using Newton's second law and considering the forces acting on each mass we can write the equations

$$m_1\ddot{y}_1 = k_2(y_2 - y_1) - k_1 y_1,$$
$$m_2\ddot{y}_2 = -k_2(y_2 - y_1).$$

Here $y_2 - y_1$ is the *relative* displacement of the two masses. The equations can be written

$$\ddot{y}_1 = -\frac{(k_1 + k_2)}{m_1} y_1 + \frac{k_2}{m_1} y_2,$$

$$\ddot{y}_2 = \frac{k_2}{m_2} y_1 - \frac{k_2}{m_2} y_2,$$

or in matrix form $\ddot{\mathbf{y}} = A\mathbf{y}$, where

$$\mathbf{y} = \begin{pmatrix} y_1 \\ y_2 \end{pmatrix} \quad \text{and} \quad A = \begin{pmatrix} -\dfrac{(k_1 + k_2)}{m_1} & \dfrac{k_2}{m_1} \\ \dfrac{k_2}{m_2} & -\dfrac{k_2}{m_2} \end{pmatrix}.$$

This system differs from the ones we have considered in this section in that second derivatives are involved instead of first. An educated guess, based on what we did in Section 2.1, can reduce this problem to one that has already been considered. We make the substitutions $\mathbf{y} = \mathbf{x} \exp(\omega t)$, $\ddot{\mathbf{y}} = \omega^2 \mathbf{x} \exp(\omega t)$, so that $\ddot{\mathbf{y}} = A\mathbf{y}$ becomes $A\mathbf{x} = \omega^2 \mathbf{x}$. Thus we need to find the eigenvalues and corresponding eigenvectors of the matrix A in order to obtain a solution of the system. The details of this are left for the exercises (see Exercise 15).

KEY WORDS AND PHRASES

Wronskian   **fundamental set of solutions**

EXERCISES 4.5

▶ 1. Given the matrix

$$A = \begin{pmatrix} 1 & 1 & 1 \\ 2 & 1 & -1 \\ -8 & -5 & -3 \end{pmatrix},$$

find its eigenvalues and corresponding eigenvectors.

2. Find the inverse of the matrix

$$P = \begin{pmatrix} -4 & -3 & 0 \\ 5 & 4 & 1 \\ 7 & 2 & -1 \end{pmatrix}.$$

3. In Eq. (4.5–3) use the initial value $\mathbf{x}(0) = (2, -12, 24)^T$ to find the values of c_1, c_2, and c_3.

4. Show that $\exp(2t)$ and $t \exp(2t)$ are linearly independent for all t. (*Hint*: Differentiate $c_1 \exp(2t) + c_2 t \exp(2t) = 0$ to obtain a second equation.)

5. In Example 4.5–2 show that $\mathbf{b} = (3, 1, -\frac{3}{2})^T$ is a solution of

$$(A - 2I)\mathbf{b} = (3, 1, -2)^T.$$

6. Show that the set of vectors

$$\{(4, 1, -3), (3, 1, -2), (3, 1, -\tfrac{3}{2})\}$$

is a linearly independent set.

7. Carry out the computational details in Example 4.5–3.

8. Verify that both Eqs. (4.5–8) and (4.5–9) satisfy the system of Example 4.5–3.

9. Obtain Eq. (4.5–8) from Eq. (4.5–9).

10. In Example 4.5–4 find $\mathbf{a}$ and $\mathbf{b}$.

11. Obtain the complementary solution for the problem of Example 4.5–4.

12. Find the general solution of the system

$$\dot{\mathbf{x}} = \begin{pmatrix} -1 & 1 \\ -5 & 3 \end{pmatrix} \mathbf{x}.$$

(Compare Example 4.5–5.)

13. Carry out the computational details in Example 4.5–5.

14. In Example 4.5–6 find $\dot{u}_1$, $\dot{u}_2$, u_1, and u_2.

15. Find the eigenvalues and corresponding eigenvectors for the two-degree-of-freedom system shown in Fig. 4.5–1 for the special case $k_1 = k_2 = k$ and $m_1 = m_2 = m$.

▶▶ **16.** Find the general solution of each of the following homogeneous systems.

   a) $\dot{\mathbf{x}} = \begin{pmatrix} 3 & 2 \\ 1 & 2 \end{pmatrix}\mathbf{x}$
     b) $\dot{\mathbf{x}} = \begin{pmatrix} 4 & -4 \\ 12 & 17 \end{pmatrix}\mathbf{x}$
     c) $\dot{\mathbf{x}} = \begin{pmatrix} 1 & -2 \\ -4 & -1 \end{pmatrix}\mathbf{x}$

   d) $\dot{\mathbf{x}} = \begin{pmatrix} 1 & 0 & 0 \\ 2 & 1 & -2 \\ 3 & 2 & 1 \end{pmatrix}\mathbf{x}$
   e) $\dot{\mathbf{x}} = \begin{pmatrix} 1 & 1 & 1 \\ 2 & 1 & -1 \\ -8 & -5 & -3 \end{pmatrix}\mathbf{x}$

17. Solve each of the following initial-value problems.

   a) $\dot{\mathbf{x}} = \begin{pmatrix} 1 & 1 \\ 0 & 1 \end{pmatrix}\mathbf{x}, \quad \mathbf{x}(0) = (1, 0)^T$

   b) $\dot{\mathbf{x}} = \begin{pmatrix} 4 & 1 \\ -8 & 8 \end{pmatrix}\mathbf{x}, \quad \mathbf{x}\left(\frac{\pi}{4}\right) = (0, 1)^T$

   c) $\dot{\mathbf{x}} = \begin{pmatrix} 1 & 3 \\ 1 & -1 \end{pmatrix}\mathbf{x}, \quad \mathbf{x}(0) = (4, 12)^T$

   d) $\dot{\mathbf{x}} = \begin{pmatrix} 6 & -7 \\ 1 & -2 \end{pmatrix}\mathbf{x}, \quad \mathbf{x}(0) = (-2, 1)^T$

   e) $\dot{\mathbf{x}} = \begin{pmatrix} 1 & -1 & 4 \\ 3 & 2 & -1 \\ 2 & 1 & -1 \end{pmatrix}\mathbf{x}, \quad \mathbf{x}(0) = (3, -5, 0)^T$

18. Solve each of the following nonhomogeneous systems.

   a) $\dot{\mathbf{x}} = \begin{pmatrix} 1 & 3 \\ 1 & -1 \end{pmatrix}\mathbf{x} + \begin{pmatrix} \sin t \\ -\cos t \end{pmatrix}$
     b) $\dot{\mathbf{x}} = \begin{pmatrix} 2 & -1 \\ 3 & -2 \end{pmatrix}\mathbf{x} + \begin{pmatrix} 1 \\ 2 \end{pmatrix}t$

   c) $\dot{\mathbf{x}} = \begin{pmatrix} 2 & 1 \\ 4 & -1 \end{pmatrix}\mathbf{x} + \begin{pmatrix} 1 \\ -1 \end{pmatrix}e^t$
     d) $\dot{\mathbf{x}} = \begin{pmatrix} -4 & 2 \\ 2 & 1 \end{pmatrix}\mathbf{x} + \begin{pmatrix} t \\ 3 \end{pmatrix}e^{2t}$

19. In the two-degree-of-freedom mass–spring system shown in Fig. 4.5–1 make the substitutions $\dot{y}_1 = y_3$ and $\dot{y}_2 = y_4$, and then write the equations in the form $\dot{\mathbf{y}} = A\mathbf{y}$.

20. In Exercise 19 make the substitutions $m_1 = m_2 = 1$, $k_1 = 2$ and $k_2 = 3$, and then obtain the general solution.

21. Obtain the solution to each of the following systems.

   a) $\dot{\mathbf{x}} = \begin{pmatrix} 2 & 3 \\ 1 & 4 \end{pmatrix}\mathbf{x}$
     b) $\dot{\mathbf{x}} = \begin{pmatrix} 6 & -7 \\ 1 & -2 \end{pmatrix}\mathbf{x}$

   c) $\dot{\mathbf{x}} = \begin{pmatrix} 2 & -1 \\ 3 & -2 \end{pmatrix}\mathbf{x}$
     d) $\dot{\mathbf{x}} = \begin{pmatrix} -3 & 8 \\ 2 & 3 \end{pmatrix}\mathbf{x}$

22. Find the general solution of each of the following systems.

a) $\dot{\mathbf{x}} = \begin{pmatrix} -1 & 0 & 1 \\ 0 & -1 & 1 \\ 1 & 1 & 0 \end{pmatrix} \mathbf{x}$

b) $\dot{\mathbf{x}} = \begin{pmatrix} 1 & -1 & 4 \\ 3 & 2 & -1 \\ 2 & 1 & -1 \end{pmatrix} \mathbf{x}$

23. Solve each of the following systems.

a) $\dot{\mathbf{x}} = \begin{pmatrix} -2 & -3 \\ 3 & 4 \end{pmatrix} \mathbf{x}$

b) $\dot{\mathbf{x}} = \begin{pmatrix} 3 & -2 \\ 5 & -3 \end{pmatrix} \mathbf{x}$

c) $\dot{\mathbf{x}} = \begin{pmatrix} 1 & -1 \\ 1 & 3 \end{pmatrix} \mathbf{x}$

d) $\dot{\mathbf{x}} = \begin{pmatrix} 4 & 2 \\ -8 & -4 \end{pmatrix} \mathbf{x}$

24. Obtain the general solution of each of the following systems.

a) $\dot{\mathbf{x}} = \begin{pmatrix} 1 & 1 & 1 \\ 2 & 1 & -1 \\ 0 & -1 & 1 \end{pmatrix} \mathbf{x}$

b) $\dot{\mathbf{x}} = \begin{pmatrix} 1 & 1 & 1 \\ 2 & 1 & -1 \\ -3 & 2 & 4 \end{pmatrix} \mathbf{x}$

▶▶▶ **25.** Show that the system

$$t\dot{\mathbf{x}} = \begin{pmatrix} a_{11} & a_{12} \\ a_{21} & a_{22} \end{pmatrix} \mathbf{x} = A\mathbf{x}$$

has a nontrivial solution $\mathbf{x} = \mathbf{c}t^{\lambda}$, $t > 0$, whenever λ is an eigenvalue of A.

26. Use the result of Exercise 25 to obtain the general solution for $t > 0$ of each of the following systems.

a) $t\dot{\mathbf{x}} = \begin{pmatrix} 5 & -1 \\ 3 & 1 \end{pmatrix} \mathbf{x}$

b) $t\dot{\mathbf{x}} = \begin{pmatrix} 2 & 1 \\ -2 & -2 \end{pmatrix} \mathbf{x}$

27. A particle moves in $\mathbb{R}^2$ in accordance with

$$\dot{\mathbf{x}} = \begin{pmatrix} 0 & a \\ -a & 0 \end{pmatrix} \mathbf{x},$$

where a is a positive constant.

a) Obtain the parametric equations of its path.

b) Show that the path is a circle with center at the origin.

28. A particle moves in $\mathbb{R}^3$ in accordance with

$$\dot{\mathbf{x}} = \begin{pmatrix} 0 & -1 & -2 \\ 1 & 0 & 1 \\ 2 & -1 & 0 \end{pmatrix} \mathbf{x}.$$

Find the parametric equations of its path.

29. a) Write the general solution of Example 4.5–3 in the form

$$\mathbf{x} = (k_2 - k_1)\mathbf{u}_1 + i(k_1 + k_2)\mathbf{u}_2,$$

where $\mathbf{u}_1$ and $\mathbf{u}_2$ are real.

b) Show that $\mathbf{u}_1$ and $\mathbf{u}_2$ each satisfy the given equation.

c) Show that $\mathbf{u}_1$ and $\mathbf{u}_2$ are linearly independent solutions of the given system.

d) Conclude that Eq. (4.5–8) with c_1 and c_2 real is the general solution of the system.

4.6 NUMERICAL METHODS

In Exercise 26 of Section 4.2 we hinted at the types of systems of linear algebraic equations one is likely to meet in practice. The entries in the coefficient matrix may be approximate numbers, numbers that have been obtained from data acquired in an experiment. When these approximate numbers are used in numerous computations, as they are when a matrix is reduced to echelon form, small errors may be magnified. It then becomes a matter of concern whether the final result is meaningful. It is the province of the numerical analyst to study problems of this kind in depth. One of our objectives in this section is to point out some pitfalls in using numerical methods. The information presented here should be of special interest to the applied mathematician since he or she often obtains solutions to problems from someone using a computer.

Systems of linear algebraic equations

We begin by examining the various methods of solving a system of linear algebraic equations. The time-honored *Cramer's rule* appears to be the most simple to use if it is applicable (that is, if the system of equations has a unique solution). This method requires that determinants be evaluated and herein lies the big disadvantage. Evaluating a determinant by expanding it in terms of cofactors is extremely inefficient. It has been estimated* that it requires about 70,000,000 multiplications and divisions to solve a system of ten equations by Cramer's rule. Clearly, large systems of equations would require an excessive amount of computer time. One way to cut down on this time is to use elementary row operations and thus transform the matrices involved to triangular form. The determinant of a triangular matrix is the product of the diagonal terms, so Cramer's rule becomes practical when matrices are in triangular form.

If a system is to be reduced to triangular form, however, it is just as simple to use *Gaussian elimination*. This method is not limited to those systems having unique solutions and is quite easy to program since only the three elementary row operations are used together with some logic. The method requires back substitution and it would appear that the *Gauss–Jordan reduction* method, which gives the results explicitly, would be more efficient. It is estimated, however, that for large n (n being the number of variables) Gauss–Jordan reduction requires about 50 percent more operations than does Gaussian elimination.[†]

---

* See Curtis F. Gerald, *Applied Numerical Analysis*, 2d ed., p. 99 (Reading, Mass.: Addison-Wesley, 1978).
[†] See Anthony Ralston and Philip Rabinowitz, *A First Course in Numerical Analysis*, 2d ed., p. 418 (New York: McGraw-Hill, 1978).

But even Gaussian elimination can be subject to errors. In some cases the coefficients in a system of equations are such that the results are particularly sensitive to roundoff. When this happens, we say the system is **ill-conditioned**, which means roughly that the matrix is "almost" singular, that is, its determinant is "nearly" zero. We will not attempt to define the exact meaning of the words in quotation marks. For a more complete discussion one of the books listed in the references at the end of this chapter may be consulted.

The Gaussian elimination method produces a row echelon form that requires that each nonzero row have unity as its first nonzero element. This, in turn, requires a division of the elements of a row by a nonzero number and, if this number is nearly zero, such division can produce meaningless results. There is a useful strategy called **pivoting**, which avoids such small divisors by rearranging the equations in such a way that the coefficient of largest magnitude appears on the diagonal at each step.

Another source of error is due to the fact that a set of equations may involve relationships between quantities that are measured in widely different units. Thus some of the equations may have very large coefficients and others may have very small ones. This can be remedied by **scaling**, that is, dividing each row by the magnitude of its largest coefficient.

Matrix inversion is another method that can be used for solving a system of equations if the coefficient matrix is nonsingular. If the matrix is "nearly" singular, then problems previously mentioned can occur. Inverting a matrix by using Gauss–Jordan reduction carries with it the disadvantages of that method. In this connection we should mention that computing the inverse by Gaussian elimination and by Gauss–Jordan reduction requires roughly the same number of operations. This is a somewhat unexpected result in view of the greater efficiency of Gaussian elimination for solving a system of equations. The following comparative figures have been given by Steinberg.* All figures refer to solving a system of ten equations in ten unknowns, $A\mathbf{x} = \mathbf{b}$.

	Number of additions	Number of multiplications
a) Compute $|A|$.	285	339
b) Gaussian elimination	375	430
c) Gauss–Jordan reduction	450	595
d) Compute A^{-1} by Gaussian elimination.	810	1000

* David I. Steinberg, *Computational Matrix Algebra*, p. 103 (New York: McGraw-Hill, 1974).

The methods for solving a system of equations discussed so far fall into a category that can be called "*direct* methods." The name implies that we arrive at a result after a definite number of arithmetic and logical operations. An entirely different approach is available in one of the "*iterative* methods." In using one of these, an estimate is successively refined until it is determined either that it qualifies as a solution or that the method fails to converge.

The most simple iterative procedure is called the **Jacobi iteration method**,* first published in 1846. We shall describe this method and then give an example. Given a set of equations, we rearrange them in such a way that the elements on the diagonal have magnitudes as large as possible relative to the magnitudes of other coefficients in the same row (this is called "pivoting"). We assume that the equations have been previously "scaled." Next we solve the first of the n equations for x_1, the second for x_2, etc. Thus the ith variable becomes

$$x_i = \frac{1}{a_{ii}} \left(b_i - \sum_{\substack{j=1 \\ j \neq i}}^{n} a_{ij} x_j \right), \qquad i = 1, 2, \ldots, n. \qquad \textbf{(4.6-1)}$$

Then we choose (guess) initial values of the variables, call them $x_1^{(0)}$, $x_2^{(0)}, \ldots, x_n^{(0)}$, substitute these into the right-hand side of Eq. (4.6–1), and compute new values $x_1^{(1)}, x_2^{(1)}, \ldots, x_n^{(1)}$. Continuing in this fashion we obtain a new set of values for the x_i from Eq. (4.6–1). When the $(k+1)$st set agrees with the kth set to within some predetermined error, the process may be stopped. We may rewrite Eq. (4.6–1) so that it has the appearance of an iterative formula:

$$x_i^{(k+1)} = \frac{1}{a_{ii}} \left(b_i - \sum_{\substack{j=1 \\ j \neq i}}^{n} a_{ij} x_j^{(k)} \right), \qquad i = 1, 2, \ldots, n. \qquad \textbf{(4.6-2)}$$

EXAMPLE 4.6–1 Solve the system

$$
\begin{aligned}
8x_1 + x_2 - x_3 &= 8 \\
2x_1 + x_2 + 9x_3 &= 12 \\
x_1 - 7x_2 + 2x_3 &= -4,
\end{aligned}
$$

by the Jacobi method to three decimal places.

Solution. After pivoting and putting the equations in the form (4.6–1) we have (see Exercise 1)

$$
\begin{aligned}
x_1 &= 1 && - 0.125x_2 + 0.125x_3 \\
x_2 &= 0.571 + 0.143x_1 && + 0.286x_3 \\
x_3 &= 1.333 - 0.222x_1 - 0.111x_2.
\end{aligned}
$$

---

* After Carl G. J. Jacobi (1804–1851), a German mathematician.

The most simple initial guess is $x_1^{(0)} = x_2^{(0)} = x_3^{(0)} = 0$. Results are given in the following table (see Exercise 2).

$k =$	0	1	2	3	4	5	6	7
$x_1^{(k)}$	0	1.000	1.095	0.995	0.993	1.002	1.001	1.000
$x_2^{(k)}$	0	0.571	1.095	1.026	0.990	0.998	1.001	1.000
$x_3^{(k)}$	0	1.333	1.048	0.969	1.000	1.004	1.001	1.000

The example above appears in Gerald's *Applied Numerical Analysis* (op. cit., p. 144). Steinberg (op. cit., p. 114) gives a *sufficient* condition for convergence of the Jacobi method, namely, that $\mathbf{x}^{(k+1)} = \mathbf{b} + A\mathbf{x}^{(k)}$ will converge if

$$\max_{\text{for } i = 1, 2, \ldots, n} \left\{ \sum_{j=1}^{n} |a_{ij}| \right\} < 1. \tag{4.6–3}$$

In other words, the sum of the absolute values of the elements in row i must be less than 1 for each value of i to ensure convergence.

The Jacobi method is also known as the *method of simultaneous displacements* since every element in the solution vector is computed before these values are used in the next iteration. An obvious improvement would be to use each x_i as soon as it is available in order to compute the remaining elements of the solution vector. Such a method is called the *method of successive displacements* or, more commonly, the *Gauss–Seidel*[*] method. This method usually converges faster than the Jacobi method. It has been shown (Ralston and Rabinowitz, op. cit., p. 445) that the Gauss–Seidel method converges independently of the initial guess if the matrix of coefficients is **positive definite**, that is, one whose eigenvalues are all positive. This last, in turn, is true if the absolute value of the diagonal element in each row exceeds the sum of the absolute values of the other elements, that is, if

$$|a_{ii}| > \sum_{\substack{j=1 \\ j \neq i}}^{n} |a_{ij}|, \qquad i = 1, 2, \ldots, n.$$

More generally, Faddeeva[†] has proved that a *necessary and sufficient* condition that an iterative process converge with any initial vector and any constant vector $\mathbf{b}$ (where $A\mathbf{x} = \mathbf{b}$) is that all eigenvalues of A be less than unity in absolute value. She also gives a method for improving the iterative process.

---

[*] After P. L. Seidel (1821–1896), who published it in 1874 on the basis of an idea due to Gauss.

[†] V. N. Faddeeva, *Computational Methods of Linear Algebra*, p. 118 (New York: Dover, 1959).

Eigenvalues and eigenvectors

In Section 4.4 we found the eigenvalues and corresponding eigenvectors of a square matrix A. In all cases the matrices were no larger than 3×3 and the eigenvalues were integers. We thus imposed restrictions (for pedagogical reasons) that seldom exist in practice. We shall now look at some features of the eigenvalue problem that will enable us to strengthen our computational techniques.

Of special significance is the sum of the diagonal elements of a matrix A. This sum is called the **trace** of A, denoted by

$$\text{tr}\,(A) = \sum_{i=1}^{n} a_{ii}.$$

It can be shown* that if $\lambda_1, \lambda_2, \ldots, \lambda_n$ are the eigenvalues of A (not necessarily distinct), then

$$\text{tr}\,(A) = \sum_{i=1}^{n} \lambda_i \quad \text{and} \quad |A| = \prod_{i=1}^{n} \lambda_i, \qquad \textbf{(4.6–4)}$$

where π denotes "product." Knowing that the sum of the eigenvalues is equal to the trace and that their product is equal to the determinant is helpful in finding the roots of the characteristic equation of a matrix.

We have already mentioned the preponderance of *symmetric* matrices in applied mathematics. Their properties are given in the following theorem, which we state without proof.

Theorem 4.6–1 Let A be a real, $n \times n$ symmetric matrix. Then:

a) the eigenvalues and eigenvectors of A are real;

b) eigenvectors belonging to distinct eigenvalues are orthogonal;

c) A is diagonalizable, that is, there exists a matrix P such that $P^T A P = \text{diag}\,(\lambda_1, \lambda_2, \ldots, \lambda_n)$, where the λ_i are eigenvalues of A.

An estimate of the *largest* eigenvalue is of value, as we shall presently see. It can be shown[†] that if λ_1 is that eigenvalue of A which has largest absolute value, then

$$|\lambda_1| \leq \max_{i=1,\,2,\,\ldots,\,n} \left\{ \sum_{j=1}^{n} |a_{ij}| \right\}.$$

This criterion is valuable for obtaining an initial estimate of the eigenvalue of largest magnitude whenever an iterative process is used.

---

* See Peter Lancaster, *Theory of Matrices*, p. 55 (New York: Academic Press, 1969).
† See Steinberg, op. cit., p. 253.

A scheme for obtaining the eigenvalue of largest magnitude (also called the **dominant** eigenvalue) known as the **power method** is originally (1929) due to von Mises.* If λ_1 is the eigenvalue sought, x_1 is the corresponding eigenvector, and $v^{(0)}$ is an arbitrary vector, then we can write the equations

$$A v^{(0)} = c_1 v^{(1)},$$
$$A v^{(1)} = c_2 v^{(2)},$$
$$\vdots$$
$$A v^{(m)} = c_{m+1} v^{(m+1)}.$$

Usually $v^{(0)}$ is chosen with unity in the ith place and zeros everywhere else. Then c_1 is such that $v^{(1)}$ also has unity in the ith place. The iterative process can be terminated when $v^{(m)} = v^{(m+1)}$ to some desired degree of accuracy, and then c_{m+1} is an approximation to λ_1. An example will illustrate the method.

EXAMPLE 4.6–2 Find the dominant eigenvalue of the symmetric matrix

$$A = \begin{pmatrix} 5 & -2 & 0 \\ -2 & 3 & -1 \\ 0 & -1 & 1 \end{pmatrix},$$

carrying the computations to two decimals.

Solution. Assume that $v^{(0)} = (1, 0, 0)^T$. Then $A v^{(0)} = (5, -2, 0)^T = 5(1, -0.4, 0)^T = 5 v^{(1)}$, $A v^{(1)} = (5.8, -3.2, 0.4)^T = 5.8(1, -0.5517, 0.0690)^T = 5.8 v^{(2)}$, and so on. The table below shows the values of the c_i and $v^{(i)}$. Hence the desired eigenvalue is 6.29 and the corresponding eigenvector is $(1, -0.65, 0.12)^T$. ■

$i =$	0	1	2	3	4	5	6	7	8
c_i	—	5.	5.8	6.104	6.220	6.264	6.280	6.286	6.288
$v^{(i)}$	$\begin{pmatrix}1\\0\\0\end{pmatrix}$	$\begin{pmatrix}1\\-0.4\\0\end{pmatrix}$	$\begin{pmatrix}1\\-0.552\\0.069\end{pmatrix}$	$\begin{pmatrix}1\\-0.610\\0.102\end{pmatrix}$	$\begin{pmatrix}1\\-0.632\\0.115\end{pmatrix}$	$\begin{pmatrix}1\\-0.640\\0.119\end{pmatrix}$	$\begin{pmatrix}1\\-0.643\\0.121\end{pmatrix}$	$\begin{pmatrix}1\\-0.644\\0.122\end{pmatrix}$	$\begin{pmatrix}1\\-0.645\\0.122\end{pmatrix}$

The power method has certain disadvantages. There may be trouble with the iterative process when two eigenvalues are very close but not identically equal. This situation can be altered by adding a fixed number to each eigenvalue. This does not change the eigenvectors,

---

*Richard E. von Mises (1883–1953), a German applied mathematician.

but improves the convergence of the process. For example, if the eigenvalues are -4.9, 2, and 5, then adding 3 to each produces -1.9, 5, and 8, so that the two that were close in magnitude are no longer close. The justification for this arbitrary addition of a number to the eigenvalues lies in the fact that if $A\mathbf{x} = \lambda\mathbf{x}$, then $A\mathbf{x} = (\lambda + c)\mathbf{x} - c\mathbf{x}$ or $(A + cI)\mathbf{x} = (\lambda + c)\mathbf{x}$ and this is the same as the original eigenvalue problem with the matrix and eigenvalues changed in a simple manner. This scheme can also be used to find an intermediate eigenvalue (Exercise 17). Choosing $\mathbf{v}^{(0)}$ to be as close as possible to the correct eigenvector will also speed convergence, but this may not be possible.

In some problems, such as the analysis of the buckling of a structure, the eigenvalue of *largest* magnitude is of interest and we can obtain this as outlined above. In vibration problems, however, the eigenvalue of *smallest* magnitude is of interest. This can be found by applying the foregoing method to A^{-1} since $A\mathbf{x} = \lambda\mathbf{x}$ is equivalent to $A^{-1}\mathbf{x} = (1/\lambda)\mathbf{x}$. In other words, the eigenvalue of largest magnitude belonging to A^{-1} is identical to the eigenvalue of smallest magnitude belonging to A. This, of course, is applicable only if A is nonsingular.

Other eigenvalues (and eigenvectors) may be obtained by using a property of real symmetric matrices, namely, that eigenvectors belonging to distinct eigenvalues are orthogonal. This property permits us to state a new problem involving a matrix that does not have the eigenvector already found.*

KEY WORDS AND PHRASES

ill-conditioning	positive definite matrix
pivoting	trace of a matrix
scaling	dominant eigenvalues
Jacobi iteration method	power method

EXERCISES 4.6

▶ 1. Use pivoting and scaling to put the system of Example 4.6–1 into the form (4.6–1).

2. Carry out the computations in Example 4.6–1 to obtain the results shown in the table.

3. Obtain the successive eigenvectors shown in the table for Example 4.6–2.

---

* Details of the process can be found in Bruce W. Arden and Kenneth N. Astill, *Numerical Algorithms: Origins and Applications*, p. 206ff. (Reading, Mass.: Addison-Wesley, 1970), and in other works listed at the end of this chapter.

▶▶ 4. Solve the system

$$2x_1 + x_2 \qquad\quad = 2$$
$$x_1 + 4x_2 + 2x_3 = 0$$
$$2x_2 + 4x_3 = 0$$

by using

a) Gaussian elimination;

b) Jacobi's method (four iterations);

c) the Gauss–Seidel method (four iterations).

Note: In parts (b) and (c) begin with $x_1 = x_2 = x_3 = 0$.

5. Solve the system

$$5x_1 - 4x_2 \qquad\quad = 2$$
$$-4x_1 + 10x_2 - 5x_3 = 0$$
$$- 5x_2 + 6x_3 = -1$$

by using

a) matrix inversion;

b) Gaussian elimination;

c) the Gauss–Seidel method.

6. Solve the following system by using Gaussian elimination and carrying all computations to three decimals.

$$3x_1 + 2x_2 - x_3 + 2x_4 = -2$$
$$x_1 + 4x_2 \qquad + 2x_4 = 1$$
$$2x_1 + x_2 + 2x_3 - x_4 = 3$$
$$x_1 + x_2 - x_3 + 3x_4 = 4$$

7. Solve the following system by using Gauss–Jordan reduction and carrying all computations to three decimals.

$$x_1 + 2x_2 - 2x_3 = -3$$
$$2x_1 - 4x_2 + 4x_3 = 0$$
$$8x_1 - 6x_2 + 2x_3 = 4$$

8. Given the following system

$$\begin{pmatrix} 3 & 2 & 100 \\ -1 & 3 & 100 \\ 1 & 2 & -1 \end{pmatrix} \begin{pmatrix} x_1 \\ x_2 \\ x_3 \end{pmatrix} = \begin{pmatrix} 105 \\ 102 \\ 2 \end{pmatrix}.$$

a) Solve by using Gaussian elimination carrying two decimals.

b) *Scale* the equations by dividing each row by the magnitude of the largest coefficient, then solving as in part (a).

9. Given the following system

$$\begin{pmatrix} -0.002 & 4.000 & 4.000 \\ -2.000 & 2.906 & -5.387 \\ 3.000 & -4.031 & -3.112 \end{pmatrix} \mathbf{x} = \begin{pmatrix} 7.998 \\ -4.481 \\ -4.143 \end{pmatrix}.$$

a) Solve by using Gaussian elimination using four-digit arithmetic, that is, carrying a total of four digits in each number.

b) Use *pivoting*, that is, rewrite the equations in the order third, first, second, and repeat part (a).

c) Explain the discrepancy between the two sets of results.

10. Given the following system

$$\begin{pmatrix} 3.02 & -1.05 & 2.53 \\ 4.33 & 0.56 & -1.78 \\ -0.83 & -0.54 & 1.47 \end{pmatrix} \mathbf{x} = \begin{pmatrix} -1.61 \\ 7.23 \\ -3.38 \end{pmatrix}.$$

a) Find the value of the determinant of the coefficient matrix.

b) Solve the system by using Gaussian elimination and three-digit arithmetic. (*Note*: It may be necessary to carry six decimals in order to have three *significant* digits—for example, 0.000362.)

c) The *ill-conditioning* exhibited here can be improved by using greater precision in the arithmetic operations. Repeat part (b) using six-digit arithmetic.

11. Consider the system of Exercise 10, which has an exact solution: $\mathbf{x} = (1, 2, -1)^T$. Show that the erroneous solution

$$\mathbf{x} = (0.880, -2.35, -2.66)^T$$

"almost" satisfies the given system, thus demonstrating an interesting phenomenon of an ill-conditioned system.

12. For the system $A\mathbf{x} = \mathbf{b}$, where

$$A = \begin{pmatrix} 2.38 & -1.42 & 3.24 \\ 1.36 & 2.54 & -1.62 \\ -1.82 & 3.65 & 1.81 \end{pmatrix} \quad \text{and} \quad \mathbf{b} = \begin{pmatrix} 1.11 \\ 1.97 \\ 2.42 \end{pmatrix},$$

find each of the following to three decimals.

a) $|A|$

b) $\mathbf{x}$ using Cramer's rule

c) A^{-1}

d) $\mathbf{x}$ using the Gauss–Seidel method

13. For the system $A\mathbf{x} = \mathbf{b}$, where

$$A = \begin{pmatrix} 5 & -1 & 1 \\ 2 & 4 & 0 \\ 1 & 1 & 5 \end{pmatrix} \quad \text{and} \quad \mathbf{b} = \begin{pmatrix} 10 \\ 12 \\ -1 \end{pmatrix},$$

find $\mathbf{x}$ to three decimals using:

a) Gaussian elimination;

b) Gauss–Jordan reduction;

c) the Gauss–Seidel method beginning with the initial guess $(2, 2, -1)^T$.

14. For the system $A\mathbf{x} = \mathbf{b}$, where

$$A = \begin{pmatrix} 2.51 & 1.48 & 4.53 \\ 1.48 & 0.93 & -1.30 \\ 2.68 & 3.04 & -1.48 \end{pmatrix} \quad \text{and} \quad \mathbf{b} = \begin{pmatrix} 0.05 \\ 1.03 \\ -0.53 \end{pmatrix},$$

find $\mathbf{x}$ to three decimals by:

a) Jacobi iteration;

b) the Gauss–Seidel method.

15. a) Compute the dominant eigenvalue of the matrix

$$A = \begin{pmatrix} 2 & 2 & 2 \\ \frac{2}{3} & \frac{5}{3} & \frac{5}{3} \\ 1 & \frac{5}{2} & \frac{11}{2} \end{pmatrix}.$$

Begin with $(1, 0, 0)^T$ and use three decimals.

b) What is the eigenvector corresponding to the eigenvalue in part (a)?

c) Repeat part (a), this time starting with $(0, 1, 0)^T$.

16. For the matrix

$$A = \begin{pmatrix} 4 & 1 & 0 \\ 0 & 2 & 1 \\ 0 & 0 & -1 \end{pmatrix},$$

find each of the following to three decimals.

a) The eigenvalues and corresponding eigenvectors using the procedure given in Section 4.4

b) The dominant eigenvalue using $(1, 0, 0)^T$ as a starting value

c) A^{-1}

d) The eigenvalue of smallest magnitude using A^{-1} in part (c).

17. For the matrix

$$A = \begin{pmatrix} 1 & 1 & 2 \\ 0 & 1 & 3 \\ 1 & 1 & 1 \end{pmatrix},$$

find the dominant eigenvalue and the corresponding eigenvector. Use the power method and three decimals.

18. In Exercise 17 find A^{-1} and then obtain the eigenvalue of smallest magnitude for A as well as the corresponding eigenvector.

▶▶▶ 19. Find all eigenvalues and corresponding eigenvectors of the matrix

$$\begin{pmatrix} 4 & 2 & 2 \\ 2 & 5 & 1 \\ 2 & 1 & 6 \end{pmatrix}.$$

Use three decimals in the computations.

20. Find the dominant eigenvalue of the matrix

$$\begin{pmatrix} 9 & 10 & 8 \\ 10 & 5 & -1 \\ 8 & -1 & 3 \end{pmatrix},$$

and also the eigenvalue of smallest magnitude.

21. Consider the coefficient matrices of the systems in Exercises 4, 5, 6, and 7. Which of these matrices are positive definite, that is, having all eigenvalues positive?

22. Without actually computing the eigenvalues, obtain as much information as possible about the eigenvalues of the coefficient matrices of the systems in Exercises 4, 5, 6, and 7.

23. Apply the criterion of equation (4.6–3) to the matrix of Example 4.6–1; then explain why the method converges.

24. Referring to Exercise 17, after finding the dominant eigenvalue, subtract that value from each of the diagonal elements and use the power method. The result should be $\lambda = -0.834$, the intermediate eigenvalue, after the subtraction has been offset by a corresponding addition.

25. Describe two other ways in which the intermediate eigenvalue of Exercise 24 can be found, assuming that the solutions to Exercises 17 and 18 are known.

26. A classical example of *ill-conditioning* is given by the Hilbert matrix,[*] which is an infinite[†] matrix. The 4×4 Hilbert matrix is

$$H = \begin{pmatrix} 1 & \frac{1}{2} & \frac{1}{3} & \frac{1}{4} \\ \frac{1}{2} & \frac{1}{3} & \frac{1}{4} & \frac{1}{5} \\ \frac{1}{3} & \frac{1}{4} & \frac{1}{5} & \frac{1}{6} \\ \frac{1}{4} & \frac{1}{5} & \frac{1}{6} & \frac{1}{7} \end{pmatrix}.$$

a) Show that H is "nearly" singular.

b) Find the solution to $H\mathbf{x} = \mathbf{b}$, where $\mathbf{b} = (2.083, 1.283, 0.950, 0.760)^T$, using three significant digits.

c) What can you say about the nature of H^{-1}?

4.7 ADDITIONAL TOPICS

When studying linear algebra it is easy to get caught up in the specifics. For example, we determine whether a *set of vectors* is linearly independent, or whether two vectors are orthogonal, or whether a space spanned by a *set of n vectors* is an *n*-dimensional vector space, etc. The domain of linear algebra, however, is much broader than these examples indicate and in this section we shall attempt to show how the terminology and concepts of this chapter can be extended to cover many topics

---

[*] After David Hilbert (1862–1943), a German mathematician.
[†] An $n \times n$ matrix with n infinite.

in mathematics. This broadening or *generalization* will provide a deeper understanding of mathematics as a whole and will, consequently, simplify further study of the subject. Moreover, the present section will serve to review much of this chapter.

In what follows we will assume that the term "scalar" refers to an element of a *field*. The set of rational numbers, the set of real numbers, and the set of complex numbers, together with the usual definitions of addition and multiplication and the properties pertaining to these operations, are familiar examples of fields. Thus we can speak of a linear combination of vectors in $\mathbb{R}^4$ over the field of real numbers, meaning an expression of the form

$$c_1\mathbf{u} + c_2\mathbf{v} + c_3\mathbf{w},$$

where the c_i are real numbers and $\mathbf{u}$, $\mathbf{v}$, and $\mathbf{w}$ are (row or column) vectors in $\mathbb{R}^4$. For example, $\mathbf{u} = (u_1, u_2, u_3, u_4)$, where the u_i are real numbers. In a similar manner we have a linear combination of $m \times n$ matrices

$$c_1A + c_2B + c_3C + c_4D,$$

where A, B, C, and D are $m \times n$ matrices. We also have

$$c_1f(x) + c_2g(x) + c_3h(x)$$

and

$$c_1P_n(x) + c_2Q_n(x),$$

where f, g, and h are functions of a real variable, continuous on $a \leq x \leq b$, and $P_n(x)$ and $Q_n(x)$ are polynomials in x of degree n. Unless otherwise stated, when we refer to scalars we will mean elements of the field of *real* numbers.

We can now define an *abstract vector space*.

Definition 4.7–1 A set of elements (or objects) V is an abstract vector space if, for any elements $\mathbf{a}$, $\mathbf{b}$, and $\mathbf{c}$ of V, any scalars α, β, and operations $\oplus$ and $\odot$, the following properties hold.

i) $\mathbf{a} \oplus \mathbf{b}$ is an element of V, and

  1. $\mathbf{a} \oplus \mathbf{b} = \mathbf{b} \oplus \mathbf{a}$;
  2. $\mathbf{a} \oplus (\mathbf{b} \oplus \mathbf{c}) = (\mathbf{a} \oplus \mathbf{b}) \oplus \mathbf{c}$;
  3. there exists a unique element $\mathbf{0}$ in V such that $\mathbf{a} \oplus \mathbf{0} = \mathbf{a}$ for every element $\mathbf{a}$ in V;
  4. for every element $\mathbf{a}$ in V there exists a unique element $-\mathbf{a}$ in V such that $\mathbf{a} \oplus (-\mathbf{a}) = \mathbf{0}$.

ii) $\alpha \odot \mathbf{a}$ is an element of V, and

  1. $\alpha \odot \mathbf{a} = \mathbf{a} \odot \alpha$;
  2. $\alpha \odot (\mathbf{a} \oplus \mathbf{b}) = \alpha \odot \mathbf{a} \oplus \alpha \odot \mathbf{b}$;

3. $(\alpha + \beta) \odot \mathbf{a} = \alpha \odot \mathbf{a} \oplus \beta \odot \mathbf{a}$;

4. $(\alpha\beta) \odot \mathbf{a} = \alpha \odot (\beta \odot \mathbf{a})$;

5. $1 \odot \mathbf{a} = \mathbf{a}$ and $0 \odot \mathbf{a} = \mathbf{0}$.

Note first the similarity between Definition 4.7–1 and Definition 4.3–1. Note also that addition of scalars is still denoted by $+$ (for example, $\alpha + \beta$) and multiplication of scalars is still noted by juxtaposition (for example, $\alpha\beta$). The elements of V, however, are not restricted to vectors and for this reason the operation of addition of these elements is denoted by $\oplus$ and scalar multiplication is denoted by $\odot$. If the elements of V are vectors, then Definition 4.7–1 is the same as Definition 4.3–1 and we write $\mathbf{a} + \mathbf{b}$ and $\alpha\mathbf{a}$ since these operations have been defined. In fact, if the elements of V are matrices, then also $\mathbf{a} \oplus \mathbf{b}$ becomes $A + B$ and $\alpha \odot a$ becomes αA since these operations have been defined. In the following example we consider three sets V whose elements are neither vectors nor matrices.

EXAMPLE 4.7–1

a) Let V be the set of all polynomials of degree two (quadratic polynomials) $y = ax^2 + bx + c$ on $[0, 1]$. Then, if $y_1 = a_1 x^2 + b_1 x + c_1$ and $y_2 = a_2 x^2 + b_2 x + c_2$, we define $y_1 \oplus y_2$ to be $(a_1 + a_2)x^2 + (b_1 + b_2)x + (c_1 + c_2)$ and $\alpha \odot y_1$ to be $\alpha a_1 x^2 + \alpha b_1 x + \alpha c_1$. We leave it as an exercise (Exercise 1) to show that V is a vector space. Observe that we can designate y as (a, b, c) so that conceptually V has the same structure as $\mathbb{R}^3$, a real three-dimensional vector space. We say that V and $\mathbb{R}^3$ are **isomorphic**, meaning that the results of operations in one vector space can be uniquely related to the results of operations in the second vector space.

b) Let V consist of all solutions of the second-order, linear, homogeneous differential equation $y'' + y = 0$ with $\oplus$ and $\odot$ defined in a natural way. Since the solutions of this differential equation are functions, we define

$$f(x) \oplus g(x) = (f + g)(x)$$

and

$$\alpha \odot f(x) = (\alpha f)(x).$$

In other words, for each value of x in the domain of definition of f and g, we evaluate $f(x)$ and $g(x)$ and add these values to obtain the value of $f + g$ at that point. The V defined as above satisfies the definition of a vector space (Exercise 2).

c) The set of all continuous functions on $[0, 1]$ can be denoted by $C[0, 1]$. With $\oplus$ and $\odot$ defined as in part (b) this set is a vector space (Exercise 3). We use the results from analysis that the sum of continuous functions is a continuous function and a finite real number times a continuous function is a continuous function. It is understood that all sums and products are taken over the domain of definition—in this case, $[0, 1]$.

d) Consider the set of all vectors in $\mathbb{R}^3$ with $\oplus$ defined as

$$\mathbf{a} \oplus \mathbf{b} = (a_1, a_2, a_3) \oplus (b_1, b_2, b_3) = (a_1 + b_1, a_2 + b_2, 0)$$

and $\odot$ defined as

$$\alpha \odot \mathbf{a} = \alpha \odot (a_1, a_2, a_3) = (\alpha a_1, \alpha a_2, 0).$$

This set of vectors does not form a vector space because, for one thing, the condition

$$1 \odot \mathbf{a} = \mathbf{a}$$

is not satisfied. ■

Our generalization of a vector space shows that the elements of the space need not be vectors in the geometric sense of the word. Thus, given a vector space, we can talk about a *basis* of the space and the *dimension* of the space. For example, consider the vector space whose elements are quadratic polynomials in x:

$$ax^2 + bx + c.$$

A basis for a space whose elements are quadratic polynomials would have to be a set of functions having the property that linear combinations of these functions would generate the quadratic polynomials of the space. Since the scalar coefficients a, b, and c distinguish one quadratic polynomial from another, it is clear that the functions x^2, x, and 1 provide a basis for the space. This last statement implies that (a) *every* quadratic polynomial can be expressed as a linear combination of the basis elements, and (b) the basis elements are linearly independent (Exercises 5 and 6). It now follows that the vector space whose elements are quadratic polynomials has dimension three. (Why?)

We must postpone the question of what constitutes a basis for the set $C[a, b]$. This question will be answered when we study Fourier series in Chapter 7. For the present we state that the basis of the vector space of functions continuous on $[a, b]$ contains an infinite number of elements, that is to say, the vector space is infinite-dimensional.

For *finite*-dimensional vector spaces there is the following powerful theorem, which we state without proof.

Theorem 4.7–1 Every n-dimensional abstract vector space is isomorphic to $\mathbb{R}^n$.

Two vector spaces are isomorphic if there is a one-to-one correspondence between their elements that preserves the operations $\oplus$ and $\odot$. In other words, if V and W are isomorphic vector spaces with

$\mathbf{x}, \mathbf{y}$ as elements of V and $\mathbf{x}'$, $\mathbf{y}'$ the *corresponding* elements of W, then we also have the correspondences

$$\mathbf{x} \oplus \mathbf{y} \leftrightarrow \mathbf{x}' \oplus' \mathbf{y}' \quad \text{and} \quad \alpha \odot \mathbf{x} \leftrightarrow \alpha \odot' \mathbf{x}'.$$

Note that the primes refer to elements *and* operations in W. It may happen that the distinction between $\oplus$ and $\oplus'$ may be a fine one. For example, if the elements of V are four-component vectors, then $\oplus$ refers to addition of these vectors, which is accomplished by adding corresponding components. If W has 2×2 matrices as its elements, then $\oplus'$ refers to addition of these matrices, which is also accomplished by adding corresponding elements of the matrices. More complicated situations are outside the province of this text. In Section 5.4 we will consider sets consisting of orthogonal basis vectors. We will also present a method of obtaining a set of orthogonal vectors from a set of linearly independent ones.

A useful term in linear algebra is given in the following definition.

Definition 4.7–2 The **rank** of a matrix is the number of nonzero rows in its row echelon form.

Using this concept we can state the following.

1. A system of n linear algebraic equations in n unknowns has a unique solution if and only if the rank of the coefficient matrix is n.
2. The equation $AX = B$, where A is $n \times n$ is consistent if and only if the matrix A and the augmented matrix $A \,|\, B$ have the same rank.
3. The equation $AX = B$, where A is $n \times n$ and has rank $r < n$ has an infinite number of solutions, which can be expressed using $n - r$ of the unknowns as arbitrary.
4. If A is an $m \times n$ matrix of rank r, then

$$n = r + \dim (\ker A).$$

By the kernel of A (ker A) we mean the kernel of the linear transformation whose matrix is A.

It is also possible to define the rank of a matrix by using the concept of determinants. This is done in the following definition.

Definition 4.7–3 Let A be an $m \times n$ matrix. Then the rank of A is the order of the largest nonsingular square submatrix of A. (*Note*: An $s \times s$ matrix is said to be of *order s*.)

Since a matrix is nonsingular if and only if its determinant is different from zero, finding the rank of a matrix is the same as finding the largest submatrix that has a nonzero determinant. In most cases, however, Definition 4.7–2 is easier to apply.

Next we continue the discussion of determinants begun in Section 4.2 and present another method for finding the inverse of a nonsingular matrix.

Definition 4.7–4 Let A be an $n \times n$ matrix, $A = (a_{ij})$. The **cofactor** of a_{ij} is $(-1)^{i+j}c_{ij}$, where c_{ij} is the determinant of the $(n-1) \times (n-1)$ submatrix obtained from A by deleting its ith row and jth column. The matrix (c_{ji}), that is, the transpose of (c_{ij}) is called the **adjoint** of A and is denoted by adj A.

EXAMPLE 4.7–2   Find adj A given that

$$A = \begin{pmatrix} 1 & 2 & 3 \\ 2 & 1 & 2 \\ -2 & 1 & -1 \end{pmatrix}.$$

Solution. We have

$$\text{adj } A = \begin{pmatrix} -3 & -2 & 4 \\ 5 & 5 & -5 \\ 1 & 4 & -3 \end{pmatrix}^T.$$

Note that the element in the c_{21} position is obtained from a_{21} in A as follows:

$$c_{21} = (-1)^{2+1} \begin{vmatrix} 2 & 3 \\ 1 & -1 \end{vmatrix} = -(-5) = 5. \quad \blacksquare$$

Sometimes the adjoint of a matrix A is useful in finding A^{-1}, as given in the following theorem whose proof we omit.

Theorem 4.7–2 For any $n \times n$ matrix A we have

$$A(\text{adj } A) = (\text{adj } A)A = I|A|.$$

Hence if A is nonsingular, then

$$A^{-1} = \frac{1}{|A|} (\text{adj } A).$$

EXAMPLE 4.7–3   Find A^{-1} for the matrix of A in Example 4.7–2.

Solution. We had

$$\text{adj } A = \begin{pmatrix} -3 & 5 & 1 \\ -2 & 5 & 4 \\ 4 & -5 & -3 \end{pmatrix}.$$

Since $|A| = 5$ (Exercise 7) we have, by Theorem 4.7–2,

$$A^{-1} = \frac{1}{5} \begin{pmatrix} -3 & 5 & 1 \\ -2 & 5 & 4 \\ 4 & -5 & -3 \end{pmatrix}. \quad \blacksquare$$

Theorem 4.7–2 can be applied to prove the next theorem.

Theorem 4.7–3 (*Cramer's rule*). Given the $n \times n$ system $AX = B$. The unique solution of this system is given by

$$x_k = \frac{|A_k|}{|A|}, \qquad k = 1, 2, \ldots, n,$$

provided that $|A| \neq 0$, where $|A_k|$ is the determinant of the matrix obtained from A by replacing its kth column by the elements of B.

PROOF. Since $|A| \neq 0$, A^{-1} exists, and multiplying $AX = B$ on the left by A^{-1} produces $X = A^{-1}B$. Using Theorem 4.7–2, we can write

$$X = \frac{1}{|A|} (\text{adj } A)B. \tag{4.7–1}$$

Row k of the left member of this equation is x_k. Row k of adj A consists of the cofactors $c_{1k}, c_{2k}, \ldots, c_{nk}$; hence Eq. (4.7–1) can be written

$$x_k = \frac{c_{1k}b_1 + c_{2k}b_2 + \cdots + c_{nk}b_n}{|A|}, \qquad k = 1, 2, \ldots, n. \tag{4.7–2}$$

On the other hand, the matrix A_k constructed from A by replacing the kth column of A by the elements of B has the property that the cofactor of b_j in A_k is the same as the cofactor c_{jk} of a_{jk} in A. Hence the numerator of the right member of Eq. (4.7–2) corresponds to the expansion of A_k by the elements of column k. Thus Eq. (4.7–2) can be written

$$x_k = \frac{|A_k|}{|A|}, \qquad k = 1, 2, \ldots, n$$

as required, completing the proof. $\square$

We conclude this chapter with an explanation of **change of basis**. Although our discussion will be confined to $\mathbb{R}^3$ for simplicity, it will apply to vector spaces in general.

Consider the natural basis for $\mathbb{R}^3$ consisting of the vectors

$$\mathbf{e}_1 = (1, 0, 0), \qquad \mathbf{e}_2 = (0, 1, 0), \qquad \mathbf{e}_3 = (0, 0, 1).$$

In Section 4.3 we showed that the vectors

$$\mathbf{w}_1 = (1, 0, 0), \qquad \mathbf{w}_2 = (1, 1, 0), \qquad \mathbf{w}_3 = (1, 1, 1)$$

are also a basis for $\mathbb{R}^3$. Thus an arbitrary vector in $\mathbb{R}^3$ can be expressed in either basis. To prevent confusion, we will use the subscripts e and w

to indicate the particular basis being used. For example, $(2, 3, 4)_e$ will mean $2\mathbf{e}_1 + 3\mathbf{e}_2 + 4\mathbf{e}_3$, and $(2, 3, 4)_w$ will mean $2\mathbf{w}_1 + 3\mathbf{w}_2 + 4\mathbf{w}_3$. It should be clear that these are two *different* vectors in $\mathbb{R}^3$.

If we know how the basis vectors are transformed, then we know the transformation from one basis to the other in general. Accordingly,

$$(1, 0, 0)_w = (1, 0, 0)_e,$$
$$(0, 1, 0)_w = (1, 0, 0)_e + (0, 1, 0)_e, \qquad \textbf{(4.7-3)}$$
$$(0, 0, 1)_w = (1, 0, 0)_e + (0, 1, 0)_e + (0, 0, 1)_e,$$

expresses the vectors in the w-basis in terms of the vectors in the e-basis. In matrix form this transformation can be written as

$$T\mathbf{x}_w = \mathbf{x}_e, \qquad \textbf{(4.7-4)}$$

where

$$T = \begin{pmatrix} 1 & 1 & 1 \\ 0 & 1 & 1 \\ 0 & 0 & 1 \end{pmatrix}.$$

Note that T is the transpose of the matrix of coefficients in Eqs. (4.7–3).

EXAMPLE 4.7–4 Transform the vector $(2, 3, 4)_w$ into the e-basis.

Solution. Proceeding directly, we have

$$(2, 3, 4)_w = 2(1, 0, 0)_e + 3(1, 1, 0)_e + 4(1, 1, 1)_e$$
$$= (9, 7, 4)_e.$$

Using the matrix T gives us

$$\begin{pmatrix} 1 & 1 & 1 \\ 0 & 1 & 1 \\ 0 & 0 & 1 \end{pmatrix} \begin{pmatrix} 2 \\ 3 \\ 4 \end{pmatrix}_w = \begin{pmatrix} 9 \\ 7 \\ 4 \end{pmatrix}_e. \quad \blacksquare$$

The matrix T is necessarily nonsingular; hence Eq. (4.7–4) can also be written as

$$\mathbf{x}_w = T^{-1}\mathbf{x}_e, \qquad \textbf{(4.7-5)}$$

where (Exercise 7a)

$$T^{-1} = \begin{pmatrix} 1 & -1 & 0 \\ 0 & 1 & -1 \\ 0 & 0 & 1 \end{pmatrix}.$$

In other words, T and T^{-1} allow us to go back and forth between representations in the two bases.

In Exercise 14 of Section 4.3 it was shown that the set

$$\mathbf{v}_1 = (1, 1, 1), \qquad \mathbf{v}_2 = (1, 2, 3), \qquad \mathbf{v}_3 = (2, 2, 0)$$

is a basis for $\mathbb{R}^3$. It can be shown (Exercise 7b) that

$$S\mathbf{x}_v = \mathbf{x}_e,$$

where

$$S = \begin{pmatrix} 1 & 1 & 2 \\ 1 & 2 & 2 \\ 1 & 3 & 0 \end{pmatrix}$$

and (Exercise 7c)

$$\mathbf{x}_v = S^{-1}\mathbf{x}_e,$$

where

$$S^{-1} = \begin{pmatrix} 3 & -3 & 1 \\ -1 & 1 & 0 \\ -\frac{1}{2} & 1 & -\frac{1}{2} \end{pmatrix}.$$

Now it is a simple matter to go from a representation in one basis to another via the natural basis. We have

$$\mathbf{x}_v = S^{-1}\mathbf{x}_e = S^{-1}(T\mathbf{x}_w) = (S^{-1}T)\mathbf{x}_w$$

and

$$\mathbf{x}_w = T^{-1}\mathbf{x}_e = T^{-1}(S\mathbf{x}_v) = (T^{-1}S)\mathbf{x}_v.$$

It also follows that if a linear transformation has matrix A in the natural basis, then it will have matrix $T^{-1}AT$ in the w-basis and $S^{-1}AS$ in the v-basis (Exercise 7d). These ideas are illustrated in the following example.

EXAMPLE 4.7–5   Consider the linear transformation

$$L(x_1, x_2, x_3) = (x_1 + x_2, x_2 + x_3, x_3 + x_1)$$

from $\mathbb{R}^3$ to $\mathbb{R}^3$. Convert the matrix of this transformation to matrices in the w- and v-bases. Then transform $(9, 7, 4)_e$ into each of the other bases.

Solution. By applying the given transformation to each of the basis vectors in the natural basis, we obtain the matrix of the transformation,

$$A = \begin{pmatrix} 1 & 1 & 0 \\ 0 & 1 & 1 \\ 1 & 0 & 1 \end{pmatrix}_e.$$

Then (Exercise 7e)

$$T^{-1}AT = \begin{pmatrix} 1 & 1 & 0 \\ -1 & 0 & 0 \\ 1 & 1 & 2 \end{pmatrix}_w$$

and

$$S^{-1}AS = \begin{pmatrix} 2 & -2 & 8 \\ 0 & 2 & -2 \\ 0 & \frac{3}{2} & 0 \end{pmatrix}_v.$$

Hence,

$$A \begin{pmatrix} 9 \\ 7 \\ 4 \end{pmatrix}_e = \begin{pmatrix} 16 \\ 11 \\ 13 \end{pmatrix}_e,$$

$$T^{-1}AT \begin{pmatrix} 2 \\ 3 \\ 4 \end{pmatrix}_w = \begin{pmatrix} 5 \\ -2 \\ 13 \end{pmatrix}_w,$$

$$S^{-1}AS \begin{pmatrix} 10 \\ -2 \\ \frac{1}{2} \end{pmatrix}_v = \begin{pmatrix} 28 \\ -5 \\ -3 \end{pmatrix}_v. \quad \blacksquare$$

KEY WORDS AND PHRASES

abstract vector space
isomorphism
rank

cofactor
adjoint
change of basis

EXERCISES 4.7

▶ 1. Show that the set V and the operations in Example 4.7–1(a) satisfy Definition 4.7–1 and thus V is a vector space.

2. Show that the set V and the operations in Example 4.7–1(b) satisfy Definition 4.7–1.

3. Show that the set $C[0, 1]$ and the operations in Example 4.7–1(c) satisfy Definition 4.7–1.

4. In Example 4.7–1(d) it was stated that one condition of Definition 4.7–1 was not satisfied. Determine what other conditions of the definition are not satisfied.

5. Show that x^2, x, and 1 are linearly independent.

6. Prove that every quadratic polynomial can be expressed as a linear combination of x^2, x, and 1.

7. Given that

$$A = \begin{pmatrix} 1 & 2 & 3 \\ 2 & 1 & 2 \\ -2 & 1 & -1 \end{pmatrix},$$

find $|A|$.

8. **a)** Obtain the matrix T^{-1} in Eq. (4.7–5).

   b) By transforming the basis vectors of the v-basis into the natural basis, obtain the matrix S.

   c) Referring to part (b), find S^{-1}.

   d) Show that Ax_e can be written as ATx_w, which becomes $T^{-1}ATx_w$ in the w-basis. Show also that A is transformed into $S^{-1}AS$ in the v-basis.

e) Compute $T^{-1}AT$ and $S^{-1}AS$ in Example 4.7–5.

f) In Example 4.7–5 show that

$$(16, 11, 13)_e = (5, -2, 13)_w = (28, -5, -3)_v.$$

▶▶ 9. Determine which of the following are vector spaces. If the given set is not a vector space, state which condition or conditions of Definition 4.7–1 are not satisfied.

a) The set of all $m \times n$ matrices with the usual definitions of matrix addition and scalar multiplication.

b) The set consisting of only the $m \times n$ zero matrix.

c) The set of all solutions of $AX = O$, where A is an $n \times n$ matrix and the usual definitions of $\oplus$ and $\odot$ apply.

d) $C[a, b]$, the space of functions continuous on $[a, b]$.

e) $C^{(n)}[a, b]$, the space of functions that are n times differentiable on $[a, b]$.

f) The space of polynomials

$$P_n(x) = a_0 + a_1x + a_2x^2 + \cdots + a_nx^n$$

for any nonnegative integer n.

g) The set of all solutions of a given linear homogeneous differential equation,

$$a_0(x)\frac{d^ny}{dx^n} + a_1(x)\frac{d^{n-1}y}{dx^{n-1}} + \cdots + a_n(x)y = 0.$$

10. a) Show that the set

$$\left\{ \begin{pmatrix} 1 & 0 & 0 \\ 0 & 0 & 0 \end{pmatrix}, \begin{pmatrix} 0 & 1 & 0 \\ 0 & 0 & 0 \end{pmatrix}, \begin{pmatrix} 0 & 0 & 1 \\ 0 & 0 & 0 \end{pmatrix}, \begin{pmatrix} 0 & 0 & 0 \\ 1 & 0 & 0 \end{pmatrix}, \begin{pmatrix} 0 & 0 & 0 \\ 0 & 1 & 0 \end{pmatrix}, \begin{pmatrix} 0 & 0 & 0 \\ 0 & 0 & 1 \end{pmatrix} \right\}$$

is a basis for the vector space whose elements are 2×3 matrices.

b) In what ways can this vector space be related to $\mathbb{R}^6$?

11. Find a basis for the vector space given in Example 4.7–1(b).

12. Determine the rank of each of the following matrices.

a) $\begin{pmatrix} -1 & -2 & 0 & 1 \\ 2 & 3 & 4 & 5 \end{pmatrix}$

b) $\begin{pmatrix} 1 & 0 & -2 \\ -1 & 3 & 5 \\ 1 & 5 & 3 \end{pmatrix}$

c) $\begin{pmatrix} 1 & 0 & 1 & 1 \\ -1 & -1 & 0 & 1 \\ 2 & -1 & 3 & 4 \\ 0 & 1 & -1 & -2 \end{pmatrix}$

d) $\begin{pmatrix} 5 & -6 & -1 & 1 \\ 0 & 2 & -3 & 2 \\ 1 & 2 & -1 & 4 \\ -1 & 0 & 2 & 1 \end{pmatrix}$

13. Find adjoint A, $|A|$, and A^{-1} for each of the following matrices.

a) $A = \begin{pmatrix} 1 & 2 & 3 \\ 0 & -4 & 2 \\ -1 & 5 & 4 \end{pmatrix}$

b) $A = \begin{pmatrix} 0 & -1 & 1 & 4 \\ 3 & 2 & -2 & 1 \\ 0 & 4 & 0 & 1 \\ 1 & 0 & -1 & 1 \end{pmatrix}$

14. Use Theorem 4.7–2 to find the inverse of each of the following matrices if the inverse exists.

a) $\begin{pmatrix} 1 & 3 & 5 \\ 2 & 4 & 4 \\ 1 & -1 & 1 \end{pmatrix}$ **b)** $\begin{pmatrix} 5 & -6 & -1 & 1 \\ 0 & 2 & -3 & 2 \\ 1 & 2 & -1 & 4 \\ -1 & 0 & 2 & 1 \end{pmatrix}$

15. a) Show that the vectors

$$\mathbf{w}_1 = (1, -1, 0), \qquad \mathbf{w}_2 = (1, 0, -1), \qquad \mathbf{w}_3 = (0, 1, 1)$$

form a basis for $\mathbb{R}^3$.

b) Obtain the matrix T in $T\mathbf{x}_w = \mathbf{x}_e$.

c) Compute T^{-1}.

d) Transform the matrix

$$A = \begin{pmatrix} 1 & 1 & 0 \\ 0 & 1 & 1 \\ 1 & 0 & 1 \end{pmatrix}$$

into the w-basis.

16. a) Show that the vectors

$$\mathbf{v}_1 = (1, 0, 1), \qquad \mathbf{v}_2 = (0, -1, -1), \qquad \mathbf{v}_3 = (1, 1, 0)$$

form a basis for $\mathbb{R}^3$.

b) Obtain the matrix S in $S\mathbf{x}_v = \mathbf{x}_e$.

c) Compute S^{-1}.

d) Transform the matrix

$$B = \begin{pmatrix} -1 & 1 & 0 \\ 0 & 0 & 0 \\ -1 & 0 & 1 \end{pmatrix}$$

into the v-basis.

17. a) Using the w- and v-bases of Exercises 15 and 16, transform $\mathbf{x}_w$ into $\mathbf{x}_v$.

b) Transform the matrix A of Exercise 15 into the v-basis.

c) Transform the matrix B of Exercise 16 into the w-basis.

▶▶▶ **18.** Show that premultiplying a matrix A by an elementary matrix does not change the rank of A.

19. Show that the rank r of an $m \times n$ matrix A cannot exceed the maximum of m and n, that is, $r \leq \max(m, n)$.

20. Prove that the system $AX = B$, where A is $n \times n$, has a unique solution if and only if the rank of A is n.

21. Prove that the system $AX = B$, where A is $n \times n$, is consistent if and only if the rank of A is the same as the rank of the augmented matrix $A \vdots B$.

22. Prove that if A is $m \times n$ of rank r, then

$$n = r + \dim(\ker A).$$

23. Prove that $|EA| = |E| \cdot |A|$ given that E is any elementary matrix.

24. If A and B are $n \times n$ matrices, then $|AB| = |A| \cdot |B|$. (*Hint*: Consider the cases A singular and A nonsingular separately. For the latter, show that $A \sim I$; then use Exercise 23.)

25. Prove that the adjoint of a singular matrix is singular. (*Hint*: Use the fact that $|AB| = |A| \cdot |B|$ from Exercise 24.)

26. a) Show that

$$\begin{pmatrix} 2 & 0 & 0 \\ 0 & 1 & 0 \end{pmatrix} \begin{pmatrix} a \\ b \\ c \end{pmatrix} = \begin{pmatrix} 2a \\ b \end{pmatrix}$$

defines a linear transformation. (Refer to Section 4.3.)

b) If $(a, b, c)^T$ represents $ax^2 + bx + c$, explain why the transformation in part (a) can be called "differentiation."

c) Find the kernel of the transformation.

d) Describe the codomain of the transformation.

e) Find a basis for the range of the transformation. Is the transformation onto? Is the transformation one-to-one?

27. a) Find a basis for the vector space whose elements are cubic polynomials, $ax^3 + bx^2 + cx + d$.

b) What is the matrix that transforms an arbitrary element of this vector space into the space of derivatives of these elements?

28. Show that a change of basis in a linear transformation results in matrices that are similar. (See Definition 4.4–1.)

29. Consider the matrix

$$A = \begin{pmatrix} 2 & 1 & 2 \\ 1 & 2 & 2 \\ 2 & 2 & 3 \end{pmatrix}.$$

Show that A maps the set of all Pythagorean triples into itself. An ordered triple (x, y, z) is called a Pythagorean triple if it has the property $x^2 + y^2 = z^2$. Show that $A(x, y, z)^T$ also has this property.

REFERENCES

Acton, F. S., *Numerical Methods That Work*. New York: Harper and Row, 1970.

Arden, B. W., and K. N. Astill, *Numerical Algorithms: Origins and Applications*. Reading, Mass.: Addison-Wesley, 1970.

Dorn, W. S., and D. D. McCracken, *Numerical Methods with Fortran IV Case Studies*. New York: Wiley, 1972.

Eisenman, R. L., *Matrix Vector Analysis*. New York: McGraw-Hill, 1963. *Contains an interesting section entitled "Applications of Matrices."*

Faddeeva, V. N., *Computational Methods of Linear Algebra*. (Translated from the Russian by Curtis D. Benster). New York: Dover, 1959.

Gerald, C. F., *Applied Numerical Analysis*, 2d ed. Reading, Mass.: Addison-Wesley, 1978.

Jennings, A., *Matrix Computation for Engineers and Scientists*. New York: Wiley, 1977.
Contains a thorough treatment of all the topics discussed in Section 4.6.

Lancaster, P., *Theory of Matrices*. New York: Academic Press, 1969.
In spite of the title, this book contains much information of value to the applied *mathematician*.

Noble, B., *Applied Linear Algebra*. Englewood Cliffs, N.J.: Prentice-Hall, 1969.

Ralston, A., and P. Rabinowitz, *A First Course in Numerical Analysis*, 2d ed. New York: McGraw-Hill, 1978.

Rice, J. R., *Matrix Computation and Mathematical Software*. New York: McGraw-Hill, 1981.

Steinberg, D. I., *Computational Matrix Algebra*. New York: McGraw-Hill, 1974.

Williams, G., *Computational Linear Algebra with Models*, 2d ed. Boston: Allyn & Bacon, 1978.
Contains many real applications and provides computer-based exercises.

A number of programs written in BASIC and described briefly below are available from CONDUIT (The University of Iowa), P.O. Box 338, Iowa City, IA 52240.

Matrix Operations: Addition, multiplication, transposition, inversion.

Systems of Linear Equations: Uses elementary row operations to solve a system not to exceed 10×20.

Independence and Bases: Determines whether vectors are linearly dependent or independent.

Eigenvalues and Eigenvectors: Elementary row operations are used to reduce $A - \lambda I$ to echelon form.

Characteristic Polynomials: The roots of the characteristic equation are found by Newton's method.

5 *vector calculus*

5.1 VECTOR ALGEBRA

In this chapter we will discuss vectors in $\mathbb{R}^3$, that is, vectors of the form

$$\mathbf{v} = (v_1, v_2, v_3),$$

where the v_i are real numbers. We will continue to represent vectors by boldface letters and scalars (real numbers) by lower-case letters. In written work the notations $\vec{v}$, $\bar{v}$, or $\underline{v}$ are more convenient.

In this section we will review briefly the part of vector algebra pertaining to vector addition, and multiplication of a vector by a scalar. We will then introduce dot, cross, and triple products and discuss the characteristics and applications of these.

Although we will continue to designate a vector by an ordered triple (or ordered pair when the space is $\mathbb{R}^2$) of real numbers, we will often find it convenient to use a notation that relates a vector to a rectangular (or cartesian) coordinate system. We will use a **right-hand** (or dextral) **coordinate system**, such as the familiar x-, y-, and z-axes of analytic geometry. In a right-hand system we may imagine the positive z-axis as being the axis of a right-hand screw jack. If the positive x-axis is the handle of the screw jack, then rotating the handle from the positive x-axis toward the positive y-axis will advance the screw in the direction of the positive z-axis (Fig. 5.1–1).

If we define $\mathbf{i}$, $\mathbf{j}$, and $\mathbf{k}$ as *unit vectors* (that is, vectors of length one) along the x-, y-, and z-axis, respectively, then we can write

$$\mathbf{v} = (v_1, v_2, v_3) = v_1\mathbf{i} + v_2\mathbf{j} + v_3\mathbf{k}.$$

This vector is shown in Fig. 5.1–2. Note that the *initial point* (also called the *tail*) of the vector is at the origin, and the *terminal point* (also called the *head*) of the vector is at the point whose coordinates are given by (v_1, v_2, v_3). The word "vector" has its origin* in the Latin word meaning "bearer" or "carrier," hence we think of a vector as "carrying" the initial point *to* the terminal point along a straight line

---

* The word was introduced by Sir William Rowan Hamilton (1805–1865), an Irish mathematician, in 1847.

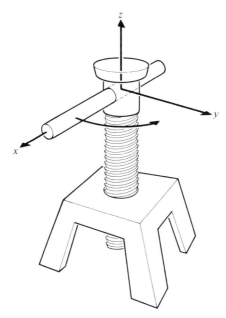

FIGURE 5.1–1 A screw jack.

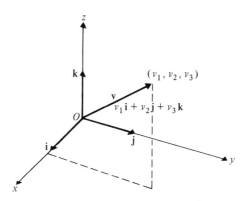

FIGURE 5.1–2 The vector **v** in $\mathbb{R}^3$.

connecting the two points. Thus a nonzero vector has *two* character-istics: its *length* and its *direction*.

By using the Pythagorean theorem we find the length of the vector **v**, designated by $|\mathbf{v}|$ or by v, to be

$$v = |\mathbf{v}| = \sqrt{v_1^2 + v_2^2 + v_3^2}, \tag{5.1–1}$$

which is commonly called the *distance formula*. We will discuss the direction of **v** later in this section. We remark that we will be dealing in this chapter mainly with **free vectors**, meaning that two vectors **u** and **v** will be called *equal* (**u** = **v**) if and only if they have the same length *and* the same direction (Fig. 5.1–3). In other words, vectors are free to be moved in space at will provided their lengths and directions remain the same. This is in contrast to **bound vectors** (also called *fixed* or *position* vectors) used in mechanics where, if a vector represents an applied force, it makes a great deal of difference where the point of application of that force is. Examples of free and bound vectors will be found at the end of this section. The **zero vector**, designated by **0**, has zero length, that is, $|\mathbf{0}| = 0$ and any direction whatever or an unspecified direction. Thus the length of a vector is a nonnegative real number.

FIGURE 5.1–3 Equal vectors.

Since a vector has two characteristics—a length and a direction—vectors are commonly used to represent physical quantities such as force, velocity, acceleration, directed displacement, electric and magnetic intensity, magnetic induction, and many others. Scalars, on the other hand, are used to represent length, temperature, mass, time, age, density, etc., since each of these can be described by a single real number.

Following are definitions of the two basic operations in vector algebra. These definitions are completely analogous to those given in Section 4.1.

Definition 5.1–1 If $\mathbf{u} = u_1\mathbf{i} + u_2\mathbf{j} + u_3\mathbf{k}$ and $\mathbf{v} = v_1\mathbf{i} + v_2\mathbf{j} + v_3\mathbf{k}$ are any two vectors in $\mathbb{R}^3$, then

$$\mathbf{u} + \mathbf{v} = (u_1 + v_1)\mathbf{i} + (u_2 + v_2)\mathbf{j} + (u_3 + v_3)\mathbf{k}.$$

For any scalar c,
$$c\mathbf{v} = (cv_1)\mathbf{i} + (cv_2)\mathbf{j} + (cv_3)\mathbf{k}.$$

Certain other properties follow from these definitions. These are stated in the following theorem, the proof of which is left for the exercises (Exercise 1). See Fig. 5.1–4 for a diagrammatic representation of the associative property in $\mathbb{R}^2$.

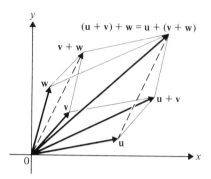

FIGURE 5.1–4 The associative property of addition in $\mathbb{R}^2$.

Theorem 5.1–1 If $\mathbf{u}$, $\mathbf{v}$, and $\mathbf{w}$ are any three vectors in $\mathbb{R}^3$ and c and d are any scalars, then:

a) $\mathbf{u} + \mathbf{v} = \mathbf{v} + \mathbf{u}$ (commutative property);

b) $\mathbf{u} + (\mathbf{v} + \mathbf{w}) = (\mathbf{u} + \mathbf{v}) + \mathbf{w}$ (associative property);

c) $\mathbf{u} + \mathbf{0} = \mathbf{u}$ (additive identity);

d) for every $\mathbf{u}$ there exists a unique vector $-\mathbf{u}$ such that $\mathbf{u} + (-\mathbf{u}) = \mathbf{0}$ (additive inverse);

e) $c(d\mathbf{v}) = (cd)\mathbf{v}$;

f) $(c + d)\mathbf{v} = c\mathbf{v} + d\mathbf{v}$;

g) $0\mathbf{v} = \mathbf{0}$, $1\mathbf{v} = \mathbf{v}$, $(-1)\mathbf{v} = -\mathbf{v}$;

h) $c(\mathbf{u} + \mathbf{v}) = c\mathbf{u} + c\mathbf{v}$.

We are now ready to discuss the matter of direction of a vector. This problem will be divided into two parts: first, the direction of vectors in $\mathbb{R}^1$ and $\mathbb{R}^2$ and then vectors in $\mathbb{R}^3$.

Vectors in $\mathbb{R}^1$ can be considered to have their initial and terminal points on the x-axis. Thus the vector *from* $x = -2$ *to* $x = 3$ can be written $(3 - (-2))\mathbf{i} = 5\mathbf{i}$. In general, the vector from $x = x_1$ to $x = x_2$ is $(x_2 - x_1)\mathbf{i}$ and the concept is the same as that of a directed line segment in analytic geometry. Hence all vectors have the direction of $\mathbf{i}$ or $-\mathbf{i}$. When dealing with vectors along the y-axis we would use $\mathbf{j}$ and $-\mathbf{j}$, etc.

In $\mathbb{R}^2$ vectors lie in the xy-plane and every such vector can be uniquely expressed as a linear combination of $\mathbf{i}$ and $\mathbf{j}$ (Exercise 2). In Fig. 5.1–5 we show the vector $\mathbf{v}$ from (x_1, y_1) to (x_2, y_2). This vector can be expressed as

$$\mathbf{v} = (x_2 - x_1)\mathbf{i} + (y_2 - y_1)\mathbf{j} \qquad (5.1\text{--}2)$$

using the definition of vector addition as given in Definition 5.1–1. The length of $\mathbf{v}$ is

$$v = |\mathbf{v}| = \sqrt{(x_2 - x_1)^2 + (y_2 - y_1)^2},$$

which is the distance formula familiar from analytic geometry.

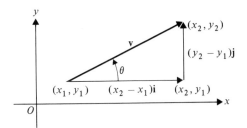

FIGURE 5.1–5 A vector in $\mathbb{R}^2$.

The direction of $\mathbf{v}$ in Eq. (5.1–2) can be given in terms of θ, the angle from the positive x-axis to the vector. We have

$$\tan \theta = \frac{y_2 - y_1}{x_2 - x_1}, \qquad x_1 \neq x_2, \qquad (5.1\text{--}3)$$

and, in general, for

$$\mathbf{v} = v_1 \mathbf{i} + v_2 \mathbf{j}$$

we have

$$\tan \theta = v_2/v_1 \quad \text{or} \quad \theta = \arctan (v_2/v_1).$$

In $\mathbb{R}^3$ the situation is somewhat different since we need to consider the angles a vector makes with all three axes. In Fig. 5.1–6 we have a vector with initial point (x_1, y_1, z_1) and terminal point (x_2, y_2, z_2). The angles α, β, and γ from the positive x-, y-, and z-axis, respectively, to the vector are called the **direction angles** of the vector. These angles are not as convenient for computation as the cosines of these angles, which are called the **direction cosines** of the vector. It can be shown (Exercise 3) by drawing appropriate right triangles that the

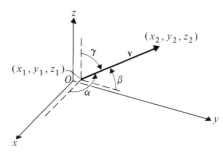

FIGURE 5.1–6 Direction angles.

direction cosines are given by

$$\cos \alpha = \frac{x_2 - x_1}{v}, \qquad \cos \beta = \frac{y_2 - y_1}{v}, \qquad \cos \gamma = \frac{z_2 - z_1}{v}. \quad \textbf{(5.1–4)}$$

These three direction cosines are not independent since it can be shown (Exercise 4) that

$$\cos^2 \alpha + \cos^2 \beta + \cos^2 \gamma = 1. \quad \textbf{(5.1–5)}$$

From Eqs. (5.1–4) we see that the right members all have the same denominators, namely, v. Thus the direction of the vector can be specified by giving the ordered triple

$$(x_2 - x_1, y_2 - y_1, z_2 - z_1). \quad \textbf{(5.1–6)}$$

These numbers are called **direction numbers** of the vector and are the ones most often used to describe the direction of a vector.

EXAMPLE 5.1–1 Consider the vector **v** from $(-2, 1, 3)$ to $(3, -2, 5)$. Find v and also the direction numbers, direction cosines, and direction angles of this vector. Specify a unit vector in the direction of **v**.

Solution. The direction numbers are given by $(3 - (-2), -2 - 1, 5 - 3)$ or $(5, -3, 2)$. Thus

$$v = |\mathbf{v}| = \sqrt{(5)^2 + (-3)^2 + (2)^2} = \sqrt{38} \doteq 6.16441.$$

The direction cosines are

$$\cos \alpha = \frac{5}{\sqrt{38}} = \frac{5\sqrt{38}}{38}, \qquad \cos \beta = \frac{-3\sqrt{38}}{38}, \qquad \cos \gamma = \frac{\sqrt{38}}{19},$$

or, in decimal form,

$$\cos \alpha \doteq 0.81111, \qquad \cos \beta \doteq -0.48666, \qquad \cos \gamma \doteq 0.32444.$$

The direction angles, expressed in radians, are

$$\alpha \doteq 0.62475, \qquad \beta \doteq 2.07906, \qquad \gamma \doteq 1.24037.$$

The unit vector in the direction of **v** is given by

$$\frac{\mathbf{v}}{v} = \frac{\sqrt{38}}{38}\,(5\mathbf{i} - 3\mathbf{j} + 2\mathbf{k}). \quad \blacksquare$$

This example shows why using direction *numbers* is the most simple way of expressing the direction of a vector in $\mathbb{R}^3$. Moreover, the preceding discussion of direction of vectors in $\mathbb{R}^3$ can be generalized to vectors in $\mathbb{R}^n$, $n \geq 1$ (see Exercise 37).

Dot product

Definition 5.1–2 The **dot product** of two vectors **u** and **v** is a scalar denoted by $\mathbf{u} \cdot \mathbf{v}$, which is given by

$$\mathbf{u} \cdot \mathbf{v} = uv \cos \theta, \tag{5.1–7}$$

where θ is the angle (positive or negative) through which, if **v** is rotated, it will have the same direction as **u**. In general, we will use the angle of smallest magnitude.

The dot product is also known as the *scalar product* (since the result is a scalar) and as the *inner product*, a term that will be explained in connection with the cross product, which will be defined later. We can see from Eq. (5.1–7) that the dot product of two nonzero vectors is zero if and only if the vectors have *perpendicular* directions. Also clear from Eq. (5.1–7) is the fact that the dot product of *unit* vectors is equal to the cosine of the angle between them. We will use this fact later to obtain the angles between curves and surfaces.

Some useful properties of the dot product are given in the following theorem. Proofs are left for the exercises (see Exercise 5).

Theorem 5.1–2 If **u**, **v**, and **w** are vectors in $\mathbb{R}^3$ and c is a scalar, then:

a) $\mathbf{u} \cdot \mathbf{v} = \mathbf{v} \cdot \mathbf{u}$ (commutative property);

b) $(c\mathbf{u}) \cdot \mathbf{v} = c(\mathbf{u} \cdot \mathbf{v})$;

c) $\mathbf{u} \cdot (c\mathbf{v}) = c(\mathbf{u} \cdot \mathbf{v})$;

d) $(\mathbf{u} + \mathbf{v}) \cdot \mathbf{w} = \mathbf{u} \cdot \mathbf{w} + \mathbf{v} \cdot \mathbf{w}$ (right distributive property);

e) $\mathbf{u} \cdot (\mathbf{v} + \mathbf{w}) = \mathbf{u} \cdot \mathbf{v} + \mathbf{u} \cdot \mathbf{w}$ (left distributive property);

f) $|\mathbf{v}|^2 = v^2 = \mathbf{v} \cdot \mathbf{v}$.

Since the unit basis vectors have the properties

$$\mathbf{i} \cdot \mathbf{i} = \mathbf{j} \cdot \mathbf{j} = \mathbf{k} \cdot \mathbf{k} = 1$$

and

$$\mathbf{i} \cdot \mathbf{j} = \mathbf{j} \cdot \mathbf{k} = \mathbf{i} \cdot \mathbf{k} = 0,$$

we can see that, for an arbitrary vector

$$\mathbf{v} = v_1\mathbf{i} + v_2\mathbf{j} + v_3\mathbf{k},$$

the components v_i are given by

$$v_1 = \mathbf{v} \cdot \mathbf{i}, \qquad v_2 = \mathbf{v} \cdot \mathbf{j}, \qquad v_3 = \mathbf{v} \cdot \mathbf{k}. \tag{5.1-8}$$

An alternative definition of dot product, namely

$$\mathbf{u} \cdot \mathbf{v} = u_1v_1 + u_2v_2 + u_3v_3, \tag{5.1-9}$$

where $\mathbf{u} = u_1\mathbf{i} + u_2\mathbf{j} + u_3\mathbf{k}$, is often more useful for computational purposes than the one given in Definition 5.1–2. (See Exercise 6.)

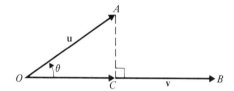

FIGURE 5.1–7 Projection of one vector on another.

One application of the dot product deserves special mention. In Fig. 5.1–7 $\mathbf{u}$ and $\mathbf{v}$ are vectors from the origin to the points A and B, respectively. The vector $\overrightarrow{OC}$ is the *projection* of $\mathbf{u}$ on $\mathbf{v}$; it has length $u \cos \theta$ and the direction of $\mathbf{v}$. Hence we write

$$\overrightarrow{OC} = \frac{\mathbf{u} \cdot \mathbf{v}}{v} \mathbf{v}, \tag{5.1-10}$$

that is, the projection of $\mathbf{u}$ on $\mathbf{v}$ is a vector in the direction of $\mathbf{v}$ whose length is the dot product of $\mathbf{u}$ and the *unit vector* in the direction of $\mathbf{v}$. Projections are useful in physics for finding the **resultant** of a number of forces (Exercise 7). Other applications of the dot product to geometry will be given later in this section.

Cross product

Definition 5.1–3 The **cross product** of two vectors $\mathbf{a}$ and $\mathbf{b}$, denoted by $\mathbf{a} \times \mathbf{b}$, is given by

$$\mathbf{a} \times \mathbf{b} = |\mathbf{a}|\,|\mathbf{b}|\,(\sin \theta)\,\mathbf{c}, \tag{5.1-11}$$

where a rotation from $\mathbf{a}$ to $\mathbf{b}$ through an angle θ $(0 \le \theta \le \pi)$ will advance a right-hand screw along the unit vector $\mathbf{c}$ in the direction of $\mathbf{c}$ that is perpendicular to the plane determined by $\mathbf{a}$ and $\mathbf{b}$ (Fig. 5.1–8).

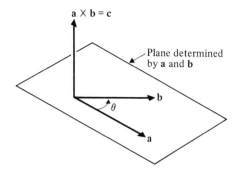

FIGURE 5.1–8 The cross product.

The cross product $\mathbf{a} \times \mathbf{b}$ is also known as the *vector product* (since the result is a vector) and the *outer product* (since the result is *not in* the plane of the vectors $\mathbf{a}$ and $\mathbf{b}$). The terms "outer" (for a cross product) and "inner" (for a dot product) were first used in 1840 by Hermann G. Grassmann (1809–1877), a German mathematician who was one of the pioneers in the development of vector analysis.

It is evident from Eq. (5.1–11) that the cross product of two nonzero vectors is the zero vector if and only if the vectors are *parallel*. We also have

$$\mathbf{i} \times \mathbf{j} = \mathbf{k},$$
$$\mathbf{j} \times \mathbf{k} = \mathbf{i},$$
$$\mathbf{k} \times \mathbf{i} = \mathbf{j}$$

for the unit basis vectors. Moreover, from the definition,

$$\mathbf{j} \times \mathbf{i} = -\mathbf{k},$$
$$\mathbf{k} \times \mathbf{j} = -\mathbf{i},$$
$$\mathbf{i} \times \mathbf{k} = -\mathbf{j},$$

and, in general,

$$\mathbf{a} \times \mathbf{b} = -(\mathbf{b} \times \mathbf{a}),$$

which shows that the cross product is **anticommutative**. This and other properties of the cross product are summarized in the following theorem.

Theorem 5.1–3 If $\mathbf{u}$, $\mathbf{v}$, and $\mathbf{w}$ are vectors in $\mathbb{R}^3$ and c is a scalar, then:

a) $\mathbf{u} \times \mathbf{v} = -(\mathbf{v} \times \mathbf{u})$ (anticommutative property);

b) $(\mathbf{u} + \mathbf{v}) \times \mathbf{w} = (\mathbf{u} \times \mathbf{w}) + (\mathbf{v} \times \mathbf{w})$ (right distributive property);

c) $\mathbf{u} \times (\mathbf{v} + \mathbf{w}) = (\mathbf{u} \times \mathbf{v}) + (\mathbf{u} \times \mathbf{w})$ (left distributive property);

d) $\mathbf{u} \times c\mathbf{v} = c(\mathbf{u} \times \mathbf{v})$;

e) $c\mathbf{u} \times \mathbf{v} = c(\mathbf{u} \times \mathbf{v})$.

The proofs of the above are left for the exercises (Exercise 8). We call attention to the fact that the cross product is *not* associative, that is, $(\mathbf{u} \times \mathbf{v}) \times \mathbf{w} \neq \mathbf{u} \times (\mathbf{v} \times \mathbf{w})$ (Exercise 9). For this reason parentheses are essential in the *triple cross product*, $\mathbf{u} \times \mathbf{v} \times \mathbf{w}$.

Using the distributive property in the last theorem, it can be shown (Exercise 10) that if

$$\mathbf{u} = u_1\mathbf{i} + u_2\mathbf{j} + u_3\mathbf{k}$$

and

$$\mathbf{v} = v_1\mathbf{i} + v_2\mathbf{j} + v_3\mathbf{k},$$

then

$$\mathbf{u} \times \mathbf{v} = \begin{vmatrix} \mathbf{i} & \mathbf{j} & \mathbf{k} \\ u_1 & u_2 & u_3 \\ v_1 & v_2 & v_3 \end{vmatrix}. \tag{5.1-12}$$

This last is a *symbolic* determinant and is to be interpreted as the vector resulting from expanding the determinant in terms of elements of the first row. In other words,

$$\mathbf{u} \times \mathbf{v} = (u_2v_3 - v_2u_3)\mathbf{i} + (v_1u_3 - u_1v_3)\mathbf{j} + (u_1v_2 - v_1u_2)\mathbf{k}.$$

It is easy to see now why $\mathbf{u} \times \mathbf{v} = -(\mathbf{v} \times \mathbf{u})$. (Explain.)

EXAMPLE 5.1–2 Find the area of the parallelogram whose two sides are the vectors $\mathbf{a} = 2\mathbf{i} + \mathbf{j} + 7\mathbf{k}$ and $\mathbf{b} = \mathbf{i} + \mathbf{j} - 4\mathbf{k}$.

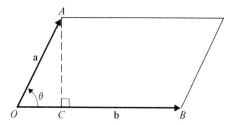

FIGURE 5.1–9 The area of a parallelogram.

Solution. Figure 5.1–9 shows the altitude AC of the parallelogram. Hence the area can be found by multiplying the length of AC by b, the length of the base. Since $AC = a \sin \theta$, the area can be written as $|\mathbf{a} \times \mathbf{b}|$. But

$$\mathbf{a} \times \mathbf{b} = \begin{vmatrix} \mathbf{i} & \mathbf{j} & \mathbf{k} \\ 2 & 1 & 7 \\ 1 & 1 & -4 \end{vmatrix} = -11\mathbf{i} + 15\mathbf{j} + \mathbf{k},$$

so that the area of the parallelogram is

$$|\mathbf{a} \times \mathbf{b}| = \sqrt{(-11)^2 + (15)^2 + (1)^2} = \sqrt{347} \doteq 18.628. \quad \blacksquare$$

Scalar triple product

When three vectors are multiplied using both the dot and vector products, we have the triple product

$$\mathbf{u} \times \mathbf{v} \cdot \mathbf{w}.$$

Parentheses are not necessary here because the grouping $(\mathbf{v} \cdot \mathbf{w})$ would make the triple product meaningless since we have not defined the cross product of a vector and a scalar. Thus the parentheses can be placed only around $\mathbf{u} \times \mathbf{v}$ and the triple product becomes $(\mathbf{u} \times \mathbf{v}) \cdot \mathbf{w}$, which is a scalar. This scalar is called the **scalar triple product** or the *box product* of $\mathbf{u}$, $\mathbf{v}$, and $\mathbf{w}$ and is denoted by $[\mathbf{u}, \mathbf{v}, \mathbf{w}]$. The latter term comes from the fact that if vectors $\mathbf{u}$, $\mathbf{v}$, and $\mathbf{w}$ are drawn so that they have common initial points, the vectors can be considered to form the edges of a parallelepiped. (Fig. 5.1–10). Except perhaps for sign, $\mathbf{u} \times \mathbf{v} \cdot \mathbf{w}$ or $[\mathbf{u}, \mathbf{v}, \mathbf{w}]$ gives the volume of the parallelepiped (box). (See Exercise 11.) It then follows that three nonzero vectors are coplanar (lie in the same plane) if and only if their scalar triple product is zero (Exercise 12). This last is more easily seen by writing the scalar triple product in the equivalent form,

$$\mathbf{u} \times \mathbf{v} \cdot \mathbf{w} = [\mathbf{u}, \mathbf{v}, \mathbf{w}] = \begin{vmatrix} u_1 & u_2 & u_3 \\ v_1 & v_2 & v_3 \\ w_1 & w_2 & w_3 \end{vmatrix} \qquad \textbf{(5.1–13)}$$

(see Exercise 13).

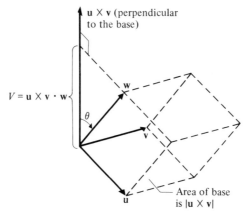

FIGURE 5.1–10 The volume of a parallelepiped.

Vector equation of a line

We may apply the preceding ideas to analytic geometry in order to obtain vector equations of lines and planes. For example, consider the line determined by the points P_1 and P_2 in $\mathbb{R}^3$. Knowing the coordinates of these points is equivalent to knowing the position vectors $\mathbf{r}_1$ and $\mathbf{r}_2$ of these points with respect to some origin as shown in Fig. 5.1–11. If $P(x, y, z)$ is an *arbitrary* point on the line with position vector $\mathbf{r}$, then the vector from P_1 to P_2 is $\mathbf{r}_2 - \mathbf{r}_1$ and we can write for some scalar t,

$$\mathbf{r} = \mathbf{r}_1 + t(\mathbf{r}_2 - \mathbf{r}_1), \qquad (5.1\text{–}14)$$

which is a **vector equation** of the line. Observe that if $t = 0$, then $\mathbf{r} = \mathbf{r}_1$; and if $t = 1$, then $\mathbf{r} = \mathbf{r}_2$. For $0 < t < 1$, $\mathbf{r}$ is the position vector of points P that are between P_1 and P_2. (See Exercise 14.)

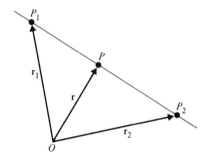

FIGURE 5.1–11 The vector equation of a line.

It is relatively simple to convert a vector equation of the line (5.1–14) to other forms more familiar from analytic geometry. In terms of the unit basis vectors,

$$\mathbf{r}_1 = x_1\mathbf{i} + y_1\mathbf{j} + z_1\mathbf{k}, \qquad \mathbf{r}_2 = x_2\mathbf{i} + y_2\mathbf{j} + z_2\mathbf{k}, \qquad \mathbf{r} = x\mathbf{i} + y\mathbf{j} + z\mathbf{k},$$

and Eq. (5.1–14) becomes

$$x\mathbf{i} + y\mathbf{j} + z\mathbf{k} = x_1\mathbf{i} + y_1\mathbf{j} + z_1\mathbf{k} + t((x_2 - x_1)\mathbf{i} + (y_2 - y_1)\mathbf{j} + (z_2 - z_1)\mathbf{k})$$

or

$$x = x_1 + t(x_2 - x_1),$$
$$y = y_1 + t(y_2 - y_1),$$
$$z = z_1 + t(z_2 - z_1).$$

But the direction numbers of the line are $x_2 - x_1, y_2 - y_1, z_2 - z_1$; we call these numbers a, b, c, respectively. Then

$$x = x_1 + ta,$$
$$y = y_1 + tb, \qquad \textbf{(5.1–15)}$$
$$z = z_1 + tc,$$

which are **parametric equations** of the line; that is, the coordinates of an arbitrary point (x, y, z) are expressed in terms of a *parameter t*. Solving each equation in the last set for t produces

$$\frac{x - x_1}{a} = \frac{y - y_1}{b} = \frac{z - z_1}{c} = t, \qquad \textbf{(5.1–16)}$$

which are equations of the line in *symmetric form*, provided $a \neq 0$, $b \neq 0$, $c \neq 0$. If $abc = 0$, then use Eqs. (5.1–15).

Vector equation of a plane

A plane is uniquely determined by the direction of a vector **normal** (or perpendicular) to the plane and a point in the plane. Referring to Fig. 5.1–12, let P_1 with position vector $\mathbf{r}_1$ be a point in the plane whose equation is required; let $\mathbf{n} = a\mathbf{i} + b\mathbf{j} + c\mathbf{k}$ be normal to the plane; and let $P(x, y, z)$ be an arbitrary point in the plane with position vector $\mathbf{r}$. Then the vector from P_1 to P is $\mathbf{r} - \mathbf{r}_1$ and

$$(\mathbf{r} - \mathbf{r}_1) \cdot \mathbf{n} = 0, \qquad \textbf{(5.1–17)}$$

which is a *vector equation* of the plane. It expresses the fact that the normal to a plane intersecting the plane at a point is perpendicular to every line in the plane through this point.

In terms of coordinate basis vectors,

$$(x\mathbf{i} + y\mathbf{j} + z\mathbf{k} - (x_1\mathbf{i} + y_1\mathbf{j} + z_1\mathbf{k})) \cdot (a\mathbf{i} + b\mathbf{j} + c\mathbf{k}) = 0$$

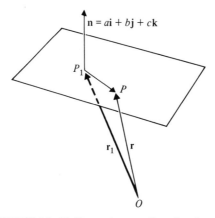

FIGURE 5.1–12 The vector equation of a plane.

or

$$a(x - x_1) + b(y - y_1) + c(z - z_1) = 0, \qquad \text{(5.1–18)}$$

which is an equation of the plane in rectangular coordinates. From Eq. (5.1–18) we can see that the plane $2x - 3y + 5z = 7$ has a normal whose direction numbers are $(2, -3, 5)$.

Applications

We conclude this section by giving a number of examples that illustrate some of the applications of vector algebra.

EXAMPLE 5.1–3 (*Work*). When a *constant* force **F** is applied to an object as it moves through a distance d *in the direction of the force*, we know from physics that the work W done by the force is defined as the product $W = Fd$. If the directions of the force and motion are different, however, then we must use the component of the force in the direction of the motion. Hence, in general, work is given by

$$W = \mathbf{F} \cdot \mathbf{d}, \qquad \text{(5.1–19)}$$

where **d** is the (directed) displacement. This formula shows that the work done is a maximum when the motion is in the same direction as the force (why?) and is zero when the motion is perpendicular to the force. ■

EXAMPLE 5.1–4 (*Angular velocity*). Let P be an arbitrary point on a rigid body that is rotating about an axis with constant angular velocity ω (measured, say, in radians/sec). If we take the origin to be on the axis of rotation as shown in Fig. 5.1–13, then **r** is the position vector of P

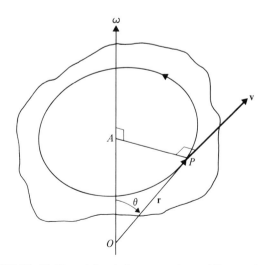

FIGURE 5.1–13 The relation between angular and linear velocities.

and the length of $\overline{AP}$ is given by $r \sin \theta$. If **v** is the linear velocity of the point (measured, say, in cm/sec), then the length of **v** is $\omega(\overline{AP}) = \omega r \sin \theta$. Thus if $\boldsymbol{\omega}$ is a vector along the axis of rotation of length ω and such that $\boldsymbol{\omega}$, **r**, and **v** form a right-hand system (Fig. 5.1–14), then

$$\mathbf{v} = \boldsymbol{\omega} \times \mathbf{r}. \qquad (5.1\text{–}20)$$

We call $\boldsymbol{\omega}$ the **angular velocity vector**. Equation (5.1–20) gives the relationship between the angular velocity vector $\boldsymbol{\omega}$ and the **linear velocity vector v**. This relationship holds even when $\boldsymbol{\omega}$ does not have constant direction (as in the case of a spinning top), provided only that the axis of rotation passes through the origin. ■

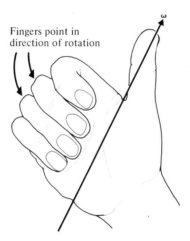

Fingers point in direction of rotation

FIGURE 5.1–14 The right-hand rule.

EXAMPLE 5.1–5 (*Vector moment*). Let **r** be the position vector of a point P with respect to an origin O, and let **F** represent a force *acting at* P. If both O and P are points of a rigid body, then **F** will tend to produce a rotation of the body about an axis through O with this axis being normal to the plane determined by **r** and **F**. The product of r and the component of **F** that is normal to **r** ($F \sin \theta$ in Fig. 5.1–15) is called the (scalar) **moment** of the force **F** about O in mechanics. The vector **M** given by

$$\mathbf{M} = \mathbf{r} \times \mathbf{F} \qquad (5.1\text{–}21)$$

is called the **vector moment** of **F** about the point O. We can see from Eq. (5.1–21) that, so far as the moment is concerned, the result is the same if **F** acts at any point along the line through O and P. ■

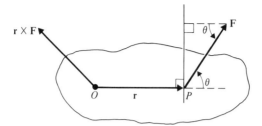

FIGURE 5.1–15 The vector moment of a force.

In connection with the last example we state without proof an important theorem from statics.

Theorem 5.1–4 (*Central axis theorem*). If **F** is the resultant (see Exercise 8) of a set of forces $\mathbf{F}_i$, $i = 1, 2, \ldots, n$, then there exists an axis such that, with respect to any point on this axis as origin, the resultant moment **M** is parallel to **F**.

EXAMPLE 5.1–6 (*Couples*). Two forces equal in magnitude but opposite in direction is called a **couple**. If the forces act at a point, then the result is a *zero couple*; hence we will consider couples such that the forces act at two distinct points having position vectors $\mathbf{r}_1$ and $\mathbf{r}_2$. The resultant moment of a couple with respect to the origin (Fig. 5.1–16) is

$$\mathbf{M} = (\mathbf{r}_1 - \mathbf{r}_2) \times \mathbf{F}. \quad \blacksquare \qquad (5.1\text{--}22)$$

It can be shown that any *system* of forces is equivalent to a force at an arbitrary point together with a couple. Further, a force acting

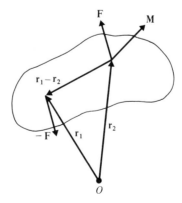

FIGURE 5.1–16 The moment of a couple.

along a line together with a couple whose moment is along a line parallel to the force is called a **wrench**. Thus it follows from the central axis theorem that a system of forces is equivalent to a wrench.

A body is said to be in **static equilibrium** if (a) the resultant force of the system of forces acting on the body is zero, and (b) the resultant moment of the system of forces with respect to an arbitrary point is zero.

KEY WORDS AND PHRASES

right-hand coordinate system
free and bound vectors
zero vector
direction angles
direction cosines
direction numbers
dot product
resultant
cross product
anticommutative

scalar triple product
normal
work
angular velocity
linear velocity
moment
couple
wrench
static equilibrium

EXERCISES 5.1

Note: Unless otherwise stated, assume that all vectors in the following exercises belong to $\mathbb{R}^3$.

▶ 1. Prove Theorem 5.1–1 using Definition 5.1–1 and the field properties of real numbers.

2. Prove that an arbitrary vector **v** in $\mathbb{R}^2$ can be *uniquely* expressed as a linear combination of **i** and **j**.

3. Show that the vector from (x_1, y_1, z_1) to (x_2, y_2, z_2) has direction cosines given by Eqs. (5.1–4).

4. Show that the direction cosines of a vector satisfy

$$\cos^2 \alpha + \cos^2 \beta + \cos^2 \gamma = 1.$$

5. Prove Theorem 5.1–2. (*Hint*: In part (d) use projections of **u**, **v**, and **u** + **v** on **w**.)

6. Establish Eq. (5.1–9). (*Hint*: Use Theorem 5.1–2.)

7. If **u** and **v** are two forces in $\mathbb{R}^3$ acting on a particle, then **u** + **v** is the single force that would produce the same effect and is called the **resultant** of **u** and **v**. Find the resultant of the forces

$$\mathbf{f}_1 = 2\mathbf{i} - 3\mathbf{j} + 5\mathbf{k} \quad \text{and} \quad \mathbf{f}_2 = \mathbf{i} + 3\mathbf{k}.$$

8. Prove Theorem 5.1–3. (*Hint*: Prove part (c) from geometrical considerations.)

9. Give a counterexample that shows that the cross product does not have the associative property.

10. If $\mathbf{u} = u_1\mathbf{i} + u_2\mathbf{j} + u_3\mathbf{k}$ and $\mathbf{v} = v_1\mathbf{i} + v_2\mathbf{j} + v_3\mathbf{k}$, show that

$$\mathbf{u} \times \mathbf{v} = \begin{vmatrix} \mathbf{i} & \mathbf{j} & \mathbf{k} \\ u_1 & u_2 & u_3 \\ v_1 & v_2 & v_3 \end{vmatrix}.$$

11. **a)** If $\mathbf{u}$, $\mathbf{v}$, and $\mathbf{w}$ have their initial points at the origin, draw a figure of a parallelepiped having these vectors as edges. (*Note*: A parallelepiped is a prism whose six faces are parallelograms.)

   b) Explain why the scalar triple product $\mathbf{u} \times \mathbf{v} \cdot \mathbf{w}$ represents the volume of the parallelepiped except perhaps for sign.

   c) Find the volume of the parallelepiped whose sides are

$$\mathbf{u} = \mathbf{i} + \mathbf{j} + 4\mathbf{k}, \qquad \mathbf{v} = 3\mathbf{i} + \mathbf{j} + \mathbf{k}, \quad \text{and} \quad \mathbf{w} = 2\mathbf{i} + \mathbf{k}.$$

12. Prove the following: Three nonzero vectors are coplanar if and only if their scalar triple product is zero. (*Note*: The word "coplanar" can be replaced by "linearly dependent.")

13. Show that the scalar triple product can be written in the form (5.1–13).

14. Give the values of t in Eq. (5.1–14) for the points described below.

   a) The midpoint of the line segment P_1P_2.

   b) The points that trisect the line segment P_1P_2.

   c) The point P_3 such that P_2 is the midpoint of the line segment P_1P_3.

   d) The point P_4 such that P_1 is the midpoint of the line segment P_4P_2.

▶▶ 15. If $\mathbf{a} = \mathbf{i} - 2\mathbf{j} + \mathbf{k}$ and $\mathbf{b} = 3\mathbf{i} + \mathbf{j} - \mathbf{k}$, find:

   a) $\mathbf{a} \cdot \mathbf{b}$;   **b)** $\mathbf{a} \times \mathbf{b}$;   **c)** $|\mathbf{a} \times \mathbf{b}|$.

16. Is it possible for a vector to make angles of $\pi/4$ with each of the three coordinate axes? Explain.

17. **a)** Find the direction angles and direction cosines of a vector that makes equal angles with the three coordinate axes.

   b) Specify a unit vector in the direction described in part (a).

18. Obtain a formula for finding the area of any triangle by using the cross product. (*Hint*: See Example 5.1–2.)

19. Show that the vector $\mathbf{v} = v_1\mathbf{i} + v_2\mathbf{j} + v_3\mathbf{k}$ has length

$$v = \sqrt{v_1^2 + v_2^2 + v_3^2}.$$

20. If $\mathbf{a} = 2\mathbf{j} - \mathbf{k}$, $\mathbf{b} = \mathbf{i} + 3\mathbf{j}$, and $\mathbf{c} = \mathbf{k}$, find:

   a) $\mathbf{a} \cdot \mathbf{b} \times \mathbf{c}$;   **b)** $\mathbf{a} \times (\mathbf{b} \times \mathbf{c})$;   **c)** $(\mathbf{a} \times \mathbf{b}) \times \mathbf{c}$.

21. Under what conditions do the following hold? (*Note*: The symbol $\Rightarrow$ means "implies.")

   a) $\mathbf{a} \times \mathbf{b} = \mathbf{a} \times \mathbf{c} \Rightarrow \mathbf{b} = \mathbf{c}$ **b)** $\mathbf{a} \cdot \mathbf{b} = \mathbf{a} \cdot \mathbf{c} \Rightarrow \mathbf{b} = \mathbf{c}$

22. A plane contains the point $(1, -2, 0)$ and the vectors $\mathbf{a} = 2\mathbf{i} + \mathbf{k}$ and $\mathbf{b} = \mathbf{i} - \mathbf{j} - 2\mathbf{k}$. Obtain its equation:

    a) in vector form;

    b) in parametric form;

    c) in the form $Ax + By + Cz = D$.

23. A plane contains the point $(1, 0, 3)$ and is normal to the vector $2\mathbf{i} - \mathbf{j} + \mathbf{k}$. Find its equation:

    a) in vector form;

    b) in parametric form,

    c) in the form $Ax + By + Cz = D$.

24. A line contains the points $(4, 2, -3)$ and $(1, 2, -1)$. Obtain its equation:

    a) in vector form;

    b) in symmetric form;

    c) in parametric form.

25. For any three vectors $\mathbf{a}$, $\mathbf{b}$, and $\mathbf{c}$, show that

    $$[\mathbf{a}, \mathbf{b}, \mathbf{c}] = [\mathbf{b}, \mathbf{c}, \mathbf{a}] = [\mathbf{c}, \mathbf{a}, \mathbf{b}] = -[\mathbf{b}, \mathbf{a}, \mathbf{c}] = -[\mathbf{c}, \mathbf{b}, \mathbf{a}] = -[\mathbf{a}, \mathbf{c}, \mathbf{b}].$$

26. Show that the vector equation of the line (5.1–14) can be written as

    $$\mathbf{r} = \mathbf{r}_2 + s(\mathbf{r}_1 - \mathbf{r}_2),$$

    where s is an arbitrary scalar.

27. Find the projection of $\mathbf{i} - 2\mathbf{j} + \mathbf{k}$ on the vector $3\mathbf{i} + \mathbf{j} - \mathbf{k}$. Explain your result.

28. What is the projection of a vector on itself?

29. If a and b are the lengths of any two sides of a triangle and if θ is the angle between these sides, obtain the following formula for the area:

    $$A = \tfrac{1}{2} ab \sin \theta.$$

    (*Hint*: See Example 5.1–2.)

30. Determine the vector equation of the plane containing the points $(0, 2, 1)$, $(1, 2, -1)$, and $(1, 0, 3)$.

31. Determine the direction cosines of the line of Exercise 24.

32. Find the work done by a force $\mathbf{F} = 2\mathbf{i} - 3\mathbf{j} + 12\mathbf{k}$ in moving a particle through a displacement $\mathbf{d} = \mathbf{i} + 4\mathbf{j} - 6\mathbf{k}$.

33. Find two unit vectors perpendicular to both $\mathbf{a} = 3\mathbf{i} + 2\mathbf{j} - 3\mathbf{k}$ and $\mathbf{b} = \mathbf{i} - 2\mathbf{j} + \mathbf{k}$.

34. Determine whether the vectors

    $$\mathbf{a} = \mathbf{i} + 3\mathbf{k}, \quad \mathbf{b} = \mathbf{i} + \mathbf{j}, \quad \text{and} \quad \mathbf{c} = \mathbf{i} + 2\mathbf{j} + \mathbf{k}$$

    form a right-hand system.

▶▶▶ 35. **a)** If $\mathbf{n}$ is a unit normal to a plane that is at a (perpendicular) distance p from the origin, show that the equation of the plane can be written as

    $$\mathbf{r} \cdot \mathbf{n} = p.$$

b) If the direction angles of the normal are α, β, and γ, show that the equation of the plane can be written as

$$x \cos \alpha + y \cos \beta + z \cos \gamma = p.$$

c) Find the distance of the plane $2x - 3y + 4z = 10$ from the origin.

36. Prove the "triangle inequality,"

$$|\mathbf{a} + \mathbf{b}| \leq |\mathbf{a}| + |\mathbf{b}|,$$

for arbitrary vectors $\mathbf{a}$ and $\mathbf{b}$.

37. Use the triangle inequality (Exercise 36) to prove

$$|\mathbf{a} - \mathbf{b}| \geq \big||\mathbf{a}| - |\mathbf{b}|\big|.$$

(*Note*: The right-hand member of the above inequality expresses the absolute value of the difference of two real numbers.)

38. The direction of a vector in $\mathbb{R}^3$ was given in terms of direction angles, direction cosines, and direction numbers. Explain how these three terms can be defined in each of the following.

a) $\mathbb{R}^1$ **b)** $\mathbb{R}^2$ **c)** $\mathbb{R}^3$

39. Establish Lagrange's identity:

$$(\mathbf{a} \times \mathbf{b}) \cdot (\mathbf{c} \times \mathbf{d}) = \begin{vmatrix} \mathbf{a} \cdot \mathbf{c} & \mathbf{a} \cdot \mathbf{d} \\ \mathbf{b} \cdot \mathbf{c} & \mathbf{b} \cdot \mathbf{d} \end{vmatrix}.$$

40. Let $\mathbf{a}$ and $\mathbf{b}$ be unit vectors in $\mathbb{R}^2$, making angles α and β, respectively, with the positive x-axis. Deduce the familiar formulas for $\cos(\alpha - \beta)$ and $\sin(\alpha - \beta)$ from $\mathbf{a} \cdot \mathbf{b}$ and $\mathbf{a} \times \mathbf{b}$.

41. Establish the **Cauchy–Schwarz* inequality**:

$$(\mathbf{a} \cdot \mathbf{b})^2 \leq a^2 b^2.$$

42. Use vector algebra to prove the law of cosines,

$$c^2 = a^2 + b^2 - 2ab \cos \theta,$$

where a, b, c, and θ are shown in Fig. 5.1–17. (*Hint*: Write $\mathbf{c} = \mathbf{b} - \mathbf{a}$; then find the dot product of each member of the equation with itself.)

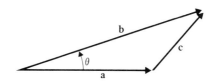

FIGURE 5.1–17 The law of cosines (Exercise 42).

---

* Hermann A. Schwarz (1843–1921), a German mathematician.

43. Establish the formula

$$\mathbf{a} \times (\mathbf{b} \times \mathbf{c}) = (\mathbf{c} \cdot \mathbf{a})\mathbf{b} - (\mathbf{b} \cdot \mathbf{a})\mathbf{c}.$$

(*Note*: A mnemonic aid for this is "cab-back.")

5.2 VECTOR DIFFERENTIATION

In the preceding section we dealt with *constant* vectors exclusively. Because of this we were limited in the use we could make of vectors. For example, we could study *lines* in $\mathbb{R}^3$ since these had constant direction. We could also study *planes* since they had normals that were constant vectors. In the applications we were limited to considering the work done by a constant force acting in a constant direction and to problems involving constant angular velocity and moment.

In this section we will encounter vectors that are *functions* of real variables. This will permit us to broaden our field of knowledge so that we will be able to consider vector fields, motion along a curved path, angles between curves, curvature, torsion, etc. These topics, in turn, will lead us into differential geometry, fluid dynamics, thermodynamics, field theory, and others.

Consider the vector

$$\mathbf{v}(t) = x(t)\mathbf{i} + y(t)\mathbf{j} + z(t)\mathbf{k}, \tag{5.2–1}$$

where $x(t)$, $y(t)$, and $z(t)$ are **scalar functions** of the real variable t. In other words, given a real number t, $x(t)$ defines a single real number; the same is true for $y(t)$ and $z(t)$. These scalar functions are familiar from the single-variable calculus and can be thought of as mappings from $\mathbb{R}$ to $\mathbb{R}$, where $\mathbb{R}$ is the set of real numbers. We use the modifier "scalar" merely to distinguish them from "vector" functions.

Equation (5.2–1) defines a **vector function** $\mathbf{v}(t)$. This function is a mapping from $\mathbb{R}$ to $\mathbb{R}^3$. It is single-valued because the components $x(t), y(t)$, and $z(t)$ have this characteristic. We will also assume that the scalar components of $\mathbf{v}(t)$ are *continuous* on some interval, such as (a, b), although we may not belabor this fact by continually referring to this interval. In brief, we assume that $\mathbf{v}(t)$ is a continuous function of t on some interval.

We define the *derivative* of $\mathbf{v}(t)$ with respect to t in a natural way as follows:

$$\frac{d\mathbf{v}(t)}{dt} = \lim_{\Delta t \to 0} \frac{\mathbf{v}(t + \Delta t) - \mathbf{v}(t)}{\Delta t}. \tag{5.2–2}$$

As in the definition of the derivative of a continuous function in the real-variable calculus, we may designate the vector $\mathbf{v}(t + \Delta t) - \mathbf{v}(t)$ by $\Delta \mathbf{v}$. Then, whenever $\Delta \mathbf{v}/\Delta t$ approaches a limit as $\Delta t \to 0$, we call this limit the derivative of $\mathbf{v}(t)$ with respect to t.

From the definition in Eq. (5.2–2) it is possible to obtain various properties of vector differentiation. These are listed in the following theorem and we give a proof of part (b), leaving the proofs of the other parts for the exercises.

Theorem 5.2–1 Let $\mathbf{u}(t)$, $\mathbf{v}(t)$, and $\mathbf{w}(t)$ be differentiable vector functions of t and let f be a differentiable scalar function of t. Then:

a) $\dfrac{d}{dt}(\mathbf{u} + \mathbf{v}) = \dfrac{d\mathbf{u}}{dt} + \dfrac{d\mathbf{v}}{dt}$;

b) $\dfrac{d}{dt}(\mathbf{u} \cdot \mathbf{v}) = \mathbf{u} \cdot \dfrac{d\mathbf{v}}{dt} + \mathbf{v} \cdot \dfrac{d\mathbf{u}}{dt}$;

c) $\dfrac{d}{dt}(f\mathbf{u}) = f\dfrac{d\mathbf{u}}{dt} + \dfrac{df}{dt}\mathbf{u}$;

d) $\dfrac{d}{dt}(\mathbf{u} \times \mathbf{v}) = \dfrac{d\mathbf{u}}{dt} \times \mathbf{v} + \mathbf{u} \times \dfrac{d\mathbf{v}}{dt}$;

e) $\dfrac{d}{dt}(\mathbf{u} \cdot \mathbf{v} \times \mathbf{w}) = \dfrac{d\mathbf{u}}{dt} \cdot \mathbf{v} \times \mathbf{w} + \mathbf{u} \cdot \dfrac{d\mathbf{v}}{dt} \times \mathbf{w} + \mathbf{u} \cdot \mathbf{v} \times \dfrac{d\mathbf{w}}{dt}$.

PROOF OF PART (b). Consider $\mathbf{u}(t) \cdot \mathbf{v}(t) = g(t)$, so that

$$g(t + \Delta t) - g(t) = \mathbf{u}(t + \Delta t) \cdot \mathbf{v}(t + \Delta t) - \mathbf{u}(t) \cdot \mathbf{v}(t).$$

But

$$\mathbf{u}(t + \Delta t) = \mathbf{u}(t) + \Delta\mathbf{u} \quad \text{and} \quad \mathbf{v}(t + \Delta t) = \mathbf{v}(t) + \Delta\mathbf{v};$$

hence

$$g(t + \Delta t) - g(t) = (\mathbf{u} + \Delta\mathbf{u}) \cdot (\mathbf{v} + \Delta\mathbf{v}) - \mathbf{u} \cdot \mathbf{v}$$
$$= \mathbf{u} \cdot \mathbf{v} + \Delta\mathbf{u} \cdot \mathbf{v} + \mathbf{u} \cdot \Delta\mathbf{v} + \Delta\mathbf{u} \cdot \Delta\mathbf{v} - \mathbf{u} \cdot \mathbf{v}$$

and

$$\frac{g(t + \Delta t) - g(t)}{\Delta t} = \frac{\Delta\mathbf{u}}{\Delta t} \cdot \mathbf{v} + \mathbf{u} \cdot \frac{\Delta\mathbf{v}}{\Delta t} + \frac{\Delta\mathbf{u}}{\Delta t} \cdot \Delta\mathbf{v}.$$

Now, taking limits as $t \to 0$, the left member becomes $dg(t)/dt$ whereas the right member has the form of the right member of (b) since $\Delta\mathbf{v} \to \mathbf{0}$ also. □

Differential geometry

Recall from calculus that if $y = f(x)$ is a differentiable function, then dy/dx is the slope of the tangent to the graph of the function. We next inquire into the significance of the derivative of a vector function. The system of equations

$$x = x(t),$$
$$y = y(t), \tag{5.2–3}$$
$$z = z(t),$$

where x, y, and z are functions of a parameter t (not necessarily "time"), can be considered parametric equations of a *path* in space. Such a path is called a *space curve* or *arc*. We observe that the word "space" is used even if we are talking about a *plane* curve and that the word "curve" is used even if we are talking about a line segment. The path described by Eq. (5.2–3) can also be expressed in vector form by

$$\mathbf{r}(t) = x(t)\mathbf{i} + y(t)\mathbf{j} + z(t)\mathbf{k}. \qquad \textbf{(5.2–4)}$$

When we do this, we consider $\mathbf{r}(t)$ to be the position vector (with respect to some origin) of the point having coordinates $(x(t), y(t), z(t))$ or (x, y, z). We shall consistently use $\mathbf{r}(t)$ in this way.

It will be convenient to work with arcs that are *smooth* according to the following definition.

Definition 5.2–1 Consider an arc described by $\mathbf{r}(t)$ as in Eq. (5.2–4). Let the point corresponding to $t = t_0$ be denoted P_0 and that corresponding to $t = t_1$ be denoted P_1. Then the arc P_0P_1 is said to be **smooth** if the following conditions are satisfied:

a) $d\mathbf{r}/dt$ is a continuous function of t for $t_0 \leq t \leq t_1$;

b) distinct points of the arc correspond to distinct values of t in the interval $t_0 \leq t \leq t_1$;

c) $d\mathbf{r}/dt \neq \mathbf{0}$ for any value of t in the interval $t_0 \leq t \leq t_1$.

Note that (a) ensures that there are no sudden changes of direction, (b) is another way of saying that the curve does not cross itself, and (c) implies that the point is never at rest.

In Fig. 5.2–1 we show a smooth curve C that is *oriented* in such a way that an arbitrary point P on C moves to Q with increasing t. If P has position vector $\mathbf{r}(t)$ and Q has position vector $\mathbf{r}(t + \Delta t)$, then the

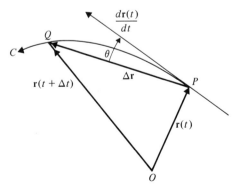

FIGURE 5.2–1 Vector differentiation.

vector from P to Q is $\mathbf{r}(t + \Delta t) - \mathbf{r}(t)$, which lies on the *secant* line PQ. Dividing the vector difference by the scalar Δt does not change the direction of this vector. Hence it is reasonable to define $d\mathbf{r}(t)/dt$ as the **vector tangent** to C at P by virtue of Eq. (5.2–2). Denoting the vector from P to Q by $\Delta \mathbf{r}$, we have

$$\Delta \mathbf{r} \cdot d\mathbf{r}/dt = |\Delta \mathbf{r}| \, |d\mathbf{r}/dt| \cos \theta$$

or

$$\cos \theta = \frac{\Delta \mathbf{r} \cdot d\mathbf{r}/dt}{|\Delta \mathbf{r}| \, |d\mathbf{r}/dt|}. \tag{5.2–5}$$

Definition 5.2–1 ensures that $\Delta \mathbf{r} \to \mathbf{0}$ if and only if $\Delta t \to 0$ and dividing the numerator and denominator in Eq. (5.2–5) by Δt produces

$$\cos \theta = \frac{(\Delta \mathbf{r}/\Delta t) \cdot (d\mathbf{r}/dt)}{|\Delta \mathbf{r}/\Delta t| \, |d\mathbf{r}/dt|}$$

so that the limit of $\cos \theta$ as $\Delta t \to 0$ is one. This justifies the definition of the tangent to a curve.

The **unit tangent vector**, denoted by $\mathbf{T}$, may be obtained from $d\mathbf{r}/dt$ by dividing the latter by $|d\mathbf{r}/dt|$. Stated another way,

$$\mathbf{T} = \frac{(dx/dt)\mathbf{i} + (dy/dt)\mathbf{j} + (dz/dt)\mathbf{k}}{\sqrt{(dx/dt)^2 + (dy/dt)^2 + (dz/dt)^2}} = \frac{d\mathbf{r}/dt}{|d\mathbf{r}/dt|},$$

which gives the unit tangent vector to a space curve in terms of a parameter t. If t represents time, then $d\mathbf{r}/dt = \mathbf{v}$ is called the **velocity** of the particle moving on C. The scalar $|d\mathbf{r}/dt| = v$ is the **speed**. Since the acceleration is the rate of change of velocity, the **acceleration vector** is given by

$$\mathbf{a} = d\mathbf{v}/dt = d^2\mathbf{r}/dt^2.$$

Thus the position, velocity, and acceleration of a particle moving along a path C are all functions of t so that $d\mathbf{r}/dt$ and $d^2\mathbf{r}/dt^2$ express the *instantaneous* velocity and acceleration, respectively, of the particle whose position vector is $\mathbf{r}(t)$.

Sometimes it is more convenient to use *arc length s* as a parameter since a number of important formulas can be more easily derived in this way. Recall from the calculus that the differential of arc length ds is given by

$$ds^2 = dx^2 + dy^2 + dz^2.$$

Hence the unit tangent vector can be expressed as

$$\mathbf{T} = d\mathbf{r}/ds \tag{5.2–6}$$

and the velocity becomes

$$\mathbf{v} = \frac{d\mathbf{r}}{dt} = \frac{d\mathbf{r}}{ds}\frac{ds}{dt} = \frac{ds}{dt}\mathbf{T} = v\mathbf{T} \tag{5.2–7}$$

using the chain rule and Eq. (5.2–6). Now the acceleration is

$$\mathbf{a} = \frac{d\mathbf{v}}{dt} = \frac{d(v\mathbf{T})}{dt} = \frac{dv}{dt}\mathbf{T} + v\frac{d\mathbf{T}}{dt} = \frac{dv}{dt}\mathbf{T} + v^2\frac{d\mathbf{T}}{ds}, \qquad \text{(5.2–8)}$$

using the fact that $d\mathbf{T}/dt = (d\mathbf{T}/ds)(ds/dt) = v(d\mathbf{T}/ds)$. But, since $\mathbf{T}$ is a unit vector, $\mathbf{T} \cdot \mathbf{T} = 1$ and differentiating this with respect to s leads to $\mathbf{T} \cdot (d\mathbf{T}/ds) = 0$, that is, $\mathbf{T}$ is normal to $d\mathbf{T}/ds$ (Exercise 5). A unit vector in the direction of $d\mathbf{T}/ds$ is denoted by $\mathbf{N}$ and called the **normal** (or *principal normal*) to the curve C. Hence

$$\frac{d\mathbf{T}}{ds} = \kappa\mathbf{N}, \qquad \text{(5.2–9)}$$

where the direction of $\mathbf{N}$ is chosen in such a way that κ is nonnegative. We call $|dT/ds| = 1/\rho = \kappa$ the **curvature** of C and ρ the **radius of curvature**. In terms of these we can rewrite Eq. (5.2–8) as

$$\mathbf{a} = \frac{dv}{dt}\mathbf{T} + \frac{v^2}{\rho}\mathbf{N}. \qquad \text{(5.2–10)}$$

The scalar quantities dv/dt and v^2/ρ are called the *tangential* and *normal components* of the acceleration, respectively. These are of practical interest since it is often useful to decompose (or resolve) the acceleration into two directions, one in the direction of motion (along $\mathbf{T}$) and the other normal to this direction (along $\mathbf{N}$).

Figure 5.2–2 shows the orientation of $\mathbf{N}$ as well as the significance of ρ. We call the point N the **center of curvature**. A circle of radius ρ drawn with N as center will fit (or match) the curve at P more closely than any other circle. This circle is called the **circle of curvature** (or osculating circle). The vector $\mathbf{N}$ is oriented in such a way that it is parallel to the vector from P to N. In the case where C is a straight line, κ is zero, $\mathbf{N}$ may have either of two directions, and ρ is infinite.

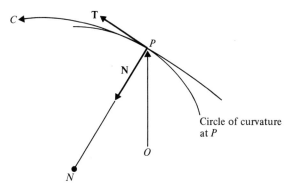

FIGURE 5.2–2 Space curve showing **N** and **T**.

The vectors **T** and **N** determine a plane that is called the **osculating plane**.

We define a third unit vector,

$$\mathbf{B} = \mathbf{T} \times \mathbf{N}, \tag{5.2-11}$$

called the **binormal**. Since the vectors **T**, **N**, and **B** form a right-hand (in that order) coordinate system (Fig. 5.2-3) of vectors at each point of the curve C, they are called a *moving trihedral*.* The plane determined by **N** and **B** is called the **normal plane** whereas that determined by **T** and **B** is the **rectifying plane**. These planes, together with the osculating plane, are also known as the *fundamental planes*.

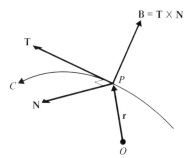

FIGURE 5.2-3 A moving trihedral.

From $\mathbf{B} \cdot (d\mathbf{B}/ds) = 0$ (see Exercise 5) we see that $d\mathbf{B}/ds$ is normal to **B**, hence has the form (why?)

$$\frac{d\mathbf{B}}{ds} = c_1 \mathbf{T} + c_2 \mathbf{N}$$

for appropriate scalars c_1 and c_2. But since $\mathbf{B} \cdot \mathbf{T} = 0$, we see that $\mathbf{B} \cdot (d\mathbf{T}/ds) + \mathbf{T} \cdot (d\mathbf{B}/ds) = 0$, that is,

$$\mathbf{T} \cdot (d\mathbf{B}/ds) = -\mathbf{B} \cdot (d\mathbf{T}/ds) = -\mathbf{B} \cdot \kappa \mathbf{N},$$

by Eq. (5.2-9). Since $\mathbf{B} \cdot \mathbf{N} = 0$, $d\mathbf{B}/ds$ is normal to **T**, hence $c_1 = 0$ and $d\mathbf{B}/ds$ has the same direction as **N**. We define

$$\frac{d\mathbf{B}}{ds} = -\tau \mathbf{N}, \tag{5.2-12}$$

---

* From the term "trihedron," a figure formed by three planes intersecting at a point.

where the scalar τ is called the **torsion** of the curve C at P. The torsion is positive if the binormal twists in the direction from $\mathbf{N}$ to $\mathbf{B}$ around the tangent as P moves along the curve in the direction of increasing s. (*Note*: Some authors define the torsion as the negative of the torsion defined above.) The reciprocal of the torsion, $\sigma = 1/\tau$, is called the **radius of torsion**. Loosely speaking, the torsion is a measure of the tendency of a curve to twist out of its osculating plane. The torsion of a plane curve is zero (Exercise 8).

We have obtained formulas for $d\mathbf{T}/ds$ and $d\mathbf{B}/ds$ in terms of $\mathbf{T}$, $\mathbf{N}$, and $\mathbf{B}$. To complete this list we add a formula for $d\mathbf{N}/ds$. Since the three unit vectors are mutually orthogonal, they are linearly independent and we can write

$$\frac{d\mathbf{N}}{ds} = a\mathbf{T} + b\mathbf{N} + c\mathbf{B}$$

for appropriate scalars a, b, and c. But $\mathbf{N} \cdot (d\mathbf{N}/ds) = 0$ (why?), hence $b = 0$. From $\mathbf{T} \cdot \mathbf{N} = 0$ we obtain $\mathbf{T} \cdot (d\mathbf{N}/ds) + \mathbf{N} \cdot (d\mathbf{T}/ds) = 0$ or $\mathbf{T} \cdot (d\mathbf{N}/ds) + \kappa = 0$ (why?), hence $a = -\kappa$. From $\mathbf{B} \cdot \mathbf{N} = 0$ we have in similar fashion $c = \tau$, so that

$$\frac{d\mathbf{N}}{ds} = -\kappa\mathbf{T} + \tau\mathbf{B}.$$

The three formulas

$$d\mathbf{T}/ds = \kappa\mathbf{N},$$
$$d\mathbf{N}/ds = -\kappa\mathbf{T} + \tau\mathbf{B}, \qquad \text{(5.2–13)}$$
$$d\mathbf{B}/ds = -\tau\mathbf{N}$$

are called the **Frenet–Serret formulas**.[*] They are fundamental in that portion of *differential geometry* (geometry in the neighborhood of a point) that deals with space (or twisted) curves.

EXAMPLE 5.2–1 Obtain the three unit vectors of the moving trihedral as well as the curvature and torsion for the right circular helix (see Fig. 5.2–4)

$$\mathbf{r}(t) = a \cos t\mathbf{i} + a \sin t\mathbf{j} + bt\mathbf{k}, \qquad t \geq 0. \qquad \text{(5.2–14)}$$

Solution. We can obtain this space curve by winding a string around a circular cylinder of radius a. The resulting curve is the same as the thread of a right-hand screw of pitch $2\pi b$. We have

$$\frac{d\mathbf{r}}{dt} = -a \sin t\mathbf{i} + a \cos t\mathbf{j} + b\mathbf{k};$$

---

[*] After the French mathematicians Jean-Frédéric Frenet (1816–1900) and Joseph A. Serret (1819–1885), who published them in 1852 and 1851, respectively.

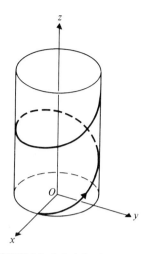

FIGURE 5.2–4 A right circular helix.

hence, by Eq. (5.2–6),

$$\mathbf{T} = \frac{d\mathbf{r}}{ds} = \frac{\dfrac{d\mathbf{r}}{dt}}{\left|\dfrac{d\mathbf{r}}{dt}\right|} = (a^2 + b^2)^{-1/2}(-a\sin t\mathbf{i} + a\cos t\mathbf{j} + b\mathbf{k}).$$

From $d\mathbf{T}/ds = \kappa\mathbf{N}$ in Eq. (5.2–13) we have

$$\frac{d\mathbf{T}}{ds} = \frac{d\mathbf{T}}{dt}\frac{ds}{dt} = (a^2 + b^2)^{-1}(-a\cos t\mathbf{i} - a\sin t\mathbf{j}) = \kappa\mathbf{N}$$

so that

$$\mathbf{N} = -\cos t\mathbf{i} - \sin t\mathbf{j}$$

and $\kappa = a/(a^2 + b^2)$. Now

$$\mathbf{B} = \mathbf{T} \times \mathbf{N} = (a^2 + b^2)^{-1/2}\begin{vmatrix} \mathbf{i} & \mathbf{j} & \mathbf{k} \\ -a\sin t & a\cos t & b \\ -\cos t & -\sin t & 0 \end{vmatrix}$$

$$= (a^2 + b^2)^{-1/2}(b\sin t\mathbf{i} - b\cos t\mathbf{j} + a\mathbf{k}).$$

From $d\mathbf{B}/ds = -\tau\mathbf{N}$ in Eq. (5.2–13) we have

$$\frac{d\mathbf{B}}{ds} = (a^2 + b^2)^{-1}(b\cos t\mathbf{i} + b\sin t\mathbf{j}) = -\tau\mathbf{N}$$

so that $\tau = b/(a^2 + b^2)$. Observe that both the curvature and torsion are constant. ■

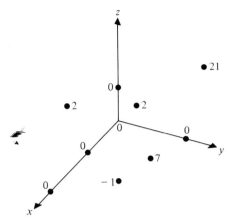

FIGURE 5.2–5 A continuous scalar field
$f(x, y, z) = y(x^2 + z)$.

Scalar and vector fields

A scalar function $f(x, y, z)$ defined over some domain D is also called
a *scalar point function*. The name implies that at every point P of D
the function $f(x, y, z)$ defines a scalar given by $f(P)$. The totality of
points P and scalars $f(P)$ is also called a **scalar field**. Thus $f(x, y, z) =$
$x^2y + yz$ is a scalar field in $\mathbb{R}^3$ since $f(0, 0, 0) = 0$, $f(1, 2, -1) = 0$,
$f(-2, 1, 3) = 7$, etc. (Fig. 5.2–5). This field is defined at *every* point and
is an example of a *continuous scalar field*. A map of a forest showing
the heights of trees is an example of a *discrete scalar field* (Fig. 5.2–6).

A function $\mathbf{F}(x, y, z)$ is called a *vector point function* since at points
P of a domain D, $\mathbf{F}(P)$ defines a vector. The totality of points P and

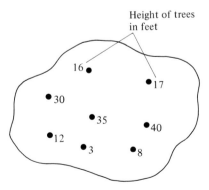

Height of trees
in feet

FIGURE 5.2–6 A discrete scalar field.

vectors $\mathbf{F}(P)$ is called a **vector field**. An example of such a field, defined in $\mathbb{R}^3$, is given by

$$\mathbf{F}(x, y, z) = xy\mathbf{i} + \cos yz\mathbf{j} + xe^z\mathbf{k}.$$

Other examples of vector fields are the velocity of a moving fluid, gravitational force fields, and electrostatic fields due to a number of charged bodies. On a global scale, currents in the Pacific Ocean, shown in Fig. 5.2–7, constitute a vector field.

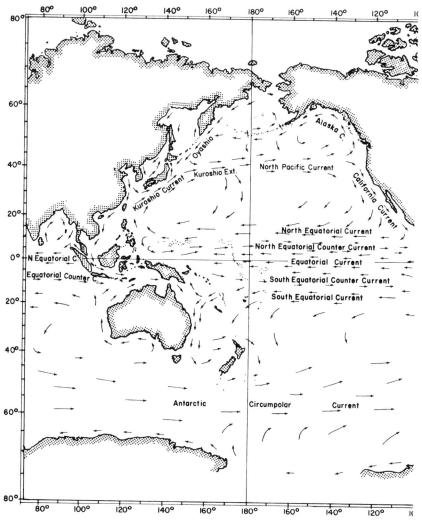

FIGURE 5.2–7 Currents in the Pacific Ocean. (From William C. Patzert et al., "Aircraft monitoring of ocean thermal structure and currents," *Nav. Res. Rev.* XXXI (September 1978), p. 1. Reproduced by permission.)

Scalar and vector fields may be defined as functions of time as well as position. The motion of a fluid, for example, is completely determined when the velocity vector is known as a function of time and position. In case a scalar field is *time-independent*, it is also referred to as a **steady** field. We will discuss other types of fields in Section 5.7.

KEY WORDS AND PHRASES

scalar function	curvature
vector function	radius of curvature
smooth arc	osculating plane
tangent vector	binormal
velocity	torsion
speed	Frenet–Serret formulas
acceleration vector	scalar field
normal	vector field

EXERCISES 5.2

▶ 1. Prove Theorem 5.2–1(a).

2. Prove Theorem 5.2–1(c).

3. Prove Theorem 5.2–1(d). (*Hint*: The order of the factors must be preserved since the cross product is not commutative.)

4. Prove Theorem 5.2–1(e).

5. Show that if r is a vector having *constant* (nonzero) length, then the directions r and dr/ds are perpendicular.

6. Show that $r + \rho N$ determines the center of curvature of a curve C at a point having position vector r.

7. Show that the acceleration vector of a curve lies in its osculating plane.

8. Prove that the torsion of a plane curve is zero.

9. Prove that the curvature of a straight line is zero.

10. Carry out the details in the text to show that

$$d\mathbf{N}/ds = -\kappa \mathbf{T} + \tau \mathbf{B}.$$

11. For the right circular helix

$$\mathbf{r}(t) = a \cos t\mathbf{i} + a \sin t\mathbf{j} + bt\mathbf{k}, \qquad t \geq 0,$$

find:

a) the vector equation of the tangent line;
(*Note*: The *equation* of the tangent *line* is not the same as the tangent *vector* **T**.)

b) the vector equation of the normal line;

c) the vector equation of the osculating plane.

▶▶ 12. Find the speed and the normal and tangential components of the acceleration of the curve

$$\mathbf{r}(t) = t^2\mathbf{i} + \frac{t^3}{3}\mathbf{j} + \frac{t^4}{4}\mathbf{k}$$

at the point where $t = 1$.

13. Find **T**, **N**, and **B** for the path

$$\mathbf{r}(t) = 3t\mathbf{i} - t^2\mathbf{j} + t^3\mathbf{k}$$

at $t = 2$.

14. Find the vectors **T**, **N**, and **B** for the space curve

$$\mathbf{r}(t) = (\sin t - t \cos t)\mathbf{i} + (\cos t + t \sin t)\mathbf{j} + t^2\mathbf{k}.$$

15. Find the normal and tangential components of the acceleration of a particle traveling in a path given in Exercise 14.

16. Find the curvature of the path given in Exercise 14.

17. Denoting differentiation with respect to t by a dot, show that for a space
curve $\mathbf{r} = \mathbf{r}(t)$:

a) $\kappa = \dfrac{|\dot{\mathbf{r}} \times \ddot{\mathbf{r}}|}{(\dot{\mathbf{r}} \cdot \dot{\mathbf{r}})^{3/2}};$ **b)** $\tau = \dfrac{(\dot{\mathbf{r}} \times \ddot{\mathbf{r}}) \cdot \dddot{\mathbf{r}}}{|\dot{\mathbf{r}} \times \ddot{\mathbf{r}}|^2}.$

18. Use the formulas given in Exercise 17 to compute the curvature and torsion of the right circular helix (5.2–14).

19. Find the curvature and torsion for the space curve

$$\mathbf{r}(t) = (3t - t^3)\mathbf{i} + 3t^2\mathbf{j} + (3t + t^3)\mathbf{k}.$$

20. Find the curvature and torsion of the *twisted cubic*

$$\mathbf{r}(t) = t\mathbf{i} + 3t^2\mathbf{j} + t^3\mathbf{k}.$$

21. Evaluate the scalar field

$$f(x, y, z) = xy + yz + xz$$

at each of the following points.

a) $(1, -2, 3)$

b) $(-2, 0, 4)$

c) Along the line $x = y = z$

22. Give three physical examples of a scalar field.

23. Obtain the equation of the tangent line of the twisted cubic (compare
Exercise 20) at the point (2, 12, 8).

24. Graph the vector field

$$\mathbf{F}(x, y) = \frac{-y\mathbf{i} + x\mathbf{j}}{\sqrt{x^2 + y^2}}.$$

(*Hint*: Compute some values of **F** on the circles $x^2 + y^2 = a^2$ for various
constants a.)

25. Graph the vector field

$$\mathbf{F}(x, y) = \frac{x\mathbf{i} + y\mathbf{j}}{\sqrt{x^2 + y^2}}.$$

26. a) Graph the vector field

$$\mathbf{F}(x, y, z) = x\mathbf{i} + y\mathbf{j} + z\mathbf{k}.$$

b) Does the graph suggest why this is called a "source" field? Explain.

27. Give three physical examples of a vector field.

▶▶▶ **28.** The equations

$$\kappa = f(s), \qquad \tau = g(s)$$

are called the *natural* or *intrinsic* equations of a curve. Obtain the natural equations of the cycloid

$$\mathbf{r}(t) = a(t - \sin t)\mathbf{i} + a(1 - \cos t)\mathbf{j}.$$

29. The vector

$$\mathbf{D} = \tau\mathbf{T} + \kappa\mathbf{B}$$

is called the *Darboux* vector* of a space curve. Show that equations (5.2–13) can be written

$$d\mathbf{T}/ds = \mathbf{D} \times \mathbf{T},$$
$$d\mathbf{N}/ds = \mathbf{D} \times \mathbf{N},$$
$$d\mathbf{B}/ds = \mathbf{D} \times \mathbf{B}.$$

30. Prove that for any space curve,

$$(d\mathbf{T}/ds) \cdot (d\mathbf{B}/ds) = -\kappa\tau.$$

31. Prove that the normal component of the acceleration is zero if and only if the path is a straight line.

32. Prove that if the motion of a particle along a path is such that the speed is constant, then the tangential component of the acceleration is zero, and conversely.

33. Prove that if a particle has position vector along a curve C given by

$$\mathbf{r}(t) = x(t)\mathbf{i} + y(t)\mathbf{j} + z(t)\mathbf{k},$$

where $x(t)$, $y(t)$, and $z(t)$ are quadratic functions of t (that is, of the form $at^2 + bt + c$), then C is a plane curve.

34. Referring to Exercise 33, prove that if $x(t)$, $y(t)$, and $z(t)$ are linear functions of t, then C is a straight line.

35. A plane curve can be written $y = f(x)$ or $y = f(t)$ or, in vector form,

$$\mathbf{r} = t\mathbf{i} + y\mathbf{j}.$$

Prove that the radius of curvature, familiar from calculus, is given by

$$\rho = \frac{(1 + (y')^2)^{3/2}}{y''}.$$

---

* Jean Gaston Darboux (1842–1917) was a French mathematician who did important work in differential geometry.

5.3 THE DEL OPERATOR: GRADIENT, DIVERGENCE, AND CURL

In the preceding section we dealt with vector functions of position that were stated in terms of a parameter t (time) or in terms of coordinates x, y, and z. In the former case we saw that the position vector $\mathbf{r}(t)$ led to ordinary derivatives $d\mathbf{r}(t)/dt$ and $d^2\mathbf{r}(t)/dt^2$, which were the velocity and acceleration vectors, respectively. When the position vector is in the form

$$\mathbf{r}(x, y, z) = a(x, y, z)\mathbf{i} + b(x, y, z)\mathbf{j} + c(c, y, z)\mathbf{k}, \qquad (5.3\text{--}1)$$

however, then we must deal with *partial derivatives*. Since $a(x, y, z)$, $b(x, y, z)$, and $c(x, y, z)$ in Eq. (5.3–1) are *scalar* functions, we define the partial derivatives of $\mathbf{r}$ in terms of the partial derivatives of these functions. For example,

$$\frac{\partial \mathbf{r}}{\partial x} = \lim_{\Delta x \to 0} \frac{\mathbf{r}(x + \Delta x, y, z) - \mathbf{r}(x, y, z)}{\Delta x}$$

$$= \lim_{\Delta x \to 0} \frac{a(x + \Delta x, y, z)\mathbf{i} + b(x + \Delta x, y, z)\mathbf{j} + c(x + \Delta x, y, z)\mathbf{k} - \mathbf{r}}{\Delta x},$$

and similarly for $\partial \mathbf{r}/\partial y$ and $\partial \mathbf{r}/\partial z$. Thus the differentiability of a vector function of several variables is intrinsically tied to the differentiability of the component scalar functions.

In this section we introduce a *partial differential vector operator* that has wide application in mathematics and physics. We define

$$\nabla \equiv \frac{\partial}{\partial x}\mathbf{i} + \frac{\partial}{\partial y}\mathbf{j} + \frac{\partial}{\partial z}\mathbf{k}, \qquad (5.3\text{--}2)$$

called the **del operator**. The name comes from the fact that the symbol ∇ is the capital Greek letter "delta" upside down. Other names have been used, such as "atled" ("delta" spelled backwards) and "nabla" (an Assyrian harp, which the symbol resembles), but have not gained widespread popularity. The operator as defined in (5.3–2) was first used by Hamilton in a paper published in 1846.

Gradient

We observe that the symbol ∇ defined in (5.3–2) is, in common with other symbols such as $\int$, $\sqrt{\ }$, and $\mathscr{L}$, meaningless by itself. We use the del operator first in connection with a scalar point function or scalar field as given in the next definition.

Definition 5.3–1 Let $f(x, y, z)$ be a differentiable real-valued function. Then the vector function defined by

$$\nabla f = \frac{\partial f}{\partial x}\mathbf{i} + \frac{\partial f}{\partial y}\mathbf{j} + \frac{\partial f}{\partial z}\mathbf{k} \qquad (5.3–3)$$

is called the **gradient** of f. The notation **grad** f is also used for the gradient.

One use of the gradient is found in describing the **directional derivative** of a function at a point. Let a curve C in space be described by the position vector

$$\mathbf{r}(s) = x(s)\mathbf{i} + y(s)\mathbf{j} + z(s)\mathbf{k},$$

using arc length s as parameter. Then

$$\mathbf{T} = \frac{d\mathbf{r}(s)}{ds} = \frac{dx}{ds}\mathbf{i} + \frac{dy}{ds}\mathbf{j} + \frac{dz}{ds}\mathbf{k}$$

is a unit tangent vector to C (compare Eq. 5.2–6). Hence,

$$\nabla f \cdot \mathbf{T} = \frac{\partial f}{\partial x}\frac{dx}{ds} + \frac{\partial f}{\partial y}\frac{dy}{ds} + \frac{\partial f}{\partial z}\frac{dz}{ds} = \frac{\partial}{\partial s}f(x, y, z),$$

which is called the *directional derivative* of f in the direction defined by $\mathbf{T}$. At a given point the directional derivative is a constant and represents the rate of change of a function in a given direction. Observe that $\partial f/\partial x$, $\partial f/\partial y$, and $\partial f/\partial z$ are special cases of the directional derivative (Exercise 1).

EXAMPLE 5.3–1 Compute the directional derivative of the function $f(x, y, z) = x^2 + yz$ at the point $(3, 1, -5)$ in the direction of the vector $\mathbf{a} = \mathbf{i} + 2\mathbf{j} - \mathbf{k}$.

Solution. We have
$$\nabla f = 2x\mathbf{i} + z\mathbf{j} + y\mathbf{k}$$
$$= 6\mathbf{i} - 5\mathbf{j} + \mathbf{k}$$

at $(3, 1, -5)$. Hence,

$$df/ds = \nabla f \cdot \mathbf{a}/a = (6\mathbf{i} - 5\mathbf{j} + \mathbf{k}) \cdot (\mathbf{i} + 2\mathbf{j} - \mathbf{k})/\sqrt{6}$$
$$= -5/\sqrt{6} = -5\sqrt{6}/6 \doteq -2.04. \quad \blacksquare$$

We interpret the above result by saying that *at the point* $(3, 1, -5)$ the function f is *decreasing* approximately two units per unit when we move in the direction of $\mathbf{a}$. Observe that it is essential to use a *unit* vector in the given direction when computing the directional derivative.

The directional derivative of a differentiable function $f(x, y, z)$ at a point P_0 in the direction of the vector $\mathbf{a}$ is given by

$$\frac{df}{ds} = (\nabla f)_0 \cdot \frac{\mathbf{a}}{a}, \qquad (5.3\text{--}4)$$

where $(\nabla f)_0$ is the value of ∇f at P_0. It follows from the definition of the dot product that df/ds attains its maximum value in the direction of the gradient and that this maximum value is, in fact, $|\nabla f|$ (Exercise 3).

Consider the function $w = f(x, y, z)$, which is differentiable in some domain D. The points at which the function is constant, that is, $f(x, y, z) = w_0$, are called **level surfaces** of the function. If w represents temperature, then the level surfaces are called *isothermal surfaces*; if w represents electric potential, then the level surfaces are called *equipotential surfaces*. We next show that the gradient of f is normal to a level surface.

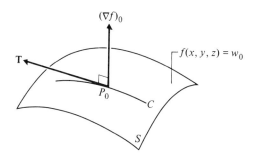

FIGURE 5.3–1 The gradient is normal to the surface.

In Fig. 5.3–1 let S represent a surface whose equation is given by $f(x, y, z) = w_0$ and let P_0 with coordinates (x_0, y_0, z_0) be a point on this surface. If C is *any* curve *on* S that has unit tangent $\mathbf{T}$ at P_0, then the directional derivative of f in the direction of $\mathbf{T}$ is given by

$$\frac{df}{ds} = (\nabla f)_0 \cdot \mathbf{T} = \nabla w_0 \cdot \mathbf{T} = 0,$$

since f is everywhere constant on S. From the last expression we see that $(\nabla f)_0$ is normal to the curve C at P_0. But we would obtain the same result for *every* curve C that passes through P_0 and *lies on* S. This is another way of saying that $(\nabla f)_0$ is *normal to the surface* S at P_0. In case the gradient at a point is the zero vector, we can still make the last statement since the zero vector may be considered to be normal to *every* direction.

EXAMPLE 5.3–2 Obtain the equations of the tangent plane and normal line to the surface of the paraboloid of revolution $x^2 + y^2 - z = 1$ at the point $(1, 2, 4)$.

Solution. If we consider $f(x, y, z) \equiv x^2 + y^2 - z$, then

$$\nabla f = 2x\mathbf{i} + 2y\mathbf{j} - \mathbf{k}$$
$$= 2\mathbf{i} + 4\mathbf{j} - \mathbf{k}$$

at $(1, 2, 4)$. Hence the direction numbers of the normal are given by $(2, 4, -1)$. The tangent plane then has equation $2x + 4y - z = d$ and knowing that $(1, 2, 4)$ is a point in the plane allows us to find d and write $2x + 4y - z = 6$ as the equation of the tangent plane. The normal line in symmetric form is

$$\frac{x - 1}{2} = \frac{y - 2}{4} = \frac{z - 4}{-1}. \quad \blacksquare$$

Note that the direction numbers $(2, 4, -1)$ apply to the *outward-pointing* normal at the given point (Fig. 5.3–2). By defining $f(x, y, z) \equiv -x^2 - y^2 + z$ we could obtain the direction numbers of the *inward-pointing* normal.

We summarize some of the properties of the gradient in the following theorem, the proof of which is left for the exercises (see Exercise 4).

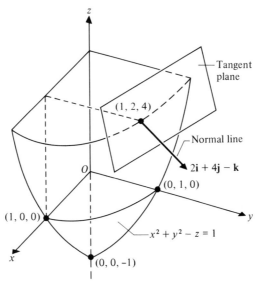

FIGURE 5.3–2

Theorem 5.3–1 If c is a constant and f and g are differentiable scalar functions, then:

a) $\mathbf{V}(f + g) = \mathbf{V}f + \mathbf{V}g$;

b) $\mathbf{V}(cf) = c\mathbf{V}f$;

c) $\mathbf{V}(fg) = f\mathbf{V}g + g\mathbf{V}f$.

We observe that properties (a) and (b) above imply that the del operator is a *linear operator*.

Divergence

Another way in which we can use the del operator is in applying it to a *vector* function. Since the operator has the form of a vector we must use either dot or cross multiplication. Given the vector function (or vector field)

$$\mathbf{v}(x, y, z) = v_1(x, y, z)\mathbf{i} + v_2(x, y, z)\mathbf{j} + v_3(x, y, z)\mathbf{k},$$

we define

$$\mathbf{V} \cdot \mathbf{v} \equiv \frac{\partial v_1}{\partial x} + \frac{\partial v_2}{\partial y} + \frac{\partial v_3}{\partial z}. \qquad (5.3\text{–}5)$$

This *scalar* is called the **divergence** of $\mathbf{v}$ and is also written **div v**. If $\mathbf{v}$ is the velocity field of a fluid, then **div v** is a measure of the fluid's *compressibility* (that is, the rate of change of its *density*); hence for an **incompressible fluid** we have $\mathbf{V} \cdot \mathbf{v} = 0$. In Sections 5.6 and 5.7 we will expand on the physical interpretation of the divergence but for now, we summarize some of its properties in the following theorem. We give a proof of part (c) and leave the proofs of the others for the exercises (see Exercise 5).

Theorem 5.3–2 Let $\mathbf{u}(x, y, z)$ and $\mathbf{v}(x, y, z)$ be differentiable vector functions, $f(x, y, z)$ a differentiable scalar function, and c a constant. Then:

a) $\mathbf{V} \cdot (\mathbf{u} + \mathbf{v}) = \mathbf{V} \cdot \mathbf{u} + \mathbf{V} \cdot \mathbf{v}$;

b) $\mathbf{V} \cdot (c\mathbf{v}) = c(\mathbf{V} \cdot \mathbf{v})$;

c) $\mathbf{V} \cdot (f\mathbf{v}) = f(\mathbf{V} \cdot \mathbf{v}) + \mathbf{V}f \cdot \mathbf{v}$.

PROOF OF PART (c). Let

$$\mathbf{v} = v_1(x, y, z)\mathbf{i} + v_2(x, y, z)\mathbf{j} + v_3(x, y, z)\mathbf{k}.$$

Then

$$f\mathbf{v} = fv_1\mathbf{i} + fv_2\mathbf{j} + fv_3\mathbf{k}$$

and

$$\mathbf{V} \cdot (f\mathbf{v}) = \frac{\partial(fv_1)}{\partial x} + \frac{\partial(fv_2)}{\partial y} + \frac{\partial(fv_3)}{\partial z}$$

$$= f\left(\frac{\partial v_1}{\partial x} + \frac{\partial v_2}{\partial y} + \frac{\partial v_3}{\partial z}\right) + v_1\frac{\partial f}{\partial x} + v_2\frac{\partial f}{\partial y} + v_3\frac{\partial f}{\partial z}$$

$$= f(\mathbf{V} \cdot \mathbf{v}) + \mathbf{v} \cdot (\mathbf{V}f),$$

from which the result follows. $\square$

EXAMPLE 5.3–3   A force field is given by $\mathbf{F} = \mathbf{r}/r^3$. This is called an *inverse square force* (Exercise 16). Find $\mathbf{V} \cdot \mathbf{F}$.

Solution. We have

$$\mathbf{F} = \frac{x\mathbf{i} + y\mathbf{j} + z\mathbf{k}}{(x^2 + y^2 + z^2)^{3/2}};$$

hence

$$\frac{\partial}{\partial x}(x(x^2 + y^2 + z^2)^{-3/2}) = (y^2 + z^2 - 2x^2)(x^2 + y^2 + z^2)^{-5/2}$$

with similar expressions for $\partial/\partial y$ and $\partial/\partial z$. Thus

$$\mathbf{V} \cdot \mathbf{F} = \frac{y^2 + z^2 - 2x^2 + x^2 + z^2 - 2y^2 + x^2 + y^2 - 2z^2}{(x^2 + y^2 + z^2)^{5/2}} = 0. \quad \blacksquare$$

A continuously differentiable vector field $\mathbf{F}$ is called **solenoidal** if $\mathbf{V} \cdot \mathbf{F} = 0$. The origin of the term is found in electromagnetic theory, where $\mathbf{B}$ is the magnetic induction vector and $\mathbf{V} \cdot \mathbf{B} = 0$ is one of Maxwell's* equations.

The gradient of a scalar field is a vector field, hence we can define the divergence of the gradient, called the **Laplacian**[†] of f,

$$\mathbf{V} \cdot \mathbf{V}f = \mathbf{V}^2f = \frac{\partial^2 f}{\partial x^2} + \frac{\partial^2 f}{\partial y^2} + \frac{\partial^2 f}{\partial z^2}. \tag{5.3–6}$$

The operator $\mathbf{V}^2$ is called the *Laplacian* and f is said to satisfy **Laplace's equation** in case $\mathbf{V}^2f = 0$.

Curl

A third way that the del operator can be used is to form the cross product $\mathbf{V} \times \mathbf{v}$. If

$$\mathbf{v}(x, y, z) = v_1(x, y, z)\mathbf{i} + v_2(x, y, z)\mathbf{j} + v_3(x, y, z)\mathbf{k}$$

---

* James Clerk Maxwell (1831–1879), a Scottish physicist.
[†] After Pierre Simon Laplace (1749–1827), a French mathematician.

as before, then we define

$$\mathbf{V} \times \mathbf{v} \equiv \begin{vmatrix} \mathbf{i} & \mathbf{j} & \mathbf{k} \\ \dfrac{\partial}{\partial x} & \dfrac{\partial}{\partial y} & \dfrac{\partial}{\partial z} \\ v_1 & v_2 & v_3 \end{vmatrix}. \qquad (5.3\text{-}7)$$

This determinant is a *symbolic* one and is to be interpreted as the vector resulting when the determinant is expanded in terms of the elements of the first row. In other words,

$$\mathbf{V} \times \mathbf{v} = \left(\frac{\partial v_3}{\partial y} - \frac{\partial v_2}{\partial z} \right) \mathbf{i} + \left(\frac{\partial v_1}{\partial z} - \frac{\partial v_3}{\partial x} \right) \mathbf{j} + \left(\frac{\partial v_2}{\partial x} - \frac{\partial v_1}{\partial y} \right) \mathbf{k}.$$

This vector is called the **curl** of **v** or the *rotation* of **v** and is also written **curl v** or **rot v**.*

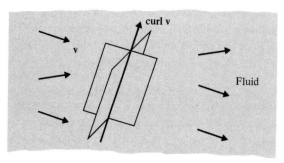

FIGURE 5.3–3 The curl of a vector field.

Although the word "curl" was first used by Maxwell in connection with electrodynamics, an example from fluid dynamics may help to clarify the notion. Imagine that we have a moving viscous fluid (like oil) whose motion is defined by the vector field **v**. We also have a small paddle wheel mounted on an axis and free to turn (Fig. 5.3–3). We place the paddle wheel into the moving fluid so that the wheel turns because of the motion of the fluid. Now we change the orientation of the axis so that the angular velocity ω of the wheel is a *maximum*. When this happens, the axis will be pointed in the direction of **curl v** and $|\mathbf{curl\ v}| = 2\omega$. This is how $\mathbf{V} \times \mathbf{v}$ is related to rotation. When a vector field **v** that is continuously differentiable in some domain D

---

* **Curl v** is also called the **vorticity** vector.

has the property that its curl vanishes ($\mathbf{V} \times \mathbf{v} = \mathbf{0}$) at every point of D, we call $\mathbf{v}$ an **irrotational field**. Such a field is also called a **conservative field**, a term whose significance will become apparent in Section 5.7.

The following theorem lists some of the properties of the curl with the proofs of all but part (g) being left for the exercises.

Theorem 5.3–3 Let $\mathbf{u}(x, y, z)$ and $\mathbf{v}(x, y, z)$ be differentiable vector functions, $f(x, y, z)$ a differentiable scalar function, and c a constant. Then:

a) $\mathbf{V} \times (\mathbf{u} + \mathbf{v}) = \mathbf{V} \times \mathbf{u} + \mathbf{V} \times \mathbf{v}$;

b) $\mathbf{V} \times (c\mathbf{v}) = c(\mathbf{V} \times \mathbf{v})$;

c) $\mathbf{V} \times (f\mathbf{v}) = f(\mathbf{V} \times \mathbf{v}) + (\mathbf{V}f) \times \mathbf{v}$;

d) $\mathbf{V} \times (\mathbf{V}f) = \mathbf{0}$;

e) $\mathbf{V} \cdot \mathbf{V} \times \mathbf{v} = 0$;

f) $\mathbf{V} \times \mathbf{r} = \mathbf{0}$;

g) $\mathbf{V} \cdot (\mathbf{u} \times \mathbf{v}) = \mathbf{v} \cdot (\mathbf{V} \times \mathbf{u}) - \mathbf{u} \cdot (\mathbf{V} \times \mathbf{v})$.

Note that additional assumptions (besides differentiability) may be necessary to prove some parts of this theorem (see Exercise 28).

PROOF OF PART (g).

$$\mathbf{V} \cdot (\mathbf{u} \times \mathbf{v}) = \mathbf{i} \cdot \frac{\partial}{\partial x}(\mathbf{u} \times \mathbf{v}) + \mathbf{j} \cdot \frac{\partial}{\partial y}(\mathbf{u} \times \mathbf{v}) + \mathbf{k} \cdot \frac{\partial}{\partial z}(\mathbf{u} \times \mathbf{v})$$

$$= \mathbf{i} \cdot \frac{\partial \mathbf{u}}{\partial x} \times \mathbf{v} + \mathbf{j} \cdot \frac{\partial \mathbf{u}}{\partial y} \times \mathbf{v} + \mathbf{k} \cdot \frac{\partial \mathbf{u}}{\partial z} \times \mathbf{v}$$

$$+ \mathbf{i} \cdot \left(\mathbf{u} \times \frac{\partial \mathbf{v}}{\partial x}\right) + \mathbf{j} \cdot \left(\mathbf{u} \times \frac{\partial \mathbf{v}}{\partial y}\right) + \mathbf{k} \cdot \left(\mathbf{u} \times \frac{\partial \mathbf{v}}{\partial z}\right)$$

$$= \mathbf{v} \cdot \left[\left(\mathbf{i} \times \frac{\partial \mathbf{u}}{\partial x}\right) + \left(\mathbf{j} \times \frac{\partial \mathbf{u}}{\partial y}\right) + \left(\mathbf{k} \times \frac{\partial \mathbf{u}}{\partial z}\right)\right]$$

$$- \mathbf{u} \cdot \left[\left(\mathbf{i} \times \frac{\partial \mathbf{v}}{\partial x}\right) + \left(\mathbf{j} \times \frac{\partial \mathbf{v}}{\partial y}\right) + \left(\mathbf{k} \times \frac{\partial \mathbf{v}}{\partial z}\right)\right]$$

$$= \mathbf{v} \cdot \mathbf{V} \times \mathbf{u} - \mathbf{u} \cdot \mathbf{V} \times \mathbf{v}. \quad \square$$

We emphasize that all the definitions in this section that involved the del operator, namely, Eqs. (5.3–3), (5.3–5), (5.3–6), and (5.3–7), were given in a *rectangular coordinate system*. These definitions must be changed if other coordinate systems are used. Some applied problems are more tractable in one coordinate system than another, so it is necessary to be familiar with other than rectangular coordinates. In the following section we will examine other coordinate systems.

KEY WORDS AND PHRASES

gradient
directional derivative
level surface
divergence
incompressible fluid
solenoidal

Laplacian
Laplace's equation
curl
irrotational field
conservative field
vorticity

EXERCISES 5.3

1. Explain why $\partial f/\partial x$, $\partial f/\partial y$, and $\partial f/\partial z$ are special cases of the directional derivative.

2. Find the maximum and minimum values of df/ds in Example 5.3–1.

3. **a)** Show that the maximum value of the directional derivative df/ds is in the direction of ∇f.

   b) Show that $(df/ds)_{\max} = |\nabla f|$.

   c) What is the *minimum* value of df/ds?

4. Prove Theorem 5.3–1.

5. Prove Theorem 5.3–2(a) and 5.3–2(b).

6. Consider the parabola $y = x^2$ in the xy-plane. Using the gradient and the derivative, show that the gradient specifies a direction that is normal to the curve at every point.

7. Graph some of the level curves of the functions $z = f(x, y)$, where $f(x, y)$ is as shown.

   a) $f(x, y) = x^2 + y^2$ **b)** $f(x, y) = x^2 - y^2$ **c)** $f(x, y) = x + y$

   d) $f(x, y) = ax^2 + by^2$, $a \neq b$ **e)** $f(x, y) = xy$

8. Describe the level surfaces of the functions $w = f(x, y, z)$, where $f(x, y, z)$ is as shown.

   a) $f(x, y, z) = x^2 + y^2 + z^2$

   b) $f(x, y, z) = x^2 + y^2 - z^2$

   c) $f(x, y, z) = ax + by + cz$ for constants a, b, and c

9. Obtain the equations of the tangent plane and normal line to the hyperboloid $x^2 + y^2 - z^2 = 18$ at the point $(3, 5, -4)$.

10. Find the equations of the tangent plane and normal line for the cone $z^2 = x^2 + y^2$ at the point $(3, 4, -5)$.

11. Find the directional derivatives of the following functions, $f = f(x, y, z)$, at the given points in the directions specified by **a**.

    a) $f = x^3 - xy^2 - z$; $(1, 1, 0)$; $\mathbf{a} = 2\mathbf{i} - 3\mathbf{j} + 6\mathbf{k}$

    b) $f = \log (x^2 + y^2 + z^2)^{1/2}$; $(3, 4, 12)$; $\mathbf{a} = 3\mathbf{i} + 6\mathbf{j} - 2\mathbf{k}$

    c) $f = e^x \cos (yz)$; $(0, 0, 0)$; $\mathbf{a} = 2\mathbf{i} + \mathbf{j} - 2\mathbf{k}$

12. Find the angle between the surfaces $x^2 + yz = 3$ and $x \log z + xy^2 = -4$ at their common point $(-1, 2, 1)$. (*Note*: The angle between two surfaces at a common point is defined as the angle between their normals at the point.)

13. If $f(x, y, z) = (x^2 + y^2 + z^2)^{-3/2}$, find each of the following at the point $(1, 2, 3)$.

    a) ∇f **b)** $|\nabla f|$

    c) the direction cosines of ∇f

14. Obtain **grad** $|\mathbf{r}|$. (*Hint*: $\mathbf{r} = x\mathbf{i} + y\mathbf{j} + z\mathbf{k}$.)

15. Compute the divergence of the vector field

$$\mathbf{v}(x, y, z) = x \cos y\mathbf{i} + y^2 z\mathbf{j} - xyz\mathbf{k}.$$

16. Explain why the force field

$$\mathbf{F} = \mathbf{r}/r^3$$

    is called an "inverse *square* force."

17. If $f(r)$ is an arbitrary function of $r = |\mathbf{r}|$ and $\mathbf{v} = \mathbf{r}/r$, show that

$$\nabla \cdot (f(r)\mathbf{v}) = \frac{1}{r^2} \frac{d}{dr} (r^2 f(r)).$$

18. Find the divergence of each of the following vector fields.

    a) $\mathbf{v} = x^2 \cos z\mathbf{i} + y \log z\mathbf{j} - yz\mathbf{k}$

    b) $\mathbf{u} = \sin (xy)\mathbf{i} + \cos (yz)\mathbf{j} + \sin (2xz)\mathbf{k}$

    c) $\mathbf{v} = y \sin (yz)\mathbf{i} - 2z \cos (2xz)\mathbf{j} - x \cos (xy)\mathbf{k}$

    d) $\mathbf{u} = xyz(x\mathbf{i} + y\mathbf{j} + z\mathbf{k})$

19. Show that the Laplacian operator is a linear operator.

20. Compute the Laplacian of

$$f(x, y, z) = (xyz)(x + y + z).$$

21. Compute the Laplacian of **grad** r (compare Exercise 14).

22. Compute the Laplacian of each of the following functions.

    a) $f(x, y, z) = e^x \cos (yz)$

    b) $f(x, y, z) = \log (x^2 + y^2 + z^2)^{1/2}$

    c) $f(x, y, z) = x^3 - xy^2 - z$

23. Show that each of the following functions satisfies Laplace's equation.

    a) $f = e^x \cos y - z$ **b)** $f = 2z^3 - 3(x^2 + y^2)z$

    c) $f = \log (x^2 + y^2)^{1/2}$ **d)** $f = \cos 5z \exp (3x + 4y)$

24. Determine which of the following fields are irrotational.

    a) $\mathbf{r} = x\mathbf{i} + y\mathbf{i} + z\mathbf{k}$ **b)** $\mathbf{v} = 2xyz\mathbf{i} + x^2 z\mathbf{j} + x^2 y\mathbf{k}$

    c) $\mathbf{u} = yz\mathbf{i} + xz\mathbf{j} + xy\mathbf{k}$

    d) $\mathbf{v} = (x + y \sin z)\mathbf{i} + x \sin z\mathbf{j} + xy \cos z\mathbf{k}$

▶▶▶ **25.** Show that $\mathbf{V} \times (\mathbf{V}f) = 0$.

26. Show that $\mathbf{V} \cdot (\mathbf{V} \times \mathbf{v}) = 0$.

27. Show that the del dot and del cross operators are linear operators.

28. Prove Theorem 5.3–3. Note that to prove part (d) it is necessary to assume that f has continuous derivatives since it is under these conditions that the second mixed partial derivatives are equal.

5.4 ORTHOGONAL COORDINATE SYSTEMS

Up to this point most of our work with vectors has been limited to a *fixed* cartesian (or rectangular) coordinate system. An exception to this was the moving trihedral defined in Section 5.2 for a space curve. Otherwise we referred vectors to an orthogonal coordinate system in which the *constant* unit vectors $\mathbf{i}$, $\mathbf{j}$, and $\mathbf{k}$ formed a right-hand system. In Section 5.3 the del operator was defined in terms of these vectors.

Cylindrical coordinates

We will now extend some of the concepts in this chapter so that eventually we will be able to solve a greater variety of applied problems in mathematics. Whenever a physical problem is such that circular symmetry is present, it is usually more advantageous to use *polar coordinates* in the plane or **cylindrical coordinates*** in space. These are defined as follows:

$$x = \rho \cos \phi,$$
$$y = \rho \sin \phi, \qquad \qquad \textbf{(5.4–1)}$$
$$z = z,$$

where $\rho \geq 0, 0 \leq \phi < 2\pi, -\infty < z < \infty$.

In Fig. 5.4–1 we show the relationship between rectangular coordinates (x, y, z) and cylindrical coordinates (ρ, ϕ, z). We observe that restricting the ranges of the variables as we have done means that Eqs. (5.4–1) provide a one-to-one correspondence between ordered triples in the two coordinate systems.

A point whose position vector $\mathbf{r}$ in rectangular coordinates is

$$\mathbf{r} = x\mathbf{i} + y\mathbf{j} + z\mathbf{k}$$

has a position vector in cylindrical coordinates given by

$$\mathbf{r} = \rho \cos \phi \mathbf{i} + \rho \sin \phi \mathbf{j} + z\mathbf{k}.$$

---

* Also called *circular* cylindrical coordinates when it is necessary to distinguish them from elliptic cylindrical coordinates, parabolic cylindrical coordinates, etc.

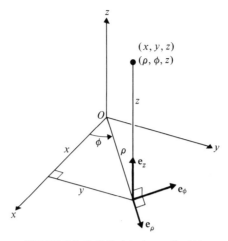

FIGURE 5.4–1 Cylindrical coordinates.

Thus unit vectors in the ρ, ϕ, and z directions as shown in Fig. 5.4–1 are defined by

$$\mathbf{e}_\rho = \partial \mathbf{r}/\partial \rho = \cos \phi \mathbf{i} + \sin \phi \mathbf{j},$$

$$\mathbf{e}_\phi = \frac{\partial \mathbf{r}/\partial \phi}{|\partial \mathbf{r}/\partial \phi|} = -\sin \phi \mathbf{i} + \cos \phi \mathbf{j}, \qquad (5.4\text{–}2)$$

$$\mathbf{e}_z = \partial \mathbf{r}/\partial z = \mathbf{k},$$

respectively. Clearly, these vectors are mutually orthogonal and form a right-hand system in the order given, that is, $\mathbf{e}_\rho \times \mathbf{e}_\phi = \mathbf{e}_z$ (Exercise 1). Observe, however, that $\mathbf{e}_\rho$ and $\mathbf{e}_\phi$ are *not constant* unit vectors but are functions of ϕ.

By solving Eqs. (5.4–2) for $\mathbf{i}$, $\mathbf{j}$, and $\mathbf{k}$ we obtain (Exercise 2)

$$\mathbf{i} = \cos \phi \mathbf{e}_\rho - \sin \phi \mathbf{e}_\phi,$$

$$\mathbf{j} = \sin \phi \mathbf{e}_\rho + \cos \phi \mathbf{e}_\phi, \qquad (5.4\text{–}3)$$

$$\mathbf{k} = \mathbf{e}_z;$$

hence we can express any of the quantities discussed in previous sections in cylindrical coordinates. For example, using a dot to designate differentiation with respect to $\mathbf{t}$, we have

$$\mathbf{r} = \rho \mathbf{e}_\rho + z \mathbf{e}_z,$$

$$\mathbf{v} = \dot{\mathbf{r}} = \dot{\rho} \mathbf{e}_\rho + \rho \dot{\mathbf{e}}_\rho + \dot{z} \mathbf{e}_z.$$

But

$$\dot{\mathbf{e}}_\rho = -\dot{\phi} \sin \phi \mathbf{i} + \dot{\phi} \cos \phi \mathbf{j} = \dot{\phi} \mathbf{e}_\phi \qquad (5.4\text{–}4)$$

from Eq. (5.4–2); hence,

$$\mathbf{v} = \dot{\rho} \mathbf{e}_\rho + \rho \dot{\phi} \mathbf{e}_\phi + \dot{z} \mathbf{e}_z. \qquad (5.4\text{–}5)$$

Continuing in this way, we can also show that

$$\mathbf{a} = (\ddot{\rho} - \rho\dot{\phi}^2)\mathbf{e}_\rho + (\rho\ddot{\phi} + 2\dot{\rho}\dot{\phi})\mathbf{e}_\phi + \ddot{z}\mathbf{e}_z \qquad \textbf{(5.4-6)}$$

(Exercise 3).

We can express the del operator in terms of cylindrical coordinates by first recalling the chain rule:

$$\frac{\partial f}{\partial x} = \frac{\partial f}{\partial \rho}\frac{\partial \rho}{\partial x} + \frac{\partial f}{\partial \phi}\frac{\partial \phi}{\partial x}.$$

Then, since

$$\rho = (x^2 + y^2)^{1/2} \quad \text{and} \quad \phi = \arctan \frac{y}{x},$$

we have

$$\frac{\partial \rho}{\partial x} = \cos \phi \quad \text{and} \quad \frac{\partial \phi}{\partial x} = -\frac{\sin \phi}{\rho}.$$

Thus,

$$\frac{\partial f}{\partial x}\mathbf{i} = \left(\frac{\partial f}{\partial \rho}\cos \phi - \frac{\partial f}{\partial \phi}\frac{\sin \phi}{\rho}\right)(\cos \phi \mathbf{e}_\rho - \sin \phi \mathbf{e}_\phi),$$

using Eq. (5.4–3). Similarly (Exercise 4),

$$\frac{\partial f}{\partial y}\mathbf{j} = \left(\frac{\partial f}{\partial \rho}\sin \phi + \frac{\partial f}{\partial \phi}\frac{\cos \phi}{\rho}\right)(\sin \phi \mathbf{e}_\rho + \cos \phi \mathbf{e}_\phi)$$

so that (Exercise 5)

$$\nabla \equiv \frac{\partial}{\partial \rho}\mathbf{e}_\rho + \frac{1}{\rho}\frac{\partial}{\partial \phi}\mathbf{e}_\phi + \frac{\partial}{\partial z}\mathbf{e}_z. \qquad \textbf{(5.4-7)}$$

EXAMPLE 5.4–1 Obtain the unit normal to the surface of the sphere $x^2 + y^2 + z^2 = a^2$ in cylindrical coordinates.

Solution. The sphere has equation $\rho^2 + z^2 = a^2$ in cylindrical coordinates; hence, taking $f(\rho, \phi, z) = \rho^2 + z^2 = a^2$, we have

$$\nabla f = 2\rho\mathbf{e}_\rho + 2z\mathbf{e}_z.$$

The unit normal is then

$$\mathbf{n} = \frac{\nabla f}{|\nabla f|} = \frac{2\rho\mathbf{e}_\rho + 2z\mathbf{e}_z}{\sqrt{4\rho^2 + 4z^2}} = \frac{1}{a}(\rho\mathbf{e}_\rho + z\mathbf{e}_z). \qquad \blacksquare$$

Note carefully that from the above example it *appears* that length is obtained in the same way in cylindrical coordinates as in rectangular coordinates. This is not so, since in rectangular coordinates we have

$$ds^2 = dx^2 + dy^2 + dz^2,$$

whereas from Eqs. (5.4–1) we obtain

$$dx = -\rho \sin \phi \, d\phi + \cos \phi \, d\rho,$$
$$dy = \rho \cos \phi \, d\phi + \sin \phi \, d\rho,$$
$$dz = dz,$$

so that (Exercise 6)

$$ds^2 = d\rho^2 + \rho^2 d\phi^2 + dz^2. \tag{5.4–8}$$

Observe from Eq. (5.4–8) that the elements of arc length in the $\mathbf{e}_\rho$, $\mathbf{e}_\phi$, and $\mathbf{e}_z$ directions are given by $d\rho$, $\rho d\phi$, and dz, respectively. The coefficients,

$$h_\rho = 1, \qquad h_\phi = \rho, \qquad h_z = 1, \tag{5.4–9}$$

are called **scale factors** and play an important role in transforming from one system of coordinates to the other. For example, the rectangular volume element dV becomes

$$dV = h_\rho h_\phi h_z \, d\rho \, d\phi \, dz = \rho d\rho \, d\phi \, dz. \tag{5.4–10}$$

Note that the scale factors can also be expressed as

$$h_\rho = \left| \frac{\partial \mathbf{r}}{\partial \rho} \right|, \qquad h_\phi = \left| \frac{\partial \mathbf{r}}{\partial \phi} \right|, \qquad h_z = \left| \frac{\partial \mathbf{r}}{\partial z} \right|.$$

Finding the divergence in cylindrical coordinates is a bit more involved than finding the gradient. Given the vector function

$$\mathbf{v} = v_1(\rho, \phi, z)\mathbf{e}_\rho + v_2(\rho, \phi, z)\mathbf{e}_\phi + v_3(\rho, \phi, z)\mathbf{e}_z,$$

we compute $\mathbf{V} \cdot v_1 \mathbf{e}_\rho$ and leave the rest of the computations for the exercises. We have

$$
\begin{aligned}
\mathbf{V} \cdot (v_1 \mathbf{e}_\rho) &= \mathbf{V} \cdot v_1 (\mathbf{e}_\phi \times \mathbf{e}_z) && \text{(why?)} \\
&= \mathbf{V} \cdot v_1 (\rho \nabla \phi \times \nabla z) && \text{(by Eq. 5.4–7)} \\
&= \mathbf{V} \cdot (v_1 \rho)(\nabla \phi \times \nabla z) && \text{(why?)} \\
&= \mathbf{V}(v_1 \rho) \cdot (\nabla \phi \times \nabla z) && \\
&\quad + (v_1 \rho)\mathbf{V} \cdot (\nabla \phi \times \nabla z) && \text{(by Theorem 5.3–2c)} \\
&= \mathbf{V}(v_1 \rho) \cdot (\nabla \phi \times \nabla z) && \text{(by Exercise 26 of Section 5.3)} \\
&= \mathbf{V}(v_1 \rho) \cdot \left(\frac{1}{\rho} \mathbf{e}_\phi \times \mathbf{e}_z \right) && \text{(why?)} \\
&= \mathbf{V}(v_1 \rho) \cdot \frac{1}{\rho} \mathbf{e}_\rho && \text{(why?)} \\
&= \frac{1}{\rho} \frac{\partial}{\partial \rho}(v_1 \rho). && \text{(using Eq. 5.4–7 again)}
\end{aligned}
$$

Similarly (Exercise 7),

$$\mathbf{V} \cdot (v_2 \mathbf{e}_\phi) = \frac{1}{\rho} \frac{\partial v_2}{\partial \phi} \quad \text{and} \quad \mathbf{V} \cdot (v_3 \mathbf{e}_z) = \frac{1}{\rho} \frac{\partial}{\partial z}(v_3 \rho),$$

so that, in cylindrical coordinates,

$$\mathbf{V} \cdot \mathbf{v} = \frac{1}{\rho}\left(\frac{\partial(v_1 \rho)}{\partial \rho} + \frac{\partial v_2}{\partial \phi} + \frac{\partial(v_3 \rho)}{\partial z} \right). \tag{5.4-11}$$

The curl in cylindrical coordinates can be obtained in a manner similar to the above. Thus

$$
\begin{aligned}
\mathbf{V} \times v_3 \mathbf{e}_z &= \mathbf{V} \times v_3 \mathbf{V}z = v_3 \mathbf{V} \times \mathbf{V}z + \mathbf{V}v_3 \times \mathbf{V}z \\
&= \mathbf{V}v_3 \times \mathbf{V}z \\
&= \left(\frac{\partial v_3}{\partial \rho} \mathbf{e}_\rho + \frac{1}{\rho} \frac{\partial v_3}{\partial \phi} \mathbf{e}_\phi + \frac{\partial v_3}{\partial z} \mathbf{e}_z \right) \times \mathbf{e}_z \\
&= -\frac{\partial v_3}{\partial \rho} \mathbf{e}_\phi + \frac{1}{\rho} \frac{\partial v_3}{\partial \phi} \mathbf{e}_\rho \\
&= \frac{1}{\rho}\left(\frac{\partial v_3}{\partial \phi} \mathbf{e}_\rho - \frac{\partial v_3}{\partial \rho} \rho \mathbf{e}_\phi \right).
\end{aligned}
$$

Combining the above with two similar expressions for $\mathbf{V} \times v_1 \mathbf{e}_\rho$ and $\mathbf{V} \times v_2 \mathbf{e}_\phi$ (Exercise 8) results in

$$\mathbf{V} \times \mathbf{v} = \frac{1}{\rho}\begin{vmatrix} \mathbf{e}_\rho & \rho\mathbf{e}_\phi & \mathbf{e}_z \\ \dfrac{\partial}{\partial \rho} & \dfrac{\partial}{\partial \phi} & \dfrac{\partial}{\partial z} \\ v_1 & \rho v_2 & v_3 \end{vmatrix}. \tag{5.4-12}$$

Spherical coordinates

Another useful orthogonal coordinate system is the **spherical coordinate system**. This is defined as follows:

$$
\begin{aligned}
x &= r \sin \theta \cos \phi, \\
y &= r \sin \theta \sin \phi, \\
z &= r \cos \theta,
\end{aligned}
\tag{5.4-13}
$$

where $r \geq 0$, $0 \leq 0 \leq \pi$, and $0 \leq \phi < 2\pi$. Figure 5.4–2 shows* the relationship between rectangular coordinates (x, y, z) and spherical coordinates (r, θ, ϕ). The position vector of an arbitrary point is given by

---

* Some authors interchange ϕ and θ, some use ρ instead of r, and some do both. It is essential, therefore, to note carefully a particular author's notation for spherical coordinates.

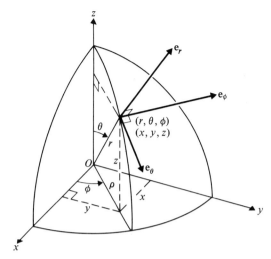

FIGURE 5.4–2 Spherical coordinates.

$\mathbf{r} = x\mathbf{i} + y\mathbf{j} + z\mathbf{k} = r \sin \theta \cos \phi \mathbf{i} + r \sin \theta \sin \phi \mathbf{j} + r \cos \theta \mathbf{k}$, where $r = |\mathbf{r}|$ in accordance with our previous notation. Scale factors are obtained from $\mathbf{r}$ by differentiation:

$$h_r = \left| \frac{\partial \mathbf{r}}{\partial r} \right| = 1,$$

$$h_\theta = \left| \frac{\partial \mathbf{r}}{\partial \theta} \right| = r, \qquad\qquad (5.4\text{–}14)$$

$$h_\phi = \left| \frac{\partial \mathbf{r}}{\partial \phi} \right| = r \sin \theta.$$

Unit vectors (Exercise 9) in the r, θ, and ϕ directions are

$$\mathbf{e}_r = \sin \theta \cos \phi \mathbf{i} + \sin \theta \sin \phi \mathbf{j} + \cos \theta \mathbf{k},$$
$$\mathbf{e}_\theta = \cos \theta \cos \phi \mathbf{i} + \cos \theta \sin \phi \mathbf{j} - \sin \theta \mathbf{k}, \qquad (5.4\text{–}15)$$
$$\mathbf{e}_\phi = -\sin \phi \mathbf{i} + \cos \phi \mathbf{j}.$$

Observe that $\mathbf{e}_\phi$ is the same as for cylindrical coordinates since ϕ is measured in the same way. The unit vectors form a right-hand orthogonal system with $\mathbf{e}_r \times \mathbf{e}_\theta = \mathbf{e}_\phi$ (Exercise 10).

We leave it for the reader (Exercise 12) to show that

$$ds^2 = (dr)^2 + r^2(d\theta)^2 + r^2 \sin^2 \theta (d\phi)^2 \qquad (5.4\text{–}16)$$

and

$$dV = r^2 \sin \theta \, dr \, d\theta \, d\phi. \qquad (5.4\text{–}17)$$

It can be further shown (Exercises 13, 14, and 25) that

$$\mathbf{V} \equiv \frac{\partial}{\partial r} \mathbf{e}_r + \frac{1}{r} \frac{\partial}{\partial \theta} \mathbf{e}_\theta + \frac{1}{r \sin \theta} \frac{\partial}{\partial \phi} \mathbf{e}_\phi, \tag{5.4-18}$$

$$\mathbf{V} \cdot \mathbf{v} = \frac{1}{r^2 \sin \theta} \left(\frac{\partial}{\partial r} (v_1 r^2 \sin \theta) + \frac{\partial}{\partial \theta} (v_2 r \sin \theta) + \frac{\partial}{\partial \phi} (r v_3) \right), \tag{5.4-19}$$

and

$$\mathbf{V} \times \mathbf{v} = \frac{1}{r^2 \sin \theta} \begin{vmatrix} \mathbf{e}_r & r\mathbf{e}_\theta & r \sin \theta \mathbf{e}_\phi \\ \dfrac{\partial}{\partial r} & \dfrac{\partial}{\partial \theta} & \dfrac{\partial}{\partial \phi} \\ v_1 & r v_2 & r \sin \theta v_3 \end{vmatrix} \tag{5.4-20}$$

where $\mathbf{v} = v_1 \mathbf{e}_r + v_2 \mathbf{e}_\theta + v_3 \mathbf{e}_\phi$.

The Laplacian in cylindrical and spherical coordinates plays an important role in applied mathematics. We will investigate this topic in Chapter 9. It should be apparent from the material in this section that transforming the gradient, divergence, and curl to **curvilinear coordinate systems** is not a simple matter. Thus defining these concepts in a rectangular coordinate framework may not have been a good idea. It appears that definitions that do not tie these ideas to a particular coordinate system would be more desirable. In Section 5.6 we will give such "coordinate-free" definitions of gradient, divergence, and curl.

Gram–Schmidt orthogonalization

In addition to "standard" orthogonal coordinate systems such as the cartesian, cylindrical, and spherical coordinate systems, it is sometimes convenient to construct special orthogonal coordinate systems. This can always be done provided we begin with a set of *linearly independent vectors*. From such a set it is possible to obtain a set of mutually orthogonal, unit vectors called an **orthonormal set**. We illustrate with an example from $\mathbb{R}^3$.

EXAMPLE 5.4–2   Construct an orthonormal set from the set

$$S_1 = \{\mathbf{w}_1 = (1, 1, 1), \mathbf{w}_2 = (1, 2, 1), \mathbf{w}_3 = (0, 1, 1)\}.$$

Solution. The vectors in the set S_1 are not mutually orthogonal (why?) nor are they unit vectors (why?). They are, however, linearly independent (Exercise 15). Beginning* with $\mathbf{w}_1$, rename it $\mathbf{v}_1$.

---

* We may begin with *any* of the vectors in S_1.

Now consider the space spanned by $\mathbf{w}_1$ and $\mathbf{w}_2$. An arbitrary vector $\mathbf{v}_2$ in this space can be written as

$$\mathbf{v}_2 = a\mathbf{w}_1 + b\mathbf{w}_2,$$

with a and b appropriate scalars. In other words,

$$\mathbf{v}_2 = a(1, 1, 1) + b(1, 2, 1) = (a + b, a + 2b, a + b).$$

We seek that particular vector $\mathbf{v}_2$ that is orthogonal to $\mathbf{v}_1$, that is, we determine a and b so that

$$\mathbf{v}_2 \cdot \mathbf{v}_1 = 3a + 4b = 0.$$

To simplify things we choose $b = 3$ so that $a = -4$ and $\mathbf{v}_2 = (-1, 2, -1)$. Finally, we consider the space spanned by $\mathbf{v}_1$, $\mathbf{v}_2$, and $\mathbf{w}_3$. An arbitrary vector in this space has the form

$$\mathbf{v}_3 = c\mathbf{v}_1 + d\mathbf{v}_2 + f\mathbf{w}_3$$
$$= (c - d, c + 2d + f, c - d + f).$$

For $\mathbf{v}_3$ to be orthogonal to both $\mathbf{v}_1$ and $\mathbf{v}_2$, we must have

$$\mathbf{v}_3 \cdot \mathbf{v}_1 = 3c + 2f = 0$$

and

$$\mathbf{v}_3 \cdot \mathbf{v}_2 = 6d + f = 0.$$

Thus $d = -\frac{1}{6}f$ and $c = -\frac{2}{3}f$, so that choosing $f = 6$ produces $c = -4$, $d = -1$, and $\mathbf{v}_3 = (-3, 0, 3)$. The set

$$S_2 = \{\mathbf{v}_1 = (1, 1, 1), \mathbf{v}_2 = (-1, 2, -1), \mathbf{v}_3 = (-3, 0, 3)\}$$

is an **orthogonal set**, hence forms an orthogonal basis for $\mathbb{R}^3$ (Fig. 5.4–3). Dividing each vector of the set S_2 by its length yields the

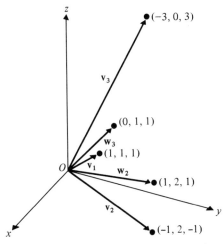

FIGURE 5.4–3 An orthogonal basis in $\mathbb{R}^3$.

orthonormal set

$$S_3 = \left\{ \mathbf{u}_1 = \frac{\sqrt{3}}{3}(1, 1, 1),\ \mathbf{u}_2 = \frac{\sqrt{6}}{6}(-1, 2, -1),\ \mathbf{u}_3 = \frac{3\sqrt{2}}{2}(-3, 0, 3) \right\}. \quad \blacksquare$$

A generalization of the process used in the last example is known as the **Gram–Schmidt* orthogonalization process**. In any vector space in which the concept of orthogonality has been defined, we can construct an orthonormal set of vectors from a linearly independent set.

KEY WORDS AND PHRASES

cylindrical coordinates
scale factors
spherical coordinate system
orthogonal vectors

orthonormal vectors
Gram–Schmidt orthogonalization
  process

EXERCISES 5.4

▶

1. Show that the vectors $\mathbf{e}_\rho$, $\mathbf{e}_\phi$, and $\mathbf{e}_z$ in Eqs. (5.4–2) are mutually orthogonal and form a right-hand system with $\mathbf{e}_\rho \times \mathbf{e}_\phi = \mathbf{e}_z$.

2. Solve the system (5.4–2) for $\mathbf{i}$, $\mathbf{j}$, and $\mathbf{k}$. (*Note*: Cramer's rule or Gaussian elimination can be used here.)

3. **a)** Show that $\dot{\mathbf{e}}_\phi = -\dot{\phi}\mathbf{e}_\rho$.

   b) Starting with Eq. (5.4–5) and using Eq. (5.4–4) and part (a), obtain Eq. (5.4–6).

4. Using the chain rule and Eq. (5.4–3), obtain $(\partial f/\partial y)\mathbf{j}$ in cylindrical coordinates.

5. Obtain Eq. (5.4–7) by filling in the details given in the text.

6. Show that the element of arc length in cylindrical coordinates is given by Eq. (5.4–8).

7. If $\mathbf{v} = v_1 \mathbf{e}_\rho + v_2 \mathbf{e}_\phi + v_3 \mathbf{e}_z$ in cylindrical coordinates, then obtain $\nabla \cdot (v_2 \mathbf{e}_\phi)$ and $\nabla \cdot (v_3 \mathbf{e}_z)$ as given in the text.

8. Follow the procedure in the text to find each of the following.

   a) $\nabla \times v_1 \mathbf{e}_\rho$

   b) $\nabla \times v_2 \mathbf{e}_\phi$

   c) $\nabla \times \mathbf{v}$ as given in Eq. (5.4–11)

---

* Jorgen P. Gram (1850–1916), a Danish actuary and mathematician; Erhard Schmidt (1876–1959), a German mathematician.

9. Show that the vectors (5.4–15) are unit vectors.

10. Show that the vectors (5.4–15) satisfy $\mathbf{e}_r \times \mathbf{e}_\theta = \mathbf{e}_\phi$.

11. Solve the system (5.4–15) for $\mathbf{i}$, $\mathbf{j}$, and $\mathbf{k}$.

12. Show that for spherical coordinates:
    a) $ds^2 = (dr)^2 + r^2(d\theta)^2 + r^2 \sin^2\theta(d\phi)^2$;
    b) $dV = r^2 \sin\theta\, dr\, d\theta\, d\phi$.

13. Obtain Eq. (5.4–18) in a manner similar to the way in which Eq. (5.4–7) was obtained in the text.

14. Obtain Eq. (5.4–19) following the method used to find the divergence in cylindrical coordinates.

15. Verify that the vectors (1, 1, 1), (1, 2, 1), and (0, 1, 1) are linearly independent. (Compare Example 5.4–2.)

▶▶ 16. Translate $x^2 + y^2 + z^2 = 4z$ into cylindrical and spherical coordinates.

17. Translate $z^2 - \rho^2$ into rectangular and spherical coordinates.

18. **Parabolic cylindrical coordinates** are defined by

$$x = \tfrac{1}{2}(u^2 - v^2), \qquad y = uv, \qquad z = z,$$

where $-\infty < u < \infty$, $v \geq 0$, and $-\infty < z < \infty$. Find the unit vectors in this system and show that the system is orthogonal (Fig. 5.4–4).

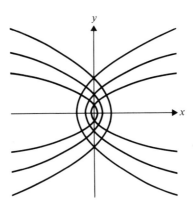

FIGURE 5.4–4 Parabolic cylindrical coordinates.

19. Consider the plane coordinate transformation

$$x = a \cosh u \cos v, \qquad y = a \sinh u \sin v,$$

where $u \geq 0$, $0 \leq v < 2\pi$. Obtain the scale factors. Is this coordinate system orthogonal? Explain. (*Note*: These are **elliptic coordinates**; the

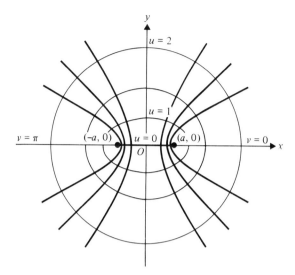

FIGURE 5.4–5 Elliptic coordinates.

curves $u = $ constant and $v = $ constant are confocal ellipses and hyperbolas, respectively, in Fig. 5.4–5.)

20. Analyze the **prolate spheroidal coordinate system** defined by

$$x = a \sinh u \sin v \cos \theta,$$
$$y = a \sinh u \sin v \sin \theta,$$
$$z = a \cosh u \cos v,$$

where $u \geq 0$, $0 \leq v \leq \pi$, and $0 \leq \theta < 2\pi$ by obtaining the scale factors. Is this system orthogonal?

21. Paraboloidal coordinates are given by

$$x = uv \cos \theta,$$
$$y = uv \sin \theta,$$
$$z = \tfrac{1}{2}(u^2 - v^2),$$

where $u \geq 0$, $v \geq 0$, $0 \leq \theta < 2\pi$. Obtain the gradient, ds^2, and dV in terms of u, v, and θ. Show that this is an orthogonal system.

22. Consider the vectors $\mathbf{w}_1 = (3, 1)$ and $\mathbf{w}_2 = (1, 3)$.

a) Verify that $\mathbf{w}_1$ and $\mathbf{w}_2$ are linearly independent.

b) Obtain an orthogonal set

$$\{\mathbf{v}_1 = \mathbf{w}_1, \mathbf{v}_2\}$$

spanning the same space as $\mathbf{w}_1$ and $\mathbf{w}_2$.

c) Obtain an orthogonal set

$$\{\mathbf{v}_1 = \mathbf{w}_2, \mathbf{v}_2\}$$

spanning the same space as $\mathbf{w}_1$ and $\mathbf{w}_2$.

d) Obtain orthonormal sets from the sets in parts (b) and (c).

e) Graph $\mathbf{w}_1$, $\mathbf{w}_2$, and $\mathbf{v}_2$.

23. Consider the set of vectors

$$S_1 = \{\mathbf{w}_1 = (1, 3, -1), \mathbf{w}_2 = (0, 1, 0), \mathbf{w}_3 = (-1, 2, 2)\}.$$

a) Show that S_1 is a set of basis vectors for $\mathbb{R}^3$.

b) Obtain the set

$$S_2 = \{\mathbf{v}_1 = \mathbf{w}_1, \mathbf{v}_2, \mathbf{v}_3\}$$

in which $\mathbf{v}_i \cdot \mathbf{v}_j = 0$ for $i \neq j$.

c) Obtain the set

$$S_3 = \{\mathbf{u}_1, \mathbf{u}_2, \mathbf{u}_3\}$$

from S_2 such that

$$\mathbf{u}_i \cdot \mathbf{u}_j = \begin{cases} 0, & \text{for } i \neq j, \\ 1, & \text{for } i = j. \end{cases}$$

d) Express each of the vectors in S_1 as a linear combination of the vectors in S_3.

24. Obtain an orthonormal set from the set

$$\{\mathbf{w}_1 = (1, 2, 3), \mathbf{w}_2 = (3, 2, 1), \mathbf{w}_3 = (2, 3, 1)\}$$

by:

a) normalizing $\mathbf{w}_1$;

b) normalizing $\mathbf{w}_2$;

c) normalizing $\mathbf{w}_3$.

▶▶▶ **25.** Express cylindrical coordinates ρ, ϕ, and z in terms of rectangular coordinates x, y, and z.

26. The formulas (see Exercise 25) for transforming x-, y-, and z-coordinates to cylindrical coordinates are not valid for points on the z-axis. Explain why this is so. These points are called *singular points* of the transformation.

27. Find the singular points of the transformation from rectangular to spherical coordinates. (Compare Exercise 26.)

28. If $\mathbf{r} = x\mathbf{i} + y\mathbf{j} + z\mathbf{k}$ is the position vector of a point, then obtain the velocity vector in spherical coordinates.

29. Follow the method used in the text for obtaining the curl in cylindrical coordinates to derive Eq. (5.4–20).

30. Two sets of vectors $\{\mathbf{e}_1, \mathbf{e}_2, \mathbf{e}_3\}$ and $\{\mathbf{a}_1, \mathbf{a}_2, \mathbf{a}_3\}$ are said to form a **reciprocal set of vectors** if

$$\mathbf{e}_i \cdot \mathbf{a}_j = \begin{cases} 1, & \text{when } i = j, \\ 0, & \text{when } i \neq j. \end{cases}$$

a) Show that

$$\{\mathbf{V}\rho, \mathbf{V}\phi, \mathbf{V}z\} \quad \text{and} \quad \left\{\frac{\partial \mathbf{r}}{\partial \rho}, \frac{\partial \mathbf{r}}{\partial \phi}, \frac{\partial \mathbf{r}}{\partial z}\right\}$$

are reciprocal sets of vectors.

b) Prove: Two sets of vectors $\{\mathbf{e}_i\}$ and $\{\mathbf{a}_j\}$ are reciprocal sets if and only if

$$\mathbf{a}_1 = \frac{\mathbf{e}_2 \times \mathbf{e}_3}{[\mathbf{e}_1, \mathbf{e}_2, \mathbf{e}_3]}, \qquad \mathbf{a}_2 = \frac{\mathbf{e}_3 \times \mathbf{e}_1}{[\mathbf{e}_1, \mathbf{e}_2, \mathbf{e}_3]}, \qquad \mathbf{a}_3 = \frac{\mathbf{e}_1 \times \mathbf{e}_2}{[\mathbf{e}_1, \mathbf{e}_2, \mathbf{e}_3]},$$

where $[\mathbf{e}_1, \mathbf{e}_2, \mathbf{e}_3] \neq 0$.

c) Show that a result analogous to part (a) holds for spherical coordinates.

31. Show that if $\mathbf{w}_1$ and $\mathbf{w}_2$ are two linearly independent vectors in $\mathbb{R}^2$, then $\mathbf{v}_1$ and $\mathbf{v}_2$ constitute an orthogonal basis for $\mathbb{R}^2$, where $\mathbf{v}_1 = \mathbf{w}_1$ and

$$\mathbf{v}_2 = -(\mathbf{w}_2 \cdot \mathbf{w}_1)\mathbf{w}_1 + (\mathbf{w}_1 \cdot \mathbf{w}_1)\mathbf{w}_2.$$

32. Referring to Exercise 31, obtain the orthonormal basis from $\mathbf{v}_1$ and $\mathbf{v}_2$.

5.5 LINE, SURFACE, AND VOLUME INTEGRALS

Before presenting three important integral theorems in Section 5.6 and their applications in Section 5.7, we consider a generalization of the Riemann* integral in this section. The Riemann integral is the one familiar from calculus courses where the word "Riemann" is usually omitted and it is called simply a *definite integral*.

Accordingly, we begin with a brief review of the definite integral,

$$\int_a^b f(x)\, dx.$$

This review is not meant to be rigorous. A more rigorous treatment can be found in George B. Thomas and R. L. Finney's *Calculus and Analytic Geometry* 5th ed., p. 222 (Reading, Mass.: Addison-Wesley, 1979), or, at a more advanced level, in Walter Rudin's, *Principles of Mathematical Analysis*, 3d ed., Ch. 6 (New York: McGraw-Hill, 1976).

Suppose f is a bounded[†] real function defined on $[a, b]$. We divide the closed interval $[a, b]$ into a finite number, say n, of subdivisions of length $\Delta x_i = x_{i+1} - x_i$, which are not necessarily all equal. Figure 5.5–1 shows such a subdivision using the points $x_0 = a, x_1, x_2, \ldots, x_{n-1}$, $x_n = b$. In each subdivision we choose a point, call it x_i^* (shown by an asterisk in the figure). We compute $f(x_i^*)$ and *multiply* it by Δx_i, the length of the ith interval. Note that x_i^* may be *anywhere* in the sub-

---

* Georg F. B. Riemann (1826–1866), a German mathematician.
† A function f is said to be *bounded* on $[a, b]$ if there is a real number M such that $|f(x)| \leq M$ for all x satisfying $a \leq x \leq b$.

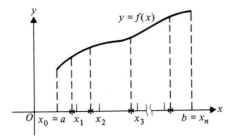

FIGURE 5.5–1 A subdivision of the interval [a, b].

division including the end points. Next we *add* all the products so obtained and write the sum:

$$\sum_{i=1}^{n} f(x_i^*)\,\Delta x_i.$$

Finally, we take the *limit* of this sum as $n \to \infty$ and, at the same time, as the length of the largest subdivision approaches zero. If this limit exists and is independent of the way in which the x_i were chosen in [a, b] and the way in which the x_i^* were chosen in each subdivision, we define this limit as *the* definite integral of $f(x)$ from a to b and write

$$\int_a^b f(x)\,dx = \lim_{n \to \infty} \sum_{i=1}^{n} f(x_i^*)\,\Delta x_i.$$

The fundamental theorem of the integral calculus then states that the definite integral is a function of its limits and is given by

$$\int_a^b f(x)\,dx = F(b) - F(a),$$

where $dF(x)/dx = f(x)$.

Line integrals

A useful generalization of the definite integral is the line integral. Let $f(x, y)$ be a scalar point function defined in some region that contains a piecewise smooth, oriented curve C (compare Definition 5.2–1). If the curve C has initial point P_0 and terminal point P_1, then the portion of C between these points can be divided into arcs, say n in number, by points

$$P_0(x_0, y_0), (x_1, y_1), \ldots, (x_{n-1}, y_{n-1}), P_1(x_n, y_n).$$

Call the arc lengths of these subdivisions Δs_i, $i = 1, 2, \ldots, n$ and choose a point on each one; call it (x_i^*, y_i^*) as shown in Fig. 5.5–2. Then evaluate

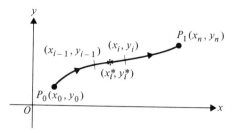

FIGURE 5.5–2 A subdivision of a smooth, oriented arc.

$f(x, y)$ at (x_i^*, y_i^*), multiply by Δs_i, and add these products to obtain the sum:

$$\sum_{i=1}^{n} f(x_i^*, y_i^*)\, \Delta s_i.$$

If this sum has a limit as $n \to \infty$ and, at the same time, as the maximum of the arc lengths approaches zero, and if this limit is independent of the way C is subdivided and the way in which the points (x_i^*, y_i^*) are chosen, then we define

$$\int_C f(x, y)\, ds = \lim_{n \to \infty} \sum_{i=1}^{n} f(x_i^*, y_i^*)\, \Delta s_i. \tag{5.5–1}$$

We call the above integral a **line integral** (although it is defined in connection with a *curve*). In order to evaluate a line integral we need to know (1) the integrand $f(x, y)$, (2) the curve C, (3) the initial point P_0 on the curve, and (4) the terminal point P_1 on the curve. In the *special case* where $f(x, y) = 1$, the line integral defines the *length* of the curve C from P_0 to P_1. Other interpretations of the line integral will be discussed later in this section.

Observe the similarity between the definitions of a definite integral and a line integral. The x-axis has been replaced by a curve and the integrand has been replaced by a function of two variables. If, however, the curve C is given in parametric form, $x = x(t)$, $y = y(t)$, then

$$ds = \sqrt{\left(\frac{dx}{dt}\right)^2 + \left(\frac{dy}{dt}\right)^2}\; dt \tag{5.5–2}$$

and the line integral becomes a definite integral.

EXAMPLE 5.5–1 Evaluate

$$\int_C (x^2 y + 2)\, ds$$

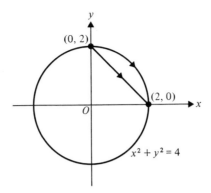

FIGURE 5.5–3 Integration paths for
Examples 5.5–1 and 5.5–2.

along the curve $C: x = 2 \cos t, y = 2 \sin t$, in a clockwise direction from
(0, 2) to (2, 0) (see Fig. 5.5–3).

Solution. The integrand in terms of t is

$$x^2 y + 2 = 8 \cos^2 t \sin t + 2,$$

and the limits are $t = \pi/2$ to $t = 0$ (why?) while

$$ds = \sqrt{(-2 \sin t)^2 + (2 \cos t)^2} \; dt = 2 \; dt.$$

Hence,

$$\int_C (x^2 y + 2) \; ds = 4 \int_{\pi/2}^{0} (4 \cos^2 t \sin t + 1) \; dt = -\frac{2}{3} (3\pi + 8) \doteq -11.62. \quad \blacksquare$$

In the following example we will change the path to a straight line
in order to illustrate some other phases of the computation involved
in evaluating a line integral.

EXAMPLE 5.5–2   Repeat Example 5.5–1 changing the path to the straight
line between the given points.

Solution. The equation of the line is $x + y = 2$ and from $y = 2 - x$ we
have

$$ds = \sqrt{dx^2 + dy^2} = \sqrt{1 + (dy/dx)^2} = \sqrt{2} \; dx,$$

so that the line integral becomes (note the limits of integration!)

$$\sqrt{2} \int_0^2 (x^2(x - 2) + 2) \; dx = \frac{-16\sqrt{2}}{3} \doteq -7.54. \quad \blacksquare$$

The above examples show clearly that the value of a line integral depends, in general, on the curve C. Later we will consider some special cases where the line integral will be independent of the path between two points.

In Section 5.1 we discussed the work W done by a constant force $\mathbf{F}$ moving a particle in a constant direction $\mathbf{d}$ (compare Example 5.1–3). In this case we had $W = \mathbf{F} \cdot \mathbf{d}$. We now extend this definition to the case where the force is a vector function,

$$\mathbf{F}(x, y) = M(x, y)\mathbf{i} + N(x, y)\mathbf{j},$$

and the particle acted upon is moving along a curve C. We define the work done as*

$$W = \int_C \mathbf{F} \cdot d\mathbf{r} = \int_C M(x, y)\, dx + N(x, y)\, dy, \qquad \textbf{(5.5–3)}$$

where, as usual, $\mathbf{r} = x\mathbf{i} + y\mathbf{j}$ is the position vector of an arbitrary point (x, y) on C and $d\mathbf{r} = dx\mathbf{i} + dy\mathbf{j}$.

EXAMPLE 5.5–3 A force field is given by

$$\mathbf{F}(x, y) = x \sin y\mathbf{i} + \cos y\mathbf{j}.$$

Find the work done in moving a particle in this field along straight lines from $(0, 0)$ to $(0, 1)$, then to $(1, 1)$.

Solution. Using Eq. (5.5–3) we need to evaluate

$$\int_C x \sin y\, dx + \cos y\, dy.$$

Along the path from $(0, 0)$ to $(0, 1)$ we have $x = 0$, $dx = 0$, hence the integral reduces to

$$\int_0^1 \cos y\, dy.$$

Along the path from $(0, 1)$ to $(1, 1)$ we have $y = 1$, $dy = 0$, and the integral reduces to

$$(\sin 1) \int_0^1 x\, dx.$$

Thus the work done is

$$W = \int_0^1 \cos y\, dy + (\sin 1) \int_0^1 x\, dx = \frac{3}{2}(\sin 1) \doteq 1.26. \quad \blacksquare$$

---

* Note that this is the same as

$$\int_C \mathbf{F} \cdot \mathbf{T}\, ds,$$

where $\mathbf{T}$ is the unit tangent vector to C (see Eq. 5.2–6).

Equation (5.5–3) can be extended in a natural way to $\mathbb{R}^3$ to find the work done when a particle moves in a force field **F** along a space curve C. If

$$\mathbf{F}(x, y, z) = F_1(x, y, z)\mathbf{i} + F_2(x, y, z)\mathbf{j} + F_3(x, y, z)\mathbf{k},$$

then the work done is given by

$$W = \int_C \mathbf{F} \cdot d\mathbf{r} = \int_C F_1\, dx + F_2\, dy + F_3\, dz.$$

If C is a *closed* curve, then we also use the notation

$$\oint_C \mathbf{F} \cdot d\mathbf{r} \qquad\qquad (5.5\text{–}4)$$

for the above line integral. This last, in fact, appears in a number of applications and has been given a name. We call the line integral (5.5–4) around a closed curve C the **circulation** of **F** about C. As we shall see later, the circulation is a *number* that provides a measure of the tendency of **F** to change along C. In case $\mathbf{F} \cdot d\mathbf{r}$ is the differential of a scalar field f, that is,

$$df = \frac{\partial f}{\partial x}\, dx + \frac{\partial f}{\partial y}\, dy + \frac{\partial f}{\partial z}\, dz = \nabla f \cdot d\mathbf{r} = \mathbf{F} \cdot d\mathbf{r}, \qquad (5.5\text{–}5)$$

then the circulation is zero, **F** is a *conservative field*, and the work done in such a field is independent of the path and depends only on the initial and terminal points. Note that Eq. (5.5–5) expresses the fact that $\mathbf{F} \cdot d\mathbf{r}$ is an *exact differential* (compare Section 1.2). When Eq. (5.5–5) holds, then we can find f, given **F**, and we have more simply

$$\int_C \mathbf{F} \cdot d\mathbf{r} = \int_{P_0}^{P_1} df = f(P_1) - f(P_0)$$

by the fundamental theorem of the integral calculus. We present two methods of obtaining f whenever **F** is given and $\nabla f = \mathbf{F}$ holds.

EXAMPLE 5.5–4 If $\mathbf{F} = (3x^2 - y^2)\mathbf{i} - 2xy\mathbf{j} - \mathbf{k}$, find f (if it exists) such that $\nabla f = \mathbf{F}$.

Solution. The parenthetical phrase is intended to show that not every vector field **F** can be expressed as the gradient of some scalar field f. *If* it can, then we have

$$\frac{\partial f}{\partial x}\mathbf{i} + \frac{\partial f}{\partial y}\mathbf{j} + \frac{\partial f}{\partial z}\mathbf{k} = (3x^2 - y^2)\mathbf{i} - 2xy\mathbf{j} - \mathbf{k}.$$

The fact that **i**, **j**, and **k** are mutually orthogonal leads to (why?)

$$\frac{\partial f}{\partial x} = 3x^2 - y^2, \qquad \frac{\partial f}{\partial y} = -2xy, \qquad \frac{\partial f}{\partial z} = -1.$$

These are three first-order partial differential equations. The first was obtained by differentiating f with respect to x while holding y and z

constant, hence we integrate the same way. Then

$$f = x^3 - xy^2 + \phi_1(y, z)$$

where instead of an arbitrary *constant* of integration we now have an arbitrary *function* $\phi_1(y, z)$. Proof of the correctness of this result is shown by differentiating it partially with respect to x and obtaining $3x^2 - y^2$. Using the other two partial differential equations, we obtain, in addition to

$$f = x^3 - xy^2 + \phi_1(y, z),$$

the expressions

$$f = -xy^2 + \phi_2(x, z)$$

and

$$f = -z + \phi_3(x, y).$$

Now, the only way that all three of the above expressions can be identical is if we take $\phi_1(y, z) = -z$, $\phi_2(x, z) = x^3 - z$, and $\phi_3(x, y) = x^3 - xy^2$. Thus

$$f(x, y, z) = x^3 - xy^2 - z + C,$$

where C is a constant, is the most general scalar function such that $\nabla f = \mathbf{F}$. ■

EXAMPLE 5.5–5   If $\mathbf{F} = e^x(\cos yz\mathbf{i} - z \sin yz\mathbf{j} - y \sin yz\mathbf{k})$, find f such that $\mathbf{F} = \nabla f$.

Solution. As in the previous example we begin with

$$\frac{\partial f}{\partial x} = e^x \cos yz,$$

which implies that

$$f = e^x \cos yz + \phi(y, z).$$

Differentiating with respect to y, we have

$$\frac{\partial f}{\partial y} = -ze^x \sin yz + \frac{\partial \phi}{\partial y} = -ze^x \sin yz.$$

(Why?) Hence

$$\frac{\partial \phi}{\partial y} = 0 \quad \text{and} \quad \phi(y, z) = g(z).$$

(Why?) Updating the expression for f gives us

$$f = e^x \cos yz + g(z).$$

Differentiating with respect to z gives us

$$\frac{\partial f}{\partial z} = -ye^x \sin yz + g'(z) = -ye^x \sin yz.$$

(Why?) Hence

$$g'(z) = 0 \quad \text{and} \quad g(z) = C, \quad \text{a constant.}$$

Thus, updating f again, we have

$$f(x, y, z) = e^x \cos yz + C. \quad \blacksquare$$

From Exercise 25 of Section 5.3 (namely, $\nabla \times (\nabla f) = 0$) it follows that if a vector field $\mathbf{F}$ is the gradient of some scalar field f, then $\nabla \times \mathbf{F} = 0$. Fields that are irrotational or, equivalently, that are gradients of some scalar function are called *gradient fields* or **conservative fields**. The latter comes from the fact that in a conservative field the total energy (sum of kinetic and potential energies) of a particle is constant (that is, it is "conserved"). This last statement is also called the *law of conservation of energy*. It now follows that the work done in moving a particle from one point to another in a conservative field is independent of the path between the two points. We state and prove this in the following theorem.

Theorem 5.5–1 Let $\mathbf{F}$ be a continuous vector field in some region D in $\mathbb{R}^3$. Then $\mathbf{F}$ is conservative if and only if the line integral between two points P_0 and P of D,

$$\int_C \mathbf{F} \cdot d\mathbf{r},$$

is independent of the smooth curve C in D connecting the two points.

PROOF. We first prove that if $\mathbf{F}$ is conservative, then the line integral is independent of the path C, that is, depends only on the points P_0 and P. Since $\mathbf{F}$ is conservative, $\mathbf{F} = \nabla f$, and along any smooth arc C,

$$\int_{P_0}^{P} \mathbf{F} \cdot d\mathbf{r} = \int_{P_0}^{P} \frac{\partial f}{\partial x} \, dx + \frac{\partial f}{\partial y} \, dy + \frac{\partial f}{\partial z} \, dz$$

$$= \int_{P_0}^{P} df = f(P) - f(P_0),$$

which shows that the line integral is independent of the path C.

Next, we choose an arbitrary but fixed point in D, call it $P_0(x_0, y_0, z_0)$. Let $P(x, y, z)$ be any other point in D and let C_1 be a smooth arc connecting P_0 and P (see Fig. 5.5–4). Then *define*

$$f(x, y, z) = \int_{P_0}^{P} \mathbf{F} \cdot d\mathbf{r}$$

to be the line integral from P_0 to P along the arc C_1. Since we are now making the assumption that this last line integral is independent of the path between P_0 and P, the above definition of $f(x, y, z)$ is not ambiguous even though we haven't specified C_1. Now recall that

$$\frac{\partial f}{\partial x} = \lim_{\Delta x \to 0} \frac{f(x + \Delta x, y, z) - f(x, y, z)}{\Delta x}$$

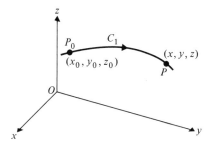

FIGURE 5.5–4

and denote the straight-line segment from (x, y, z) to $(x + \Delta x, y, z)$ by C_2 (see Exercise 1). Then the integral from P to $(x + \Delta x, y, z)$ along C_2 can be written

$$\int_{(x, y, z)}^{(x + \Delta x, y, z)} \mathbf{F} \cdot d\mathbf{r}.$$

Thus the integral from P_0 to $(x + \Delta x, y, z)$ via P, that is, along C_1 and C_2 becomes

$$f(x + \Delta x, y, z) = f(x, y, z) + \int_{(x, y, z)}^{(x + \Delta x, y, z)} \mathbf{F} \cdot d\mathbf{r}.$$

This last integral, however, is just (why?)

$$\int_{(x, y, z)}^{(x + \Delta x, y, z)} F_1 \, dx,$$

where $\mathbf{F} = F_1(x, y, z)\mathbf{i} + F_2(x, y, z)\mathbf{j} + F_3(x, y, z)\mathbf{k}$. Hence,

$$\frac{f(x + \Delta x, y, z) - f(x, y, z)}{\Delta x} = \frac{1}{\Delta x} \int_{(x, y, z)}^{(x + \Delta x, y, z)} F_1 \, dx,$$

which is just the *mean value* of F_1 along C_2 and approaches $F_1(x, y, z)$ as $\Delta x \to 0$. But this is the same as saying (why?)

$$\frac{\partial f}{\partial x} = F_1(x, y, z).$$

By holding x and z constant, then x and y constant, we can show in a similar manner (Exercise 2) that

$$\frac{\partial f}{\partial y} = F_2(x, y, z) \quad \text{and} \quad \frac{\partial f}{\partial z} = F_3(x, y, z)$$

and, finally, that $\mathbf{F} = \nabla f$. $\quad\square$

It follows as a simple corollary (Exercise 3) of the last theorem that the work done in a conservative force field around a *closed* curve is zero. We can thus list the following statements, each of which implies the others:

i) **F** is a conservative vector field;

ii) $\mathbf{V} \times \mathbf{F} = \mathbf{0}$;

iii) $\int_C \mathbf{F} \cdot d\mathbf{r}$ is independent of the path C;

iv) $\oint_C \mathbf{F} \cdot d\mathbf{r} = 0$;

v) $\mathbf{F} = \mathbf{V}f$.

From a practical standpoint, (ii) is the most useful for determining whether a vector field is conservative.

EXAMPLE 5.5–6   Find the work done in moving a particle from $(a, 0, 0)$ to $(0, a, \pi b/2)$ along the right circular helix $\mathbf{r}(t) = a \cos t\mathbf{i} + a \sin t\mathbf{j} + bt\mathbf{k}$ in the force field $\mathbf{F} = (4xy - 3x^2z^2)\mathbf{i} + 2x^2\mathbf{j} - 2x^3z\mathbf{k}$.

Solution. We first note (Exercise 4) that **F** is a conservative field everywhere; hence the work done is independent of the path. Thus we may choose a path consisting of three straight-line segments from $(a, 0, 0)$ to $(0, 0, 0)$ to $(0, a, 0)$ to $(0, a, \pi b/2)$, as shown in Fig. 5.5–5. Along the first of these, $y = 0$, $z = 0$, and $W_1 = 0$. Along the second line segment, $x = 0$ and $z = 0$ so that $W_2 = 0$. Along the third line segment, $x = 0$, $y = a$, and $W_3 = 0$. Hence the total work done is zero. ∎

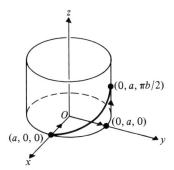

FIGURE 5.5–5 Integration path for Example 5.5–6.

Surface integrals

The two-dimensional analog of a line integral is a surface integral. Let S be a *smooth* surface, meaning a surface for which it is possible to define a unit normal vector **n** at every point of S in such a way that **n**

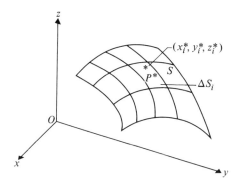

FIGURE 5.5–6 A subdivision of a smooth surface.

varies continuously on S. Let $f(x, y, z)$ be a scalar function that is defined and continuous on S. Now let the surface S be divided into n parts by curves on the surface. Call the surface area of one of these ΔS_i, as shown in Fig. 5.5–6. In this "patch" choose a point $P^*(x_i^*, y_i^*, z_i^*)$ and evaluate $f(x, y, z)$ at this point. Form the product $f(x_i^*, y_i^*, z_i^*)\,\Delta S_i$ and the sum of all such products,

$$\sum_{i=1}^{n} f(x_i^*, y_i^*, z_i^*)\,\Delta S_i. \tag{5.5–6}$$

Now let $n \to \infty$, let the "patch" having maximum area approach zero, and, at the same time, let the maximal dimension of each "patch" approach zero. If the sum (5.5–6) has a limit that is independent of the manner in which S is subdivided and independent of the way in which P^* is chosen in each subdivision, then we define

$$\iint_{S} f(x, y, z)\, dS = \lim_{n \to \infty} \sum_{i=1}^{n} f(x_i^*, y_i^*, z_i^*)\,\Delta S_i. \tag{5.5–7}$$

The double integral is called the **surface integral** of f over the surface S. The notation

$$\oiint_{S} f(x, y, z)\, dS$$

is used to denote the surface integral of f over a *closed* surface S.

In case $f(x, y, z) \equiv 1$, then the integral in (5.5–7) represents the *surface area* of S. Otherwise, the *number* obtained when evaluating a surface integral may have some other meaning, as we will see in the next section. For the present we give some examples to show how a surface integral may be computed.

EXAMPLE 5.5-7   If $\mathbf{F} = 6z\mathbf{i} + (2x + y)\mathbf{j} - x\mathbf{k}$, evaluate

$$\iint_S \mathbf{F} \cdot \mathbf{n} \, dS$$

over the surface bounded by the cylinder $x^2 + z^2 = 9$, $x = 0$, $y = 0$, $z = 0$, and $y = 6$.

Solution. The surface is shown in Fig. 5.5–7. Also shown is a typical patch dS on the surface S and the unit normal $\mathbf{n}$ at an arbitrary point of dS. Using the gradient we obtain $\mathbf{n}$ as follows. Let $f(x, y, z) = x^2 + z^2 - 9 = 0$ represent the surface S. Then $\nabla f = 2x\mathbf{i} + 2z\mathbf{k}$ and

$$\mathbf{n} = \frac{2x\mathbf{i} + 2z\mathbf{k}}{\sqrt{4x^2 + 4z^2}} = \frac{x\mathbf{i} + z\mathbf{k}}{3},$$

using the fact that $x^2 + z^2 = 9$ everywhere on S. Hence the integrand becomes

$$\mathbf{F} \cdot \mathbf{n} = \frac{6xz - xz}{3} = \frac{5}{3} xz.$$

We next project the patch dS into the xy-plane, forming the rectangle shown in Fig. 5.5–7. (An alternative way of looking at this is to begin with the rectangle dx by dy and project it *up* to intersect S in the patch dS. After all, it is immaterial how these patches are obtained since the value of the surface integral is independent of this.) We now observe that, in general, $dS \geq dy \, dx$. The angle γ, however, can be used to

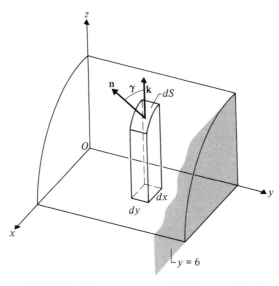

FIGURE 5.5-7 The projection of dS (Example 5.5–7).

transform this inequality into an equality, for we have (Exercise 6)

$$dS|\cos \gamma| = dy\, dx$$

or

$$dS = \frac{dy\, dx}{|\cos \gamma|} = \frac{dy\, dx}{|\mathbf{n} \cdot \mathbf{k}|}. \tag{5.5–8}$$

(If the element of surface area dS is projected into one of the other coordinate planes, then we have

$$dS = \frac{dy\, dz}{|\mathbf{n} \cdot \mathbf{i}|} \quad \text{or} \quad dS = \frac{dx\, dz}{|\mathbf{n} \cdot \mathbf{j}|} \tag{5.5–9}$$

in terms of the projections in the yz- or xz-planes, respectively.) Note that γ is one of the direction angles of $\mathbf{n}$. In the present example we have $|\mathbf{n} \cdot \mathbf{k}| = |z|/3 = z/3$, hence the surface integral becomes

$$5 \int_0^3 \int_0^6 x\, dy\, dx = 5 \int_0^3 xy \Big|_0^6 dx = 5 \int_0^3 6x\, dx = 15x^2 \Big|_0^3 = 135.$$

Observe that the limits of integration are constant in this example since $dy\, dx\, (=dA)$ covers a *rectangle* in the integration. ∎

EXAMPLE 5.5–8   If $\mathbf{F} = x\mathbf{i} + y\mathbf{j} + (z^2 - 1)\mathbf{k}$, find

$$\iint_S \mathbf{F} \cdot \mathbf{n}\, dS$$

over the entire closed surface bounded by the cylinder $x^2 + y^2 = a^2$ and the planes $z = 0$, $z = b$, where $\mathbf{n}$ is the unit outward normal. The surfaces are shown in Fig. 5.5–8.

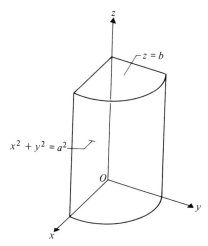

FIGURE 5.5–8 Surfaces for Example 5.5–8.

Solution. There are two key words in the statement of the problem, namely, "entire" and "outward." The first of these indicates that we are dealing with a *piecewise* smooth surface consisting of the lateral portion, the top, and the bottom. We consider these in that order.

Lateral surface

$$\mathbf{n} = \frac{\nabla f}{|\nabla f|} = \frac{2x\mathbf{i} + 2y\mathbf{j}}{\sqrt{4x^2 + 4y^2}} = \frac{x\mathbf{i} + y\mathbf{j}}{a},$$

$$\mathbf{F} \cdot \mathbf{n} = \frac{x^2 + y^2}{a} = \frac{a^2}{a} = a,$$

$$\iint_S \mathbf{F} \cdot \mathbf{n}\, dS = a \iint_S dS = a(2\pi ab) = 2\pi a^2 b$$

Top surface

$$\mathbf{n} = \mathbf{k}, \qquad \mathbf{F} \cdot \mathbf{n} = z^2 - 1 = b^2 - 1,$$

$$\iint_S \mathbf{F} \cdot \mathbf{n}\, dS = (b^2 - 1) \iint_S dS = (b^2 - 1)\pi a^2$$

Bottom surface

$$\mathbf{n} = -\mathbf{k}, \qquad \mathbf{F} \cdot \mathbf{n} = 1 - z^2 = 1,$$

$$\iint_S \mathbf{F} \cdot \mathbf{n}\, dS = \iint_S dS = \pi a^2$$

Entire surface

$$\iint_S \mathbf{F} \cdot \mathbf{n}\, dS = 2\pi a^2 b + (b^2 - 1)\pi a^2 + \pi a^2 = \pi a^2 b(b + 2) \quad \blacksquare$$

The last example should be studied carefully. Observe that when the surface integral was evaluated over the lateral surface, the integrand reduced to a constant. This was fortunate since otherwise we would have had to consider what to use for the unit *outward* normal. See Exercise 9 for a variation of the last example, in which the signs of *x* and *y* must be considered. Another approach is given in the following example.

EXAMPLE 5.5–9 Evaluate the surface integral of z^2 over the sphere of radius *a* centered at the origin.

Solution. One can easily be convinced that the use of rectangular coordinates is not the best approach here. Referring to Section 5.4, we see that the patch dS in spherical coordinates can be approximated by the product of two arcs, namely, $r \sin \theta\, d\phi$ and $r\, d\theta$ (Exercise 10).

On a sphere of radius a we have $r = a$, so that

$$dS = a^2 \sin \theta \, d\theta \, d\phi$$

and

$$z^2 = r^2 \cos^2 \theta = a^2 \cos^2 \theta.$$

Hence the surface integral becomes

$$\iint_S z^2 \, dS = a^4 \int_0^{2\pi} \int_0^{\pi} \sin \theta \cos^2 \theta \, d\theta \, d\phi = \frac{4\pi a^4}{3}. \quad \blacksquare$$

Volume integrals

It should now be a relatively simple step to define a **volume integral** (Exercise 11),

$$\iiint_V f(x, y, z) \, dV.$$

We remark again that this integral gives the *volume* of V only if $f(x, y, z) \equiv 1$. Note also that the volume element can be expressed in six different ways ($dx \, dy \, dz$, $dx \, dz \, dy$, etc.) in rectangular coordinates leading to various formulations of a volume integral. Cylindrical and spherical coordinates can be used where appropriate and often serve to simplify the evaluation of an integral. These variations are included in the exercises.

EXAMPLE 5.5–10 Find the volume integral of $\mathbf{V} \cdot \mathbf{F}$ over the volume enclosed by $z = 4 - x^2$, $y = 0$, $y = 3$, $x = 0$, and $z = 0$, if the vector field $\mathbf{F} = (2x^2 - 3z)\mathbf{i} - 2xy\mathbf{j} - 4z\mathbf{k}$.

Solution. The integrand is

$$\mathbf{V} \cdot \mathbf{F} = 2x - 4$$

and the volume is shown in Fig. 5.5–9. Also shown is the volume element dV, which is a rectangular parallelepiped with sides dx, dy, and dz. The element is shown as it varies in the z-direction since we have chosen to integrate first with respect to z (Exercise 13). Then

$$\iiint_V \mathbf{V} \cdot \mathbf{F} \, dV = \int_0^2 \int_0^3 \int_0^{4-x^2} (2x - 4) \, dz \, dy \, dx = -40. \quad \blacksquare$$

Recall from calculus that care must be used in setting up multiple integrals. The volume element must be placed in an *arbitrary* position within the volume. A common error that is made in the problem of the last example is to use 0 and 4 as the limits for the integration in the z-direction. A figure—even a rough sketch—is a great aid in preventing errors of this kind.

We remark that the notations

$$\int_S f(x, y, z) \, dS \quad \text{and} \quad \int_V f(x, y, z) \, dV$$

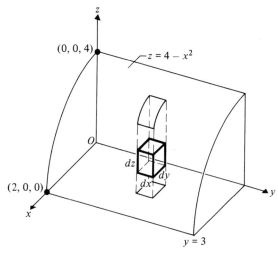

FIGURE 5.5–9 Volume for Example 5.5–10.

are also used for surface and volume integrals, respectively. Another frequently used notation is

$$\iint_S \mathbf{F} \cdot d\mathbf{S} \quad \text{for} \quad \iint_S \mathbf{F} \cdot \mathbf{n} \, dS.$$

In other words, the vector $d\mathbf{S}$ has the direction of the normal to the surface patch and magnitude dS, the area of the patch. In the next two sections we will discuss various relations between the integrals studied in this section.

KEY WORDS AND PHRASES

line integral   conservative field
independent of path   surface integral
circulation   volume integral

EXERCISES 5.5

▶   **1.** Show that if D is an open region and $P(x, y, z)$ lies in D, then $(x + \Delta x, y, z)$ also lies in D for sufficiently small values of Δx.

 2. In the proof of Theorem 5.5–1 provide the details to show that

$$\frac{\partial f}{\partial y} = F_2(x, y, z) \quad \text{and} \quad \frac{\partial f}{\partial z} = F_3(x, y, z).$$

3. Prove that the work done around every closed path in a conservative vector field is zero.

4. Show that the vector field

$$\mathbf{F} = (4xy - 3x^2z^2)\mathbf{i} + 2x^2\mathbf{j} - 2x^3z\mathbf{k}$$

is conservative.

5. Solve the problem of Example 5.5–6 by taking a path consisting of three straight-line segments from $(a, 0, 0)$ to $(a, 0, \pi b/2)$ to $(a, a, \pi b/2)$ to $(0, a, \pi b/2)$.

6. Explain why $|\cos \gamma|$ is used to relate dS and $dy\,dx$ in Example 5.5–7.

7. Solve the problem of Example 5.5–7 by:

   a) expressing the surface integral as $\iint 5x\,dx\,dy$ and using appropriate limits of integration;

   b) projecting dS into the yz-plane.

8. Carry out the details of the integrations in Example 5.5–8.

9. If $\mathbf{F} = y\mathbf{i} + x\mathbf{j} + (z^2 - 1)\mathbf{k}$, find

$$\iint\limits_{S} \mathbf{F} \cdot \mathbf{n}\, dS$$

over the *entire* closed surface bounded by the cylinder $x^2 + y^2 = a^2$ and the planes $x = 0$ $(y \geq 0)$, $y = 0$ $(x \leq 0)$, $z = 0$, $z = b$, where $\mathbf{n}$ is the unit outward normal.

10. Show that in spherical coordinates the element of area dS can be approximated by the product of a "vertical" arc $r\,d\theta$ and a "horizontal" arc $r \sin \theta\, d\phi$.

11. Give a definition of a volume integral patterned after the definition given in the text of a surface integral.

12. Carry out the details of the integrations in Example 5.5–9.

13. Solve the problem in Example 5.5–9 by integrating first

   a) in the y-direction;   **b)** in the x-direction.

▶▶ 14. Evaluate the line integral of Example 5.5–1 along each of the following paths.

   a) The circle: $x = 2 \cos t$, $y = 2 \sin t$ counterclockwise from $(2, 0)$ to $(0, 2)$

   b) Straight-line segments from $(2, 0)$ to $(0, 0)$ to $(0, 2)$

   c) Straight-line segments from $(2, 0)$ to $(2, 2)$ to $(0, 2)$

   d) From $(0, 2)$ to $(2, 0)$ along the line $y = 2 - x$

15. Evaluate the work done in the force field of Example 5.5–3 on a particle moving on each of the given paths.

   a) $(0, 0)$ to $(1, 0)$, then to $(1, 1)$

   b) $(0, 0)$ to $(1, 1)$ along a straight-line path

   c) $(0, 0)$ to $(1, 1)$ along the curve $y = x^2$

   d) $(0, 0)$ to $(1, 1)$ along the curve $x = y^2$

16. Which of the following vector fields are conservative? For those that are, find a scalar function f such that $\mathbf{F} = \nabla f$.

a) $\mathbf{F} = (yze^{xyz} + z)\mathbf{i} + xze^{xyz}\mathbf{j} + (xye^{xyz} + x)\mathbf{k}$

b) $\mathbf{F} = (3x - 2y)\mathbf{i} + (2x - 3y)\mathbf{j} + z\mathbf{k}$

c) $\mathbf{F} = \dfrac{-z}{y^2 + z^2}\mathbf{j} + \dfrac{yx}{y^2 + z^2}\mathbf{k}$

d) $\mathbf{F} = \dfrac{-y}{x^2 + y^2}\mathbf{i} + \dfrac{x}{x^2 + y^2}\mathbf{j}$

17. Find the work done on a particle moving in the circle: $x = \cos t$, $y = \sin t$, $0 \le t < 2\pi$ for each of the force fields shown.

   a) $\mathbf{F} = -y\mathbf{i} + x\mathbf{j}$ b) $\mathbf{F} = x\mathbf{i} + y\mathbf{j}$

   c) $\mathbf{F} = (x - y)\mathbf{i} + (x + y)\mathbf{j}$

   d) Can you determine *on the basis of your answers* which of the above fields are irrotational? Why?

18. Evaluate

$$\int_C \mathbf{F} \cdot d\mathbf{r}$$

   for the field $\mathbf{F}$ and path C given.

   a) $\mathbf{F} = \mathbf{V}(xy^2z^3)$; C: the ellipse formed by the plane $z = 2x + 3y$ cutting the right circular cylinder $x^2 + y^2 = 16$ from $(4, 0, 8)$ to $(0, 4, 12)$

   b) The same as part (a) except starting at $(4, 0, 8)$ and going completely around the ellipse in a counterclockwise direction as viewed from the positive z-axis looking toward the origin.

   c) $\mathbf{F} = \mathbf{V} \times \mathbf{r}$; straight-line segment from the origin to $(1, 2, 3)$

   d) $\mathbf{F} = \mathbf{V}(xy^2z^3)$; straight-line segment from $(1, 1, 1)$ to $(2, 1, -1)$

19. Compute the surface integral of $\sin z$ over the portion of the plane $z = x + y$ for which $x \ge 0$, $y \ge 0$, and $z \le \pi$.

20. Evaluate

$$\iint_S \mathbf{V} \times \mathbf{F} \cdot \mathbf{n}\, dS$$

   over the portion of the paraboloid $z = 4 - x^2 - y^2$ that lies above the plane $z = 0$, where $\mathbf{F} = (z - y)\mathbf{i} + (z + x)\mathbf{j} - (x + y)\mathbf{k}$.

21. Evaluate

$$\iint_S \mathbf{V} \times \mathbf{F} \cdot d\mathbf{S}$$

   over the hemisphere $z = \sqrt{4 - x^2 - y^2}$, $0 \le x^2 + y^2 < 4$, that lies above the xy-plane, where $F = y\mathbf{i} - x\mathbf{j}$.

22. Evaluate

$$\iiint_V \mathbf{V} \cdot \mathbf{F}\, dV,$$

   where $\mathbf{F} = (x + y)\mathbf{i} + (y + z)\mathbf{j} + (x + z)\mathbf{k}$ over each of the given volumes.

   a) $x^2 + y^2 \le 4 - z$, $z \ge 0$ b) $x^2 + y^2 \le 9$, $0 \le z \le 5$

   c) $x^2 + y^2 + z^2 \le a^2$

23. Evaluate $x + 2y$ over the closed region (volume) bounded by the parabolic cylinder $z = 4 - x^2$ and the planes $x = 0$, $y = 0$, $y = 3$, and $z = 0$.

24. Evaluate

$$\int_V \mathbf{V} \cdot \mathbf{F} \, dV,$$

where $\mathbf{F} = (2x^2 - 3z)\mathbf{i} - 2xy\mathbf{j} - 4z\mathbf{k}$ over the closed region in the first octant bounded by the plane $2x + 2y + z = 4$.

25. Show that it is impossible to find a scalar function f such that $\mathbf{grad} \, f = xy^2\mathbf{i} + x^3y\mathbf{j}$.

26. Consider the vector field

$$\mathbf{F} = 2xyz\mathbf{i} + x^2z\mathbf{j} + x^2y\mathbf{k}.$$

a) Show that $\mathbf{F}$ is conservative.

b) Find a scalar f such that $\mathbf{V}f = \mathbf{F}$.

c) Find the work done in moving a particle through this field from $(0, 1, 4)$ to $(1, -3, 2)$.

27. Distinguish between a surface integral and the area of a surface.

28. Distinguish between a volume integral and the volume enclosed by a closed surface.

5.6 INTEGRAL THEOREMS

In Section 5.5 we discussed line, surface, and volume integrals. This section features three important theorems that state that under certain conditions a line integral can be replaced by a surface integral and a surface integral by a volume integral. These facts will enable us to give coordinate-free definitions of gradient, divergence, and curl and also contribute to the discussion of applications in Section 5.7.

Although the theorems in question were stated and proved as recently as the nineteenth century, there is some uncertainty today about their authorship. Because of this the theorems bear a variety of names. We will use the names that seem to be most common but caution the reader to keep an open mind when consulting other texts.

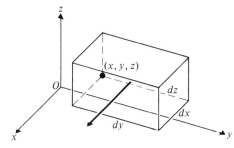

FIGURE 5.6–1 Volume element.

Divergence theorem

In Section 5.3 we discussed the relation between the gradient and the normal to a surface and the relation between the curl and the vorticity of a flow. We did not, however, assign any physical significance to the divergence. Accordingly, consider a volume element shown in Fig. 5.6–1. Let this small volume represent the volume of a *compressible* fluid with velocity $\mathbf{v}(x, y, z)$ and variable mass density $\rho(x, y, z)$ and consider the flow in the positive x-direction. The mass per unit time flowing *into* the element in this direction is approximately*

$$\rho v_1(x, y, z) \, dy \, dz,$$

where $\mathbf{v}(x, y, z) = v_1\mathbf{i} + v_2\mathbf{j} + v_3\mathbf{k}$. The rate of flow *out* of the element (still in the positive x-direction) is

$$\left(\rho v_1(x, y, z) + \frac{\partial}{\partial x}(\rho v_1) \, dx\right) dy \, dz.$$

The net rate of flow, that is, the difference between flow out and flow in, is

$$\frac{\partial}{\partial x}(\rho v_1) \, dx \, dy \, dz.$$

By considering the flow in the positive y- and z-directions, we get terms similar to the last one so that the *total* net rate of flow out of the element per unit time is

$$\left(\frac{\partial}{\partial x}(\rho v_1) + \frac{\partial}{\partial y}(\rho v_2) + \frac{\partial}{\partial z}(\rho v_3)\right) dx \, dy \, dz$$

or

$$\mathbf{V} \cdot (\rho\mathbf{v}) \, dx \, dy \, dz.$$

Therefore the total net flow of the fluid *out* of the element per unit volume per unit time is $\mathbf{V} \cdot \rho\mathbf{v}$. For this reason the name *divergence* is an appropriate one for the operator "$\mathbf{V} \cdot$." It should be mentioned that Maxwell originally used the term *convergence* and the present operator is the negative of this. Because of the law of conservation of mass[†] the net flow out must be balanced and we have the **equation of continuity** in hydrodynamics,

$$\mathbf{V} \cdot (\rho\mathbf{v}) + \frac{\partial \rho}{\partial t} = 0. \tag{5.6–1}$$

If now we assume the volume element dV of the previous discussion as being part of a volume V *enclosed* by the surface S, then

---

* The values given here and in the remainder of the discussion can be considered to be *average* values over the faces in question, which reduce to the ones shown for an arbitrarily small volume element.

[†] Another way of saying this is that the volume element contains no **sources** or **sinks**, that is, points where fluid is created or destroyed, respectively.

the *total* net flow per unit time can be represented by the volume integral

$$\iiint_V \mathbf{V} \cdot \rho \mathbf{v} \, dV.$$

On the other hand, the surface integral of the normal component of the velocity field, namely,

$$\iint_S \rho \mathbf{v} \cdot \mathbf{n} \, dS,$$

should give the same quantity. Thus the statement of Theorem 5.6–1 below should come as no surprise. We have, of course, neglected the mathematical niceties in the foregoing. Neither will we prove the integral theorems in this section. Proofs that are not overly complicated can be given only for certain simplified special cases. Many books on advanced calculus consider more general cases but only a few advanced texts go beyond this. An excellent treatment of the subject at an intermediate level can be found in Robert C. James's *Advanced Calculus*, Chapter 9 (Belmont, Calif.: Wadsworth, 1966).

The following theorem is also called *Gauss's theorem* but it is not to be confused with *Gauss's law*, which will be discussed in the following section.

Theorem 5.6–1 (*Divergence theorem*). Let V be a bounded region in $\mathbb{R}^3$ whose boundary S consists of a piecewise smooth surface. Let $\mathbf{n}$ be the outward normal to S and $\mathbf{F}(x, y, z)$ a vector field whose components have continuous first-order partial derivatives at all points of V and S. Then

$$\iiint_V \nabla \cdot \mathbf{F} \, dV = \iint_S \mathbf{F} \cdot \mathbf{n} \, dS. \tag{5.6–2}$$

See Fig. 5.6–2.

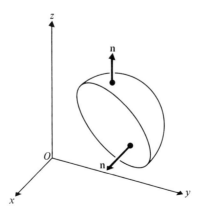

FIGURE 5.6–2 The divergence theorem.

DISCUSSION (in lieu of a proof). The words "outward normal" require some comment. It is assumed that we can distinguish one side of the surface S from the other, that is, the surface is **orientable**. This cannot be done for all surfaces. For example, the Möbius* strip is a *nonorientable* surface. We do not give a rigorous definition of an orientable surface since this would require the use of a branch of mathematics called *manifold theory.*[†]

The requirement that S consist of a piecewise smooth surface is to ensure that S is a **simple** surface, that is, one that does not intersect itself. This means that there cannot be more than one normal vector **n** at any point. Moreover, each smooth surface is bounded by smooth curves and **n** exists and varies continuously with position on the surface. If the surface is described by the equation $f(x, y, z) = 0$, then this last requirement is equivalent to requiring that the partial derivatives $\partial f/\partial x$, $\partial f/\partial y$, and $\partial f/\partial z$ exist and be continuous on the surface. Fortunately, the above restrictions are met by nearly all surfaces that are of practical interest.

EXAMPLE 5.6–1 Compute

$$\iint_S \mathbf{F} \cdot \mathbf{n} \, dS$$

over the closed surface bounded by the right circular cylinder $x^2 + y^2 = 9$ and the planes $z = 0$ and $z = 5$, where $\mathbf{F} = x\mathbf{i} + y\mathbf{j} + (z^2 - 1)\mathbf{k}$, (a) directly, (b) by using the divergence theorem.

---

* A Möbius strip (after August F. Möbius, 1790–1868, a German mathematician) can be formed by taking the rectangular strip with vertices $(0, 0)$, $(0, 1)$, $(12, 1)$, and $(12, 0)$ and twisting it so that $(12, 0)$ coincides with $(0, 1)$ and $(12, 1)$ with $(0, 0)$ (Fig. 5.6–3). By pasting the edges together in this fashion a continuous surface can be obtained that has only *one* side. On this surface a seemingly "outward" normal can be moved around continuously and end up in the *opposite* direction.

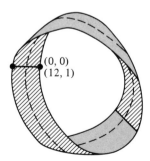

FIGURE 5.6–3 The Möbius strip.

[†] See, for example, M. Spivak, *Calculus on Manifolds* (Reading, Mass.: W. A. Benjamin, 1965).

Solution. We may obtain the result for part (a) from Example 5.5–8 by putting $a = 3$ and $b = 5$. The value of the surface integral is $\pi a^2 b(b + 2) = 315\pi$. For part (b) we have $\mathbf{V} \cdot \mathbf{F} = 2(z + 1)$ and

$$\iiint_V \mathbf{V} \cdot \mathbf{F} \, dV = 2 \int_0^5 \int_0^{2\pi} \int_0^3 (z + 1)\rho \, d\rho \, d\phi \, dz = 315\pi,$$

using cylindrical coordinates (see Eq. 5.4–10). ∎

The last example is somewhat typical in that often it is simpler to evaluate a volume integral than a surface integral. In a given situation, however, one may not be able to determine a priori which integral is easier to compute whenever the divergence theorem is applicable.

We use the divergence theorem to give another definition of divergence. Consider a "small" volume ΔV with boundary ΔS satisfying the hypotheses of Theorem 5.6–1 (the divergence theorem). If ΔV is sufficiently small, then we may take $\mathbf{V} \cdot \mathbf{F}$ to be approximately constant throughout ΔV. More precisely, we can state that

$$\lim_{\Delta V \to 0} \frac{\displaystyle\iiint_{\Delta V} \mathbf{V} \cdot \mathbf{F} \, dV}{\Delta V} = \mathbf{V} \cdot \mathbf{F},$$

where the volume integral is taken over ΔV. Now, using Theorem 5.6–1, we can write

$$\mathbf{V} \cdot \mathbf{F} = \lim_{\Delta V \to 0} \frac{1}{\Delta V} \iint_{\Delta S} \mathbf{F} \cdot \mathbf{n} \, dS$$

or, more simply,

$$\mathbf{V} \cdot \mathbf{F} = \lim_{V \to 0} \frac{1}{V} \iint_S \mathbf{F} \cdot \mathbf{n} \, dS, \tag{5.6–3}$$

provided the limit exists. This is a *coordinate-free* definition of *divergence* from which the one in Eq. (5.3–5) can be obtained. From Eq. (5.6–3) we can see the justification for saying that the divergence of a vector field $\mathbf{F}$ gives, at any point P, the *flux output* per unit volume at P. Here we use a common definition* of flux, namely, that the flux of $\mathbf{F}$ through a closed surface S is the numerical value of

$$\iint_S \mathbf{F} \cdot \mathbf{n} \, dS.$$

Thus the divergence theorem states that the flux of a vector field $\mathbf{F}$ through a closed surface S is a measure of the divergence of $\mathbf{F}$ in V, the interior of S.

---

* If $\mathbf{F}$ is the velocity field of a fluid, then the flux can be interpreted as the total volume flowing over the surface S in unit time.

The next two theorems follow from the divergence theorem and, in turn, will lead to coordinate-free definitions of the gradient and curl.

Theorem 5.6–2 Let V and S be defined as in Theorem 5.6–1 and let $f(x, y, z)$ be a differentiable scalar function. Then

$$\iiint_V \nabla f \, dV = \iint_S f\mathbf{n} \, dS. \tag{5.6–4}$$

PROOF. Let $\mathbf{F} = f\mathbf{A}$, where $\mathbf{A}$ is a *constant* vector. Then Eq. (5.6–2) becomes

$$\iiint_V \nabla \cdot f\mathbf{A} \, dV = \iint_S f\mathbf{A} \cdot \mathbf{n} \, dS. \tag{5.6–5}$$

But $\nabla \cdot f\mathbf{A} = \nabla f \cdot \mathbf{A} = \mathbf{A} \cdot \nabla f$ by Theorem 5.3–2(c) and $f\mathbf{A} \cdot \mathbf{n} = \mathbf{A} \cdot f\mathbf{n}$. (Why?) Hence Eq. (5.6–5) can be written as

$$\iiint_V (\mathbf{A} \cdot \nabla f) \, dV = \iint_S \mathbf{A} \cdot f\mathbf{n} \, dS,$$

or

$$\mathbf{A} \cdot \iiint_V \nabla f \, dV = \mathbf{A} \cdot \iint_S f\mathbf{n} \, dS,$$

or

$$\mathbf{A} \cdot \left(\iiint_V \nabla f \, dV - \iint_S f\mathbf{n} \, dS \right) = 0.$$

But $|\mathbf{A}| \neq 0$ and the direction of $\mathbf{A}$ is arbitrary so that the cosine of the angle between the vectors in the last zero dot product cannot *always* be zero. Hence the quantity in parentheses is zero and the desired result follows. $\square$

Theorem 5.6–3 Let V, S, and $\mathbf{F}$ be defined as in Theorem 5.6–1. Then

$$\iiint_V \nabla \times \mathbf{F} \, dV = \iint_S \mathbf{n} \times \mathbf{F} \, dS. \tag{5.6–6}$$

PROOF. Let $\mathbf{B} = \mathbf{F} \times \mathbf{A}$, where $\mathbf{A}$ is a constant vector, and apply the divergence theorem to $\mathbf{B}$. We have

$$\iiint_V \nabla \cdot (\mathbf{F} \times \mathbf{A}) \, dV = \iint_S (\mathbf{F} \times \mathbf{A}) \cdot \mathbf{n} \, dS.$$

But $\nabla \cdot (\mathbf{F} \times \mathbf{A}) = \mathbf{A} \cdot \nabla \times \mathbf{F}$ by Theorem 5.3–3(c). Also

$$(\mathbf{F} \times \mathbf{A}) \cdot \mathbf{n} = \mathbf{F} \cdot (\mathbf{A} \times \mathbf{n}) = (\mathbf{A} \times \mathbf{n}) \cdot \mathbf{F} = \mathbf{A} \cdot (\mathbf{n} \times \mathbf{F})$$

by Exercise 24 in Section 5.1. Hence,

$$\iiint_V \mathbf{A} \cdot (\nabla \times \mathbf{F}) \, dV = \iint_S \mathbf{A} \cdot (\mathbf{n} \times \mathbf{F}) \, dS$$

or

$$\mathbf{A} \cdot \iiint_V (\mathbf{\nabla} \times \mathbf{F}) \, dV = \mathbf{A} \cdot \iint_S (\mathbf{n} \times \mathbf{F}) \, dS$$

and the desired result follows by the same reasoning used to prove Theorem 5.6–2. $\square$

From the last two theorems we can obtain coordinate-free definitions of gradient and curl similar to the definition of divergence in Eq. (5.6–3). We have

$$\mathbf{\nabla}f = \lim_{V \to 0} \frac{1}{V} \iint_S f\mathbf{n} \, dS \qquad (5.6\text{–}7)$$

and

$$\mathbf{\nabla} \times \mathbf{F} = \lim_{V \to 0} \frac{1}{V} \iint_S (\mathbf{n} \times \mathbf{F}) \, dS, \qquad (5.6\text{–}8)$$

whenever the above limits exist.

Note that both Eqs. (5.6–7) and (5.6–8) are examples of vector integration in which $\mathbf{i}$, $\mathbf{j}$, and $\mathbf{k}$ are constant, hence can be brought outside the integrals. For example,

$$\int_0^1 (t\mathbf{i} + t^2\mathbf{j} + t^3\mathbf{k}) \, dt = \frac{t^2}{2}\mathbf{i} + \frac{t^3}{3}\mathbf{j} + \frac{t^4}{4}\mathbf{k} \bigg|_0^1$$

$$= \frac{1}{2}\mathbf{i} + \frac{1}{3}\mathbf{j} + \frac{1}{4}\mathbf{k}.$$

Stokes's theorem

A second important integral theorem is called *Stokes's theorem*, after George G. Stokes (1819–1903), the Anglo-Irish mathematical physicist who submitted the theorem in 1854 as an examination question (!) in the Smith's Prize Exam at Cambridge University.*

Theorem 5.6–4 (*Stokes's theorem*). Let S be a piecewise smooth orientable surface with unit normal $\mathbf{n}$ and having as boundary the simple closed piecewise smooth curve C. Let $\mathbf{F}(x, y, z)$ be a vector field whose components have continuous first-order partial derivatives on S. Then

$$\iint_S (\mathbf{\nabla} \times \mathbf{F}) \cdot \mathbf{n} \, dS = \int_C \mathbf{F} \cdot d\mathbf{r}, \qquad (5.6\text{–}9)$$

provided the orientations of C and S are compatible. That is, a person traversing C with his head pointing in the direction of the outward normal has the surface S on his left. (See Fig. 5.6–4.)

* See Victor J. Katz, "The History of Stokes' Theorem," *Math Mag.* **52** (May 1979): 146–156.

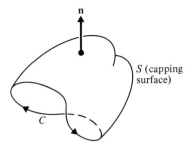

FIGURE 5.6–4 Stokes's theorem.

DISCUSSION. The comments about orientable surfaces in the discussion of the divergence theorem apply equally well here. In words, Stokes's theorem states that the normal component of the curl over a surface that "caps" a closed space curve is equal to the circulation around the curve. Note that the surface S is *not closed* in this theorem but that its open boundary constitutes the curve C. Such a surface may be called a "capping surface." One can imagine that S is the surface of a duffel bag and C represents the drawstring.

EXAMPLE 5.6–2   Use Stokes's theorem to evaluate the normal component of $\mathbf{V} \times \mathbf{F}$ over the surface bounded by the planes $4x + 6y + 3z = 12$, $x = 0$, and $y = 0$ above the xy-plane, where $\mathbf{F} = y\mathbf{i} + xz\mathbf{j} + (x^2 - yz)\mathbf{k}$. (See Fig. 5.6–5.)

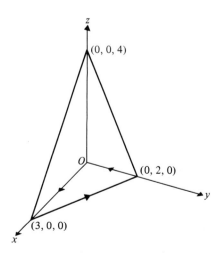

FIGURE 5.6–5 The path for Example 5.6–2.

Solution. The hypotheses of Stokes's theorem (Theorem 5.6–4) are satisfied and the curve C is the triangle with vertices $(0, 0, 0)$, $(3, 0, 0)$, and $(0, 2, 0)$ oriented in that order. We have $\mathbf{F} \cdot d\mathbf{r} = y\, dx$ since $z = 0$ and $dz = 0$ on C. Hence the circulation of F on C is given by

$$\int_0^3 0\, dx - \frac{3}{2} \int_0^2 y\, dy + \int_2^0 y(0) = -3.$$

Note: The hypotenuse of the triangular path has equation given by $2x + 3y = 6$.

Using an approach similar to the one used to obtain a coordinate-free definition of divergence, we have

$$(\mathbf{V} \times \mathbf{F}) \cdot \mathbf{n} = \lim_{\Delta S \to 0} \frac{1}{\Delta S} \oint_C \mathbf{F} \cdot d\mathbf{r}. \qquad \textbf{(5.6–10)}$$

From Eq. (5.6–10) we can deduce that the normal component of the curl of $\mathbf{F}$ at point P is the circulation per unit area of $\mathbf{F}$ around P in a plane normal to $\mathbf{n}$.

Green's theorem in the plane

If the space curve C is a *plane* closed curve, in particular a simple closed curve in the xy-plane, then Stokes's theorem reduces to the following one.

Theorem 5.6–5 (*Green's* theorem in the plane*). Consider a closed region R in the xy-plane having as boundary the simple, closed, piecewise smooth curve C. Let $M(x, y)$ and $N(x, y)$ be continuous functions of x and y that have continuous first-order partial derivatives in R. Then

$$\oint_C (M(x, y)\, dx + N(x, y)\, dy) = \iint_R \left(\frac{\partial N}{\partial x} - \frac{\partial M}{\partial y} \right) dx\, dy, \qquad \textbf{(5.6–11)}$$

where C is traversed in the positive direction (the region R is on the *left* of a person walking on C). See Fig. 5.6–6.

EXAMPLE 5.6–3 Evaluate

$$\oint_C \mathbf{F} \cdot d\mathbf{r},$$

where C is the positive closed path enclosing the region between the circles $x^2 + y^2 = 4$ and $x^2 + y^2 = 16$ if $F = xy\mathbf{i} - x\mathbf{j}$.

---

* After George Green (1793–1841), an English mathematician.

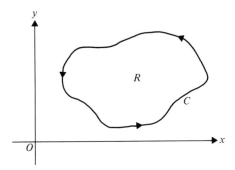

FIGURE 5.6-6 Green's theorem in the plane.

Solution. Here the curve C encloses an annular region as shown in Fig. 5.6–7. Note that the direction on one circle is opposite to that on the other. Note also that the two circular paths are connected by a straight-line segment $\overline{AB}$ and that this portion of the path is traversed twice in opposite directions. By parameterizing the outer circle we obtain

$$C_1: \quad \mathbf{r}_1(t) = 4\cos t\,\mathbf{i} + 4\sin t\,\mathbf{j}, \qquad 0 \le t < 2\pi,$$

and similarly for the inner circle

$$C_2: \quad \mathbf{r}_2(t) = 2\cos t\,\mathbf{i} + 2\sin t\,\mathbf{j}, \qquad 0 \le t < 2\pi.$$

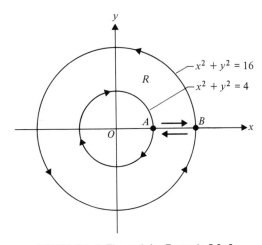

FIGURE 5.6-7 The path for Example 5.6–3.

Hence the required line integral becomes

$$\int_0^{2\pi} (8 \cos t \sin^2 t + 4 \cos^2 t)\, dt + \int_0^{2\pi} (-64 \cos t \sin^2 t - 16 \cos^2 t)\, dt$$

$$= \int_0^{2\pi} (-56 \cos t \sin^2 t - 12 \cos^2 t)\, dt$$

$$= -56 \frac{\sin^3 t}{3} - 12 \left(\frac{t}{2} + \frac{\sin 2t}{4} \right) \Big|_0^{2\pi}$$

$$= -12\pi \doteq -37.7.$$

On the other hand, using Theorem 5.6–5 we have

$$M(x, y) = xy, \qquad \frac{\partial M}{\partial y} = x,$$

$$N(x, y) = -x, \qquad \frac{\partial N}{\partial x} = -1,$$

so that

$$\iint_R \left(\frac{\partial N}{\partial x} - \frac{\partial M}{\partial y} \right) dx\, dy = \iint_R (-1 - x)\, dx\, dy$$

$$= \iint_R (-1 - \rho \cos \phi)\rho\, d\rho\, d\phi$$

in polar coordinates. Then

$$\int_0^{2\pi} \int_2^4 (-1 - \rho \cos \phi)\rho\, d\rho\, d\phi = -12\pi \doteq -37.7$$

as before, illustrating the use of Green's theorem in the plane. ■

The following is a useful and interesting corollary of Theorem 5.6–5.

Corollary 5.6–6 The area bounded by a simple closed curve C is given by

$$A = \frac{1}{2} \oint_C (x\, dy - y\, dx). \tag{5.6–12}$$

PROOF. In Theorem 5.6–5 let $M(x, y) = -y$ and $N(x, y) = x$, which leads to

$$\oint_C (x\, dy - y\, dx) = 2 \iint_R dx\, dy$$

from which the desired result follows. □

EXAMPLE 5.6–4 Find the area bounded by the four-cusped hypocycloid $x^{2/3} + y^{2/3} = a^{2/3}$, $a > 0$.

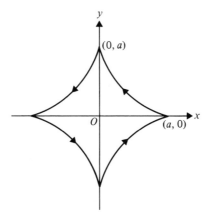

FIGURE 5.6–8 A four-cusped hypocycloid.

Solution. The curve is shown in Fig. 5.6–8. It is a closed curve traced by a point on a circle of radius $a/4$ as it rolls on the inside of a circle of radius a without slipping. The given equation can be parameterized by writing

$$x = a \cos^3 \theta, \qquad y = a \sin^3 \theta;$$

hence, using Corollary 5.6–6, we have

$$A = \frac{1}{2} \int_0^{2\pi} (3a^2 \sin^2 \theta \cos^4 \theta \, d\theta + 3a^2 \sin^4 \theta \cos^2 \theta \, d\theta)$$

$$= \frac{3a^2}{2} \int_0^{2\pi} \sin^2 \theta \cos^2 \theta \, d\theta = \frac{3a^2}{8} \int_0^{2\pi} \sin^2 2\theta \, d\theta$$

$$= \frac{3a^2}{8} \left(\frac{\theta}{2} - \frac{\sin 4\theta}{8} \right) \Big|_0^{2\pi} = \frac{3\pi a^2}{8}. \quad \blacksquare$$

In Section 5.7 we will present a number of applications to illustrate the uses to which the concepts of vector analysis can be put. In particular, the integral theorems of this section will be helpful in expressing some important results.

KEY WORDS AND PHRASES

equation of continuity	orientable surface
source	coordinate-free definition
sink	Stokes's theorem
divergence theorem	Green's theorem in the plane

EXERCISES 5.6

▶ 1. Show that it follows from the equation of continuity (5.6–1) that for an incompressible fluid, $\mathbf{V} \cdot \mathbf{v} = 0$.

2. Furnish the details for the computation in part (b) of Example 5.6–1.

3. Explain why the definition of divergence in Eq. (5.6–3) is called "co-ordinate-free."

4. Solve the problem stated in Example 5.6–2 by evaluating the normal component of the curl over the given capping surface, that is, without using Stokes's theorem.

▶▶ In Exercises 5–11, use the divergence theorem whenever it is applicable.

5. Evaluate the surface integral of z^2 over the sphere of radius a centered at the origin. (Compare Example 5.5–9.)

6. Find the volume integral of $\mathbf{V} \cdot \mathbf{F}$ over the volume enclosed by $z = 4 - x^2$, $y = 0$, $y = 3$, $x = 0$, and $z = 0$, where $\mathbf{F} = (2x^2 - 3z)\mathbf{i} - 2xy\mathbf{j} - 4z\mathbf{k}$. (Compare Example 5.5–10.)

7. Evaluate

$$\int_V \mathbf{V} \cdot \mathbf{F} \, dV,$$

where $\mathbf{F} = (2x^2 - 3z)\mathbf{i} - 2xy\mathbf{j} - 4z\mathbf{k}$ over the closed region in the first octant bounded by the plane $2x + 2y + z = 4$. (Compare Exercise 24 in Section 5.5.)

8. Evaluate

$$\int_V \mathbf{V} \cdot \mathbf{F} \, dV,$$

where $\mathbf{F} = (x + y)\mathbf{i} + (y + z)\mathbf{j} + (x + z)\mathbf{k}$ over the given volume.
   a) $x^2 + y^2 \leq 4 - z$, $z \geq 0$
   b) $x^2 + y^2 \leq 9$, $0 \leq z \leq 5$
   c) $x^2 + y^2 + z^2 \leq a^2$

(Compare Exercise 22 in Section 5.5.)

9. a) Show that

$$\iiint_V \mathbf{V} \cdot \mathbf{r} \, dV = 3V.$$

   b) Express the result of part (a) in terms of a surface integral.

10. Evaluate

$$\iint_S \mathbf{F} \cdot \mathbf{n} \, dS$$

over the entire surface of the cube bounded by the planes $x = \pm 1$, $y = \pm 1$, $z = \pm 1$, where $\mathbf{F}$ is each of the following.
   a) $\mathbf{F} = \mathbf{V}(xyz)$
   b) $\mathbf{F} = \mathbf{r}$
   c) $\mathbf{F} = x^2\mathbf{i} + y^2\mathbf{j} + z^2\mathbf{k}$

11. Compute

$$\iint\limits_{S} \mathbf{F} \cdot \mathbf{n} \, dS$$

over the surface of the tetrahedron bounded by the planes $x = 0$, $y = 0$, $z = 0$, and $x + y + z = 1$, where $\mathbf{F} = xy\mathbf{i} + z^2\mathbf{j} + 2yz\mathbf{k}$.

12. Explain how the divergence theorem could be used in Exercise 11 if the surface $y = 0$ were omitted from S.

13. Verify Eq. (5.6–6) for $\mathbf{F} = (2x^2 - 3z)\mathbf{i} - 2xy\mathbf{j} - 4x\mathbf{k}$, where V is the region bounded by $x = 0$, $y = 0$, $z = 0$, and $2x + 2y + z = 4$.

In Exercises 14–18, use Stokes's theorem whenever it is applicable.

14. Evaluate

$$\iint\limits_{S} \mathbf{V} \times \mathbf{F} \cdot \mathbf{n} \, dS$$

over the portion of the paraboloid $z = 4 - x^2 - y^2$ that lies above the plane $z = 0$, where $\mathbf{F} = (z - y)\mathbf{i} + (z + x)\mathbf{j} - (x + y)\mathbf{k}$. (Compare Exercise 20 in Section 5.5.)

15. Evaluate

$$\iint\limits_{S} \mathbf{V} \times \mathbf{F} \cdot d\mathbf{S}$$

over the hemisphere $z = \sqrt{4 - x^2 - y^2}$, $0 \le x^2 + y^2 < 4$, which lies above the xy-plane, where $\mathbf{F} = y\mathbf{i} - x\mathbf{j}$.

16. Find the surface integral of the normal component of the curl of the vector field

$$\mathbf{F} = (y^2 + z^2)\mathbf{i} + (x^2 + z^2)\mathbf{j} + (x^2 + y^2)\mathbf{k}$$

over the given surface.

a) $z = \sqrt{1 - x^2}$, $y = -1$, $y = 1$, $z > 0$

b) The surface of the upper half of the cube with one vertex at $(1, 1, 1)$, center at the origin, and edges parallel to the axes

c) The surface of a tetrahedron, excluding the face in the yz-plane, with vertices at $(0, 0, 0)$, $(1, 0, 0)$, $(0, 1, 0)$, and $(0, 0, 1)$

17. Compute the circulation of $\mathbf{F} = x\mathbf{i} + (2z - x)\mathbf{j} + y^2\mathbf{k}$ around the circle $x^2 + y^2 = 4$, $z = 2$, where the capping surface is given by each of the following.

a) $x^2 + y^2 = 2z$

b) $x^2 + y^2 + z^2 - 4z = 0$

c) $x^2 + y^2 = 4$ and $x^2 + y^2 = 4$, $z = 0$

18. Compute

$$\iint\limits_{S} \mathbf{V} \times \mathbf{F} \cdot \mathbf{n} \, dS$$

given that $\mathbf{F} = (y - z + 2)\mathbf{i} + (yz + 4)\mathbf{j} - xz\mathbf{k}$ and S is the surface of the open box formed by the planes $x = 0$, $y = 0$, $z = 0$, $x = 3$, $y = 3$, $z = 3$, with the top being open.

19. Use Stokes's theorem to evaluate the circulation of

$$\mathbf{F} = yz \exp(xy)\mathbf{i} + xz(1 + \exp(xy))\mathbf{j} + \exp(xy)\mathbf{k}$$

around the curve of intersection $x^2 + y^2 = 4$ and $2x + 3y + z = 5$ in a counterclockwise direction as viewed from the positive z-axis.

20. Evaluate the line integral

$$\oint_C (2y\,dx - 2x\,dy + xz^2\,dz)$$

around the circle $x^2 + y^2 = 1$, $z = 5$, by (a) direct computation, and (b) using Stokes's theorem.

21. Show that, in addition to the formula for area given by Eq. (5.6–12), the formulas

$$A = \oint_C x\,dy = -\oint_C y\,dx$$

and infinitely many others are also applicable.

22. Use the formula

$$A = -\oint_C y\,dx$$

to show that the area of a triangle with vertices at (x_1, y_1), (x_2, y_2), and (x_3, y_3) can be written as $\frac{1}{2}((x_2 - x_1)(y_3 - y_1) - (y_2 - y_1)(x_3 - x_1))$. Note that this result can also be expressed as

$$A = \frac{1}{2}\begin{vmatrix} x_1 & y_1 & 1 \\ x_2 & y_2 & 1 \\ x_3 & y_3 & 1 \end{vmatrix}.$$

23. The equation of an ellipse can be written in parametric form,

$$\mathbf{r}(\theta) = a\cos\theta\mathbf{i} + b\sin\theta\mathbf{j},$$

where $ab > 0$. Find the area of an elliptical region by means of a line integral.

24. Find the area bounded by one arch of the **cycloid*** and the x-axis. Use the parametric equations of the cycloid obtained from the vector equation,

$$\mathbf{r}(\theta) = a(\theta - \sin\theta)\mathbf{i} + a(1 - \cos\theta)\mathbf{j}, \qquad a > 0.$$

25. Use a line integral to find the area of the annular region between the circles $x^2 + y^2 = 4$ and $x^2 + y^2 = 16$. (*Hint:* Use polar coordinates.)

26. Find the circulation of $\mathbf{F} = -y\mathbf{i} + x\mathbf{j}$ around the square bounded by $x = 1$, $y = 1$, $x = 0$, and $y = 0$ in a counterclockwise direction (a) directly, and (b) by using Theorem 5.6–5.

27. Find the circulation of $\mathbf{F} = (\sin y)\mathbf{i} + (x\cos y)\mathbf{j}$ around the triangle with vertices at $(-1, 0)$, $(1, 0)$, and $(0, 1)$ in a counterclockwise direction (a) directly, and (b) by using Theorem 5.6–5.

▶▶▶ **28.** Explain why Green's theorem in the plane does not apply if

$$M = \frac{-y}{x^2 + y^2} \quad \text{and} \quad N = \frac{x}{x^2 + y^2}$$

and C is the circle $x^2 + y^2 = a^2$.

---

* A cycloid is the curve traced by a circle of radius a as it rolls along the x-axis without slipping.

29. Show that Green's theorem in the plane *does* apply when the functions of Exercise 28 are used and R is the annular region bounded by $x^2 + y^2 = a^2$ and $x^2 + y^2 = \varepsilon$ for arbitrarily small ε such that $0 < \varepsilon < a$.

30. Show that the vector form of Green's theorem in the plane can be written

$$\oint_C \mathbf{F} \cdot d\mathbf{r} = \iint_R \nabla \times \mathbf{F} \cdot \mathbf{k} \, dA.$$

31. Prove that the normal component of the curl of a differentiable vector field integrated over a closed surface is zero.

32. Use Stokes's theorem to show that if **curl F** $= 0$ in a simply connected region, then **F** is conservative.

33. Show that

$$\iint_S \nabla \times \mathbf{F} \cdot \mathbf{n} \, dS$$

has the same value for all oriented surfaces that cap a positively oriented curve C provided that the directions of **n** and C are compatible.

34. Let S be a capping surface bounded by a curve C as in Stokes's theorem. If **a** is a constant vector, show that

$$\iint_S \mathbf{a} \cdot \mathbf{n} \, dS = \frac{1}{2} \oint_C (\mathbf{a} \times \mathbf{r}) \cdot d\mathbf{r}.$$

5.7 APPLICATIONS

Vectors provide a means of presenting a large amount of information succinctly. When we write **v**, for example, we are implying a magnitude and a direction that could involve n mutually orthogonal basis vectors. The use of the del operator permits the writing of relations between physical quantities with a minimum amount of notation. A classic example of this is given by Maxwell's equations in electromagnetic theory, which will be discussed later in this section. Also included will be some applications from potential theory, electrostatics, kinematics, and thermodynamics. These are not intended to be exhaustive by any means but will indicate the advantages of vector notation and the usefulness of the concepts discussed in the previous sections.

Central force fields

When a vector field **F** is such that at each point, $\mathbf{F}(x, y, z)$ is a vector directed toward a *fixed point*, the field is called a **central field**. A simple example of such a field is given by

$$\mathbf{F}(x, y, z) = \frac{-x\mathbf{i} - y\mathbf{j} - z\mathbf{k}}{\sqrt{x^2 + y^2 + z^2}} = \frac{-\mathbf{r}}{r}.$$

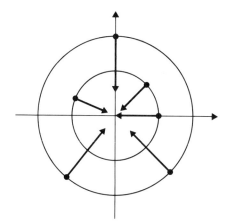

FIGURE 5.7–1 A central field.

This field, which is defined everywhere except at the origin, consists of unit vectors at every point directed toward the origin (see Fig. 5.7–1). More generally, any force of the form

$$\mathbf{F} = f(r)\mathbf{e}_r, \qquad (5.7\text{–}1)$$

where $\mathbf{e}_r$ is a unit vector in the radial direction (compare Section 5.4) and f is a scalar function, is a **central force**.

It can be shown by direct calculation (Exercise 1) that $\nabla \times \mathbf{F} = \mathbf{0}$ for the force given in Eq. (5.7–1). Hence central fields are conservative and the work done in moving a particle in such a field is independent of the path. This last statement is equivalent to saying that every central field $\mathbf{F}$ can be expressed as the negative gradient of a scalar function ϕ, called a scalar **potential**,

$$\mathbf{F} = -\nabla\phi.$$

We next give some examples of central forces.

EXAMPLE 5.7–1 (*Centrifugal force*). A particle on a circular disk rotating with constant angular speed ω experiences a radial outward force given by (Fig. 5.7–2)

$$\mathbf{F}_C = \omega^2 r \mathbf{e}_r.$$

This force is called a **centrifugal* force**. The work done in moving a particle in this field from zero to r is

$$\int_0^r \mathbf{F}_C \cdot d\mathbf{r} = \phi_C(0) - \phi_C(r),$$

---

* Literally, "fleeing the center."

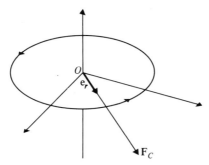

FIGURE 5.7–2 Centrifugal force.

where $\phi_C(r)$ is the scalar potential of $\mathbf{F}_C$. By taking $\phi_C(0) = 0$ arbitrarily and observing that in the above integral we obtain a contribution to ϕ only when $d\mathbf{r}$ is radial since $\mathbf{F}_C$ is radial we have

$$\phi_C(r) = -\int_0^r \mathbf{F}_C \cdot d\mathbf{r} = -\int_0^r \omega^2 r \, dr = -\tfrac{1}{2}\omega^2 r^2,$$

the *potential* of a centrifugal force. ∎

EXAMPLE 5.7–2   (*Gravitational force*). According to **Newton's universal law of gravitation**, two bodies of mass m_1 and m_2 a distance r apart attract each other with a gravitational force of magnitude

$$G\,\frac{m_1 m_2}{r^2},$$

where G is a constant, called the constant of gravitation. In SI units* the value of G is approximately 6.67×10^{-11} N-m^2/kg^2, a value that is determined experimentally. According to Newton's third law of motion, the gravitational force acts along the line joining the two bodies, hence, if we take the origin at the center of the earth, the gravitational force is

$$\mathbf{F}_G = -G\,\frac{m_1 m_2}{r^2}\,\mathbf{e}_r. \qquad (5.7\text{–}2)$$

This is also a central force directed radially inward. We define the gravitational potential at (x, y, z) as the work done in bringing a unit mass from infinity to the point. As in the previous example,

$$\phi_G(r) = -\int_\infty^r \mathbf{F}_G \cdot d\mathbf{r} = \int_r^\infty \mathbf{F}_G \cdot d\mathbf{r} = -G\,\frac{m_1 m_2}{r},$$

called the gravitational *potential*. Note that defining the gravitational

* Système International d'Unites.

potential in this way assigns zero to $\phi_G(\infty)$. This acknowledges that only *potential difference* can be defined since from $\mathbf{F} = -\nabla\phi$, we can determine ϕ only to within an additive constant. (See Examples 5.5–4 and 5.5–5.) ∎

Central forces thus have an important role in natural phenomena. It can be shown* that it follows from the laws of Newtonian mechanics that the orbit in which a body travels due to a central force field is necessarily a *plane* curve (Exercise 12). In particular, the curve is an ellipse if it is closed and usually a hyperbola if it is not. There is a very small probability that the path is a parabola in the latter case. Kepler's[†] first law states that planetary orbits are plane, and the second law says that a planet's position vector sweeps across equal areas in equal times. Since the origin of the position vectors is taken as the fixed point of the central-force gravitational field, Kepler's laws are directly derivable from the fact that planetary motion takes place in a central-force field.[‡]

Electrostatics

The study of electric fields that are due to stationary electric charges is called *electrostatics*. Fundamental in this subject is Coulomb's law,[§] which states that two point charges q_1 and q_2 attract (or repel) each other with a force that is proportional to the product q_1q_2 and inversely proportional to the square of the distance between the charges. Calling the force $\mathbf{F}_E$, we have

$$\mathbf{F}_E = k\frac{q_1q_2}{r^3}\mathbf{r}, \tag{5.7-3}$$

where $\mathbf{r}$ is the vector from charge q_1 to charge q_2 and k is a constant. The approximate value of k (called the Coulomb constant) is 9×10^9 N-m²/C², where C is the SI unit of charge called the *coulomb*. In place of k the constant $1/(4\pi\varepsilon_0)$, where ε_0 is the permittivity of free space,[‖] is frequently used. Note the close resemblance between $\mathbf{F}_E$ in Eq. (5.7–3) and the gravitational field $\mathbf{F}_G$ in Eq. (5.7–2).

* See John B. Fraleigh, *Calculus: A Linear Approach*, vol. 2, p. 720ff. (Reading, Mass.: Addison-Wesley, 1972).
† Johannes Kepler (1571–1630), a German astronomer.
‡ See Jerry B. Marion, *Classical Dynamics of Particles and Systems*, 2d ed., Chapter 8 (New York: Academic Press, 1970).
§ Charles A. de Coulomb (1736–1806), a French physicist, experimentally confirmed this relationship.
‖ "Free space" can be defined as a vacuum devoid of matter or, equivalently, infinitely removed from matter. In free space the electric field $\mathbf{E}$ and the electric displacement $\mathbf{D}$ are related by $\mathbf{D} = \varepsilon_0\mathbf{E}$, the permittivity ε_0 having dimensions of farads/m in the MKS system.

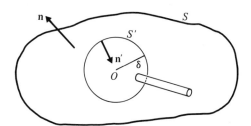

FIGURE 5.7–3 Gauss's law (Example 5.7–3).

If a point electric charge q is placed at the origin, then this charge will produce an electric field given by

$$\mathbf{E} = \frac{q}{4\pi\varepsilon_0 r^2}\, \mathbf{e}_r. \tag{5.7–4}$$

The total flux of $\mathbf{E}$ over a surface S is given by the surface integral

$$\iint_S \mathbf{E} \cdot \mathbf{n}\, dS.$$

EXAMPLE 5.7–3   (*Gauss's law**). For a point charge q at the origin we have

$$\iint_S \mathbf{E} \cdot \mathbf{n}\, dS = \begin{cases} q/\varepsilon_0, & \text{if } q \text{ is inside } S, \\ 0, & \text{if } q \text{ is outside } S. \end{cases}$$

For the second part we can use the divergence theorem since the integrands do not include the origin (Exercise 2). For the first part we may surround the origin with a sphere S' of radius δ contained within the surface S (Fig. 5.7–3). The surface S and the sphere S' may be joined by a hollow tube to form a single simply connected closed surface. The radius of the tube can be made increasingly small so there will be no contribution from its surface. Thus the divergence theorem applies to the volume between S' and S and we can also use the second part of Gauss's law to obtain

$$\iint_S \mathbf{E} \cdot \mathbf{n}\, dS - \iint_{S'} \mathbf{E} \cdot \mathbf{n}'\, dS' = 0.$$

The negative sign for the second integral is due to the fact that $\mathbf{n}'$, the normal to the sphere S', points inward. Moreover, on S' the vectors $\mathbf{E}$ and $\mathbf{n}'$ are parallel to $\mathbf{e}_r$, hence the second integral has the value

$$\frac{q}{4\pi\varepsilon_0 \delta^2}\,(4\pi\delta^2) = \frac{q}{\varepsilon_0}$$

and the desired result has been established. ∎

---

* Published in 1867.

We observe that if a volume V contains a distributed charge, that is, if q is replaced by a total charge having charge density $\rho(x, y, z)$ enclosed by the surface S, then

$$\iint_S \mathbf{E} \cdot \mathbf{n} \, dS = \iiint_V \frac{\rho}{\varepsilon_0} \, dV.$$

The divergence theorem now permits us to write

$$\iiint_V \mathbf{V} \cdot \mathbf{E} \, dv = \iiint_V \frac{\rho}{\varepsilon_0} \, dV$$

and, since V is arbitrary, the integrands must be equal, that is,

$$\mathbf{V} \cdot \mathbf{E} = \frac{\rho}{\varepsilon_0}. \tag{5.7-5}$$

This last is one of Maxwell's four fundamental equations of electromagnetism. It should be noted that this equation could have served as the starting point and Gauss's law could have been obtained from it by reversing the above argument (Exercise 3). If we replace $\mathbf{E}$ by $-\mathbf{V}\phi$ (since $\mathbf{E}$ is conservative), then Eq. (5.7–5) becomes

$$\mathbf{V}^2 \phi = \frac{-\rho}{\varepsilon_0}, \tag{5.7-6}$$

which is known as **Poisson's* equation**. In case a region is charge-free, that is, $\rho = 0$, we have the equation

$$\mathbf{V}^2 \phi = 0, \tag{5.7-7}$$

known as **Laplace's equation**, which will be studied in some detail in Chapters 8 and 9. The scalar function $\phi(x, y, z)$ appearing in Eqs. (5.7–6) and (5.7–7) is called the *electric potential*.[†] It is possible to define the electric field completely in terms of this potential. When this is done, it is customary to define the potential at infinity to be zero. The negative sign in $\mathbf{E} = -\mathbf{V}\phi$ indicates that $\mathbf{E}$ is in the direction of *decreasing* potential, which agrees with usual convention.

Maxwell's equations

From Eq. (5.7–5) we can see that in a region in which there are no charges, we have $\mathbf{V} \cdot \mathbf{E} = 0$. This "incompressibility condition" is equivalent to saying that the region contains no "sources" or "sinks," which in this case are points where electric charges are created or destroyed. One might expect that the *magnetic induction vector* **B**

---

* Siméon D. Poisson (1781–1840), a French mathematical physicist.
[†] Note that in some texts $V(x, y, z)$ is used to denote this quantity.

would behave in a similar manner to **E**. There are no free magnetic poles, however, hence we always have

$$\mathbf{V} \cdot \mathbf{B} = 0. \tag{5.7-8}$$

This is the second of Maxwell's equations. The term "solenoidal" is applied to **B** since a magnetic field that has zero divergence everywhere may be generated by a solenoid.

We have seen that the electric field vector **E** is conservative, hence

$$\oint_C \mathbf{E} \cdot d\mathbf{r} = 0$$

around any simple closed path C. If the path is linked by a *changing* magnetic field, however, then **Faraday's* law of induction** states that a current is generated in the closed circuit. If Ψ is the magnetic flux linking the closed circuit, then

$$\oint_C \mathbf{E} \cdot d\mathbf{r} = -\frac{\partial \Psi}{\partial t},$$

where

$$\Psi = \iint_S \mathbf{B} \cdot \mathbf{n} \, dS.$$

Hence for any closed path whatever,

$$\oint_C \mathbf{E} \cdot d\mathbf{r} = -\frac{\partial}{\partial t} \iint_S \mathbf{B} \cdot \mathbf{n} \, dS$$

$$= -\iint_S \frac{\partial \mathbf{B}}{\partial t} \cdot \mathbf{n} \, dS$$

since spatial and time coordinates are independent so that differentiation and integration can be interchanged. Applying Stokes's theorem, we find

$$\iint_S (\mathbf{V} \times \mathbf{E}) \cdot \mathbf{n} \, dS + \iint_S \frac{\partial \mathbf{B}}{\partial t} \cdot \mathbf{n} \, dS = 0$$

or

$$\iint_S \left(\mathbf{V} \times \mathbf{E} + \frac{\partial \mathbf{B}}{\partial t} \right) \cdot \mathbf{n} \, dS = 0.$$

Since the last equality holds for an arbitrary surface, the integrand must be zero, that is,

$$\mathbf{V} \times \mathbf{E} = -\frac{\partial \mathbf{B}}{\partial t}. \tag{5.7-9}$$

---

* Michael Faraday (1791–1867), a British physicist. Actually, the law was also discovered independently by Joseph Henry (1797–1878), an American physicist.

This is the third of Maxwell's equations and is called the **law of electromagnetic induction**.

We summarize the three equations discussed above and add to these a fourth. Following are Maxwell's equations:

$$\mathbf{V} \cdot \mathbf{E} = 0, \tag{5.7-5a}$$

$$\mathbf{V} \cdot \mathbf{B} = 0, \tag{5.7-8}$$

$$\mathbf{V} \times \mathbf{E} = -\frac{\partial \mathbf{B}}{\partial t}, \tag{5.7-9}$$

$$\mathbf{V} \times \mathbf{B} = \varepsilon_0 \mu_0 \frac{\partial \mathbf{E}}{\partial t}. \tag{5.7-10}$$

These equations hold in a vacuum containing no electric charges. They are valid for nonisotropic and nonhomogeneous media. The constant μ_0 in Eq. (5.7–10) is the permeability of free space. Maxwell's equations are not independent since Eq. (5.7–8) can be deduced from Eq. (5.7–9), and Eq. (5.7–5a) can be deduced from Eq. (5.7–10).*

If plane electromagnetic waves are propagated in free space, the vector $\mathbf{E} \times \mathbf{B}$ gives the direction of propagation. It is customary to write

$$\mathbf{S} = \mathbf{E} \times \mathbf{H}, \tag{5.7-11}$$

where $\mathbf{B} = \mu\mathbf{H}$, $\mathbf{B}$ being magnetic induction and $\mathbf{H}$ magnetic field intensity, the magnetic analog to $\mathbf{E}$. The scalar μ is the permeability of the medium. The vector $\mathbf{S}$ is the **Poynting**[†] **vector**. The integral of this vector over a closed surface gives the total outward flow of energy per unit of time.

Coriolis[‡] acceleration

Consider a circular disk rotating with constant angular velocity ω about the z-axis that passes through the center of the disk. Then a particle at the point $P(x, y, 0)$ on the disk has position vector (see Fig. 5.7–4)

$$\mathbf{r}(t) = x(t)\mathbf{i} + y(t)\mathbf{j}.$$

The vectors $\mathbf{i}$ and $\mathbf{j}$ are orthogonal unit vectors that rotate with the disk. Hence the velocity vector $\mathbf{v}$ is given by

$$\mathbf{v}(t) = \frac{dx}{dt}\mathbf{i} + \frac{dy}{dt}\mathbf{j} + x\frac{d\mathbf{i}}{dt} + y\frac{d\mathbf{j}}{dt}.$$

---

* For details see Paul Lorrain and Dale R. Corson, *Electromagnetic Fields and Waves*, 2d ed., Chapter 10 (San Francisco: W. H. Freeman, 1970).

 [†] John H. Poynting (1852–1914), an English physicist.

 [‡] Gaspard G. de Coriolis (1792–1843), a French physicist, published a paper on this topic in 1835.

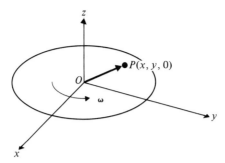

FIGURE 5.7–4 Coriolis acceleration.

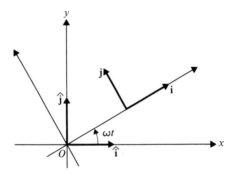

FIGURE 5.7–5 Rotation of axes.

The rotating vectors **i** and **j** can be expressed in terms of fixed vectors*
î and **ĵ** as follows:

$$\mathbf{i} = \cos \omega t \hat{\mathbf{i}} + \sin \omega t \hat{\mathbf{j}},$$
$$\mathbf{j} = -\sin \omega t \hat{\mathbf{i}} + \cos \omega t \hat{\mathbf{j}}. \qquad (5.7\text{–}12)$$

Note that Eqs. (5.7–12) describe a change of coordinate systems due to
rotation of axes (see Fig. 5.7–5).

From relations (5.7–12) we find

$$\frac{d\mathbf{i}}{dt} = \omega \mathbf{j} \quad \text{and} \quad \frac{d\mathbf{j}}{dt} = -\omega \mathbf{i}$$

so that the velocity becomes

$$\mathbf{v}(t) = (\dot{x} - \omega y)\mathbf{i} + (\dot{y} + \omega x)\mathbf{j}.$$

---

* Such fixed (or *effectively* fixed) vectors constitute a reference frame, which is
called an *inertial frame*, and is defined as one in which Newton's laws are valid.

Differentiating again gives the acceleration (Exercise 4):

$$\mathbf{a} = (\ddot{x}\mathbf{i} + \ddot{y}\mathbf{j}) - \omega^2(x\mathbf{i} + y\mathbf{j}) + 2\omega(-\dot{y}\mathbf{i} + \dot{x}\mathbf{j}). \qquad (5.7\text{--}13)$$

There are three terms in the acceleration shown in Eq. (5.7–13). The first term, $\ddot{x}\mathbf{i} + \ddot{y}\mathbf{j}$, is the one we would expect if there were no rotation and, if ω were zero, this term *would* be the only one present. The second, $-\omega^2(x\mathbf{i} + y\mathbf{j}) = -\omega^2\mathbf{r}$, is the **centripetal acceleration** due to the rotation; it is directed toward the center and is thus the negative of the centrifugal acceleration. The third term, $2\omega(-\dot{y}\mathbf{i} + \dot{x}\mathbf{j})$, is the **Coriolis acceleration**. Since the velocity of the particle relative to the rotating reference system is $\dot{x}\mathbf{i} + \dot{y}\mathbf{j}$, it can be seen that the Coriolis acceleration is at right angles to this velocity. Moreover, it can be shown (Exercise 5) that the Coriolis acceleration is in the direction of the rotation.

We point out that the **Coriolis force** is not really a force in the usual sense of the word. It is an artificial or pseudo force by which we attempt to extend Newton's equations to a noninertial system. Nevertheless, the Coriolis force accounts for the deviations of projectile motion and falling bodies and also explains the operation of the Foucault pendulum.* Moreover, winds would normally blow radially inward toward a low-pressure center. Because of the deflection due to Coriolis forces, however, winds form a counterclockwise pattern (as seen from above) in the northern hemisphere and a clockwise pattern in the southern hemisphere. These rotations can be seen very clearly in satellite photographs of hurricanes.

Heat conduction

We consider a material that is *thermally isotropic*, that is, one whose density, specific heat, and thermal conductivity are independent of direction. We assume also the following facts regarding thermal conductivity, all of which are based on experimental evidence.

1. Heat flows from a region of higher temperature to a region of lower temperature.
2. Heat flows through an area at a rate proportional to the temperature gradient normal to the area.
3. The change in heat of a body as its temperature changes is proportional to its mass and to the change in temperature.

Now consider a region V bounded by a closed surface S. If we denote the temperature by $u(x, y, z, t)$, the rate at which heat flows

---

* Devised in 1851 by Jean L. Foucault (1819–1868) to demonstrate the rotation of the earth mechanically. The pendulum led to his invention of the gyroscope in 1852.

outward from the region through a surface element dS is given by

$$-\frac{dQ_1}{dt} = -k\frac{du}{dn}\,dS = -k(\nabla u \cdot \mathbf{n})\,dS,$$

where $\mathbf{n}$ is the unit *outward* normal to dS and k is a proportionality constant called the *thermal conductivity* of the material. The derivative du/dn is the **normal derivative,*** that is, the rate of change of u in the direction of $\mathbf{n}$. Hence the net rate of heat flow *into* V is given by

$$k\oiint_S \nabla u \cdot \mathbf{n}\,dS,$$

the surface integral being taken over the entire surface of S. This inward flow must be balanced[†] by the amount of heat lost by the body, which is given by

$$\iiint_V c\rho\,\frac{\partial u}{\partial t}\,dV,$$

where c is the specific heat and ρ is the density of the material.

Equating the two integrals, we have

$$k\oiint_S \nabla u \cdot \mathbf{n}\,dS = \iiint_V c\rho\,\frac{\partial u}{\partial t}\,dV$$

and, applying the divergence theorem gives us

$$k\iiint_V \nabla^2 u\,dV = c\rho\iiint_V \frac{\partial u}{\partial t}\,dV.$$

This last can be written

$$\iiint_V \left(k\nabla^2 u - c\rho\,\frac{\partial u}{\partial t}\right)dV = 0$$

and, since the above holds for an *arbitrary* region V, the integrand must be zero and we have

$$k\nabla^2 u = c\rho\,\frac{\partial u}{\partial t}$$

or

$$h^2\nabla^2 u = \frac{\partial u}{\partial t}, \tag{5.7-14}$$

where $h^2 = k/c\rho$ is the *thermal diffusivity* of the material. Equation (5.7–14) is known as the **heat equation** or the **diffusion equation**. It plays an important role in applied mathematics and will be studied further in Section 8.3.

---

* Also called the directional derivative.
† This assumes that there are no heat sources or sinks in V.

KEY WORDS AND PHRASES

central field
potential
centrifugal force
gravitational force
Newton's universal law
   of gravitation
Coulomb's law
Poisson's equation

Laplace's equation
Faraday's law of induction
Poynting vector
centripetal acceleration
Coriolis acceleration
normal derivative
diffusion equation

EXERCISES 5.7

▶ 1. a) Graph the central field

$$\mathbf{F}(x, y, z) = -\mathbf{r}/r.$$

   b) Show that the field in part (a) is conservative.

   c) Show that

$$\mathbf{F} = f(r)\mathbf{e}_r,$$

   where $\mathbf{e}_r$ is a unit vector in the radial direction and f is a differentiable scalar function is conservative. (*Hint*: See Eq. 5.4–20.)

2. Prove Gauss's law (Example 5.7–3) for the case where the surface S does not enclose the origin.

3. Starting with Eq. (5.7–5) derive Gauss's law (Example 5.7–3).

4. Obtain the expression for acceleration in Eq. (5.7–13).

5. Show that the Coriolis acceleration in the example given in the text is in the direction of rotation.

6. In obtaining Eq. (5.7–14) show that if there is a source of heat continuously distributed throughout V given by $f(x, y, z, t)$, then the diffusion equation becomes

$$h^2 \nabla^2 u + f/c\rho = \frac{\partial u}{\partial t}.$$

7. Show that the equation obtained in Exercise 6 becomes Poisson's equation if u is independent of t and Laplace's equation if, additionally, there are no heat sources present.

8. In the diffusion equation (5.7–14) u is given in °C and t in seconds. Determine the units of ρ, k, and h^2 if c is given in calories/(g°C).

9. In obtaining Eq. (5.7–14) it was assumed that k was constant. If k is a function of position, show that the diffusion equation becomes

$$\nabla \cdot k\nabla u = c\rho \frac{\partial u}{\partial t}.$$

▶▶ 10. a) Graph the central field

$$\mathbf{F} = -\omega^2 \mathbf{r},$$

   where ω is a constant.

b) Is this field conservative?

c) Show that the potential of this field is $\frac{1}{2}\omega^2 r^2$ by assuming that $\phi(0) = 0$. (*Note*: This is the potential of a **simple harmonic oscillator**.)

11. a) Show that the earth's gravitational field can be written

$$\mathbf{F}_G = \frac{k\mathbf{r}}{r^3},$$

where k is a constant. (Compare Eq. 5.7–2.)

b) Show that $\mathbf{V} \cdot \mathbf{F}_G = 0$.

12. Show that a planet (or satellite) moving in a gravitational field necessarily moves on a plane curve. (*Hint*: The torsion of a space curve is given by

$$\tau = \frac{(\dot{\mathbf{r}} \times \ddot{\mathbf{r}}) \cdot \dddot{\mathbf{r}}}{|\dot{\mathbf{r}} \times \ddot{\mathbf{r}}|^2},$$

which is zero for a plane curve. (Compare Exercises 8 and 17 of Section 5.2.)

13. If $\mathbf{u}$ and $\mathbf{v}$ are each irrotational, show that $\mathbf{u} \times \mathbf{v}$ is solenoidal.

14. Show that

$$\mathbf{V} \cdot \mathbf{V}f(r) = \frac{2}{r}\frac{df}{dr} + \frac{d^2f}{dr^2}$$

for an arbitrary differentiable scalar function $f(r)$, where $r = \sqrt{x^2 + y^2 + z^2}$.

15. Show that any scalar function that satisfies Laplace's equation is necessarily both solenoidal and irrotational.

16. a) Show that a steady (independent of t) electric field $\mathbf{E}$ satisfies Laplace's equation. (Compare Exercise 18.)

b) Show that the equation obtained in part (a) can be written as three separate equations.

▶▶▶ **17.** Show that in rectangular coordinates,

$$\mathbf{V} \times (\mathbf{V} \times \mathbf{A}) = \mathbf{V}(\mathbf{V} \cdot \mathbf{A}) - \mathbf{V}^2\mathbf{A}$$

for an arbitrary vector $\mathbf{A}$.

18. Eliminate $\mathbf{B}$ from Eqs. (5.7–9) and (5.7–10) to obtain the *electromagnetic vector wave equation*,

$$\mathbf{V} \cdot \mathbf{V}\mathbf{E} = \varepsilon_0\mu_0 \frac{\partial^2\mathbf{E}}{\partial t^2}.$$

(*Hint*: Apply the curl operator to Eq. (5.7–9) and $\partial/\partial t$ to Eq. (5.7–10), interchange spatial and time derivatives, and use Eq. (5.7–5a) and Exercise 17.)

19. Prove that if $\mathbf{F} = \mathbf{V} \times \mathbf{V}$, then $\mathbf{F}$ is solenoidal. (*Note*: The vector $\mathbf{V}$ is called the *vector potential* of the field $\mathbf{F}$.)

20. Show that for arbitrary vectors $\mathbf{A}$ and $\mathbf{B}$,

$$\mathbf{V} \cdot (\mathbf{A} \times \mathbf{B}) = \mathbf{B} \cdot (\mathbf{V} \times \mathbf{A}) - \mathbf{A} \cdot (\mathbf{V} \times \mathbf{B}).$$

21. Explain the direction in which water rotates as it drains out of an outlet in the northern hemisphere. If $\mathbf{v}$ is the velocity of the stream, what can you conclude about **curl v**?

22. If we assume that the earth is revolving around a fixed axis in space (a reasonable assumption on the basis of the small effects of other motions), what is the direction of the Coriolis force for a particle in the northern hemisphere moving (a) due south? (b) due east?

23. Give two nontrivial examples of nonconservative vector fields in $\mathbb{R}^3$.

REFERENCES

Following are some suggestions for further reading. As indicated, topics of this chapter may be covered in greater detail and/or to a greater depth than was possible in this text.

Crowe, M. J., *A History of Vector Analysis*. Notre Dame, Ind.: University of Notre Dame Press, 1967.
This source book provides historical details of the early users of vector analysis and the development of the notation.

Davis, H. F., *Introduction to Vector Analysis*. Boston: Allyn and Bacon, 1961.
An elementary textbook requiring a minimum of background and one that is very readable. The last chapter contains an introduction to tensor analysis.

Fraleigh, J. B., *Calculus: A Linear Approach*, vol. 2. Reading, Mass.: Addison-Wesley, 1972.
An elementary treatment of the vector analysis of plane curves, line, surface, and volume integrals, and Stokes's theorem. Uses the language of differential forms.

Landau, L. D., and E. M. Lifshitz, *Mechanics*, 3d ed. Elmsford, N.Y.: Pergamon Press, 1979. *A scholarly, mathematical approach.*

Lorrain, P., and D. R. Corson, *Electromagnetic Fields and Waves*, 2d ed. San Francisco: W. H. Freeman, 1970.
This book should be useful to engineers and scientists who wish to review the subject and to students who wish to study more or less independently. It contains an introductory chapter on vectors that makes the book self-contained. Two chapters on relativity are included in line with the authors' objective that the reader obtain a working knowledge of the basic concepts of electromagnetism.

Marion, J. B., *Classical Dynamics of Particles and Systems*, 2d ed. New York: Academic Press, 1970.
This is an unusually complete and readable book. An introductory chapter on vector analysis and matrices is followed by such topics as Newtonian mechanics, linear and nonlinear oscillations, Lagrangian and Hamiltonian dynamics, central force motion, and the special theory of relativity. Historical footnotes abound throughout the book.

Miller, F., Jr., *College Physics*. New York: Harcourt Brace Jovanovich, 1972.
Chapter 7 contains an elementary explanation of inertial and noninertial frames of reference. Chapter 8 contains a discussion of centripetal and centrifugal forces with several illustrative examples.

Pollard, H., *Celestial Mechanics*; A Carus Mathematical Monograph. Math. Assoc. of America, 1976.
A complete mathematical treatment of the central force problem is given in Chapter 1.

Schey, H. M., *Div, Grad, Curl, and All That*. New York: W. W. Norton, 1973.
This is an informal text on vector calculus presented in the context of simple electrostatics.

Segel, L. A., *Mathematics Applied to Continuum Mechanics*. New York: Mac-millan, 1977.
  This text contains more advanced applications to viscous fluids, static and dynamic problems in elasticity, and water waves.

Sposito, G., *An Introduction to Classical Dynamics*. New York: John Wiley, 1976.
  The applications of classical dynamics to geophysics, atmospheric physics, space physics, physical oceanography, and environmental physics are represented significantly in the text and in the problems.

Symon, K. R., *Mechanics*, 3d ed. Reading Mass.: Addison-Wesley, 1971.
  A complete text at an intermediate level.

Tipler, P. A., *Physics*. Worth, 1976.
  Beginning on p. 163 is an article, "Global Winds" by Richard Goody, that deals with the role of Coriolis forces in meteorology.

Williamson, R. E., R. H. Crowell, and H. F. Trotter, *Calculus of Vector Functions*, 3d ed. Englewood Cliffs, N.J.: Prentice-Hall, 1972.
  Provides detailed coverage of vector calculus with applications to vector field theory.

6

partial differential equations

6.1 FIRST-ORDER EQUATIONS

Ordinary differential equations play a very important role in applied mathematics. A wide variety of problems from engineering and physics can be formulated in terms of ordinary differential equations. We have presented some of these in Sections 1.5 and 2.5 as examples of the kinds of problems that arise. But, just as it is not always possible to simplify a problem by neglecting friction, air resistance, Coriolis force, etc., we cannot always neglect the presence of other independent variables. Often, time must necessarily be considered in addition to one or more spatial variables.

Whenever more than a *single* independent variable must be taken into account, it may be possible to formulate a problem in terms of a **partial differential equation**. This chapter, as well as Chapters 8 and 9, will be concerned with obtaining solutions to such equations. Much of the terminology introduced in Chapters 1 and 2 in connection with ordinary differential equations is extended in a natural way to partial differential equations. For example, the *order* of an equation is the same as the highest ordered derivative that appears.

The major emphasis in this text is on partial differential equations of *second* order. In this section, however, we will discuss *first-order* equations because they have application in aerodynamics, hydro-dynamics, thermodynamics, calculus of variations, and probability. We have, in fact, solved some first-order partial differential equations in Section 5.5 in the process of obtaining a potential f satisfying $\nabla f = \mathbf{v}$, where $\mathbf{v}$ was a vector field.

The most general first-order partial differential equation in two independent variables has the form

$$F(x, y, z, z_x, z_y) = 0, \tag{6.1–1}$$

where $z(x, y)$ is to be found and where*

$$\frac{\partial z}{\partial x} = z_x \quad \text{and} \quad \frac{\partial z}{\partial y} = z_y.$$

---

* The notation $p = z_x$, $q = z_y$ is also in use.

We will assume that the given function F is continuous in each of its five arguments. Equation (6.1–1) is said to be *linear* if F is linear in z, z_x, and z_y. Thus a linear equation can be written in the form

$$a_0(x, y)z(x, y) + a_1(x, y)z_x + a_2(x, y)z_y = b(x, y), \qquad \textbf{(6.1–2)}$$

whereas *quasilinear* equations have the form

$$P(x, y, z)z_x + Q(x, y, z)z_y = R(x, y, z). \qquad \textbf{(6.1–3)}$$

We remark that some authors call Eq. (6.1–3) a linear equation. This may be confusing since an equation like

$$\sin z\, z_x + z^2 z_y = z^3(x + y)$$

would then be considered linear. Recall that, in ordinary differential equations, the dependent variable cannot occur with any exponent except one and products of the dependent variable and its derivatives are not allowed. In a quasilinear equation the first derivatives z_x and z_y must appear linearly but there is no other restriction.

It was Joseph Lagrange (1736–1813), a French physicist, who showed that solutions of equations of the form given in (6.1–3) can be obtained by solving certain ordinary differential equations. For this reason Eq. (6.1–3) is sometimes called *Lagrange's equation*. (*Note*: This is not to be confused with the Newtonian equations which, in a more general form, are called *Lagrange's equations*.) We present the following theorem without proof.[*]

Theorem 6.1–1 The general solution of the first-order quasilinear partial differential equation

$$P(x, y, z)z_x + Q(x, y, z)z_y = R(x, y, z)$$

is given by $F(u, v) = 0$, where F is arbitrary, and $u(x, y, z) = c_1$ and $v(x, y, z) = c_2$ form a general solution of the two *ordinary* differential equations

$$\frac{dx}{P} = \frac{dy}{Q} = \frac{dz}{R}. \qquad \textbf{(6.1–4)}$$

Note that (6.1–4) is equivalent to (why?)

$$\frac{dy}{dx} = \frac{Q}{P} \quad \text{and} \quad \frac{dz}{dx} = \frac{R}{P} \qquad \textbf{(6.1–5)}$$

in which the independent variable is x. The general solutions of Eqs. (6.1–5) can be expressed in the form $y = y(x, c_1, c_2)$ and $z = z(x, c_1, c_2)$ from which $u(x, y, z) = c_1$ and $v(x, y, z) = c_2$ can be obtained. The details are left for the exercises (see Exercise 2).

---

[*] A proof can be found in Ian N. Sneddon, *Elements of Partial Differential Equations*, p. 50 (New York: McGraw-Hill, 1957).

EXAMPLE 6.1–1 Obtain the general solution of

$$zz_x + yz_y = x.$$

Solution. Comparing this problem with the notation used in Theorem 6.1–1, we have $P = z$, $Q = y$, and $R = x$, functions that are everywhere continuously differentiable. Then

$$\frac{dx}{z} = \frac{dy}{y} = \frac{dz}{x},$$

leading to $x^2 - z^2 = c_1$. Some ingenuity may be required in order to obtain an expression involving y. In the present case, we try the combination $dx + dz$. Then

$$dx + dz = \frac{z}{y}\,dy + \frac{x}{y}\,dy = \frac{x + z}{y}\,dy$$

and, separating variables, we have

$$\frac{dx + dz}{x + z} = \frac{dy}{y},$$

leading to $(x + z)/y = c_2$ (Exercise 3). Hence the general solution according to Theorem 6.1–1 is

$$F\left(x^2 - z^2, \frac{x + z}{y}\right) = 0. \quad \blacksquare \qquad\qquad \textbf{(6.1–6)}$$

Observe an important characteristic of the above general solution: It contains an arbitrary *function* instead of an arbitrary *constant*. Note also that an equivalent way of writing this general solution is

$$x^2 - z^2 = f\left(\frac{x + z}{y}\right) \quad \text{or} \quad \frac{x + z}{y} = g(x^2 - z^2),$$

where f and g are arbitrary functions. Because of the presence of arbitrary functions in the general solutions of partial differential equations, the *form* of the solutions can be extremely varied. Consequently, the applications will be concentrated on finding *particular* solutions that satisfy prescribed initial and boundary conditions.

The equations (6.1–4) can be put into the form

$$A(x, y, z)\,dx + B(x, y, z)\,dy + C(x, y, z)\,dz = 0.$$

This is called a **Pfaffian* differential equation** in three variables. The equation in two variables has been discussed in Section 1.2. Necessary and sufficient conditions for the existence of a solution of a Pfaffian differential equation are given in Sneddon's *Elements of Partial Differential Equations* (op. cit., p. 18ff).

---

* After Johann F. Pfaff (1765–1825), a German mathematician, who published a result on the theory of differential forms in 1815.

We next consider the problem of determining the particular form of the function $F(u, v)$ of Theorem 6.1–1. If c_1 and c_2 are known, then $u(x, y, z) = c_1$ and $v(x, y, z) = c_2$ represent surfaces and their *curve of intersection* satisfies both of these equations. This curve is called a **characteristic curve** and the totality of such curves is an **integral surface**. Hence, if we are given a curve,

$$C: \quad x = x(t), \quad y = y(t), \quad z = z(t),$$

expressed in terms of a parameter t, then $u(x, y, z) = c_1$ becomes $u(x(t), y(t), z(t)) = c_1$ and $v(x, y, z) = c_2$ becomes $v(x(t), y(t), z(t)) = c_2$. By eliminating t between these two equations we can find the equation of the integral surface. It should be emphasized that the curve C is *not* a characteristic curve but it does lie in the integral surface.

EXAMPLE 6.1–2 Determine the integral surface of

$$yz_x + xz_y = 0$$

that passes through the curve

$$C: \quad x = 0, \quad y = t, \quad z = t^4.$$

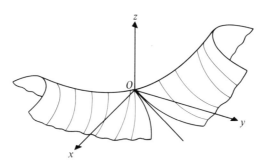

FIGURE 6.1–1 Integral surface (Example 6.1–2).

Solution. Using Theorem 6.1–1 we have

$$\frac{dx}{y} = \frac{dy}{x} = \frac{dz}{0};$$

hence* $x^2 - y^2 = c_1$ and $z = c_2$ so that the general solution becomes $F(x^2 - y^2, z) = 0$ or $z = f(x^2 - y^2)$. Then, from the curve C, $t^4 = f(-t^2)$ from which $t^2 = f(t)$ and $z = (x^2 - y^2)^2$ (see Fig. 6.1–1). ∎

---

* $dz/0$ is to be interpreted as $dz = 0$, hence $z = $ a constant. From $dx/y = dy/x$ we have $x\,dx - y\,dy = 0$ and it follows that $x^2 - y^2 = $ a constant.

We remark that the above solution is *unique*. If, however, the curve C is defined as the hyperbola $x^2 - y^2 = 1$, $z = 1$, then C is a characteristic curve and there are infinitely many solutions (Exercise 4). On the other hand, there are *no* solutions if C is defined as

$$C: \quad x = t, \quad y = \sqrt{t^2 - 1}, \quad z = t$$

(Exercise 5). In this case C is not a characteristic curve but its projection in the xy-plane coincides with the projection of a characteristic curve.

A method of solving linear, first-order, partial differential equations parallels the method used for ordinary differential equations. There the general solution was the sum of the solution of the homogeneous equation and a particular solution. The following example illustrates this method as it is applied to a partial differential equation.

EXAMPLE 6.1–3 Solve the equation

$$z_x + xz = x^3 + 3xy.$$

Solution. We begin with the homogeneous equation

$$z_x + xz = 0 \quad \text{or} \quad \frac{\partial z}{\partial x} = -xz.$$

Since $\partial z/\partial x$ implies that y is being held constant, this last equation can be treated as an ordinary differential equation except in one respect. The solution is

$$z = f(y) \exp(-x^2/2),$$

where $f(y)$ is an arbitrary function of y. Finding a particular solution may be more involved. We can use the superposition principle and make some appropriate estimates. For example, in order to obtain the term x^3, assume

$$z_1 = Ax^3 + Bx^2 + Cx + D$$

and this leads to $z_1 = x^2 - 2$. To obtain the term $3xy$ assume

$$z_2 = Ex^2y + Fxy + Gy$$

and this leads to $z_2 = 3y$. Thus the general solution is

$$z = f(y) \exp(-x^2/2) + x^2 - 2 + 3y. \quad \blacksquare$$

In many other problems involving first-order linear equations, solutions may be obtained almost by inspection. This is particularly true if either z_x or z_y is missing from the equation (see Example 6.1–5). The following examples show methods of solution in various cases.

EXAMPLE 6.1–4   Solve the equation

$$xz_x - yz_y = 0.$$

Solution. Using Eqs. (6.1–5), we have

$$\frac{dy}{dx} = -\frac{y}{x} \quad \text{and} \quad \frac{dz}{dx} = 0,$$

which have solutions (why?) $xy = c_1$ and $z = c_2$, respectively. Hence the general solution of the given equation is $z = f(xy)$, where f is an arbitrary differentiable function.   ■

EXAMPLE 6.1–5   Solve the equation

$$z_x = x^2 + y^2.$$

Solution. Since z is not present, we have immediately by integration

$$z = \frac{x^3}{3} + xy^2 + f(y). \quad ■$$

EXAMPLE 6.1–6   Solve the equation

$$xz_x + yz_y = \log x, \qquad x > 0, \qquad y > 0.$$

Solution. Although this equation can be solved by using Theorem 6.1–1 (Exercise 7), we will use another method. Make a change of independent variable, putting $u = \log x$ and $v = \log y$. Then

$$\frac{\partial z}{\partial x} = \frac{\partial z}{\partial u} \frac{du}{dx} = \frac{\partial z}{\partial u} \frac{1}{x}$$

and, similarly,

$$\frac{\partial z}{\partial y} = \frac{\partial z}{\partial v} \frac{1}{y}$$

so that the given equation becomes

$$\frac{\partial z}{\partial u} + \frac{\partial z}{\partial v} = u.$$

This can be solved to obtain (Exercise 8)

$$u = v + c_1 \quad \text{and} \quad \frac{u^2}{2} = z + c_2,$$

so that

$$z = \frac{u^2}{2} + f(u - v)$$

or

$$z = \frac{1}{2}(\log x)^2 + f\left(\frac{x}{y}\right). \quad ■$$

A useful application of first-order quasilinear partial differential equations is found in determining a family of surfaces orthogonal to a given family. Let $f(x, y, z) = c$ be a family of surfaces.* The direction numbers of the normal to this family at a point (x, y, z) are

$$(f_x, f_y, f_z).$$

Denote these direction numbers by (P, Q, R) to simplify the notation. Now suppose $z = g(x, y)$ is a surface that intersects each surface of the given family orthogonally. The direction numbers of the normal to $z = g(x, y)$ at an arbitrary point are $(z_x, z_y, -1)$ and for orthogonality we must have[†]

$$Pz_x + Qz_y = R.$$

Thus, by Theorem 6.1–1, the surfaces orthogonal to $f(x, y, z) = c$ are those generated by the characteristic curves of the equations,

$$\frac{dx}{P} = \frac{dy}{Q} = \frac{dz}{R}. \qquad \qquad \textbf{(6.1–7)}$$

An application of orthogonal surfaces can be found in potential theory. Let $f(x, y, z) = c$ represent a family of *equipotential surfaces*, that is, for each value of c the given surface has a certain potential. Then the characteristic curves of the equations

$$\frac{dx}{f_x} = \frac{dy}{f_y} = \frac{dz}{f_z}$$

represent the *lines of force*.

EXAMPLE 6.1–7   Find the surfaces orthogonal to the family

$$f(x, y, z) = \frac{z(x + y)}{z + 1} = c.$$

Solution. We have

$$f_x = \frac{z}{z + 1}, \qquad f_y = \frac{z}{z + 1}, \qquad f_z = \frac{x + y}{(z + 1)^2};$$

hence Eqs. (6.1–7) become (Exercise 9)

$$\frac{dx}{z(z + 1)} = \frac{dy}{z(z + 1)} = \frac{dz}{x + y}.$$

---

  * Such a family is also called a *one-parameter* family because of the presence of the parameter c in the equation.
  [†] Note that this equation expresses the fact that the dot product of $P\mathbf{i} + Q\mathbf{j} + R\mathbf{k}$ and $z_x\mathbf{i} + z_y\mathbf{j} - \mathbf{k}$ is zero.

From the first two equations we have $x = y + c_1$ or $c_1 = x - y$. From the last two we have $(2y + c_1)\,dy = (z^2 + z)\,dz$, hence it follows that $y^2 + c_1 y = z^3/3 + z^2/2 + c'_2$. This last can be written as

$$c_2 = 6xy - 2z^3 - 3z^2.$$

Thus the orthogonal surfaces are

$$F(x - y, 6xy - 2z^3 - 3z^2)$$

or

$$6xy - 2z^3 - 3z^2 = g(x - y). \quad \blacksquare$$

KEY WORDS AND PHRASES

quasilinear equation   characteristic curve
Pfaffian differential equation   integral surface

EXERCISES 6.1

▶ 1. Show that Eqs. (6.1–3) can be written as

$$\frac{dx}{dy} = \frac{P}{Q} \quad \text{and} \quad \frac{dz}{dy} = \frac{R}{Q}$$

   in which the independent variable is y (compare Eqs. 6.1–5).

2. Show that the solutions $y = y(x, c_1, c_2)$ and $z = z(x, c_1, c_2)$ of Eqs. (6.1–5) can be put into the form $u(x, y, z) = c_1$ and $v(x, y, z) = c_2$.

3. Solve the equations

$$\frac{dx}{z} = \frac{dy}{y} = \frac{dz}{x}$$

   of Example 6.1–1.

4. Show that there are infinitely many solutions of $yz_x + xz_y = 0$ that pass through the curve

$$C: \quad x = t, \quad y = \sqrt{t^2 - 1}, \quad z = 1.$$

   (Compare Example 6.1–2.)

5. Show that there are no solutions of $yz_x + xz_y = 0$ that pass through the curve

$$C: \quad x = t, \quad y = \sqrt{t^2 - 1}, \quad z = t.$$

   (Compare Example 6.1–2.)

6. Carry out the details in Example 6.1–3.

7. Solve the equation

$$xz_x + yz_y = \log x$$

   by using Theorem 6.1–1. (Compare Example 6.1–6.)

8. Solve the equation

$$\frac{\partial z}{\partial u} + \frac{\partial z}{\partial v} = u$$

by using Theorem 6.1–1. (Compare Example 6.1–6.)

9. Carry out the details in Example 6.1–7.

▶▶ 10. Obtain the general solution of each of the following equations.

    a) $(x + z)z_x + (y + z)z_y = 0$ (*Hint*: Equations 6.1–4 become $dx/(x + z) = dy/(y + z) = dz/0$, hence $dz = 0$ and $z = c_1$.)

    b) $xzz_x + yzz_y = -(x^2 + y^2)$

    c) $z_x + xz_y = z$

    d) $yz_x + z_y = z$

    e) $(y + x)z_x + (y - x)z_y = z$

    f) $x^2 z_x + y^2 z_y = (x + y)z$ (*Hint*: Observe that $dx/x + dy/y - dz/z = 0$.)

    g) $(xz + y)z_x - (x + yz)z_y = x^2 - y^2$ (*Hint*: Compute $y\,dx + x\,dy - dz$.)

    h) $xz_x + yz_y = 0$

    i) $xz_x + yz_y = z$

11. Solve each of the following equations.

    a) $z_y + 2yz = 0$ **b)** $z_x - 2xyz = 0$

    c) $z_y = \sin(y/x)$ **d)** $z_y = x^2 + y^2$

    e) $z_x - 2z_y = x^2$

12. Obtain the solution of the equation $z_x = z_y$ that passes through the curve

$$C: \quad x = t^2 + 1, \quad y = 2t, \quad z = (t + 1)^4.$$

13. **a)** Find the general solution of the equation

$$z_x + zz_y = 1.$$

    b) Obtain the solution that passes through the curve

$$C: \quad x = t, \quad y = t, \quad z = 2.$$

14. Solve each of the following equations, obtaining the general solution.

    a) $z(yz_y - xz_x) = y^2 - x^2$

    b) $(x^2 - y^2 - z^2)z_x + 2xyz_y = 2xz$

    c) $(y + 1)z_x + (x + 1)z_y = z$

    d) $yzz_x + xzz_y = x + y$

15. In each case find the integral surface that contains the given curve.

    a) $(y + xz)z_x + (x + yz)z_y = z^2 - 1$; $C: x = t, y = 2, z = t^2$

    b) $(y - z)z_x + (z - x)z_y = x - y$; $C: x = t, y = 2t, z = 0$

    c) $yz_x - xz_y = 2xyz$; $C: x = y = z = t$

    d) $x^2 z_x + y^2 z_y = z^2$; $C: x = t, y = 2t, z = 1$

16. Below are the general solutions of certain partial differential equations. By differentiating each of these and eliminating the arbitrary functions, obtain a differential equation of lowest order.

a) $z = e^{-x}F(x + 2y)$

b) $z = yf(x)$

c) $x + z = yf(x^2 - z^2)$

d) $z(x - y) = xy \log\left(\dfrac{x}{y}\right) + f\left(\dfrac{x - y}{xy}\right)$

▶▶▶ **17.** Show that the general solution of

$$2xz_x - yz_y = 0$$

is $z = f(xy^2)$.

18. If $f(x, y)$ satisfies the equation $z_x = z_y$, prove that f is a function of $x + y$.

19. Use the method of Example 6.1–6 to solve each of the following equations.

a) $2xz_x - 3yz_y = 0$

b) $xz_x - 2yz_y = x^2y$

c) $3xz_x - yz_y + 4z = x^2 \cos x$

20. Show that the surfaces

$$z(x + y) = c(z + 1) \quad \text{and} \quad 6xy - 2z^3 - 3z^2 = g(x - y),$$

where c is an arbitrary constant and g is an arbitrary differentiable function, are orthogonal wherever they intersect (compare Example 6.1–7).

21. a) Solve the equation

$$\frac{\partial z}{\partial x} = \frac{\partial z}{\partial y}.$$

b) Read the solution given in *Amer. Math. Monthly* **85**, no. 10 (Dec. 1978), p. 829 to Problem no. 5871 proposed and solved by P. R. Chernoff.

22. Solve the homogeneous equation

$$az_x + bz_y = 0,$$

where a and b are constants.

23. a) Obtain the surfaces orthogonal to the family $x^2 + y^2 = C$.

b) Sketch some of the orthogonal surfaces.

6.2 HIGHER-ORDER EQUATIONS

In studying partial differential equations of order higher than one it will be convenient to use the subscript notation. For example, we will write

$$\frac{\partial^2 u}{\partial x^2} = u_{xx}, \qquad \frac{\partial^2 u}{\partial y^2} = u_{yy}, \qquad \frac{\partial^2 u}{\partial x \, \partial y} = u_{yx} = \frac{\partial}{\partial x}\left(\frac{\partial u}{\partial y}\right), \quad \text{etc.}$$

Since we will be concerned to a great extent with *second-order* partial differential equations, we consider the most general equation of this type,

$$Au_{xx} + Bu_{xy} + Cu_{yy} + Du_x + Eu_y + Fu = G. \tag{6.2–1}$$

If the functions $A, B, \ldots, G$ depend on x and y (but *not* on u or its derivatives), the equation is *linear*. If $G \equiv 0$, then the equation is *homogeneous*; otherwise it is *nonhomogeneous*.

By a *solution* of Eq. (6.2–1) we mean a function $u(x, y)$ that satisfies the equation identically. A **general solution** is one that contains two arbitrary, independent *functions* (in contrast with two arbitrary *constants* in the case of a second-order ordinary differential equation). A **particular solution** is one that can be obtained from the general solution by a specific choice of the arbitrary functions.

Linear equations of the form (6.2–1) are categorized as to type in an interesting way. Recall from analytic geometry that the most general *second-degree* equation in two variables is

$$Ax^2 + Bxy + Cy^2 + Dx + Ey + F = 0,$$

where $A, B, \ldots, F$ are constants. Equations of this type represent conic sections (possibly degenerate ones) as follows:

$$\text{an ellipse if } B^2 - 4AC < 0;$$
$$\text{a parabola if } B^2 - 4AC = 0;$$
$$\text{a hyperbola if } B^2 - 4AC > 0.$$

Similarly, *linear* partial differential equations (6.2–1) are called **elliptic**, **parabolic**, or **hyperbolic** according as $B^2 - 4AC$ is negative, zero, or positive, respectively. Since A, B, and C are functions of x and y in Eq. (6.2–1), it is quite possible for a partial differential equation to be of **mixed** type. For example,

$$xu_{xx} + yu_{yy} + 2yu_x - xu_y = 0$$

is elliptic if $xy > 0$, hyperbolic if $xy < 0$, and parabolic if $xy = 0$. If x and y are *spatial* coordinates, then the equation is elliptic in the first and third quadrants, hyperbolic in the second and fourth quadrants, and parabolic on the coordinate axes. The theory of equations of **mixed type** was originated by Tricomi* in 1923. In the study of transonic flow the so-called Tricomi equation,

$$y\Psi_{xx} + \Psi_{yy} = 0,$$

occurs. This equation is elliptic for $y > 0$ and hyperbolic for $y < 0$.

We shall see later in this section that the classification of a second-order partial differential equation into various types is important since unique, stable solutions are obtained for each type under different kinds of boundary conditions.

Although x and y are generally used to designate spatial coordinates, this need not always be the case. One of these could be a time

---

* Francesco G. Tricomi (1897–   ), an Italian mathematician.

coordinate, for example. The only restriction, in fact, is that x and y be independent variables. If we have spatial independent variables, then, by specifying the values of the dependent variable on some boundary—called **boundary conditions**—we can evaluate the arbitrary functions in the general solution and obtain a particular solution. If, on the other hand, one of the independent variables is time, then we must also specify **initial conditions** in order to obtain a particular solution. Problems in which boundary conditions or both boundary and initial conditions are specified are called **boundary-value problems**. These will be treated in detail in Chapters 8 and 9.

Next we give examples to illustrate the three types of second-order equations. Particular solutions to these will be obtained in ensuing sections.

EXAMPLE 6.2–1   Elliptic boundary-value problem.

$$\text{P.D.E.: } u_{xx} + u_{yy} = 0, \qquad a < x < b, \qquad c < y < d;$$
$$\text{B.C.: } u(a, y) = u_0, \qquad u(b, y) = u_1, \qquad c < y < d,$$
$$u(x, c) = u_2, \qquad u(x, d) = u_3, \qquad a < x < b;$$

where a, b, c, d, u_0, u_1, u_2, u_3 are constants. This elliptic equation is called **Laplace's equation*** and we shall study it in greater detail in Section 6.3 and again in Chapters 8 and 9. The boundary conditions (B.C.) show that $a \le x \le b$ and $c \le y \le d$, hence the values of the unknown function $u(x, y)$ are known on a rectangle in the xy-plane (Fig. 6.2–1). The partial differential equation (P.D.E.) shown indicates that $u(x, y)$ is to satisfy the equation everywhere within the *open* rectangular region. ∎

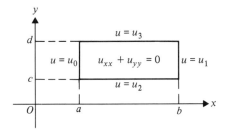

FIGURE 6.2–1 Laplace's equation (Example 6.2–1).

_____

* After Pierre S. de Laplace (1749–1827), a French astronomer and mathematician.

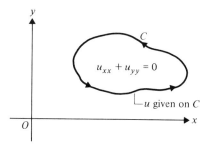

FIGURE 6.2–2 Dirichlet problem in two dimensions.

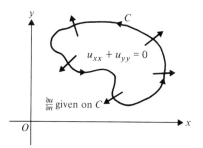

FIGURE 6.2–3 Neumann problem in two dimensions.

Boundary-value problems of this type, that is, those in which u satisfies Laplace's equation in an open region and takes on prescribed values on the boundary of the region (Fig. 6.2–2), are called **Dirichlet problems**.* If the above boundary conditions are replaced by

$$u_x(a, y) = u_0, \qquad u_x(b, y) = u_1, \qquad c < y < d,$$
$$u_y(x, c) = u_2, \qquad u_y(x, d) = u_3, \qquad a < x < b,$$

then the derivatives u_x and u_y are prescribed and the problem is called a **Neumann problem**.† In this case we say the **normal derivative**, $\partial u / \partial n$, that is, the rate of change of u in a direction normal to the boundary, is specified (Fig. 6.2–3). Boundary-value problems may, of course, be of *mixed* type as we shall see in Section 8.3. Other notations

---

* After Peter G. L. Dirichlet (1805–1859), a German mathematician.
† After Carl G. Neumann (1832–1925), a German mathematician.

for $u_x(a, y) = u_0$ are in use. Some of these are the following:

$$\frac{\partial u(x, y)}{\partial x}\bigg|_{x=a} = u_0,$$

$$\lim_{x \to a^+} u_x(x, y) = u_0,$$

$$u_x(a^+, y) = u_0.$$

EXAMPLE 6.2–2 Hyperbolic boundary-value problem.

P.D.E.: $u_{tt} = a^2 u_{xx}$, $t > 0$, $0 < x < L$;
B.C.: $u(0, t) = u(L, t) = 0$, $t > 0$;
I.C.: $u(x, 0) = u_0$, $u_t(x, 0) = u_0'$, $0 < x < L$.

Here a is a constant and u is a displacement and we have both boundary
conditions and initial conditions (I.C.) prescribed. This time u is to
satisfy the given equation for positive values of t and for all x in the
open interval $0 < x < L$ (Fig. 6.2–4). The boundary conditions state
that for positive t, $u = 0$ when $x = 0$ and when $x = L$, whereas the
initial conditions give the values of the initial displacement and initial
velocity as u_0 and u_0', respectively. The differential equation must be
dimensionally correct, hence the constant a has the dimension of
velocity (Exercise 1). ■

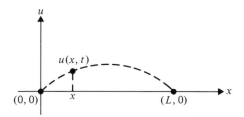

FIGURE 6.2–4 One-dimensional wave
equation (Example 6.2–2).

The partial differential equation in the last example is the **wave
equation** in one dimension. We will derive the equation and obtain
its general solution in Section 6.4. In Section 8.4 we will consider the
semi-infinite $(0 < x < \infty)$ and infinite $(-\infty < x < \infty)$ cases using the
Fourier transform.

EXAMPLE 6.2–3 Parabolic boundary-value problem.

P.D.E.: $u_t = k u_{xx}$, $t > 0$, $0 < x < L$, $k > 0$;
B.C.: $u(0, t) = a$, $u(L, t) = b$, $t > 0$;
I.C.: $u(x, 0) = f(x)$, $0 < x < L$.

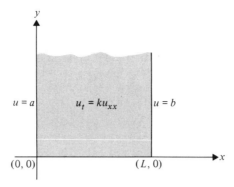

FIGURE 6.2-5 One-dimensional diffusion equation (Example 6.2-3).

Again we have $u(x, t)$ satisfying a partial differential equation on an open interval for all positive t. The boundary conditions prescribe the value of u at the end points of the given interval (Fig. 6.2–5), whereas the initial condition specifies values of u at time $t = 0$. In this example the partial differential equation is the one-dimensional* **diffusion equation**. We will solve it in Section 8.3 over finite regions and in Section 8.4 for infinite and semi-infinite regions. ■

The existence of arbitrary *functions* in the general solution of a partial differential equation means that the totality of functions that satisfy such an equation is very large. For example, the functions

$$\arctan \frac{y}{x}, \quad e^x \sin y, \quad \log \sqrt{x^2 + y^2}, \quad \sin x \sinh y$$

are quite varied, yet each satisfies Laplace's equation (Exercise 2). In applications involving partial differential equations, however, we have certain information about a physical system that allows us to find particular solutions. Most of our study of boundary-value problems will be concentrated on this point.

As in the case of ordinary differential equations and first-order partial differential equations, the most simple second-order partial differential equations to solve are the homogeneous ones with *constant* coefficients. Consider such an equation in which the first derivatives are absent, namely,

$$au_{xx} + bu_{xy} + cu_{yy} = 0 \qquad (6.2\text{–}2)$$

and let[†] $u = f(y + rx)$, where r is a constant. Then $u_x = rf'(y + rx)$ and $u_{xx} = r^2 f''(y + rx)$, where the prime denotes the derivative of the

---

* We will see later that $0 < y < \infty$ implies that u varies only in the x-direction.
[†] This is actually a "guess" prompted by the form of Eq. (6.2–2).

function f with respect to its argument, $y + rx$. Substitution into Eq. (6.2–2) produces

$$(ar^2 + br + c)f''(y + rx) = 0$$

and the characteristic equation

$$ar^2 + br + c = 0. \qquad\qquad \textbf{(6.2–3)}$$

If the roots of Eq. (6.2–3) are real and distinct, say r_1 and r_2, then the *general solution* of Eq. (6.2–2) can be written

$$u(x, y) = f(y + r_1 x) + g(y + r_2 x),$$

where f and g are twice differentiable, but otherwise arbitrary functions. (See Exercises 3 and 4.)

EXAMPLE 6.2–4 Solve the hyperbolic equation

$$a^2 u_{xx} - b^2 u_{yy} = 0,$$

where a and b are real constants.

Solution. The characteristic equation is

$$a^2 r^2 - b^2 = 0$$

and the general solution is

$$u(x, y) = f\left(y + \frac{b}{a}x\right) + g\left(y - \frac{b}{a}x\right). \quad \blacksquare$$

Modifications of the above method can be found in the exercises. It should be mentioned that if the characteristic equation has two equal roots, then the general solution has the form

$$u(x, y) = f(y + rx) + xg(y + rx). \qquad\qquad \textbf{(6.2–4)}$$

Exercises 18–20 deal with the case where the characteristic equation has complex roots.

KEY WORDS AND PHRASES

general solution
particular solution
elliptic, parabolic, and hyperbolic
    partial differential equations
boundary and initial conditions
boundary-value problem

Laplace's equation
Dirichlet and Neumann problems
normal derivative
wave equation
diffusion equation

EXERCISES 6.2

▶ 1. Show that the constant a in the wave equation has the dimension of velocity.

2. Verify that the functions

   a) $\arctan \dfrac{y}{x}$,   **b)** $e^x \sin y$,

   c) $\log \sqrt{x^2 + y^2}$,   **d)** $\sin x \sinh y$,

   all satisfy Laplace's equation.

3. Verify that

$$u(x, y) = f(y + r_1 x) + g(y + r_2 x),$$

   where r_1 and r_2 satisfy

$$ar^2 + br + c = 0,$$

   is the general solution of Eq. (6.2–2).

4. Referring to Exercise 3, verify that

$$u(x, y) = c_1 f(y + r_1 x) + c_2 g(y + r_2 x)$$

   also satisfies Eq. (6.2–2) for arbitrary constants c_1 and c_2. How does this solution compare with the general solution?

▶▶ 5. Classify each of the following partial differential equations as elliptic, hyperbolic, or parabolic. Consider the appropriate values of the independent variables in each case.

   a) $u_{xx} + 4u_{xy} + 3u_{yy} + 4u_x - 3u = xy$

   b) $xu_{xx} + u_{yy} - 2x^2 u_y = 0$

   c) $u_{xy} - u_x = x \sin y$

   d) $(y^2 - 1)u_{xx} - 2xy u_{xy} + (x^2 - 1)u_{yy} + e^x u_x + u_y = 0$

6. If f and g are twice-differentiable but otherwise *arbitrary* functions, verify that $f(x + at)$, $g(x - at)$, and $f(x + at) + g(x - at)$ are solutions of $u_{tt} = a^2 u_{xx}$. (*Hint*: By the chain rule,

$$g_t(x - at) = \frac{dg(x - at)}{d(x - at)} \frac{\partial(x - at)}{\partial t} = -ag'(x - at).$$

7. Verify that

$$u = (c_1 \cos \lambda x + c_2 \sin \lambda x)(c_3 \sin \lambda a t + c_4 \cos \lambda a t)$$

   is a solution of the wave equation $u_{tt} = a^2 u_{xx}$, where c_1, c_2, c_3, c_4, and λ are constants.

8. Show that the function $u(x, t)$ in Exercise 7 becomes $c_5 \cos \lambda a t \sin \lambda x$ when the boundary and initial conditions

$$u(0, t) = 0 \quad \text{and} \quad u_t(x, 0) = 0$$

   are imposed.

9. Verify that $u = \exp(-k\lambda^2 t)(c_1 \cos \lambda x + c_2 \sin \lambda x)$ is a solution of the diffusion equation $u_t = k u_{xx}$, where c_1, c_2, and λ are constants.

10. Verify that

$$u(x, t) = \frac{1}{2a} \int_{x-at}^{x+at} g(s)\, ds$$

is a solution of the wave equation $u_{tt} = a^2 u_{xx}$ satisfying the conditions $u(x, 0) = 0$, $u_t(x, 0) = g(x)$. (*Hint*: Use the Leibniz rule for differentiating an integral; compare Exercise 23 in Section 1.3.)

11. For each of the following partial differential equations (i) give the order and (ii) state whether or not the equation is linear; if it is not linear, explain why.

   a) $x u_x + y u_y = u$ b) $u(u_{xx}) + (u_y)^2 = 0$

   c) $u_{xx} - u_{xy} - 2 u_{yy} = 1$ d) $u_{xx} - 2 u_y = 2x - e^u$

   e) $(u_x)^2 - x(u_{xy}) = \sin y$

12. a) Show that if $a = 0$ in Eq. (6.2–2), the method given in the text fails. Show, however, that in this case the substitution $u = f(x + ry)$ will produce the general solution.

   b) Obtain the general solution to $u_{xy} - 3 u_{yy} = 0$.

   c) Use the substitution $u = f(x + ry)$ to solve $u_{xx} + u_{xy} - 6 u_{yy} = 0$.

13. Verify that the general solution of Eq. (6.2–2) is given by Eq. (6.2–4) when the characteristic equation has equal roots.

14. It can be shown* that if $u = F(x, y)$, then the mixed second partial derivatives u_{xy} and u_{yx} are equal whenever they exist and are continuous. Show that the mixed second partial derivatives are equal for each of the following functions.

   a) $u = e^x \cos y$ b) $u = \arctan \dfrac{y}{x}$

   c) $u = e^{xy} \tan xy$ d) $u = \sqrt{\dfrac{x+y}{x-y}}$

15. Obtain a solution of

$$u_{xx} + u_{xy} - 6 u_{yy} = 0.$$

16. Obtain the general solution of each of the following equations.

   a) $u_{xx} + u_{xy} - 6 u_{yy} = 0$ b) $u_{xx} - 9 u_{yy} = 0$

   c) $u_{xx} + 4 u_{yy} = 0$ d) $6 u_{xx} + u_{xy} - 2 u_{yy} = 0$

17. Classify each of the following equations as to whether it is elliptic, hyperbolic, or parabolic.

   a) $x^2 u_{xx} + 2xy u_{xy} + y^2 u_{yy} = 4x^2$

   b) $u_{xx} - (2 \sin x) u_{xy} - (\cos^2 x) u_{yy} - (\cos x) u_y = 0$

   c) $u_{xx} + u_{yy} + u_{xy} + au = 0$ (a a constant)

   d) $x^2 u_{xx} - y^2 u_{yy} = xy$

   e) $4 u_{xx} - 8 u_{xy} + 4 u_{yy} = 3$

   * See Robert C. James, *Advanced Calculus*, p. 298 (Belmont, Calif.: Wadsworth, 1966).

▶▶▶ 18. Verify that

$$u = f_1(x + iy) + f_2(x - iy)$$

is a solution of $u_{xx} + u_{yy} = 0$.

19. Generalize the result of Exercise 18 to the case where the characteristic equation (6.2–3) has complex roots.

20. Show that

$$u = f_1(y - ix) + xf_2(y - ix) + f_3(y + ix) + xf_4(y + ix)$$

is a solution of

$$u_{xxxx} + 2u_{yyxx} + u_{yyyy} = 0.$$

21. The method given in the text can be extended to homogeneous partial differential equations of order four. Obtain the general solutions of each of the following equations.

a) $\dfrac{\partial^4 u}{\partial x^4} + 2\dfrac{\partial^4 u}{\partial x^2 \, \partial y^2} + \dfrac{\partial^4 u}{\partial y^4} = 0$*

b) $\dfrac{\partial^4 u}{\partial x^4} - \dfrac{\partial^4 u}{\partial y^4} = 0$

c) $\dfrac{\partial^4 u}{\partial x^4} - 2\dfrac{\partial^4 u}{\partial x^2 \, \partial y^2} + \dfrac{\partial^4 u}{\partial y^4} = 0$

22. Show that if ψ_1 and ψ_2 are any two harmonic functions of x and y, then any function ϕ of the form

$$\phi(x, y) = x\psi_1(x, y) + \psi_2(x, y)$$

satisfies the biharmonic equation. (*Note*: A harmonic function is one that satisfies Laplace's equation; the biharmonic equation is given in Exercise 21(a).)

6.3 SEPARATION OF VARIABLES

Laplace's equation,

$$u_{xx} + u_{yy} + u_{zz} = 0, \tag{6.3–1}$$

is one of the classical partial differential equations of mathematical physics. We present it above in the three-dimensional case and we will examine the two-dimensional case later in this section. The importance of this equation is due to its occurrence in so many branches of science.

In the study of *electrostatics* it is shown that the electric intensity vector **E** due to a collection of stationary charges is given by

$$\mathbf{E} = -\nabla\phi = -(\phi_x\mathbf{i} + \phi_y\mathbf{j} + \phi_z\mathbf{k}),$$

where ϕ is a scalar point function called the *electric potential*. In the above, $\nabla\phi$ is the *gradient* of ϕ and **i**, **j**, and **k** are unit vectors along the

---

* This equation, called the *biharmonic equation*, occurs in the study of elasticity and hydrodynamics.

x-, y-, and z-axis, respectively. Further, Gauss's law states that

$$\mathbf{V} \cdot \mathbf{E} = \mathbf{V} \cdot (-\mathbf{V}\phi) = -(\phi_{xx} + \phi_{yy} + \phi_{zz}) = 4\pi\rho(x, y, z),$$

where $\rho(x, y, z)$ is the charge density. The constant 4π appears because it represents the surface area of a unit sphere and Gauss's law is generally stated in the form of a surface integral. (See Section 5.7.) Thus the potential ϕ satisfies the equation

$$\phi_{xx} + \phi_{yy} + \phi_{zz} = -4\pi\rho(x, y, z) \qquad \textbf{(6.3–2)}$$

which is known as *Poisson's equation*. In a region that is free of charges, $\rho(x, y, z) = 0$ and Eq. (6.3–2) reduces to Laplace's equation,

$$\phi_{xx} + \phi_{yy} + \phi_{zz} = 0. \qquad \textbf{(6.3–3)}$$

In this case it is assumed that the electric potential is due to charges located outside of or on the boundary of the charge-free region. It is significant that *every* potential function that is derived from any electrostatic distribution whatever must satisfy Eq. (6.3–3) in free space.

In an analogous manner, in *magnetostatics* the *magnetic potential* due to the presence of poles satisfies Eq. (6.3–3) in regions free of poles. Similarly, the *gravitational potential* due to the presence of matter satisfies Eq. (6.3–3) in regions devoid of matter. In *aerodynamics* and *hydrodynamics* the *velocity potential* ϕ has the property $\mathbf{V}\phi = \mathbf{v}$, where $\mathbf{v}$ is the velocity vector field. For an idealized fluid—one that is incompressible and irrotational—the velocity potential satisfies Eq. (6.3–3) in those portions of the fluid that contain no sources or sinks. Hence Laplace's equation plays an important role in *potential theory*. For this reason it is often called the **potential equation** and functions satisfying it are called **potential functions** as well as **harmonic functions**.

Laplace's equation is also prominent in other branches of science. If a membrane of constant density is stretched uniformly over a supporting frame, if no external forces are applied to the membrane except at the frame, and if the frame is given a displacement normal to the plane of the membrane (the xy-plane), then the *static transverse displacement* of the membrane $z(x, y)$ satisfies the two-dimensional Laplace equation. In this case it is also assumed that z and its derivatives are so small that higher powers of z, z_x, and z_y can be neglected. We shall see in Section 8.3 that the steady-state (independent of time) *temperature* in a substance having constant thermal conductivity and containing no heat sources or sinks also satisfies Laplace's equation.

It should be apparent from the foregoing that the occurrence of Laplace's equation in such diverse fields merits the attention we will pay to its solution. We begin with a simple boundary-value problem involving Laplace's equation in two variables.

EXAMPLE 6.3–1 Solve the following boundary-value problem.

$$P.D.E.: \ u_{xx} + u_{yy} = 0, \qquad 0 < x < \pi, \qquad 0 < y < b;$$
$$B.C.: \ u(0, y) = g_1(y), \qquad u(\pi, y) = g_2(y), \qquad 0 < y < b,$$
$$u(x, 0) = f(x), \qquad u(x, b) = f_1(x), \qquad 0 < x < \pi.$$

DISCUSSION. We need to find a function $u(x, y)$ that satisfies the P.D.E. in the open rectangular region $0 < x < \pi$, $0 < y < b$, and that takes on the prescribed values $f(x)$, $f_1(x)$, $g_1(y)$, $g_2(y)$ on the boundary of the region (the rectangle) as shown in Fig. 6.3–1. In other words, the functions f and g are assumed to be known so that u is assigned a unique value at every point of the rectangle. There must consequently be continuity at the four corners. This means that we must have

$$\lim_{x \to \pi^-} u(x, 0) = \lim_{x \to \pi^-} f(x) = \lim_{y \to 0^+} u(\pi, y) = \lim_{y \to 0^+} g_2(y)$$

with similar limits holding at the other three corners.

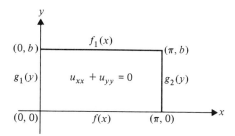

FIGURE 6.3–1 Boundary-value problem (Example 6.3–1).

The problem as stated can be considerably simplified because we can apply the *principle of superposition* since Laplace's equation is a linear, homogeneous equation. Thus if u_1, u_2, u_3, and u_4 all satisfy Laplace's equation and the following boundary conditions shown* in Fig. 6.3–2,

i) $u_1(0, y) = g_1(y), \quad u_1(\pi, y) = u_1(x, 0) = u_1(x, b) = 0,$

ii) $u_2(\pi, y) = g_2(y), \quad u_2(0, y) = u_2(x, 0) = u_2(x, b) = 0,$

iii) $u_3(x, 0) = f(x), \quad u_3(0, y) = u_3(\pi, y) = u_3(x, b) = 0,$

iv) $u_4(x, b) = f_1(x), \quad u_4(0, y) = u_4(\pi, y) = u_4(x, 0) = 0,$

---

* The symbol ∇^2 is also used for the Laplacian, that is, $\nabla^2 u = u_{xx} + u_{yy}$ in two dimensions.

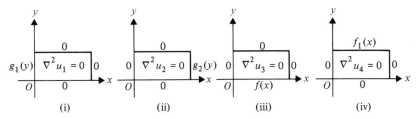

FIGURE 6.3–2 Applying the principle of superposition (Example 6.3–1).

then $u = u_1 + u_2 + u_3 + u_4$ will be the solution of the problem in Example 6.3–1. This, in turn, suggests that we need to be concerned only with solving *one* of these problems since the other three are quite similar. Accordingly, we rephrase the problem of Example 6.3–1. ■

EXAMPLE 6.3–2 Solve the following boundary-value problem.

$$\text{P.D.E.: } u_{xx} + u_{yy} = 0, \qquad 0 < x < \pi, \qquad 0 < y < b;$$
$$\text{B.C.: } u(0, y) = u(\pi, y) = 0, \qquad 0 < y < b,$$
$$u(x, b) = 0, \qquad u(x, 0) = f(x), \qquad 0 < x < \pi.$$

Solution (partial). Now we have three *homogeneous* boundary conditions and u is prescribed on an open interval by $f(x)$, as shown in Fig. 6.3–2(iii). We will discuss the nature of $f(x)$ and the values $f(0)$ and $f(\pi)$ in Chapter 7. The significance of the value $x = \pi$ will also become apparent there.

We may not know a priori which of the infinite variety of functions $u(x, y)$ satisfying Laplace's equation will also satisfy the given boundary conditions. We do know, however, that x and y are *independent* variables; hence a method of solution known as **separation of variables** seems to be a logical choice. Moreover, we shall see later that this method is particularly useful when most of the boundary conditions are homogeneous. Finally, the method leads to *ordinary, homogeneous* differential equations with constant coefficients with which we are familiar (compare Section 2.1 from which we will draw freely in the remainder of this chapter). Accordingly, separation of variables promises to be effective.

Assume that $u(x, y)$ can be expressed as a *product* of two functions, one a function of x alone, the other a function of y alone. Then

$$u(x, y) = X(x)Y(y) \tag{6.3–4}$$

and

$$u_{xx} = X''Y, \qquad u_{yy} = XY'',$$

where the primes denote *ordinary* derivatives, the differentiation being performed with respect to the arguments of X and Y. Then, substituting into the P.D.E., we have

$$X''Y + XY'' = 0.$$

Next, we recognize that, although $u(x, y) = 0$ satisfies the P.D.E., we wish to rule out this **trivial solution**. This, in turn, means that neither $X(x)$ nor $Y(y)$ can be identically zero so that we can divide the last equation by the product XY. Thus

$$\frac{X''}{X} = -\frac{Y''}{Y} \qquad\qquad \textbf{(6.3–5)}$$

and the variables have been *separated*, since the left member of Eq. (6.3–5) is a function of x alone and the right member is a function of y alone.

Varying x in Eq. (6.3–5) will change the left member but not the right; hence the equality can hold, in general, only if both members are constant, that is,

$$\frac{X''}{X} = -\frac{Y''}{Y} = k. \qquad\qquad \textbf{(6.3–6)}$$

In order to determine the nature of the constant k, we examine the following two-point boundary-value problem:

$$X'' - kX = 0, \qquad X(0) = X(\pi) = 0, \qquad\qquad \textbf{(6.3–7)}$$

where the conditions $u(0, y) = 0$ and $u(\pi, y) = 0$ have been translated into $X(0) = 0$ and $X(\pi) = 0$, respectively, by using Eq. (6.3–4). (Note that this can be done only with *homogeneous* conditions.) We now distinguish the three possible cases: $k = 0$, $k > 0$, and $k < 0$.

Case I $k = 0$

The general solution of $X'' = 0$ is $X(x) = c_1 x + c_2$ and the conditions $X(0) = 0$ and $X(\pi) = 0$ lead to $c_2 = 0$ and $c_1 = 0$, respectively. Hence this case leads to the trivial solution so we discard the possibility that $k = 0$.

Case II $k > 0$

We have the general solution

$$X(x) = c_3 e^{\sqrt{k}x} + c_4 e^{-\sqrt{k}x},$$

and the condition $X(0) = 0$ produces $c_3 + c_4 = 0$ so the solution becomes

$$X(x) = c_3(e^{\sqrt{k}x} - e^{-\sqrt{k}x}).$$

(We will consistently follow the procedure of *updating* a solution whenever new information about its character is obtained.) Next, the condition $X(\pi) = 0$ yields

$$c_3(e^{\sqrt{k}\pi} - e^{-\sqrt{k}\pi}) = 0.$$

But the quantity in parentheses cannot be zero unless $k = 0$ (Exercise 1), which is ruled out in this case. Hence $c_3 = 0$ and again we have the trivial solution so that we must discard the possibility that $k > 0$.

Case III $k < 0$, say, $k = -\lambda^2$

Then the general solution is

$$X(x) = c_5 \cos \lambda x + c_6 \sin \lambda x,$$

while the condition $X(0) = 0$ leads to $c_5 = 0$ and the updated solution

$$X(x) = c_6 \sin \lambda x.$$

Applying the condition $X(\pi) = 0$ leads either to $c_6 = 0$, which we reject since it produces the trivial solution, or to $\sin \lambda \pi = 0$. Thus λ must be a nonzero integer and we take

$$\lambda = n, \qquad n = 1, 2, 3, \ldots, \tag{6.3-8}$$

resulting in the updated solutions

$$X_n(x) = c_n \sin nx, \qquad n = 1, 2, 3, \ldots. \tag{6.3-9}$$

We have subscripted the function $X(x)$ and the arbitrary constant in Eq. (6.3–9) to emphasize the fact that the two-point boundary-value problem (6.3–7) has an *infinite* number of solutions depending on n. Values of n^2, that is, the values 1, 4, 9, ... are called **eigenvalues** of (6.3–7) and the corresponding functions (6.3–9) are called **eigenfunctions**. Without loss of generality we can take $c_n \equiv 1$ since eigenfunctions (like the eigenvectors of Section 4.4) are known only to within a constant factor. In other words, the eigenfunctions in Eq. (6.3–9) may be written as

$$X_n(x) = \sin nx, \qquad n = 1, 2, 3, \ldots.$$

We can now obtain, for each n, the function $Y_n(y)$ corresponding to X_n. From Eq. (6.3–6) we see that Y_n must be a solution to the problem

$$Y_n'' - n^2 Y_n = 0, \qquad Y_n(b) = 0, \qquad n = 1, 2, 3, \ldots. \tag{6.3-10}$$

We have translated the condition $u(x, b) = 0$ to $Y_n(b) = 0$ but the condition $u(x, 0) = f(x)$ cannot be changed to a condition on $Y_n(y)$ since $f(x)$ is not zero. The solutions of the differential equations in (6.3–10) are

$$Y_n(y) = d_n e^{ny} + f_n e^{-ny}$$

and the condition $Y_n(b) = 0$ implies that

$$d_n e^{nb} + f_n e^{-nb} = 0.$$

Thus

$$d_n = -f_n e^{-2nb}$$

and the updated solutions become

$$Y_n(y) = f_n(-e^{-2nb}e^{ny} + e^{-ny})$$
$$= f_n e^{-nb}(-e^{-nb}e^{ny} + e^{nb}e^{-ny})$$
$$= f_n e^{-nb}(e^{n(b-y)} - e^{-n(b-y)})$$
$$= 2f_n e^{-nb} \sinh n(b-y).$$

But $2f_n e^{-nb}$ are arbitrary constants, call them g_n; hence the solutions of (6.3–10) are

$$Y_n(y) = g_n \sinh n(b-y). \qquad \textbf{(6.3–11)}$$

Going back to Eq. (6.3–4), we have

$$u_n(x, y) = B_n \sin nx \sinh n(b-y), \qquad n = 1, 2, 3, \ldots \quad \textbf{(6.3–12)}$$

for arbitrary constants B_n ($=c_n g_n$). Each of these functions satisfies the given P.D.E. and also the three homogeneous boundary conditions.

It remains to satisfy the nonhomogeneous boundary condition, $u(x, 0) = f(x)$. It is clear from the expression

$$u_n(x, 0) = B_n \sinh nb \sin nx = f(x) \qquad \textbf{(6.3–13)}$$

that it will not be possible to satisfy this final condition with any one of our solutions $u_n(x, y)$ unless $f(x) = C_n \sin nx$ for some constant C_n. We will continue the discussion of Example 6.3–2 in Section 8.1. In Chapter 7 we will see what conditions must be placed on $f(x)$ and later, how the solutions (6.3–12) must be modified in order that they satisfy a nonhomogeneous boundary condition. ∎

It should be emphasized that the method of separation of variables is not guaranteed to solve *every* linear partial differential equation (Exercise 4). On the other hand, the method is well suited to obtaining *particular* rather than *general* solutions and can be applied to first-order equations, as shown in the following example.

EXAMPLE 6.3–3   Obtain the solution of

$$2xu_x - yu_y = 0$$

that passes through the curve (Fig. 6.3–3)

$$C: \quad x = t, \quad y = t, \quad u = t^3.$$

Assume that the solution can be written as

$$u(x, y) = X(x)Y(y).$$

Solution. Differentiating and substituting into the given P.D.E. yields

$$2xX'(x)Y(y) - yX(x)Y'(y) = 0$$

FIGURE 6.3–3 The curve
C of Example 6.3–3.

or

$$\frac{2xX'}{X} = \frac{yY'}{Y} = k.$$

Solving these two first-order, separable, ordinary differential equations produces

$$X = c_1 x^{k/2} \quad \text{and} \quad Y = c_2 y^k;$$

hence

$$u = c_3 (xy^2)^{k/2}.$$

Now, applying the given condition, we have $c_3 = 1$ and $k = 2$, so that the required solution is $u = xy^2$. (Compare Exercise 17 of Section 6.1). ■

We will use the method of separation of variables often in solving boundary-value problems phrased in various systems of coordinates in Chapters 8 and 9.

KEY WORDS AND PHRASES

potential equation
potential function
harmonic function
separation of variables

trivial solution
eigenvalue
eigenfunction

EXERCISES 6.3

▶ **1.** Show that

$$\exp(\sqrt{k}\pi) - \exp(-\sqrt{k}\pi) = 0$$

implies that $k = 0$. (*Hint*: Recall the definition of sinh $\sqrt{k}\pi$.)

2. Show that choosing λ to be a negative integer in Eq. (6.3–8) will not change the eigenfunctions given in Eq. (6.3–9).

3. Verify that the functions of Eq. (6.3–12) satisfy the P.D.E. and the homogeneous boundary conditions of Example 6.3–2.

4. Explain the difficulty encountered in attempting to solve the equation

$$u_{xx} - u_{xy} + 2u_{yy} = 0$$

by the method of separation of variables.

5. Carry out the details in Example 6.3–3.

▶▶ 6. Show that each of the following functions is a potential function.

    a) $u = c/r$, where $r = \sqrt{x^2 + y^2 + z^2}$ and c is a constant

    b) $u = c \log r + k$, r as above, c and k constants

    c) $u = \arctan \dfrac{2xy}{x^2 - y^2}$

7. Obtain the particular solution to the following boundary-value problem. Note that this problem can be solved completely.

$$\text{P.D.E.: } u_{xx} + u_{yy} = 0, \quad 0 < x < \pi, \quad 0 < y < b;$$
$$\text{B.C.: } u(0, y) = u(\pi, y) = 0, \quad 0 < y < b,$$
$$u(x, b) = 0, \quad u(x, 0) = 3 \sin x, \quad 0 < x < \pi.$$

8. In Exercise 7 use $b = 2$ and compute the value of $u(x, y)$ at each of the given points.

    a) $(\pi/2, 0)$ **b)** $(\pi, 1)$

    c) $(\pi/2, 2)$ **d)** $(\pi/2, 1)$

Use the method of separation of variables given in this section to obtain two *ordinary* differential equations in each of Exercises 9–11. Do not attempt to solve the resulting equations.

9. $u_t = ku_{xx}$, where k is a constant

10. $u_{tt} = c^2 u_{xx}$, where c is a constant

11. $u_t = ku_{xx} + au$, where a and k are constants

In Exercises 12–14, follow the procedure shown in Example 6.3–2 to obtain solutions, carrying the work as far as possible.

12. Solve the following boundary-value problem.

$$\text{P.D.E.: } u_{xx} + u_{yy} = 0, \quad 0 < x < \pi, \quad 0 < y < b;$$
$$\text{B.C.: } u(0, y) = g_1(y), \quad u(\pi, y) = 0, \quad 0 < y < b,$$
$$u(x, 0) = u(x, b) = 0, \quad 0 < x < \pi.$$

13. Solve the following boundary-value problems.

$$\text{P.D.E.: } u_{xx} + u_{yy} = 0, \quad 0 < x < \pi, \quad 0 < y < b;$$
$$\text{B.C.: } u(\pi, y) = g_2(y), \quad u(0, y) = 0, \quad 0 < y < b,$$
$$u(x, 0) = u(x, b) = 0, \quad 0 < x < \pi.$$

14. Solve the following boundary-value problem.

$$\text{P.D.E.: } u_{xx} + u_{yy} = 0, \quad 0 < x < \pi, \quad 0 < y < b;$$
$$\text{B.C.: } u(x, b) = f_1(x), \quad u(x, 0) = 0, \quad 0 < x < \pi,$$
$$u(0, y) = u(\pi, y) = 0, \quad 0 < y < b.$$

▶▶▶ **15.** By differentiating the left member of Eq. (6.3–5) partially with respect to y, show that both members must be constant.

16. Repeat Case II of Example 6.3–2 by starting with the general solution

$$X(x) = C_3 \cosh \sqrt{k}x + C_4 \sinh \sqrt{k}x.$$

6.4 THE VIBRATING-STRING EQUATION

Although our main objective is to present various methods of solving problems, it may be instructive to see the genesis of a boundary-value problem. Accordingly, this section is devoted to a discussion of a physical problem, the simplifying assumptions that are necessary in order to derive a simple partial differential equation, and, finally, obtaining a particular solution.

Consider a string (a more accurate word would be "wire" since what we are about to describe can be likened to the vibrations of a guitar "string") of length L fastened at two points. It will be convenient to take the x-axis as the position of the string when no external forces are acting. Let $x = 0$ and $x = L$ represent the points at which the string is fastened. When the string is caused to vibrate, a point on the string will, at time t, assume a position with coordinates (x, y). We will be interested in obtaining the equation satisfied by y, as a function of x and t. In other words, if $y(x, t)$ is the vertical displacement of the string at a distance x from the fixed left end at time t, then what is the partial differential equation satisfied by $y(x, t)$?

We first make some assumptions that will simplify the derivation. These are listed and discussed below.

Vibrating string: Simplifying assumptions

1. The string is homogeneous, that is, the cross section and density are constant throughout its length.

2. Each point of the string moves along a line perpendicular to the x-axis.

3. The maximum deflection is "small" in comparison to the length L. This loosely stated assumption can be more clearly understood by saying that y should be on the order of a few millimeters for a string one meter in length.

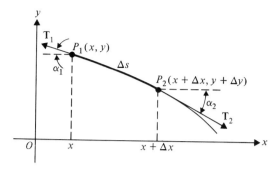

FIGURE 6.4–1 A portion of the vibrating string.

4. The string is perfectly flexible and under uniform (constant) tension throughout its length.

5. External forces, such as air resistance and weight of the string, are ignored.

Next, consider a portion of the string shown greatly magnified in Fig. 6.4–1. The coordinates of two neighboring points P_1 and P_2 are (x, y) and $(x + \Delta x, y + \Delta y)$, respectively. Denote the tension at the two points by T_1 and T_2, respectively. These two tension forces necessarily act along the tangents to the curve at the two points, the tangents making angles α_1 and α_2 with the horizontal as shown. Let the length of the portion of string considered be Δs and denote the density of the string per unit length by δ. The horizontal components of $\mathbf{T}_1$ and $\mathbf{T}_2$ must be equal; otherwise assumption (2) would be violated. Thus*

$$- T_1 \cos \alpha_1 + T_2 \cos \alpha_2 = 0$$

or

$$T_1 \cos \alpha_1 = T_2 \cos \alpha_2 = T_0, \quad \text{a constant.}$$

The *net* vertical component of tension on the element Δs is

$$T_1 \sin \alpha_1 - T_2 \sin \alpha_2$$

or

$$T_0(\tan \alpha_1 - \tan \alpha_2) = T_0 \left(-\frac{\partial y(x, t)}{\partial x} + \frac{\partial y(x + \Delta x, t)}{\partial x} \right).$$

(Recall from calculus that the derivative at a point is defined as the tangent of the slope angle which, in turn, is the angle measured from the positive x-axis to the tangent line in a counterclockwise sense.) By Newton's second law the sum of the forces on the element Δs must

---

* Using the notation given in Chapter 5: $T_1 = |\mathbf{T}_1|$ and $T_2 = |\mathbf{T}_2|$.

be zero for equilibrium. Hence

$$T_0 \left(-\frac{\partial y(x, t)}{\partial x} + \frac{\partial y(x + \Delta x, t)}{\partial x} \right) = \delta \Delta s \frac{\partial^2 y(\bar{x}, t)}{\partial t^2}, \qquad \text{(6.4–1)}$$

where $\bar{x}$ represents the coordinate of the center of mass of the element Δs. Because of assumption (3), $\Delta s \doteq \Delta x$ so that dividing both members of Eq. (6.4–1) by Δx and then taking the limit of the expression as $\Delta x \to 0$, we obtain

$$T_0 \frac{\partial^2 y}{\partial x^2} = \delta \frac{\partial^2 y}{\partial t^2}$$

or

$$\frac{\partial^2 y}{\partial t^2} = a^2 \frac{\partial^2 y}{\partial x^2}, \qquad a^2 = \frac{T_0}{\delta}. \qquad \text{(6.4–2)}$$

Equation (6.4–2) is the vibrating-string equation or the wave equation in one (spatial) dimension (compare Example 6.2–2). Since the equation is hyperbolic, its general solution is obtained as shown in Example 6.2–4 and is given by

$$y(x, t) = \phi(x + at) + \psi(x - at), \qquad \text{(6.4–3)}$$

where ϕ and ψ are twice-differentiable but otherwise arbitrary functions.

If now we impose the initial conditions

$$y(x, 0) = f(x), \qquad y_t(x, 0) = 0, \qquad 0 < x < L, \qquad \text{(6.4–4)}$$

then we will gain some further insight into the functions ϕ and ψ. From Eq. (6.4–3) we have

$$y_t(x, t) = a\phi'(x + at) - a\psi'(x - at),$$

where primes indicate differentiation with respect to the arguments. For example, $\phi'(x + at) = d\phi(x + at)/d(x + at)$ so that

$$\frac{\partial \phi(x + at)}{\partial t} = \frac{d\phi(x + at)}{d(x + at)} \frac{\partial(x + at)}{\partial t} = a\phi'(x + at).$$

Then

$$a\phi'(x) - a\psi'(x) = 0,$$

which shows that $\phi'(x) = \psi'(x)$, that is, ϕ and ψ differ by at most a constant, $\phi(x) = \psi(x) + C$. Hence

$$y(x, 0) = \phi(x) + \psi(x) = 2\psi(x) + C = f(x)$$

or

$$\psi(x) = \frac{1}{2}(f(x) - C)$$

and

$$\phi(x) = \frac{1}{2}(f(x) + C).$$

Thus the solution to Eq. (6.4–2) with the initial conditions shown in Eqs. (6.4–4) is

$$y(x, t) = \frac{1}{2}(f(x + at) + f(x - at)). \qquad \textbf{(6.4–5)}$$

We now observe that Eq. (6.4–5) was obtained without using the fact that the string was fastened at $x = 0$ and $x = L$. In fact, the derivation of Eq. (6.4–2) was also independent of these boundary conditions. This implies that Eq. (6.4–2) holds equally well for a string of infinite length, provided, of course, that the simplifying assumptions apply throughout the length of the string. Thus we can summarize the foregoing in the next example.

EXAMPLE 6.4–1

$$\text{P.D.E.: } y_{tt} = a^2 y_{xx}, \qquad -\infty < x < \infty, \qquad t > 0;$$
$$\text{I.C.: } y(x, 0) = f(x), \qquad -\infty < x < \infty,$$
$$y_t(x, 0) = 0, \qquad -\infty < x < \infty,$$

has solutions

$$y(x, t) = \frac{1}{2}(f(x + at) + f(x - at)). \qquad \blacksquare \qquad \textbf{(6.4–6)}$$

The particular solution (6.4–6) was obtained by specifying an initial displacement and *zero* initial velocity. We next consider the case in which the initial displacement is zero and the initial velocity is prescribed.

EXAMPLE 6.4–2 Solve the following boundary-value problem.

$$\text{P.D.E.: } y_{tt} = a^2 y_{xx}, \qquad -\infty < x < \infty, \qquad t > 0;$$
$$\text{I.C.: } y(x, 0) = 0, \qquad -\infty < x < \infty,$$
$$y_t(x, 0) = g(x), \qquad -\infty < x < \infty.$$

Solution. Our point of departure is again the general solution given in Eq. (6.4–3),

$$y(x, t) = \phi(x + at) + \psi(x - at).$$

This time the condition $y(x, 0) = 0$ implies $\phi(x) = -\psi(x)$, so that

$$y(x, t) = \phi(x + at) - \phi(x - at)$$

and

$$y_t(x, t) = a\phi'(x + at) + a\phi'(x - at).$$

Hence $y_t(x, 0) = g(x)$ produces

$$\phi'(x) = \frac{1}{2a} g(x)$$

and

$$\phi(x) = \frac{1}{2a} \int_0^x g(s) \, ds,$$

using the fundamental theorem of the integral calculus. The general solution is thus transformed into

$$y(x, t) = \frac{1}{2a} \left(\int_0^{x+at} g(s) \, ds - \int_0^{x-at} g(s) \, ds \right)$$

$$= \frac{1}{2a} \int_{x-at}^{x+at} g(s) \, ds. \quad \blacksquare \qquad\qquad (6.4\text{–}7)$$

Using superposition (Exercise 3), we may obtain the solution to the following example.

EXAMPLE 6.4–3

$$\text{P.D.E.:} \quad y_{tt} = a^2 y_{xx}, \qquad -\infty < x < \infty, \qquad t > 0;$$
$$\text{I.C.:} \quad y(x, 0) = f(x), \qquad -\infty < x < \infty,$$
$$\qquad\quad y_t(x, 0) = g(x), \qquad -\infty < x < \infty,$$

has solution

$$y(x, t) = \frac{1}{2} (f(x + at) + f(x - at)) + \frac{1}{2a} \int_{x-at}^{x+at} g(s) \, ds. \quad \blacksquare \quad (6.4\text{–}8)$$

This last is called **D'Alembert's solution.**[*]
In order to gain further insight into the solution given in Eq. (6.4–8), consider an initial displacement $f(x)$ shown in Fig. 6.4–2. If the string has zero initial velocity and is given an initial displacement as shown by the rectangular pulse, then the figure shows the string at time $t = 0$. The position of the pulse at a later time can be calculated. For example, at time $t = 1/2a$ we would graph $y = \frac{1}{2}(f(x + \frac{1}{2}) + f(x - \frac{1}{2}))$ by using the following values to obtain the result graphed in Fig. 6.4–3:

x	0	$\pm\frac{1}{2}$	± 1	$\pm\frac{3}{2}$	± 2
y	4	4	2	2	0

---

[*] After Jean-Le Rond D'Alembert (1717–1783), a French mathematician, whose major contributions were in mechanics.

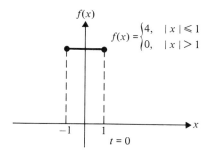

FIGURE 6.4–2 Initial displacement.

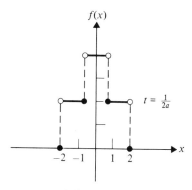

FIGURE 6.4–3

Using $t = 1/a$, $3/2a$, and $2/a$, we obtain graphs that show the propagation of the initial displacement of the string in two directions. This is shown in Fig. 6.4–4.

We conclude this section by showing an analogy to the one-dimensional wave equation in atomic physics. We have

$$y_{tt}(x, t) = c^2 y_{xx}(x, t), \qquad (6.4\text{–}9)$$

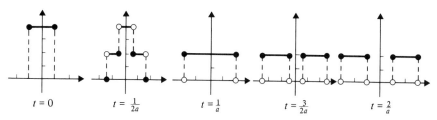

FIGURE 6.4–4 The time history of the vibrating string.

where c is the speed of light (approximately 3.0×10^{10} cm/sec). If we write

$$y(x, t) = \phi(x)e^{i\omega t},$$

which is one way of using the method of separation of variables, Eq. (6.4–9) becomes

$$\phi_{xx} + \frac{\omega^2}{c^2}\phi = 0.$$

The relation between the de Broglie wavelength, λ, and the angular frequency, ω, is given by $\lambda = 2\pi c/\omega$ so that the last equation can be written

$$\phi_{xx} + \left(\frac{2\pi}{\lambda}\right)^2 \phi = 0.$$

In the case of a single helium atom, or other elementary particle, the momentum p of the particle, the wavelength, λ, and Planck's constant, h, are related by $p = h/\lambda$. Thus the wave equation for a particle in a one-dimensional "box" becomes

$$\phi_{xx} + \frac{4\pi^2 p^2}{h^2}\phi = 0.$$

Finally, the momentum of such a free particle is related to the particle's mass, m, and energy, E, by $p^2/2m = E$, so that we have

$$\phi_{xx} + \frac{8\pi^2 mE}{h^2}\phi = 0. \qquad (6.4\text{–}10)$$

Equation (6.4–10) is the one-dimensional form of the **Schrödinger wave equation**.* The function ϕ is the **wave function**, which has no direct physical meaning.

KEY WORDS AND PHRASES

vibrating string   wave function
D'Alembert's solution   Schrödinger wave equation

EXERCISES 6.4

▶ **1. a)** Referring to Fig. 6.4–1, show that

$$\mathbf{T}_1 = -T_1 \cos \alpha_1 \mathbf{i} + T_1 \sin \alpha_1 \mathbf{j},$$

which resolves the tension at P_1 into horizontal and vertical components.

* After Erwin Schrödinger (1887–1961), an Austrian physicist.

b) Resolve $\mathbf{T}_2$ in a similar fashion.

c) By equating the horizontal components of $\mathbf{T}_1$ and $\mathbf{T}_2$, show that

$$T_1 \cos \alpha_1 = T_2 \cos \alpha_2 = T_0, \quad \text{a constant.}$$

d) Show that the net vertical component of tension on the element Δs is

$$T_1 \sin \alpha_1 - T_2 \sin \alpha_2.$$

2. Carry out the details needed to arrive at Eq. (6.4–2). (*Hint:* Recall from calculus the definition of $\partial y/\partial x$,

$$\frac{\partial y}{\partial x} = \lim_{\Delta x \to 0} \frac{y(x + \Delta x, t) - y(x, t)}{\Delta x};$$

then replace the y's by y_x.)

3. Obtain D'Alembert's solution to the wave equation in Example 6.4–3 without using the principle of superposition; that is, apply the initial conditions to the general solution.

▶▶ 4. Solve Eq. (6.4–2) with the initial conditions $y(x, 0) = \sin x$, $y_t(x, 0) = 0$.

5. Solve Eq. (6.4–2) with the initial conditions $y(x, 0) = 0$, $y_t(x, 0) = \cos x$.

6. Solve Eq. (6.4–2) with the initial conditions $y(x, 0) = \sin x$, $y_t(x, 0) = \cos x$.

7. If a string is given an initial displacement defined by

$$f(x) = \begin{cases} a(ax + 1), & -\dfrac{1}{a} \le x \le 0, \\[2mm] a(1 - ax), & 0 \le x \le \dfrac{1}{a}, \\[2mm] 0, & \text{otherwise,} \end{cases}$$

show the graph of the solution $y(x, t)$ of Eq. (6.4–2) for

a) $t = 0$ **b)** $t = \dfrac{1}{2a}$

c) $t = \dfrac{1}{a}$ **d)** $t = \dfrac{3}{2a}$.

8. Verify that the initial displacement $f(x)$ of Exercise 7 travels from $x = 0$ to $x = \infty$ and from $x = 0$ to $x = -\infty$ with speed a.

9. Solve Eq. (6.4–2) with the initial conditions $y(x, 0) = 0$, $y_t(x, 0) = f(x)$ defined in Exercise 7.

▶▶▶ 10. Verify that if $f(x)$ is a twice-differentiable function and $g(x)$ is a differentiable function, then Eq. (6.4–8) satisfies Eq. (6.4–2).

11. Explain why the result obtained in Eq. (6.4–7) is independent of the constant used as the lower limit in the previous step.

12. In the general solution of the wave equation, show that

$$y(x, t) = -y\left(L - x, t + \frac{L}{a}\right)$$

for some positive constant L. Interpret this result physically.

13. A variation of the method of separation of variables mentioned in this section consists of making the substitution

$$y(x, t) = X(x)e^{i\omega t}.$$

Solve the following problem by this method.

P.D.E.: $y_{tt} = a^2 y_{xx}, \quad 0 < x < L, \quad t > 0;$

B.C.: $\left.\begin{array}{l} y(0, t) = 0, \\ y(L, t) = 0, \end{array}\right\} t > 0;$

I.C.: $y(x, 0) = 3 \sin \dfrac{2\pi x}{L}, \quad 0 < x < L,$

$y_t(x, 0) = 0, \quad 0 < x < L.$

Interpret the problem physically and explain why the separation constant appears to be missing.

14. Solve Exercise 13 with the initial conditions changed to

$$y(x, 0) = 0, \quad 0 < x < L,$$

$$y_t(x, 0) = 2 \cos \frac{3\pi x}{L}, \quad 0 < x < L.$$

15. If an external force per unit length of magnitude F is applied to the string, show that the resulting partial differential equation becomes

$$y_{tt}(x, t) = a^2 y_{xx}(x, t) + F/\delta.$$

16. If the external force in Exercise 15 consists of the weight of the string, show that the equation there becomes

$$y_{tt}(x, t) = a^2 y_{xx}(x, t) - g,$$

where g is the acceleration due to gravity.

17. Obtain the static (independent of time) displacements $y(x)$ of points of a string of length L that hangs at rest under its own weight. Show that the string hangs in a parabolic arc and find the maximum displacement.

18. If a string is vibrating in a medium that provides a damping coefficient b ($b > 0$), show that the partial differential equation becomes

$$y_{tt}(x, t) = a^2 y_{xx}(x, t) - \frac{b}{\delta} y_t(x, t).$$

19. In the one-dimensional wave equation, make the substitution $\tau = at$, and obtain

$$y_{\tau\tau} = y_{xx}.$$

7

fourier series and fourier integrals

7.1 FOURIER COEFFICIENTS

In Example 6.3–2 we attempted to solve a boundary-value problem involving Laplace's equation in two dimensions. The function $u(x, y)$ was to satisfy this equation for $0 < x < \pi$, $0 < y < b$; be equal to zero on the three sides of the rectangle $x = 0$, $x = \pi$, and $y = b$; and be equal to a prescribed function $f(x)$ on the fourth side, $y = 0$. By using the method of separation of variables we obtained the functions

$$u_n(x, y) = B_n \sinh n(b - y) \sin nx, \qquad n = 1, 2, 3, \ldots, \quad \textbf{(7.1–1)}$$

which meet all the requirements except that they do not reduce to $f(x)$ when $y = 0$. In this section we will resolve this remaining difficulty.

The situation before us is one that was faced by mathematicians in the eighteenth century. A problem from astronomy at that time involved the expansion of the reciprocal of the distance between two planets in a series of cosines of multiples of the angle between the radius vectors. As early as 1749 and 1754 D'Alembert and Euler published papers in which the expansion of a function in a series of cosines was discussed. In 1811 Fourier* developed the idea to the point where it became generally useful. Fourier was involved in studying the mathematical theory of heat conduction in the course of which he had to solve a problem similar to the one in Example 6.3–2. Further historical details may be found in H. S. Carslaw's *Introduction to the Theory of Fourier's Series and Integrals*, 3d ed. (New York: Dover, 1930) and in I. Grattan-Guinness's *Joseph Fourier* 1768–1830 (Cambridge, Mass.: The MIT Press, 1972).

Since Laplace's equation is a linear homogeneous partial differential equation, we know that any linear combination of solutions is a solution. Moreover, such a linear combination also satisfies the

---

* Jean B. J. Fourier (1768–1830), a French physicist and mathematician, better known today as Joseph Fourier.

homogeneous boundary conditions. In other words, the sum

$$u(x, y) = \sum_{n=1}^{N} B_n \sinh n(b - y) \sin nx$$

accomplishes everything that the functions in Eq. (7.1–1) do for any *finite N*. It is natural, consequently, to ask if the *infinite* series

$$u(x, y) = \sum_{n=1}^{\infty} B_n \sinh n(b - y) \sin nx \qquad \textbf{(7.1–2)}$$

will do the same. Clearly, (7.1–2) *does* satisfy the homogeneous boundary conditions since $u(x, y)$ reduces to zero when $x = 0$, $x = \pi$, and $y = b$. Whether $u(x, y)$ satisfies Laplace's equation, however, is another matter that can be answered only if we know the nature of the constants B_n and if we know under what conditions an infinite series can be differentiated term by term. We consider the constants first and postpone the second question.

From Eq. (7.1–2) we have, on applying the remaining boundary condition of Example 6.3–2,

$$u(x, 0) = f(x) = \sum_{n=1}^{\infty} B_n \sinh nb \sin nx$$

or

$$f(x) = \sum_{n=1}^{\infty} b_n \sin nx, \qquad \textbf{(7.1–3)}$$

where we have replaced the arbitrary constants $B_n \sinh nb$ by b_n for notational simplicity. So far we have said nothing about the function $f(x)$ except that it is defined for all x in the interval $0 < x < \pi$. We now extend the interval of definition and place some restriction on $f(x)$. Let

$$f(x) = \begin{cases} -f(-x), & \text{for } -\pi < x < 0, \\ 0, & \text{for } x = -\pi, x = 0, x = \pi, \end{cases}$$

and suppose that $f(x)$ is continuous* for $0 < x < \pi$, as shown in Fig. 7.1–1. Writing Eq. (7.1–3) in the equivalent form

$$f(x) = b_1 \sin x + b_2 \sin 2x + b_3 \sin 3x + \cdots,$$

we seek a means of finding any one of the coefficients b_i. Suppose we want the value of b_2. We multiply each term of the preceding equation

---

* This assumption will be relaxed later.

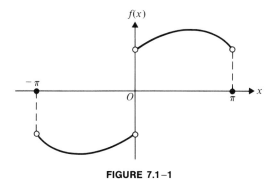

FIGURE 7.1–1

by $\sin 2x$ and integrate from $x = -\pi$ to $x = \pi$. Thus

$$\int_{-\pi}^{\pi} f(x) \sin 2x \, dx = b_1 \int_{-\pi}^{\pi} \sin x \sin 2x \, dx$$

$$+ b_2 \int_{-\pi}^{\pi} \sin^2 2x \, dx$$

$$+ b_3 \int_{-\pi}^{\pi} \sin 3x \sin 2x \, dx + \cdots. \quad \textbf{(7.1–4)}$$

But (Exercise 1)

$$\int_{-\pi}^{\pi} \sin nx \sin mx \, dx = 0, \quad \text{if } n \neq m \quad \textbf{(7.1–5)}$$

and

$$\int_{-\pi}^{\pi} \sin^2 nx \, dx = \pi. \quad \textbf{(7.1–6)}$$

Hence each integral on the right in Eq. (7.1–4) vanishes except the one involving b_2 and we have

$$b_2 = \frac{1}{\pi} \int_{-\pi}^{\pi} f(x) \sin 2x \, dx.$$

An identical procedure can be used to find any b_i so that we can write, in general,

$$b_n = \frac{1}{\pi} \int_{-\pi}^{\pi} f(x) \sin nx \, dx, \quad n = 1, 2, 3, \ldots. \quad \textbf{(7.1–7)}$$

We may simplify Eq. (7.1–7) somewhat further. Both $f(x)$ and $\sin nx$ are *odd functions*, which means that they have the property

$$F(-x) = -F(x),$$

that is, they are symmetrical about the origin. [Other examples of odd functions are x, x^3, x^5, $\tan x$, $\csc x$, and $\sinh x$ (Fig. 7.1–2).] But the product of two odd functions is an *even function*, which has the

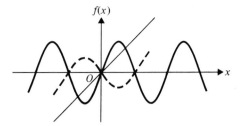

FIGURE 7.1-2 Odd functions.

property

$$F(-x) = F(x)$$

or symmetry about the line $x = 0$. [Examples of even functions are 1, x^2, x^4, $\cos x$, $\sec x$, and $\cosh x$ (Fig. 7.1–3).] Therefore,

$$b_n = \frac{2}{\pi} \int_0^\pi f(x) \sin nx \, dx, \qquad n = 1, 2, 3, \ldots \qquad \textbf{(7.1–8)}$$

and, recalling that

$$B_n = \frac{b_n}{\sinh nb},$$

we can write Eq. (7.1–2) as

$$u(x, y) = \sum_{n=1}^\infty \frac{b_n}{\sinh nb} \sinh n(b - y) \sin nx$$

or

$$u(x, y) = \frac{2}{\pi} \sum_{n=1}^\infty \frac{\sinh n(b - y) \sin nx}{\sinh nb} \int_0^\pi f(s) \sin ns \, ds. \qquad \textbf{(7.1–9)}$$

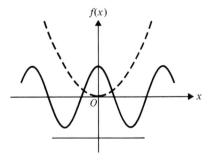

FIGURE 7.1-3 Even functions.

In this last expression we have changed the dummy variable of integration to s to prevent confusion with the independent variable x in $u(x, y)$.

We claim that (7.1–9) is the solution of the Dirichlet problem of Example 6.3–2. Admittedly, there are a number of unanswered questions at this point. Some of these are the following:

1. In view of the property expressed in Eq. (7.1–5) can the interval $0 < x < \pi$ be extended to $0 < x < a$ for some constant a?

2. Under what conditions can an infinite series be integrated term by term? (This was done in obtaining the value of b_n in Eq. 7.1–8.)

3. Does $f(x)$ have to be continuous for $0 < x < \pi$ as we assumed or is some lesser restriction sufficient?

4. Does $f(x)$ have to be defined so that $f(-\pi) = f(0) = f(\pi) = 0$ as we have done?

5. Under what conditions can an infinite series be differentiated term by term? [This must be done to verify that (7.1–9) satisfies Laplace's equation.]

We shall consider these and other questions in the remainder of this chapter.

A function $f(x)$ defined on $[-\pi, \pi]$ has a **Fourier series representation** if we can write

$$f(x) = \frac{1}{2} a_0 + \sum_{n=1}^{\infty} (a_n \cos nx + b_n \sin nx) \qquad \textbf{(7.1–10)}$$

with

$$a_n = \frac{1}{\pi} \int_{-\pi}^{\pi} f(s) \cos ns \, ds, \qquad n = 0, 1, 2, \ldots \qquad \textbf{(7.1–11)}$$

and

$$b_n = \frac{1}{\pi} \int_{-\pi}^{\pi} f(s) \sin ns \, ds, \qquad n = 1, 2, 3, \ldots, \qquad \textbf{(7.1–12)}$$

provided the last two integrals converge. The coefficients a_n and b_n are called **Fourier coefficients** (or Euler–Fourier coefficients) for the interval $[-\pi, \pi]$. The "equality" in Eq. (7.1–10) has a special significance that will be brought out later. Values of a_n and b_n are obtained by using the relations (Exercise 2)

$$\int_{-\pi}^{\pi} \sin nx \cos mx \, dx = 0 \qquad \textbf{(7.1–13)}$$

and

$$\int_{-\pi}^{\pi} \cos nx \cos mx \, dx = 0, \quad \text{if } n \neq m, \qquad \textbf{(7.1–14)}$$

in addition to the one in Eq. (7.1–5). These three are called **orthogonality relations** in accordance with the following definition.

Definition 7.1–1 The set of functions

$$\{\phi_i(x), \quad i = 1, 2, 3, \ldots\}$$

is **orthogonal** on (a, b) with weight function $w(x)$ if

$$\int_a^b \phi_n(x)\phi_m(x)w(x)\, dx = 0, \qquad n \neq m,$$

and

$$\int_a^b (\phi_n(x))^2 w(x)\, dx \neq 0.$$

Orthogonality of *functions* as given in Definition 7.1–1 is a generalization of orthogonality of *vectors*. Note that the *sum* of products in the scalar (or dot) multiplication of vectors has been replaced by the *integral* of products. Although the weight function $w(x)$ will be equal to unity in this section, it is included since it will take on some other values in later discussions. So far we have considered the following orthogonal sets of functions.

1. The set

$$\{\sin nx, \quad n = 1, 2, 3, \ldots\}$$

is orthogonal on $(-\pi, \pi)$ with weight function $w(x) = 1$ by virtue of Eqs. (7.1–5) and (7.1–6).

2. The set

$$\{\cos nx, \quad n = 0, 1, 2, \ldots\}$$

is orthogonal on $(-\pi, \pi)$ with weight function $w(x) = 1$ by virtue of Eq. (7.1–14) and the fact that (Exercise 2)

$$\int_{-\pi}^{\pi} \cos^2 nx\, dx = \pi, \qquad n = 1, 2, 3, \ldots \qquad \textbf{(7.1–15)}$$

and

$$\int_{-\pi}^{\pi} dx = 2\pi. \qquad \textbf{(7.1–16)}$$

3. The set

$$\{1, \cos x, \sin x, \cos 2x, \sin 2x, \ldots, \cos nx, \sin nx, \ldots\}$$

is orthogonal on $(-\pi, \pi)$ with weight function $w(x) = 1$ by virtue of Eqs. (7.1–5), (7.1–6), and (7.1–13)–(7.1–16).

Orthogonal functions play an important role in applied mathematics. We have already used the orthogonality property of the sine functions to complete the solution to the problem of Example 6.3–2 and we shall use similar methods for solving many other boundary-value problems.

A set of functions having an additional property, given in the following definition, will be of even greater value in our work.

Definition 7.1–2 The set of functions

$$\{\phi_i(x), \quad i = 1, 2, 3, \ldots\}$$

is **orthonormal** on (a, b) with weight function $w(x)$ if

$$\int_a^b \phi_n(x)\phi_m(x)w(x)\,dx = 0, \qquad n \neq m$$

and

$$\int_a^b (\phi_n(x))^2 w(x)\,dx = 1.$$

Thus orthonormal functions have the same properties as orthogonal functions but, in addition, they have been *normalized*. Again we have an analogy to *unit* (or normalized) vectors. The two relations in Definition 7.1–2 can be more simply expressed by using a symbol called the **Kronecker delta**,* defined by

$$\delta_{mn} = \begin{cases} 0, & \text{if } m \neq n, \\ 1, & \text{if } m = n. \end{cases}$$

If this symbol is used, then the functions in an orthonormal set have the property

$$\int_a^b \phi_n(x)\phi_m(x)w(x)\,dx = \delta_{mn}. \qquad (7.1\text{–}17)$$

The set of functions

$$\left\{ \frac{1}{\sqrt{2\pi}}, \frac{\cos x}{\sqrt{\pi}}, \frac{\sin x}{\sqrt{\pi}}, \frac{\cos 2x}{\sqrt{\pi}}, \frac{\sin 2x}{\sqrt{\pi}}, \ldots \right\} \qquad (7.1\text{–}18)$$

form an orthonormal set on $(-\pi, \pi)$ with weight function $w(x) = 1$.

Next we observe that if a function $f(x)$ is to have a valid Fourier series representation as given by (7.1–10) on an interval extending beyond $[-\pi, \pi]$, then it must be **periodic**. This is due to the fact that the individual terms of the series are, themselves, periodic. Recall that a function is *periodic of period p* if

$$f(x + p) = f(x)$$

for all values of x. Each term in (7.1–10) is periodic of period 2π. (Note that a constant function $f(x) = k$ satisfies the above definition of periodicity for any value of p.) Thus a function must be periodic of period 2π in order for the Fourier representation to hold outside of $[-\pi, \pi]$. This requirement was not a restriction in Example 6.3–2 since there we were interested only in representing the function on $(0, \pi)$ and it was unimportant to what values the series might converge outside this interval.

A second property that the function $f(x)$ must have is that it must be **piecewise smooth**. A function $f(x)$ is said to be piecewise smooth if $f(x)$ and $f'(x)$ are *both* piecewise continuous. In Definition 3.1–1 we

---

* After Leopold Kronecker (1823–1891), a German mathematician.

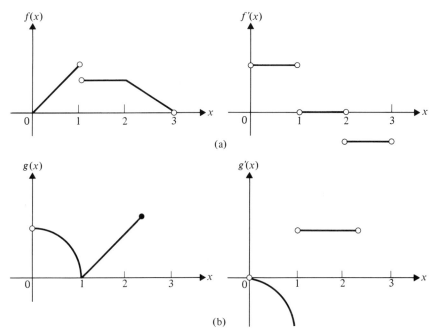

FIGURE 7.1–4 (a) $f(x)$ and $f'(x)$ are both piecewise continuous; (b) $g(x)$ is piecewise continuous but $g'(x)$ is not.

gave the definition of a piecewise continuous function. See also Fig. 7.1–4(a) for an example of a piecewise smooth function.

Sufficient (but not necessary) conditions for obtaining a Fourier series representation of a function are given in Theorem 7.1–1. These conditions are called the **Dirichlet conditions** since they appeared in papers of Dirichlet published in 1829 (*Journal für Math.*, **4**) and 1837 (*Dove's Repertorium der Physik*, **1**, p. 152).

Theorem 7.1–1 If $f(x)$ is a periodic* function of period 2π and piecewise smooth for $-\pi \le x \le \pi$, then the Fourier series (7.1–10) of $f(x)$ converges to $f(x)$ at all points where $f(x)$ is continuous and converges to the average of the left- and right-hand limits of $f(x)$ where $f(x)$ is discontinuous.

It should be apparent now why we defined $f(-\pi)$, $f(0)$, and $f(\pi)$ as zero in Example 6.3–2. These are the average values of the left- and right-hand limits of the function at the points of discontinuity, as shown in Fig. 7.1–5. No matter *how* $f(x)$ is defined at these points the

---

* By requiring that $f(x)$ be periodic, we are assured that the conclusion of the theorem applies to the infinite interval $-\infty < x < \infty$, rather than to just those values of x belonging to $[-\pi, \pi]$.

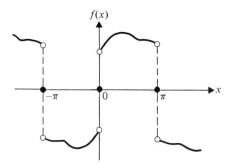

FIGURE 7.1-5 Average values at points of discontinuity.

Fourier series will converge to zero there according to Theorem 7.1–1. At the present time the convergence problem for a Fourier series is still unsolved. The Dirichlet conditions are known not to be necessary, and necessary conditions that have been obtained are known not to be sufficient. Suffice it to say that the conditions* stated in Theorem 7.1–1 are satisfied by a large number of the functions that we encounter in applied mathematics; hence we will not delve any deeper into the theory. We illustrate how Fourier series representations are obtained in the following examples.

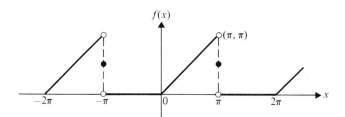

FIGURE 7.1-6 The function of Example 7.1–1.

EXAMPLE 7.1–1 Find the Fourier series representation of the function

$$f(x) = \begin{cases} 0, & \text{for } -\pi < x < 0, \\ x, & \text{for } 0 < x < \pi, \end{cases}$$

$$f(x + 2\pi) = f(x).$$

Solution. A sketch of the function is shown in Fig. 7.1–6. Note that the function satisfies the Dirichlet conditions of Theorem 7.1–1. Using

* The condition that $f(x)$ be periodic will be eliminated in Section 7.3.

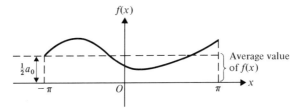

FIGURE 7.1–7 $\int_{-\pi}^{\pi} f(x)\, dx = \pi a_0$

Eq. (7.1–11) we have

$$a_n = \frac{1}{\pi} \int_{-\pi}^{\pi} f(s) \cos ns \, ds = \frac{1}{\pi} \int_{0}^{\pi} s \cos ns \, ds$$

$$= \frac{1}{\pi} \left(\frac{s}{n} \sin ns + \frac{1}{n^2} \cos ns \right) \Bigg|_{0}^{\pi}$$

$$= \frac{1}{\pi n^2} (\cos n\pi - 1)$$

$$= \frac{1}{\pi n^2} ((-1)^n - 1) = \begin{cases} 0, & \text{if } n \text{ is even,} \\ -\dfrac{2}{\pi n^2}, & \text{if } n \text{ is odd.} \end{cases}$$

The integration can be done by parts or tables may be used. We have also used the relation $\cos n\pi = (-1)^n$ in the computation. Note that the above scheme for obtaining a_n is not valid for $n = 0$. (Why?) We can, however, put $n = 0$ *before* performing the integration so that

$$a_0 = \frac{1}{\pi} \int_{0}^{\pi} s \, ds = \frac{\pi}{2}.$$

The constant term in the Fourier series, namely, $\frac{1}{2}a_0$, is the *average value* of the *function* being represented (Fig. 7.1–7) over the given interval.* Continuing by using Eq. (7.1–12), we have

$$b_n = \frac{1}{\pi} \int_{-\pi}^{\pi} f(s) \sin ns \, ds = \frac{1}{\pi} \int_{0}^{\pi} s \sin ns \, ds$$

$$= \frac{1}{\pi} \left(-\frac{s}{n} \cos ns + \frac{1}{n^2} \sin ns \right) \Bigg|_{0}^{\pi}$$

$$= \frac{1}{n} (-\cos n\pi) = \frac{(-1)^{n+1}}{n}.$$

---

* Recall from calculus that the average (or mean) value of $f(x)$ on the interval $a \le x \le b$ is given by

$$\frac{1}{b-a} \int_{a}^{b} f(x) \, dx.$$

Thus the function can be represented as

$$f(x) = \frac{\pi}{4} - \frac{2}{\pi}\left(\frac{\cos x}{1^2} + \frac{\cos 3x}{3^2} + \frac{\cos 5x}{5^2} + \cdots\right)$$
$$+ \left(\frac{\sin x}{1} - \frac{\sin 2x}{2} + \frac{\sin 3x}{3} - + \cdots\right)$$

or, if we use summation notation,

$$f(x) = \frac{\pi}{4} + \sum_{n=1}^{\infty}\left(-\frac{2}{\pi}\frac{\cos(2n-1)x}{(2n-1)^2} + \frac{(-1)^{n+1}}{n}\sin nx\right). \quad \blacksquare \quad \textbf{(7.1–19)}$$

A remark is necessary now about the use of an equality in (7.1–19). For those values of x for which $f(x)$ is continuous, the equality is appropriate in the sense that the series converges to the function. In other words, the more terms we add, the closer the sum will be to the functional value at that point. This pointwise convergence does not hold, however, at points of discontinuity since there the convergence is to the average of the left- and right-hand limits. Hence we will use the symbol $\sim$ rather than $=$ for the Fourier series representation of a function. Accordingly, Eq. (7.1–19) becomes

$$f(x) \sim \frac{\pi}{4} + \sum_{n=1}^{\infty}, \quad \text{etc.,}$$

and we read the symbol $\sim$ as "has the representation" in the sense of Theorem 7.1–1.

Although we cannot graph the result in Eq. (7.1–19), we can graph approximations to an infinite series in the form of *finite* partial sums. Let the sum of the first N terms of an infinite series be denoted by S_N. In the present case,

$$S_N = \frac{\pi}{4} + \sum_{n=1}^{N}\left(-\frac{2}{\pi}\frac{\cos(2n-1)x}{(2n-1)^2} + \frac{(-1)^{n+1}}{n}\sin nx\right). \quad \textbf{(7.1–20)}$$

Graphs of Eq. (7.1–20) are shown in Fig. 7.1–8 for $N = 5$, 10, and 20. Note the fairly good resemblance of S_5 to the function being represented.

In Fig. 7.1–8 the **overshoot** at $x = \pi^-$ and the **undershoot** at $x = \pi^+$ is a characteristic of Fourier (and other) series at points of discontinuity and is known as the **Gibbs phenomenon**.* The overshoot and undershoot both generally amount to about 18 percent of the distance between the functional values at a discontinuity. This

---

* After Josiah W. Gibbs (1839–1903), an American mathematical physicist. Gibbs pointed this out in 1899 (*Nature*, **59**, p. 606), and in 1906 Maxime Bôcher (1867–1918) gave a mathematical explanation (*Annals of Math.*, **2**(7), p. 81).

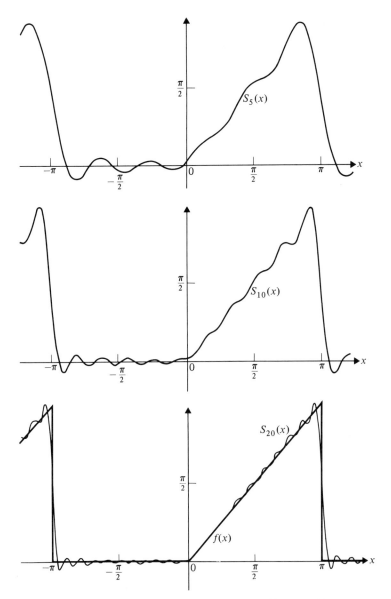

FIGURE 7.1–8 The graph of Eq. (7.1–20) for $N = 5, 10, 20$.

phenomenon persists even though a large number of terms are summed.* A detailed, computer-generated diagram of the Gibbs phenomenon is shown in Fig. 7.1–9. Here the function $f(x) = 1$ is being

* See also David Shelupsky, "Derivation of the Gibbs Phenomenon," *Amer. Math. Monthly* **87**, no. 3 (March 1980), pp. 210–212.

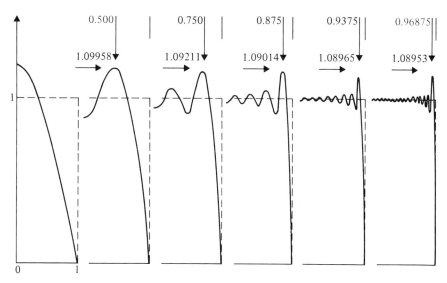

FIGURE 7.1–9 The Gibbs phenomenon. (From Kurt B. Wolf, *Integral Transforms in Science and Engineering*, New York: Plenum Press, 1979, by permission of the author and publisher.)

represented by S_N with $N = 2$, 4, 8, 16, 32, and 64. With increasing N the highest peak moves closer to the discontinuity at $x = 1$ but the overshoot remains approximately 1.09.

EXAMPLE 7.1–2 Obtain the Fourier series representation of the function

$$f(x) = \begin{cases} x + 2, & \text{for } -2 < x < 0, \\ 1, & \text{for } 0 < x < 2, \end{cases}$$

$$f(x + 4) = f(x).$$

Solution. This function, shown in Fig. 7.1–10, meets all the requirements of Theorem 7.1–1 except that the period is 4 instead of 2π. We

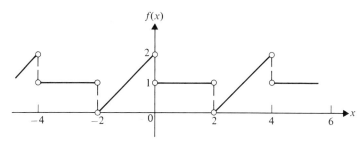

FIGURE 7.1–10

can easily remedy this, however, by making a scale according to the proportion

$$\frac{x}{\pi} = \frac{t}{2}.$$

Then $x = \pi t/2$, $dx = \pi \, dt/2$, so that Eqs. (7.1–11) and (7.1–12) become

$$a_n = \frac{1}{2} \int_{-2}^{2} f(s) \cos \frac{n\pi s}{2} \, ds, \qquad n = 0, 1, 2, \ldots$$

and

$$b_n = \frac{1}{2} \int_{-2}^{2} f(s) \sin \frac{n\pi s}{2} \, ds, \qquad n = 1, 2, \ldots,$$

respectively. Hence,

$$a_n = \frac{1}{2} \int_{-2}^{0} (s + 2) \cos \frac{n\pi s}{2} \, ds + \frac{1}{2} \int_{0}^{2} \cos \frac{n\pi s}{2} \, ds$$

$$= \frac{4}{n^2 \pi^2} \quad \text{if } n \text{ is odd, 0 otherwise};$$

$$a_0 = 2;$$

$$b_n = \frac{1}{2} \int_{-2}^{0} (s + 2) \sin \frac{n\pi s}{2} \, ds + \frac{1}{2} \int_{0}^{2} \sin \frac{n\pi s}{2} \, ds$$

$$= -\frac{2}{n\pi} \quad \text{if } n \text{ is even, 0 otherwise.}$$

Thus

$$f(x) \sim 1 + \frac{4}{\pi^2} \sum_{m=1}^{\infty} \frac{1}{(2m-1)^2} \cos \frac{(2m-1)\pi x}{2} - \frac{1}{\pi} \sum_{m=1}^{\infty} \frac{\sin m\pi x}{m}. \qquad \blacksquare \quad (7.1\text{--}21)$$

We have discussed the convergence of a Fourier series representation of a function (Theorem 7.1–1). We have attempted to show convergence diagrammatically by graphing partial sums S_N in Figs. 7.1–8 and 7.1–9. Now we will look into the matter of convergence in greater detail.

As before, let $S_N(x)$ represent the partial sums of the Fourier series representation of a function $f(x)$ on $[-\pi, \pi]$. The expression

$$\int_{-\pi}^{\pi} (f(x) - S_N(x))^2 \, dx$$

can be expected to be "small" since it is a measure of the error of the representation. If, for a certain class of functions f,

$$\lim_{N \to \infty} \int_{-\pi}^{\pi} (f(x) - S_N(x))^2 \, dx = 0, \qquad (7.1\text{--}22)$$

then we say the orthonormal functions (7.1–18) form a **complete set** with respect to the given class. An equivalent way of stating Eq. (7.1–22) is

$$\text{l.i.m.}_{N \to \infty} S_N(x) = f(x), \tag{7.1–23}$$

where "l.i.m." is read "limit in the mean" and we say that $S_N(x)$ **converges in the mean** to $f(x)$. It can be shown* that the set of orthonormal trigonometric functions (7.1–18) is *complete* with respect to the class of all *piecewise smooth functions*. It should be clear that completeness is an important property of a set of orthonormal functions and we will refer to this property in Sections 8.5, 9.3, and 9.4.

We conclude this section by observing that some unexpected useful information can also be gained from the Fourier series representation in (7.1–21). If we put $x = 0$ we obtain

$$f(0) = 1 + \frac{4}{\pi^2} \sum_{m=1}^{\infty} \frac{1}{(2m-1)^2}.$$

We know, however, from Theorem 7.1–1 that $f(0) = \frac{3}{2}$. (Why?) Hence

$$\sum_{m=1}^{\infty} \frac{1}{(2m-1)^2} = \frac{\pi^2}{8},$$

a result that can be verified by other means.

KEY WORDS AND PHRASES

Fourier series representation	Dirichlet conditions
Fourier coefficients	average value of a function on
orthogonality relation	an interval
orthogonal and orthonormal sets	the Gibbs phenomenon
of functions	complete orthonormal set
Kronecker delta	convergence in the mean
piecewise smooth function	

EXERCISES 7.1

▶ **1. a)** Show that

$$\int_{-\pi}^{\pi} \sin nx \sin mx \, dx = 0,$$

if $n \neq m$.

b) Show that

$$\int_{-\pi}^{\pi} \sin^2 nx \, dx = \pi.$$

---

*See Hans Sagan, *Boundary and Eigenvalue Problems in Mathematical Physics*, Ch. 4 (New York: Wiley, 1961).

2. Verify each of the following relations.

a) $\int_{-\pi}^{\pi} \sin nx \cos mx \, dx = 0$

b) $\int_{-\pi}^{\pi} \cos nx \cos mx \, dx = 0, \quad$ if $n \neq m$

c) $\int_{-\pi}^{\pi} \cos^2 nx \, dx = \pi, \quad n = 1, 2, \ldots$

3. Verify that the set of functions in (7.1–18) is an orthonormal set on $(-\pi, \pi)$ with weight function $w(x) = 1$.

4. Verify that the function $f(x)$ of Example 7.1–2 is piecewise smooth on the interval $-2 < x < 2$.

5. Carry out the details of the computations in Example 7.1–2.

▶▶ **6.** Verify that each of the following functions is an odd function by showing that each has the property $F(-x) = -F(x)$.

a) x^3 b) $\tan x$

c) $\csc x$ d) $\sinh x$

7. Verify that each of the following functions is an even function by showing that each has the property $F(-x) = F(x)$.

a) 1 b) $\cos x$ c) $\sec x$

d) $\cosh x$ e) $(x - a)^2$

8. Show that each of the following functions is neither odd nor even.

a) $ax^2 + bx + c$ b) $\log x$

c) e^x d) $x^2/(1 + x)$

9. Obtain the following multiplication table pertaining to multiplication of odd and even functions.

×	Odd	Even
Odd	Even	Odd
Even	Odd	Even

(*Note*: The entry in the first line and second column of the table shows that the product of an odd function and an even function is an odd function.)

10. Show that the function $f(x) = c$, where c is a constant, is periodic of period p for any value of p.

11. The smallest value of p for which $f(x + p) = f(x)$ is called the *fundamental period* of the function $f(x)$. Find the fundamental period of each of the following functions.

a) $\sin \dfrac{1}{2} x$ b) $\cos 2x$ c) $\cos 3\pi x$

d) $\sin \pi x$ e) $\cos \dfrac{\pi}{2} x$

12. a) If $f(x) = 1$ in Eq. (7.1–9), evaluate the integral and find $u(x, y)$.

      b) Assuming that the infinite series can be differentiated term by term, verify that the function $u(x, y)$ found in part (a) satisfies Laplace's equation.

13. If $f(x) = x$ in Eq. (7.1–9), evaluate the integral and find $u(x, y)$.

14. Using the result of Exercise 13, putting $b = 2$, and using tables and/or a calculator, compute $u(\pi/2, 1)$ to three decimals.

▶▶▶ 15. a) Prove that

$$\sum_{m=1}^{\infty} \frac{1}{(2m-1)^2} = \frac{\pi^2}{8}$$

by computing $f(2)$ from Eq. (7.1–21).

b) Compute the first ten terms of the series in part (a) and compare the result with the correct value.

16. Prove that if a function is periodic with period p, then it is also periodic of period np, $n = \pm 1, \pm 2, \ldots$.

17. If $f(x) = x - x^2$ when $0 < x < 1$ and $f(x)$ is periodic of period 1, show that $f(x)$ is an even function.

18. Show that, although most functions are neither odd nor even, every function defined on $(-c, c)$ can be expressed as the sum of an even function and an odd function by using the identity

$$f(x) \equiv \tfrac{1}{2}(f(x) + f(-x)) + \tfrac{1}{2}(f(x) - f(-x)).$$

19. Show that the set of trigonometric functions

$$\left\{ \frac{1}{\sqrt{2\pi}}, \frac{\cos nx}{\sqrt{\pi}}, n = 1, 2, \ldots \right\}$$

is *not* complete with respect to the class of *continuous* functions on $[-\pi, \pi]$ by showing that $f(x) = \sin x$ cannot be represented by a series of functions in the given set.

7.2 SINE, COSINE, AND EXPONENTIAL SERIES

In many of the applications of Fourier series a function $f(x)$ is defined on an interval $0 < x < L$. We may then represent this function either as a series consisting only of sine terms or as one consisting only of cosine terms. This is accomplished by making either an odd or an even **periodic extension** of the given function, respectively.

Generalizing the formulas of Example 7.1–2, we have (Exercises 1, 2, and 3)

$$f(x) \sim \frac{1}{2} a_0 + \sum_{n=1}^{\infty} a_n \cos \frac{n\pi x}{L} + b_n \sin \frac{n\pi x}{L}, \qquad \text{(7.2–1)}$$

where

$$a_n = \frac{1}{L} \int_{-L}^{L} f(s) \cos \frac{n\pi s}{L} \, ds, \qquad n = 0, 1, 2, \ldots \qquad \text{(7.2–2)}$$

and

$$b_n = \frac{1}{L} \int_{-L}^{L} f(s) \sin \frac{n\pi s}{L} \, ds, \qquad n = 1, 2, \ldots . \qquad \text{(7.2–3)}$$

Note that these three reduce to Eqs. (7.1–10), (7.1–11), and (7.1–12) when $L = \pi$ as they should.

Cosine series

If $f(x)$ is defined on $0 < x < L$ and is to be represented by a series of cosines, we make an **even periodic extension** of $f(x)$, as shown in Fig. 7.2–1. The resulting function, defined by

$$f(x) = \begin{cases} f(-x), & 0 < x < L, \\ f(x), & -L < x < 0, \end{cases}$$

$$f(x + 2L) = f(x), \qquad f(0) = \lim_{\varepsilon \to 0^+} f(\varepsilon), \qquad f(L) = \lim_{\varepsilon \to 0} f(L - \varepsilon),$$

is an even periodic function; hence $b_n \equiv 0$, $n = 1, 2, \ldots$, and the formula (7.2–2) can be simplified to

$$a_n = \frac{2}{L} \int_0^L f(s) \cos \frac{n\pi s}{L} \, ds, \qquad n = 0, 1, 2, \ldots . \qquad \textbf{(7.2–4)}$$

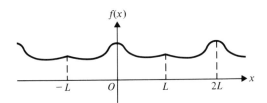

$f(x)$

FIGURE 7.2–1 Even periodic extension.

EXAMPLE 7.2–1 Obtain the Fourier cosine series representation of the function $f(x) = x$, $0 \le x \le \pi$.

Solution. The even periodic extension of the function is shown in Fig. 7.2–2. Using Eq. (7.2–4) we have

$$a_n = \frac{2}{\pi} \int_0^\pi s \cos ns \, ds,$$

$$a_n = \begin{cases} 0, & \text{if } n \text{ is even,} \\ -\dfrac{4}{\pi n^2}, & \text{if } n \text{ is odd,} \end{cases}$$

$$a_0 = \frac{2}{\pi} \int_0^\pi s \, ds = \pi.$$

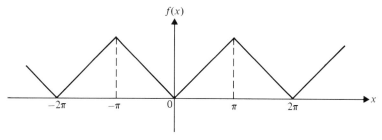

FIGURE 7.2–2

Hence the required series can be written (Exercise 5)

$$f(x) \sim \frac{\pi}{2} - \frac{4}{\pi} \sum_{m=1}^{\infty} \frac{\cos (2m-1)x}{(2m-1)^2}. \quad \blacksquare \qquad (7.2\text{–}5)$$

Sine series

If $f(x)$ is defined on $0 < x < L$ and is to be represented by a series of sines, we make an **odd periodic extension** of $f(x)$, as shown in Fig. 7.2–3. The resulting function, defined by

$$f(x) = \begin{cases} f(x), & 0 < x < L, \\ -f(-x), & -L < x < 0, \end{cases}$$

$$f(x + 2L) = f(x), \qquad f(0) = f(L) = 0,$$

is an odd periodic function, hence $a_n \equiv 0$, $n = 0, 1, 2, \ldots$, and the formula (7.2–3) can be simplified to

$$b_n = \frac{2}{L} \int_0^L f(s) \sin \frac{n\pi s}{L} \, ds, \qquad n = 1, 2, 3, \ldots . \qquad (7.2\text{–}6)$$

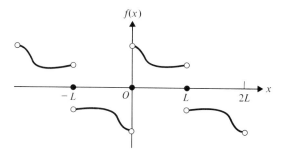

FIGURE 7.2–3 Odd periodic extension.

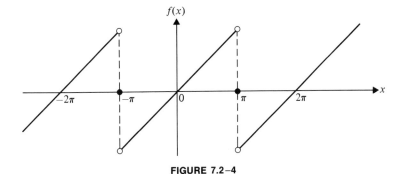

FIGURE 7.2–4

EXAMPLE 7.2–2 Obtain the Fourier sine series representation of the function $f(x) = x$, $0 \leq x < \pi$, $f(\pi) = 0$.

Solution. The odd periodic extension of the function is shown in Fig. 7.2–4. Using Eq. (7.2–6) we have

$$b_n = \frac{2}{\pi} \int_0^\pi s \sin ns \, ds = \frac{2}{n} (-1)^{n+1}$$

so that the required series can be written (Exercise 7)

$$f(x) \sim 2 \sum_{n=1}^{\infty} \frac{(-1)^{n+1} \sin nx}{n}. \quad \blacksquare \qquad (7.2\text{–}7)$$

It is to be noted that although the expressions (7.1–19), (7.2–5), and (7.2–7) above are all different, they all represent the function $f(x) = x$ on the interval $0 < x < \pi$. This versatility of Fourier series representation makes the technique a valuable one in applied mathematics. We have already seen that some boundary-value problems require that a given function be represented by a series of sines (compare Eq. 7.1–3) and we shall see in Chapters 8 and 9 that other problems may require a series of cosines. The exercises at the end of this section are designed to illustrate both of these representations.

Exponential series

Another useful form of Fourier series is the **exponential form**. This is obtained from the standard form (see Eq. 7.1–10),

$$f(x) \sim \frac{1}{2} a_0 + \sum_{n=1}^{\infty} a_n \cos nx + b_n \sin nx,$$

by using Euler's formula

$$e^{inx} = \cos nx + i \sin nx.$$

We have

$$\cos nx = \frac{1}{2}(e^{inx} + e^{-inx}),$$

$$\sin nx = \frac{1}{2i}(e^{inx} - e^{-inx}),$$

so that we can write

$$f(x) \sim \frac{1}{2}a_0 + \frac{1}{2}\sum_{n=1}^{\infty} a_n(e^{inx} + e^{-inx}) - ib_n(e^{inx} - e^{-inx})$$

$$= \frac{1}{2}a_0 + \frac{1}{2}\sum_{n=1}^{\infty} (a_n - ib_n)e^{inx} + (a_n + ib_n)e^{-inx}.$$

If we now define **complex Fourier coefficients** by

$$c_0 = \tfrac{1}{2}a_0, \quad c_n = \tfrac{1}{2}(a_n - ib_n), \quad c_{-n} = \tfrac{1}{2}(a_n + ib_n), \quad n = 1, 2, \ldots, \quad \textbf{(7.2–8)}$$

then we have

$$f(x) \sim \sum_{n=-\infty}^{\infty} c_n e^{inx}, \qquad \textbf{(7.2–9)}$$

which is the **complex form** (or exponential form) of Fourier series. It is commonly used in physics and engineering because of its notational simplicity.

In the above development of the exponential form we have assumed that $f(x)$ had period 2π but this was done only to simplify the notation. For period $2L$ we would write

$$f(x) \sim \sum_{n=-\infty}^{\infty} c_n e^{in\pi x/L}$$

with corresponding changes in the c_n. Details of these changes, as well as the derivation of the formula

$$c_n = \frac{1}{2\pi} \int_{-\pi}^{\pi} f(s)e^{-ins}\, ds, \qquad n = 0, \pm 1, \pm 2, \ldots, \qquad \textbf{(7.2–10)}$$

are left to the exercises.

Differentiation of Fourier series

In Example 7.2–1 we obtained the representation

$$f(x) \sim \frac{\pi}{2} - \frac{4}{\pi}\sum_{m=1}^{\infty} \frac{\cos(2m-1)x}{(2m-1)^2} \qquad \textbf{(7.2–5)}$$

of the function $f(x) = x$ for $0 \leq x \leq \pi$. If we differentiate both members of (7.2–5), we obtain

$$f'(x) \sim \frac{4}{\pi} \sum_{m=1}^{\infty} \frac{\sin (2m-1)x}{2m-1}.$$

It can be shown (Exercise 24) that the above is the representation of $f'(x) = 1$ on $0 < x < \pi$ and that $f'(0) = f'(\pi) = 0$ as expected. On the other hand, if we begin with (7.2–7), then differentiation produces

$$f'(x) \sim 2 \sum_{n=1}^{\infty} (-1)^{n+1} \cos nx,$$

which converges *nowhere* since the limit of the nth term does not approach zero as $n \to \infty$, a necessary condition for convergence. Clearly, the conditions under which a Fourier series representation may be differentiated term by term must be examined. Sufficient conditions are stated in the following theorem, which we state without proof.

Theorem 7.2–1 Let $f(x)$ have a Fourier series representation for $-\pi \leq x \leq \pi$ given by

$$f(x) \sim \frac{1}{2} a_0 + \sum_{n=1}^{\infty} (a_n \cos nx + b_n \sin nx)$$

with

$$a_n = \frac{1}{\pi} \int_{-\pi}^{\pi} f(s) \cos ns \, ds, \qquad n = 0, 1, 2, \ldots$$

and

$$b_n = \frac{1}{\pi} \int_{-\pi}^{\pi} f(s) \sin ns \, ds, \qquad n = 1, 2, \ldots .$$

Then the Fourier series representation is differentiable at each point where $f''(x)$ exists, provided that $f(x)$ is continuous and $f'(x)$ is piecewise continuous on $-\pi \leq x \leq \pi$ and that $f(-\pi) = f(\pi)$. Moreover,

$$f'(x) \sim \sum_{n=1}^{\infty} n(-a_n \sin nx + b_n \cos nx), \qquad -\pi < x < \pi.$$

Integration of a Fourier series

Integration of a Fourier series is a much simpler matter. This is to be expected, since integration is a "smoothing" process that tends to eliminate discontinuities, whereas the process of differentiation has the opposite effect. The following theorem applies to integration of a Fourier series.

Theorem 7.2–2 Let $f(x)$ be piecewise continuous on $-\pi < x < \pi$ and have a Fourier series representation

$$f(x) \sim \frac{1}{2} a_0 + \sum_{n=1}^{\infty} (a_n \cos nx + b_n \sin nx),$$

with a_n and b_n as before. Then

$$\int_{-\pi}^{x} f(s) \, ds = \frac{1}{2} a_0(x + \pi) + \sum_{n=1}^{\infty} \frac{1}{n} (a_n \sin nx - b_n(\cos nx - \cos n\pi))$$

for $-\pi \le x \le \pi$.

KEY WORDS AND PHRASES

odd and even periodic extension   complex Fourier coefficients

EXERCISES 7.2

▶ **1.** Verify each of the following orthogonality relations.

 a) $\int_{-L}^{L} \sin \frac{m\pi x}{L} \cos \frac{n\pi x}{L} \, dx = 0$

 b) $\int_{-L}^{L} \cos \frac{m\pi x}{L} \cos \frac{n\pi x}{L} \, dx = 0, \quad \text{if } n \ne m$

 c) $\int_{-L}^{L} \sin \frac{m\pi x}{L} \sin \frac{n\pi x}{L} \, dx = 0, \quad \text{if } n \ne m$

2. Show that

$$\int_{-L}^{L} \sin^2 \frac{n\pi x}{L} \, dx = \int_{-L}^{L} \cos^2 \frac{n\pi x}{L} \, dx = L, \qquad n = 1, 2, \ldots .$$

3. Using the results of Exercises 1 and 2, what can you say about the set

$$\left\{ \frac{1}{\sqrt{2L}}, \frac{\cos \frac{n\pi x}{L}}{\sqrt{L}}, \frac{\sin \frac{n\pi x}{L}}{\sqrt{L}} \right\}, \qquad n = 1, 2, \ldots ?$$

4. Explain how the function of Example 7.2–1 satisfies the definition of an even periodic extension given in the text.

5. Carry out the details necessary to obtain the result (7.2–5) in Example 7.2–1.

6. Explain how the function of Example 7.2–2 satisfies the definition of an odd periodic extension given in the text.

7. Carry out the details necessary to obtain the result (7.2–7) in Example 7.2–2.

8. a) Show that

$$\int_{-\pi}^{\pi} e^{inx} e^{-imx}\, dx = \begin{cases} 0, & \text{if } n \neq m, \\ 2\pi, & \text{if } n = m. \end{cases}$$

b) Use the result of part (a) to obtain a set that is orthonormal on $[-\pi, \pi]$. (*Note*: The type of orthogonality exhibited here is called **Hermitian orthogonality**.* If the set of complex-valued functions of the real variable x,

$$\{\phi_n(x)\}, \qquad n = 1, 2, \ldots,$$

has the property

$$\int_a^b \phi_n(x)\overline{\phi_m(x)}\, dx = 0, \quad \text{for } n \neq m,$$

then the set is orthogonal (in the Hermitian sense) on (a, b) with weight function 1. The bar indicates the *complex conjugate* (see Section 10.1). In case the functions are *real-valued*, the above reduces to the orthogonality defined in Definition 7.1–1.

▶▶ In Exercises 9–14, obtain (a) the Fourier sine and (b) the Fourier cosine representations of the given function.

9. $f(x) = 2 - x, \quad 0 < x \le 2$

10. $f(x) = a, \quad 0 < x \le 3, \quad a > 0$

11. $f(x) = x^2, \quad 0 \le x < 1$

12. $f(x) = e^x, \quad 0 < x < 2$

13. $f(x) = \sin \pi x, \quad 0 \le x < 1$

14. $f(x) = \cos x, \quad 0 < x \le \pi/2$

15. Represent the function of Exercise 11 by a Fourier series that contains both sines and cosines. (*Hint*: Define a periodic extension that is neither odd nor even. There are an infinite number of ways of doing this.)

16. Find a Fourier sine series representation of the function $x - 1$ on the interval $1 < x < 2$.

17. Find a Fourier cosine series representation of the function $x - 1$ on the interval $1 < x < 2$.

18. a) Find the Fourier series representation of the function

$$f(x) = \begin{cases} x + 1, & -1 \le x \le 0, \\ -x + 1, & 0 \le x \le 1, \end{cases}$$

$$f(x + 2) = f(x).$$

b) Use the result of part (a) to show that

$$1 + \frac{1}{3^2} + \frac{1}{5^2} + \cdots = \frac{\pi^2}{8}.$$

19. Find the Fourier sine series representation of the function $f(x) = 1$, $0 < x < \pi, f(0) = f(\pi) = 0$.

---

* After Charles Hermite (1822–1901), a French mathematician.

20. Obtain each of the following series representations.

a) $\dfrac{\pi}{4} = 1 - \dfrac{1}{3} + \dfrac{1}{5} - \dfrac{1}{7} + - \cdots$

b) $\dfrac{\pi}{4} = \dfrac{\sqrt{2}}{2}\left(1 + \dfrac{1}{3} - \dfrac{1}{5} - \dfrac{1}{7} + + - - \cdots\right)$

(*Hint*: Use the result of Exercise 19 with appropriate values for x.)

21. Consider the following triangular function:

$$f(x) = \begin{cases} 2 - x, & 1 < x < 2, \\ x - 2, & 2 < x < 3. \end{cases}$$

a) Obtain a Fourier sine series representation of this function.

b) Obtain a Fourier cosine series representation of this function.

22. Given the function

$$f(x) = \begin{cases} 1, & 0 < x < 2, \\ -1, & -2 < x < 0, \end{cases}$$

$f(x + 4) = f(x)$, $f(0) = f(2) = 0$, write the exponential form of the Fourier series representation of this function. (*Hint*: See Exercise 30 for the coefficients.)

23. Given the function

$$f(x) = 1, \qquad -\infty < x < \infty,$$

write the exponential form of the Fourier series representation of this function.

24. Show that when $f(x)$ given in (7.2–5) is differentiated, the result converges to $f'(x)$ for $0 < x < \pi$ and to zero for $x = 0$ and $x = \pi$. Graph $f'(x)$ and compare it with the graph obtained from the $f(x)$ shown in the text.

25. Explain how $f(x)$ of (7.2–5) meets the conditions of Theorem 7.2–1.

26. In what ways does (7.2–7) fail to meet the conditions of Theorem 7.2–1?

27. Obtain a Fourier series representation for $f(x) = x$ on $0 \leq x < 1$ by integrating the representations of x^2 in Exercise 11. For what values of x is your result valid?

▶▶▶ **28.** Show that the set

$$\left\{\sqrt{\dfrac{2}{L}}\sin\dfrac{n\pi x}{L}\right\}, \qquad n = 1, 2, \ldots$$

is orthonormal on $[0, L]$. *Hint*: Use the fact that

$$\sin\dfrac{n\pi x}{L} = \dfrac{e^{in\pi x/L} - e^{-in\pi x/L}}{2i}.$$

29. Derive the formula (7.2–10) from Eqs. (7.1–11), (7.1–12), and (7.2–8).

30. Show that for a function of period $2L$ the complex Fourier coefficients are given by

$$c_n = \dfrac{1}{2L}\int_{-L}^{L} f(s)e^{-in\pi s/L}\,ds, \qquad n = 0, 1, 2, \ldots.$$

31. Step functions occur in modeling off–on controls in mechanical systems. Such a function is given by

$$f(x) = (-1)^n h, \qquad n = 0, \pm 1, \pm 2, \ldots, n < x < n + 1.$$

a) Sketch the function.

b) Obtain the Fourier series representation of this function.

c) Sketch the first term of the Fourier series representation.

d) Sketch the first two terms of the Fourier series representation.

32. Given the function

$$f(x) = 2 - x, \qquad 0 < x < 2,$$

define a function whose Fourier sine series representation will converge to $f(x)$ for all values of x. (*Note*: The answer is not unique.)

7.3 FOURIER INTEGRALS AND TRANSFORMS

The following extension of Fourier series to the Fourier integral is intended to be a *plausible* rather than a *mathematical* development. Our justification for this lies in the fact that a rigorous treatment such as that found in I. N. Sneddon's *Fourier Transforms* (New York: McGraw-Hill, 1951) would lead us too far afield.

A periodic function $f(x)$ that satisfies the Dirichlet conditions, Theorem 7.1–1, may be represented by a Fourier series. The Dirichlet conditions are *sufficient* but not *necessary*. The representation is in the sense that the series will converge to the average or mean value at points where $f(x)$ has a jump discontinuity.

As an example, the function defined by

$$f(x) = e^{-|x|}, \qquad -L < x < L,$$
$$f(x + 2L) = f(x)$$

$$(7.3{-}1)$$

has a Fourier series representation

$$f(x) \sim \frac{1}{2} a_0 + \sum_{n=1}^{\infty} \left(a_n \cos \frac{n\pi x}{L} + b_n \sin \frac{n\pi x}{L} \right), \qquad (7.3{-}2)$$

where

$$a_n = \frac{1}{L} \int_{-L}^{L} f(s) \cos \frac{n\pi s}{L} \, ds, \qquad n = 0, 1, 2, \ldots,$$

$$b_n = \frac{1}{L} \int_{-L}^{L} f(s) \sin \frac{n\pi s}{L} \, ds, \qquad n = 1, 2, \ldots .$$

We are using s as the dummy variable of integration here. For the example given in Eqs. (7.3–1) we would have $b_n = 0$ for all n and a simpler formula for a_n since $f(x)$ is an even function.

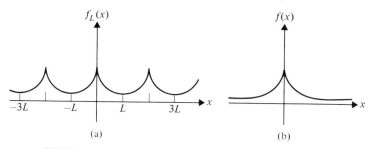

FIGURE 7.3–1 (a) Periodic function. (b) Nonperiodic function.

If we denote the function defined by Eqs. (7.3–1) as $f_L(x)$ to high-light the fact that this is a periodic function with half-period equal to L, then we can make an obvious observation. The Fourier series representation given by (7.3–2) is valid no matter how large L is so long as it remains finite. This then leads us naturally to consider the function $f(x)$ defined as

$$f(x) = \lim_{L \to \infty} f_L(x).$$

The two functions $f(x)$ and $f_L(x)$ are shown in Fig. 7.3–1. Now $f(x)$ is no longer a periodic function but is piecewise smooth. We will impose one more condition on $f(x)$, namely, that it be absolutely integrable on the real line, that is, that the improper integral

$$\int_{-\infty}^{\infty} |f(x)|\, dx$$

be finite. In the case of our sample function we have

$$\int_{-\infty}^{\infty} |e^{-|x|}|\, dx = 2 \int_0^{\infty} e^{-x}\, dx = 2 \lim_{L \to \infty} \int_0^L e^{-x}\, dx.$$

We can easily verify that the value of the integral is 2 (Exercise 1). Following the usual custom we write

$$\int_0^{\infty} f(x)\, dx \quad \text{instead of} \quad \lim_{L \to \infty} \int_0^L f(x)\, dx.$$

Now we make the substitution $\alpha_n = n\pi/L$ and replace a_n and b_n by their values in (7.3–2). Then

$$f_L(x) \sim \frac{1}{2L} \int_{-L}^{L} f_L(s)\, ds$$

$$+ \frac{1}{L} \sum_{n=1}^{\infty} \left(\cos\left(\alpha_n x\right) \int_{-L}^{L} f_L(s) \cos\left(\alpha_n s\right) ds \right.$$

$$+ \left. \sin\left(\alpha_n x\right) \int_{-L}^{L} f_L(s) \sin\left(\alpha_n s\right) ds \right). \tag{7.3–3}$$

Since

$$\Delta\alpha = \alpha_{n+1} - \alpha_n = \frac{(n+1)\pi}{L} - \frac{n\pi}{L} = \frac{\pi}{L},$$

we may write (7.3–3) as

$$f_L(x) \sim \frac{1}{2L} \int_{-L}^{L} f_L(s)\, ds + \frac{1}{\pi} \sum_{n=1}^{\infty} \left(\cos(\alpha_n x)\, \Delta\alpha \int_{-L}^{L} f_L(s) \cos(\alpha_n s)\, ds \right.$$

$$\left. + \sin(\alpha_n x)\, \Delta\alpha \int_{-L}^{L} f_L(s) \sin(\alpha_n s)\, ds \right). \tag{7.3–4}$$

This last representation of the periodic, piecewise smooth function $f_L(x)$ is valid for any *finite* L.

We now let L approach ∞. Then the first integral on the right of (7.3–4) approaches zero because $f(x)$ is absolutely integrable.* Moreover, it seems plausible that the infinite series becomes an integral from 0 to ∞. Hence

$$f(x) = \frac{1}{\pi} \int_{0}^{\infty} \left(\cos(\alpha x) \int_{-\infty}^{\infty} f(s) \cos(\alpha s)\, ds \right.$$

$$\left. + \sin(\alpha x) \int_{-\infty}^{\infty} f(s) \sin(\alpha s)\, ds \right) d\alpha, \tag{7.3–5}$$

which is the **Fourier integral representation** of $f(x)$. Equation (7.3–5) is often written in the form

$$f(x) = \int_{0}^{\infty} (A(\alpha) \cos \alpha x + B(\alpha) \sin \alpha x)\, d\alpha, \qquad -\infty < x < \infty, \tag{7.3–6}$$

where

$$A(\alpha) = \frac{1}{\pi} \int_{-\infty}^{\infty} f(s) \cos \alpha s\, ds$$

and $\tag{7.3–7}$

$$B(\alpha) = \frac{1}{\pi} \int_{-\infty}^{\infty} f(s) \sin \alpha s\, ds.$$

Note the close similarity between the coefficients in the Fourier integral and in the Fourier series.

Sufficient conditions for the validity of Eq. (7.3–5) can be stated in the form of the following theorem.

Theorem 7.3–1 If $f(x)$ is piecewise smooth and absolutely integrable on the real line, then $f(x)$ can be represented by a Fourier integral. At a point where $f(x)$ is discontinuous, the value of the Fourier integral is the average of the left- and right-hand limits of $f(x)$ at that point.

---

* Recall that an improper integral is integrable if it is absolutely integrable.

The Fourier integral can be written in a more compact form. Going back to Eq. (7.3–5), we recognize that $\cos(\alpha x)$ and $\sin(\alpha x)$ do not depend on s, hence these terms may be put *inside* the integrals. Then we have

$$f(x) = \frac{1}{\pi} \int_0^\infty \int_{-\infty}^\infty f(s)(\cos \alpha x \cos \alpha s + \sin \alpha x \sin \alpha s) \, ds \, d\alpha \quad \textbf{(7.3–8)}$$

$$= \frac{1}{\pi} \int_0^\infty \int_{-\infty}^\infty f(s) \cos \alpha(s - x) \, ds \, d\alpha. \quad \textbf{(7.3–9)}$$

If $f(x)$ is an *even* function, then $f(s) \sin \alpha s$ is an odd function of s and Eq. (7.3–8) reduces to

$$f(x) = \frac{2}{\pi} \int_0^\infty \int_0^\infty f(s) \cos \alpha x \cos \alpha s \, ds \, d\alpha. \quad \textbf{(7.3–10)}$$

If $f(x)$ is an *odd* function, then $f(s) \cos \alpha s$ is an odd function of s and Eq. (7.3–8) reduces to

$$f(x) = \frac{2}{\pi} \int_0^\infty \int_0^\infty f(s) \sin \alpha x \sin \alpha s \, ds \, d\alpha. \quad \textbf{(7.3–11)}$$

From Eq. (7.3–9) we see that the inner integral is an even function of α; hence we can write

$$f(x) = \frac{1}{2\pi} \int_{-\infty}^\infty \int_{-\infty}^\infty f(s) \cos \alpha(s - x) \, ds \, d\alpha. \quad \textbf{(7.3–12)}$$

On the other hand,

$$\frac{i}{2\pi} \int_{-\infty}^\infty \int_{-\infty}^\infty f(s) \sin \alpha(s - x) \, ds \, d\alpha = 0 \quad \textbf{(7.3–13)}$$

because the inner integral is an odd function of α. Thus, adding Eqs. (7.3–12) and (7.3–13), we obtain

$$f(x) = \frac{1}{2\pi} \int_{-\infty}^\infty \int_{-\infty}^\infty f(s) e^{i\alpha(s - x)} \, ds \, d\alpha, \quad \textbf{(7.3–14)}$$

which is the **complex form** of the Fourier integral.

Fourier transform

The **Fourier transform** is obtained from Eq. (7.3–14) as follows. Rewrite Eq. (7.3–14) as an *iterated integral*,

$$f(x) = \frac{1}{2\pi} \int_{-\infty}^\infty f(s) e^{i\alpha s} \, ds \int_{-\infty}^\infty e^{-i\alpha x} \, d\alpha.$$

Next, note that the variable s is a *dummy variable* and may be replaced by any other, say x, since the first integral is a function of α. Thus we have the *pair* of equations

$$\bar{f}(\alpha) = \int_{-\infty}^{\infty} f(x)e^{i\alpha x}\,dx,$$

$$f(x) = \frac{1}{2\pi} \int_{-\infty}^{\infty} \bar{f}(\alpha)e^{-i\alpha x}\,d\alpha.$$

(7.3–15)

Such a pair is called a **Fourier transform pair**. We call $\bar{f}(\alpha)$ the **Fourier transform** of $f(x)$. The second equation in (7.3–15) defines $f(x)$, which is the **inverse Fourier transform** of $\bar{f}(\alpha)$. An example of a simple function and its Fourier transform is shown in Fig. 7.3–2.

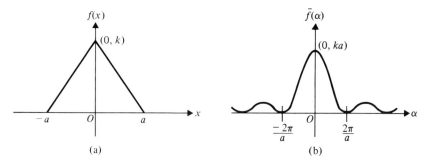

(a)   (b)

FIGURE 7.3–2 A Fourier transform pair.

Before we give an example of the use of the Fourier transform, a few words of caution are in order. Some authors define a Fourier transform pair as

$$\bar{f}(\alpha) = \frac{1}{\sqrt{2\pi}} \int_{-\infty}^{\infty} f(x)^{i\alpha x}\,dx,$$

$$f(x) = \frac{1}{\sqrt{2\pi}} \int_{-\infty}^{\infty} \bar{f}(\alpha)e^{-i\alpha x}\,d\alpha$$

in order to keep as much symmetry as possible in the pair. Others interchange the negative signs in the exponents, whereas still others use combinations of these two variations. Because of the negative sign in *one* of the exponents, true symmetry is impossible to achieve. The important thing is that any given pair should reduce to Eq. (7.3–14) or its equivalent. Thus in applications it doesn't matter if the factor $1/2\pi$ is inserted when the function is transformed or when the inverse transform is found. You should keep this fact in mind when consulting tables of Fourier transforms.

Transforms are common in mathematics since they are useful in converting many problems into more easily solvable ones. For example, logarithms are useful for converting multiplication into addition, the Laplace transform is useful for converting ordinary differential equations into algebraic equations, and the Fourier transform will be useful for converting partial differential equations into ordinary differential equations. In each case an inverse process is necessary.

In preparation for the next example we compute the Fourier transforms of du/dx and d^2u/dx^2. We *assume* that u and du/dx both approach zero as $x \to \pm\infty$, that u satisfies the Dirichlet conditions, and that u is absolutely integrable on the real line. We have, after integrating by parts,

$$\int_{-\infty}^{\infty} \frac{du}{dx} e^{i\alpha x}\, dx = ue^{i\alpha x}\Big|_{-\infty}^{\infty} - i\alpha \int_{-\infty}^{\infty} ue^{i\alpha x}\, dx$$

$$= -i\alpha\bar{u}(\alpha), \tag{7.3-16}$$

using the assumptions. For the second derivative we have, after integrating by parts,

$$\int_{-\infty}^{\infty} \frac{d^2u}{dx^2} e^{i\alpha x}\, dx = \frac{du}{dx} e^{i\alpha x}\Big|_{-\infty}^{\infty} - \alpha i \int_{-\infty}^{\infty} \frac{du}{dx} e^{i\alpha x}\, dx$$

$$= -i\alpha(-i\alpha\bar{u}(\alpha)) = -\alpha^2\bar{u}(\alpha), \tag{7.3-17}$$

using the result of Eq. (7.3–16).

The results in Eqs. (7.3–16) and (7.3–17) can also be written as

$$\mathscr{F}\left(\frac{du}{dx}\right) = -i\alpha\mathscr{F}(u) = -i\alpha\bar{u}(\alpha)$$

and

$$\mathscr{F}\left(\frac{d^2u}{dx^2}\right) = -\alpha^2\mathscr{F}(u) = -\alpha^2\bar{u}(\alpha),$$

respectively. Following our notation in Chapter 3 where we used $\mathscr{L}(f(x))$ for the *Laplace transform* of $f(x)$, we use $\mathscr{F}(f(x))$ for the *Fourier transform* of $f(x)$. We further distinguish between the two transforms by writing $\bar{f}(\alpha)$ for the Fourier transform of $f(x)$ in contrast with $F(s)$, the Laplace transform of $f(x)$.

In addition to the Fourier transform pair defined in (7.3–15), we also have the Fourier *sine* and *cosine* transform pairs, which will be defined in Section 7.4. We will present applications of these there. We conclude this section with an example.

EXAMPLE 7.3–1 Solve the following one-dimensional heat equation:

$$u_t(x, t) = u_{xx}(x, t), \qquad -\infty < x < \infty, \qquad 0 < t,$$

given that $u(x, 0) = f(x)$ and $|u(x, t)| < \infty$.

Solution. We assume that $u(x, t)$ and $u_x(x, t)$ approach zero as $x \to \pm \infty$ and that $f(x)$ is piecewise smooth and absolutely integrable. Then $f(x)$ has a Fourier transform

$$\bar{f}(\alpha) = \int_{-\infty}^{\infty} f(x)e^{i\alpha x} \, dx$$

$$= \int_{-\infty}^{\infty} f(s)e^{i\alpha s} \, ds,$$

if we change the dummy variable of integration in order to prevent confusion later. The Fourier transform of $u_t(x, t)$ is

$$\int_{-\infty}^{\infty} \frac{\partial u(x, t)}{\partial t} e^{i\alpha x} \, dx = \frac{\partial}{\partial t} \int_{-\infty}^{\infty} u(x, t)e^{i\alpha x} \, dx$$

$$= \frac{\partial \bar{u}(\alpha, t)}{\partial t} = \frac{d\bar{u}(\alpha, t)}{dt},$$

where we have assumed that differentiation and integration may be interchanged and where we have noted that α plays the role of a parameter in obtaining the transform.

Transforming the given partial differential equation and the initial condition produces

$$\frac{d\bar{u}}{dt} + \alpha^2 \bar{u} = 0, \qquad \bar{u}(\alpha, 0) = \bar{f}(\alpha).$$

The solution to this problem is readily found (Exercise 4) to be

$$\bar{u}(\alpha, t) = \bar{f}(\alpha)e^{-\alpha^2 t}.$$

Using the inversion formula (7.3–15) to get back to $u(x, t)$, we have

$$u(x, t) = \frac{1}{2\pi} \int_{-\infty}^{\infty} \bar{u}(\alpha, t)e^{-i\alpha x} \, d\alpha = \frac{1}{2\pi} \int_{-\infty}^{\infty} \bar{f}(\alpha)e^{-\alpha^2 t}e^{-i\alpha x} \, d\alpha$$

$$= \frac{1}{2\pi} \int_{-\infty}^{\infty} \int_{-\infty}^{\infty} f(s)e^{i\alpha s}e^{-\alpha^2 t}e^{-i\alpha x} \, d\alpha \, ds$$

$$= \frac{1}{2\pi} \int_{-\infty}^{\infty} \int_{-\infty}^{\infty} f(s)e^{i\alpha(s-x)}e^{-\alpha^2 t} \, d\alpha \, ds.$$

Now

$$e^{-\alpha^2 t}e^{i\alpha(s-x)} = e^{-\alpha^2 t}(\cos \alpha(s-x) + i \sin \alpha(s-x))$$

and the first term in this sum is an even function of α, whereas the second term is an odd function of α. Hence

$$u(x, t) = \frac{1}{\pi} \int_{-\infty}^{\infty} \int_{0}^{\infty} f(s) \cos \alpha(s-x)e^{-\alpha^2 t} \, d\alpha \, ds. \quad \blacksquare \qquad \text{(7.3–18)}$$

We observe that the above solution may be simplified if $f(x)$ is known. In that case the order of integration in Eq. (7.3–18) can be reversed and the integration with respect to s can be accomplished. Examples of this procedure can be found in the exercises. From a practical viewpoint the evaluation of $u(x, t)$ from Eq. (7.3–18) for various values of x and t may be done numerically. The use of the **fast Fourier transform** (FFT) technique is recommended in this case.*

KEY WORDS AND PHRASES

Fourier transform pair   inverse Fourier transform

EXERCISES 7.3

▶ 1. Show that
$$f(x) = e^{-|x|}$$
is absolutely integrable on the real line.

2. If $f(x)$ is absolutely integrable on the real line and
$$\lim_{L \to \infty} f_L(x) = f(x),$$
show that
$$\lim_{L \to \infty} \frac{1}{2L} \int_{-L}^{L} f_L(x)\, dx = 0.$$

3. Carry out the details of the integrations by parts required to obtain Eqs. (7.3–16) and (7.3–17).

4. Obtain the solution
$$\bar{u}(\alpha, t) = \bar{f}(\alpha)e^{-\alpha^2 t}$$
in Example 7.3–1.

▶▶ 5. Determine which of the following functions is absolutely integrable on the real line.

 a) $f(x) = |1 - x|, \quad -1 \le x \le 1$

 b) $f(x) = \sin \pi x,$

 c) $f(x) = x^{1/3}$

6. From Eqs. (7.3–6) and (7.3–7) show that if
$$f(x) = \begin{cases} 1, & \text{for } |x| < 1, \\ 0, & \text{for } |x| > 1, \\ \frac{1}{2}, & \text{for } |x| = 1, \end{cases}$$

then $f(x)$ satisfies the conditions of Theorem 7.3–1, hence has a valid Fourier integral representation for all x given by
$$f(x) = \frac{2}{\pi} \int_0^\infty \frac{\sin \alpha \cos \alpha x}{\alpha}\, d\alpha, \quad -\infty < x < \infty.$$

* See Anthony Ralston and P. Rabinowitz, *A First Course in Numerical Analysis*, 2d ed., p. 263ff. (New York: McGraw-Hill, 1978).

7. Use the result of Exercise 6 to show that

$$\int_0^\infty \frac{\sin 2\alpha}{\alpha} \, d\alpha = \frac{\pi}{2}.$$

8. Obtain the result

$$\int_0^\infty \frac{\sin ax}{x} \, dx = \frac{\pi}{2}, \quad \text{if } a > 0.$$

(*Hint*: Make a change of variable in Exercise 7.)

9. Given the function

$$f(x) = \begin{cases} e^{-x}, & \text{for } x > 0, \\ 0, & \text{for } x < 0, \\ \frac{1}{2}, & \text{for } x = 0, \end{cases}$$

show that $f(x)$ satisfies the conditions of Theorem 7.3–1, hence has a valid Fourier integral representation for all x given by

$$f(x) = \frac{1}{\pi} \int_0^\infty \frac{\cos \alpha x + \alpha \sin \alpha x}{1 + \alpha^2} \, d\alpha, \quad -\infty < x < \infty.$$

10. Obtain the result

$$\int_0^\infty \frac{1}{1 + \alpha^2} \, d\alpha = \frac{\pi}{2}$$

from Exercise 9.

11. Obtain the Fourier integral representation of the function $e^{-|x|}$. For what values of x is the representation valid?

12. Prove that the function

$$f(x) = \begin{cases} 0, & \text{for } x \le 0, \\ \sin x, & \text{for } 0 \le x \le \pi, \\ 0, & \text{for } x \ge \pi, \end{cases}$$

has a Fourier integral representation given by

$$f(x) = \frac{1}{\pi} \int_0^\infty \frac{\cos \alpha x + \cos \alpha(\pi - x)}{1 - \alpha^2} \, d\alpha, \quad -\infty < x < \infty.$$

13. Use the result of Exercise 12 to show that

$$\int_0^\infty \frac{\cos (\pi \alpha / 2)}{1 - \alpha^2} \, d\alpha = \frac{\pi}{2}.$$

▶▶▶ 14. Obtain the Fourier integral representation of the function

$$f(x) = \frac{1}{1 + x^2}.$$

(*Hint*: $\sin \alpha x / (1 + x^2)$ is an odd function of x; then use the result of Exercise 11 with α and x interchanged.)

15. Find the Fourier integral representation of the function

$$f(x) = \frac{\sin x}{x}.$$

16. Prove that the function of Exercise 15 is *not* absolutely integrable. Nevertheless, the Fourier integral representation there is valid. Does this violate Theorem 7.3–1? Explain.

17. Prove that the Fourier transform is a linear transformation.

18. Find $f(x)$ and its transform $\bar{f}(\alpha)$ for the example shown in Fig. 7.3–2.

7.4 APPLICATIONS

In this section we will give some examples showing how Fourier transforms can be used to solve boundary-value problems. Transform methods are especially useful when the region of interest is infinite or semi-infinite. For the latter we define special forms of the Fourier transform.

If $f(x)$ is an **odd** function, then we define the **Fourier sine transform** pair,

$$\mathscr{F}_s(f(x)) = \bar{f}_s(\alpha) = \int_0^\infty f(x) \sin \alpha x \, dx,$$

$$f(x) = \frac{2}{\pi} \int_0^\infty \bar{f}_s(\alpha) \sin \alpha x \, d\alpha. \tag{7.4-1}$$

We use integration by parts to compute the Fourier sine transform of d^2u/dx^2 to obtain

$$\int_0^\infty \frac{d^2u}{dx^2} \sin \alpha x \, dx = \frac{du}{dx} \sin \alpha x \Big|_0^\infty - \alpha \int_0^\infty \frac{du}{dx} \cos \alpha x \, dx$$

$$= -\alpha \int_0^\infty \frac{du}{dx} \cos \alpha x \, dx$$

$$= -\alpha u \cos \alpha x \Big|_0^\infty - \alpha^2 \int_0^\infty u \sin \alpha x \, dx.$$

Hence

$$\mathscr{F}_s\left\{\frac{d^2u}{dx^2}\right\} = \alpha u(0) - \alpha^2 \bar{u}_s(\alpha). \tag{7.4-2}$$

In the above computation we have made the assumptions that u and du/dx approach zero as $x \to \infty$ and that

$$\int_0^\infty |u| \, dx$$

is finite.

If $f(x)$ is an **even** function, then we define the **Fourier cosine transform** pair

$$\mathscr{F}_c(f(x)) = \bar{f}_c(\alpha) = \int_0^\infty f(x) \cos \alpha x \, dx,$$

$$f(x) = \frac{2}{\pi} \int_0^\infty \bar{f}_c(\alpha) \cos \alpha x \, d\alpha. \tag{7.4-3}$$

In the same manner as before we can compute the Fourier cosine transform of d^2u/dx^2 to obtain (Exercise 1)

$$\mathscr{F}_c\left\{\frac{d^2u}{dx^2}\right\} = -u'(0) - \alpha^2\bar{u}_c(\alpha) \tag{7.4-4}$$

under the previous assumptions. When the meaning is clear we will omit the subscripts "s" and "c" in order to simplify the notation.

EXAMPLE 7.4–1 Solve the following boundary-value problem (Fig. 7.4–1).

P.D.E.: $u_t = ku_{xx}$, $0 < x < \infty$, $0 < t$;
B.C.: $u(0, t) = 0$, $0 < t$;
I.C.: $u(x, 0) = f(x)$, $0 < x < \infty$.

Solution. Since $0 < x < \infty$ we can make either an odd or an even extension of the given function $f(x)$. In other words, it "appears" that we may solve the problem by means of either the Fourier sine or the cosine transform. We note, however, that in the boundary condition we are given the value of $u(x, t)$ at $x = 0$. For this reason we choose the *sine* transform. We assume that u and du/dt approach zero as $x \to \infty$ and that both $u(x, t)$ and $f(x)$ are absolutely integrable on $0 < x < \infty$. Then

$$u(x, t) = \frac{2}{\pi}\int_0^\infty \bar{u}(\alpha, t) \sin \alpha x \, d\alpha,$$

and, if we use Eq. (7.4–2), the partial differential equation is transformed into

$$\frac{d\bar{u}}{dt} + \alpha^2 k\bar{u} = 0, \qquad \bar{u}(0) = \bar{f}(\alpha).$$

(See also Example 7.3–1.) Hence

$$\bar{u} = \bar{f}(\alpha)e^{-\alpha^2 kt}$$

and, using the inverse sine transform, we have

$$u(x, t) = \frac{2}{\pi}\int_0^\infty \bar{f}(\alpha)e^{-\alpha^2 kt} \sin \alpha x \, d\alpha.$$

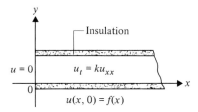

FIGURE 7.4–1

But

$$\bar{f}(\alpha) = \int_0^\infty f(x) \sin \alpha x \, dx$$

$$= \int_0^\infty f(s) \sin \alpha s \, ds.$$

Thus

$$u(x, t) = \frac{2}{\pi} \int_0^\infty f(s) \sin \alpha s \, ds \int_0^\infty e^{-\alpha^2 k t} \sin \alpha x \, d\alpha. \quad \blacksquare$$

EXAMPLE 7.4–2 Solve the following boundary-value problem.

P.D.E.: $u_{tt} = a^2 u_{xx}, \quad -\infty < x < \infty, \quad 0 < t;$

I.C.: $u(x, 0) = f(x), \quad -\infty < x < \infty,$

$u_t(x, 0) = g(x), \quad -\infty < x < \infty.$

Solution. This problem can be solved by a method similar to the one we used in solving the heat equation in Example 7.3–1. We assume that u, $f(x)$, and $g(x)$ all have Fourier transforms. Then the partial differential equation is transformed into

$$\frac{d^2 \bar{u}}{dt^2} + \alpha^2 a^2 \bar{u} = 0$$

with initial conditions

$$\bar{u}(0) = \bar{f}(\alpha) \quad \text{and} \quad \bar{u}'(0) = \bar{g}(\alpha).$$

Hence

$$\bar{u}(\alpha, t) = c_1(\alpha) \cos \alpha a t + c_2(\alpha) \sin \alpha a t,$$

with $c_1(\alpha) = \bar{f}(\alpha)$. Then

$$\bar{u}'(\alpha, t) = -\alpha a \bar{f}(\alpha) \sin \alpha a t + c_2(\alpha) \alpha a \cos \alpha a t$$

and the condition

$$\bar{u}'(0) = \bar{g}(\alpha)$$

implies that

$$c_2(\alpha) = \bar{g}(\alpha)/\alpha a.$$

Thus

$$\bar{u}(\alpha, t) = \bar{f}(\alpha) \cos \alpha a t + \frac{\bar{g}(\alpha) \sin \alpha a t}{\alpha a}$$

and

$$u(x, t) = \frac{1}{2\pi} \int_{-\infty}^\infty \left(\bar{f}(\alpha) \cos \alpha a t + \frac{\bar{g}(\alpha) \sin \alpha a t}{\alpha a} \right) e^{-i\alpha x} \, d\alpha.$$

The last expression can be put into a more easily recognizable form if we make use of the connection between the hyperbolic functions of complex quantities and the circular functions, namely,

$$\sin x = -\tfrac{1}{2} i (e^{ix} - e^{-ix})$$

and
$$\cos x = \tfrac{1}{2}(e^{ix} + e^{-ix}).$$

Using these, we can write the solution to the problem as
$$u(x, t) = \frac{1}{2\pi} \int_{-\infty}^{\infty} \bar{f}(\alpha) \frac{e^{-i\alpha(x-at)} + e^{-i\alpha(x+at)}}{2} \, d\alpha$$

$$+ \frac{1}{2\pi} \int_{-\infty}^{\infty} \bar{g}(\alpha) \frac{e^{-i\alpha(x-at)} - e^{-i\alpha(x+at)}}{2\alpha a i} \, d\alpha.$$

But
$$f(x - at) = \frac{1}{2\pi} \int_{-\infty}^{\infty} \bar{f}(\alpha) e^{-i\alpha(x-at)} \, d\alpha,$$

so that the first integral in the solution is
$$\tfrac{1}{2}(f(x - at) + f(x + at)).$$

Further, from
$$g(x) = \frac{1}{2\pi} \int_{-\infty}^{\infty} \bar{g}(\alpha) e^{-i\alpha x} \, d\alpha,$$

we obtain for arbitrary c and d
$$\int_{c}^{d} g(x) \, dx = \frac{1}{2\pi} \int_{-\infty}^{\infty} \bar{g}(\alpha) \, d\alpha \int_{c}^{d} e^{-i\alpha x} \, dx$$

$$= \frac{1}{2\pi} \int_{-\infty}^{\infty} \bar{g}(\alpha) \, d\alpha \left(\frac{e^{-i\alpha x}}{-i\alpha} \right)\Big|_{c}^{d}$$

$$= \frac{1}{2\pi} \int_{-\infty}^{\infty} \bar{g}(\alpha) \frac{e^{-i\alpha c} - e^{-i\alpha d}}{i\alpha} \, d\alpha,$$

assuming that we can change the order of integration. Hence
$$\frac{1}{2\pi} \int_{-\infty}^{\infty} \bar{g}(\alpha) \frac{e^{-i\alpha(x-at)} - e^{-i\alpha(x+at)}}{2ai\alpha} \, d\alpha = \frac{1}{2a} \int_{x-at}^{x+at} g(s) \, ds$$

and the final solution to our problem takes on the familiar form
$$u(x, t) = \frac{1}{2}(f(x + at) + f(x - at)) + \frac{1}{2a} \int_{x-at}^{x+at} g(s) \, ds \quad \textbf{(7.4–5)}$$

(see Eq. 6.4–8). ∎

An important special application of the Fourier transform will be given next. Suppose that we have a rectangular pulse of duration $2c$ defined by
$$f(t) = \begin{cases} 1, & \text{when } |t| < c, \\ 0, & \text{when } |t| > c, \\ \tfrac{1}{2}, & \text{when } |t| = c. \end{cases}$$

This pulse is shown in Fig. 7.4–2.

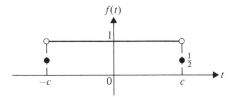

FIGURE 7.4–2 Rectangular pulse.

Since $f(t)$ is piecewise smooth and absolutely integrable, we can compute its Fourier transform (also called the **spectrum**) as follows:

$$\bar{f}(\alpha) = \int_{-c}^{c} f(t)e^{i\alpha t}\,dt = \int_{-c}^{c} e^{i\alpha t}\,dt$$

$$= \frac{e^{i\alpha t}}{i\alpha}\bigg|_{-c}^{c} = \frac{1}{i\alpha}(e^{i\alpha c} - e^{-i\alpha c}).$$

Then, using the relation

$$\frac{e^{ix} - e^{-ix}}{2i} = \sin x,$$

we obtain

$$\bar{f}(\alpha) = \frac{2\sin \alpha c}{\alpha}.$$

The graph of $\bar{f}(\alpha)$, the spectrum of $f(t)$, is shown in Fig. 7.4–3.

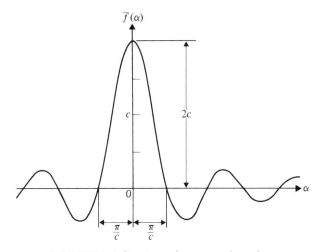

FIGURE 7.4–3 Spectrum of a rectangular pulse.

Note that we can evaluate $\bar{f}(0)$ by means of the well-known limit

$$\lim_{\theta \to 0} \frac{\sin \theta}{\theta} = 1.$$

Hence $\bar{f}(0) = 2c$, as shown in Fig. 7.4–3. As $c \to \infty$, the pulse $f(t)$ becomes increasingly broad in time. On the other hand, the central peak of $\bar{f}(\alpha)$ becomes increasingly higher and narrower. Thus most of the energy of the pulse lies within this central peak of width $2\pi/c$. Therefore, the longer the pulse, the more narrow is the **spectral bandwidth** into which its energy is concentrated.

If we think of α as angular frequency ($\alpha = 2\pi f$) and let $\Delta\alpha$ be the angular frequency separating the maximum of $\bar{f}(\alpha)$ at $\alpha = 0$ from the first zero at $\alpha = \pi/c$, then $\Delta\alpha = \pi/c$. However, c represents the corresponding duration of the pulse in time, call it Δt; hence $(\Delta\alpha)(\Delta t) = \pi$, or $2\pi \, \Delta f \, \Delta t = \pi$ or $\Delta f \, \Delta t = \frac{1}{2}$. Thus there is a *constant* relation between the time duration of the pulse and its frequency bandwidth. In other words, the shape of a pulse in the time domain and the shape of its amplitude spectrum in the frequency domain are not independent. A relation of this kind forms the basis of the Heisenberg* uncertainty principle in quantum mechanics manifested in the uncertainty relation between position and momentum measurements.[†]

A visual formulation of the foregoing relation between the time and frequency domains can be obtained. Imagine that all the harmonic (sinusoidal) components indicated by $\bar{f}(\alpha)$ are graphed as functions of *time*. Now, if the ordinates are added, the result is the single step $f(t)$. Stated in another way, since each harmonic component extends from $t = -\infty$ to $t = \infty$, the linear superposition of harmonic components results in a complete cancellation for $|t| > c$, in unity for $|t| < c$, and in one-half for $|t| = c$. Still another, and more mathematical, way of expressing the foregoing is in the form

$$f(t) = \frac{1}{2\pi} \int_{-\infty}^{\infty} \bar{f}(\alpha) e^{-i\alpha t} \, d\alpha.$$

We mention one important offshoot of the rectangular pulse example. If we let c become infinite, then the energy of the pulse is confined to a *zero* bandwidth. Under this limiting condition $\bar{f}(\alpha)$ becomes the so-called **Dirac delta function** $\delta(\alpha)$, which has the unusual properties

$$\delta(\alpha) = 0 \quad \text{if } \alpha \neq 0$$

and

$$\int_{-\infty}^{\infty} \delta(\alpha) \, d\alpha = 1.$$

---

* Werner (Carl) Heisenberg (1901–1976), a German physicist.
[†] See Kurt B. Wolf, *Integral Transforms in Science and Engineering*, Chapter 7 (New York: Plenum Press, 1979).

The Dirac delta function plays an important role in ordinary differential equations when the forcing functions are impulses (see Section 3.5).

EXAMPLE 7.4–3   Find a harmonic* function $u(x, y)$ in the xy-plane that is bounded for $y \geq 0$ and for which

$$\lim_{y \to 0^+} u(x, y) = f(x).$$

Solution. We solve Laplace's equation,

$$u_{xx} + u_{yy} = 0,$$

by separation of variables. This results in the two ordinary differential equations

$$X'' + \alpha^2 X = 0$$

and

$$Y'' - \alpha^2 Y = 0,$$

with solutions

$$X(x) = c_1 \cos \alpha x + c_2 \sin \alpha x$$

and

$$Y(y) = c_3 e^{\alpha y} + c_4 e^{-\alpha y},$$

respectively. We choose $c_3 = 0$ since the solution must be bounded for $y \geq 0$. Thus the solution has the form

$$u(x, y) = \int_0^\infty e^{-\alpha y}(A \cos \alpha x + B \sin \alpha x) \, d\alpha$$

and we can satisfy the given boundary condition by considering A and B to be functions of α. Then

$$u(x, y) = \int_0^\infty e^{-\alpha y}(A(\alpha) \cos \alpha x + B(\alpha) \sin \alpha x) \, d\alpha$$

and

$$f(x) = \int_0^\infty (A(\alpha) \cos \alpha x + B(\alpha) \sin \alpha x) \, d\alpha.$$

If we refer to Eq. (7.3–5), the last equation can be satisfied if we take

$$A(\alpha) \cos \alpha x + B(\alpha) \sin \alpha x = \frac{1}{\pi} \int_{-\infty}^\infty f(s) \cos \alpha(s - x) \, ds.$$

Then

$$u(x, y) = \frac{1}{\pi} \int_0^\infty \left(\int_{-\infty}^\infty f(s) \cos \alpha(s - x) \, ds \right) e^{-\alpha y} \, d\alpha.$$

---

* A harmonic (or potential) function is one that satisfies Laplace's equation.

Reversing the order of integration, we have

$$u(x, y) = \frac{1}{\pi} \int_{-\infty}^{\infty} \left(\int_{0}^{\infty} e^{-\alpha y} \cos \alpha(s - x) \, d\alpha \right) f(s) \, ds.$$

Thus, evaluating the inner integral (Exercise 5) gives us

$$u(x, y) = \frac{1}{\pi} \int_{-\infty}^{\infty} \frac{yf(s)}{y^2 + (s - x)^2} \, ds. \quad \blacksquare$$

Additional examples of the use of transform methods for solving boundary-value problems will be given in Chapter 8. We conclude this section with a medical application of Fourier analysis.

The use of high-frequency sound waves has become common in medical diagnosis. Sound waves do not appear to cause any side effects and, unlike x-rays, can be safely used to examine the position and development of a fetus. The human heart can be examined by sending short bursts of sound through the chest wall and recording the echoes. Since various portions of the heart have different acoustic impedances, an **echocardiogram** is useful in diagnosis. If $f(t)$ denotes the time-varying amplitude of an echocardiogram waveform, then the *finite* Fourier series representation of $f(t)$ is given by (see Exercise 7)

$$f(t) = \sqrt{C_0} + \sum_{n=1}^{N} \sqrt{C_n} \sin (n\omega_0 t + \theta_n) + e(t).$$

Here ω_0, the fundamental frequency of the series, is 2π times the reciprocal of the heart period. The term $e(t)$ denotes the error resulting from using a finite value of N. The terms C_n and θ_n are the "power" and "phase angle," respectively, of the nth harmonic. Raeside, Chu, and Chandraratna have shown* that a knowledge of only C_0 and C_1 was sufficient to distinguish between those patients with a normal heart and those with three different cardiac conditions.

KEY WORDS AND PHRASES

Fourier sine transform	spectral bandwidth
Fourier cosine transform	Dirac delta function
spectrum	echocardiogram

---

* D. E. Raeside, W. K. Chu, and P. A. N. Chandraratna, "Medical Application of Fourier Analysis," *SIAM Review* **20**, no. 4 (1978), pp. 850–854.

EXERCISES 7.4

▶ 1. Furnish the necessary details for obtaining Eq. (7.4–4).

2. What is the solution of the problem in Example 7.4–1 if

$$f(x) = e^{-x}?$$

3. Solve Example 7.4–1 if the boundary condition $u(0, t) = 0$ is replaced by $u_x(0, t) = 0$, that is, if the end at $x = 0$ is insulated.

4. Solve Example 7.4–3 by using the Fourier transform.

5. In Example 7.4–3 evaluate

$$\int_0^\infty e^{-\alpha y} \cos \alpha(s - x) \, d\alpha$$

and verify the result obtained in the text.

6. Show that the Fourier series for $f(t)$ in the medical application can be written as shown in the text. (*Hint*: Use the trigonometric identity for $\sin(A + B)$.)

▶▶ 7. Find the temperature $u(x, t)$ in a semi-infinite rod, initially at temperature zero, given that one end is kept at constant temperature u_0. State any assumptions that must be made to solve this problem by using the Fourier sine transform.

8. A semi-infinite rod is insulated at one end and has an initial temperature distribution given by e^{-ax}, $a > 0$. Find the temperature $u(x, t)$ by:

a) using separation of variables;

b) using the Fourier cosine transform.

c) Verify that the results in parts (a) and (b) are the same.

9. A semi-infinite rod has one end at temperature zero and an initial temperature distribution given by $f(x)$. Solve this problem given that

$$f(x) = \begin{cases} u_0, & \text{for } 0 < x < L, \\ 0, & \text{otherwise.} \end{cases}$$

(*Hint*: Compare Example 7.4–1.)

10. a) Obtain the spectrum of the function

$$f(t) = \begin{cases} \cos t, & \text{for } -\pi/2 \le t \le \pi/2, \\ 0, & \text{elsewhere.} \end{cases}$$

b) Graph the function and its spectrum.

▶▶▶ 11. Use the Fourier cosine transform to show that the steady-state temperatures in the semi-infinite slab $y > 0$, when the temperature on the edge $y = 0$ is kept at unity over the interval $|x| < c$ and at zero outside this interval, is given by

$$u(x, y) = \frac{1}{\pi}\left(\arctan\left(\frac{c + x}{y}\right) + \arctan\left(\frac{c - x}{y}\right)\right).$$

(*Hint*: The result

$$\int_0^\infty e^{-ax} x^{-1} \sin bx \, dx = \arctan \frac{b}{a},$$

$a > 0$, $b > 0$, may be useful.)

12. Express the solution to Exercise 9 in terms of the **error function**:

$$\text{erf } x = \frac{2}{\sqrt{\pi}} \int_0^x e^{-s^2} \, ds.$$

(*Hint*: Use the result

$$\frac{\sin \alpha x}{\alpha} = \int_0^x \cos \alpha s \, ds$$

and the fact that the Fourier transform of

$$\frac{1}{\sqrt{4\pi a}} e^{-x^2/4a}$$

is $e^{-a\alpha^2}$ for $a > 0$.)

13. An important application of the Fourier transform is the resolution of a *finite* sinusoidal wave into an *infinite* one. If $\sin \omega_0 t$ is clipped so that N cycles remain, then

$$f(t) = \begin{cases} \sin \omega_0 t, & \text{for } |t| < \dfrac{N\pi}{\omega_0}, \\ 0, & \text{elsewhere.} \end{cases}$$

Use the Fourier sine transform to show that

$$\bar{f}_s(\alpha) = \frac{2}{\pi} \int_0^{N\pi/\omega_0} \sin \omega_0 t \sin \alpha t \, dt.$$

Compute $\bar{f}_s(\omega_0)$ and explain the significance of this value.

14. Let $f(x)$ be absolutely integrable on $-\infty < x < \infty$ and have Fourier transform $\bar{f}(\alpha)$. Show the following for $a > 0$.

a) $\dfrac{1}{a} f(x/a)$ has transform $\bar{f}(a\alpha)$.   b) $f(ax)$ has transform $\dfrac{1}{a} \bar{f}(\alpha/a)$.

c) $\bar{f}(x)$ has transform $2\pi f(-\alpha)$.

15. Let $f(x)$ be absolutely integrable on $-\infty < x < \infty$ and have Fourier transform $\bar{f}(\alpha)$. Show the following for any real b.

a) $f(x - b)$ has transform $e^{i\alpha b} \bar{f}(\alpha)$.

b) $\frac{1}{2}(f(x - b) + f(x + b))$ has transform $(\cos \alpha b) \bar{f}(\alpha)$.

16. The transport equation* describing the neutron distribution from a plane source at $x = 0$ in an infinite medium is

$$\theta \frac{\partial \Psi(x, \theta)}{\partial x} + S\psi(x, \theta) = \frac{1}{2} nS \int_{-1}^1 \Psi(x, \theta) \, d\theta + \delta(x)/4\pi.$$

---

*See J. H. Tait, *An Introduction to Neutron Transport Theory*, p. 30 (New York: American Elsevier, 1965).

Here Ψ is the flux of neutrons per cm^2 per sec, θ is the cosine of a solid angle, n is the number of nuclei per cm^3, $\delta(x)$ is the Dirac delta function, and S is the sum of cross sections for elastic scattering, inelastic scattering, and fission. The above equation can be solved by means of the Fourier transform, as outlined in the following steps.

a) Multiply the equation by $\exp(-i\alpha x)$ and integrate over x to obtain

$$(S + i\alpha\theta)\overline{\Psi}(\alpha, \theta) = \frac{1}{2} nS \int_{-1}^{1} \overline{\Psi}(\alpha, \theta)\, d\theta + \frac{1}{4\pi}.$$

b) The right-hand side of the equation in part (a) is independent of θ, call it $F(\alpha)$. Show that

$$F(\alpha) = \frac{1}{2} nSF(\alpha)(1/i\alpha) \log \left(\frac{S + i\alpha}{S - i\alpha}\right) + \frac{1}{4\pi}.$$

c) Using the fact that

$$\frac{1}{2i} \log \left(\frac{S + i\alpha}{S - i\alpha}\right) = \arctan(\alpha/S)$$

is real, obtain

$$F(\alpha) = \frac{1}{4\pi} \left(1 - \frac{nS}{\alpha} \arctan \frac{\alpha}{S}\right)^{-1}.$$

d) Invert $\overline{\Psi}(\alpha, \theta)$ to obtain

$$\Psi(x, \theta) = \frac{1}{8\pi^2} \int_{-\infty}^{\infty} \exp(i\alpha x)(S + i\alpha\theta)^{-1} \left(1 - \frac{nS}{2i\alpha} \log \frac{S + i\alpha}{S - i\alpha}\right)^{-1} d\alpha.$$

17. Obtain the Fourier transform of the product $xy(x)$. (*Hint*: Start with

$$\overline{y}(\alpha) = \int_{-\infty}^{\infty} y(x)e^{i\alpha x}\, dx$$

and differentiate both members with respect to α.)

REFERENCE

Erdelyi, A. (ed.), *Tables of Integral Transforms*, vol. I. New York: McGraw-Hill, 1954.
  An excellent collection of Fourier sine, cosine, and exponential transforms and Laplace transforms.

8
boundary-value problems
in rectangular coordinates

8.1 LAPLACE'S EQUATION

In Section 6.3 we indicated that one of the most common second-order partial differential equations is Laplace's equation,

$$u_{xx} + u_{yy} + u_{zz} = 0. \tag{8.1-1}$$

Up to this point we have dealt with this equation in one and two dimensions. Now we will present an example of a boundary-value problem in three dimensions.

EXAMPLE 8.1–1   Solve the following boundary-value problem.

P.D.E.: $u_{xx} + u_{yy} + u_{zz} = 0$, $0 < x < a$, $0 < y < b$, $0 < z < c$;

B.C.: $u(0, y, z) = u(a, y, z) = 0$, $0 < y < b$, $0 < z < c$,

$u(x, 0, z) = u(x, b, z) = 0$, $0 < x < a$, $0 < z < c$,

$u(x, y, c) = 0$, $u(x, y, 0) = f(x, y)$, $0 < x < a$, $0 < y < b$.

Solution. This problem could arise in attempting to find a potential function inside a rectangular parallelepiped whose four lateral faces and top are at potential zero and the potential on the bottom is a given function of x and y (see Fig. 8.1–1). We shall later specify the properties that this function must have.

We solve the problem by the method of separation of variables. Let

$$u(x, y, z) = X(x)Y(y)Z(z),$$

differentiate, and substitute into Eq. (8.1–1) to obtain

$$X''YZ + XY''Z + XYZ'' = 0.$$

The primes denote *ordinary* derivatives with respect to the arguments of the functions. As usual, we are interested in finding a nontrivial solution so that the last equation can be divided by the product XYZ. Then

$$\frac{Y''}{Y} + \frac{Z''}{Z} = -\frac{X''}{X} = \lambda, \tag{8.1-2}$$

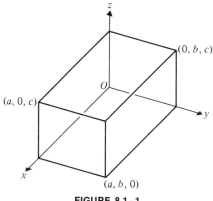

FIGURE 8.1–1

where λ is a separation constant whose exact nature will be determined by the boundary conditions. Note that although the separation of variables is not complete, the left-hand side of Eq. (8.1–2) is free of x, whereas the remaining term contains only x. This condition can exist only if both terms are constant as shown.

The two-point boundary-value problem in X can be written as

$$X'' + \lambda X = 0, \qquad X(0) = 0, \qquad X(a) = 0. \tag{8.1–3}$$

We leave it for the exercises (see Exercise 1) to show that $\lambda = 0$ and $\lambda < 0$ lead to trivial solutions. Thus

$$X(x) = c_1 \cos\left(\sqrt{\lambda} x\right) + c_2 \sin\left(\sqrt{\lambda} x\right)$$

and the condition $X(0) = 0$ implies that $c_1 = 0$, whereas $X(a) = 0$ implies that $\sqrt{\lambda} = n\pi/a$, $n = 1, 2, \ldots$. Hence the eigenvalues of Eq. (8.1–3) are

$$\lambda = n^2 \pi^2 / a^2, \qquad n = 1, 2, \ldots,$$

and the corresponding eigenfunctions are

$$X_n(x) = \sin\left(\frac{n\pi}{a} x\right), \qquad n = 1, 2, \ldots .$$

We have suppressed the arbitrary constant since *any* constant multiple of the above eigenfunction is also a solution of the homogeneous boundary-value problem in (8.1–3).

A second separation of variables now produces

$$\frac{Z''}{Z} - \frac{n^2 \pi^2}{a^2} = -\frac{Y''}{Y} = \mu.$$

The two-point boundary-value problem in Y has exactly the same form as (8.1–3); hence

$$\mu = m^2\pi^2/b^2, \qquad m = 1, 2, \ldots$$

and the eigenfunctions are

$$Y_m(y) = \sin\left(\frac{m\pi}{b} y\right), \qquad m = 1, 2, \ldots.$$

Although both separation constants are functions of the positive integers, they are *independent* of each other.

The problem in Z can be written as

$$Z'' - \pi^2\left(\frac{n^2}{a^2} + \frac{m^2}{b^2}\right)Z = 0, \qquad Z(c) = 0,$$

or, putting

$$\omega_{mn}^2 = \pi^2\left(\frac{n^2}{a^2} + \frac{m^2}{b^2}\right),$$

gives us

$$Z'' - \omega_{mn}^2 Z = 0, \qquad Z(c) = 0.$$

The solution to this problem for specified values of m and n is (see Exercise 2)

$$Z_{mn}(z) = B_{mn} \sinh \omega_{mn}(c - z),$$

where B_{mn} is a constant that depends on m and n. This constant will be evaluated when we apply the last (nonhomogeneous) boundary condition.

Since m and n are independent, we must take a linear combination of a linear combination of products for our final solution. This results in the **double infinite series**

$$u(x, y, z) = \sum_{n=1}^{\infty} \sum_{m=1}^{\infty} B_{mn} \sinh \omega_{mn}(c - z) \sin\left(\frac{m\pi}{b} y\right)\sin\left(\frac{n\pi}{a} x\right), \quad \text{(8.1–4)}$$

which is to be interpreted as follows: For *each* value of n, m takes on values $m = 1, 2, 3, \ldots$, which gives the *double* series.

Applying the final boundary condition, we have

$$\sum_{n=1}^{\infty}\left(\sum_{m=1}^{\infty} B_{mn} \sinh (c\omega_{mn}) \sin\left(\frac{m\pi}{b} y\right)\right) \sin\left(\frac{n\pi}{a} x\right) = f(x, y) \quad \text{(8.1–5)}$$

and the parentheses show that for each m we must have

$$\sum_{n=1}^{\infty} B_{mn} \sinh (c\omega_{mn}) \sin\left(\frac{m\pi}{b} y\right) = \frac{2}{a}\int_0^a f(s, y) \sin\left(\frac{n\pi s}{a}\right)ds. \quad \text{(8.1–6)}$$

In other words, for *each fixed* value of y ($0 < y < b$), Eq. (8.1–5) shows that $f(x, y)$ must be represented as a Fourier sine series in x. On the other hand, the right-hand side of Eq. (8.1–6) is a function of y for each n, call it $F_n(y)$, and we can write

$$\sum_{n=1}^{\infty} B_{mn} \sinh(c\omega_{mn}) \sin\left(\frac{m\pi}{b}y\right) = F_n(y).$$

This shows that $F_n(y)$ is represented by a Fourier sine series in y so that the coefficients are given by

$$B_{mn} \sinh(c\omega_{mn}) = \frac{2}{b} \int_0^b F_n(t) \sin\left(\frac{m\pi}{b}t\right) dt.$$

Thus

$$B_{mn} = \frac{2}{b \sinh(c\omega_{mn})} \int_0^b F_n(t) \sin\left(\frac{m\pi}{b}t\right) dt$$

$$= \frac{4}{ab \sinh(c\omega_{mn})} \int_0^b \int_0^a f(s, t) \sin\left(\frac{n\pi}{a}s\right) \sin\left(\frac{m\pi}{b}t\right) ds\, dt. \quad \textbf{(8.1–7)}$$

Hence the solution to the problem is given by Eq. (8.1–4) with the B_{mn} given in Eq. (8.1–7) and ω_{mn} defined by

$$\omega_{mn} = \pi \sqrt{\frac{n^2}{a^2} + \frac{m^2}{b^2}}.$$

We observe that the function $f(x, y)$ of Example 8.1–1 must satisfy the Dirichlet conditions with respect to both variables; that is, for fixed $y = y_0$ ($0 < y_0 < b$), $f(x, y_0)$ must be a piecewise smooth function of x ($0 < x < a$). Similarly, for fixed $x = x_0$ ($0 < x_0 < a$), $f(x_0, y)$ must be a piecewise smooth function of y ($0 < y < b$). ■

In the next example we will solve Laplace's equation over a semi-infinite domain.

EXAMPLE 8.1–2 Find the potential $V(x, y)$ at any point of a plate bounded by $x = 0$ and $y = b$ if $V(0, y) = V(x, b) = 0$ and $V(x, 0) = f(x)$.

Solution. We state the problem in mathematical terms, being careful to give the ranges of the variables (Fig. 8.1–2).

$$\text{P.D.E.: } V_{xx} + V_{yy} = 0, \qquad 0 < x < \infty, \qquad 0 < y < b;$$
$$\text{B.C.: } V(0, y) = 0, \qquad 0 < y < b,$$
$$V(x, b) = 0, \qquad 0 < x < \infty,$$
$$V(x, 0) = f(x), \qquad 0 < x < \infty.$$

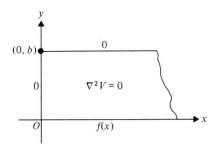

FIGURE 8.1–2

We use the Fourier sine transform and transform x because $V(0, y) = 0$ *and* $0 < x < \infty$. Then

$$\bar{V}(\alpha, y) = \int_0^\infty V(x, y) \sin \alpha x \, dx$$

and, if we use Eq. (7.4–2), the partial differential equation becomes

$$-\alpha^2 \bar{V}(\alpha, y) + \frac{d^2 \bar{V}(\alpha, y)}{dy^2} = 0.$$

The sine transform of $V(x, b) = 0$ is $\bar{V}(\alpha, b) = 0$ and that of $V(x, 0) = f(x)$ is $\bar{V}(\alpha, 0) = \bar{f}(\alpha)$. In order to make these transformations we must assume that

$$\lim_{x \to \infty} V(x, y) \quad \text{and} \quad \lim_{x \to \infty} V_x(x, y)$$

are both zero and that $f(x)$ is absolutely integrable on the real half-line, $0 < x < \infty$. (Why?)

The solution of the second-order, homogeneous, ordinary differential equation with constant coefficients is

$$\bar{V}(\alpha, y) = C_1(\alpha) \cosh (\alpha y) + C_2(\alpha) \sinh (\alpha y).$$

The condition $\bar{V}(\alpha, b) = 0$ yields

$$C_1 = -C_2 \frac{\sinh (\alpha b)}{\cosh (\alpha b)}$$

so that the updated solution becomes (Exercise 5)

$$\bar{V}(\alpha, y) = -C_2(\alpha) \frac{\sinh (\alpha b)}{\cosh (\alpha b)} \cosh (\alpha y) + C_2(\alpha) \sinh (\alpha y)$$

$$= \frac{C_2(\alpha) \sinh \alpha(y - b)}{\cosh (\alpha b)}.$$

From the condition $\bar{V}(\alpha, 0) = \bar{f}(\alpha)$, we now obtain

$$C_2(\alpha) = \frac{-\bar{f}(\alpha) \cosh(\alpha b)}{\sinh(\alpha b)}$$

and the updated solution,

$$\bar{V}(\alpha, y) = \frac{\bar{f}(\alpha) \sinh \alpha(b - y)}{\sinh(\alpha b)}.$$

Using the inverse transform (7.4–1), we get

$$V(x, y) = \frac{2}{\pi} \int_0^\infty \frac{\bar{f}(\alpha) \sinh \alpha(b - y)}{\sinh(\alpha b)} \sin(\alpha x)\, d\alpha$$

$$= \frac{2}{\pi} \int_0^\infty \int_0^\infty f(s) \sin \alpha s \, \frac{\sinh \alpha(b - y)}{\sinh(\alpha b)} \sin(\alpha x)\, ds\, d\alpha.$$

We note that no further simplification is possible unless $f(x)$ is known. (See Exercise 6.) ∎

A function that satisfies Laplace's equation is said to be a **harmonic function**. Harmonic functions have some special properties, which we state in the following theorems.

Theorem 8.1–1 If a function f is harmonic in a bounded region and is zero everywhere on the boundary of the region, then f is identically zero throughout the region.

Theorem 8.1–2 If a function f is harmonic in a bounded region and its normal derivative $\partial f/\partial n$ is zero everywhere on the boundary of the region, then f is a constant in the region.

A **Dirichlet problem** is defined as the problem of finding a function that is harmonic in a region with prescribed values on the boundary of the region.

Theorem 8.1–3 If a Dirichlet problem for a bounded region has a solution, then that solution is unique.

A **Neumann problem** is defined as the problem of finding a function f that is harmonic in a region with prescribed values of the normal derivative $\partial f/\partial n$ on the boundary of the region.

Theorem 8.1–4 If a Neumann problem for a bounded region has a solution, then that solution is unique except possibly for an additive constant.

Dirichlet and Neumann problems arise naturally from the heat-conduction (or diffusion) equation when the steady-state solution is of interest. In two dimensions the heat-conduction equation is

$$u_t = k(u_{xx} + u_{yy}),$$

where u is the temperature and k is a constant called the diffusivity. If we seek the steady-state temperature, that is, the temperature after a long period of time has elapsed, then u is independent of t and the above equation reduces to the two-dimensional Laplace equation. A similar situation holds in three dimensions.

EXAMPLE 8.1–3 Find the steady-state temperatures in a rectangular plate of length a and width b if the edges $x = 0$ and $x = a$ are perfectly insulated, the edge $y = b$ is kept at temperature zero, and the edge $y = 0$ has a temperature distribution given by $(10/a)(a - x)$. See Fig. 8.1–3.

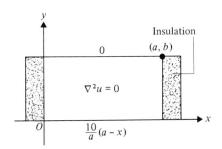

FIGURE 8.1–3

Solution. Recall that, according to Newton's law of cooling (see Section 1.5), the rate of change of temperature at an interface measured normal to the interface is proportional to the difference in temperatures of the two media. "Perfect insulation" implies that this normal derivative must be zero. Accordingly, we can state the following boundary-value problem:

P.D.E.: $u_{xx} + u_{yy} = 0$, $\quad 0 < x < a$, $\quad 0 < y < b$;

B.C.: $\left.\begin{array}{l} u_x(0, y) = 0, \\ u_x(a, y) = 0, \end{array}\right\} \quad 0 < y < b,$

$\left.\begin{array}{l} u(x, b) = 0, \\ u(x, 0) = \dfrac{10}{a}(a - x), \end{array}\right\} \quad 0 < x < a.$

Separation of variables gives a two-point boundary-value problem:

$$X'' + \lambda^2 X = 0, \qquad X'(0) = 0, \qquad X'(a) = 0.$$

The eigenfunctions are (Exercises 9, 10, and 11)

$$X_n(x) = \cos\left(\frac{n\pi}{a}x\right), \qquad n = 0, 1, 2, \ldots .$$

We also have

$$Y_n'' - \frac{n^2\pi^2}{a^2}Y_n = 0, \qquad Y_n(b) = 0,$$

with solutions (Exercise 12)

$$Y_n(y) = \frac{c_n}{\cosh\left(\dfrac{n\pi b}{a}\right)} \sinh \frac{n\pi}{a}(y - b), \qquad n = 1, 2, \ldots .$$

Now that all the homogeneous boundary conditions are satisfied, we can form a linear combination of the eigenfunction products. Then (Exercise 13)

$$u(x, y) = c_0(y - b) + \sum_{n=1}^{\infty} c_n \cos\left(\frac{n\pi}{a}x\right) \frac{\sinh \dfrac{n\pi}{a}(y - b)}{\cosh\left(\dfrac{n\pi b}{a}\right)}.$$

Applying the fourth boundary condition, we get

$$-bc_0 + \sum_{n=1}^{\infty} c_n \frac{\left(-\sinh\left(\dfrac{n\pi b}{a}\right)\right)}{\cosh\left(\dfrac{n\pi b}{a}\right)} \cos\left(\frac{n\pi}{a}x\right) = \frac{10}{a}(a - x)$$

or

$$-bc_0 + \sum_{n=1}^{\infty} a_n \cos\left(\frac{n\pi}{a}x\right) = \frac{10}{a}(a - x).$$

This shows that we can make an even periodic extension of the function $(10/a)(a - x)$ in order to represent it as a Fourier *cosine* series (Fig. 8.1–4). Then,

$$a_0 = \frac{2}{a}\int_0^a \frac{10}{a}(a - s)\,ds = 10;$$

hence, $-bc_0 = \frac{1}{2}a_0 = 5$ from which $c_0 = -5/b$. We also have (Exercise 14)

$$a_n = \frac{2}{a}\int_0^a \frac{10}{a}(a - s)\cos\left(\frac{n\pi}{a}s\right)ds$$

$$= \begin{cases} 0, & \text{if } n \text{ is even,} \\[2mm] \dfrac{40}{n^2\pi^2}, & \text{if } n \text{ is odd.} \end{cases}$$

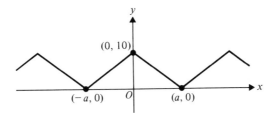

FIGURE 8.1–4 Even periodic extension
(Example 8.1–3).

The solution can now be written as

$$u(x, y) = \frac{5}{b}(b - y) + \frac{40}{\pi^2} \sum_{n=1}^{\infty} \frac{\cos \dfrac{(2n - 1)\pi x}{a} \sinh \dfrac{(2n - 1)\pi}{a}(b - y)}{(2n - 1)^2 \sinh \dfrac{(2n - 1)\pi b}{a}}. \quad \blacksquare$$

We emphasize that even though three of the four boundary conditions in the last example are homogeneous, the algebra involved in obtaining the solution can become quite complicated. For this reason the principle of superposition is recommended when more than one nonhomogeneous boundary condition is present (see Section 6.3).

KEY WORDS AND PHRASES

double infinite series   **Dirichlet problem**
harmonic function   **Neumann problem**

EXERCISES 8.1

▶ **1.** Show that if $\lambda = 0$ or $\lambda < 0$, then the only solution to (8.1–3) is $X(x) = 0$.

2. Show that the solution to

$$Z'' - \omega_{mn}^2 Z = 0, \qquad Z(c) = 0,$$

is

$$B_{mn} \sinh \omega_{mn}(c - z),$$

where B_{mn} is an arbitrary constant whose value depends on both m and n.

3. Solve the problem of Example 8.1–1 if $a = b = c = \pi$.

4. Find B_{mn} in Example 8.1–1 if

$$f(x, y) = xy.$$

5. Obtain the result

$$\bar{V}(\alpha, y) = \frac{c_2(\alpha) \sinh \alpha(y - b)}{\cosh (\alpha b)}$$

from the previous step in Example 8.1–2.

6. Obtain the solution to Example 8.1–2 if $f(x) = e^{-x}$. Does this function meet the necessary requirements?

7. Obtain the solution to Example 8.1–2 if

$$f(x) = \begin{cases} \sin x, & \text{for } 0 < x < \pi, \\ 0, & \text{otherwise.} \end{cases}$$

8. Solve the problem of Example 8.1–2 if $b = 1$ and the condition $V(0, y) = 0$ is replaced by $V_x(0, y) = 0$.

9. Show that $X_0 = 1$ is an eigenfunction corresponding to the eigenvalue $\lambda = 0$ in Example 8.1–3.

10. In Example 8.1–3 show that

$$X'' - \lambda^2 X = 0, \qquad X'(0) = X'(a) = 0,$$

has only the trivial solution.

11. Obtain the eigenfunctions

$$X_n(x) = \cos \left(\frac{n\pi}{a} x \right), \qquad n = 0, 1, 2, \ldots$$

in Example 8.1–3.

12. Obtain the functions

$$Y_n(y) = \frac{c_n}{\cosh \left(\dfrac{n\pi b}{a} \right)} \sinh \frac{n\pi}{a} (y - b), \qquad n = 1, 2, \ldots$$

in Example 8.1–3.

13. In Example 8.1–3 show that

$$Y_0(y) = y - b$$

is the solution corresponding to the eigenvalue $n = 0$.

14. Show that

$$a_{2n-1} = \frac{40}{(2n - 1)^2 \pi^2}, \qquad n = 1, 2, \ldots$$

in Example 8.1–3.

15. Solve the problem of Example 8.1–3 if $a = b = \pi$ and $u(x, 0) = \sin x$.

▶▶ 16. Solve the following boundary-value problem (Fig. 8.1–5).

$$\begin{aligned} \text{P.D.E.:} \quad & u_{xx} + u_{yy} = 0, \quad 0 < x < 1, \quad y > 0; \\ \text{B.C.:} \quad & u(1, y) = 0, \quad y > 0, \\ & u(0, y) = e^{-ay}, \quad a > 0, \quad y > 0, \\ & u_y(x, 0) = 0, \quad 0 < x < 1. \end{aligned}$$

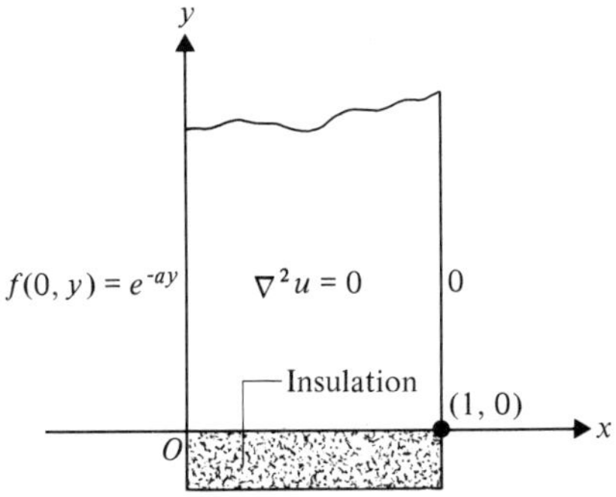

$f(0, y) = e^{-ay}$ $\nabla^2 u = 0$ 0

— Insulation

$(1, 0)$

FIGURE 8.1–5

17. Solve Laplace's equation over the infinite strip, $0 < y < b$, given that the potential is zero when $y = b$ and $f(x)$ when $y = 0$ (Fig. 8.1–6). State any other conditions that must be satisfied. (*Hint*: Use the Fourier transform.)

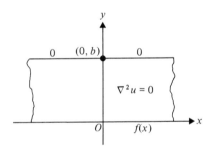

0    $(0, b)$ 0

$\nabla^2 u = 0$

$f(x)$

FIGURE 8.1–6

18. A rectangular plate has its edges $y = 0$ and $y = b$ perfectly insulated, while its edges $x = 0$ and $x = a$ are kept at temperature zero. Use *separation of variables* to find the steady-state temperatures in the plate. Is the result in agreement with what you would expect in the light of Theorems 8.1–1 and 8.1–2?

19. Solve the problem of Example 8.1–1 if on the face $x = 0$, $u(0, y, z) = \sin(\pi y/b)\sin(\pi z/c)$ and on all other faces $u = 0$.

20. In Example 8.1–1 put $a = 1$, $b = 2$, and compute ω_{11}, ω_{12}, ω_{21}, ω_{22}.

21. Let $u(x, y) = f(\phi)$, where $\phi(x, y)$ is a nonconstant, harmonic function. Under what conditions is u harmonic?

22. Show that the separable solutions of Example 8.1–1 can be written in the form

$$\exp(\pm i\alpha x)\exp(\pm i\beta y)\exp(\pm \gamma z).$$

23. Rephrase Theorem 8.1–1, changing "zero" to "a constant c." Explain how this change can be justified.

8.2 THE WAVE EQUATION

In Section 6.4 we derived the vibrating-string equation after making a number of simplifying assumptions. This derivation can be extended in a natural way (Exercise 1) to a **vibrating membrane** (such as a drumhead, for example) to obtain the two-dimensional wave equation

$$u_{tt} = c^2(u_{xx} + u_{yy}). \tag{8.2–1}$$

We are calling the constant c here rather than a as in Section 6.4. It is defined as Tg/w, where T is the (constant) tension per unit length, w is the weight per unit area, and g is the acceleration due to gravity. A boundary-value problem involving this equation in rectangular coordinates will generally have four boundary conditions and two initial conditions specified. We illustrate with an example.

EXAMPLE 8.2–1 Solve the following problem.

P.D.E.: $u_{tt} = c^2(u_{xx} + u_{yy})$, $0 < x < a$, $0 < y < b$, $t > 0$;

B.C.: $u(0, y, t) = u(a, y, t) = 0$, $0 < y < b$, $t > 0$,

$u(x, 0, t) = u(x, b, t) = 0$, $0 < x < a$, $t > 0$;

I.C.: $u_t(x, y, 0) = 0$, $u(x, y, 0) = f(x, y)$, $0 < x < a$, $0 < y < b$.

Solution. Using separation of variables in a different form from that used previously, let

$$u(x, y, t) = \Phi(x, y)T(t)$$

and substitute into the P.D.E. Then

$$\ddot{T}\Phi = c^2 T(\Phi_{xx} + \Phi_{yy}),$$

the dot denoting differentiation with respect to t. Dividing by $c^2\Phi T$ yields the desired separation,

$$\frac{\ddot{T}}{c^2 T} = \frac{(\Phi_{xx} + \Phi_{yy})}{\Phi} = -\lambda^2, \tag{8.2–2}$$

where $-\lambda^2$ is a negative constant.

The physical interpretation of this problem (see Fig. 8.2–1) is the following: A rectangular membrane is fastened along its four edges to a frame that is given an initial displacement in the z-direction defined by $f(x, y)$. Since no damping or other external forces are acting, we

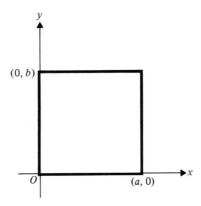

FIGURE 8.2–1

expect that every interior point of the membrane will vibrate indefinitely. For this reason the separation constant is chosen to be *negative*. In other words, we were motivated in choosing the above product form of $u(x, y, t)$ by the knowledge that u must be *periodic* in t.

From Eq. (8.2–2) we have

$$\ddot{T} + c^2 \lambda^2 T = 0, \qquad \dot{T}(0) = 0,$$

using the homogeneous initial condition. The solution to this problem is (Exercise 2)

$$T(t) = \cos (c\lambda t),$$

where λ is still to be determined.

In Eq. (8.2–2) we make the substitution $\Phi(x, y) = X(x) Y(y)$ to obtain

$$\frac{X''}{X} + \frac{Y''}{Y} = -\lambda^2.$$

But this is exactly the problem (including the boundary conditions) that we solved in Example 8.1–1. Hence, solutions of the present problem are products of

$$X_n(x) = \sin \left(\frac{n\pi}{a} x \right), \qquad n = 1, 2, \ldots,$$

$$Y_m(y) = \sin \left(\frac{m\pi}{b} y \right), \qquad m = 1, 2, \ldots,$$

$$T_{mn}(t) = \cos (c\omega_{mn} t),$$

where the m and n are independent and

$$\omega_{mn}^2 = \pi^2 \left(\frac{n^2}{a^2} + \frac{m^2}{b^2} \right).$$

A linear combination of these products summed over m and n is an expression that still contains arbitrary constants. These constants can be found by imposing the final nonhomogeneous initial condition as in Example 8.1–1. Thus,

$$u(x, y, t) = \sum_{m=1}^{\infty} \sum_{n=1}^{\infty} B_{mn} \sin\left(\frac{m\pi}{b} y\right) \sin\left(\frac{n\pi}{a} x\right) \cos\left(c\omega_{mn}t\right)$$

with

$$B_{mn} = \frac{4}{ab} \int_0^b \sin\left(\frac{m\pi}{b} y\right) \int_0^a f(x, y) \sin\left(\frac{n\pi}{a} x\right) dx \, dy.$$

We observe that $f(x, y)$, $f_x(x, y)$, and $f_y(x, y)$ should all be continuous on $0 < x < a$, $0 < y < b$, and vanish on the boundaries of the rectangle. Note that the angular frequency of the vibrating membrane ($c\omega_{mn}$) depends on both m and n and does not change by integral multiples of some fixed basic frequency. Consequently, the vibrating membrane does not produce a musical note as the vibrating string does. (See Exercise 5.) ■

Longitudinal waves

The waves produced in a vibrating string are **transverse waves** (Fig. 8.2–2), that is, the direction of motion of the individual particles of the string is perpendicular to the direction of propagation of the waves (compare Section 6.4). In a solid metal bar, however, elastic waves that are **longitudinal waves** can occur, that is, the direction of motion of the individual particles is the same as the direction of propagation of the waves (Fig. 8.2–3).

Consider a bar of uniform cross section and density extending along the x-axis from the origin to the point $(0, L)$. We assume that the bar is perfectly elastic, meaning that if external forces are applied at the ends so that elongation takes place, tensile forces in the direction of the x-axis will result. If the external forces are now removed, the bar will vibrate longitudinally in accordance with the laws of elasticity.

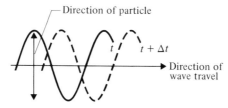

FIGURE 8.2–2 Transverse wave.

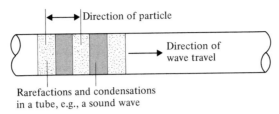

Rarefactions and condensations
in a tube, e.g., a sound wave

FIGURE 8.2–3 Longitudinal wave.

Suppose the bar has density ρ (mass per unit volume), cross section A, and Young's modulus of elasticity E, and let a cross section at x be displaced an amount u as shown in Fig. 8.2–4. From the definition of Young's modulus E, the force on the cross section at x is given by

$$EA \frac{\partial u}{\partial x}$$

since $\partial u/\partial x$ represents the elongation per unit length. On the other hand, the force on this cross section of length Δx is also given by

$$\rho A \, \Delta x \, \frac{\partial^2 u}{\partial t^2},$$

where $\partial^2 u/\partial t^2$ is evaluated at a point between x and $x + \Delta x$, say at the center of mass of the element. The net force per unit length is

$$\frac{EA}{\Delta x} \left(\frac{\partial u(x + \Delta x, t)}{\partial x} - \frac{\partial u(x, t)}{\partial x} \right)$$

and, on equating the two forces and taking the limit as $\Delta x \to 0$, we have

$$\frac{\partial^2 u}{\partial t^2} = \frac{E}{\rho} \frac{\partial^2 u}{\partial x^2} = c^2 \frac{\partial^2 u}{\partial x^2}. \qquad \textbf{(8.2–3)}$$

Thus the small longitudinal vibrations of an elastic rod satisfy the one-dimensional wave equation (Exercise 7).

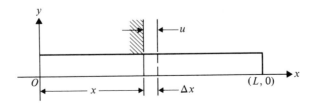

FIGURE 8.2–4 Longitudinal waves in a bar.

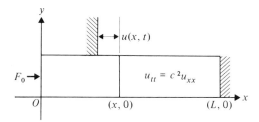

FIGURE 8.2–5

EXAMPLE 8.2–2 The end $x = L$ of a long thin bar is kept fixed and a constant compressive force of F_0 units per unit area is applied to the end $x = 0$ (Fig. 8.2–5). If the bar is initially at rest and unstrained, find the longitudinal displacement of an arbitrary cross section at any time t.

Solution. We need to solve the following boundary-value problem:

P.D.E.: $u_{tt} = c^2 u_{xx}$, $0 < x < L$, $t > 0$, $c^2 = E/\rho$;

  B.C.: $u(L, t) = 0$, $t > 0$ (the end $x = L$ is kept fixed),

           $Eu_x(0, t) = F_0$, $t > 0$ (a constant force is applied at $x = 0$);

  I.C.: $u(x, 0) = 0$, $0 < x < L$ (the bar is initially unstrained),

         $u_t(x, 0) = 0$, $0 < x < L$ (the bar is initially at rest).

We have indicated in the above formulation of the problem how the *physical facts* are translated into *mathematical terms*. Since $u(x, t)$ represents the displacement of the bar, $u(L, t) = 0$ shows that the displacement is zero when $x = L$, that is, the end $x = L$ is kept fixed. As stated earlier, the force on a cross section at x is given by

$$EA \frac{\partial u}{\partial x}$$

and at the end $x = 0$ this force is given as $F_0 A$. From this it follows that

$$Eu_x(0, t) = F_0.$$

Since the bar is initially unstrained, it cannot have any displacement at $t = 0$, that is, $u(x, 0) = 0$. Finally, the bar is initially at rest, meaning that at time $t = 0$, the velocity u_t is zero. The ability to translate physical phenomena into mathematical language is an invaluable aid in problem solving.

Although there are three homogeneous conditions, separation of variables will produce only the trivial solution (Exercise 8). We can obtain a nontrivial solution by making the change of variable

$$u(x, t) = U(x, t) + \phi(x),$$

where $\phi(x)$ is to be determined. Then the problem becomes the following one:

P.D.E.: $U_{tt} = c^2(U_{xx} + \phi''(x))$, $0 < x < L$, $t > 0$;

B.C.: $\left.\begin{array}{l} U(L, t) + \phi(L) = 0, \\ EU_x(0, t) + E\phi'(0) = F_0, \end{array}\right\}$ $t > 0$;

I.C.: $\left.\begin{array}{l} U(x, 0) + \phi(x) = 0, \\ U_t(x, 0) = 0, \end{array}\right\}$ $0 < x < L$.

If we now set

$$\phi''(x) = 0, \qquad \phi(L) = 0, \qquad \phi'(0) = F_0/E,$$

then $\phi(x) = (F_0/E)(x - L)$ (Exercise 9). Thus the problem is transformed into the following familiar one:

P.D.E.: $U_{tt} = c^2 U_{xx}$, $0 < x < L$, $t > 0$;

B.C.: $\left.\begin{array}{l} U(L, t) = 0, \\ U_x(0, t) = 0, \end{array}\right\}$ $t > 0$;

I.C.: $\left.\begin{array}{l} U(x, 0) = \dfrac{F_0}{E}(L - x), \\ U_t(x, 0) = 0, \end{array}\right\}$ $0 < x < L$.

This problem has solution (see Eq. 6.4–6 in Example 6.4–1)

$$U(x, t) = \frac{F_0}{2E}(\Phi(x + ct) + \Phi(x - ct)),$$

where $\Phi(x)$ is the odd periodic extension of $x - L$ and is defined for $-\infty < x < \infty$ (see Fig. 8.2–6). The solution to the original problem is given by

$$u(x, t) = \frac{F_0}{E}(x - L) + \frac{F_0}{2E}(\Phi(x + ct) + \Phi(x - ct)). \qquad \textbf{(8.2–4)}$$

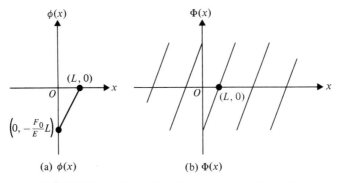

(a) $\phi(x)$ (b) $\Phi(x)$

FIGURE 8.2–6 $\phi(x)$ and its odd periodic extension $\Phi(x)$.

We leave it for the exercises (see Exercises 10 and 11) to show that the above solution satisfies the P.D.E. and all conditions. ■

In the problem of Example 8.2–2 we showed how a nonhomogeneous boundary condition can be transferred from one independent variable to another by a change in the dependent variable. Other examples of this technique will be given in Section 8.3.

Another case in which the one-dimensional wave equation applies will be considered next. Let θ be the angular displacement of a cross section of a uniform circular shaft from some equilibrium position. If θ is small, then, according to the theory of elasticity, θ satisfies the differential equation

$$\theta_{tt} = c^2 \theta_{xx}, \tag{8.2–5}$$

with $c^2 = Gg/\rho$, where G is the shear modulus, ρ is the density (mass per unit volume), and g is the acceleration due to gravity.

EXAMPLE 8.2–3   Suppose that a shaft of circular cross section and length L is clamped at one end while the other end is twisted, then freed. As the shaft oscillates, the free end is clamped at time $t = t_0$. At this instant the angular velocity is $\omega_0 x/L$ and the angle θ is zero. State all conditions and solve the boundary-value problem.

Solution. We take coordinates so that the shaft lies along the x-axis with the free end at $x = L$. Then we have the following.

$$\text{P.D.E.: } \theta_{tt} = c^2 \theta_{xx}, \quad 0 < x < L, \quad t > 0;$$
$$\text{B.C.: } \theta(0, t) = 0, \quad t > 0 \quad \text{(clamped at one end)},$$
$$\theta(L, t) = 0, \quad t > t_0 \quad \text{(free end clamped at } t = t_0\text{)};$$
$$\text{I.C.: } \left.\begin{array}{l} \theta(x, t_0) = 0, \\ \theta_t(x, t_0) = \omega_0 x/L, \end{array}\right\} \quad 0 < x < L.$$

We can simplify the problem considerably by taking $t_0 = 0$ so that the time coordinate is shifted by an amount t_0. Then, using separation of variables, we find

$$\frac{\ddot{T}}{c^2 T} = \frac{X''}{X} = -\lambda^2.$$

The separation constant must be negative since we are dealing with periodic oscillations in time. Thus the two-point boundary-value problem

$$X'' + \lambda^2 X = 0, \qquad X(0) = 0, \qquad X(L) = 0,$$

has solutions

$$X_n(x) = \sin\left(\frac{n\pi}{L}x\right), \qquad n = 1, 2, \ldots,$$

whereas the initial-value problem

$$\ddot{T}_n + \frac{n^2\pi^2c^2}{L^2}\,T_n = 0, \qquad T_n(0) = 0,$$

has solutions

$$T_n(t) = \sin\left(\frac{n\pi c}{L}\,t\right), \qquad n = 1, 2, \dots .$$

Hence

$$\theta(x, t) = \sum_{n=1}^{\infty} b_n \sin\left(\frac{n\pi}{L}\,x\right)\sin\left(\frac{n\pi c}{L}\,t\right)$$

with (Exercise 14)

$$b_n = \frac{2}{L}\int_0^L \frac{\omega_0 s}{L}\sin\left(\frac{n\pi}{L}\,s\right)ds$$

$$= \frac{2\omega_0 L}{c\pi^2}\frac{(-1)^{n-1}}{n^2}. \quad \blacksquare$$

KEY WORDS AND PHRASES

vibrating membrane   **longitudinal waves**
transverse waves

EXERCISES 8.2

▶ 1. State the simplifying assumptions that must be made in order to derive the two-dimensional wave equation (8.2–1). (*Hint*: See Section 6.4.)

2. Find the solutions of

$$\ddot{T} + c^2\lambda^2 T = 0, \qquad \dot{T}(0) = 0.$$

3. Carry out the remaining details in Example 8.2–1.

4. Solve the problem of Example 8.2–1 given that $f(x, y) = xy(a - x)(b - y)$. Show that this function satisfies the conditions in the example.

5. If

$$u(x, y, 0) = k \sin\frac{\pi x}{a}\sin\frac{\pi y}{b},$$

where k is a constant, in Example 8.2–1, obtain the solution. Note that under this condition the vibrating membrane *does* produce a musical tone. What is the frequency of the tone?

6. The frequencies ω_{mn} in Example 8.2–1 are called **characteristic frequencies**. List the first six characteristic frequencies of the vibrating rectangular membrane, that is, $\omega_{11}, \omega_{12}, \omega_{21}, \omega_{22}, \omega_{13}, \omega_{31}$, if $a = b = \pi$.

7. Show that E/ρ in Eq. (8.2–3) has dimensions of velocity.

8. Use separation of variables in Example 8.2–2 to show that the equation in t has only the trivial solution regardless of the choice of the (real) separation constant.

9. Solve the two-point boundary-value problem

$$\phi''(x) = 0, \qquad \phi(L) = 0, \qquad \phi'(0) = F_0/E.$$

10. Verify that Eq. (8.2–4) satisfies the one-dimensional wave equation.

11. Verify that Eq. (8.2–4) satisfies the boundary and initial conditions of the problem in Example 8.2–2. (*Hint*: Keep in mind that Φ is an *odd* function.)

12. Find the eigenvalues and eigenfunctions of the two-point boundary-value problem

$$X'' + \lambda^2 X = 0, \qquad X(0) = 0, \qquad X(L) = 0.$$

13. Solve

$$\ddot{T} + \frac{n^2\pi^2 c^2}{L^2} T = 0, \qquad T(0) = 0.$$

14. Obtain the coefficients b_n in the problem of Example 8.2–3.

▶▶ 15. **a)** Show that the constant c in Eq. (8.2–1) has dimensions of velocity.

b) Deduce the dimensions of E in Eq. (8.2–3) and of G in Eq. (8.2–5).

16. If the initial conditions of the problem of Example 8.2–1 are replaced by

$$u_t(x, y, 0) = g(x, y), \qquad u(x, y, 0) = 0,$$

what is the solution $u(x, y, t)$?

17. Use the principle of superposition to solve the problem of Example 8.2–1 if the initial conditions there are replaced by $u(x, y, 0) = f(x, y), u_t(x, y, 0) = g(x, y)$.

18. **a)** Referring to Exercise 5, find

$$u_t\left(\frac{a}{2}, \frac{b}{2}, t\right).$$

b) Interpret the result in part (a) physically.

19. Solve Example 8.2–3 if $\theta_t(x, t_0) = k$, a constant, all other conditions remaining the same.

20. Solve the vibrating-string equation for the case of the plucked string of length L, that is, for initial conditions

$$u(x, 0) = \begin{cases} \dfrac{2h}{L} x, & \text{for } 0 \le x \le \dfrac{L}{2}, \\[2mm] \dfrac{2h}{L}(L - x), & \text{for } \dfrac{L}{2} \le x \le L, \end{cases}$$

$$u_t(x, 0) = 0.$$

(See Fig. 8.2–7.)

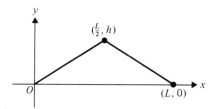

FIGURE 8.2-7

8.3 THE DIFFUSION EQUATION

A classic example of a second-order partial differential equation of parabolic type is the diffusion equation. It has the form

$$u_t = k\nabla^2 u, \tag{8.3-1}$$

where ∇^2 is the Laplacian operator,

$$\frac{\partial^2}{\partial x^2} + \frac{\partial^2}{\partial y^2} + \frac{\partial^2}{\partial z^2}.$$

We show the operator in three dimensions and in rectangular coordinates here (compare Section 5.4).

Since heat may be considered as a "fluid" inside matter, the diffusion equation plays an important role in problems of heat conduction. In this application u represents the temperature in a body, whereas the constant k, called the **thermal diffusivity**, is defined as

$$k = \frac{K}{\sigma\rho},$$

where K is the *thermal conductivity*, σ is the *specific heat*, and ρ is the *density* (mass per unit volume). At times we will use $k = 1$ since this special value can be achieved by making a scale change in t. (See Exercise 1.)

Let S be the smooth surface of a solid body. Assume that heat is transferred within the body by *conduction* and "flows" from regions of higher temperature to regions of lower temperature. At a point of the surface S we define the *flux* of heat Φ as the quantity of heat per unit area, per unit time, that is being conducted across S at the point. The flux Φ is proportional to the directional derivative of the temperature u in a direction *normal* to the surface S, that is,

$$\Phi = -K\frac{\partial u}{\partial n}, \tag{8.3-2}$$

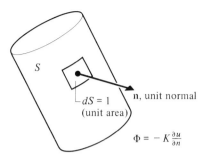

FIGURE 8.3-1 Flux.

where the proportionality constant K (>0) is the thermal conductivity and $\partial u / \partial n$ is the rate of change of temperature along the *outward* pointing normal (Fig. 8.3–1). The SI units of flux are calories/cm²/sec.

If heat transfer takes place between a body and the surrounding medium, then Newton's law of cooling (see Section 1.5) states that the flux is proportional to the temperature difference between the body and the surrounding medium. Use of the word "cooling" implies that the body is at a higher temperature than the surrounding medium, but the reverse may be true as well.

In Examples 8.3–1, 8.3–2, and 8.3–3 we will illustrate how various boundary conditions can be treated. In these examples we will be considering a slender rod insulated so that heat conduction takes place in only the x-direction. We will thus be dealing with the *one-dimensional heat equation* in rectangular coordinates. The equation also applies to a semi-infinite slab whose edges are kept at constant temperatures or insulated.

EXAMPLE 8.3–1 Solve the following problem.

$$\text{P.D.E.: } u_t = u_{xx}, \quad 0 < x < L, \quad t > 0;$$

$$\text{B.C.: } \left. \begin{array}{l} -u_x(0, t) = 0, \\ u(L, t) = 0, \end{array} \right\} \quad t > 0;$$

$$\text{I.C.: } u(x, 0) = f(x), \quad 0 < x < L.$$

This is the problem of heat conduction in a bar whose left end is perfectly insulated, whose right end is kept at temperature zero, and along which there is an initial temperature distribution defined by $f(x)$. See Fig. 8.3–2.

Solution. Using separation of variables, let $u(x, t) = X(x)T(t)$ and substitute into the partial differential equation. Then

$$X(x)T'(t) = X''(x)T(t)$$

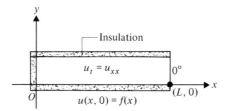

FIGURE 8.3–2

and, dividing by $X(x)T(t)$, we get

$$\frac{T'(t)}{T(t)} = \frac{X''(x)}{X(x)} = -\lambda^2 \quad \text{(Exercise 2)}.$$

The two-point boundary-value problem in $X(x)$ is

$$X'' + \lambda^2 X = 0, \qquad X'(0) = 0, \qquad X(L) = 0,$$

with solutions (Exercise 3)

$$X_n(x) = \cos \frac{(2n-1)}{L} \frac{\pi}{2} x, \quad n = 1, 2, \dots .$$

Since

$$T'_n(t) + \frac{(2n-1)^2 \pi^2}{4L^2} T_n(t) = 0$$

has solutions

$$T_n(t) = \exp\left(\frac{-\pi^2 (2n-1)^2}{4L^2} t\right),$$

we take for $u(x, t)$ the linear combination

$$u(x, t) = \sum_{n=1}^{\infty} a_{2n-1} \exp\left(\frac{-\pi^2 (2n-1)^2}{4L^2} t\right) \cos \frac{(2n-1)\pi}{2L} x. \quad \textbf{(8.3–3)}$$

Then, applying the I.C., we have

$$\sum_{n=1}^{\infty} a_{2n-1} \cos \frac{(2n-1)\pi}{2L} x = f(x).$$

If $f(x)$ meets the required conditions, then it can be expanded in a Fourier cosine series so that

$$a_{2n-1} = \frac{2}{L} \int_0^L f(s) \cos \frac{(2n-1)\pi}{2L} s \, ds, \qquad n = 1, 2, \dots . \quad \textbf{(8.3–4)}$$

The complete solution is given by Eq. (8.3–3) with a_{2n-1} defined in Eq. (8.3–4). ∎

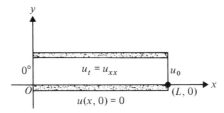

FIGURE 8.3–3

The next example will illustrate how a **nonhomogeneous bound-ary condition** can be made homogeneous.

EXAMPLE 8.3–2 Solve the following problem. See Fig. 8.3–3.

$$\text{P.D.E.: } u_t = u_{xx}, \quad 0 < x < L, \quad t > 0;$$

$$\left.\begin{array}{l} \text{B.C.: } u(0, t) = 0, \\ \qquad\ \ u(L, t) = u_0, \text{ a constant,} \end{array}\right\} \quad t > 0;$$

$$\text{I.C.: } u(x, 0) = 0, \quad 0 < x < L.$$

Solution. The method of separation of variables fails here because it leads to the trivial solution (Exercise 6). If we make a change of variable, however, the difficulty can be overcome. Accordingly, let

$$u(x, t) = U(x, t) + \phi(x),$$

so that the problem becomes the following one.

$$\text{P.D.E.: } U_t = U_{xx} + \phi''(x), \quad 0 < x < L, \quad t > 0;$$

$$\left.\begin{array}{l} \text{B.C.: } U(0, t) + \phi(0) = 0, \\ \qquad\ \ U(L, t) + \phi(L) = u_0, \end{array}\right\} \quad t > 0;$$

$$\text{I.C.: } U(x, 0) + \phi(x) = 0, \quad 0 < x < L.$$

Now, if we choose $\phi(x)$ so that it satisfies

$$\phi''(x) = 0, \qquad \phi(0) = 0, \qquad \phi(L) = u_0, \qquad \textbf{(8.3–5)}$$

then the resulting problem in $U(x, t)$ is one that can be solved by the method of separation of variables. The details are left for the exercises. (See Exercises 7 and 8.) ∎

The technique used in solving the problem of Example 8.3–2 is an illustration of a powerful method in mathematics, that is, reducing a problem to one whose solution is known. Other examples of this technique occur in the integral calculus where integration by substitution is used and in ordinary differential equations where certain second-order equations are reduced to first-order ones. See also Exercises 19 and 20.

We observe that an alternative way of looking at Example 8.3–2 is the following. One end of a bar is kept at temperature zero, the other at temperature u_0, a constant. The entire bar is initially at temperature zero. It is apparent from physical considerations that as $t \to \infty$ the temperature $u(x)$ for $0 < x \le L$ will approach u_0. But this is just another way of saying that u_0 is the solution of the **steady-state** (independent of time) problem. This problem can be stated as

$$u''(x) = 0, \qquad u(0) = 0, \qquad u(L) = u_0.$$

Thus the function $\phi(x)$ in (8.3–5) is the solution to the steady-state problem. When this solution is added to the **transient solution**, the result is the complete solution obtained in Exercise 8.

EXAMPLE 8.3–3 A cylindrical rod of length L is initially at temperature $A \sin x \; (A > 0)$. Its left end is kept at temperature zero, while heat is transferred from the right end into the surroundings at temperature zero (Fig. 8.3–4). Find $u(x, t)$.

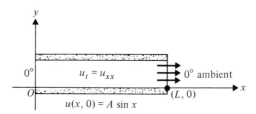

FIGURE 8.3–4

Solution. Here the problem to be solved can be expressed as follows:

P.D.E.: $u_t = u_{xx}, \quad 0 < x < L, \quad t > 0;$

B.C.: $\left. \begin{aligned} &u(0, t) = 0, \\ &u_x(L, t) = hu(L, t), \quad h > 0, \end{aligned} \right\} \quad t > 0;$

I.C.: $u(x, 0) = A \sin x, \quad 0 < x < L.$

Using separation of variables, we have

$$\frac{X''}{X} = \frac{T'}{T} = -\lambda^2,$$

hence

$$X(x) = c_1 \cos \lambda x + c_2 \sin \lambda x.$$

The first boundary condition implies that $c_1 = 0$, whereas the second results in

$$\tan \lambda L = \lambda/h. \tag{8.3–6}$$

Thus a solution can be written as

$$u(x, t) = c_2 e^{-\lambda^2 t} \sin \lambda x$$

and, if we apply the initial condition, this becomes

$$u(x, 0) = c_2 \sin \lambda x = A \sin x,$$

which can be satisfied by choosing $\lambda = 1$ and $c_2 = A$. Finally, we have

$$u(x, t) = A e^{-t} \sin x \qquad (8.3-7)$$

as the solution. Verification of this result is left for the exercises. (See Exercise 10.) ∎

It should be obvious that the *form* of the initial condition in the last example had a lot to do with the ease of obtaining the solution (8.3–7). If the initial condition had been some other function, we would not have been able to express it as an infinite series of orthonormal functions since such functions did not appear to be available. We will pursue this phase of problem solving in Section 8.5.

We conclude this section by mentioning that the one-dimensional Fermi* age equation for the **diffusion of neutrons** in a medium such as graphite is

$$\frac{\partial^2 q(x, \tau)}{\partial x^2} = \frac{\partial q(x, \tau)}{\partial \tau}.$$

Here q represents the number of neutrons that "slow down" (that is, fall below some given energy level) per second per unit volume. The Fermi age, τ, is a measure of the energy loss.

KEY WORDS AND PHRASES

thermal diffusivity   diffusion of neutrons
nonhomogeneous boundary
    condition

EXERCISES 8.3

▶ 1. Show that a time scale change given by $\tau = kt$ will transform the diffusion equation (8.3–1) into

$$u_\tau = \nabla^2 u.$$

2. In Example 8.3–1 show that:

a) choosing $\lambda = 0$ leads to the trivial solution;

b) choosing $+\lambda^2$ for the separation constant leads to the trivial solution.

---

* Enrico Fermi (1901–1954), an Italian nuclear physicist, who supervised the construction of the first nuclear reactor.

3. Find the eigenvalues and eigenfunctions of the problem

$$X'' + \lambda^2 X = 0, \qquad X'(0) = 0, \qquad X(L) = 0.$$

4. Explain why there is no a_0-term in the solution to Example 8.3–1.

5. Verify the solution to Example 8.3–1 completely.

6. Apply the method of separation of variables to the problem of Example 8.3–2 and show that the result is the trivial solution. (*Hint*: Examine the problem in $T(t)$.)

7. Show that the solution of Eq. (8.3–5) is

$$\phi(x) = \frac{u_0}{L} x.$$

8. Using the result in Exercise 7, solve the problem of Example 8.3–2.

9. In Example 8.3–3 show that:

   a) $\lambda = 0$ leads to the trivial solution;

   b) the separation constant $+\lambda^2$ leads to the trivial solution. [*Hint*: Show that $f(\lambda) = \tanh \lambda L - (\lambda/h)$ is zero for $\lambda = 0$ but nowhere else by examining $f'(\lambda)$.]

10. Verify the result in Eq. (8.3–7).

▶▶ 11. Solve the following problem.

$$\text{P.D.E.: } u_t = u_{xx}, \quad 0 < x < L, \quad t > 0;$$

$$\text{B.C.: } \left.\begin{array}{l} u_x(L, t) = 0, \\ u(0, t) = 0, \end{array}\right\} \quad t > 0;$$

$$\text{I.C.: } u(x, 0) = f(x), \quad 0 < x < L.$$

12. Solve the problem of Exercise 11 given that

$$f(x) = \begin{cases} 0, & 0 \le x < \dfrac{L}{2}, \\ L - x, & \dfrac{L}{2} \le x \le L. \end{cases}$$

13. A thin cylindrical bar of length L has both ends perfectly insulated and has an initial temperature distribution given by $f(x)$. (See Fig. 8.3–5.) Find the temperature at any point in the bar at any time t.

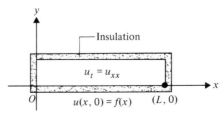

FIGURE 8.3–5

14. If heat is generated uniformly throughout a semi-infinite slab of width L at a constant rate C, then the one-dimensional heat equation has the form

$$u_t = u_{xx} + C, \qquad C > 0.$$

Solve this equation under the assumptions that the edges $x = 0$ and $x = L$ of the slab are kept at temperature zero, the faces of the slab are insulated, and the initial temperature distribution is $f(x)$. (*Hint:* Make a change of variable by putting $u(x, t) = U(x, t) + \phi(x)$.)

15. In Exercise 14 put $f(x) = 0$ and obtain the solution. (*Note:* This problem is a simplification of one that occurs in the manufacture of plywood, where heat is supplied by means of high-frequency heating.)

16. Solve the following problem.

$$\text{P.D.E.:}\ u_t = u_{xx}, \quad 0 < x < 2, \quad t > 0;$$

$$\left.\begin{array}{l} \text{B.C.:}\ u_x(0, t) = 0, \\ \phantom{\text{B.C.:}\ }u(2, t) = 0, \end{array}\right\}\ t > 0;$$

$$\text{I.C.:}\ u(x, 0) = ax, \quad a > 0, \quad 0 < x < 2.$$

17. Solve the following problem.

$$\text{P.D.E.:}\ u_t = u_{xx}, \quad 0 < x < 1, \quad t > 0;$$

$$\left.\begin{array}{l} \text{B.C.:}\ u(0, t) = 0, \\ \phantom{\text{B.C.:}\ }u_x(1, t) = 0, \end{array}\right\}\ t > 0;$$

$$\text{I.C.:}\ u(x, 0) = u_0 x, \quad u_0 > 0, \quad 0 < x < 1.$$

18. If $f(x)$ in Example 8.3–1 is defined as

$$f(x) = \begin{cases} 0, & 0 \le x < \dfrac{L}{2}, \\[2mm] L, & \dfrac{L}{2} \le x \le L, \end{cases}$$

obtain the result

$$\begin{aligned} u(x, t) = \frac{2L}{\pi} \Bigg(& \exp\left(\frac{-\pi^2 t}{4L^2}\right)(2 - \sqrt{2}) \cos\left(\frac{\pi x}{2L}\right) \\ & - \exp\left(\frac{-3\pi^2 t}{4L^2}\right)\left(\frac{2 + \sqrt{2}}{3}\right) \cos\left(\frac{3\pi x}{2L}\right) \\ & + \exp\left(\frac{-5\pi^2 t}{4L^2}\right)\left(\frac{2 + \sqrt{2}}{5}\right)\Bigg) \cos\left(\frac{5\pi x}{2L}\right) + \cdots . \end{aligned}$$

19. The faces $x = 0$, $x = a$, $y = 0$, and $y = b$ of a semi-infinite, solid, rectangular parallelepiped are kept at temperature zero and there is an initial temperature distribution on the bottom face given by $f(x, y)$. Find the temperature in the interior at any time t. (*Hint:* Compare Example 8.1–1.)

20. A rod one unit in length is perfectly insulated along its length so that heat conduction can take place in only the x-direction. The left end is kept at temperature zero and the right end is insulated. If the temperature is given initially by the function ax^2 (a is a positive constant), find the temperature in the rod at any time t. (*Hint:* See Exercise 17.)

▶▶▶ 21. A rod of length L is insulated *at both ends* only and the initial temperature distribution is given by $f(x)$. If there is a linear surface heat transfer between the rod and its surroundings, then the equation

$$v_t(x, t) = kv_{xx}(x, t) - hv(x, t),$$

where h is a positive constant, applies (Fig. 8.3–6). Using the substitution

$$v(x, t) = \exp(-ht)\, u(x, t),$$

reduce this problem to one that has been previously solved. (Compare Exercise 13.)

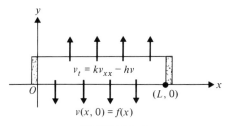

$$v_t = kv_{xx} - hv$$

$$(L, 0)$$

$$v(x, 0) = f(x)$$

FIGURE 8.3–6

22. Solve Exercise 21, given that the ends of the rod are kept at temperature zero, rather than insulated.

8.4 TRANSFORM METHODS

Whenever a boundary-value problem is to be solved over an infinite or a semi-infinite domain, transform methods provide a good approach. In Sections 7.3 and 7.4 we presented some examples using the Fourier transform. We shall give further examples in this section, using both the Fourier transform and the Laplace transform.

EXAMPLE 8.4–1   Find the bounded, harmonic function $v(x, y)$ in the semi-infinite strip $0 < x < c$, $y > 0$ that satisfies the following conditions:

a) $v(0, y) = 0$

b) $v_y(x, 0) = 0$

c) $v_x(c, y) = f(y)$.

Interpret this problem physically.

Solution. Laplace's equation

$$v_{xx} + v_{yy} = 0, \qquad 0 < x < c, \qquad y > 0$$

is to be solved. We use the Fourier *cosine* transform and transform the variable y because of boundary condition (b). Thus (refer to Eq. 7.4–4)

$$\frac{d^2 \bar{v}(x, \alpha)}{dx^2} - \alpha^2 \bar{v}(x, \alpha) = 0$$

(a′) $\bar{v}(0, \alpha) = 0,$ (c′) $\dfrac{d\bar{v}(c, \alpha)}{dx} = \bar{f}(\alpha),$

where

$$\bar{f}(\alpha) = \int_0^\infty f(y) \cos \alpha y \; dy.$$

The general solution of the second-order ordinary differential equation is

$$\bar{v}(x, \alpha) = c_1(\alpha) \cosh (\alpha x) + c_2(\alpha) \sinh (\alpha x),$$

with $c_1(\alpha) = 0$ by condition (a′), whereas

$$c_2(\alpha) = \frac{\bar{f}(\alpha)}{\alpha \cosh (\alpha c)}$$

by condition (c′). Hence,

$$\bar{v}(x, \alpha) = \frac{\bar{f}(\alpha) \sinh (\alpha x)}{\alpha \cosh (\alpha c)}$$

and the inverse transform is

$$v(x, y) = \frac{2}{\pi} \int_0^\infty \frac{\bar{f}(\alpha) \sinh (\alpha x)}{\alpha \cosh (\alpha c)} \cos (\alpha y) \; d\alpha$$

$$= \frac{2}{\pi} \int_0^\infty \frac{\sinh (\alpha x) \cos (\alpha y)}{\alpha \cosh (\alpha c)} \; d\alpha \int_0^\infty f(s) \cos (\alpha s) \; ds.$$

Referring to Fig. 8.4–1, we can interpret the problem as finding the steady-state temperatures in a semi-infinite slab of width c, given that the left edge is kept at temperature zero, the bottom edge is perfectly insulated, and the flux across the right edge is a given function $f(y)$. ∎

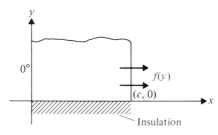

0° $f(y)$ $(c, 0)$ Insulation

FIGURE 8.4–1

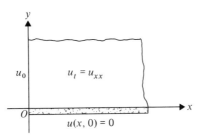

FIGURE 8.4-2

In the next example we shall use the Fourier sine transform.

EXAMPLE 8.4-2 Solve the following problem (Fig. 8.4-2).

$$\text{P.D.E.: } u_t = u_{xx}, \quad x > 0, \quad t > 0;$$
$$\text{B.C.: } u(0, t) = u_0, \quad t > 0;$$
$$\text{I.C.: } u(x, 0) = 0, \quad x > 0.$$

Solution. We use the Fourier sine transform to transform u_{xx}. Thus (refer to Eq. 7.4–2),

$$\frac{d\bar{u}(\alpha, t)}{dt} = \alpha u_0 - \alpha^2 \bar{u}(\alpha, t)$$

and this first-order, nonhomogeneous, ordinary differential equation has solution

$$\bar{u}(\alpha, t) = \frac{u_0}{\alpha} (1 - \exp(-\alpha^2 t)),$$

using the initial condition $\bar{u}(\alpha, 0) = 0$ (Exercise 3). The inverse Fourier sine transform of $\bar{u}(\alpha, t)$ is

$$u(x, t) = \frac{2u_0}{\pi} \int_0^\infty (1 - \exp(-\alpha^2 t)) \sin \alpha x \, \frac{d\alpha}{\alpha}.$$

But, from Exercise 8 in Section 7.3,

$$\int_0^\infty \frac{\sin \alpha x \, d\alpha}{\alpha} = \frac{\pi}{2} \quad \text{if } x > 0.$$

Hence,

$$u(x, t) = u_0 - \frac{2u_0}{\pi} \int_0^\infty \exp(-\alpha^2 t) \sin \alpha x \, \frac{d\alpha}{\alpha}$$

and, using the result

$$\frac{\sin \alpha x}{\alpha} = \int_0^x \cos \alpha s \, ds,$$

we have

$$u(x, t) = u_0 - \frac{2u_0}{\pi} \int_0^\infty \exp(-\alpha^2 t)\, d\alpha \int_0^x \cos \alpha s\, ds$$

$$= u_0 - \frac{2u_0}{\pi} \int_0^x ds \int_0^\infty \exp(-\alpha^2 t) \cos \alpha s\, d\alpha.$$

The integral with respect to α appears in tables of definite integrals, hence,

$$u(x, t) = u_0 - \frac{u_0}{\sqrt{\pi}} \int_0^x \frac{\exp(-s^2/4t)\, ds}{\sqrt{t}}.$$

Now the substitution $v^2 = s^2/4t$ transforms the result into

$$u(x, t) = u_0 - \frac{2u_0}{\sqrt{\pi}} \int_0^{x/2\sqrt{t}} \exp(-v^2)\, dv$$

$$= u_0 - u_0 \operatorname{erf}(x/2\sqrt{t})$$
$$= u_0(1 - \operatorname{erf}(x/2\sqrt{t}))$$

$$= u_0 \operatorname{erfc}\left(\frac{x}{2\sqrt{t}}\right),$$

using the definition of erfc, the **complementary error function**:

$$\operatorname{erfc} x = 1 - \operatorname{erf} x = \frac{2}{\sqrt{\pi}} \int_x^\infty e^{-s^2}\, ds,$$

where (see Exercise 13 in Section 7.4) the **error function**, erf x, is defined by

$$\operatorname{erf} x = \frac{2}{\sqrt{\pi}} \int_0^x e^{-s^2}\, ds. \quad \blacksquare$$

We will solve the problem of Example 8.4–2 again, this time using the **Laplace transform**. Recall from Section 3.1 the definition of the Laplace transform of $u(t)$,

$$U(s) = \int_0^\infty e^{-st} u(t)\, dt, \qquad s > 0.$$

Hence,

$$\int_0^\infty e^{-st} \frac{\partial u(x, t)}{\partial t}\, dt = e^{-st} u(x, t)\Big|_0^\infty + s \int_0^\infty e^{-st} u(x, t)\, dt$$

$$= sU(x, s),$$

given that $u(x, 0) = 0$ so that the partial differential equation, $u_t = u_{xx}$, becomes

$$\frac{d^2 U(x, s)}{dx^2} - sU(x, s) = 0.$$

with

$$U(0, s) = \int_0^\infty e^{-st} u_0 \, dt = \frac{u_0}{s}.$$

Now, the solution of this second-order, homogeneous, ordinary differential equation is

$$U(x, s) = c_1(s)e^{\sqrt{s}x} + c_2(s)e^{-\sqrt{s}x}.$$

We take $c_1(s)$ to be zero so that $U(x, s)$ will remain bounded for $x > 0$ and, applying the condition $U(0, s) = u_0/s$, we have

$$U(x, s) = \frac{u_0}{s} e^{-\sqrt{s}x}.$$

Using the table of Laplace transforms in Table 1, we find

$$u(x, t) = u_0 \, \text{erfc} \, (x/2\sqrt{t})$$

as before. ∎

The last example illustrates the effect produced by a nonhomogeneous boundary condition. Such a condition resulted in a *nonhomogeneous* differential equation in the transformed variable. In Section 8.3 we presented another method for dealing with a nonhomogeneous boundary condition. (See Example 8.3–2.) The following example will show how easily the Laplace transform can handle some nonhomogeneous boundary conditions.

EXAMPLE 8.4–3 Solve the following problem using the Laplace transform (Fig. 8.4–3).

$$\text{P.D.E.:} \quad u_t = u_{xx}, \quad 0 < x < d, \quad t > 0;$$

$$\text{B.C.:} \quad \left. \begin{array}{l} u(0, t) = a, \\ u(d, t) = a, \end{array} \right\} \quad t > 0;$$

$$\text{I.C.:} \quad u(x, 0) = a + b \sin\left(\frac{\pi}{d} x\right), \quad 0 < x < d,$$

where a and b are constants.

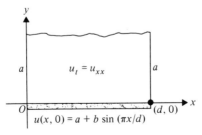

FIGURE 8.4–3

Solution. Transforming the equation and the boundary conditions results in the problem

$$\frac{d^2 U(x, s)}{dx^2} - sU(x, s) = -a - b \sin\left(\frac{\pi}{d} x\right),$$

$$U(0, s) = \frac{a}{s},$$

$$U(d, s) = \frac{a}{s}.$$

This is an ordinary differential equation with complementary solution (see Section 2.1):

$$U_c(x, s) = c_1(s)e^{\sqrt{s}x} + c_2(s)e^{-\sqrt{s}x}.$$

A particular integral,

$$U_p(x, s) = \frac{a}{s} + \frac{bd^2}{d^2 s + \pi^2} \sin\left(\frac{\pi}{d} x\right),$$

can be found by the method of undetermined coefficients (see Section 2.3). The sum of $U_c(x, s)$ and $U_p(x, s)$ is the general solution and, evaluating $c_1(s)$ and $c_2(s)$ with the aid of the conditions $U(0, s) = U(d, s) = a/s$, we have

$$U(x, s) = \frac{a}{s} + \frac{bd^2}{d^2 s + \pi^2} \sin\left(\frac{\pi}{d} x\right).$$

In obtaining the inverse of this, we note that $\sin(\pi x/d)$ can be treated as a constant. Thus

$$u(x, t) = a + b \sin\left(\frac{\pi}{d} x\right) \exp\left(-\pi^2 t/d^2\right).$$

The details are left for the exercises. (See Exercise 5.) ■

EXAMPLE 8.4–4 Solve the following problem using the Laplace transform.

P.D.E.: $u_{tt} = u_{xx}$, $\quad 0 < x < c$, $\quad t > 0$;

B.C.: $\left.\begin{array}{l} u(0, t) = 0, \\ u(c, t) = 0, \end{array}\right\}$ $t > 0$;

I.C.: $u(x, 0) = b \sin\left(\frac{\pi}{c} x\right),$

$\left.\begin{array}{l} \\ u_t(x, 0) = -b \sin\left(\frac{\pi}{c} x\right), \end{array}\right\}$ $0 < x < c.$

Solution. Transforming the equation and the boundary conditions yields

$$\frac{d^2U(x, s)}{dx^2} = s^2U(x, s) - bs \sin\left(\frac{\pi}{c}x\right) + b \sin\left(\frac{\pi}{c}x\right),$$

$$U(0, s) = U(c, s) = 0,$$

which has solution (Exercise 7)

$$U(x, s) = \frac{c^2b(s - 1)}{c^2s^2 + \pi^2} \sin\left(\frac{\pi}{c}x\right).$$

Hence,

$$u(x, t) = b \sin\left(\frac{\pi}{c}x\right)\left(\cos\left(\frac{\pi}{c}t\right) - \frac{c}{\pi}\sin\left(\frac{\pi}{c}t\right)\right). \quad \blacksquare$$

In the next example we will *simplify* the conditions in Example 8.4–4 and show that, in spite of this, the problem is intractable by the Laplace transformation method.

EXAMPLE 8.4–5 Solve the following problem using the Laplace transform (Fig. 8.4–4).

$$\text{P.D.E.:} \quad u_t = u_{xx}, \quad 0 < x < 1, \quad t > 0;$$

$$\text{B.C.:} \quad \left.\begin{array}{l} u(0, t) = 1, \\ u(1, t) = 1, \end{array}\right\} \quad t > 0;$$

$$\text{I.C.:} \quad u(x, 0) = 0, \quad 0 < x < 1.$$

The transformed problem is

$$\frac{d^2U(x, s)}{dx^2} = sU(x, s),$$

$$U(0, s) = \frac{1}{s}, \qquad U(1, s) = \frac{1}{s},$$

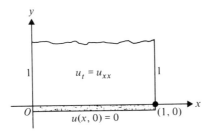

FIGURE 8.4–4

which looks innocent enough. Its solution, however, is found to be (Exercise 9)

$$U(x, s) = \frac{1}{s} \cosh \sqrt{s}x + \frac{(1 - \cosh \sqrt{s}) \sinh \sqrt{s}x}{s \sinh \sqrt{s}}$$

$$= \frac{\sinh \sqrt{s}x + \sinh \sqrt{s}(1 - x)}{s \sinh \sqrt{s}}. \quad \blacksquare$$

Even with reasonably extensive tables of Laplace transforms at our disposal, we could not expect to find the function $U(x, s)$ of Example 8.4–5. What is needed here is a *general* method for obtaining the Laplace inverse of a function such as the method we used in Section 7.3 for obtaining the inverse of a Fourier transform. Finding the inverse Laplace transform involves complex variables and, in particular, contour integration. We will treat this topic in Section 10.5. See Exercise 10, however, for a method of solution of the present problem.

KEY WORDS AND PHRASES

error function   complementary error function

EXERCISES 8.4

▶ **1.** Carry out the details in Example 8.4–1.

2. Solve Example 8.4–1 given that $f(y) = e^{-y}$.

3. Solve the initial-value problem

$$\frac{d\bar{u}(t)}{dt} + \alpha^2 \bar{u}(t) = \alpha u_0, \qquad \bar{u}(0) = 0,$$

where α and u_0 are constants. (Compare Example 8.4–2.)

4. Fill in the details required to arrive at the solution to Example 8.4–2.

5. Given that

$$\frac{d^2 U(x, s)}{dx^2} - sU(x, s) = -a - b \sin \left(\frac{\pi}{d} x\right),$$

$$U(0, s) = U(d, s) = \frac{a}{s}.$$

a) Obtain the complementary solution $U_c(x, s)$.

b) Obtain a particular integral by the method of undetermined coefficients.

c) Obtain the complete solution

$$U(x, s) = \frac{a}{s} + \frac{bd^2}{d^2 s + \pi^2} \sin \left(\frac{\pi}{d} x\right).$$

d) Find the inverse Laplace transform of the function $U(x, s)$ obtained in part (c).

6. Verify completely the solution to Example 8.4–3.

7. Solve the following two-point boundary-value problem.

$$\frac{d^2 U(x, s)}{dx^2} - s^2 U(x, s) = b(1 - s) \sin \frac{\pi}{c} x,$$

$$U(0, s) = U(c, s) = 0.$$

(Compare Example 8.4–4.)

8. Obtain the inverse Laplace transform of the function $U(x, s)$ of Example 8.4–4.

9. Solve the two-point boundary-value problem in Example 8.4–5 for $U(x, s)$.

10. In the problem of Example 8.4–5, make the substitution

$$u(x, t) = U(x, t) + \phi(x).$$

Then solve the problem by the method of separation of variables.

▶▶ 11. Solve the following problem using a Fourier transform (Fig. 8.4–5).

$$\text{P.D.E.:} \quad u_t = u_{xx}, \quad x > 0, \quad t > 0;$$
$$\text{B.C.:} \quad u_x(0, t) = u_0, \quad t > 0;$$
$$\text{I.C.:} \quad u(x, 0) = 0, \quad x > 0.$$

Interpret this problem physically.

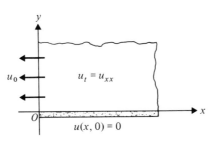

FIGURE 8.4–5

12. Solve Exercise 11 using the Laplace transform.

13. **a)** Find the bounded, harmonic function $v(x, y)$ over the semi-infinite strip $0 < x < c$, $y > 0$ that satisfies the following conditions:

i) $v(0, y) = 0$;

ii) $v_y(x, 0) = 0$;

iii) $v_x(c, y) = f(y)$.

b) Interpret this problem physically.

14. a) Find the bounded, harmonic function $v(x, y)$ over the semi-infinite strip $0 < y < b$, $x > 0$ that satisfies the following conditions:

    i) $v_y(x, 0) = 0$;

    ii) $v_x(0, y) = 0$;

    iii) $v(x, b) = f(x)$.

b) Interpret this problem physically.

15. Solve the following problem by using the Laplace transform.

$$\text{P.D.E.: } u_{tt} = u_{xx} + \sin \frac{\pi x}{c} \sin \omega t, \quad 0 < x < c, \quad t > 0;$$

$$\text{B.C.: } u(0, t) = 0, \quad u(c, t) = 0, \quad t > 0;$$

$$\text{I.C.: } u(x, 0) = 0, \quad u_t(x, 0) = 0, \quad 0 < x < c.$$

16. Show that

$$\int_0^\infty \frac{\exp(-ax)}{\sqrt{x}} = \sqrt{\frac{\pi}{a}}, \quad a > 0.$$

17. a) Graph $f(x) = \exp(-x^2)$.

b) Graph $f(x) = \operatorname{erf} x$. (*Hint*: Values of erf x can be found in tables.)

c) Graph $f(x) = \operatorname{erfc} x$.

d) Graph $u(x, t) = u_0 \operatorname{erfc} (x/2\sqrt{t})$ for various fixed values of t; take $u_0 = 1$ for convenience.

8.5 STURM–LIOUVILLE PROBLEMS

A boundary-value problem that has the form

$$\frac{d}{dx}\left(r(x)\frac{dy}{dx}\right) + (q(x) + \lambda w(x))y = 0, \tag{8.5–1}$$

$$\begin{aligned} a_1 y(a) + a_2 y'(a) = 0, \\ b_1 y(b) + b_2 y'(b) = 0, \end{aligned} \tag{8.5–2}$$

is called a **Sturm–Liouville system** if the above constants and functions have certain properties that will be described later. The system is named after Jacques C. F. Sturm (1803–1855), a Swiss who taught at the Sorbonne, and Joseph Liouville (1809–1882), a professor at the Collège de France. These two mathematicians, together with Augustin-Louis Cauchy (1789–1857), a colleague of Liouville, had discussed the conditions under which a function may be expanded into an infinite series of orthogonal functions. A **Sturm–Liouville problem** consists of finding the values of λ and the corresponding values of y that satisfy the system. Since systems described by Eqs. (8.5–1) and (8.5–2) occur in a wide variety of applications, Sturm–Liouville theory has been extensively developed.

In Eq. (8.5–1) we require that $r(x)$ be a real function that is continuous and has a continuous derivative over the interval of interest

$a \leq x \leq b$. We also require that $q(x)$ be real and continuous and $w(x)$ be positive on the same interval. It may happen that $w(x) = 0$ at isolated points in the interval. We further require that λ be a constant (independent of x) and that a_1 and a_2 be not both zero and that b_1 and b_2 be not both zero in Eq. (8.5–2). If $r(x) > 0$ for $a \leq x \leq b$, then the Sturm–Liouville problem is called **regular**.

To solve a regular Sturm–Liouville problem means to find values of λ (called **eigenvalues**, or characteristic values) and to find the corresponding values of y (called **eigenfunctions**, or characteristic functions). Our task is made easier by the following facts, which we shall not prove.

1. All eigenvalues are real.
2. There are an infinite number of eigenvalues and they can be ordered $\lambda_1 < \lambda_2 < \lambda_3 \cdots$.
3. To each eigenvalue there corresponds an eigenfunction.
4. Eigenfunctions belonging to different eigenvalues are linearly independent.
5. The eigenfunctions constitute an orthogonal set on the interval $a < x < b$ with respect to the weight function $w(x)$.
6. $\lim_{n \to \infty} \lambda_n = \infty$.

Next we give some examples of the foregoing. In each case we will identify the terms in the examples and compare them to the terms in Eqs. (8.5–1) and (8.5–2).

EXAMPLE 8.5–1 Solve the following system:

$$y'' + \lambda y = 0, \qquad y(0) = 0, \qquad y(\pi) = 0.$$

Solution. Here $r(x) = 1$, $q(x) = 0$, $w(x) = 1$, $a = 0$, $b = \pi$, $a_1 = b_1 = 1$, and $a_2 = b_2 = 0$. The solution to the differential equation is

$$y = c_1 \cos (\sqrt{\lambda}x) + c_2 \sin (\sqrt{\lambda}x)$$

with $\lambda > 0$. If $\lambda \leq 0$, then the system has only the trivial solution $y = 0$ (Exercise 1). This is not of interest, since *every* Sturm–Liouville system has a trivial solution. Note that we admit zero as an eigenvalue but not as an eigenfunction.

The condition $y(0) = 0$ implies that $c_1 = 0$; hence the updated solution becomes

$$y = c_2 \sin (\sqrt{\lambda}x).$$

The second condition $y(\pi) = 0$ implies that either $c_2 = 0$ (which would lead to the trivial solution) or $\sqrt{\lambda}\pi = n\pi$, that is, $\lambda = n^2$, $n = 1, 2, 3, \ldots$. Thus the eigenvalues of the system are $\lambda_1 = 1$, $\lambda_2 = 4$, $\lambda_3 = 9, \ldots$. The corresponding eigenfunctions are

$$y_1(x) = \sin x, \qquad y_2(x) = \sin 2x, \qquad y_3(x) = \sin 3x, \ldots$$

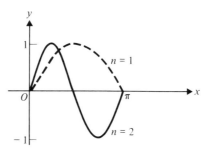

FIGURE 8.5–1 Eigenfunctions (Example 8.5–1).

and, in general (Fig. 8.5–1),

$$y_n(x) = \sin nx, \qquad n = 1, 2, 3, \ldots,$$

where the arbitrary constants have been set equal to one.

It can now be readily verified that these eigenfunctions form an orthogonal set on the interval $0 < x < \pi$ with weight function $w(x) = 1$. We have

$$\int_0^\pi \sin nx \sin mx \, dx = \left. \frac{\sin (n-m)x}{2(n-m)} - \frac{\sin (n+m)x}{2(n+m)} \right|_0^\pi = 0,$$

if $m \neq n$. The orthogonality of these eigenfunctions is of prime importance in Fourier series expansions, as we have seen in Section 7.2. ■

EXAMPLE 8.5–2 Solve the following system:

$$y'' + \lambda y = 0, \qquad y'(0) = 0, \qquad y'(\pi) = 0.$$

Solution. Now we have $r(x) = w(x) = 1$, $q(x) = 0$, $a = 0$, $b = \pi$, $a_1 = b_1 = 0$, and $a_2 = b_2 = 1$. The solution to the differential equation is again

$$y = c_1 \cos (\sqrt{\lambda}x) + c_2 \sin (\sqrt{\lambda}x), \qquad \lambda \geq 0.$$

From this we have

$$y' = -c_1\sqrt{\lambda} \sin (\sqrt{\lambda}x) + c_2\sqrt{\lambda} \cos (\sqrt{\lambda}x).$$

The first condition $y'(0) = 0$ implies that $c_2 = 0$ so the solution becomes

$$y = c_1 \cos (\sqrt{\lambda}x) \quad \text{and} \quad y' = -c_1\sqrt{\lambda} \sin (\sqrt{\lambda}x).$$

The second condition $y'(\pi) = 0$ implies that $\sqrt{\lambda} = n$, $\lambda = n^2$, $n = 0, 1, 2, \ldots$. Hence the eigenvalues are $\lambda_0 = 0$, $\lambda_1 = 1$, $\lambda_2 = 4$, $\lambda_3 = 9, \ldots$, and the corresponding eigenfunctions are $y_0 = 1, y_1 = \cos x, y_2 = \cos 2x$, $y_3 = \cos 3x, \ldots$, where again we have set the arbitrary constants equal to unity.

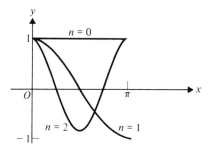

FIGURE 8.5-2 Eigenfunctions (Example 8.5–2).

We can verify the orthogonality of the eigenfunctions as follows:

$$\int_0^\pi \cos nx \cos mx \, dx = \frac{\sin (n - m)x}{2(n - m)} + \frac{\sin (n + m)x}{2(n + m)} \Big|_0^\pi = 0$$

if $n \neq m$. Hence the set (Fig. 8.5–2)

$$\{1, \cos x, \cos 2x, \cos 3x, \ldots\}$$

is an orthogonal set on the interval $0 < x < \pi$ with weight function $w(x) = 1$. This property is also useful in Fourier series expansions. ■

In the last example $\lambda = 0$ was an eigenvalue. It can be shown (Exercise 2) that $\lambda < 0$ leads to the trivial solution.

EXAMPLE 8.5–3 Solve the following system:

$$y'' + \lambda y = 0, \qquad y(0) + y'(0) = 0, \qquad y(1) = 0.$$

Solution. Here we have $r(x) = w(x) = 1$, $q(x) = 0$, $a = 0$, $b = 1$, $a_1 = a_2 = b_1 = 1$, and $b_2 = 0$. If $\lambda < 0$, we have a trivial solution (Exercise 3). If $\lambda = 0$, then $y = k_1 + k_2 x$ and the conditions applied to this function show that an eigenfunction belonging to the eigenvalue $\lambda = 0$ is $1 - x$.

If $\lambda > 0$, we have as the solution of the differential equation

$$y = c_1 \cos (\sqrt{\lambda} x) + c_2 \sin (\sqrt{\lambda} x).$$

The condition $y(0) + y'(0) = 0$ implies that $c_1 + c_2\sqrt{\lambda} = 0$, that is, $c_1 = -c_2\sqrt{\lambda}$. Then the condition $y(1) = 0$ implies that $\sqrt{\lambda} = \tan \sqrt{\lambda}$. Thus the eigenvalues are the squares of the solutions of the transcendental equation $t = \tan t$. This equation cannot be solved by algebraic methods so we graph the curves $u = t$ and $u = \tan t$ and observe the values of t where the two curves intersect. Figure 8.5–3 shows the first two eigenvalues $\lambda_1 \doteq (4.5)^2$ and $\lambda_2 \doteq (7.7)^2$, together with the eigenvalue $\lambda = 0$ that has already been found.

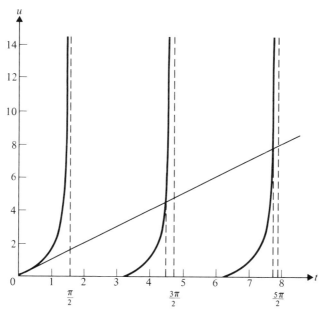

FIGURE 8.5–3 Simultaneous solutions of $u = t$ and $u = \tan t$.

We can see from the figure that there are an infinite number of eigenvalues and that they are approaching the square of the odd multiples of $\pi/2$. In other words,

$$\lambda_n \doteq \frac{(2n + 1)^2 \pi^2}{4},$$

the approximation getting better as n increases through the positive integers. A computer is useful in obtaining the eigenvalues. For example, the following computational scheme shows how to obtain λ_1 to three-decimal accuracy with the aid of a hand-held calculator.

t	$\tan t$	
4.5	4.637	
4.4	3.096	
4.45	3.723	
4.49	4.422	
4.495	4.527	
4.4935	4.495	
4.4934	4.493	Hence $\lambda_1 \doteq 20.187$.

The values of $\tan t$ are a guide to what should be chosen for the next value of t. The eigenvalues of this problem also appear in published tables.

The eigenfunctions corresponding to the eigenvalues λ_n, $n = 1, 2, \ldots$, are

$$y_n(x) = \sin(\sqrt{\lambda_n}x) - \sqrt{\lambda_n}\cos(\sqrt{\lambda_n}x).$$

The Sturm–Liouville theory guarantees the orthogonality of these eigenfunctions on the interval $0 < x < 1$. In other words,

$$\int_0^1 (\sin(\sqrt{\lambda_n}x) - \sqrt{\lambda_n}\cos(\sqrt{\lambda_n}x))(\sin(\sqrt{\lambda_m}x) - \sqrt{\lambda_m}\cos(\sqrt{\lambda_m}x))\,dx = 0$$

if $m \neq n$. As an *example* we evaluate the above integral for $n = 1$ and $m = 2$, that is, $\lambda_1 = 4.493$ and $\lambda_2 = 7.725$. We have

$$\int_0^1 (\sin 4.493x - 4.493\cos 4.493x)(\sin 7.725x - 7.725\cos 7.725x)\,dx$$

$$= -0.002.$$

The approximate values of the eigenvalues used as well as roundoff error prevented the result from being exactly zero. ■

Up to this point all the examples presented involved a very simple second-order differential equation. We can show, however, that *every* second-order, linear, homogeneous differential equation can be transformed into the form shown in (8.5–1). Consider

$$A(x)y'' + B(x)y' + C(x)y = 0, \tag{8.5-3}$$

with $A'(x) \neq B(x)$. Note that if this restriction is *not* imposed, then Eq. (8.5–3) will have the required form except for the constant λ. If we multiply Eq. (8.5–3) by

$$\frac{1}{A(x)}\exp\left(\int^x \frac{B(t)\,dt}{A(t)}\right) = \mu(x)/A(x),$$

we can write the result as

$$\frac{d}{dx}\left(\mu(x)\frac{dy}{dx}\right) + \frac{C(x)}{A(x)}\mu(x)y = 0,$$

which is the form required for a Sturm–Liouville system. The parameter λ is, of course, an essential part of the system.

We simplify the notation in what follows by defining a differential operator L,

$$L \equiv \frac{d}{dx}\left(r(x)\frac{d}{dx}\right) + q(x), \tag{8.5-4}$$

that is,

$$Ly = \frac{d}{dx}\left(r(x)\frac{dy}{dx}\right) + q(x)y = (ry')' + qy,$$

the primes denoting differentiation with respect to x. With this notation Eq. (8.5–1) can be written

$$Ly = -\lambda wy, \tag{8.5-5}$$

which shows more clearly why λ is called an eigenvalue of the eigenfunction y. (Compare Section 4.4.)

If y_1 and y_2 are functions that are twice differentiable on $[a, b]$, then

$$
\begin{aligned}
y_1 L y_2 - y_2 L y_1 &= y_1(r y_2')' - y_2(r y_1')' \\
&= y_1(r y_2'' + r' y_2') - y_2(r y_1'' + r' y_1') \\
&= r'(y_1 y_2' - y_2 y_1') + r(y_1 y_2'' - y_2 y_1'') \\
&= (r(y_1 y_2' - y_2 y_1'))'.
\end{aligned}
$$

On the other hand, if λ_1 is an eigenvalue belonging to y_1 while λ_2 is an eigenvalue belonging to y_2, then, from Eq. (8.5–5),

$$
y_1 L y_2 - y_2 L y_1 = (\lambda_1 - \lambda_2) w y_1 y_2.
$$

Thus, equating the two quantities, we get

$$
(\lambda_1 - \lambda_2) w y_1 y_2 = (r(y_1 y_2' - y_2 y_1'))'
$$

and, integrating from a to b gives us

$$
(\lambda_1 - \lambda_2) \int_a^b w(x) y_1(x) y_2(x)\, dx = r(y_1 y_2' - y_2 y_1')\Big|_a^b. \qquad \textbf{(8.5–6)}
$$

We now see that if $\lambda_1 \neq \lambda_2$, then $y_1(x)$ and $y_2(x)$ are orthogonal on the interval (a, b) with weight function $w(x)$, provided the boundary conditions are such that the right side of Eq. (8.5–6) vanishes. We leave it as an exercise to show that the boundary conditions of Example 8.5–3 make $y_n y_m' - y_m y_n'$ vanish at $x = 0$ and at $x = 1$. (See Exercise 4.)

In more advanced texts* it is shown that the eigenfunctions of a regular Sturm–Liouville system form a **complete set** (see Eq. 7.1–22) with respect to piecewise smooth functions on the interval (a, b). Recall that this means that if the set

$$
\{\phi_n(x),\ n = 1, 2, \ldots\}
$$

consists of normalized eigenfunctions of Eq. (8.5–1) and if f is piecewise smooth on (a, b), then

$$
\lim_{N \to \infty} \int_a^b \left(f(x) - \sum_{n=1}^N c_n \phi_n(x) \right)^2 w(x)\, dx = 0. \qquad \textbf{(8.5–7)}
$$

In other words, f can be represented by a series of eigenfunctions

$$
f(x) \sim \sum_{n=1}^\infty c_n \phi_n(x)
$$

with

$$
c_n = \int_a^b f(s) \phi_n(s) w(s)\, ds, \qquad n = 1, 2, 3, \ldots
$$

---

* See, for example, Arthur E. Danese, *Advanced Calculus: An Introduction to Applied Mathematics*, vol. 1, ch. 17 (Boston: Allyn and Bacon, 1965).

and such that Eq. (8.5–7) holds. We have already seen applications stemming from Examples 8.5–1, 8.5–2, and 8.5–3 in Chapter 7.

Up to this point we have been dealing with *regular* problems. A Sturm–Liouville problem is said to be *singular* if (1) $r(a) = 0$ and $a_1 = a_2 = 0$, or (2) $r(b) = 0$ and $b_1 = b_2 = 0$. For these cases it is still true that the eigenfunctions are orthogonal (see Exercise 5). Singular problems also arise when $r(x)$ or $w(x)$ vanishes at $x = a$ or $x = b$, when $q(x)$ is discontinuous at these points, or when the interval, $a \leq x \leq b$, is unbounded.

KEY WORDS AND PHRASES

Sturm–Liouville systems and problems
regular Sturm–Liouville problem

eigenvalue
eigenfunction

EXERCISES 8.5

▶ 1. Verify that the problem of Example 8.5–1 has only the trivial solution if $\lambda \leq 0$.

2. Show that $\lambda < 0$ in Example 8.5–2 leads to the trivial solution.

3. Verify that $\lambda < 0$ leads to the trivial solution in the problem of Example 8.5–3.

4. Show that the eigenfunctions $y_n(x)$ and boundary conditions of Example 8.5–3 are such that $y_n y_m' - y_m y_n'$ vanishes at $x = 0$ and at $x = 1$.

5. **a)** In Eqs. (8.5–1) and (8.5–2) if $r(a) = a_1 = a_2 = 0$, explain why the eigenfunctions are orthogonal.

   b) In Eqs. (8.5–1) and (8.5–2) if $r(b) = b_1 = b_2 = 0$, explain why the eigenfunctions are orthogonal. (*Hint*: Examine the right-hand side of Eq. 8.5–6.)

▶▶ 6. Obtain the eigenvalues and corresponding eigenfunctions of the regular Sturm–Liouville system,

$$y'' + \lambda y = 0, \qquad y'(0) = 0, \qquad y(\pi) = 0.$$

In Exercises 7–11, find the eigenvalues and corresponding eigenfunctions of each system.

7. $y'' + \lambda y = 0, \quad y'(-\pi) = 0, \quad y'(\pi) = 0$

8. $y'' + \lambda y = 0, \quad y(0) = 0, \quad y'(\pi) = 0$

9. $y'' + (1 + \lambda)y = 0, \quad y(0) = 0, \quad y(\pi) = 0$

10. $y'' + 2y' + (1 - \lambda)y = 0, \quad y(0) = 0, \quad y(1) = 0$

11. $y'' + 2y' + (1 - \lambda)y = 0$, $\quad y'(0) = 0$, $\quad y'(\pi) = 0$

12. Transform the differential equation of Exercise 11 into the form (8.5–1). (*Hint*: Find $\mu(x)$ as shown in the text.)

13. State the orthogonality relation for the eigenfunctions in Exercises 8–11, that is, give the weight function and the interval of orthogonality for each one.

14. In Eq. (8.5–1) why do we require that $r(x) > 0$ for $a \leq x \leq b$ for a regular Sturm–Liouville problem?

▶▶▶ 15. Given the problem

$$y'' + \lambda y = 0, \qquad y(-c) = y(c), \qquad y'(-c) = y'(c).$$

The boundary conditions here are called **periodic boundary conditions**.

a) Show that $\lambda = 0$ is an eigenvalue with corresponding eigenfunction $y_0(x) = 1$.

b) Show that $\lambda < 0$ leads to the trivial solution.

c) For $\lambda > 0$ obtain the eigenvalues $\lambda_n = (n\pi/c)^2$, $n = 1, 2, \ldots$ with corresponding eigenfunctions

$$y_n(x) = a_n \cos\left(\frac{n\pi}{c} x\right) + b_n \sin\left(\frac{n\pi}{c} x\right),$$

where a_n and b_n are not both zero but are otherwise arbitrary.

d) Where have we made use of the fact (proved elsewhere) that the eigenfunctions are orthogonal on the interval $-c < x < c$ with weight function $w(x) = 1$?

16. Consider
$$Ly(x) = A(x)y'' + B(x)y' + C(x)y,$$

where $A(x)$, $B(x)$, and $C(x)$ are real functions of x such that $A(x)$ is twice continuously differentiable, $B(x)$ is continuously differentiable, and $C(x)$ is continuous on $[a, b]$. Moreover, $A(x)$ does not vanish on (a, b). Then

$$L^*y(x) = \frac{d^2}{dx^2}(A(x)y(x)) - \frac{d}{dx}(B(x)y(x)) + C(x)y(x)$$

is called the **adjoint**[†] of $y(x)$.

a) Show that
$$L^*y = Ay'' + (2A' - B)y' + (A'' - B' + C)y.$$

b) Show that L is **self-adjoint**, that is, $L = L^*$ if and only if $A' = B$.

c) Show that the Sturm–Liouville equation (8.5–1) is a self-adjoint equation.

17. Show that the differential operator, L, defined in Eq. (8.5–4) is a linear operator.

[†] This is another use of the word "adjoint" different from the one given in Section 4.7.

18. Laguerre's equation,

$$xy'' + (1 - x)y' + \lambda y = 0,$$

is of importance in quantum mechanics.

a) Transform the equation into the form (8.5–1).

b) What is the weight function?

19. Hermite's differential equation,

$$y'' - 2xy' + 2\lambda y = 0,$$

arises in the quantum mechanical theory of the linear oscillator.

a) Transform the equation into the form (8.5–1).

b) What is the weight function?

20. Chebyshev's differential equation,

$$(1 - x^2)y'' - xy' + n^2 y = 0,$$

is found in mathematical physics.

a) Transform the equation into self-adjoint form.

b) What is the weight function?

9

boundary-value problems in other coordinate systems

9.1 BOUNDARY-VALUE PROBLEMS IN CIRCULAR REGIONS

Whenever a region has circular symmetry there is usually an advantage in using **polar coordinates**. The relationship between rectangular and polar coordinates is shown in Fig. 9.1–1 and the equations accompanying the figure.

From these relations we can compute the partial derivatives:

$$\frac{\partial x}{\partial \rho} = \cos \phi, \qquad \frac{\partial y}{\partial \rho} = \sin \phi,$$

$$\frac{\partial x}{\partial \phi} = -\rho \sin \phi, \qquad \frac{\partial y}{\partial \phi} = \rho \cos \phi.$$

If $u(x, y)$ is a twice-differentiable function of x and y, then, by the chain rule,

$$\frac{\partial u}{\partial \rho} = \frac{\partial u}{\partial x}\frac{\partial x}{\partial \rho} + \frac{\partial u}{\partial y}\frac{\partial y}{\partial \rho} = \frac{\partial u}{\partial x}\cos \phi + \frac{\partial u}{\partial y}\sin \phi. \qquad \textbf{(9.1–1)}$$

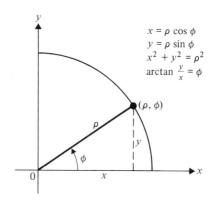

$$x = \rho \cos \phi$$
$$y = \rho \sin \phi$$
$$x^2 + y^2 = \rho^2$$
$$\arctan \frac{y}{x} = \phi$$

FIGURE 9.1–1 Polar coordinates.

Hence,

$$\frac{\partial^2 u}{\partial \rho^2} = \frac{\partial}{\partial \rho}\left(\frac{\partial u}{\partial x}\right)\cos\phi + \frac{\partial u}{\partial x}\frac{\partial}{\partial \rho}(\cos\phi) + \frac{\partial}{\partial \rho}\left(\frac{\partial u}{\partial y}\right)\sin\phi + \frac{\partial u}{\partial y}\frac{\partial}{\partial \rho}(\sin\phi)$$

$$= \frac{\partial}{\partial \rho}\left(\frac{\partial u}{\partial x}\right)\cos\phi + \frac{\partial}{\partial \rho}\left(\frac{\partial u}{\partial y}\right)\sin\phi,$$

since the other two terms are zero.

In order to evaluate a term like $\partial/\partial\rho(\partial u/\partial x)$, we observe that Eq. (9.1–1) applies not only to $u(x, y)$ but to *any* differentiable function of x and y. We can, in fact, write Eq. (9.1–1) symbolically as

$$\frac{\partial(\ \ )}{\partial \rho} = \frac{\partial(\ \ )}{\partial x}\cos\phi + \frac{\partial(\ \ )}{\partial y}\sin\phi.$$

Thus

$$\frac{\partial u_x}{\partial \rho} = u_{xx}\cos\phi + u_{xy}\sin\phi$$

and

$$\frac{\partial u_y}{\partial \rho} = u_{yx}\cos\phi + u_{yy}\sin\phi.$$

Therefore,

$$\frac{\partial^2 u}{\partial \rho^2} = u_{xx}\cos^2\phi + 2u_{xy}\sin\phi\cos\phi + u_{yy}\sin^2\phi,$$

if we assume that $u_{xy} = u_{yx}$, which is the case for the functions with which we are dealing.

In a similar manner we find

$$\frac{1}{\rho^2}\frac{\partial^2 u}{\partial \phi^2} = u_{xx}\sin^2\phi - 2u_{xy}\sin\phi\cos\phi + u_{yy}\cos^2\phi$$

$$- \frac{1}{\rho}u_x\cos\phi - \frac{1}{\rho}u_y\sin\phi$$

and, since

$$\frac{1}{\rho}\frac{\partial u}{\partial \rho} = \frac{1}{\rho}u_x\cos\phi + \frac{1}{\rho}u_y\sin\phi,$$

we have

$$\frac{\partial^2 u}{\partial \rho^2} + \frac{1}{\rho^2}\frac{\partial^2 u}{\partial \phi^2} + \frac{1}{\rho}\frac{\partial u}{\partial \rho} = \frac{\partial^2 u}{\partial x^2} + \frac{\partial^2 u}{\partial y^2}. \qquad \textbf{(9.1–2)}$$

In Eq. (9.1–2) we have the Laplacian in two dimensions in both polar and rectangular coordinates. If u is *independent of* ϕ, then we have the simpler result,

$$\frac{\partial^2 u}{\partial x^2} + \frac{\partial^2 u}{\partial y^2} = \frac{\partial^2 u}{\partial \rho^2} + \frac{1}{\rho}\frac{\partial u}{\partial \rho} = \frac{1}{\rho}\frac{\partial}{\partial \rho}\left(\rho\frac{\partial u}{\partial \rho}\right). \qquad \textbf{(9.1–3)}$$

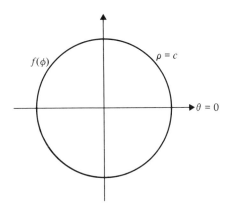

FIGURE 9.1–2 Temperatures in a disk.

EXAMPLE 9.1–1 Find the steady-state, bounded temperatures in a circular metallic disk of radius c, if the temperatures on the edge are given by $f(\phi)$. See Fig. 9.1–2.

Solution. We first observe that $f(\phi)$ must be a piecewise smooth function that is periodic of period 2π. We also note that $u(-\pi, \phi) = u(\pi, \phi)$ and $u_\rho(-\pi, \phi) = u_\rho(\pi, \phi)$, these restrictions being necessary so that the temperatures will be uniquely determined. Then we have the following problem:

$$\text{P.D.E.:}\ \frac{1}{\rho}\frac{\partial}{\partial\rho}(\rho u_\rho) + \frac{1}{\rho^2} u_{\phi\phi} = 0, \quad 0 < \rho < c, \quad -\pi < \phi \leq \pi;$$

$$\text{B.C.:}\ u(c, \phi) = f(\phi), \quad -\pi < \phi \leq \pi.$$

It appears that we do not have enough boundary conditions to solve the problem but the fact that the temperatures must be *bounded* will provide additional information. We use the method of separation of variables and assume that we can write

$$u(\rho, \phi) = R(\rho)\Phi(\phi).$$

Then

$$\frac{\Phi}{\rho}\frac{d}{d\rho}\left(\rho\frac{dR}{d\rho}\right) + \frac{R}{\rho^2}\frac{d^2\Phi}{d\phi^2} = 0$$

or, dividing by $R\Phi/\rho^2$, we have

$$\frac{\rho}{R}\frac{d}{d\rho}\left(\rho\frac{dR}{d\rho}\right) = -\frac{1}{\Phi}\frac{d^2\Phi}{d\phi^2} = n^2, \quad n = 0, 1, 2, \ldots .$$

Choosing the separation constant as we have will ensure that Φ (and $u(\rho, \phi)$) will be periodic of period 2π in ϕ. This choice is dictated by the physical nature of the problem.

Thus the equation

$$\frac{d^2\Phi}{d\phi^2} + n^2\Phi = 0, \qquad n = 0, 1, 2, \ldots$$

has solutions

$$\Phi_n(\phi) = a_n \cos n\phi + b_n \sin n\phi.$$

The second ordinary differential equation can be written as

$$\rho^2 \frac{d^2 R_n}{d\rho^2} + \rho \frac{dR_n}{d\rho} - n^2 R_n = 0. \qquad (9.1\text{--}4)$$

We first take the case $n = 0$ and solve the resulting equation by reduction of order (see Section 2.1) to obtain (Exercise 2)

$$R_0(\rho) = c_1 \log \rho + c_2.$$

In order that this solution remain bounded in the neighborhood of $\rho = 0$ we must take $c_1 = 0$. (Why?) Hence the solution corresponding to $n = 0$ is a constant that we may take to be unity.

Equation (9.1–4) is a Cauchy–Euler equation (compare Section 2.4) whose solutions are (Exercise 3)

$$R_n(\rho) = A_n \rho^n + B_n \rho^{-n}, \qquad n = 1, 2, \ldots,$$

and we must take $B_n = 0$ so that $R_n(\rho)$ (and $u(\rho, \phi)$) will remain bounded at $\rho = \varepsilon > 0$, that is, in the neighborhood of $\rho = 0$. Without loss of generality we can also take $A_n = 1$. Then

$$u_n(\rho, \phi) = (a_n \cos n\phi + b_n \sin n\phi)\rho^n$$

are bounded functions that satisfy the given partial differential equation. To satisfy the boundary condition we take

$$u(\rho, \phi) = \frac{1}{2} a_0 + \sum_{n=1}^{\infty} (a_n \cos n\phi + b_n \sin n\phi)\rho^n \qquad (9.1\text{--}5)$$

and leave it as an exercise (Exercise 4) to show that the solution to the problem is given by Eq. (9.1–5) with the constants a_n and b_n defined as follows:

$$a_n = \frac{1}{\pi c^n} \int_{-\pi}^{\pi} f(s) \cos ns \, ds, \qquad n = 0, 1, 2, \ldots$$

$$\qquad\qquad\qquad\qquad\qquad\qquad\qquad\qquad (9.1\text{--}6)$$

$$b_n = \frac{1}{\pi c^n} \int_{-\pi}^{\pi} f(s) \sin ns \, ds, \qquad n = 1, 2, \ldots.$$

Since the function $f(\phi)$ is being represented by a Fourier series, this function must satisfy the conditions necessary for such representation (see Section 7.1). ■

In the next example we extend the above techniques to three dimensions, that is, we use **cylindrical coordinates**.

EXAMPLE 9.1–2 Solve the following boundary-value problem.

P.D.E.: $\nabla^2 u = 0$ in cylindrical coordinates, $b < \rho < c$,
$$-\pi < \phi \leq \pi, \quad -\infty < z < \infty;$$
B.C.: $u(b, \phi, z) = f(\phi), \quad -\pi < \phi \leq \pi, \quad -\infty < z < \infty,$
$$u(c, \phi, z) = 0, \quad -\pi < \phi \leq \pi, \quad -\infty < z < \infty.$$

Solution. The boundary conditions indicate that u is independent of z; consequently this three-dimensional problem reduces to a two-dimensional one (Fig. 9.1–3). Using separation of variables, we obtain the following ordinary differential equations and boundary conditions (compare Example 9.1–1):

$$\Phi'' + n^2\Phi = 0, \quad \Phi(-\pi) = \Phi(\pi), \quad n = 0, 1, 2, \ldots ;$$
$$\rho^2 R_n'' + \rho R_n' - n^2 R_n = 0, \quad R_n(c) = 0.$$

The solutions of these equations can be written as (Exercises 7 and 8)

$$\Phi(n\phi) = A_n \cos n\phi + B_n \sin n\phi$$

and

$$R_n(\rho) = \left(\frac{c}{\rho}\right)^n - \left(\frac{\rho}{c}\right)^n, \quad n = 1, 2, \ldots,$$

respectively. The case $n = 0$ must be taken separately (why?), resulting in $\Phi = $ constant and $R = \log(\rho/c)$.

Forming a linear combination of the products, we have

$$u(\rho, \phi) = A_0 \log\left(\frac{\rho}{c}\right) + \sum_{n=1}^{\infty} \left(\left(\frac{c}{\rho}\right)^n - \left(\frac{\rho}{c}\right)^n\right)(A_n \cos n\phi + B_n \sin n\phi).$$

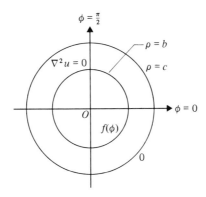

FIGURE 9.1–3

The nonhomogeneous boundary condition now produces

$$u(b, \phi) = f(\phi)$$

$$= A_0 \log\left(\frac{b}{c}\right) + \sum_{n=1}^{\infty}\left(\left(\frac{c}{b}\right)^n - \left(\frac{b}{c}\right)^n\right)(A_n \cos n\phi + B_n \sin n\phi).$$

This is a Fourier series representation of $f(\phi)$; hence, making the substitutions

$$\frac{1}{2}a_0 = A_0 \log\left(\frac{b}{c}\right), \qquad a_n = \left(\left(\frac{c}{b}\right)^n - \left(\frac{b}{c}\right)^n\right)A_n, \qquad b_n = \left(\left(\frac{c}{b}\right)^n - \left(\frac{b}{c}\right)^n\right)B_n,$$

we have the final result,

$$u(\rho, \phi) = \frac{\log\left(\dfrac{\rho}{c}\right)}{2\log\left(\dfrac{b}{c}\right)} a_0 + \sum_{n=1}^{\infty} \frac{\left(\dfrac{c}{\rho}\right)^n - \left(\dfrac{\rho}{c}\right)^n}{\left(\dfrac{c}{b}\right)^n - \left(\dfrac{b}{c}\right)^n}(a_n \cos n\phi + b_n \sin n\phi),$$

where a_0, a_n, and b_n are the Fourier coefficients in the expansion of $f(\phi)$. ∎

Before we can consider the solution of the wave and diffusion equations in cylindrical coordinates, we will need to have a knowledge of Bessel functions. This, in turn, requires that we solve second-order, linear, homogeneous, ordinary differential equations of a type not encountered in Chapter 2. Accordingly, we take up this topic in the next section.

KEY WORDS AND PHRASES

polar coordinates   **cylindrical coordinates**

EXERCISES 9.1

▶ 1. **a)** In Example 9.1–1 explain in physical terms why $u(\rho, \phi)$ must be periodic of period 2π in ϕ.

   b) Explain why the case $n = 0$ was included in Example 9.1–1.

2. Solve the equation

$$\rho^2 \frac{d^2R}{d\rho^2} + \rho \frac{dR}{d\rho} = 0$$

by reduction of order. (Compare Example 9.1–1.)

3. Solve the Cauchy–Euler equation,

$$\rho^2 \frac{d^2 R_n}{d\rho^2} + \rho \frac{dR_n}{d\rho} - n^2 R_n = 0, \qquad n = 1, 2, \dots .$$

(Compare Example 9.1–1.)

4. Supply the details in Example 9.1–1 to show that the solution is given by Eq. (9.1–5) with the constants defined as in Eq. (9.1–6).

5. Give a physical interpretation of the problem in Example 9.1–2.

6. Explain why the boundary conditions imply that u is independent of z in Example 9.1–2. Is this consistent with the physical situation?

7. In Example 9.1–2 show that the solutions of

$$\Phi'' + n^2 \Phi = 0, \qquad \Phi(-\pi) = \Phi(\pi), \qquad \Phi'(-\pi) = \Phi'(\pi), \qquad n = 0, 1, 2, \dots ,$$

are

$$\Phi(n\phi) = A_n \cos n\phi + B_n \sin n\phi.$$

(*Note*: The boundary conditions here are called "periodic boundary conditions.")

8. In Example 9.1–2 show that the solutions of

$$\rho^2 R_n'' + \rho R_n' - n^2 R_n = 0, \qquad R_n(c) = 0$$

are

$$R_n(\rho) = \left(\frac{c}{\rho}\right)^n - \left(\frac{\rho}{c}\right)^n, \qquad n = 1, 2, \dots$$

and

$$R_0(\rho) = \log(\rho/c).$$

▶▶ 9. In Example 9.1–1 obtain the solution given that $f(\phi) = u_0$, a constant. Does your result agree with the physical facts? Explain.

10. **a)** In Example 9.1–1 obtain the solution if $f(\phi)$ is given by

$$f(\phi) = \begin{cases} 0, & \text{for } -\pi < \phi < 0, \\ u_0, & \text{for } 0 < \phi < \pi. \end{cases}$$

b) Compute $u(a, 0)$, $u(a, \pi/2)$, $u(a, \pi)$, $u(a, -\pi)$, and $u(0, 0)$.

11. Change Example 9.1–2 so that the outer surface is insulated and solve the resulting problem.

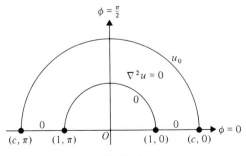

FIGURE 9.1–4

12. Find the harmonic function in the region $1 < \rho < c$, $0 < \phi < \pi$, if $u = u_0$ when $\rho = c$ and all other boundaries are kept at zero (Fig. 9.1–4).

13. Find the steady-state temperatures in the quadrant $1 < \rho < c$, $0 < \phi < \pi/2$, if the boundaries $\phi = 0$ and $\phi = \pi/2$ are kept at zero, the boundary $\rho = c$ is insulated, and the remaining boundary is kept at temperature u_0, a constant (Fig. 9.1–5).

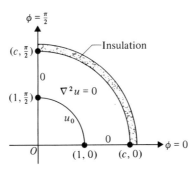

FIGURE 9.1–5

14. Find the steady-state temperatures in the region $1 < \rho < c$, $0 < \phi < \pi/2$, if the temperatures of the boundaries $\rho = 1$ and $\rho = c$ are kept at zero and $f(\phi)$, respectively, and the remaining two boundaries are insulated (Fig. 9.1–6).

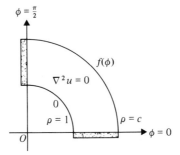

FIGURE 9.1–6

15. Find the bounded, steady-state temperatures in the unbounded region $\rho > c$ if $u(c, \phi) = f(\phi)$, $-\pi < \phi \le \pi$.

▶▶▶ **16. a)** In Example 9.1–1 evaluate $u(0, \phi)$.

    b) Does the result obtained in part (a) agree with physical fact (see Exercise 5)?

    c) Explain this apparent paradox.

17. Verify the result obtained in Example 9.1–2.

18. a) Solve the following problem relating to the static, transverse displacement of a membrane:

$$\text{D.E.:} \quad \frac{d}{d\rho}\left(\rho \frac{dz}{d\rho}\right) = 0, \quad 1 < \rho < \rho_0;$$

$$\text{B.C.:} \quad z(1) = 0, \quad z(\rho_0) = z_0.$$

    b) Interpret the problem in part (a) in physical terms.

    c) Solve the problem in part (a) given that the inner circle has radius 10 cm, the outer circle has radius 20 cm, and the outer rim is displaced 2 cm initially.

    d) Graph the solution in part (c).

19. In Example 9.1–2 examine the solution as $c \to 0$. Next, try separation of variables with $c = 0$ and carry the solution as far as possible.

9.2 SERIES SOLUTIONS OF ORDINARY DIFFERENTIAL EQUATIONS

We digress in this section in order to consider a powerful method of solving linear, ordinary differential equations. Since this method involves infinite series, it is appropriate to review some aspects of these first.

Each of the following is an example of a **series of constants**:

$$1 + 2 + 3 + 4 + \cdots + n + \cdots; \tag{9.2-1}$$

$$1 - 1 + 1 - 1 + 1 - + \cdots + (-1)^{n+1} + \cdots; \tag{9.2-2}$$

$$0 + 0 + 0 + \cdots + 0 + \cdots; \tag{9.2-3}$$

$$1 + \frac{1}{2^p} + \frac{1}{3^p} + \cdots + \frac{1}{n^p} + \cdots; \tag{9.2-4}$$

$$a(1 + r + r^2 + \cdots + r^n + \cdots); \tag{9.2-5}$$

$$1 - \frac{1}{3} + \frac{1}{5} - + \cdots + \frac{(-1)^{n+1}}{2n - 1} + \cdots. \tag{9.2-6}$$

The series in (9.2–1) is **divergent** because the **partial sums**

$$S_1 = 1, \quad S_2 = 1 + 2, \quad S_3 = 1 + 2 + 3, \quad S_4 = 1 + 2 + 3 + 4, \ldots$$

form a **sequence**,

$$\{S_1, S_2, S_3, \ldots\} = \{1, 3, 6, 10, \ldots\},$$

which has *no* **limit point**.* On the other hand, the series in (9.2–2) is divergent because its sequence of partial sums has *two* limit points, +1 and 0. The series in (9.2–3) is a trivial example of a convergent series since its sequence of partial sums has a unique limit point, namely, zero.

By a more sophisticated test (the **integral test**) it can be shown that the series of (9.2–4) is convergent for $p > 1$ and divergent for $p \leq 1$. The series of (9.2–5) is a **geometric series** with first term a and **common ratio** r. It can be shown (by the **ratio test**) that (9.2–5) converges if $|r| < 1$ and diverges if $r \geq 1$ and $a \neq 0$. The sum of (9.2–5) can be written in **closed form** as

$$a \sum_{n=0}^{\infty} r^n = \frac{a}{1 - r}, \qquad |r| < 1.$$

Finally, the series is an example of an **alternating series**, which can be proved (using a theorem of Leibniz) to be convergent because the following two conditions hold:

1. The absolute value of each term is less than or equal to the absolute value of its predecessor.

2. The limiting value of the nth term is zero as $n \to \infty$.

We remark that it is one thing to determine whether a given series converges, but quite another to determine what it converges to. It is not obvious that the sum of the series in (9.2–6) is $\pi/4$, although we have obtained this result and some others in the exercises for Section 7.2 (see Exercises 18 and 20 there).

Of greater interest to us than a series of constants will be **power series** of the form

$$a_0 + a_1(x - x_0) + a_2(x - x_0)^2 + \cdots + a_n(x - x_0)^n + \cdots. \quad \textbf{(9.2–7)}$$

Such a series is called a power series in $x - x_0$. A power series *always* converges. For example, (9.2–7) converges for $x = x_0$ but we will be interested in convergence on an *interval* such as $(x - R, x_0 + R)$. We call R the **radius of convergence** of the power series and can determine its value by using the ratio test, as shown in the following example.

EXAMPLE 9.2–1 Find the radius of convergence of the series

$$(x - 1) - \frac{(x - 1)^2}{2} + \frac{(x - 1)^3}{3} - \frac{(x - 1)^4}{4} + - \cdots.$$

---

* A point is called a *limit point* of a sequence if an infinite number of terms in the sequence are within a distance of ε of the point, where ε is an arbitrarily small positive number. A limit point need not be unique and need not be an element of the sequence. For example, 1 is the unique limit point of the sequence

$$\{\tfrac{1}{2}, \tfrac{3}{4}, \tfrac{7}{8}, \tfrac{15}{16}, \cdots\}.$$

Solution. It will be convenient to write the series using summation notation,

$$\sum_{n=1}^{\infty} \frac{(-1)^{n+1}(x-1)^n}{n}.$$

According to the ratio test a series converges whenever

$$\lim_{n \to \infty} \left| \frac{u_{n+1}}{u_n} \right| < 1,$$

where u_n represents the nth term of the series. In the present case,

$$\lim_{n \to \infty} \left| \frac{u_{n+1}}{u_n} \right| = \lim_{n \to \infty} \left| \frac{(-1)^{n+2}(x-1)^{n+1}}{n+1} \cdot \frac{n}{(-1)^{n+1}(x-1)^n} \right|$$

$$= \lim_{n \to \infty} \frac{n}{n+1} |x-1| = |x-1| < 1.$$

Hence $-1 < x - 1 < 1$ or $0 < x < 2$, which shows that the radius of convergence is 1. In many problems it is necessary to examine the endpoints of the **interval of convergence** as well and this must be done separately. It can be shown (Exercise 1) that the interval of convergence here is $0 < x \leq 2$. ∎

We will need two theorems, which we state without proof.

Theorem 9.2–1 A power series $\sum_{n=0}^{\infty} a_n(x - x_0)^n$ and its derivative $\sum_{n=1}^{\infty} na_n(x - x_0)^{n-1}$ have the same radius of convergence.

Theorem 9.2–2 Let a function $f(x)$ be represented by a power series $\sum_{n=0}^{\infty} a_n(x - x_0)^n$ in the interior of its interval of convergence. Then the function is differentiable there and its derivative is given by

$$f'(x) = \sum_{n=1}^{\infty} na_n(x - x_0)^{n-1}.$$

We are now ready to look at a simple differential equation with a view to solving it by using series. To begin with we will take x_0 to be zero. Later we will indicate why this is not always possible.

EXAMPLE 9.2–2 Find a solution of the equation $y'' - xy = 0$.

Solution. We *assume* that there is a solution of the form

$$y = \sum_{n=0}^{\infty} a_n x^n.$$

Then

$$y' = \sum_{n=1}^{\infty} na_n x^{n-1} \quad \text{and} \quad y'' = \sum_{n=2}^{\infty} n(n-1)a_n x^{n-2}.$$

Substituting these values into the given equation produces

$$\sum_{n=2}^{\infty} n(n-1)a_n x^{n-2} - \sum_{n=0}^{\infty} a_n x^{n+1} = 0.$$

In order to collect terms it would be convenient to have x^n in both summations. This can be accomplished by realizing that n is a *dummy index* of summation and can be replaced by any other letter just as we change variables in definite integrals. Accordingly, we replace *each* n in the first sum by $n + 2$ and each n in the second sum by $n - 1$. Then

$$\sum_{n=0}^{\infty} (n+2)(n+1)a_{n+2} x^n - \sum_{n=1}^{\infty} a_{n-1} x^n = 0.$$

Next we combine the two sums into one with n going from 1 to ∞, adding any terms that are left out of this sum. Thus

$$\sum_{n=1}^{\infty} ((n+2)(n+1)a_{n+2} - a_{n-1})x^n + 2a_2 = 0,$$

which is a linear combination of $1, x, x^2, \ldots$. Since the set

$$\{1, x, x^2, x^3, \ldots\}$$

is linearly independent, a linear combination of these functions can be zero if and only if each coefficient is zero. Hence

$$2a_2 = 0$$

and

$$(n+2)(n+1)a_{n+2} - a_{n-1} = 0.$$

From the first of these, $a_2 = 0$, and from the second we obtain the **recursion formula**,

$$a_{n+2} = \frac{a_{n-1}}{(n+2)(n+1)}, \qquad n = 1, 2, \ldots .$$

For $n = 1$ we have $a_3 = a_0/(3 \cdot 2)$, so that a_0 can be arbitrary. For $n = 2$ we have $a_4 = a_1/(4 \cdot 3)$, so that a_1 can be arbitrary. For $n = 3$ we have $a_5 = a_2/(5 \cdot 4) = 0$; consequently $a_2, a_5, a_8, \ldots$, are all zero. For $n = 4$ we have $a_6 = a_0/(3 \cdot 2 \cdot 6 \cdot 5)$. For $n = 5$ we have $a_7 = a_4/(7 \cdot 6) = a_1/(4 \cdot 3 \cdot 7 \cdot 6)$, and so on. The solution to the given differential equation is

$$y = a_0 + a_1 x + \frac{a_0}{6} x^3 + \frac{a_1}{12} x^4 + \frac{a_0}{180} x^6 + \frac{a_1}{504} x^7 + \cdots .$$

This last equation can also be written as

$$y = a_0\left(1 + \frac{x^3}{6} + \frac{x^6}{180} + \cdots\right) + a_1\left(x + \frac{x^4}{12} + \frac{x^7}{504} + \cdots\right),$$

which shows the two arbitrary constants we expect to find in the solution of a second-order differential equation. It can be shown (Exercise 2) that both series converge for $-\infty < x < \infty$. ■

Although many series can be written in **closed form**—for example,

$$e^x = 1 + x + \frac{x^2}{2!} + \frac{x^3}{3!} + \cdots; \qquad (9.2\text{--}8)$$

$$\sin x = x - \frac{x^3}{3!} + \frac{x^5}{5!} - \frac{x^7}{7!} + - \cdots; \qquad (9.2\text{--}9)$$

$$\cos x = 1 - \frac{x^2}{2!} + \frac{x^4}{4!} - \frac{x^6}{6!} + - \cdots; \qquad (9.2\text{--}10)$$

$$\log(1 + x) = x - \frac{x^2}{2} + \frac{x^3}{3} - \frac{x^4}{4} + - \cdots \qquad (9.2\text{--}11)$$

—this is not always possible. If a function can be represented in an open interval containing x_0 by a *convergent* series of the form $\sum_{n=0}^{\infty} a_n(x - x_0)^n$, then the function is said to be **analytic** at $x = x_0$. The functions in Eqs. (9.2–8) through (9.2–11) are all analytic at $x = 0$. If a function is analytic at every point at which it is defined, it is called an **analytic function**. All polynomials are analytic and so are rational functions except where their denominators vanish.

Now let us look at another example of a series solution of a differential equation.

EXAMPLE 9.2–3   Solve the equation

$$(x - 1)y'' - xy' + y = 0.$$

Solution. As before, assume*

$$y = \sum_0 a_n x^n, \qquad y' = \sum_1 na_n x^{n-1}, \qquad y'' = \sum_2 n(n - 1)a_n x^{n-2},$$

and substitute into the given differential equation. Then

$$\sum_2 n(n - 1)a_n x^{n-1} - \sum_2 n(n - 1)a_n x^{n-2} - \sum_1 na_n x^n + \sum_0 a_n x^n = 0.$$

---

* In the remainder of this section we will simplify the notation by omitting the upper limit on all summations.

Replace n by $n + 1$ in the first sum and replace n by $n + 2$ in the second sum so that

$$\sum_{1} (n + 1)na_{n+1}x^n - \sum_{0} (n + 2)(n + 1)a_{n+2}x^n - \sum_{1} na_n x^n + \sum_{0} a_n x^n = 0$$

or

$$\sum_{1} \left(n(n + 1)a_{n+1} - (n + 1)(n + 2)a_{n+2} - na_n + a_n\right)x^n - 2a_2 + a_0 = 0.$$

Equating the coefficients of various powers of x to zero gives us the following:

a_0 is arbitrary;

a_1 is arbitrary;

$$a_2 = \frac{1}{2}a_0;$$

$$a_{n+2} = \frac{n(n + 1)a_{n+1} + (1 - n)a_n}{(n + 1)(n + 2)}, \qquad n = 1, 2, \ldots;$$

$$a_3 = \frac{2a_2}{2 \cdot 3} = \frac{a_2}{3} = \frac{a_0}{3 \cdot 2};$$

$$a_4 = \frac{6a_3 - a_2}{3 \cdot 4} = \frac{a_3}{2} - \frac{a_2}{12} = \frac{a_0}{12} - \frac{a_0}{24} = \frac{a_0}{4!};$$

etc.

Hence

$$y = a_1 x + a_0 \left(1 + \frac{x^2}{2!} + \frac{x^3}{3!} + \frac{x^4}{4!} + \cdots\right),$$

and it can be shown (Exercise 3) that x and e^x are two linearly independent solutions of the given equation. Here the solution can be written in closed form in contrast to the solution of Example 9.2–2 (Exercise 4). ■

Unfortunately, the series method of solving ordinary differential equations is not as simple as the last two examples seem to indicate. Consider the equation

$$x^2 y'' - xy' - 3y = 0.$$

We leave it as an exercise (Exercise 5) to show that the series method with $x_0 = 0$ will produce only the trivial solution, $y = 0$. Yet the given equation is a Cauchy–Euler equation and both x^3 and $1/x$ are solutions (Exercise 6). The answer to the apparent mystery lies in the fact that the solutions of a Cauchy–Euler equation are not linearly independent on any interval that includes the origin. Recall that in Section 2.4 we solved Cauchy–Euler equations assuming that $x > 0$ *or* $x < 0$.

Consider the most general second-order, linear, homogeneous ordinary differential equation,

$$y'' + P(x)y' + Q(x)y = 0. \tag{9.2-12}$$

Those values of x, call them x_0, at which *both* $P(x)$ and $Q(x)$ are analytic are called **ordinary points** of Eq. (9.2–12). If either $P(x_0)$ or $Q(x_0)$ is not analytic, then x_0 is a **singular point** of Eq. (9.2–12). If, however, x_0 is a singular point but *both* $(x - x_0)P(x)$ and $(x - x_0)^2 Q(x)$ are analytic at $x = x_0$, then x_0 is called a **regular singular point** of Eq. (9.2–12). All other singular points are called **irregular singular points**.

EXAMPLE 9.2–4   Classify the singular points of the equation

$$(1 - x^2)y'' - 2xy' + n(n + 1)y = 0,$$

where $n = 0, 1, 2, \ldots$.

Solution. The only singular points are $x_0 = \pm 1$. If $x_0 = -1$, then

$$\frac{(x + 1)(-2x)}{1 - x^2} = \frac{2x}{x - 1} \quad \text{and} \quad \frac{(x + 1)^2 n(n + 1)}{1 - x^2} = \frac{n(n + 1)(x + 1)}{1 - x}.$$

Since both of these rational functions are analytic at $x = -1$, the latter is a regular singular point. Similarly, for $x_0 = 1$ we have

$$\frac{(x - 1)(-2x)}{1 - x^2} = \frac{2x}{x + 1} \quad \text{and} \quad \frac{(x - 1)^2 n(n + 1)}{1 - x^2} = \frac{n(n + 1)(1 - x)}{1 + x},$$

so that $x_0 = 1$ is also a regular singular point. ■

The point of all this is contained in Fuchs's theorem[*], which states that it is always possible to obtain *at least one* power series solution to a linear differential equation provided the assumed series solution is about an ordinary point or, at worst, a regular singular point.

The work of Fuchs was extended by Georg Frobenius (1849–1917), a German mathematician, who suggested that instead of assuming a series solution of the form $\sum_0 a_n x^n$, one should use the form $\sum_0 a_n x^{n+r}$. The use of this form to solve linear, ordinary differential equations is known today as the **method of Frobenius**. We illustrate with an example using the Cauchy–Euler equation referred to above.

EXAMPLE 9.2–5   Solve the equation

$$x^2 y'' - xy' - 3y = 0$$

by the method of Frobenius.

[*] After Lazarus Fuchs (1833–1902), a German mathematician.

Solution. We have

$$y = \sum_0 a_n x^{n+r},$$

$$y' = \sum_0 (n + r)a_n x^{n+r-1},$$

$$y'' = \sum_0 (n + r)(n + r - 1)a_n x^{n+r-2},$$

and, on substituting into the given equation, we have

$$\sum_0 ((n + r)(n + r - 1) - (n + r) - 3)a_n x^{n+r} = 0.$$

Since the coefficient of x^{n+r} must be zero for $n = 0, 1, 2, \ldots$, we have for $n = 0$,

$$(r^2 - 2r - 3)a_0 = 0.$$

Choosing a_0 to be arbitrary, that is, nonzero, produces

$$r^2 - 2r - 3 = 0,$$

which is called the **indicial equation**. Its roots are -1 and 3. In general,

$$a_n(n^2 + 2nr - 2n) = 0, \qquad n = 1, 2, \ldots,$$

which can be satisfied only by taking $a_n = 0$, $n = 1, 2, \ldots$. Hence we are left with the two possibilities

$$y_1(x) = a_0 x^{-1} \quad \text{and} \quad y_2(x) = b_0 x^3.$$

Note that the two constants are arbitrary since each root of the indicial equation leads to an infinite series. In this example, however, each series consists of a single term. ∎

When solving *second-order*, linear differential equations by the method of Frobenius the indicial equation is a quadratic equation and three possibilities exist. These are different from what one might expect, so we list them here.

1. If the roots of the indicial equation are *equal*, then only *one* solution can be obtained.

2. If the roots of the indicial equation differ by a number that is not an integer, then two linearly independent solutions may be obtained.

3. If the roots of the indicial equation differ by an integer, then the larger integer of the two will yield a solution, whereas the smaller may or may not produce a solution.

We conclude this section by solving two important differential equations, which will appear again in the following two sections.

EXAMPLE 9.2–6   Obtain a solution of the differential equation

$$\frac{d^2y}{dx^2} + \frac{1}{x}\frac{dy}{dx} + \left(1 - \frac{n^2}{x^2}\right)y = 0, \qquad n = 0, 1, 2, \ldots . \qquad \textbf{(9.2–13)}$$

This equation is known as **Bessel's differential equation**. It was originally obtained by Friedrich Wilhelm Bessel (1784–1846), a German mathematician, in the course of his studies of planetary motion. Since then this equation has appeared in problems of heat conduction, electromagnetic theory, and acoustics that are expressed in *cylindrical coordinates*.

Solution. Since the coefficients are not constant, we seek a series solution. Multiplying Eq. (9.2–13) by x^2, we obtain

$$x^2y'' + xy' + (x^2 - n^2)y = 0. \qquad \textbf{(9.2–14)}$$

We note that $x = 0$ is a regular singular point, hence we use the method of Frobenius. Assume that

$$y = \sum_{m=0}^{\infty} a_m x^{m+r},$$

$$y' = \sum_{m=0}^{\infty} a_m(m + r)x^{m+r-1},$$

$$y'' = \sum_{m=0}^{\infty} a_m(m + r)(m + r - 1)x^{m+r-2},$$

and substitute into Eq. (9.2–14). Then

$$\sum_{m=0}^{\infty} a_m(m + r)(m + r - 1)x^{m+r} + \sum_{m=0}^{\infty} a_m(m + r)x^{m+r}$$

$$+ \sum_{m=0}^{\infty} a_m x^{m+r+2} - n^2\sum_{m=0}^{\infty} a_m x^{m+r} = 0.$$

If we replace m by $m - 2$ in the third series, the last equation can be written as

$$\sum_{m=2}^{\infty} (a_m(m+r)(m+r-1) + a_m(m+r) + a_{m-2} - n^2 a_m)x^{m+r} + a_0 r(r-1)x^r$$

$$+ a_0 rx^r - n^2 a_0 x^r + a_1 r(r+1)x^{r+1} + a_1(r+1)x^{r+1} - n^2 a_1 x^{r+1} = 0.$$

Simplifying, we get

$$\sum_{m=2} (a_m((m + r)^2 - n^2) + a_{m-2})x^{m+r} + a_0(r^2 - n^2)x^r$$

$$+ a_1(r^2 + 2r + 1 - n^2)x^{r+1} = 0.$$

The coefficient of x^r must be zero; hence if we assume $a_0 \neq 0$, then we obtain $r = \pm n$. We choose the positive sign since n was defined as a nonnegative integer in (9.2–13). Since the coefficient of x^{r+1} must also be zero, we choose $a_1 = 0$. Then the recursion formula is obtained by setting the coefficient of x^{m+r} equal to zero. Thus,

$$a_m = \frac{-a_{m-2}}{m(m + 2n)}, \qquad m = 2, 3, \ldots .$$

The first few coefficients can be computed from this formula. They are as follows:

$$m = 2: \quad a_2 = \frac{-a_0}{2^2(n + 1)};$$

$$m = 4: \quad a_4 = \frac{-a_2}{2^3(n + 2)} = \frac{a_0}{2^4 \cdot 2(n + 1)(n + 2)};$$

$$m = 6: \quad a_6 = \frac{-a_4}{2^2 \cdot 3(n + 3)} = \frac{-a_0}{2^6 \cdot 3!(n + 1)(n + 2)(n + 3)}.$$

In general we have

$$a_{2m} = \frac{(-1)^m a_0}{2^{2m}m!(n + 1)(n + 2) \cdots (n + m)}, \qquad m = 1, 2, \ldots ,$$

and a solution to Eq. (9.2–13) can be written as

$$y_n(x) = a_0 \sum_{m=0} \frac{(-1)^m x^{2m+n}}{2^{2m}m!(n + 1)(n + 2) \cdots (n + m)}$$

$$= 2^n n! a_0 \sum_{m=0} \frac{(-1)^m}{m!(m + n)!} \left(\frac{x}{2}\right)^{2m+n}. \quad \blacksquare$$

The **Bessel function of the first kind of order** n is defined by giving a_0 the value $1/2^n n!$. Hence we have

$$J_n(x) = \sum_{m=0} \frac{(-1)^m}{m!(m + n)!} \left(\frac{x}{2}\right)^{2m+n}, \qquad n = 0, 1, 2, \ldots , \quad \textbf{(9.2–15)}$$

and this is a particular solution of Bessel's differential equation. We will consider this function in greater detail in Section 9.3.

EXAMPLE 9.2–7 Obtain a solution of the equation

$$(1 - x^2)y'' - 2xy' + n(n + 1)y = 0, \qquad n = 0, 1, 2, \ldots . \quad \textbf{(9.2–16)}$$

This equation is known as **Legendre's differential equation.***

Solution. Since $x = \pm 1$ are regular singular points (see Example 9.2–4) we may assume a power series about $x = 0$ which is an ordinary point. Accordingly, put

$$y = \sum_{m=0}^{\infty} a_m x^m, \qquad y' = \sum_{m=1}^{\infty} a_m m x^{m-1}, \qquad y'' = \sum_{m=2}^{\infty} a_m m(m - 1)x^{m-2}.$$

Substituting these values into Eq. (9.2–16) produces

$$\sum_{m=2}^{\infty} a_m m(m - 1)x^{m-2} - \sum_{m=2}^{\infty} a_m m(m - 1)x^m$$

$$- 2\sum_{m=1}^{\infty} a_m m x^m + n(n + 1)\sum_{m=0}^{\infty} a_m x^m = 0.$$

Replacing m by $m + 2$ in the first sum, we get

$$\sum_{m=0}^{\infty} a_{m+2}(m + 2)(m + 1)x^m - \sum_{m=2}^{\infty} a_m m(m - 1)x^m$$

$$- 2\sum_{m=1}^{\infty} a_m m x^m + n(n + 1)\sum_{m=0}^{\infty} a_m x^m = 0,$$

or

$$\sum_{m=2}^{\infty} (a_{m+2}(m + 2)(m + 1) - a_m m(m - 1) - 2a_m m + a_m n(n + 1))x^m$$

$$+ 2a_2 + 6a_3 x - 2a_1 x + n(n + 1)a_0 + n(n + 1)a_1 x = 0.$$

Setting the coefficient of each power of x equal to zero in the above, we have

$$2a_2 + n(n + 1)a_0 = 0, \qquad a_2 = \frac{-n(n + 1)a_0}{2}, \qquad a_0 \text{ arbitrary};$$

$$6a_3 - 2a_1 + n(n + 1)a_1 = 0, \qquad a_3 = \frac{(2 - n(n + 1))a_1}{6}, \qquad a_1 \text{ arbitrary}.$$

---

* After Adrien Marie Legendre (1752–1833), a French mathematician who is known mainly for his work in number theory, elliptic functions, and calculus of variations.

In general,

$$a_{m+2}(m+2)(m+1) - (m(m-1) + 2m - n(n+1))a_m = 0;$$

$$a_{m+2} = \frac{m(m+1) - n(n+1)}{(m+2)(m+1)} a_m;$$

$$a_{m+2} = \frac{(m-n)(m+n+1)}{(m+2)(m+1)} a_m, \qquad m = 0, 1, 2, \ldots. \qquad \textbf{(9.2–17)}$$

Equation (9.2–17) is the recurrence relation from which the coefficients can be found.

Computing the first few coefficients gives us

$$a_2 = \frac{-n(n+1)}{1 \cdot 2} a_0,$$

$$a_4 = \frac{(2-n)(n+3)}{4 \cdot 3} a_2 = \frac{n(n-2)(n+1)(n+3)}{4!} a_0,$$

$$a_6 = \frac{(4-n)(n+5)}{6 \cdot 5} a_4 = \frac{-n(n-2)(n-4)(n+1)(n+3)(n+5)}{6!} a_0,$$

$$a_3 = \frac{(1-n)(n+2)}{3 \cdot 2} a_1 = \frac{-(n-1)(n+2)}{3!} a_1,$$

$$a_5 = \frac{(3-n)(n+4)}{5 \cdot 4} a_3 = \frac{(n-1)(n-3)(n+2)(n+4)}{5!} a_1,$$

$$a_7 = \frac{(5-n)(n+6)}{7 \cdot 6} a_5 = \frac{-(n-1)(n-3)(n-5)(n+2)(n+4)(n+6)}{7!} a_1.$$

Hence a solution to Legendre's equation can be written as

$$y_n(x) = a_0 \left(1 - \frac{n(n+1)}{2!} x^2 + \frac{n(n-2)(n+1)(n+3)}{4!} x^4 \right.$$

$$\left. - \frac{n(n-2)(n-4)(n+1)(n+3)(n+5)}{6!} x^6 + - \cdots \right)$$

$$+ a_1 \left(x - \frac{(n-1)(n+2)}{3!} x^3 + \frac{(n-1)(n-3)(n+2)(n+4)}{5!} x^5 \right.$$

$$\left. - \frac{(n-1)(n-3)(n-5)(n+2)(n+4)(n+6)}{7!} x^7 + - \cdots \right).$$

$$\textbf{(9.2–18)}$$

Both series converge for $-1 < x < 1$.

If $n = 0, 2, 4, \ldots$, and a_1 is chosen to be zero, then the solutions, using Eq. (9.2–18), become

$$y_0(x) = a_0,$$
$$y_2(x) = a_0(1 - 3x^2),$$
$$y_4(x) = a_0(1 - 10x^2 + \tfrac{35}{3}x^4), \quad \text{etc.}$$

If we also impose the condition that $y_n(1) = 1$, then we can evaluate the a_0 to obtain

$$P_0(x) = 1, \qquad P_2(x) = \tfrac{1}{2}(3x^2 - 1), \qquad P_4(x) = \tfrac{1}{8}(35x^4 - 30x^2 + 3), \ldots.$$
$$(9.2–19)$$

These *polynomials* are called the **Legendre polynomials of even degree**.

If $n = 1, 3, 5, \ldots$, and a_0 is chosen to be zero, then the solutions, using Eq. (9.2–18), become

$$y_1(x) = a_1 x,$$
$$y_3(x) = a_1(x - \tfrac{5}{3}x^3),$$
$$y_5(x) = a_1(x - \tfrac{14}{3}x^3 + \tfrac{21}{5}x^5), \quad \text{etc.}$$

Again if we impose the condition that $y_n(1) = 1$, then we can evaluate the a_1 to obtain

$$P_1(x) = x, \qquad P_3(x) = \tfrac{1}{2}(5x^3 - 3x), \qquad P_5(x) = \tfrac{1}{8}(63x^5 - 70x^3 + 15x), \ldots.$$
$$(9.2–20)$$

These polynomials are called the **Legendre polynomials of odd degree.** ■

The **Legendre polynomials** will be of use in Section 9.4 since they arise in boundary-value problems expressed in *spherical coordinates*.

KEY WORDS AND PHRASES

series of constants
divergent and convergent series
sequence
partial sums
limit point
integral test
ratio test
geometric series
alternating series
power series
closed form
radius of convergence
interval of convergence

recursion formula
analytic function
singular point
ordinary point
regular and irregular
    singular point
method of Frobenius
indicial equation
Bessel's differential equation
Bessel function of the first kind
    of order n
Legendre's differential equation
Legendre polynomials

EXERCISES 9.2

▶ **1.** Show that the interval of convergence of the series in Example 9.2–1 is $0 < x \le 2$.

2. a) Show that one solution of the differential equation $y'' - xy = 0$ in Example 9.2–2 can be written as

$$y_1(x) = a_0 \sum_{n=0}^{\infty} \frac{1 \cdot 4 \cdot 7 \cdots (3n-2)}{(3n)!} x^{3n}.$$

b) Write the second solution in a similar form.

c) Find the radius of convergence of the series in part (a).

3. Verify that $y_1(x) = x$ and $y_2(x) = e^x$ are linearly independent solutions of the differential equation $(x-1)y'' - xy' + y = 0$.

4. a) Show that the solution obtained in Example 9.2–4 is equivalent to

$$y(x) = c_1 x + c_2 e^x.$$

b) For what values of x is the above solution valid?

5. Show that assuming a solution of the form $y = \sum_0 a_n x^n$ for the equation

$$x^2 y'' - xy' - 3y = 0$$

leads to the trivial solution $y = 0$.

6. Verify that x^3 and $1/x$ are linearly independent solutions of the equation in Exercise 5 on every interval not containing the origin.

▶▶ **7.** Consider the sequence

$$\left\{ \tfrac{1}{2}, \tfrac{3}{4}, \tfrac{7}{8}, \tfrac{15}{16}, \ldots \right\}.$$

a) Write the nth term of the sequence.

b) Show that 1 is the limit point of the sequence.

8. Apply the ratio test to the series (9.2–5) to show that the series converges for $|r| < 1$.

9. a) Identify the series

$$\frac{1}{10} + \frac{1}{100} + \frac{1}{1000} + \cdots.$$

b) Find the sum of the series.

10. Find the radius of convergence of each of the following power series.

a) $(x-1) + \dfrac{(x-1)^3}{3} + \dfrac{(x-1)^5}{5} + \cdots$

b) $1 + \dfrac{x^2}{2!} + \dfrac{x^4}{4!} + \dfrac{x^6}{6!} + \cdots$

c) $1 + \dfrac{(x+3)}{2} + \dfrac{(x+3)^2}{3} + \dfrac{(x+3)^3}{4} + \cdots$

d) $x + \dfrac{2!x^2}{2^2} + \dfrac{3!x^3}{3^3} + \dfrac{4!x^4}{4^4} + \cdots$

(*Hint*: Use the limit definition of e.)

e) $1 + \dfrac{(x+2)}{3} + \dfrac{(x+2)^2}{2 \cdot 3^2} + \dfrac{(x+2)^3}{3 \cdot 3^3} + \cdots$

f) $1 + \dfrac{(x-1)^2}{2!} + \dfrac{(x-1)^4}{4!} + \dfrac{(x-1)^6}{6!} + \cdots$

11. Classify the singular points of each of the following differential equations.

  a) $x^2 y'' + x y' + (x^2 - n^2) y = 0, \quad n = 0, 1, 2, \ldots$

  b) $x^3 y'' - x y' + y = 0$

  c) $x^2 y'' + (4x - 1) y' + 2y = 0$

  d) $x^3 (x - 1)^2 y'' + x^4 (x - 1)^3 y' + y = 0$

12. Use power series to solve each of the following equations.

  a) $y'' + y = 0$

  b) $y'' - y = 0$

  c) $y' - y = x^2$ (Note that the power series method is not limited to homogeneous equations.)

  d) $y' - xy = 0$ (If possible, write the solution in closed form.)

  e) $(1 + x^2) y'' + 2xy' - 2y = 0$

13. Solve each of the following differential equations by the method of Frobenius.

  a) $xy'' + y' + xy = 0$

  b) $4xy'' + 2y' + y = 0$

  c) $x^2 y'' + 2xy' - 2y = 0$

14. Solve the equation

$$xy'' + 2y' = 0$$

by two methods. (*Hint*: x is an integrating factor.)

15. Solve the equation

$$y'' - xy' - y = 0$$

by assuming a solution that is a power series in $(x - 1)$. In this case the coefficient x must also be written in terms of $x - 1$. This can be done by assuming that $x = A(x - 1) + B$ and determining the constants A and B.

▶▶▶ **16.** The differential equation

$$y'' - xy = 0$$

is known as **Airy's equation*** and its solutions are called **Airy functions**, which have applications in the theory of diffraction (Fig. 9.2–1).

  a) Obtain the solution in terms of a power series in x.

  b) Obtain the solution in terms of a power series in $(x - 1)$. (Compare Exercise 15.)

---

* After Sir George Airy (1801–1892), an English mathematician and astronomer.

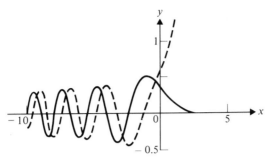

FIGURE 9.2–1 Airy functions.

17. Illustrate Theorem 9.2–1 by differentiating the series in Exercise 10 and finding the radii of convergence of the differentiated series. (Note that this does *not* constitute a *proof* of the theorem.)

18. a) Illustrate Theorem 9.2–2 by differentiating the functions and series in Eqs. (9.2–8) through (9.2–11).

b) What is the radius of convergence of the series in Eqs. (9.2–8) through (9.2–11)? (*Note:* You will have to use the fact that if the series of *absolute values* converges, then the alternating series also converges.)

19. Compare the solutions of Exercise 16 with the solutions of $y'' - y = 0$. Comment.

9.3 BESSEL FUNCTIONS

Laplace's equation in cylindrical coordinates (ρ, ϕ, z) has the form (Exercise 1)

$$\nabla^2 u = \frac{1}{\rho}\frac{\partial}{\partial\rho}\left(\rho\frac{\partial u}{\partial\rho}\right) + \frac{1}{\rho^2}\frac{\partial^2 u}{\partial\phi^2} + \frac{\partial^2 u}{\partial z^2} = 0. \tag{9.3–1}$$

We apply the method of separation of variables by assuming that u is a product of functions of ρ, ϕ, and z, that is,

$$u = R(\rho)\Phi(\phi)Z(z),$$

and substitute the appropriate derivatives into the partial differential equation. Then

$$\frac{\Phi Z}{\rho}\frac{d}{d\rho}\left(\rho\frac{dR}{d\rho}\right) + \frac{RZ}{\rho^2}\frac{d^2\Phi}{d\phi^2} + R\Phi\frac{d^2Z}{dz^2} = 0,$$

and dividing by $R\Phi Z/\rho^2$ yields

$$\frac{\rho}{R}\frac{d}{d\rho}\left(\rho\frac{dR}{d\rho}\right) + \frac{\rho^2}{Z}\frac{d^2Z}{dz^2} = -\frac{1}{\Phi}\frac{d^2\Phi}{d\phi^2}.$$

Since the left member of the last equation is independent of ϕ, the equation can be satisfied only if both members are equal to a constant. Hence

$$-\frac{1}{\Phi}\frac{d^2\Phi}{d\phi^2} = n^2, \qquad n = 0, 1, 2, \ldots,$$

and a second separation yields

$$\frac{1}{\rho R}\frac{d}{d\rho}\left(\rho\frac{dR}{d\rho}\right) - \frac{n^2}{\rho^2} = -\frac{1}{Z}\frac{d^2Z}{dz^2} = -\lambda^2.$$

We have called the first *separation constant* n^2 because this will force Φ (and u) to be periodic of period 2π in ϕ. This is the desired situation in many applied problems. We have called the second separation constant λ^2 because we do *not* want Z (and u) to be periodic in z.

The values of the separation constants n and λ are actually dictated by the nature of the boundary conditions that u must satisfy. We have chosen those values that conform to the boundary conditions most often found in problems dealing with this general topic at this level.

Thus, by separation of variables, we have reduced Laplace's equation to the following three ordinary, linear, homogeneous differential equations:

$$\frac{d^2Z}{dz^2} - \lambda^2 Z = 0, \tag{9.3-2}$$

$$\frac{d^2\Phi}{d\phi^2} + n^2\Phi = 0, \qquad n = 0, 1, 2, \ldots, \tag{9.3-3}$$

$$\frac{d^2R}{d\rho^2} + \frac{1}{\rho}\frac{dR}{d\rho} + \left(\lambda^2 - \frac{n^2}{\rho^2}\right)R = 0. \tag{9.3-4}$$

The solutions to the first two are straightforward and are given by

$$Z(\lambda z) = Ae^{\lambda z} + Be^{-\lambda z} \tag{9.3-5}$$

and

$$\Phi(n\phi) = C\cos n\phi + D\sin n\phi, \tag{9.3-6}$$

respectively. Equation (9.3–4) is a Bessel equation (see Example 9.2–6) with solutions $J_n(\lambda\rho)$ and $Y_n(\lambda\rho)$. The first of these is called the **Bessel function of the first kind of order** n and the second is the **Bessel function of the second kind of order** n.* Hence the general solution of Eq. (9.3–4) can be written as

$$R_n(\lambda\rho) = EJ_n(\lambda\rho) + FY_n(\lambda\rho). \tag{9.3-7}$$

We remark that the solutions to Laplace's equation in cylindrical coordinates are *products* of the functions shown in Eqs. (9.3–5), (9.3–6),

---

* Bessel functions of the second kind will be discussed later in this section.

and (9.3–7). A function u that satisfies $\nabla^2 u = 0$ is called a *harmonic function*; hence the products referred to are called **cylindrical harmonics**. Since $J_n(\lambda\rho)$ is defined at $\rho = 0$ whereas $Y_n(\lambda\rho)$ is not (as we shall see later), we choose the arbitrary constant F to be zero if u must be bounded at the origin. We would also choose $A = 0$ if $\lambda > 0$ and it is required that $\lim_{z\to\infty} |u|$ exist. Moreover, one of the boundary conditions may require that either C or D be zero. Thus, in practice, the cylindrical harmonics are not as formidable as they may seem.

In Section 9.5 we will look at some applications that involve Bessel functions but in this section we examine these functions in more detail. We have solved the equivalent of Eq. (9.3–4) (see Exercise 2) and obtained the solution

$$J_n(x) = \sum_{m=0}^{\infty} \frac{(-1)^m}{m!(m+n)!}\left(\frac{x}{2}\right)^{2m+n}, \qquad n = 0, 1, 2, \ldots \quad \textbf{(9.3–8)}$$

in Example 9.2–6.

We next examine $J_0(x)$ and $J_1(x)$ in some detail. From Eq. (9.3–8) we have

$$J_0(x) = 1 - \frac{x^2}{2^2} + \frac{x^4}{2^2 \cdot 4^2} - \frac{x^6}{2^2 \cdot 4^2 \cdot 6^2} + - \cdots,$$

$$J_1(x) = \frac{x}{2} - \frac{x^3}{2^2 \cdot 4} + \frac{x^5}{2^2 \cdot 4^2 \cdot 6} - \frac{x^7}{2^2 \cdot 4^2 \cdot 6^2 \cdot 8} + - \cdots.$$

Comparing these functions with the Maclaurin's series for $\cos x$ and $\sin x$,

$$\cos x = 1 - \frac{x^2}{2!} + \frac{x^4}{4!} - \frac{x^6}{6!} + - \cdots,$$

$$\sin x = x - \frac{x^3}{3!} + \frac{x^5}{5!} - \frac{x^7}{7!} + - \cdots,$$

we note a similarity between $J_0(x)$ and $\cos x$ and also between $J_1(x)$ and $\sin x$. For example (Exercise 4),

$$J_0(0) = 1, \qquad\qquad \cos 0 = 1;$$
$$J_0(-x) = J_0(x), \qquad \cos(-x) = \cos x;$$

$$J_0'(0) = 0, \qquad\qquad \frac{d}{dx}(\cos x)\Big|_{x=0} = 0;$$

$$J_1(0) = 0, \qquad\qquad \sin 0 = 0;$$
$$J_1(-x) = -J_1(x), \qquad \sin(-x) = -\sin x;$$

$$J_0'(x) = -J_1(x), \qquad \frac{d}{dx}(\cos x) = -\sin x.$$

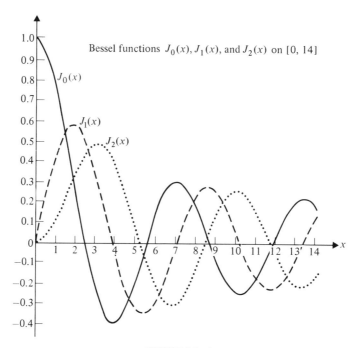

FIGURE 9.3–1

These similarities can be seen in the graphs of $J_0(x)$ and $J_1(x)$ shown in Fig. 9.3–1. Note that the Bessel functions of the first kind are almost periodic with a variable period nearly equal to 2π. In fact, the period approaches 2π as $x \to \infty$. Note also that the amplitudes are decreasing with increasing x.

In solving boundary-value problems the zeros of $J_n(x)$, that is, the roots of $J_n(x) = 0$, are of importance. Approximate values of the first few zeros of $J_0(x)$ and $J_1(x)$ are listed below.

	1st	2nd	3rd	4th	5th	6th
$J_0(x)$	2.405	5.520	8.654	11.792	14.931	18.071
$J_1(x)$	0	3.832	7.016	10.173	13.323	16.471

Another useful relation, namely,

$$\frac{d}{dx}(x^n J_n(x)) = x^n J_{n-1}(x), \qquad n = 1, 2, \ldots, \tag{9.3–9}$$

can be easily obtained (Exercise 4) from Eq. (9.3–8). In differential form Eq. (9.3–9) becomes

$$d(x^n J_n(x)) = x^n J_{n-1}(x)\, dx$$

and integrating from 0 to c $(c > 0)$, we have

$$x^n J_n(x)\Big|_0^c = \int_0^c x^n J_{n-1}(x)\,dx$$

or

$$\int_0^c x^n J_{n-1}(x)\,dx = c^n J_n(c).$$

When $n = 1$ this reduces to

$$\int_0^c x J_0(x)\,dx = c J_1(c), \qquad\qquad \textbf{(9.3–10)}$$

a result that will be useful in Section 9.5.

Orthogonality of Bessel functions

Bessel functions of the first kind satisfy an **orthogonality relation** under certain conditions and we look at this property next.

Bessel's differential equation of order n can be written as

$$x^2 \frac{d^2 u}{dx^2} + x \frac{du}{dx} + (\lambda^2 x^2 - n^2)u = 0. \qquad\qquad \textbf{(9.3–11)}$$

A particular solution of this equation is $J_n(\lambda x)$, the Bessel function of the first kind of order n. Similarly, $J_n(\mu x)$ is a particular solution of

$$x^2 \frac{d^2 v}{dx^2} + x \frac{dv}{dx} + (\mu^2 x^2 - n^2)v = 0. \qquad\qquad \textbf{(9.3–12)}$$

Now multiply Eq. (9.3–11) by v/x and Eq. (9.3–12) by u/x and subtract to obtain

$$vx \frac{d^2 u}{dx^2} + v\frac{du}{dx} + (\lambda^2 x^2 - n^2)\frac{uv}{x} - ux\frac{d^2 v}{dx^2} - u\frac{dv}{dx} - (\mu^2 x^2 - n^2)\frac{uv}{x} = 0.$$

This can be written as

$$(\lambda^2 - \mu^2)xuv = ux\frac{d^2 v}{dx^2} - vx\frac{d^2 u}{dx^2} + u\frac{dv}{dx} - v\frac{du}{dx}$$

$$= \frac{d}{dx}\left(x\left(u\frac{dv}{dx} - v\frac{du}{dx} \right) \right)$$

Hence, for any $c > 0$,

$$(\lambda^2 - \mu^2)\int_0^c xuv\,dx = \int_0^c d\left(x\left(u\frac{dv}{dx} - v\frac{du}{dx} \right) \right)$$

or, if we replace u and v by their values $J_n(\lambda x)$ and $J_n(\mu x)$, respectively,

$$(\lambda^2 - \mu^2)\int_0^c x J_n(\lambda x)J_n(\mu x)\,dx = x(\mu J_n(\lambda x)J_n'(\mu x) - \lambda J_n(\mu x)J_n'(\lambda x))\Big|_0^c$$

$$= c(\mu J_n(\lambda c)J_n'(\mu c) - \lambda J_n(\mu c)J_n'(\lambda c)).$$

Thus we have

$$\int_0^c x J_n(\lambda x) J_n(\mu x)\, dx = \frac{c}{\lambda^2 - \mu^2}\left(\mu J_n(\lambda c) J_n'(\mu c) - \lambda J_n(\mu c) J_n'(\lambda c)\right).$$

From this it follows that

$$\int_0^c x J_n(\lambda x) J_n(\mu x)\, dx = 0,$$

provided that $\lambda \neq \mu$ *and*

$$\mu J_n(\lambda c) J_n'(\mu c) - \lambda J_n(\mu c) J_n'(\lambda c) = 0. \qquad \textbf{(9.3–13)}$$

Now Eq. (9.3–13) holds if λc and μc are different roots of:

a) $J_n(x) = 0$, because then $J_n(\lambda c) = 0$ and $J_n(\mu c) = 0$;

b) $J_n'(x) = 0$, because then $J_n'(\mu c) = 0$ and $J_n'(\lambda c) = 0$;

c) $h J_n(x) + x J_n'(x) = 0$, where $h > 0$.

To see the last condition note that if λc and μc are different roots of $h J_n(x) + x J_n'(x) = 0$, then it follows that

$$h J_n(\lambda c) + \lambda c J_n'(\lambda c) = 0 \quad \text{and} \quad h J_n(\mu c) + \mu c J_n'(\mu c) = 0.$$

Multiplying the first of these by $\mu J_n'(\mu c)$ and the second by $\lambda J_n'(\lambda c)$ results in

$$h \mu J_n'(\mu c) J_n(\lambda c) + c \lambda \mu J_n'(\lambda c) J_n'(\mu c) = 0,$$
$$h \lambda J_n'(\lambda c) J_n(\mu c) + c \lambda \mu J_n'(\lambda c) J_n'(\mu c) = 0.$$

Subtracting, we have

$$h\left(\mu J_n(\lambda c) J_n'(\mu c) - \lambda J_n(\mu c) J_n'(\lambda c)\right) = 0$$

or

$$\mu J_n(\lambda c) J_n'(\mu c) - \lambda J_n(\mu c) J_n'(\lambda c) = 0,$$

which is identical to Eq. (9.3–13).

Finally observe that if $h = 0$ in condition (c) above, then condition (c) is equivalent to condition (b). Hence we may take $h \geq 0$ in condition (c).

In the foregoing we have shown that the Bessel functions $J_n(\lambda x)$ and $J_n(\mu x)$ are orthogonal on the interval $0 < x < c$ *with weight function* x provided that $\lambda \neq \mu$ and one of conditions (a), (b), or (c) above holds.

Fourier–Bessel series

We note that, although Bessel's differential equation does not meet the conditions required of a *regular* Sturm–Liouville system (8.5–1), it does fall into a special class of *singular* systems discussed briefly at the end of Section 8.5. It can also be shown* that the normalized eigenfunctions that we will obtain in this section form a **complete set** (see Eq. 7.1–22) with respect to the class of piecewise smooth functions on the interval $(0, c)$ with $c > 0$. Thus we will be able to represent piecewise smooth functions by a series of Bessel functions of the first kind, called a **Fourier–Bessel series**.

Let $c\lambda_j, j = 1, 2, 3, \ldots$, be the positive roots (zeros) of $J_n(\lambda c) = 0$. These roots can be found tabulated in various mathematical tables for computational purposes. For example, if $n = 0$, then, approximately, $\lambda_1 c = 2.405$, $\lambda_2 c = 5.520$, $\lambda_3 c = 8.654$, etc.

Now consider the representation on $(0, c)$ of the function f:

$$f(x) = A_1 J_n(\lambda_1 x) + A_2 J_n(\lambda_2 x) + A_3 J_n(\lambda_3 x) + \cdots. \qquad (9.3-14)$$

If we wish to find the value of A_2, say, then we multiply each term by $x J_n(\lambda_2 x) \, dx$ and integrate from 0 to c. Then we have

$$\int_0^c x f(x) J_n(\lambda_2 x) \, dx = A_1 \int_0^c x J_n(\lambda_1 x) J_n(\lambda_2 x) \, dx$$
$$+ A_2 \int_0^c x J_n(\lambda_2 x) J_n(\lambda_2 x) \, dx$$
$$+ A_3 \int_0^c x J_n(\lambda_3 x) J_n(\lambda_2 x) \, dx + \cdots.$$

Because of the orthogonality of the Bessel functions all integrals on the right with the exception of the second one are zero. Thus we have

$$\int_0^c x f(x) J_n(\lambda_2 x) \, dx = A_2 \int_0^c x J_n^2(\lambda_2 x) \, dx,$$

from which we obtain

$$A_2 = \frac{\int_0^c x f(x) J_n(\lambda_2 x) \, dx}{\int_0^c x J_n^2(\lambda_2 x) \, dx}.$$

Since the same procedure can be used for *any* coefficient, we have, in general,

$$A_j = \frac{\int_0^c x f(x) J_n(\lambda_j x) \, dx}{\int_0^c x J_n^2(\lambda_j x) \, dx}, \qquad j = 1, 2, 3, \ldots . \qquad (9.3-15)$$

Next we evaluate the denominator in the last expression. To do this we go back to Bessel's differential equation of order n,

$$x u'' + u' + \left(\lambda_j^2 x - \frac{n^2}{x} \right) u = 0.$$

---

* See Andrew Gray, and G. B. Mathews, *A Treatise on Bessel Functions and Their Applications to Physics*, 2d ed., p. 94 ff. (Philadelphia: Dover, 1966).

A particular solution of this equation is $u = J_n(\lambda_j x)$. Multiplying by the *integrating factor* $2u'x$ yields

$$2x^2 u'u'' + 2(u')^2 x + \left(\lambda_j^2 x - \frac{n^2}{x}\right) 2xu'u = 0$$

or

$$2xu'(xu'' + u') + (\lambda_j^2 x^2 - n^2)2uu' = 0. \qquad \textbf{(9.3–16)}$$

Using the fact that

$$\frac{d}{dx}(xu')^2 = 2xu'(xu'' + u')$$

and

$$\frac{d}{dx}(u^2) = 2uu',$$

we can write the differential equation (9.3–16) as

$$\frac{d}{dx}(xu')^2 + (\lambda_j^2 x^2 - n^2)\frac{d}{dx}(u^2) = 0.$$

Hence, integrating from 0 to c, we have

$$\int_0^c d(xu')^2 + \int_0^c (\lambda_j^2 x^2 - n^2)\, d(u^2) = 0.$$

Using integration by parts for the second integral gives us

$$w = \lambda_j^2 x^2 - n^2, \qquad\qquad dv = d(u^2),$$
$$dw = 2\lambda_j^2 x\, dx, \qquad\qquad v = u^2,$$

$$(xu')^2\Big|_0^c + u^2(\lambda_j^2 x^2 - n^2)\Big|_0^c - 2\lambda_j^2 \int_0^c xu^2\, dx = 0.$$

But $u = J_n(\lambda_j x)$ and $u' = \lambda_j J_n'(\lambda_j x)$ so the last expression becomes

$$\lambda_j^2 c^2 [J_n'(\lambda_j c)]^2 + J_n^2(\lambda_j c)(\lambda_j^2 c^2 - n^2) + n^2 J_n^2(0) = 2\lambda_j^2 \int_0^c xJ_n^2(\lambda_j x)\, dx.$$

$$\textbf{(9.3–17)}$$

Since $J_n(\lambda_j c) = 0$ and $J_n(0) = 0$ for $n = 1, 2, 3, \ldots$, Eq. (9.3–17) reduces to

$$\int_0^c xJ_n^2(\lambda_j x)\, dx = \frac{c^2}{2}[J_n'(\lambda_j c)]^2. \qquad \textbf{(9.3–18)}$$

Hence the coefficients in Eq. (9.3–15) become

$$A_j = \frac{2}{c^2 [J_n'(\lambda_j c)]^2}\int_0^c xf(x)J_n(\lambda_j x)\, dx,$$

$$j = 1, 2, 3, \ldots. \qquad \textbf{(9.3–19)}$$

Noting that x in the above is the variable in a definite integral and can thus be replaced by any other letter, we can write the Fourier–

Bessel series in Eq. (9.3–14) as

$$f(x) = \frac{2}{c^2} \sum_{j=1}^{\infty} \frac{J_n(\lambda_j x)}{[J_n'(\lambda_j c)]^2} \int_0^c s f(s) J_n(\lambda_j s) \, ds.$$

The equality in this representation of $f(x)$ is not to be taken literally. As in the case of Fourier series, this series will converge to the *mean value* of the function at points where f has a jump discontinuity. At points where f is continuous the series will converge to the functional value at that point.

Equation (9.3–18) expresses the **square of the norm** of the eigenfunction $J_n(\lambda_j x)$. Thus we have for the **norm** (indicated by the double bars)

$$\|J_n(\lambda_j x)\| = \frac{c}{\sqrt{2}} J_n'(\lambda_j c).$$

The set

$$\left\{ \frac{\sqrt{2}}{c} \frac{J_n(\lambda_1 x)}{J_n'(\lambda_1 c)}, \frac{\sqrt{2}}{c} \frac{J_n(\lambda_2 x)}{J_n'(\lambda_2 c)}, \ldots \right\}$$

is an orthonormal set on the interval $(0, c)$ with weight function x if the λ_j are such that $J_n(\lambda_j c) = 0$.

If the λ_j are such that $h J_n(\lambda_j c) + \lambda_j c J_n'(\lambda_j c) = 0$, then solving for $J_n'(\lambda_j c)$, we have

$$J_n'(\lambda_j c) = -\frac{h}{\lambda_j c} J_n(\lambda_j c). \tag{9.3–20}$$

Substituting this value into Eq. (9.3–17) produces

$$\int_0^c x J_n^2(\lambda_j x) \, dx = \frac{\lambda_j^2 c^2 - n^2 + h^2}{2\lambda_j^2} J_n^2(\lambda_j c).$$

Thus in this case we use, in place of Eq. (9.3–19), the following formula for the coefficients in the Fourier–Bessel series:

$$A_j = \frac{2\lambda_j^2}{(\lambda_j^2 c^2 - n^2 + h^2) J_n^2(\lambda_j c)} \int_0^c x f(x) J_n(\lambda_j x) \, dx,$$

$$j = 1, 2, 3, \ldots . \tag{9.3–21}$$

The formula (9.3–21) is not valid for $j = 1$ in the single case when $h = 0$ and $n = 0$. Then Eq. (9.3–20) becomes

$$J_0'(\lambda_j c) = 0,$$

that is, $\lambda_j c$ is a zero of $J_0'(x)$. The *first* zero of $J_0'(x)$ is at $x = 0$, hence $\lambda_1 = 0$, the first zero of $J_0'(\lambda c)$. But then $J_0(0) = 1$ and the coefficient of the integral in Eq. (9.3–21) can be evaluated by L'Hôpital's rule. Hence

$$A_1 = \frac{2}{c^2} \int_0^c x f(x) \, dx, \tag{9.3–22}$$

with $A_j, j = 2, 3, \ldots$, given by Eq. (9.3–21).

Bessel functions of the second kind

We conclude this section with a brief reference to the Bessel function of the *second* kind of order n. It is a second linearly independent solution of Bessel's differential equation, which can be obtained by the method of variation of parameters. The details will be omitted and we will give only $Y_0(x)$ and graphs of this function as well as $Y_1(x)$ and $Y_2(x)$ in Fig. 9.3–2. The Bessel function of the second kind of order

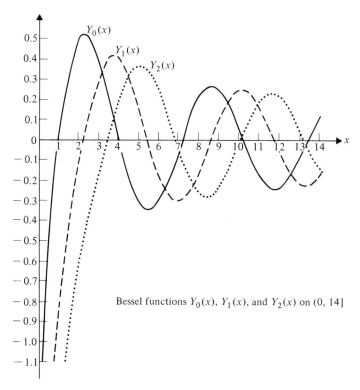

Bessel functions $Y_0(x)$, $Y_1(x)$, and $Y_2(x)$ on $(0, 14]$

FIGURE 9.3–2

zero is given by*

$$Y_0(x) = \frac{2}{\pi}\left(J_0(x)\left(\log\frac{x}{2} + \gamma\right)\right) + \frac{2}{\pi}\left(\frac{x^2}{4} - \frac{3x^4}{128} + \frac{11x^6}{13,824} + - \cdots\right),$$

(9.3–23)

where the term γ is called Euler's constant. It is an irrational number defined by

$$\gamma = \lim_{n\to\infty}\left(\sum_{k=1}^{n}\frac{1}{k} - \log n\right) \doteq 0.577215.$$

Of importance in our future work is the fact that all Bessel functions of the second kind contain the term $\log x/2$. Consequently, $Y_n(0)$ is undefined, as Fig. 9.3–2 indicates. In view of the fact that the applications involving Bessel functions in this text are such that $x > 0$, we will not be using Bessel functions of the second kind. It should be noted, however, that the latter are required in certain instances—for example, in solving problems involving electromagnetic waves in coaxial cables.

KEY WORDS AND PHRASES

Bessel function of the first and
    second kind of order n

cylindrical harmonics
norm

EXERCISES 9.3

▶ 1. Obtain Eq. (9.3–1) from Eq. (9.1–2).

2. By making the substitutions $y = R$ and $x = \lambda\rho$, show that Eq. (9.3–4) can be reduced to Eq. (9.2–13).

3. Use the ratio test to show that the series (9.3–8) representing Bessel's function of the first kind of order n converges for all x.

4. Use Eq. (9.3–8) to prove each of the following.

    a) $J_0'(0) = 0$ b) $J_1(0) = 0$

    c) $J_1(-x) = -J_1(x)$ d) $J_0'(x) = -J_1(x)$

    e) $xJ_n'(x) = -nJ_n(x) + xJ_{n-1}(x), \quad n = 1, 2, \ldots$

    f) $\dfrac{d}{dx}(x^n J_n(x)) = x^n J_{n-1}(x), \quad n = 1, 2, \ldots$

---

* Some authors call this the **Neumann function** of order zero, while others call it Weber's Bessel function of the second kind of order zero.

5. Verify that $x = 0$ is a regular singular point of Bessel's differential equation (9.3–11).

6. Obtain the general solution of each of the following differential equations *by inspection.*

a) $\dfrac{d}{dx}\left(x\dfrac{dy}{dx}\right) + xy = 0$ **b)** $4xy'' + 4y' + y = 0$

c) $\dfrac{d^2y}{dx^2} + ye^x = 0$

(*Hint*: Let $u = e^x$.)

7. If $J_0(\lambda_j) = 0$, then obtain each of the following.

a) $\displaystyle\int_0^1 J_1(\lambda_j s)\, ds = 1/\lambda_j$ **b)** $\displaystyle\int_0^{\lambda_j} J_1(s)\, ds = 1$

c) $\displaystyle\int_0^\infty J_1(\lambda_j s)\, ds = 0$

8. Obtain each of the following.

a) $\displaystyle\int^x J_0(s)J_1(s)\, ds = -\tfrac{1}{2}(J_0(x))^2$ **b)** $\displaystyle\int^x s^2 J_0(s)J_1(s)\, ds = \tfrac{1}{2}x^2(J_1(x))^2$

9. Expand each of the following functions in a Fourier–Bessel series of functions $J_0(\lambda_j x)$ on the interval $0 < x < c$, where $J_0(\lambda_j c) = 0$. (*Note*: The coefficients are given by Eq. (9.3–19) but the integral need not be evaluated in all cases.)

a) $f(x) = 1$

b) $f(x) = x^2$

(*Note*: Use the following reduction formula:
$$\int_0^x s^n J_0(s)\, ds = x^n J_1(x) + (n-1)J_0(x) - (n-1)^2 \int_0^x s^{n-2} J_0(s)\, ds,$$
$$n = 2, 3, \dots .)$$

c) $f(x) = \begin{cases} 0, & \text{for } 0 < x < 1, \\ 1/x, & \text{for } 1 \le x \le 2. \end{cases}$

(See Fig. 9.3–3.)

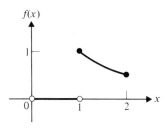

FIGURE 9.3–3

10. Bessel's differential equation is notorious for its many disguises. Show that each of the following is a Bessel's differential equation.

a) $\dfrac{dy}{dx} + ay^2 + \dfrac{1}{x}y + \dfrac{1}{a} = 0$

(This is a **Riccati equation** but the substitution

$$y = \frac{1}{az}\frac{dz}{dx}$$

transforms it into a Bessel equation.)

b) $r^2 \dfrac{d^2R}{dr^2} + 2r\dfrac{dR}{dr} + (\lambda^2 r^2 - n(n+1))R = 0$

(This equation arises when the **Helmholtz equation*** in spherical coordinates is solved by separation of variables. Make the substitution

$$R(\lambda r) = \frac{Z(\lambda r)}{(\lambda r)^{1/2}}$$

to transform it into a Bessel equation of order $n + \frac{1}{2}$.)

c) $\dfrac{d^2y}{dx^2} + \dfrac{1}{x}\dfrac{dy}{dx} + \dfrac{n}{k}y = 0$

(This is **Fourier's equation** but the substitution $x\sqrt{(n/k)} = z$ will transform it into a Bessel equation.)

11. In Bessel's differential equation of order $\frac{1}{2}$ make the substitution $y = u/\sqrt{x}$ to obtain

$$\frac{d^2u}{dx^2} + u = 0.$$

Solve this equation to obtain

$$y = c_1\frac{\sin x}{\sqrt{x}} + c_2\frac{\cos x}{\sqrt{x}},$$

and explain the qualitative nature of $J_{1/2}(x)$, that is, the decaying amplitude of this function.

12. **a)** Expand the function of Exercise 9(c) in a Fourier–Bessel series in terms of $J_1(\lambda_j x)$ where the λ_j satisfy $J_1(2\lambda_j) = 0$.

b) To what value will the series converge when $x = 1$? Explain.

13. **a)** Prove that
$$J_{1/2}(x) = \sqrt{2/\pi x}\,\sin x.$$

(*Hint*: Use the Maclaurin's series for sin x.)

b) Prove that
$$J_{-1/2}(x) = \sqrt{2/\pi x}\,\cos x.$$

14. Evaluate
$$\int^x s^n J_{n-1}(s)\,ds.$$

(*Hint*: Use Exercise 4f.)

---

* After Hermann von Helmholtz (1821–1894), a German military surgeon who turned to mathematical physics in 1871.

15. a) Prove that

$$\frac{d}{dx}(x^{-n}J_n(x)) = -x^{-n}J_{n+1}(x).$$

b) Evaluate

$$\int^x \frac{J_{n+1}(s)}{s^n}\,ds.$$

16. a) The equation

$$y'' + \frac{1}{x}y' - y = 0$$

is called the **modified Bessel equation of order zero**. Show that its solution is

$$J_0(ix) = 1 + \frac{x^2}{2^2} + \frac{x^4}{2^2 \cdot 4^2} + \frac{x^6}{2^2 \cdot 4^2 \cdot 6^2} + \cdots.$$

We also write $I_0(x) = J_0(ix)$, where $I_0(x)$ is called the **modified Bessel function of the first kind of order zero**.

b) Find the interval of convergence of $I_0(x)$.

17. Prove that

$$Y_0'(x) = -Y_1(x).$$

18. a) By dividing Bessel's differential equation (9.3–11) by x, show that it has the form of the Sturm–Liouville equation (8.5–1).

b) Using the notation of Section 8.5, show that $r(a) = 0$ and $a_1 = a_2 = 0$.

c) Show that the second boundary condition of (8.5–2) is satisfied by the Bessel function of the first kind.

9.4 LEGENDRE POLYNOMIALS

Spherical coordinates (r, ϕ, θ) may be defined as shown in Fig. 9.4–1. We have $r \geq 0$, $0 \leq \phi < 2\pi$, and $0 \leq \theta \leq \pi$. A word of caution is in order here. Some authors use ρ instead of r, some interchange ϕ and θ, and some follow both practices. It is essential, therefore, to note a particular author's definition.

In *cylindrical coordinates* (ρ, ϕ, z) the Laplacian is

$$\nabla^2 u = \frac{\partial^2 u}{\partial \rho^2} + \frac{1}{\rho}\frac{\partial u}{\partial \rho} + \frac{1}{\rho^2}\frac{\partial^2 u}{\partial \phi^2} + \frac{\partial^2 u}{\partial z^2}. \qquad (9.4\text{–}1)$$

The relation between rectangular and spherical coordinates is given by

$$x = r \sin \theta \cos \phi, \qquad y = r \sin \theta \sin \phi, \qquad z = r \cos \theta,$$

and we could obtain the Laplacian in spherical coordinates from these. It is a bit simpler and more instructive, however, to begin with

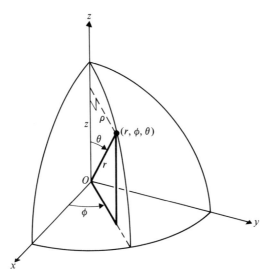

FIGURE 9.4–1 Spherical coordinates.

Eq. (9.4–1) and evaluate the four terms of that sum in spherical coordinates.

If we hold ϕ fixed, then u is a function of ρ and z and by virtue of

$$z = r \cos \theta, \qquad \rho = r \sin \theta,$$

u is a function of r and θ. Thus we have by the chain rule

$$\frac{\partial u}{\partial \rho} = \frac{\partial u}{\partial r}\frac{\partial r}{\partial \rho} + \frac{\partial u}{\partial \theta}\frac{\partial \theta}{\partial \rho}$$

$$= \frac{\rho}{r}\frac{\partial u}{\partial r} + \frac{z}{r^2}\frac{\partial u}{\partial \theta}. \qquad (9.4\text{--}2)$$

In Eq. (9.4–2) we have used the relations $z^2 + \rho^2 = r^2$ and $\rho/z = \tan \theta$ to find $\partial r/\partial \rho$ and $\partial \theta/\partial \rho$. Note that z is held constant when differentiating these relations partially with respect to ρ.

Then

$$\frac{\partial^2 u}{\partial \rho^2} = \frac{\partial}{\partial \rho}\left(\frac{\rho}{r}\frac{\partial u}{\partial r} + \frac{z}{r^2}\frac{\partial u}{\partial \theta}\right)$$

and we can use Eq. (9.4–2) since it serves as a formula for differentiating *any* function of r and θ. Symbolically,

$$\frac{\partial(\;)}{\partial \rho} = \frac{\partial(\;)}{\partial r}\frac{\rho}{r} + \frac{\partial(\;)}{\partial \theta}\frac{z}{r^2}.$$

Hence,

$$\frac{\partial^2 u}{\partial \rho^2} = \frac{\partial u_r}{\partial \rho}\frac{\rho}{r} + \frac{1}{r}u_r + u_r\rho\frac{\partial}{\partial \rho}\left(\frac{1}{r}\right) + \frac{\partial u_\theta}{\partial \rho}\frac{z}{r^2} + u_\theta z\frac{\partial}{\partial \rho}\left(\frac{1}{r^2}\right)$$

$$= \frac{\rho}{r}\left(\frac{\partial^2 u}{\partial r^2}\frac{\rho}{r} + \frac{\partial^2 u}{\partial \theta\,\partial r}\frac{z}{r^2}\right) + \frac{1}{r}u_r + u_r\rho\left(-\frac{1}{r^2}\frac{\rho}{r}\right)$$

$$+ \frac{z}{r^2}\left(\frac{\partial^2 u}{\partial r\,\partial \theta}\frac{\rho}{r} + \frac{\partial^2 u}{\partial \theta^2}\frac{z}{r^2}\right) + u_\theta z\left(-\frac{2}{r^3}\frac{\rho}{r}\right)$$

$$= \frac{\rho^2}{r^2}\frac{\partial^2 u}{\partial r^2} + \frac{2\rho z}{r^3}\frac{\partial^2 u}{\partial r\,\partial\theta} + \frac{1}{r}\frac{\partial u}{\partial r} - \frac{\rho^2}{r^3}\frac{\partial u}{\partial r} + \frac{z^2}{r^4}\frac{\partial^2 u}{\partial \theta^2} - \frac{2\rho z}{r^4}\frac{\partial u}{\partial\theta}.$$

In a similar manner we can calculate $\partial^2 u/\partial z^2$. We have

$$\frac{\partial u}{\partial z} = \frac{\partial u}{\partial r}\frac{\partial r}{\partial z} + \frac{\partial u}{\partial \theta}\frac{\partial \theta}{\partial z} = \frac{\partial u}{\partial r}\frac{z}{r} + \frac{\partial u}{\partial \theta}\left(-\frac{\rho}{r^2}\right)$$

$$\frac{\partial^2 u}{\partial z^2} = \frac{\partial}{\partial z}\left(\frac{z}{r}\frac{\partial u}{\partial r}\right) - \frac{\partial}{\partial z}\left(\frac{\rho}{r^2}\frac{\partial u}{\partial \theta}\right)$$

$$= \frac{1}{r}\frac{\partial u}{\partial r} + z\frac{\partial u}{\partial r}\left(-\frac{1}{r^2}\right)\frac{z}{r} + \frac{z}{r}\left(\frac{\partial^2 u}{\partial r^2}\frac{z}{r} - \frac{\rho}{r^2}\frac{\partial^2 u}{\partial r\,\partial\theta}\right)$$

$$- \rho\frac{\partial u}{\partial \theta}\left(-\frac{2}{r^3}\right)\frac{z}{r} - \frac{\rho}{r^2}\left(\frac{\partial^2 u}{\partial\theta\,\partial r}\frac{z}{r} + \frac{\partial^2 u}{\partial\theta^2}\left(-\frac{\rho}{r^2}\right)\right)$$

$$= \frac{1}{r}\frac{\partial u}{\partial r} - \frac{z^2}{r^3}\frac{\partial u}{\partial r} + \frac{z^2}{r^2}\frac{\partial^2 u}{\partial r^2} - \frac{2\rho z}{r^3}\frac{\partial^2 u}{\partial r\,\partial\theta} + \frac{2\rho z}{r^4}\frac{\partial u}{\partial\theta} + \frac{\rho^2}{r^4}\frac{\partial^2 u}{\partial\theta^2}.$$

Hence

$$u_{zz} + u_{\rho\rho} = \frac{\partial u}{\partial r}\left(\frac{r^2 - z^2 + r^2 - \rho^2}{r^3}\right) + \frac{\partial^2 u}{\partial r^2}\left(\frac{z^2 + \rho^2}{r^2}\right) + \frac{\partial^2 u}{\partial\theta^2}\left(\frac{z^2 + \rho^2}{r^4}\right)$$

$$= \frac{1}{r}\frac{\partial u}{\partial r} + \frac{\partial^2 u}{\partial r^2} + \frac{1}{r^2}\frac{\partial^2 u}{\partial\theta^2}.$$

Finally, adding the equivalent of the terms

$$\frac{1}{\rho}\frac{\partial u}{\partial \rho} \quad \text{and} \quad \frac{1}{\rho^2}\frac{\partial^2 u}{\partial \phi^2},$$

we have

$$\nabla^2 u = \frac{\partial^2 u}{\partial r^2} + \frac{2}{r}\frac{\partial u}{\partial r} + \frac{1}{r^2 \sin^2\theta}\frac{\partial^2 u}{\partial\phi^2} + \frac{1}{r^2}\frac{\partial^2 u}{\partial\theta^2} + \frac{\cot\theta}{r^2}\frac{\partial u}{\partial\theta}, \quad \textbf{(9.4-3)}$$

which is the Laplacian in spherical coordinates. It corresponds to Eq. (9.4–1) in cylindrical coordinates and to

$$\nabla^2 u = \frac{\partial^2 u}{\partial x^2} + \frac{\partial^2 u}{\partial y^2} + \frac{\partial^2 u}{\partial z^2}$$

in rectangular coordinates.

Solutions of Laplace's equation in spherical coordinates

Laplace's equation (also called the potential equation) in spherical coordinates can be written as

$$\nabla^2 u = \frac{\partial^2 u}{\partial r^2} + \frac{2}{r}\frac{\partial u}{\partial r} + \frac{1}{r^2 \sin^2 \theta}\frac{\partial^2 u}{\partial \phi^2} + \frac{1}{r^2}\frac{\partial^2 u}{\partial \theta^2} + \frac{\cot \theta}{r^2}\frac{\partial u}{\partial \theta} = 0.$$

An equivalent form of this equation is (Exercise 1)

$$\frac{1}{r^2}\frac{\partial}{\partial r}\left(r^2 \frac{\partial u}{\partial r}\right) + \frac{1}{r^2 \sin \theta}\frac{\partial}{\partial \theta}\left(\sin \theta \frac{\partial u}{\partial \theta}\right) + \frac{1}{r^2 \sin^2 \theta}\frac{\partial^2 u}{\partial \phi^2} = 0. \quad \textbf{(9.4–4)}$$

We seek a solution by the method of separation of variables. Assume that

$$u(r, \phi, \theta) = R(r)\Phi(\phi)\Theta(\theta)$$

and substitute into Eq. (9.4–4). Then

$$\frac{1}{r^2}\frac{d}{dr}\left(r^2 \Phi\Theta \frac{dR}{dr}\right) + \frac{1}{r^2 \sin \theta}\frac{d}{d\theta}\left(R\Phi \sin \theta \frac{d\Theta}{d\theta}\right) + \frac{1}{r^2 \sin^2 \theta}R\Theta\frac{d^2\Phi}{d\phi^2} = 0.$$

Next, divide each term by $R\Phi\Theta/r^2 \sin^2 \theta$ to obtain

$$\frac{\sin^2 \theta}{R}\frac{d}{dr}\left(r^2 \frac{dR}{dr}\right) + \frac{\sin \theta}{\Theta}\frac{d}{d\theta}\left(\sin \theta \frac{d\Theta}{d\theta}\right) = -\frac{1}{\Phi}\frac{d^2\Phi}{d\phi^2}.$$

Since the left side is independent of ϕ, we have

$$-\frac{1}{\Phi}\frac{d^2\Phi}{d\phi^2} = m^2, \qquad m = 0, 1, 2, \ldots, \qquad \textbf{(9.4–5)}$$

where the first separation constant m is chosen to be a nonnegative integer in order that the function Φ (and also u) be periodic of period 2π in ϕ. This is often necessary from physical considerations, as we shall see later.

Separating variables again produces

$$\frac{1}{R}\frac{d}{dr}\left(r^2 \frac{dR}{dr}\right) = -\left(\frac{1}{\Theta \sin \theta}\frac{d}{d\theta}\left(\sin \theta \frac{d\Theta}{d\theta}\right) - \frac{m^2}{\sin^2 \theta}\right) = \lambda,$$

where we have called the second separation constant λ. At this point nothing further is known about this quantity.

Thus we have reduced Laplace's equation to the following three second-order, linear, homogeneous ordinary differential equations:

$$\frac{d^2\Phi}{d\phi^2} + m^2\Phi = 0, \qquad \textbf{(9.4–6)}$$

$$\frac{1}{\sin \theta}\frac{d}{d\theta}\left(\sin \theta \frac{d\Theta}{d\theta}\right) + \left(\lambda - \frac{m^2}{\sin^2 \theta}\right)\Theta = 0, \qquad \textbf{(9.4–7)}$$

$$\frac{d}{dr}\left(r^2 \frac{dR}{dr}\right) - \lambda R = 0. \qquad \textbf{(9.4–8)}$$

Note that the first and third equations each contain one of the two separation constants, whereas the second contains both constants. Products of the solutions of these three equations are called **spherical harmonics**.

Equation (9.4–6) has constant coefficients, so that the solution presents no difficulty. It is

$$\Phi(m\phi) = A_m \cos m\phi + B_m \sin m\phi, \qquad m = 0, 1, 2, \ldots, \qquad \textbf{(9.4–9)}$$

where A_m and B_m are arbitrary constants that can be determined from given boundary conditions.

Next we look at Eq. (9.4–8), which can be written in the equivalent form

$$r^2 \frac{d^2 R}{dr^2} + 2r \frac{dR}{dr} - \lambda R = 0.$$

This is a Cauchy–Euler equation. It can be solved by making the substitution $R = r^k$ in which case the differential equation becomes

$$r^2 k(k-1) r^{k-2} + 2rk r^{k-1} - \lambda r^k = 0$$

or

$$(k^2 + k - \lambda) r^k = 0.$$

Hence $R = r^k$ is a solution of Eq. (9.4–8), provided that $k^2 + k - \lambda = 0$. To find a second, linearly independent solution may not be so simple. If we choose $k = n$, a *nonnegative integer*, then $\lambda = n(n+1)$ and if we further (wisely) choose $k = -(n+1)$, then *also* $\lambda = n(n+1)$. Thus* with $\lambda = n(n+1)$ Eq. (9.4–8) has the two linearly independent solutions r^n and $r^{-(n+1)}$ (Exercise 2) so that the general solution can be written as

$$R_n(r) = C_n r^n + D_n r^{-(n+1)}. \qquad \textbf{(9.4–10)}$$

In order to solve Eq. (9.4–7) we make the following substitutions:

$$x = \cos\theta, \qquad \Theta(\theta) = y(x), \qquad \frac{d}{d\theta} = \frac{dx}{d\theta}\frac{d}{dx} = -\sin\theta \frac{d}{dx}.$$

Then

$$\frac{d}{d\theta}\left(\sin\theta \frac{d\Theta}{d\theta}\right) = -\sin\theta \frac{d}{dx}\left(\sin\theta \frac{dx}{d\theta}\frac{d\Theta}{dx}\right)$$

$$= -\sin\theta \frac{d}{dx}\left(-\sin^2\theta \frac{dy}{dx}\right)$$

$$= \sqrt{1-x^2}\frac{d}{dx}\left((1-x^2)\frac{dy}{dx}\right).$$

---

* We point out that the value of λ was chosen with the intention of obtaining solutions to Eq. (9.4–7) that yield orthogonal functions. In short, some foresight is being used here.

With these substitutions Eq. (9.4–7) becomes

$$\frac{d}{dx}\left((1-x^2)\frac{dy}{dx}\right)+\left(n(n+1)-\frac{m^2}{1-x^2}\right)y=0$$

or, in equivalent form,

$$(1-x^2)\frac{d^2y}{dx^2}-2x\frac{dy}{dx}+\left(n(n+1)-\frac{m^2}{1-x^2}\right)y=0. \qquad \textbf{(9.4–11)}$$

Equation (9.4–11) is **Legendre's associated differential equation**. Its general solution, found by a series method, is

$$y_{n,m}(x)=c_{n,m}P_n^m(x)+d_{n,m}Q_n^m(x),$$

where $P_n^m(x)$ and $Q_n^m(x)$ are called **Legendre functions of the first and second kind**, respectively. The use of subscripts and superscripts indicates that these functions depend on m and n as well as on the argument x.

If $m=0$ and n is a nonnegative integer, then Eq. (9.4–11) becomes

$$(1-x^2)\frac{d^2y}{dx^2}-2x\frac{dy}{dx}+n(n+1)y=0, \qquad \textbf{(9.4–12)}$$

which is known as **Legendre's differential equation**. A particular solution is $y=P_n(x)$, the Legendre polynomial of degree n, $n=0$, $1, 2, \ldots$ (see Example 9.2–7).

A second linearly independent solution is $Q_n(x)$. Since $Q_n(x)$ has a singularity at $x=\pm1$ (as we will show later) it can be used only when $x \neq 1$ ($\theta \neq 0$) and when $x \neq -1$ ($\theta \neq \pi$).

In the case* where $m=0$ and n is a nonnegative integer, Laplace's equation in spherical coordinates (9.4–4),

$$\frac{1}{r^2}\frac{\partial}{\partial r}\left(r^2\frac{\partial u}{\partial r}\right)+\frac{1}{r^2\sin^2\theta}\frac{\partial}{\partial\theta}\left(\sin\theta\frac{\partial u}{\partial\theta}\right)=0,$$

has solutions that are products of

$$R_n(r)=C_n r^n + D_n r^{-(n+1)}$$

and

$$\Theta_n(\theta)=E_n P_n(\cos\theta)+F_n Q_n(\cos\theta), \qquad n=0, 1, 2, \ldots \ .$$

Admittedly, we have made a number of seemingly restrictive simplifying assumptions in order to get to this point in solving Laplace's equation in spherical coordinates. This was done not merely to simplify the mathematical aspects. We shall see in Section 9.5 that many of the applications will yield to our simplified approach. It should be

---

* Note that $m=0$ implies that Laplace's equation is independent of ϕ (see Eq. 9.4–9).

kept in mind, however, that the nature of the separation constants m and λ depends on the boundary conditions in any given case.

Legendre polynomials

In Example 9.2–7 we solved Legendre's differential equation (9.4–12) by the method of Frobenius and obtained the particular solutions $P_n(x)$, called **Legendre polynomials**. For reference we list the first few Legendre polynomials here (Fig. 9.4–2):

$$P_0(x) = 1, \qquad P_1(x) = x, \qquad P_2(x) = \tfrac{1}{2}(3x^2 - 1),$$
$$P_3(x) = \tfrac{1}{2}(5x^3 - 3x), \qquad P_4(x) = \tfrac{1}{8}(35x^4 - 30x^2 + 3).$$

Some properties of the Legendre polynomials that are useful in solving certain boundary-value problems are listed below. (See also Exercise 3.)

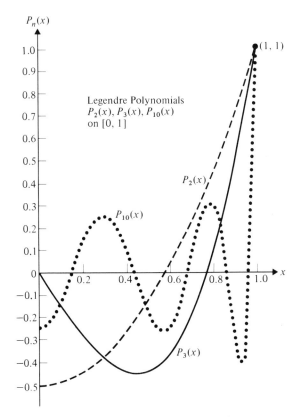

FIGURE 9.4–2 Legendre polynomials.

a) $P_{2n+1}(0) = 0$

b) $P_n(1) = 1$

c) $P_n(-1) = (-1)^n$ **(9.4–13)**

d) $P'_{n+1}(x) - xP'_n(x) = (n+1)P_n(x), \quad n = 1, 2, \ldots$

e) $xP'_n(x) - P'_{n-1}(x) = nP_n(x), \quad n = 1, 2, \ldots$

f) $P'_{n+1}(x) - P'_{n-1}(x) = (2n+1)P_n(x), \quad n = 1, 2, \ldots$

Note that property (f) is the sum of properties (d) and (e). We can prove property (d) from the **definition of the Legendre polynomials**,

$$P_n(x) = \frac{1}{2^n} \sum_{k=0}^{N} \frac{(-1)^k (2n-2k)!}{k!(n-2k)!(n-k)!} x^{n-2k}, \qquad \textbf{(9.4–14)}$$

where $N = n/2$ if n is even and $N = (n-1)/2$ if n is odd. We have

$$P_{n+1}(x) = \frac{1}{2^{n+1}} \sum_{k=0}^{N} \frac{(-1)^k (2n-2k+2)!}{k!(n-2k+1)!(n-k+1)!} x^{n-2k+1}$$

$$P'_{n+1}(x) = \frac{1}{2^{n+1}} \sum_{k=0}^{N} \frac{(-1)^k (2n-2k+2)!(n-2k+1)}{k!(n-2k+1)!(n-k+1)!} x^{n-2k}$$

$$P'_n(x) = \frac{1}{2^n} \sum_{k=0}^{N} \frac{(-1)^k (2n-2k)!(n-2k)}{k!(n-2k)!(n-k)!} x^{n-2k-1}$$

$$xP'_n(x) = \frac{1}{2^n} \sum_{k=0}^{N} \frac{(-1)^k (2n-2k)!(n-2k)}{k!(n-2k)!(n-k)!} x^{n-2k}$$

$$P'_{n+1}(x) - xP'_n(x) = \frac{1}{2^{n+1}} \sum_{k=0}^{N} \frac{(-1)^k (2n-2k+2)(2n-2k+1)(2n-2k)!}{k!(n-2k)!(n-k+1)(n-k)!} x^{n-2k}$$

$$- \frac{1}{2^n} \sum_{k=0}^{N} \frac{(-1)^k (2n-2k)!(n-2k)}{k!(n-2k)!(n-k)!} x^{n-2k}$$

$$P'_{n+1}(x) - xP'_n(x) = (2n-2k+1-n+2k) \frac{1}{2^n} \sum_{k=0}^{N} \frac{(-1)^k (2n-2k)!}{k!(n-2k)!(n-k)!} x^{n-2k}$$

$$= (n+1)P_n(x).$$

Property (e) may be shown in a similar manner (Exercise 3).

Orthogonality of Legendre polynomials

We now show under what conditions the Legendre polynomials are orthogonal. This property is essential in solving boundary-value problems, as we will see in the following section.

The Legendre polynomials $P_n(x)$ satisfy Legendre's differential equation,

$$\frac{d}{dx}((1 - x^2)P_n'(x)) + n(n + 1)P_n(x) = 0, \qquad n = 0, 1, 2, \ldots .$$

Multiplying this equation by $P_m(x)\, dx$ and integrating from -1 to 1 results in

$$\int_{-1}^{1} P_m(x)\frac{d}{dx}((1 - x^2)P_n'(x))\, dx + n(n + 1)\int_{-1}^{1} P_m(x)P_n(x)\, dx = 0.$$

$$(9.4\text{--}15)$$

The first integral may be integrated by parts, putting

$$u = P_m(x), \qquad\qquad du = P_m'(x)\, dx,$$

$$dv = \frac{d}{dx}((1 - x^2)P_n'(x))\, dx, \qquad v = (1 - x^2)P_n'(x).$$

Then

$$\int_{-1}^{1} P_m(x)\frac{d}{dx}((1 - x^2)P_n'(x))\, dx$$

$$= P_m(x)P_n'(x)(1 - x^2)\Big|_{-1}^{1} - \int_{-1}^{1}(1 - x^2)P_n'(x)P_m'(x)\, dx.$$

The first term on the right vanishes at both limits by virtue of the term $(1 - x^2)$. Thus Eq. (9.4–15) reduces to

$$-\int_{-1}^{1}(1 - x^2)P_n'(x)P_m'(x)\, dx + n(n + 1)\int_{-1}^{1} P_m(x)P_n(x)\, dx = 0.$$

In this last equation m and n have no special significance except that they are both nonnegative integers. Hence we may interchange m and n to obtain

$$-\int_{-1}^{1}(1 - x^2)P_m'(x)P_n'(x)\, dx + m(m + 1)\int_{-1}^{1} P_n(x)P_m(x)\, dx = 0.$$

Subtracting this equation from the previous one yields

$$(n - m)(n + m + 1)\int_{-1}^{1} P_m(x)P_n(x)\, dx = 0.$$

Now suppose $n \neq m$. Then $n - m \neq 0$ and $n + m + 1 = 0$ is impossible. (Why?) Hence we are left with the result,

$$\int_{-1}^{1} P_m(x)P_n(x)\, dx = 0, \qquad m \neq n. \qquad (9.4\text{--}16)$$

This shows that the set

$$\{P_0(x), P_1(x), P_2(x), \ldots\}$$

is an orthogonal set on $(-1, 1)$ with weight function one.

In applications the Legendre polynomials are often expressed in terms of θ. Let $x = \cos \theta$ and $dx = -\sin \theta \, d\theta$, and change limits accordingly. Then Eq. (9.4–16) becomes

$$\int_{\pi}^{0} P_m(\cos \theta)P_n(\cos \theta)(-\sin \theta \, d\theta) = 0, \qquad m \neq n$$

or

$$\int_{0}^{\pi} \sin \theta \, P_m(\cos \theta)P_n(\cos \theta) \, d\theta = 0, \qquad m \neq n.$$

Thus the set

$$\{P_0(\cos \theta), P_1(\cos \theta), P_2(\cos \theta), \ldots\}$$

is an orthogonal set on the interval $0 < \theta < \pi$ with weight function $\sin \theta$.

If in Eq. (9.4–16) we replace n by $2n$ and m by $2m$, then

$$\int_{-1}^{1} P_{2m}(x)P_{2n}(x) \, dx = 2 \int_{0}^{1} P_{2m}(x)P_{2n}(x) \, dx = 0, \qquad n \neq m.$$

In other words, the Legendre polynomials of *even* degree are orthogonal on the interval $0 < x < 1$ with weight function one. Similarly, the Legendre polynomials of *odd* degree are orthogonal on the interval $0 < x < 1$ with weight function one (Exercise 4).

Legendre series

The orthogonality property of the Legendre polynomials makes it possible to represent certain functions f by a **Legendre series**, that is, a series of Legendre polynomials. This representation is possible because Legendre's differential equation (9.4–12), together with an appropriate boundary condition, constitutes a singular Sturm–Liouville problem of the type discussed at the end of Section 8.5 (see Exercise 22). Moreover, it can be shown * that the normalized Legendre polynomials that we will obtain in this section form a **complete orthonormal set** with respect to piecewise smooth functions on $(-1, 1)$.

For such a function we can write

$$f(x) = A_0 P_0(x) + A_1 P_1(x) + A_2 P_2(x) + A_3 P_3(x) + \cdots.$$

In order to find A_2, for example, we multiply the above by $P_2(x) \, dx$ and integrate from -1 to 1. Then

$$\int_{-1}^{1} f(x)P_2(x) \, dx = A_0 \int_{-1}^{1} P_0(x)P_2(x) \, dx + A_1 \int_{-1}^{1} P_1(x)P_2(x) \, dx$$

$$+ A_2 \int_{-1}^{1} P_2(x)P_2(x) \, dx + A_3 \int_{-1}^{1} P_3(x)P_2(x) \, dx + \cdots.$$

---

* See Ruel V. Churchill, and J. W. Brown, *Fourier Series and Boundary Value Problems*, 3d ed., p. 234 ff. (New York: McGraw-Hill, 1978).

Because of the orthogonality property of the $P_n(x)$, each integral on the right is zero with the exception of the third. Thus

$$\int_{-1}^{1} f(x)P_2(x)\,dx = A_2 \int_{-1}^{1} (P_2(x))^2\,dx,$$

from which we get

$$A_2 = \frac{\int_{-1}^{1} f(x)P_2(x)\,dx}{\int_{-1}^{1} (P_2(x))^2\,dx}.$$

Each coefficient A_n can be found in the same way so that, in general,

$$A_n = \frac{\int_{-1}^{1} f(x)P_n(x)\,dx}{\int_{-1}^{1} (P_n(x))^2\,dx}, \qquad n = 0, 1, 2, \ldots. \qquad \textbf{(9.4–17)}$$

Next, we evaluate the denominator in the above equation. This quantity is referred to as the **square of the norm** and is denoted by $\|P_n\|^2$. In order to evaluate this, we first need a result known as **Rodrigues' formula.***

We begin with the binomial expansion of $(x^2 - 1)^n$, which can be written

$$(x^2 - 1)^n = \sum_{k=0}^{n} (-1)^k \frac{n!}{k!(n-k)!} x^{2n-2k}.$$

Differentiating this n times yields (Exercise 16)

$$\frac{d^n(x^2 - 1)^n}{dx^n} = \sum_{k=0}^{N} \frac{(-1)^k n!(2n-2k)!}{k!(n-k)!(n-2k)!} x^{n-2k}, \qquad \textbf{(9.4–18)}$$

where the last term is a constant. But $n - 2N = 0$ implies that $N = n/2$, whereas $n - 1 - 2N = 0$ implies that $N = (n-1)/2$. Then, since N is a nonnegative integer, it is defined in Eq. (9.4–18) as $n/2$ if n is even and $(n-1)/2$ if n is odd.

Recall the definition of $P_n(x)$ in summation form,

$$P_n(x) = \frac{1}{2^n} \sum_{k=0}^{N} \frac{(-1)^k(2n-2k)!}{k!(n-2k)!(n-k)!} x^{n-2k}, \qquad \textbf{(9.4–14)}$$

where N is defined as in Eq. (9.4–18). A comparison of Eqs. (9.4–14) and (9.4–18) shows that

$$P_n(x) = \frac{1}{2^n n!} \frac{d^n(x^2 - 1)^n}{dx^n}, \qquad n = 0, 1, 2, \ldots, \qquad \textbf{(9.4–19)}$$

which is **Rodrigues' formula**.

---

* After Olinde Rodrigues (1794–1851), a French economist and mathematician.

We can use Eq. (9.4–19) to evaluate $\|P_n\|^2$ as follows. We have

$$\int_{-1}^{1} (P_n(x))^2 \, dx = \int_{-1}^{1} P_n(x) \frac{1}{2^n n!} \frac{d^n}{dx^n} (x^2 - 1)^n \, dx$$

and integrating by parts with

$$u = P_n(x), \qquad dv = \frac{d^n}{dx^n} (x^2 - 1)^n \, dx,$$

$$du = P_n'(x) \, dx, \qquad v = \frac{d^{n-1}}{dx^{n-1}} (x^2 - 1)^n,$$

produces

$$\int_{-1}^{1} (P_n(x))^2 \, dx$$

$$= \frac{1}{2^n n!} \left[P_n(x) \frac{d^{n-1}}{dx^{n-1}} (x^2 - 1)^n \Big|_{-1}^{1} - \int_{-1}^{1} P_n'(x) \frac{d^{n-1}}{dx^{n-1}} (x^2 - 1)^n \, dx \right].$$

The first term on the right vanishes at both limits by virtue of the term $(x^2 - 1)$. Hence after $(n - 1)$ integrations by parts we have

$$\int_{-1}^{1} (P_n(x))^2 \, dx = \frac{(-1)^{n-1}}{2^n n!} \int_{-1}^{1} P_n^{(n-1)}(x) \frac{d}{dx} (x^2 - 1)^n \, dx.$$

One more integration produces

$$\int_{-1}^{1} (P_n(x))^2 \, dx = \frac{(-1)^n}{2^n n!} \int_{-1}^{1} P_n^{(n)}(x)(x^2 - 1)^n \, dx.$$

We now observe that

$$P_n^{(n)}(x) = \frac{(2n)!}{2^n (n!)}$$

from Eq. (9.4–14). Using the reduction formula found in most integral tables,

$$\int x^m (ax^n + b)^p \, dx = \frac{1}{m + np + 1} \left(x^{m+1}(ax^n + b)^p + npb \int x^m (ax^n + b)^{p-1} \, dx \right),$$

and integrating n times, we have

$$\int_{-1}^{1} (x^2 - 1)^n \, dx = \frac{(-1)^n 2^{2n+1}(n!)^2}{(2n + 1)!}.$$

Thus, putting everything together (Exercise 5), we have

$$\int_{-1}^{1} (P_n(x))^2 \, dx = \frac{(-1)^n}{2^n n!} \frac{(2n)!}{2^n (n!)} \frac{(-1)^n 2^{2n+1}(n!)^2}{(2n + 1)!}$$

or

$$\|P_n\|^2 = \frac{2}{2n + 1}, \qquad n = 0, 1, 2, \ldots . \qquad \text{(9.4–20)}$$

Hence the set

$$\left\{ \frac{P_0(x)}{\sqrt{2}}, \frac{P_1(x)}{\sqrt{2/3}}, \frac{P_2(x)}{\sqrt{2/5}}, \ldots \right\}$$

is an *orthonormal* set on the interval $-1 < x < 1$ with weight function 1. We can also update Eq. (9.4–17) to read

$$A_n = \frac{2n+1}{2} \int_{-1}^{1} f(x)P_n(x)\,dx, \qquad n = 0, 1, 2, \ldots . \quad \textbf{(9.4–21)}$$

Legendre series have something in common with Fourier series. If a function is defined on $(0, 1)$, it can be represented by a series containing Legendre polynomials of even degree. For this we may make an *even extension* of the function as we did to obtain Fourier cosine series in Section 7.2. We illustrate the procedure for this and for an *odd extension* in the following example.

EXAMPLE 9.4–1   Define

$$f(x) = 2(1 - x), \qquad 0 < x < 1.$$

Obtain the first two terms of the Legendre series representation of this function using (a) polynomials of even degree and (b) polynomials of odd degree.

Solution. For (a) we make an even extension and modify Eq. (9.4–21) to

$$A_{2n} = (4n + 1) \int_0^1 f(x)P_{2n}(x)\,dx, \qquad n = 0, 1, 2, \ldots . \quad \textbf{(9.4–22)}$$

Since the integrand is an even function, we can use symmetry and change the limits as shown. Then

$$A_0 = 2 \int_0^1 (1 - x)\,dx = 1,$$

$$A_2 = 5 \int_0^1 (1 - x)(3x^2 - 1)\,dx = -\frac{5}{4};$$

hence

$$f(x) \sim P_0(x) - \frac{5}{4} P_2(x) + \cdots .$$

For (b) we make an odd extension and modify Eq. (9.4–21) to

$$A_{2n+1} = (4n + 3) \int_0^1 f(x)P_{2n+1}(x)\,dx, \qquad n = 0, 1, 2, \ldots . \quad \textbf{(9.4–23)}$$

Note that the integrand is an even function again. Thus

$$A_1 = 6 \int_0^1 (1 - x)x\,dx = 1,$$

$$A_3 = 7 \int_0^1 (1 - x)(5x^3 - 3x)\,dx = -\frac{7}{4},$$

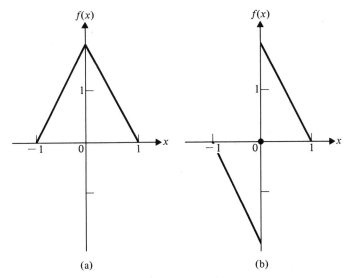

(a)   (b)

FIGURE 9.4–3 (a) Even extension. (b) Odd extension.

so that

$$f(x) \sim P_1(x) - \frac{7}{4} P_3(x) + \cdots .$$

The even and odd extensions are shown in Fig. 9.4–3. Observe that $f(0) = 0$ in (b). (Why?) ∎

Another way of solving the problems in Example 9.4–1 does not require making even and odd extensions of the given function $f(x)$. Since $f(x)$ is defined only on $(0, 1)$ and since the even and odd Legendre polynomials separately are orthogonal on this interval, we can proceed immediately to Eqs. (9.4–22) and (9.4–23) to obtain the appropriate coefficients.

Legendre functions of the second kind

We close this section with a discussion of the Legendre functions of the second kind. Linear combinations of these and the Legendre polynomials comprise the general solution of Legendre's differential equation (9.4–12).

One solution of Legendre's differential equation,

$$(1 - x^2)y'' - 2xy' + n(n + 1)y = 0, \qquad -1 < x < 1,$$

is

$$u = P_n(x), \qquad n = 0, 1, 2, \ldots,$$

the Legendre polynomials of degree n. We seek a second linearly independent solution by the method known as variation of parameters.

Assume that $y = uv$ is a solution. Then $y' = uv' + vu'$ and $y'' = uv'' + 2u'v' + vu''$, and substituting into the differential equation gives us

$$uv'' + 2u'v' + vu'' - x^2(uv'' + 2u'v' + vu'') - 2x(uv' + vu') + n(n+1)uv = 0$$

or

$$v[u'' - x^2u'' - 2xu' + n(n+1)u] + uv'' + 2u'v' - x^2uv'' - 2x^2u'v' - 2xuv' = 0.$$

But the quantity in brackets is zero because u is a solution of Legendre's differential equation. Hence we are left with

$$v''(1 - x^2)u + v'(2u' - 2x^2u' - 2xu) = 0$$

or

$$v'' + v'\left(\frac{2u'}{u} - \frac{2x}{1 - x^2}\right) = 0.$$

This last differential equation can be solved by a method called reduction of order. Let $w = v'$ and $w' = v''$. Then

$$w' + w\left(\frac{2u'}{u} - \frac{2x}{1 - x^2}\right) = 0, \tag{9.4-24}$$

a linear, first-order, homogeneous differential equation. The integrating factor is

$$\exp\left(2 \int^u \frac{t'}{t}\, dt + \int^x \frac{-2t\, dt}{1 - t^2}\right) = \exp\left(\log u^2 + \log(1 - x^2)\right)$$

$$= u^2(1 - x^2).$$

Multiplying Eq. (9.4–24) by this factor produces

$$w'u^2(1 - x^2) + w(2(1 - x^2)uu' - 2xu^2) = 0$$

or

$$d(wu^2(1 - x^2)) = 0.$$

Hence,

$$w = \frac{dv}{dx} = \frac{A}{u^2(1 - x^2)}$$

and

$$v = B + A \int \frac{dx}{u^2(1 - x^2)},$$

so that the assumed solution becomes

$$y_n(x) = vP_n(x) = B_nP_n(x) + A_nP_n(x) \int \frac{dx}{(1 - x^2)(P_n(x))^2}. \tag{9.4-25}$$

The **Legendre function of the second kind**, $Q_n(x)$, is obtained from Eq. (9.4–25) by putting $A_n = 1$ and $B_n = 0$. Thus

$$Q_0(x) = \int^x \frac{dx}{1-x^2} = \frac{1}{2} \log \left(\frac{1+x}{1-x} \right),$$

$$Q_1(x) = x \int^x \frac{dx}{x^2(1-x^2)} = \frac{x}{2} \log \left(\frac{1+x}{1-x} \right) - 1.$$

Continuing in this way, we have

$$Q_2(x) = \frac{1}{4}(3x^2 - 1) \log \left(\frac{1+x}{1-x} \right) - \frac{3}{2}x,$$

$$Q_3(x) = \frac{x}{4}(5x^2 - 3) \log \left(\frac{1+x}{1-x} \right) - \frac{5}{2}x^2 + \frac{2}{3}.$$

Graphs of these functions are shown in Fig. 9.4–4.

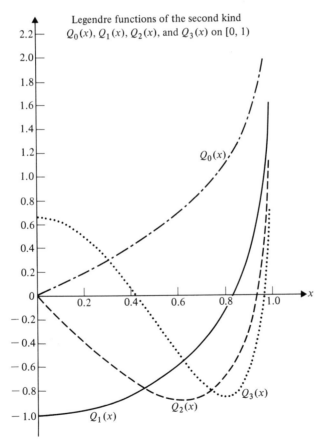

FIGURE 9.4–4

From the definition of the Legendre Q-function,

$$Q_n(x) = P_n(x) \int \frac{dx}{(1 - x^2)(P_n(x))^2}, \qquad \text{(9.4–26)}$$

it follows that

$$Q_{2n}(-x) = P_{2n}(-x) \int \frac{-dx}{(1 - x^2)(P_{2n}(-x))^2}$$

$$= -P_{2n}(x) \int \frac{dx}{(1 - x^2)(P_{2n}(x))^2} = -Q_{2n}(x)$$

and also

$$Q_{2n+1}(-x) = P_{2n+1}(-x) \int \frac{-dx}{(1 - x^2)(P_{2n+1}(-x))^2}$$

$$= P_{2n+1}(x) \int \frac{dx}{(1 - x^2)(P_{2n+1}(x))^2} = Q_{2n+1}(x).$$

These results may be combined so that we have

$$Q_n(-x) = (-1)^{n+1} Q_n(x).$$

It is for this reason that the graphs in Fig. 9.4–4 are shown only on the interval $0 \le x < 1$.

The term

$$\log \left(\frac{1 + x}{1 - x} \right)$$

in the Q-functions shows that these functions have a *singularity* at $x = \pm 1$. In spherical coordinates (r, ϕ, θ) this singularity is translated into singularities at $\theta = 0$ and $\theta = \pi$ by virtue of the relation $x = \cos \theta$.

We have obtained the Legendre functions of the second kind (Q-functions) in closed form. These functions can also be expressed as infinite series by expanding $\log (1 + x)/(1 - x)$ into a Maclaurin's series, that is,

$$\log \left(\frac{1 + x}{1 - x} \right) = 2 \left(x + \frac{x^3}{3} + \frac{x^5}{5} + \frac{x^7}{7} + \cdots \right), \qquad -1 < x < 1.$$

KEY WORDS AND PHRASES

spherical coordinates
Legendre's associated differential
    equation
Legendre functions of the first
    and second kind
Legendre's differential equation

Legendre polynomials
Legendre series
Rodrigues' formula
Legendre functions of the
    second kind

EXERCISES 9.4

▶ 1. Show that Eqs. (9.4–3) and (9.4–4) are equivalent forms of Laplace's equation in spherical coordinates.

2. Verify that r^n and $r^{-(n+1)}$ are linearly independent solutions of Eq. (9.4–8). (*Hint*: Use the Wronskian and recall the nature of n.)

3. Prove the following properties of the Legendre polynomials (compare Eq. 9.4–13).

   a) $P_{2n+1}(0) = 0$ **b)** $P_n(-1) = (-1)^n$

   c) $xP_n'(x) - P_{n-1}'(x) = nP_n(x),$ $n = 1, 2, \ldots$

   d) $P_{2n}'(0) = 0$ **e)** $P_{2n}(0) = (-1)^n \dfrac{(2n)!}{2^{2n}(n!)^2}$

4. Prove that the Legendre polynomials of odd degree are orthogonal on $0 < x < 1$ with weight function 1.

5. Carry out the details needed to arrive at Eq. (9.4–20).

6. Verify Eq. (9.4–20) for $n = 0, 1, 2, 3$, by computing the square of the norm directly from $P_n(x)$.

▶▶ 7. By direct computation show that $P_0(x)$, $P_1(x)$, and $P_2(x)$ are orthogonal on $-1 < x < 1$ with weight function 1.

8. Show that the interval of convergence of $Q_n(x)$ is $-1 < x < 1$.

9. Show that

$$\int_0^1 P_{2n}(x)\, dx = 0, \qquad n = 1, 2, \ldots\, .$$

(*Hint*: $P_0(x) = 1$.)

10. Show that

$$\int_{-1}^1 P_n'(x)P_m(x)\, dx = 1 - (-1)^{n+m},$$

where $0 \le m \le n$. (*Hint*: Use integration by parts.)

11. Show that

$$\int_{-1}^1 x(P_n(x))^2\, dx = 0.$$

12. Express each of the following polynomials in terms of Legendre polynomials.

   a) $ax + b$

   b) $ax^2 + bx + c$

   c) $ax^3 + bx^2 + cx + d$

13. Prove that

$$P_n'(1) = \frac{n}{2}(n + 1).$$

14. **a)** In Example 9.4–1 obtain the coefficients A_0, A_2, A_1, A_3.

   b) Compute A_4 and A_5.

c) Obtain three coefficients in the Legendre series representation of the function

$$f(x) = \begin{cases} 0, & -1 < x < 0, \\ 2(1 - x), & 0 < x < 1. \end{cases}$$

d) Evaluate $f(0)$ in part (c).

15. Obtain the first three nonzero coefficients in the Legendre series representation of each of the following functions.

a) $f(x) = \begin{cases} 0, & -1 < x < 0, \\ 1, & 0 < x < 1 \end{cases}$

(*Hint*: Use the result in Exercise 3e.)

b) $f(x) = \begin{cases} 0, & -1 < x < 0, \\ x, & 0 < x < 1 \end{cases}$

c) $f(x) = |x|, -1 < x < 1$

▶▶▶ 16. Verify Eq. (9.4–18). Explain why the last term in the summation is a constant.

17. Obtain the formula

$$\int_x^1 P_n(s)\, ds = \frac{1}{2n + 1}(P_{n-1}(x) - P_{n+1}(x)), \qquad n = 1, 2, \dots .$$

(*Hint*: Use Eq. 9.4–13f.)

18. If $f(x)$ is a polynomial of degree $m < n$, show that

$$\int_{-1}^1 f(x)P_n(x)\, dx = 0.$$

19. An integral representation of $P_n(x)$ is given by

$$P_n(x) = \frac{1}{\pi}\int_0^\pi (x + (x^2 - 1)^{1/2}\cos\phi)^n\, d\phi.$$

Verify this representation for $n = 0, 1, 2$.

20. **a)** Use Rodrigues' formula to obtain

$$2^n n!\, P_{n+1}(x) = (2n + 1)\frac{d^{(n-1)}u^n}{dx^{(n-1)}} + 2n\frac{d^{(n-1)}u^{n-1}}{dx^{(n-1)}},$$

where $u = x^2 - 1$.

b) In the relation obtained in part (a) make the substitution

$$2^{n-1}(n - 1)!\, P_{n-1}(x) = \frac{d^{(n-1)}u^{n-1}}{dx^{(n-1)}}$$

to obtain

$$P_{n+1}(x) - P_{n-1}(x) = \frac{2n + 1}{2^n n!}\frac{d^{(n-1)}u^n}{dx^{(n-1)}}.$$

21. Referring to Eq. (9.4–10), determine under what conditions r^n and $r^{-(n+1)}$ are linearly independent.

22. **a)** Write Legendre's differential equation (9.4–12) in the form of the Sturm–Liouville equation (8.5–1).

    b) Using the notation of Section 8.5, show that $r(b) = 0$ and that the first boundary condition of Eq. (8.5–2) holds.

9.5 APPLICATIONS

We are now ready to solve some boundary-value problems in three-space using cylindrical and spherical coordinates. In the following examples we will use many of the results developed in Sections 9.3 and 9.4.

EXAMPLE 9.5–1   Find the steady-state, bounded temperatures in the interior of a solid sphere of radius b if the temperatures on the surface are given by $f(\cos \theta)$.

Solution. Since the surface temperatures are known as a function of θ alone, the temperatures are independent of ϕ. Hence we have the following problem (see Eq. 9.4–3):

$$\text{P.D.E.:}\quad \nabla^2 u(r, \theta) = \frac{\partial^2 u}{\partial r^2} + \frac{2}{r}\frac{\partial u}{\partial r} + \frac{1}{r^2}\frac{\partial^2 u}{\partial \theta^2} + \frac{\cot \theta}{r^2}\frac{\partial u}{\partial \theta} = 0, \quad 0 < r < b,$$

$$0 < \theta < \pi;$$

$$\text{B.C.:}\quad u(b, \theta) = f(\cos \theta), \quad 0 < \theta < \pi.$$

The portion of the sphere in the first octant is shown in Fig. 9.5–1.

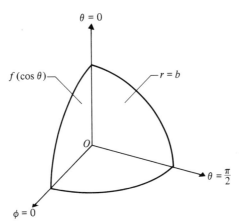

FIGURE 9.5–1

Additional boundary conditions will come from the fact that the temperatures must be bounded. Using separation of variables and assuming a product solution,

$$u(r, \theta) = R(r)\Theta(\theta),$$

we have

$$\Theta \frac{d^2 R^2}{dr^2} + \frac{2}{r} \Theta \frac{dR}{dr} + \frac{R}{r^2} \frac{d^2\Theta}{d\theta^2} + \frac{R \cot \theta}{r^2} \frac{d\Theta}{d\theta} = 0$$

or

$$\frac{r^2}{R} \frac{d^2 R}{dr^2} + \frac{2r}{R} \frac{dR}{dr} = -\frac{1}{\Theta} \frac{d^2\Theta}{d\theta^2} - \frac{\cot \theta}{\Theta} \frac{d\Theta}{d\theta} = \lambda.$$

This results in

$$r^2 R'' + 2rR' - \lambda R = 0,$$

which, for $\lambda = n(n + 1)$,* has solutions

$$R_n(r) = C_n r^n + D_n r^{-(n+1)}, \qquad n = 0, 1, 2, \ldots,$$

as given by Eq. (9.4–10). The second equation can be shown (Exercise 1) to be equivalent to Legendre's differential equation,

$$\frac{1}{\sin \theta} \frac{d}{d\theta}\left(\sin \theta \frac{d\Theta}{d\theta}\right) + n(n + 1)\Theta = 0,$$

which has the general solution given by

$$\Theta_n(\theta) = E_n P_n(\cos \theta) + F_n Q_n(\cos \theta), \qquad n = 0, 1, 2, \ldots.$$

To keep u bounded we must take $D_n = 0$ and $F_n = 0$ (Exercise 2). Thus $u(r, \theta)$ consists of products,

$$r^n P_n(\cos \theta),$$

and we take a linear combination of these in order to satisfy the remaining nonhomogeneous boundary condition. Then,[†]

$$u(r, \theta) = \sum_{n=0}^{} A_n r^n P_n(\cos \theta),$$

with the A_n to be determined. Using the given surface temperatures, we have

$$u(b, \theta) = \sum_{n=0}^{} A_n b^n P_n(\cos \theta) = f(\cos \theta),$$

---

* Recall that for these values of λ we were able to solve the θ equation (see Example 9.2–7). We are faced with the same equation in this example and, in addition, we will need to represent the given function f by a Legendre series.

[†] In this section we omit the upper limit on all summations since they all represent *infinite* series.

which shows that $f(\cos \theta)$ is to be expressed as a Legendre series. From Eq. (9.4–21),

$$A_n b^n = \frac{2n + 1}{2} \int_{-1}^{1} f(x)P_n(x)\, dx, \qquad n = 0, 1, 2, \ldots,$$

so that the solution can be written

$$u(r, \theta) = \frac{1}{2} \sum_{n=0}^{\infty} \left(\frac{r}{b}\right)^n P_n(\cos \theta)(2n + 1) \int_{-1}^{1} f(x)P_n(x)\, dx. \quad \blacksquare \quad (9.5\text{--}1)$$

EXAMPLE 9.5–2 Find the steady-state, bounded temperatures in the interior of a solid cylinder of radius c and height b, given that the temperature of the curved lateral surface is kept at zero, the base is insulated, and the top is kept at 100°.

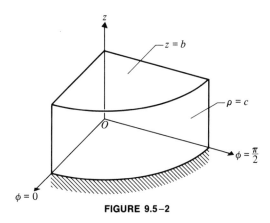

FIGURE 9.5–2

Solution. We take the axis of the cylinder along the z-axis and use cylindrical coordinates as shown in Fig. 9.5–2. Then we have the following problem (Exercise 3):

$$\text{P.D.E.:}\ u_{\rho\rho} + \frac{1}{\rho}u_{\rho} + u_{zz} = 0, \quad 0 < \rho < c, \quad 0 < z < b;$$

$$\text{B.C.:}\ u(c, z) = 0, \quad 0 < z < b,$$
$$u_z(\rho, 0) = 0, \quad 0 < \rho < c,$$
$$u(\rho, b) = 100, \quad 0 < \rho < c.$$

If we assume a product $u(\rho, z) = R(\rho)Z(z)$, the partial differential equation becomes

$$ZR'' + \frac{1}{\rho} ZR' + RZ'' = 0$$

or

$$\frac{R''}{R} + \frac{R'}{\rho R} = -\frac{Z''}{Z} = -\lambda^2.$$

We have assumed a *negative* separation constant in order that Z (and u) will *not* be periodic in z. (Why?) The resulting differential equations with their boundary conditions (only the *homogeneous* conditions may be used here) are

$$Z'' - \lambda^2 Z = 0, \qquad Z'(0) = 0,$$

and

$$R'' + \frac{1}{\rho} R' + \lambda^2 R = 0, \qquad R(c) = 0.$$

These differential equations are similar to Eqs. (9.3–2) and (9.3–4), respectively; hence we have the general solutions

$$Z(\lambda z) = Ae^{\lambda z} + Be^{-\lambda z}$$

and

$$R_0(\lambda \rho) = EJ_0(\lambda \rho) + FY_0(\lambda \rho).$$

To keep the solution bounded we take $F = 0$. The condition $R(c) = 0$ results in $J_0(\lambda c) = 0$, that is, λc is a zero of the Bessel function $J_0(x)$. Call these positive zeros, $\lambda_j c$, $j = 1, 2, \ldots$, that is, $\lambda_1 c \doteq 2.405$, $\lambda_2 c \doteq 5.520$, etc. The equation in z has solutions (Exercise 5)

$$Z(\lambda_j z) = \cosh(\lambda_j z), \qquad j = 1, 2, \ldots.$$

Thus

$$u(\rho, z) = \sum_{j=1}^{\infty} a_j \cosh(\lambda_j z) J_0(\lambda_j \rho)$$

and, applying the nonhomogeneous boundary condition, we have

$$u(\rho, b) = \sum_{j=1}^{\infty} a_j \cosh(\lambda_j b) J_0(\lambda_j \rho) = 100,$$

which shows that $f(\rho) = 100$ must be expressed as a Fourier–Bessel series on the interval $0 < \rho < c$. We use Eq. (9.3–19) to write

$$A_j = a_j \cosh(\lambda_j b) = \frac{2}{c^2 (J_0'(\lambda_j c))^2} \int_0^c 100 x J_0(\lambda_j x) \, dx$$

$$= \frac{200}{c^2 (J_0'(\lambda_j c))^2} \frac{1}{\lambda_j^2} \int_0^{\lambda_j c} s J_0(s) \, ds$$

on making the substitution $s = \lambda_j x$. Then, using Eq. (9.3–10), we have

$$A_j = a_j \cosh (\lambda_j b) = \frac{200}{\lambda_j c J_1(\lambda_j c)},$$

noting that $J_0'(x) = -J_1(x)$ and simplifying (Exercise 6). Thus

$$u(\rho, z) = \frac{200}{c} \sum_{j=1}^{\infty} \frac{J_0(\lambda_j \rho) \cosh (\lambda_j z)}{\lambda_j \cosh (\lambda_j b) J_1(\lambda_j c)}. \quad \blacksquare \qquad (9.5-2)$$

In the next example we consider the two-dimensional wave equation over a circular region.

EXAMPLE 9.5–3 Solve the following boundary-value problem.

$$\text{P.D.E.:} \quad z_{tt} = \frac{a^2}{\rho} (\rho z_\rho)_\rho, \quad 0 < \rho < c, \quad t > 0;$$

$$\text{B.C.:} \quad z(c, t) = 0, \quad t > 0;$$

$$\text{I.C.:} \quad z_t(\rho, 0) = 0, \quad 0 < \rho < c,$$

$$z(\rho, 0) = f(\rho), \quad 0 < \rho < c.$$

Solution. Here we have a homogeneous membrane (compare Example 8.2–1) of radius c fastened in a frame along its circular edge (see Fig. 9.5–3). The frame is given an initial displacement $f(\rho)$ in the z-direction and we seek the displacements at an arbitrary point of the membrane at any time t. The fact that the initial displacement is a function of ρ alone indicates that z is independent of ϕ. Using separation of variables, we arrive (Exercise 7) at the following ordinary

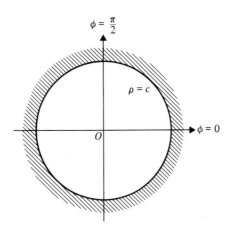

FIGURE 9.5–3

differential equations and homogeneous conditions:

$$T'' + \lambda^2 a^2 T = 0, \qquad T'(0) = 0,$$
$$\rho^2 R'' + \rho R' + \lambda^2 \rho^2 R = 0, \qquad R(c) = 0.$$

The separation constant was chosen so that T (and z) will be periodic in t consistent with physical fact. The second equation has solutions $J_0(\lambda_j \rho)$, where $J_0(\lambda_j c) = 0$, $j = 1, 2, \dots$. The first equation has solutions $\cos(\lambda_j at)$, $j = 1, 2, \dots$; hence we take a linear combination of products of these, that is,

$$z(\rho, t) = \sum_{j=1}^{\infty} A_j J_0(\lambda_j \rho) \cos(\lambda_j at).$$

To satisfy the nonhomogeneous initial condition we again use Eq. (9.3–19) to determine the A_j. Thus

$$A_j = \frac{2}{c^2 (J_1(\lambda_j c))^2} \int_0^c x f(x) J_0(\lambda_j x)\, dx, \qquad j = 1, 2, \dots$$

and

$$z(\rho, t) = \frac{2}{c^2} \sum_{j=1}^{\infty} \frac{J_0(\lambda_j \rho) \cos(\lambda_j at)}{(J_1(\lambda_j c))^2} \int_0^c x f(x) J_0(\lambda_j x)\, dx, \qquad \textbf{(9.5–3)}$$

where $\lambda_j c$ are positive roots of $J_0(x) = 0$. ∎

In the next example we investigate the two-dimensional heat equation over a circular region.

EXAMPLE 9.5–4   Solve the following boundary-value problem.

$$\text{P.D.E.:}\ \ u_t = \frac{k}{\rho}(\rho u_\rho)_\rho, \quad 0 < \rho < c, \quad t > 0;$$

$$\text{B.C.:}\ \ u_\rho(c, t) = 0, \quad t > 0;$$
$$\text{I.C.:}\ \ u(\rho, 0) = f(\rho), \quad 0 < \rho < c.$$

Solution. Here we have a homogeneous circular disk of radius c whose outer edge is insulated (see Fig. 9.5–4). We assume that the heat flow is two-dimensional, that is, the disk is thin and its top and bottom circular faces are also insulated. Moreover, the temperature is independent of ϕ since the initial temperature distribution is a function of ρ alone. We seek the temperatures in the disk at any time t. Using separation of variables, we obtain (Exercise 8) the following ordinary differential equations:

$$T' + k\lambda^2 T = 0,$$

$$R'' + \frac{1}{\rho} R' + \lambda^2 R = 0, \qquad R'(c) = 0.$$

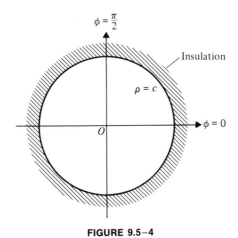

FIGURE 9.5–4

Again the separation constant was chosen to be negative, this time in order that $u_\rho(\rho, t)$ have a zero limit as $t \to \infty$ (Exercise 9).

The second equation is Bessel's differential equation of order zero having bounded solution $J_0(\lambda\rho)$. Applying the condition shown, we have

$$J_0'(\lambda c) = -\lambda J_1(\lambda c) = 0,$$

using Exercise 4(d) in Section 9.3, and showing that λc is a zero of $J_1(x) = 0$. Call these nonnegative zeros $\lambda_j c$, $j = 1, 2, \ldots$, that is, $\lambda_1 c = 0$, $\lambda_2 c \doteq 3.832$, $\lambda_3 c \doteq 7.016$, etc. Then

$$u(\rho, t) = \sum_{j=1}^{\infty} A_j \exp(-k\lambda_j^2 t) J_0(\lambda_j \rho),$$

and to satisfy the nonhomogeneous boundary condition we compute the A_j using Eq. (9.3–21) with $h = n = 0$ and Eq. (9.3–22). Hence,

$$A_1 = \frac{2}{c^2} \int_0^c x f(x)\, dx,$$

$$A_j = \frac{2}{c^2 (J_0(\lambda_j c))^2} \int_0^c x f(x) J_0(\lambda_j x)\, dx, \qquad j = 2, 3, \ldots,$$

so that the final result is

$$u(\rho, t) = A_1 + \sum_{j=2}^{\infty} A_j \exp(-k\lambda_j^2 t) J_0(\lambda_j \rho), \qquad \textbf{(9.5–4)}$$

with A_1 and A_j as defined above and with $J_1(\lambda_j c) = 0$, $j = 1, 2, 3, \ldots$. ∎

EXAMPLE 9.5–5 A solid hemisphere of radius b has its plane face perfectly insulated while the temperature of its curved surface is given by $f(\cos \theta)$. Find the steady-state, bounded temperature at any point inside.

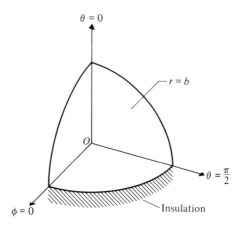

$\theta = 0$

$r = b$

O

$\theta = \dfrac{\pi}{2}$

$\phi = 0$

Insulation

FIGURE 9.5–5

Solution. Figure 9.5–5 shows one-quarter of the hemisphere. Since the temperature of the surface is independent of ϕ, we have the following problem.

P.D.E.: $\nabla^2 u = 0$ in spherical coordinates independent of ϕ with
$0 < r < b, 0 < \theta < \pi/2$;

B.C.: $u_z(r, \pi/2) = 0, \quad 0 < r < b,$
$u(b, \theta) = f(\cos \theta), \quad 0 < \theta < \pi/2.$

Since the variable z is not one of the coordinates in the spherical coordinate system, we need to alter the homogeneous boundary condition. Referring to Fig. 9.4–1, we have

$$z = r \cos \theta,$$

so that

$$\frac{\partial u}{\partial \theta} = \frac{\partial u}{\partial z} \frac{\partial z}{\partial \theta} = -r \sin \theta \frac{\partial u}{\partial z}.$$

Accordingly, at $\theta = \pi/2$ we have

$$\frac{\partial u}{\partial z} = -\frac{1}{r} \frac{\partial u}{\partial \theta}$$

and the condition $u_z = 0$ implies $u_\theta = 0$ at $\theta = \pi/2$.

Referring to Example 9.5–1, we have

$$u(r, \theta) = \sum_{n=0} A_n r^n P_n(\cos \theta),$$

which satisfies Laplace's equation. Now

$$u_\theta(r, \theta) = \sum_{n=0} A_n r^n P_n'(\cos \theta)(-\sin \theta)$$

so that

$$u_\theta(r, \pi/2) = \sum_{n=0} A_n r^n P_n'(0)(-1) = 0,$$

from which it follows that n is *even* (see Exercise 3d in Section 9.4); call it $2m$. **Updating** the solution gives us

$$u(r, \theta) = \sum_{m=0} A_{2m} r^{2m} P_{2m}(\cos \theta).$$

Now we apply the nonhomogeneous boundary condition to obtain

$$u(b, \theta) = \sum_{m=0} A_{2m} b^{2m} P_{2m}(\cos \theta) = f(\cos \theta),$$

that is, $f(\cos \theta)$ is to be represented by a series of even-degree Legendre polynomials on the interval $0 < \theta < \pi/2$. This is possible because these polynomials are orthogonal on the given interval and we can use Eq. (9.4–22) to compute the coefficients. Hence,

$$A_{2m} b^{2m} = (4m + 1) \int_0^{\pi/2} f(\cos \theta) P_{2m}(\cos \theta) \sin \theta \, d\theta$$

and the final solution becomes

$$u(r, \theta) = \sum_{m=0} (4m + 1) \left(\frac{r}{b}\right)^{2m} P_{2m}(\cos \theta) \int_0^1 f(x) P_{2m}(x) \, dx. \quad \blacksquare$$

We recommend the procedure used in the last example of updating the solution every time new information is obtained about it. Generally, it is better to apply the *homogeneous* conditions to the separate eigenfunctions, that is, before the product of eigenfunctions is formed. In the last example, however, we could safely depart from this procedure. (Why?)

We have presented a few examples to show how circular symmetry in a problem can lead to Bessel functions and spherical symmetry to Legendre polynomials. We have used separation of variables to solve the boundary-value problems since this technique allows us to transfer *homogeneous* conditions to the separate *ordinary* differential equations. We have also used our knowledge of the physical situation wherever possible to assign particular values to the separation constants.

EXERCISES 9.5

▶ **1.** In Example 9.5–1 show that

$$\frac{1}{\Theta}\frac{d^2\Theta}{d\theta^2} + \frac{\cot\theta}{\Theta}\frac{d\Theta}{d\theta} = -n(n+1)$$

is equivalent to

$$\frac{1}{\sin\theta}\frac{d}{d\theta}\left(\sin\theta\frac{d\Theta}{d\theta}\right) + n(n+1)\Theta = 0.$$

2. Show that we must take $D_n = F_n = 0$ in Example 9.5–1 even though the region in which we seek a solution does *not* include the points $r = 0$, $(b, 0)$, and (b, π). Why must these points be excluded? (*Hint*: Look at the partial differential equation being solved.)

3. Translate each of the given boundary conditions (B.C.) of Example 9.5–2 into mathematical statements.

4. Carry out the details in separating the variables in Example 9.5–2.

5. Solve

$$Z'' - \lambda_j^2 Z = 0, \qquad Z'(0) = 0.$$

(Compare Example 9.5–2.)

6. Fill in the necessary details to obtain A_j in Example 9.5–2.

7. Use separation of variables to obtain the ordinary differential equations and boundary conditions in Example 9.5–3.

8. Use separation of variables to obtain the two ordinary differential equations in Example 9.5–4.

9. In Example 9.5–4 explain why we must have

$$\lim_{t\to\infty} u_\rho(\rho, t) = 0.$$

▶▶ **10.** Obtain the solution to the problem in Example 9.5–1 given that the surface temperature is a constant 100°. Does your result agree with physical fact and with Theorem 8.1–1?

11. Obtain the solution to the problem in Example 9.5–1 if the surface temperature is given by $f(\cos\theta) = \cos\theta$. (*Hint*: Recall from Section 9.4 that $P_1(\cos\theta) = \cos\theta$.)

12. a) In Eq. (9.5–2) put $b = c = 1$ and write out the first three terms of the sum.

b) Use the result in part (a) to compute $u(0, 0)$.

c) Is the result in part (b) what you would expect? Explain.

13. Solve the problem of Example 9.5–2 given that the base and curved lateral surface are both kept at zero and the top is kept at 100°.

14. What would the result be if the separation constant in Example 9.5–3 were nonnegative? Do these results agree with physical fact? Explain.

15. In Example 9.5–3 change $f(\rho)$ to 1 and obtain the solution corresponding to Eq. (9.5–3). Is such an initial displacement physically possible? Explain.

16. According to Eq. (9.5–4),
$$\lim_{t\to\infty} u(\rho, t) = A_1.$$
Explain this result.

17. Solve the problem of Example 9.5–4 given that the outer edge of the disk is kept at temperature zero instead of being insulated and all other conditions remain the same.

18. Find the bounded, steady-state temperatures inside a solid hemisphere of radius b given that the bottom is kept at temperature zero and the remaining surface has a temperature distribution given by $f(\cos\theta)$.

19. Modify Exercise 18 so that $f(\cos\theta) = 100$ and obtain the solution to this problem.

20. Use the condition $f(\cos\theta) = 100$ in Example 9.5–5 and obtain the solution.

▶▶▶ 21. A dielectric sphere* of radius b is placed in a uniform electric field of intensity E in the z-direction. Determine the potential inside the sphere and outside the sphere. *Hint*: Both the potential inside (u) and the potential outside (U) must satisfy the potential equation. Continuity conditions are given by
$$u(b, \theta) = U(b, \theta), \qquad 0 < \theta < \pi,$$
and
$$Ku_r(b, \theta) = U_r(b, \theta), \qquad 0 < \theta < \pi, \qquad K > 0.$$
Moreover,
$$\lim_{r\to\infty} U(r, \theta) = -Ez = -Er \cos\theta.$$

22. Find the potential of a grounded, conducting sphere of radius b placed in a uniform electric field of intensity E in the z-direction. *Hint*: Again $\nabla^2 u = 0$ with
$$u(b, \theta) = 0$$
and
$$\lim_{r\to\infty} u(r, \theta) = -Ez = -Er \cos\theta.$$

23. Concentric spheres of radius a and b are held at constant potentials u_1 and u_2, respectively. Determine the potential at any point between the spheres.

9.6 NUMERICAL METHODS

Up to this point all the boundary-value problems that we have considered were such that their solutions could be obtained analytically. In actual practice, however, it is usually a rarity when a problem falls neatly into a class that can be solved by the methods we have discussed. Classical approaches may fail for one or more of the following reasons.

a) The partial differential equation is nonlinear and cannot be linearized without seriously affecting the result.

---

* This is *not* a solid sphere.

b) The boundary is irregular.

c) Boundary conditions are of mixed types.

d) Boundary values are time-dependent.

e) Materials must be considered that are not homogeneous and isotropic.

Some of the above can cause complexities that make any method except a numerical one completely impractical. Of course, numerical methods also have a number of shortcomings, as we will see later.

We present first a numerical method which is unusual in that it uses a principle from probability theory. Suppose that we need to solve a Dirichlet problem ($\nabla^2 u = 0$ with the values of u known on the boundary) in the plane over a region R that has an irregular boundary C. In particular, we seek the value of u at some given point P.

Beginning at P we start a random walk that eventually takes us to a boundary point; call it C_1. A **random walk** consists of unit steps in a direction parallel to the axes. An example of such a walk is shown in Fig. 9.6–1. Since the value of u is known at C_1, we record it; call it $u(C_1)$. We begin at P again and start a second random walk to obtain $u(C_2)$, and so on.

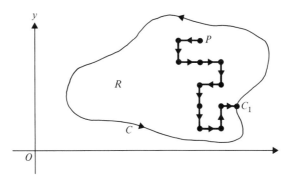

FIGURE 9.6–1 A random walk.

The theory of probability tells us that the probability of reaching a part of the boundary that is *near* to P is greater than the probability of reaching a part of the boundary that is *far* from P. But this is just another way of saying that the boundary values near to P have a greater influence in determining $u(P)$ than those farther away. This statement can be readily verified by experiment in the case where R is a metal plate and u is temperature, and we seek the steady-state temperature at a point P.

Hence, after a large number of walks (say, 1000) we would expect that the *average* value,

$$\frac{1}{1000} \sum_{i=1}^{1000} u(C_i),$$

would be close to the desired value $u(P)$. It can be shown that theoretically

$$u(P) = \lim_{n \to \infty} \frac{1}{n} \sum_{i=1}^{n} u(C_i). \tag{9.6–1}$$

The method described above is called a **Monte Carlo method** for solving a Dirichlet problem. It is especially useful if solutions at a few isolated points are desired. Usually, however, solutions at many points are required, and of special importance are curves that connect points having the same value of u. In a steady-state temperature problem, these curves are called **isothermals**.

A numerical method that yields the solution at each of a series of closely spaced points is the **finite-difference method**. It is also known as **Liebmann's method*** and the **relaxation method**. The use of this method requires covering the region R with a square grid that is as fine as necessary. Computations are made at each grid point (or node) on the basis of the fact that the value of u at each point can be expressed as a function of the values at neighboring points. This functional relationship is obtained by using a difference quotient. It can be seen, for example, that the derivative

$$\frac{du}{dx} = \lim_{h \to 0} \frac{u(x + h) - u(x)}{h}$$

can be approximated at a given point x by the difference quotient,

$$\frac{u(x + h) - u(x)}{h},$$

provided that h is sufficiently small. Similarly,

$$\frac{d^2u}{dx^2} = \lim_{h \to 0} \frac{1}{h} \left(\frac{du(x + h)}{dx} - \frac{du(x)}{dx} \right)$$

$$= \lim_{h \to 0} \frac{1}{h} \left(\frac{u(x + 2h) - u(x + h)}{h} - \frac{u(x + h) - u(x)}{h} \right)$$

can be replaced by

$$\frac{1}{h^2} (u(x + 2h) - 2u(x + h) + u(x)).$$

---

* After Karl O. H. Liebmann (1874–1939), a German mathematician.

By moving to the left a distance h, this last becomes

$$\frac{d^2u}{dx^2} \doteq \frac{1}{h^2}(u(x+h) - 2u(x) + u(x-h)).$$

If u is a function of two variables, x and y, then

$$\frac{\partial^2 u(x, y)}{\partial x^2} \doteq \frac{1}{h^2}(u(x+h, y) - 2u(x, y) + u(x-h, y))$$

and

$$\frac{\partial^2 u(x, y)}{\partial y^2} \doteq \frac{1}{h^2}(u(x, y+h) - 2u(x, y) + u(x, y-h));$$

hence the potential equation can be approximated as

$$u_{xx} + u_{yy} \doteq \frac{1}{h^2}(u(x+h, y) + u(x, y+h) + u(x-h, y)$$

$$+\, u(x, y-h) - 4u(x, y)) = 0,$$

so that

$$u(x, y) = \tfrac{1}{4}(u(x+h, y) + u(x, y+h) + u(x-h, y) + u(x, y-h))$$

$$\tag{9.6-2}$$

is a finite-difference approximation to Laplace's equation. In terms of a grid over the region R, Fig. 9.6–2 shows that the value of u at the point P is the average of the values at the four neighboring points numbered 1, 2, 3, and 4. In other words,

$$u(P) = \tfrac{1}{4}(u_1 + u_2 + u_3 + u_4).$$

This relation must hold at *every* point of the grid so that the method produces a system of linear algebraic equations that can be solved

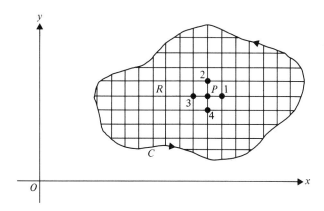

FIGURE 9.6–2 Finite-difference approximation.

with the aid of a computer. Note that some of the grid points fall on the boundary, hence are known. If the boundary lies between two grid points, then interpolation can be used. Initial values at interior points are *assumed*, then corrected as the computations progress. It should be pointed out that the better the beginning approximation, the more rapid the convergence.

Most numerical methods for partial differential equations are based on some variation of the finite-difference method. In other words, a partial differential equation is approximated by a system of linear algebraic equations. These can then be solved by a variety of numerical methods, some of which were described in Section 4.6.

It would seem that any desired degree of accuracy may be achieved by a finite-difference method simply by making the superimposed grid fine enough. This is not the case, however, since round-off and other errors tend to deteriorate the accuracy. Moreover, a fine grid means that a large number of algebraic equations need to be solved, which, in turn, may require more computer storage capacity than is available. These and other difficulties make many numerical methods less attractive than they first appear. A thorough analysis of these problems belongs in the realm of numerical analysis. An excellent elementary treatment of this subject can be found in Curtis E. Gerald's *Applied Numerical Analysis*, 2nd ed. (Reading, Mass.: Addison-Wesley, 1978). Gerald devotes three chapters to this subject, one to each of the three types of second-order, linear partial differential equations: elliptic, parabolic, and hyperbolic. Sample computer programs are given and comparisons are made with analytical methods.

KEY WORDS AND PHRASES

random walk   finite-difference method
Monte Carlo method

EXERCISES 9.6

▶ **1.** Generate a set of 100 random numbers in the following way. Using a telephone directory, start at some page and record the last two digits of each telephone number at the top and bottom of each column. Divide each number in this set repeatedly* (if necessary) by four and record the *remainder*. This will produce a set like

$$\{2, 1, 0, 3, 1, 2, 2, 0, 3, \ldots\}.$$

---

* The number 25 divided by 4 is 6, which can again be divided by 4, so the remainder is 2.

2. Consider the Dirichlet problem over a rectangular plate measuring 10 cm by 20 cm with one of the shorter sides held at temperature 100° and the other three sides held at zero. Choose coordinate axes as shown in Fig. 9.6–3. Find the temperature at the center of the plate by making 10 random walks using the numbers generated in Exercise 1. The numbers can be interpreted as follows:

   0   move 2 cm to the right,

   1   move 2 cm up,

   2   move 2 cm to the left,

   3   move 2 cm down.

When one random walk is finished, continue in the set of numbers for the second random walk, etc. If more numbers are needed, enlarge your set.

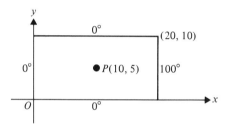

FIGURE 9.6–3

3. Divide the region of Exercise 2 into 8 subregions by lines 5 cm apart parallel to the axes so that *three* interior points will be produced. Use Eq. (9.6–2) to write a system of three algebraic equations. Solve these equations.

4. In Exercise 3 change the grid spacing from 5 cm to 2.5 cm. How many interior points (and equations) are there now?

5. Change the grid spacing of Exercise 3 to $3\frac{1}{3}$ cm and solve the resulting system of equations. (*Hint:* Use symmetry.)

6. Show that the one-dimensional wave equation and initial conditions $u_{tt} = u_{xx}$, $u(x, 0) = f(x)$, $u_t(x, 0) = g(x)$ can be approximated by the finite-difference equations

$$U(x, t + k) = 2U(x, t) - U(x, t - k) + \lambda^2(U(x + h, t) - 2U(x, t) + U(x - h, t)),$$

$$U(x, 0) = f(x),$$

$$U(x, k) = kg(x) + f(x),$$

where $\lambda = k/h$. (*Note:* It can be shown that the convergence of U to u requires that $\lambda < 1$.)

7. Show that the one-dimensional diffusion equation and initial condition

$$u_t = u_{xx}, \qquad u(x, 0) = f(x)$$

can be approximated by the finite-difference equations

$$U(x, t + k) = \lambda U(x + h, t) + (1 - 2\lambda)U(x, t) + \lambda U(x - h, t),$$
$$U(x, 0) = f(x),$$

where $\lambda = k/h^2$. (*Note*: It can be shown that the difference equation is unstable* for $\lambda > \frac{1}{2}$.)

---

* Meaning that U progressively departs from the true value u as the computation proceeds.

complex variables

10.1 THE ALGEBRA OF COMPLEX NUMBERS

Complex numbers can be very useful in solving problems in a number of areas. We have encountered examples of this usefulness in Sections 2.1 and 2.4 where certain quadratic equations had complex roots. In this chapter we will extend our knowledge of complex *numbers* to the *algebra* of complex numbers and then to complex *functions*, that is, functions of a *complex variable*.

A **complex number** is a number of the form $x + iy$, where x and y are *real* numbers and i satisfies $i^2 = -1$. We can also associate the complex number $x + iy$ with the ordered pair of real numbers (x, y). This is particularly convenient because it provides a one-to-one correspondence between complex numbers and points in the Euclidean plane. As shown in Fig. 10.1–1, the point P may be associated with the ordered pair (x, y) or with the complex number $x + iy$. We may, in fact, associate the vector from O to P with the complex number $x + iy$. The geometric interpretation of complex numbers was originated in 1806 by Argand*. For this reason the complex plane (Fig. 10.1–1) is sometimes referred to as an **Argand diagram**.

Equality of two complex numbers implies more, however, than does the equality of two vectors. Recall that in Chapter 5 we defined two vectors to be *equal* if they had the same direction and the same length. Two complex numbers $a + bi$ and $c + di$ are equal if and only if $a = c$ *and* $b = d$.

Addition of two complex numbers is accomplished in a natural way. Let $x_1 + iy_1$ and $x_2 + iy_2$ be two complex numbers. Then we define the **sum** of these as

$$(x_1 + iy_1) + (x_2 + iy_2) = (x_1 + x_2) + i(y_1 + y_2). \qquad \textbf{(10.1–1)}$$

Diagrammatically, the sum of two complex numbers, considered as vectors, corresponds to the diagonal of the parallelogram having the addends as sides (see Fig. 10.1–2). Note that this diagrammatic defi-

---

* Jean R. Argand (1768–1822), a Swiss mathematician.

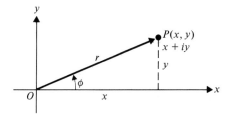

FIGURE 10.1–1 A complex plane.

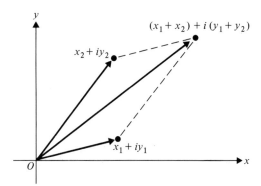

FIGURE 10.1–2 The addition of complex numbers.

nition of addition corresponds to the addition (or resultant) of forces in physics.

Since the x_i and y_i in the definition of addition (10.1–1) are *real* numbers, we immediately have the following theorem, the proof of which is left for the exercises (see Exercise 1).

Theorem 10.1–1 Let $x_1 + iy_1$, $x_2 + iy_2$, and $x_3 + iy_3$ be three complex numbers. Then:

a) $(x_1 + iy_1) + (x_2 + iy_2)$ is a complex number (closure property);

b) $(x_1 + iy_1) + (x_2 + iy_2) = (x_2 + iy_2) + (x_1 + iy_1)$ (commutative property);

c) $[(x_1 + iy_1) + (x_2 + iy_2)] + (x_3 + iy_3) = (x_1 + iy_1) + [(x_2 + iy_2) + (x_3 + iy_3)]$ (associative property);

d) $(x_1 + iy_1) + (0 + 0i) = (x_1 + iy_1)$ (existence of an additive identity).

It can be shown (Exercise 2) that the complex zero, $0 + 0i$, has the same properties for complex numbers as the real zero, 0, has for real numbers. Thus it is natural to make the correspondence,

$$0 + 0i \leftrightarrow 0.$$

We may carry this correspondence a step further and obtain a one-to-one correspondence between all real numbers and all complex numbers of the form $x + 0i$, that is,

$$x + 0i \leftrightarrow x. \tag{10.1-2}$$

Another way of interpreting (10.1–2) is to use the language of sets and say that the set of real numbers is a *subset* of the set of complex numbers. In brief, every real number is also a complex number, a view that may help to remove some of the mysticism sometimes associated with complex numbers.

So far we have discussed only *addition* of complex numbers. Before we consider subtraction, we define what is meant by multiplying a complex number by a real number. If a is a real number, then

$$a(x + iy) = ax + iay. \tag{10.1-3}$$

Here again, the vector analogy is helpful for visualizing the operation (Exercise 3).

If in Eq. (10.1–3) we put $a = -1$, we have

$$-(x + iy) = -x - iy,$$

which is the negative of the complex number $x + iy$. It now follows that

$$(x + iy) + (-x - iy) = 0 + 0i,$$

that is, $-(x + iy)$ is the *additive inverse* of $x + iy$. The use of additive inverses leads to the operation of **subtraction**, as shown in the next example.

EXAMPLE 10.1–1   Solve the equation

$$x + iy + 2 - 3i = 1 + 2i.$$

Solution. The additive inverse of $2 - 3i$ is $-2 + 3i$. Adding the latter to both members of the given equation results in

$$x + iy = -1 + 5i.$$

We can interpret this last statement as "the complex number $x + iy$ has the value $-1 + 5i$" or as "$x = -1, y = 5$." ∎

If we use polar coordinates, the point P of Fig. 10.1–1 can also be labeled $(r \cos \phi, r \sin \phi)$, hence we can write

$$x + iy = r(\cos \phi + i \sin \phi), \tag{10.1-4}$$

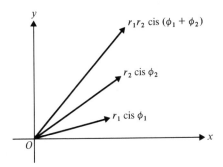

FIGURE 10.1–3 The multiplication of complex numbers.

the quantity in parentheses often being abbreviated **cis** ϕ. We observe that the angle ϕ is not unique, hence a multiple of 2π added to it would not change the polar representation.

The **polar form** of a complex number (10.1–4) is sometimes preferred because it simplifies algebraic operations involving multiplication and division. For example, if $x_1 + iy_1 = r_1$ cis ϕ_1 and $x_2 + iy_2 = r_2$ cis ϕ_2, then the following can be readily shown (Exercise 5):

$$(x_1 + iy_1)(x_2 + iy_2) = r_1 r_2 \text{ cis } (\phi_1 + \phi_2), \qquad \textbf{(10.1–5)}$$

(see Fig. 10.1–3);

$$\frac{(x_1 + iy_1)}{(x_2 + iy_2)} = \frac{r_1}{r_2} \text{ cis } (\phi_1 - \phi_2), \qquad r_2 \neq 0, \qquad \textbf{(10.1–6)}$$

(see Fig. 10.1–4);

$$(x + iy)^n = r^n \text{ cis } n\phi, \qquad n = 1, 2, \ldots . \qquad \textbf{(10.1–7)}$$

When $r = 1$ in Eq. (10.1–7), we have

$$(\cos \phi + i \sin \phi)^n = \cos n\phi + i \sin n\phi, \qquad \textbf{(10.1–8)}$$

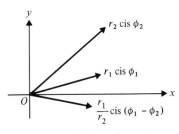

FIGURE 10.1–4 The division of complex numbers.

which is known as **de Moivre's formula**.* The formula is useful in obtaining *roots* of complex numbers.

We have been using the expression "a complex number $x + iy$." In order to simplify the notation, it is desirable to denote a complex number by a *single* letter. We will consistently use z for $x + iy$, z_1 for $x_1 + iy_1$, etc., throughout the remainder of this chapter.

EXAMPLE 10.1–2 Solve for z:

$$z^2 - i = 0.$$

Solution. Since $i = 0 + 1i$, we can write it in polar form as $\cos \pi/2 + i \sin \pi/2$. Thus we seek a complex number z such that $z^2 = \operatorname{cis} \pi/2$, that is,

$$z = \left(\operatorname{cis} \frac{\pi}{2} \right)^{1/2} = \operatorname{cis} \frac{\pi}{4} = \frac{\sqrt{2}}{2} (1 + i).$$

Observe, however, that we could also have written $z^2 = \operatorname{cis} 5\pi/2$ (why?), from which we obtain

$$z = \left(\operatorname{cis} \frac{5\pi}{2} \right)^{1/2} = \operatorname{cis} \frac{5\pi}{4} = \frac{-\sqrt{2}}{2} (1 + i),$$

as we would expect from $z = \pm \sqrt{i}$. We have thus shown that $\sqrt{i} = (\sqrt{2}/2)(1 + i)$. ∎

Example 10.1–2 can be generalized to show that if $z = r \operatorname{cis} \phi$, then

$$z^{1/n} = r^{1/n} \operatorname{cis} \left(\frac{\phi + 2k\pi}{n} \right), \qquad k = 0, 1, 2, \ldots, n - 1. \quad \textbf{(10.1–9)}$$

In Eq. (10.1–9) $r > 0$ with $r = 0$ if and only if $z = 0$. We have for $z = x + iy$,

$$r = \sqrt{x^2 + y^2} = |z|,$$

the real quantity $|z|$ being called the **absolute value** or **modulus** of the complex number z. The angle ϕ is called the **argument** or **phase** of the number z and we write $\phi = \arg z$. If ϕ is an argument of z, then so is $\phi \pm 2n\pi$, $n = 1, 2, 3, \ldots$.

We present another example, this time from ordinary differential equations.

EXAMPLE 10.1–3 Find the general solution of the fourth-order, linear, homogeneous differential equation,

$$\frac{d^4y}{dx^4} - 8 \frac{dy}{dx} = 0.$$

---

* After Abraham de Moivre (1667–1754), a French mathematician who lived in England.

Solution. The characteristic equation in terms of the variable* z is (see Section 2.1)

$$z(z^3 - 8) = 0.$$

In addition to $z = 0$, we need to find *three* values of z satisfying $z^3 - 8 = 0$. Using Eq. (10.1–9), we have

$$z^{1/3} = 8^{1/3} \text{ cis} \left(\frac{0 + 2k\pi}{3} \right), \qquad k = 0, 1, 2,$$

from which we obtain the three values 2 and $-1 \pm \sqrt{3}i$. Hence the general solution of the differential equation is

$$y(x) = c_1 + c_2 \exp(2x) + \exp(-x)(c_3 \cos\sqrt{3}x + c_4 \sin\sqrt{3}x). \quad \blacksquare$$

It is worthwhile to note that the nth roots of a complex number are *equally spaced* on a circle whose radius is the nth root of the absolute value of the number. In Example 10.1–3, the equation $z^3 - 8 = 0$ had one real root at $(2, 0)$ and complex roots at the intersections of $\phi = 2\pi/3$ and $\phi = 4\pi/3$ with the circle $|z| = 2$. These roots, as well as the results of Example 10.1–2, are shown in Fig. 10.1–5.

From algebra it is known that the complex roots of *real* equations always occur in *conjugate pairs*. Putting this another way, given that $P_n(z) = 0$ is a polynomial equation with real coefficients of which $z = x + iy$ is a root, then $\bar{z} = x - iy$ is also a root. We call $\bar{z}$ the **complex conjugate** of z; the operation of conjugation provides a notational

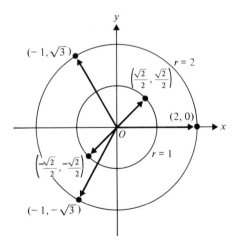

FIGURE 10.1–5 Roots of complex numbers.

---

* We are anticipating that the characteristic equation will have complex roots. If this is *not* the case, no harm has been done, as we have seen that the real numbers can be considered as complex numbers of a certain kind.

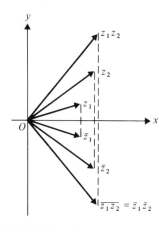

FIGURE 10.1–6 A diagram of Eq. (10.1–11).

simplicity that will be used in later sections. For the present we list some of the properties of conjugation—an operation that can be looked at geometrically as the reflection of a point in the x-axis:

$$\overline{z_1 \pm z_2} = \bar{z}_1 \pm \bar{z}_2, \qquad\qquad (10.1\text{–}10)$$

$$\overline{z_1 z_2} = \bar{z}_1 \bar{z}_2, \qquad\qquad (10.1\text{–}11)$$

(see Fig. 10.1–6);

$$\overline{\left(\frac{z_1}{z_2}\right)} = \frac{\bar{z}_1}{\bar{z}_2}, \qquad z_2 \neq 0, \qquad\qquad (10.1\text{–}12)$$

$$\text{if } \bar{z} = w, \text{ then } \bar{w} = z. \qquad\qquad (10.1\text{–}13)$$

Proofs of the above are left for the exercises (see Exercise 6).

One of the important uses of the complex conjugate stems from the fact that $z\bar{z}$ is *real* (Exercise 7). Given z_1 and $z_2 \neq 0$ (that is, $x_2 y_2 \neq 0$), we compute the **quotient** z_1/z_2 as follows:

$$\frac{z_1}{z_2} = \frac{x_1 + iy_1}{x_2 + iy_2} = \frac{(x_1 + iy_1)}{(x_2 + iy_2)} \cdot \frac{(x_2 - iy_2)}{(x_2 - iy_2)}$$

$$= \frac{(x_1 x_2 + y_1 y_2) + i(x_2 y_1 - x_1 y_2)}{x_2^2 + y_2^2}. \qquad\qquad (10.1\text{–}14)$$

Note that the last result can be written in the form $(x + iy)$. In fact, Eq. (10.1–14) can be considered a *formula* for dividing a complex number by a nonzero complex number.

Since a complex number contains two parts (real numbers without and with the symbol i) it is customary to call these the **real** (Re) and **imaginary** (Im) parts of the complex number. Notationally, if $z =$

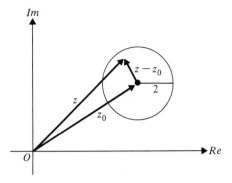

FIGURE 10.1–7

$x + iy$, then $\operatorname{Re}(z) = x$ and $\operatorname{Im}(z) = y$, both of these quantities being *real* numbers. Thus there is a certain inconsistency in saying, as we do, "The imaginary part of the complex number z is the real number y."

In closing this section we point out that because complex numbers are associated with points in a *plane* rather than points on a *line*, the relation $<$ cannot be used with complex numbers. In other words, complex numbers cannot be ordered. We can say, however, that $|z_1| < |z_2|$, $|z - z_0| \leq 2$, etc. This last defines all points z that lie in the closed circular disk of radius 2 with center at z_0 as shown in Fig. 10.1–7. Note that complex numbers may be represented by vectors as well as by points, as the figure shows.

KEY WORDS AND PHRASES

complex number   **argument**
Argand diagram   **complex conjugate**
de Moivre's formula   **real and imaginary parts**
modulus

EXERCISES 10.1

▶ 1. Prove Theorem 10.1–1. (*Hint*: Use the familiar field properties of the real numbers.)

2. Show that the complex number $0 + 0i$ has the following properties.
   a) $(x + iy) + (0 + 0i) = (x + iy)$
   b) $(x + iy)(0 + 0i) = (0 + 0i)$
   c) $(0 + 0i)/(x + iy) = (0 + 0i)$ if $xy \neq 0$

3. If $x + iy$ is a complex number with $xy \neq 0$, show diagrammatically each of the following.

a) $2(x + iy)$ **b)** $-(x + iy)$ **c)** $0.5(x + iy)$

4. Refer to Exercise 3. What can you conclude about the result of multiplying a complex number by a real number?

5. If $x + iy = r \operatorname{cis} \phi$ and $x_j + iy_j = r_j \operatorname{cis} \phi_j$, prove each of the following.

a) $(x_1 + iy_1)(x_2 + iy_2) = r_1 r_2 \operatorname{cis} (\phi_1 + \phi_2)$ **(10.1–5)**

(*Hint*: Use the trigonometric identities for $\cos (A + B)$ and $\sin (A + B)$.)

b) $\dfrac{x_1 + iy_1}{x_2 + iy_2} = \dfrac{r_1}{r_2} \operatorname{cis} (\phi_1 - \phi_2), \quad r_2 \neq 0$ **(10.1–6)**

(*Hint*: Multiply the numerator and denominator of the left member by $x_2 - iy_2$.)

c) $(x + iy)^n = r^n \operatorname{cis} n\phi, \quad n = 1, 2, \ldots$ **(10.1–7)**

(*Hint*: Use mathematical induction.)

6. If $z = x + iy$ and $\bar{z} = x - iy$, prove each of the following.

a) $\overline{z_1 \pm z_2} = \bar{z}_1 \pm \bar{z}_2$ **(10.1–10)**

b) $\overline{z_1 z_2} = \bar{z}_1 \bar{z}_2$ **(10.1–11)**

c) $\overline{\left(\dfrac{z_1}{z_2}\right)} = \dfrac{\bar{z}_1}{\bar{z}_2}, \quad z_2 \neq 0$ **(10.1–12)**

d) $\bar{w} = z$ where $w = \bar{z}$ **(10.1–13)**

7. Show that $z\bar{z}$ is real.

▶▶ 8. Perform the indicated operations and write each result in the form $x + iy$.

a) $3(2 + i) - 2i(i - 2) + 5$ **b)** $(2i - 3)(3 + i)$

c) $(1 - i) \div (2 + i)$ **d)** $(i - 1)^2 (2 + i)^3$

9. Convert each of the following numbers to polar form.

a) $1 + i$ **b)** $1 - i$ **c)** 3

d) $-i$ **e)** $-\sqrt{3} + 3i$ **f)** $1 - \sqrt{3}i$

g) $2 - 3i$

10. Find the absolute value and argument of each of the following complex numbers.

a) $1 + \sqrt{3}i$ **b)** $(4 \operatorname{cis} \pi/6)^{1/2}$ **c)** $(i - 2)^3$

d) -5 **e)** zero   **f)** $(2 - 2i)^2$

11. **a)** Solve for z:

$$z^2 + i = 0.$$

b) Use the results of part (a) and of Example 10.1–2 to find the general solution of

$$\frac{d^4 y}{dx^4} + y = 0.$$

12. Describe each of the following geometrically.

a) $|z| < 1$ **b)** $\operatorname{Re}(z) > -1$ **c)** $0 < \operatorname{Im}(z) < 1$

d) $2 < |z| < 5$ **e)** $|z + 2| < 3$

13. Using vectors to represent complex numbers, illustrate the following triangle inequalities geometrically and state each relation in words.

   a) $|z_1 + z_2| \le |z_1| + |z_2|$ **b)** $|z_1 - z_2| \ge ||z_1| - |z_2||$

14. Use the four roots of the equation $z^4 + 1 = 0$ to factor $z^4 + 1$ into *quadratic* factors with real coefficients. (*Note:* This procedure has application in finding inverse Laplace transforms. See Section 3.3.)

15. Find the trigonometric form of $1 - \cos \phi + i \sin \phi$.

16. Compute each of the following.

   a) $(-64)^{1/6}$ **b)** $\sqrt{3 + 4i}$ **c)** $\sqrt[3]{i - 1}$

17. Solve for z in each case and express the result in the form $x + iy$.

   a) $z^2 - 2iz - 5 = 0$

   b) $z^2 - (2 + 3i)z - 1 + 3i = 0$

18. **a)** Show that multiplying a complex number z by i has the effect of increasing arg z by $\pi/2$.

   b) How does this naturally lead to the definition $i^2 = -1$?

19. **a)** Write the quadratic equation that has a root equal to $2 - i$.

   b) Prove that if $z = x + iy$ is a root of a quadratic equation, then $\bar{z}$ is also a root.

20. Write each of the following in the form $x + iy$.

   a) 4 cis $\pi/3$ **b)** $3\sqrt{2}$ cis $3\pi/4$

   c) 3 cis π **d)** $\sqrt{2}$ cis $\pi/4$

21. Prove that $0 + 0i$ is the *unique* additive identity for complex numbers. (*Hint:* Assume there is a complex number $a + bi$ such that $(x + iy) + (a + bi) = (x + iy)$; then draw a conclusion about $a + bi$.)

22. Prove that the additive inverse of a complex number is unique. (Compare Exercise 21.)

23. Show that the division formula (10.1–14) gives the correct result when z_2 is real.

▶▶▶ 24. Prove the triangle inequality of Exercise 13(a) algebraically. (*Hint:* Begin with
$$|z_1 + z_2|^2 = (z_1 + z_2)(\bar{z}_1 + \bar{z}_2).)$$

25. Prove that if $z^2 = (\bar{z})^2$, then either Re$(z) = 0$ or Im$(z) = 0$.

26. Prove each of the following.

   a) $z\bar{z} = |z|^2$,   that is, $z\bar{z}$ is real for all z.

   b) $z + \bar{z} = 2$ Re(z),   that is, $z + \bar{z}$ is real for all z.

   c) $z - \bar{z} = 2i$ Im(z),   that is, Re$(z - \bar{z}) = 0$.

27. Prove that every negative real number has a negative real $(2n - 1)$th root for $n = 1, 2, \ldots$.

28. Find the eigenvalues and corresponding eigenvectors of the matrix

$$A = \begin{pmatrix} 1 & 0 & 0 \\ 0 & 0 & 1 \\ 0 & -4 & 0 \end{pmatrix}$$

10.2 ELEMENTARY FUNCTIONS

When we write $y = f(x)$ with x and y representing real numbers, we are indicating a **function of a real variable**. The allowable values of x (called the **domain** of the function) form a subset of a real line (the x-axis). Corresponding values of y (called the **range** of the function) also form a subset of a real line (the y-axis). Figure 10.2–1 shows an example of a function of a real variable.

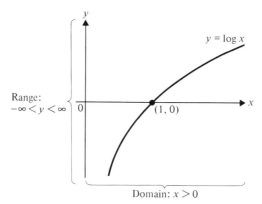

FIGURE 10.2–1 A function of a real variable.

On the other hand, the notation $w = f(z)$ represents a (complex-valued) **function of a complex variable**. Writing $z = x + iy$ and $w = u + iv$, we have

$$w = u + iv = f(z) = f(x + iy) = u(x, y) + iv(x, y),$$

so that both complex variables w and z have real and imaginary parts. As a result, the function f defines a mapping from one *plane* (the z-plane) to another (the w-plane). The **domain of definition** of the function is a portion of the xy-plane, whereas the **range** of the function is a portion of the uv-plane. We show an example of a mapping in Fig. 10.2–2 and give a specific one in the following example.

EXAMPLE 10.2–1 Analyze the mapping

$$w = \frac{1}{z}, \qquad 1 \le |z| \le 2,$$

algebraically and geometrically.

Solution. We have

$$w = \frac{1}{z} = \frac{1}{x + iy} = \frac{x - iy}{x^2 + y^2} = \frac{x}{x^2 + y^2} - i\,\frac{y}{x^2 + y^2}.$$

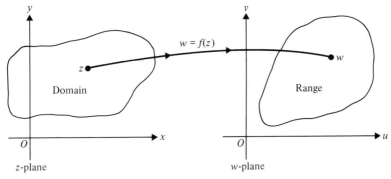

FIGURE 10.2–2 A function of a complex variable.

Hence,

$$w = u + iv = \frac{x}{x^2 + y^2} - i\,\frac{y}{x^2 + y^2}$$

so that

$$u(x, y) = \frac{x}{x^2 + y^2}$$

and

$$v(x, y) = \frac{-y}{x^2 + y^2}.$$

The domain of definition of f is a closed region consisting of all points of the xy-plane that satisfy the inequalities

$$1 \le x^2 + y^2 \le 4.$$

This closed region is shown shaded in Fig. 10.2–3(a). For each point in the shaded region in the z-plane we can use the functions $u(x, y)$ and $v(x, y)$ to compute the corresponding point in the shaded region in the w-plane. For example, the point A has coordinates $(1, 0)$; hence it maps into $A'(1, 0)$ in the w-plane. Similarly, $B(2, 0)$ maps into $B'(\frac{1}{2}, 0)$. We leave it as an exercise to show that C and D map into C' and D', respectively, as shown (Exercise 1). We can, in fact, show (Exercise 2) that

$$u^2 + v^2 = \frac{1}{x^2 + y^2}$$

so that the range of the function is the closed region in the w-plane consisting of points that satisfy

$$\tfrac{1}{4} \le u^2 + v^2 \le 1. \quad \blacksquare$$

In Section 1.3 we referred to the class of elementary functions of a real variable and indicated in a footnote in that section some members of this class. We are now ready to discuss elementary functions of a

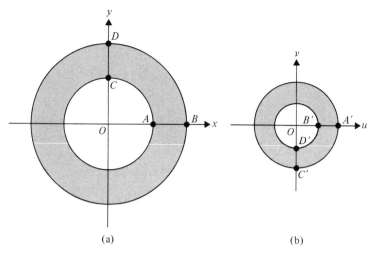

FIGURE 10.2-3 The function $1/z$ as a mapping.

complex variable in general and it seems appropriate to define the term "elementary" more rigorously. We do this in the following two definitions.

Definition 10.2-1 Let $f(z)$ and $g(z)$ be functions of a complex variable z and let α be a complex constant. Then the **elementary operations** on the functions $f(z)$ and $g(z)$ are those that result in any of the following:

$$f(z) \pm g(z), \qquad f(z) \cdot g(z), \qquad \frac{f(z)}{g(z)}, \qquad (f(z))^{\alpha}, \qquad (\alpha)^{f(z)}, \qquad \log(f(z)),$$

provided the indicated operations are defined.

Definition 10.2-2 An **elementary function** of the complex variable z is one obtained from constants (real or complex) and the variable z by means of a *finite* number of elementary operations.

The function $w = 1/z$ is a **single-valued function** and this is what is implied when we speak of a function of a complex variable. On the other hand, $w = z^{1/2}$ is a **multiple-valued function** (Exercise 3).

Definitions of a number of basic elementary functions of a complex variable stem from Euler's formula. We have, for example, the **exponential function**

$$e^z = e^{x+iy} = e^x e^{iy} = e^x(\cos y + i \sin y). \tag{10.2-1}$$

This definition is in agreement with what we know about the exponential and trigonometric functions of a real variable. For example, the Maclaurin's series for e^x, $\cos y$, $\sin y$, and e^{iy} satisfy the relation shown in Eq. (10.2-1) (Exercise 4).

We also have the following **hyperbolic functions**:

$$\sinh z = \frac{e^z - e^{-z}}{2}, \tag{10.2-2}$$

$$\cosh z = \frac{e^z + e^{-z}}{2}, \tag{10.2-3}$$

$$\tanh z = \frac{\sinh z}{\cosh z} = \frac{e^z - e^{-z}}{e^z + e^{-z}}. \tag{10.2-4}$$

From the above definitions, and with the aid of Exercise 5(c), we can see that sinh z and cosh z are periodic of complex period $2\pi i$. We can also show that tanh z is periodic of complex period πi (Exercise 6).

The **trigonometric functions** are now obtained from these definitions:

$$\sin z = \frac{e^{iz} - e^{-iz}}{2i}, \tag{10.2-5}$$

$$\cos z = \frac{e^{iz} + e^{-iz}}{2}. \tag{10.2-6}$$

We leave it for the exercises to show that the above definitions lead to familiar trigonometric identities (Exercise 7). We can also show (Exercise 8) that sin z and cos z have real period 2π. From this we can deduce (Exercise 9) that sin $z = 0$ if and only if $z = n\pi$, where n is an integer. In other words, sin z has only *real* zeros with a similar situation holding for cos z, as shown in the next example.

EXAMPLE 10.2–2 Determine all values of z that satisfy cos $z = 0$.

Solution. From Eq. (10.2–6) we have

$$\cos z = \frac{1}{2}(e^{iz} + e^{-iz}) = \frac{1}{2}(e^{i(x+iy)} + e^{-i(x+iy)})$$

$$= \frac{1}{2}(e^{-y}(\cos x + i \sin x) + e^{y}(\cos x - i \sin x))$$

$$= \cos x \left(\frac{e^y + e^{-y}}{2}\right) - i \sin x \left(\frac{e^y - e^{-y}}{2}\right)$$

$$= \cos x \cosh y - i \sin x \sinh y.$$

Next, we observe that a complex number z is zero if and only if $|z|^2 = 0$. Here,

$$|\cos z|^2 = \cos^2 x \cosh^2 y + \sin^2 x \sinh^2 y$$
$$= \cos^2 x(1 + \sinh^2 y) + (1 - \cos^2 x) \sinh^2 y$$
$$= \cos^2 x + \sinh^2 y$$
$$= 0$$

if and only if $\cos x = 0$ *and* $\sinh y = 0$. These last two equations are satisfied by

$$x = (2n + 1)\frac{\pi}{2}, \qquad n = 0, \pm 1, \pm 2, \dots \quad \text{and} \quad y = 0.$$

Thus the only zeros of $\cos z$ are *real* and are given by

$$z = (2n + 1)\frac{\pi}{2}, \qquad n = 0, \pm 1, \pm 2, \dots . \quad \blacksquare$$

We also observe an interesting relation between real and complex functions (Exercise 10). If x is real, then

$$\sin x = -i \sinh ix$$

and

$$\cos x = \cosh ix.$$

Next we examine the **inverse** of the exponential function. If $z = \exp w$, then we would like to define $w = \log z$ as the inverse of the exponential function. We have seen, however, that

$$\exp (w + 2n\pi i) = \exp w$$

for any integer n. Hence $\log z$ has infinitely many values and does not qualify as a single-valued function. In polar form,

$$z = |z|e^{i\phi} = |z|e^{i(\phi + 2n\pi)},$$

so that for $z \neq 0$,

$$\log z = \text{Log } |z| + i(\phi + 2n\pi), \tag{10.2–7}$$

where the upper-case "L" indicates the logarithm of a *real* number. We take Eq. (10.2–7) as the definition of the logarithm of a complex number with ϕ being a particular value of the argument of z. Later it will be convenient to consider the **logarithmic function**; hence we define

$$w = \text{Log } z = \text{Log } |z| + i \text{ Arg } z. \tag{10.2–8}$$

The value Log z in Eq. (10.2–8) is called the **principal value** of $\log z$ and is obtained from Eq. (10.2–7) by taking $n = 0$. In Eq. (10.2–8) we have $-\pi < \text{Arg } z \le \pi$ so that Log z is a single-valued function defined for all z except $z = 0$ and points on the negative x-axis.

If z is complex and $a \neq 0$, then we define

$$a^z = \exp (z \log a), \tag{10.2–9}$$

where

$$\log a = \text{Log } |a| + i(\arg a + 2n\pi),$$

with n an integer. The above definition shows that a^z is multivalued in general since $\log a$ is multivalued. More important, however, is the fact that a^z does not obey *all* the laws of exponents. For example, we cannot say that $(a^z)^w = a^{zw}$ (see Exercise 24) or that $(a^z)^w = (a^w)^z$ (see Exercise 25).

EXAMPLE 10.2-3 Determine all the values of $(-3)^i$.

Solution. Using Eqs. (10.2–9) and then (10.2–7), we have

$$(-3)^i = \exp(i \log(-3))$$
$$= \exp(i(\text{Log } 3 + i(\pi + 2n\pi)))$$
$$= \exp(i \text{ Log } 3) \exp(-\pi - 2n\pi), \qquad n = 0, \pm 1, \pm 2, \ldots . \ \blacksquare$$

We can carry the result of Example 10.2–3 further by using Euler's formula. For example, since

$$\exp(i \text{ Log } 3) = \cos(\text{Log } 3) + i \sin(\text{Log } 3)$$
$$\doteq 0.9998 + 0.0192i,$$

the value of $(-3)^i$ corresponding to $n = 0$ is approximately $0.0432 + 0.0008i$.

KEY WORDS AND PHRASES

function of a complex variable	exponential function
domain of definition	hyperbolic function
range of a function	trigonometric function
single- and multiple-valued	logarithmic function
function	principal value

EXERCISES 10.2

▶ 1. Verify that the points A, B, C, and D of Fig. 10.2–3(a) map into the points A', B', C', and D' of Fig. 10.2–3(b), respectively, using the function $w = 1/z$.

2. In Example 10.2–1 verify that the function $w = 1/z$ maps circles in the z-plane into circles in the w-plane. What is the relationship:

    a) between the arguments in the z-plane and w-plane?

    b) between the moduli in the two planes?

3. Show that $w = z^{1/2}$ is a two-valued function.

4. **a)** In the Maclaurin series for e^x replace x by iy to obtain the series for e^{iy}.

    b) Show that the series for e^{iy} in part (a) is the same as the one obtained by adding the series for $\cos y$ and $i \sin y$.

    c) Deduce from Eq. (10.2–1) that e^z converges for all z.

5. Show that e^z has the following properties.

    a) $e^{z_1} e^{z_2} = e^{z_1 + z_2}$

    b) $e^{z_1}/e^{z_2} = e^{z_1 - z_2}$

    c) e^z is periodic with complex period $2\pi i$

d) $e^z = 1$ if and only if $z = 2n\pi i$, where n is an integer

e) $e^{z_1} = e^{z_2}$ if and only if $z_1 = z_2 + 2n\pi i$, where n is an integer

6. Show that $\tanh z$ is periodic with complex period πi. *Hint*: $\tanh z$ can be written as

$$\frac{1 - e^{-2z}}{1 + e^{-2z}}.$$

7. Use Eqs. (10.2–5) and (10.2–6) to obtain each of the following identities.

a) $\sin^2 z + \cos^2 z = 1$

b) $\sin (z_1 \pm z_2) = \sin z_1 \cos z_2 \pm \cos z_1 \sin z_2$

c) $\cos (z_1 \pm z_2) = \cos z_1 \cos z_2 \mp \sin z_1 \sin z_2$

d) $\sin 2z = 2 \sin z \cos z$

e) $\cos 2z = \cos^2 z - \sin^2 z$

8. **a)** Show that e^{iz} is periodic with real period 2π.

b) Use part (a) to show that $\sin z$ and $\cos z$ are periodic with real period 2π.

9. Prove that $\sin z = 0$ if and only if $z = n\pi$, where n is an integer.

10. Show that:

a) $\sin x = -i \sinh ix$;   **b)** $\cos x = \cosh ix$.

11. Show that $\log z$ has the following properties.*

a) $\log z_1 z_2 = \log z_1 + \log z_2$

b) $\log \dfrac{z_1}{z_2} = \log z_1 - \log z_2$

12. Obtain values of $(-3)^i$ in Example 10.2–3 corresponding to (a) $n = 1$ and (b) $n = -1$.

▶▶ 13. Use the function $w = 1/z$ to map the unbounded region $0 < \arg z < \pi/4$ into the w-plane.

14. Use the function $w = z^2$ to map the region $1 \le |z| \le 2$, $\text{Im}(z) \ge 0$ into the w-plane. Analyze and graph the mapping as in Example 10.2–1.

15. In each case find $\log z$ and $\text{Log } z$.

a) $z = 2$ **b)** $z = i - 1$ **c)** $z = -1$

16. Show that:

a) $\cos (iy) = \cosh y$;   **b)** $\sin (iy) = i \sinh y$.

(*Hint*: Use Eqs. 10.2–5 and 10.2–6.)

17. Write each of the following functions in the form $w = u(x, y) + iv(x, y)$.

a) $w = f(z) = z^2 - 2z$ **b)** $w = f(z) = z^3$

c) $w = f(z) = \dfrac{(z + i)}{(z^2 + 1)}$ **d)** $w = f(z) = \dfrac{(z^2 - 3)}{|z - 1|}$

18. Use the definitions (10.2–5) and (10.2–6) to define each of the following trigonometric functions.

a) $\tan z$ **b)** $\cot z$

c) $\sec z$ **d)** $\csc z$

* Note that the argument of each term must be chosen appropriately.

19. Using the definitions (10.2–2) and (10.2–3) define each of the following hyperbolic functions.

    a) coth z **b)** sech z **c)** csch z

20. Express each of the following functions in the form $u(x, y) + iv(x, y)$.

    a) $\sin z$ **b)** $\cos z$ **c)** $\tan z$

    d) $\sinh z$ **e)** $\cosh z$

21. Write each of the following complex numbers in the form $a + bi$.

    a) $\sin i$ **b)** $\cos (2 - i)$

    c) $\sinh (1 - i\pi)$ **d)** $\exp (3 - i\pi/4)$

22. Determine all values of $(-2)^i$.

▶▶▶ 23. Derive each of the following identities.

    a) $\sinh (z_1 \pm z_2) = \sinh z_1 \cosh z_2 \pm \cosh z_1 \sinh z_2$

    b) $\cosh (z_1 \pm z_2) = \cosh z_1 \cosh z_2 \pm \sinh z_1 \sinh z_2$

    c) $\cosh^2 z - \sinh^2 z = 1$

24. Show that $(-1)^{2i} = \exp(-2(2n + 1)\pi)$, where n is an integer, whereas

$$(-1)^{2i} = ((-1)^2)^i = 1^i = e^{i \log 1} = e^{-2n\pi},$$

where n is an integer, thus showing that, in general, $(a^z)^w \neq a^{zw}$.

25. Show that

$$(a^z)^w = a^{wz} \exp(2n\pi i)w,$$

where n is an integer, thus showing that, in general, $(a^z)^w \neq (a^w)^z$.

26. Prove that $(\exp z)^n = \exp(nz)$ for all integers n.

27. Show that $|\exp (z)| = 1$ implies that $\operatorname{Re}(z) = 0$.

28. Consider

$$F(s) = \frac{1}{s^2 + 1},$$

the Laplace transform of some function $f(t)$. By writing

$$F(s) = \frac{A}{s + i} + \frac{B}{s - i}$$

evaluate A and B. Then use Table 3.1–1 to obtain $f(t)$.

10.3 DIFFERENTIATION

In this section we will examine derivatives of functions of a complex variable. To prepare for this we need some concepts regarding sets of points in the z-plane or the xy-plane, if you wish to think of it that way.

    The **ε-neighborhood** of a point z_0 (Fig. 10.3–1) consists of all points z that satisfy

$$|z - z_0| < \varepsilon,$$

in which ε is a given positive number. Denote a set of points in the

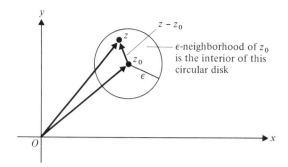

FIGURE 10.3–1 The ε-neighborhood of a point.

z-plane by S. A point z_1 is said to be an **interior point** of the set S if there is an ε-neighborhood of z_1 that contains only points of S (Fig. 10.3–2). If every point of a set is an interior point, we call the set an **open set**. A point is a **boundary point** of a set S if every ε-neighborhood of the point contains at least one point of S and at least one point that is not in S. The totality of boundary points of S is called the **boundary** of S. A set that contains all its boundary points is called a **closed set**.

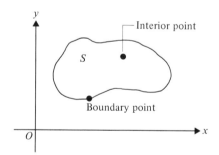

FIGURE 10.3–2 Interior and boundary points.

EXAMPLE 10.3–1 Consider each of the following sets.

a) $|z - i| < 1$

b) $1 \le |z| \le 2$

Set (a) is an open set: Every point of the set is an interior point. For example, $z = 0.01i$ is an interior point since an ε-neighborhood of the point with $\varepsilon = 0.001$ contains only points of the set (a). Geometrically, the set consists of all points within the circular region of radius 1 with center at i. Set (b) is a closed set. Its boundary points consist of all points on the unit circle and all points on the circle of radius 2, both centered at the origin (see Fig. 10.2–3a). ■

Note that a closed set must contain *all* its boundary points. Hence some sets cannot be classified as open or closed. The set $1 \le \text{Re } z \le 2$ is such a set (Exercise 1).

The foregoing concepts can be used to define a continuous function. Let f be a function that is defined in some neighborhood of z_0 although not necessarily at z_0 itself. We say that the limit of $f(z)$ as z approaches z_0 is w_0, provided that for any $\varepsilon > 0$, there exists a positive number δ, such that

$$|f(z_0) - w_0| < \varepsilon, \quad \text{whenever } 0 < |z - z_0| < \delta.$$

In brief, we write

$$\lim_{z \to z_0} f(z) = w_0.$$

If $w_0 = f(z_0)$, that is, if

$$\lim_{z \to z_0} f(z) = f(z_0),$$

then we say that f is **continuous at** z_0. If f is continuous at every point of a set S, we say that f is **continuous on** S.

The definition of limit is directly related to that for functions of a real variable studied in calculus. Hence it is not surprising that the following properties of limits hold.

Theorem 10.3–1 If $\lim_{z \to z_0} f(z) = w_1$ and $\lim_{z \to z_0} g(z) = w_2$, then:

a) $\lim_{z \to z_0} (f(z) \pm g(z)) = w_1 \pm w_2$;

b) $\lim_{z \to z_0} f(z)g(z) = w_1 w_2$;

c) $\lim_{z \to z_0} \dfrac{f(z)}{g(z)} = \dfrac{w_1}{w_2}$, provided that $w_2 \ne 0$.

It is a direct consequence of Theorem 10.3–1 that sums, differences, and products of continuous functions are also continuous, as are quotients, except for those values of z for which the denominator vanishes. Moreover, if $w = f(z) = u(x, y) + iv(x, y)$, then it follows that f is continuous if and only if $u(x, y)$ and $v(x, y)$ are both continuous (Exercise 2). Thus we can conclude, for example, that the function

$$f(z) = (2x + y) + i(x^2 y)$$

is everywhere continuous since the component functions $(2x + y)$ and $(x^2 y)$ are everywhere continuous functions of x and y.

Consider a fixed point z_0 in the domain of definition of a function f. Let z be any point of some neighborhood of z_0 and let $\Delta z = z - z_0$.

Then f is **differentiable at** z_0 with the derivative $f'(z_0)$ given by

$$f'(z_0) = \lim_{\Delta z \to 0} \frac{f(z_0 + \Delta z) - f(z_0)}{\Delta z}, \qquad (10.3\text{--}1)$$

provided the limit exists for all choices of Δz. It follows from the definition of the derivative (10.3–1) that f is continuous wherever it is differentiable.

We will be interested chiefly in a certain type of differentiable function, as defined next.

Definition 10.3–1 A complex-valued function f is said to be **analytic*** on an open set S if it is differentiable at every point of S.

Implied in the above definition is the fact that if a function f is analytic at a point z_0, then not only does $f'(z_0)$ exist but $f'(z)$ exists at every point z in some neighborhood of z_0. If a function is analytic at every point of the z-plane, it is called an **entire function**. The most simple examples of entire functions are the polynomials. Before we give other examples, we need the following important result.

Theorem 10.3–2 A necessary condition for a function $f(z) = u(x, y) + iv(x, y)$ to be differentiable at a point z_0 is that $\partial u / \partial x$, $\partial u / \partial y$, $\partial v / \partial x$, and $\partial v / \partial y$ exist at z_0 and satisfy the **Cauchy–Riemann equations,**[†]

$$\frac{\partial u}{\partial x} = \frac{\partial v}{\partial y} \quad \text{and} \quad \frac{\partial u}{\partial y} = -\frac{\partial v}{\partial x}, \qquad (10.3\text{--}2)$$

there. Moreover, then

$$f'(z) = \frac{\partial u}{\partial x} + i \frac{\partial v}{\partial x} = \frac{\partial v}{\partial y} - i \frac{\partial u}{\partial y}. \qquad (10.3\text{--}3)$$

The Cauchy–Riemann equations play an important role in the application of complex analysis to fluid flow, potential theory, and other problems in physics. A proof of Theorem 10.3–2 can be found in E. B. Saff and A. D. Snider's *Fundamentals of Complex Analysis for Mathematics, Science, and Engineering*, p. 44 (Englewood Cliffs, N.J.: Prentice-Hall, 1976). Note that the theorem gives a *necessary* condition only. Sufficient conditions for differentiability require that the first partial derivatives of u and v with respect to x and y not only exist at a point z_0 but also be continuous there.[‡]

---

* Some authors use the terms *regular* or *holomorphic* in place of analytic.
† After Augustin-Louis Cauchy (1789–1857), a French mathematician, who discovered the equations, and Georg F. B. Riemann (1826–1866), a German mathematician, who was responsible for a general theory of functions of a complex variable.
‡ See J. D. Gray and S. A. Morris, "When is a Function that Satisfies the Cauchy–Riemann Equations Analytic?", *Amer. Math. Monthly* **85**, no. 4 (April 1978), pp. 246–256.

EXAMPLE 10.3–2   The function

$$f(z) = e^z$$

is an entire function. We have

$$e^z = e^x \cos y + ie^x \sin y;$$

hence

$$u(x, y) = e^x \cos y \quad \text{and} \quad v(x, y) = e^x \sin y.$$

Thus

$$\frac{\partial u}{\partial x} = e^x \cos y = \frac{\partial v}{\partial y}$$

and

$$\frac{\partial u}{\partial y} = -e^x \sin y = -\frac{\partial v}{\partial x},$$

so that the Cauchy–Riemann equations (10.3–2) are satisfied. Moreover, $e^x \cos y$ and $e^x \sin y$ exist for all values of x and y, hence for all z. Further, by Eq. (10.3–3),

$$f'(z) = e^x \cos y + ie^x \sin y = f(z)$$

as we would expect. ■

EXAMPLE 10.3–3   The function

$$f(z) = (x^2 - y^2) - 2ixy$$

is nowhere analytic. Here we have

$$\frac{\partial u}{\partial x} = 2x, \quad \frac{\partial v}{\partial y} = -2x, \quad \frac{\partial u}{\partial y} = -2y, \quad -\frac{\partial v}{\partial x} = 2y,$$

so that the Cauchy–Riemann equations are satisfied only when $x = y = 0$, that is, when $z = 0$. Analyticity, however, requires differentiability in some *neighborhood*; hence the given function is nowhere analytic. ■

   With the aid of Theorem 10.3–2 we can now establish a number of formulas for differentiating various functions of a complex variable. For example,

$$\sin z = \sin (x + iy) = \sin x \cos (iy) + \cos x \sin (iy)$$
$$= \sin x \cosh y + i \cos x \sinh y,$$

with the aid of Exercise 16 of Section 10.2. Hence, $u(x, y) = \sin x \cosh y$ and $v(x, y) = \cos x \sinh y$ so that, using Eq. (10.3–3), we have

$$\frac{d(\sin z)}{dz} = \cos x \cosh y - i \sin x \sinh y$$

$$= \cos x \cos (iy) - \sin x \sin (iy)$$
$$= \cos (x + iy) = \cos z.$$

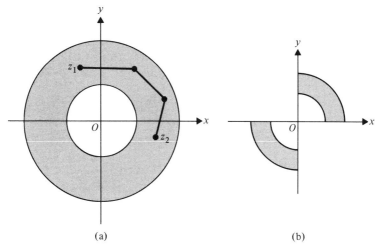

FIGURE 10.3–3 (a) Connected set. (b) Nonconnected set.

We leave it for the exercises to show that $\sin z$ and $\cos z$ are entire functions (Exercise 3).

It will be convenient to define a particular kind of open set: one that is connected. An open set S is said to be **connected** if any two points of the set can be joined by a continuous chain of a finite number of line segments, all of whose points lie entirely in S. For example, the set in Fig. 10.3–3(a) is connected, whereas that in Fig. 10.3–3(b) is not. An open connected set is called a **domain**. Thus we speak of a function being differentiable in a certain domain or a function being analytic in a certain domain.

Analytic functions have an important property to which we have already alluded in Sections 6.2 and 8.1. If $f(z) = u(x, y) + iv(x, y)$ and $f(z)$ is analytic in some domain D, then both $u(x, y)$ and $v(x, y)$ satisfy Laplace's equation in D. This follows directly from the Cauchy–Riemann equations (10.3–2). As stated in the earlier sections, the functions $u(x, y)$ and $v(x, y)$ are called harmonic functions.

EXAMPLE 10.3–4 Verify that the real and imaginary parts of the function

$$f(z) = (z + 1)^2$$

are harmonic functions.

Solution. We have

$$(z + 1)^2 = (x + iy)^2 + 2(x + iy) + 1$$
$$= (x + 1)^2 - y^2 + 2yi(x + 1).$$

Hence,

$$u(x, y) = (x + 1)^2 - y^2, \qquad v(x, y) = 2y(x + 1),$$

from which

$$u_{xx} = 2, \qquad u_{yy} = -2, \quad \text{and} \quad v_{xx} = v_{yy} = 0.$$

Thus both $u(x, y)$ and $v(x, y)$ are harmonic functions. ■

In Example 10.3–4 we showed that there is a close connection between analytic functions and harmonic functions. Both the real and imaginary parts of the analytic function $(z + 1)^2$ are harmonic. Of great practical value, however, is the fact that the families of curves,

$$u(x, y) = \text{constant} \quad \text{and} \quad v(x, y) = \text{constant},$$

are orthogonal to each other wherever they intersect and wherever their tangents are defined. In a two-dimensional electrostatic problem, for example, the curves $u = $ constant may define lines of force in which case the curves $v = $ constant define the equipotentials. In Fig. 10.3–4 we show the orthogonal families of curves obtained from Example 10.3–4.

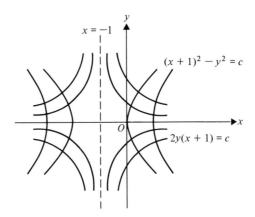

FIGURE 10.3–4 Orthogonal families of curves (Example 10.3–4).

We close this section by remarking that *every* harmonic function is the real part of some analytic function. In other words, given a harmonic function $u(x, y)$, we can find another harmonic function $v(x, y)$ such that

$$f(z) = u(x, y) + iv(x, y)$$

is an analytic function. We illustrate with an example.

EXAMPLE 10.3–5   Given the harmonic function

$$u(x, y) = \cosh x \cos y,$$

construct an analytic function $f(z)$ whose real part is $u(x, y)$.

Solution. Using the Cauchy–Riemann equations (10.3–2), we have

$$u_x = \sinh x \cos y = v_y;$$

hence

$$v(x, y) = \sinh x \sin y + \phi(x).$$

Then

$$v_x = \cosh x \sin y + \phi'(x)$$
$$= \cosh x \sin y = -u_y,$$

so that $\phi'(x) = 0$, $\phi(x) = C$, a real constant, and

$$v(x, y) = \sinh x \sin y + C.$$

Using Exercise 4, we find that the required analytic function is

$$f(z) = \cosh z + \alpha,$$

where α is a complex constant.   ■

   The function $v(x, y)$ obtained in Example 10.3–5 from the harmonic function $u(x, y)$ is called a **harmonic conjugate** of u. Note that this is another use of the word "conjugate," which is not to be confused with the term "complex conjugate" defined in Section 10.1.

KEY WORDS AND PHRASES

ε-neighborhood	entire function
interior point	Cauchy–Riemann equations
open set	connected set
boundary point	domain
boundary	harmonic function
closed set	harmonic conjugate
analytic function	

EXERCISES 10.3

▶   **1.** Identify the boundary points of the set $1 \le \operatorname{Re} z \le 2$. Is the set closed? Why?

  2. Prove that if $f(z) = u(x, y) + iv(x, y)$, then f is continuous if and only if $u(x, y)$ and $v(x, y)$ are both continuous.

3. Show that each of the following is an entire function.

   a) $\sin z$
   b) $\cos z$

4. Show that $\cosh z = \cosh x \cos y + i \sinh x \sin y$. (*Hint:* Use Eqs. 10.2–2, 10.2–3, 10.2–5, and 10.2–6.)

▶▶ 5. Describe each of the following sets. Where appropriate, state what constitutes the boundary.

   a) $0 \le \arg z \le \pi/2$
   b) $|z - z_0| < \rho$
   c) $\rho_1 \le |z| \le \rho_2$
   d) $\operatorname{Re} z \le 0$
   e) $\operatorname{Im} z > 0$

6. Describe each of the following sets and give their boundary points, if any.

   a) $|z - i| \le 1$
   b) $1 < |z| < 2$
   c) $1 \le |z| < 2$

7. Prove that each of the following functions is continuous everywhere except possibly for certain points. State the exceptions.

   a) $w = 1 + z^2$
   b) $w = |z|$
   c) $w = 1/z$

   d) $w = \operatorname{Im}(z)$
   e) $w = \bar{z}$
   f) $w = \dfrac{z + 2}{z^2 + z - 2}$

8. Explain why $f(z) = \bar{z}$ is nowhere differentiable. (*Hint:* Use Eq. 10.3–1 and show that the limit does not exist.)

9. Show that $f(z) = z$ is an entire function by computing $f'(z)$ from Eq. (10.3–1).

10. Find $f'(z)$ in each case.

   a) $f(z) = \cos z$
   b) $f(z) = \tan z$
   c) $f(z) = \sinh z$
   d) $f(z) = \cosh z$
   e) $f(z) = 1/z$
   f) $f(z) = z^{1/2}$

11. **a)** Show that the function $\operatorname{Log} z$ is analytic everywhere except for points on the nonpositive real axis.

   b) Show that

   $$\frac{d}{dz}(\operatorname{Log} z) = \frac{1}{z}$$

   for all z where $\operatorname{Log} z$ is analytic.

12. Verify that both the real and imaginary parts of the following functions are harmonic.

   a) $f(z) = 1/z$
   b) $f(z) = z^3 - i(z + 1)$
   c) $f(z) = e^{-y}(\cos x + i \sin x)$

13. Find the harmonic conjugates of each of the following functions.

   a) $x^3 - 3xy^2 + y$
   b) $x^3 - 3xy^2$
   c) $\frac{1}{2} \log (x^2 + y^2)$, $xy \neq 0$

▶▶▶ 14. Show that if $f(z) = u(x, y) + iv(x, y)$, then each of the functions $u(x, y)$ and $v(x, y)$ is harmonic wherever $f(z)$ is analytic.

15. Prove Theorem (10.1–1). *Hint:* With $f(z) = u(x, y) + iv(x, y)$, $z = x + iy$, $z_0 = x_0 + iy_0$, and $w_1 = u_1 + iv_1$, $\lim_{z \to z_0} f(z) = w_1$ implies that

$$\lim_{\substack{x \to x_0 \\ y \to y_0}} u(x, y) = u_1 \quad and \quad \lim_{\substack{x \to x_0 \\ y \to y_0}} v(x, y) = v_1.$$

16. If $f(z)$ and $g(z)$ are continuous at z_0, prove each of the following.

a) $f(z) \pm g(z)$ is continuous at z_0.   **b)** $f(z)g(z)$ is continuous at z_0.

c) $\dfrac{f(z)}{g(z)}$ is continuous at z_0, provided that $g(z_0) \neq 0$.

17. Referring to Exercise 13, determine the analytic function in each case.

10.4 MAPPING

In Example 10.2–1 we presented one mapping defined by $w = 1/z$ called an **inversion**. In this section we will look at a number of mappings and examine their characteristics with a view to solving certain boundary-value problems later by using an appropriate mapping. The next example shows how this might be done in a simple case.

EXAMPLE 10.4–1   Solve the following boundary-value problem (Fig. 10.4–1).

$$\text{P.D.E.:} \quad u_{xx} + u_{yy} = 0, \quad x^2 + y^2 > 1;$$
$$\text{B.C.:} \quad u = x + 1, \quad x^2 + y^2 = 1, \quad |u| \text{ is bounded.}$$

Solution. Here we are seeking a bounded, harmonic function in the unbounded region outside the unit circle that reduces to $x + 1$ on the unit circle. Define $z = x + iy$ and $\zeta = \xi + i\eta$ and use the inversion mapping $\zeta = 1/z$. Then the problem becomes

$$\text{P.D.E.:} \quad U_{\xi\xi} + U_{\eta\eta} = 0, \quad \xi^2 + \eta^2 < 1;$$
$$\text{B.C.:} \quad U = \xi + 1, \quad \xi^2 + \eta^2 = 1.$$

Here

$$U(\xi, \eta) = u(x(\xi, \eta), y(\xi, \eta))$$

and it is clear (see Exercise 18) that $U = \xi + 1$ is a solution (actually, the unique solution). But the inversion mapping (see Example 10.2–1)

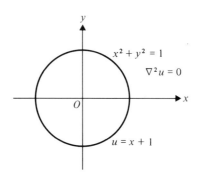

$$x^2 + y^2 = 1$$
$$\nabla^2 u = 0$$
$$u = x + 1$$

FIGURE 10.4–1

shows that

$$\xi(x, y) = \frac{x}{x^2 + y^2};$$

hence the solution to the given problem is

$$u(x, y) = \frac{x}{x^2 + y^2} + 1. \qquad \textbf{(10.4–1)}$$

We observe that $u(x, y) = x + 1$ satisfies the given P.D.E. and B.C. but it is not bounded, whereas Eq. (10.4–1) gives a bounded solution (Exercise 1). ■

 The inversion mapping is a particular case of a mapping defined by

$$w = \frac{\alpha z + \beta}{\gamma z + \delta}, \qquad \alpha\delta - \beta\gamma \neq 0 \qquad \textbf{(10.4–2)}$$

and called a **bilinear transformation*** (or mapping). The constants α, β, γ, and δ are complex constants and the restriction $\alpha\delta - \beta\gamma \neq 0$ is necessary so that w will not be a constant (Exercises 3 and 4).
 Observe that the mapping $w = 1/z$ is a special case of the bilinear transformation defined in Eq. (10.4–2). We will next look briefly at this and other special cases.

1. $\alpha = \delta = 1$ and $\gamma = 0$ produces $w = z + \beta$. This mapping is called a **translation**; it shifts a point z in the z-plane to another point $z + \beta$, as shown in Fig. 10.4–2. Note that a translation can be regarded as the result of adding two vectors.

2. $\beta = \gamma = 0$ and $\delta = 1$ produces $w = \alpha z$. This mapping produces a **magnification** and a **rotation**. We have

$$w = |\alpha| |z| \exp(i \arg(\alpha + z))$$

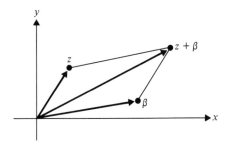

FIGURE 10.4–2 A translation.

---

* Also known as a **linear fractional transformation** and a **Möbius transformation**.

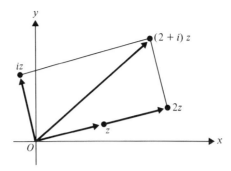

FIGURE 10.4–3 The mapping $w = (2 + i)z$.

so that $|\alpha| > 1$ produces a *stretching* while $|\alpha| < 1$ produces a *contraction*. An example of this mapping is $w = (2 + i)z$ shown in Fig. 10.4–3.

3. $\gamma = 0$ and $\delta = 1$ produces $w = \alpha z + \beta$, which is a combination of cases (1) and (2) above; that is, a rotation through the angle arg α and a magnification by the factor $|\alpha|$, followed by a translation through the vector β.

4. $\alpha = \delta = 0$ and $\beta = \gamma = 1$ is the inversion transformation already considered. This transformation carries the set of straight lines and circles into itself. (See Exercise 5 for a partial proof of this statement.)

It can be seen from these special cases that every bilinear transformation can be expressed as a finite sequence of translations, magnifications, rotations, and inversions. There is another important property, however, that bilinear transformations have. They are conformal (or angle-preserving) for all z except $z = -\delta/\gamma$ according to the following definition.

Definition 10.4–1 Let a mapping be given by $w = f(z)$. Then the mapping is **conformal** at z_0 if $w_0 = f(z_0)$ and any two smooth curves C_1 and C_2 through z_0 that intersect at an angle θ are mapped into two curves Γ_1 and Γ_2 through w_0 that also intersect at an angle θ.

Inherent in the above definition is the fact that a conformal mapping preserves both the *magnitude* and *sense* of an angle. Figure 10.4–4 shows a conformal mapping. It can be proved* that the mapping $w = f(z)$ is conformal at each point z of a domain where f is analytic and $f'(z) \neq 0$.

---

* See E. B. Saff and A. D. Snider, *Fundamentals of Complex Analysis for Mathematics, Science, and Engineering*, p. 304ff. (Englewood Cliffs, N.J.: Prentice-Hall, 1976).

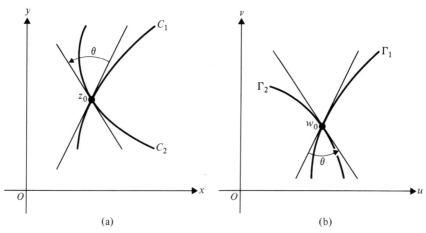

FIGURE 10.4-4 Conformal mapping: (a) z-plane; (b) w-plane.

EXAMPLE 10.4-2 Analyze the transformation

$$w = \frac{1}{2}\left(z + \frac{1}{z}\right).$$

Solution. We first observe that this is *not* a bilinear transformation. We have

$$w = u + iv = \frac{1}{2}\left(x + iy + \frac{1}{x + iy}\right)$$

$$= \frac{1}{2}\left(x + \frac{x}{x^2 + y^2}\right) + \frac{i}{2}\left(y - \frac{y}{x^2 + y^2}\right);$$

hence

$$u = \frac{1}{2}\left(x + \frac{x}{x^2 + y^2}\right) \quad \text{and} \quad v = \frac{1}{2}\left(y - \frac{y}{x^2 + y^2}\right).$$

Thus

$$u_x = \frac{1}{2}\left(1 + \frac{y^2 - x^2}{(x^2 + y^2)^2}\right) = v_y$$

and

$$u_y = \frac{-xy}{2(x^2 + y^2)^2} = -v_x.$$

This shows that f is analytic everywhere except at $z = 0$. Moreover,

$$f'(z) = \frac{1}{2}\left(1 - \frac{1}{z^2}\right) = \frac{1}{2}\left(\frac{z^2 - 1}{z^2}\right);$$

hence f is conformal except at $z = 0$ and $z = \pm 1$. In polar form the transformation can be written as

$$w = \frac{1}{2}\left(|z|e^{i\phi} + \frac{e^{-i\phi}}{|z|}\right).$$

From this we observe that if $|z| = c > 1$, then

$$w = \frac{1}{2}\left(c(\cos\phi + i\sin\phi) + \frac{1}{c}(\cos\phi - i\sin\phi)\right)$$

from which

$$u = \frac{1}{2}\left(c + \frac{1}{c}\right)\cos\phi, \qquad v = \frac{1}{2}\left(c - \frac{1}{c}\right)\sin\phi,$$

so that solving for $\cos\phi$ and $\sin\phi$, squaring, and adding gives us

$$4\left(\frac{c}{c^2 + 1}\right)^2 u^2 + 4\left(\frac{c}{c^2 - 1}\right)^2 v^2 = 1,$$

showing that circles centered at the origin are mapped into ellipses centered at the origin. On the other hand, if $y = mx$, where m is a real constant, then

$$u = \frac{1}{2}\left(x + \frac{1}{x(1 + m)^2}\right), \qquad v = \frac{m}{2}\left(x - \frac{1}{x(1 + m^2)}\right)$$

and

$$m^2 u^2 - v^2 = \frac{m^2}{1 + m^2}.$$

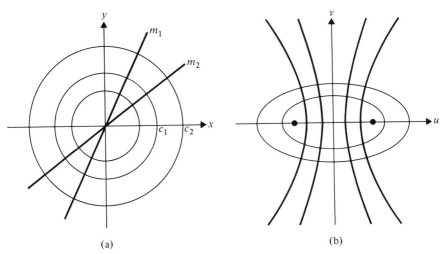

(a)   (b)

FIGURE 10.4–5 The transformation $w = \frac{1}{2}(z + 1/z)$ (Example 10.4–2): (a) z-plane; (b) w-plane.

Thus straight lines through the origin are mapped into hyperbolas centered at the origin. It can be shown that the foci of the ellipses and the hyperbolas are at $w = \pm 1$. Some of these curves are shown in Fig. 10.4–5. The transformation of this example is of interest in fluid dynamics. (See Exercise 6, however, in this connection.) ■

In the next example we indicate that an entire function can also be used advantageously in mapping.

EXAMPLE 10.4–3   Study the mapping $w = \sin z$.

Solution. Since $\sin z$ is periodic with period 2π, it is sufficient to examine the fundamental strip $-\pi/2 < x \le \pi/2$, $-\infty < y < \infty$. We have

$$w = u + iv = \sin z = \sin (x + iy)$$
$$= \sin x \cos iy + \cos x \sin iy$$
$$= \sin x \cosh y + i \cos x \sinh y,$$

so that

$$u = \sin x \cosh y, \qquad v = \cos x \sinh y. \qquad \textbf{(10.4–3)}$$

From these last equations we see that the portion of the x-axis, $0 \le x \le \pi/2$, is mapped into the portion of the u-axis, $0 \le u \le 1$; the line $x = \pi/2$, $y \ge 0$ is mapped into the portion of the u-axis, $u \ge 1$; the nonnegative y-axis is mapped into the nonnegative v-axis. Figure 10.4–6 shows the mapping with corresponding points indicated by primed and unprimed letters. In Section 10.6 we will use this mapping in the inverse sense; that is, we will map the upper halfplane into a semi-infinite strip of width π by means of the transformation $z = \sin w$. (See Exercise 7 for another example.) ■

We close this section with a mapping that preserves the magnitude but not the sense of angles. Mappings that have this property are

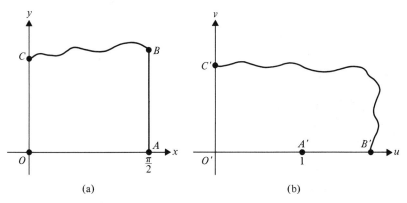

FIGURE 10.4–6 The mapping $w = \sin z$ (Example 10.4–3): (a) z-plane; (b) w-plane.

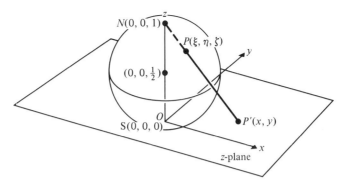

FIGURE 10.4–7 Stereographic projection.

called **isogonal** (see Exercise 8 for another example). We consider a particular isogonal mapping called a **stereographic projection**. Consider a sphere* of radius $\frac{1}{2}$ that is tangent to the z-plane at its origin. Choose a system of coordinates (ξ, η, ζ) so that the point of tangency will be at $(0, 0, 0)$, the center of the sphere at $(0, 0, \frac{1}{2})$, and the "north pole" at $(0, 0, 1)$. These points are shown in Fig. 10.4–7. The stereographic projection is defined as follows: A point $P(\xi, \eta, \zeta)$ *on the sphere* is mapped into a point $P'(x, y)$ *in the plane* by extending the line $\overrightarrow{NP}$ until it pierces the plane. This mapping is **bicontinuous**, meaning that points in a neighborhood of P on the sphere will be mapped into points in a neighborhood of P' in the plane, and conversely, points "near to" P' will also be "near to" P. In order to make the mapping and its inverse one-to-one, we must assign a point N' to the point N. Accordingly, we assign $z = \infty$ to N and call it **the point at infinity**. With the addition of this "point" we have what is called the *extended complex plane*. It is now possible to write $z \to \infty$ and to evaluate

$$f(z) = \frac{3z + 1}{z - 2}$$

at $z = \infty$ and $z = 2$. We readily find that $f(2) = \infty$ and $f(\infty) = 3$. Other problems involving the stereographic projection and the point at infinity appear in the exercises.

KEY WORDS AND PHRASES

inversion	conformal mapping
bilinear transformation	isogonal mapping
translation	stereographic projection
magnification	bicontinuous mapping
rotation	point at infinity

* Also called the **Riemann sphere**.

EXERCISES 10.4

▶ **1. a)** Show that $u(x, y) = x + 1$ satisfies the P.D.E. and B.C. of Example 10.4–1 but is not bounded.

   b) Show that $u(x, y) = x/(x^2 + y^2) + 1$ satisfies the P.D.E. and B.C. of Example 10.4–1 and is also bounded. (*Hint*: Write Eq. 10.4–1 as $[\cos (\arg z)/|z|] + 1$ and find the limit as $|z| \to \infty$.)

2. Show that a bilinear transformation can be written in the form

$$Awz + Bw + Cz + D = 0,$$

which is linear in both w and z, thus accounting for the name "bilinear."

3. Show that a bilinear transformation can be written in the form

$$w = -\frac{\alpha\delta - \beta\gamma}{\gamma} \frac{1}{\gamma z + \delta} + \frac{\alpha}{\gamma};$$

hence w is a constant if $\alpha\delta = \beta\gamma$. (*Hint*: Assume that $\gamma \neq 0$; then add and subtract $\alpha\delta/\gamma$ in the numerator of Eq. 10.4–2.)

4. If in a bilinear transformation $\gamma = 0$ (see Exercise 3), then $\delta \neq 0$ and $\alpha\delta = \beta\gamma$ again leads to w being a constant. Fill in the details for the above argument.

5. Consider the inversion mapping $w = 1/z$ referred to in this section and in Example 10.2–1.

   a) Show that straight lines through the origin are mapped into straight lines through the origin. What is the relationship between the slopes of the lines?

   b) Show that the circles $(x - a)^2 + y^2 = a^2$ are mapped into the lines $u = 1/2a$.

   c) Show that the circles $x^2 + (y - a)^2 = a^2$ are mapped into the lines $v = -1/2a$.

   d) What do the points of intersection of the circles in parts (b) and (c) map into?

6. Consider the mapping

$$w = \frac{1}{2}\left(z + \frac{1}{z}\right)$$

of Example 10.4–2.

   a) Investigate the case $|z| = 1$.

   b) Investigate the case $|z| < 1$. In particular, show that $|z| = 1/C$ maps into the same ellipse as $|z| = C$.

   c) Using the result of part (b), explain why the mapping is limited to $|z| > 1$ in practice.

7. Use Eqs. (10.4–3) to map the rectangular region $-\pi/2 \leq x \leq \pi/2, 0 \leq y \leq c$, into a semielliptic region. Label corresponding points on a figure.

8. Let f be a function analytic over a domain D. Show that the mapping $w = f(\bar{z})$ preserves the magnitude of angles but reverses their orientation.

▶▶ 9. Show that the bilinear transformation

$$w = \frac{1}{z}$$

maps the infinite strip $0 < y < \frac{1}{2}a$ into the domain

$$u^2 + (v + a)^2 > a^2, \qquad v < 0.$$

10. Show that the bilinear transformation

$$w = \frac{z}{z - 1}$$

maps the circular disk $|z| \leq 1$ into the halfplane $u \leq \frac{1}{2}$.

11. Show that the bilinear transformation

$$w = \frac{z - i}{z + i}$$

maps the circular disk $|z| < 1$ into the halfplane $u < 0$.

12. The **fixed points** (or **invariant points**) of a transformation $w = f(z)$ are given by the solutions of $z = f(z)$. Prove that a bilinear transformation can have at most *two* fixed points.

13. Find the fixed points of each of the following transformations.

a) $w = \dfrac{1}{z}$ **b)** $w = \dfrac{z}{z - 1}$ **c)** $w = z + \dfrac{1}{z}$

14. A value z_0 for which $f'(z_0) = 0$ is called a **critical point** of the transformation $w = f(z)$. Find the critical points of each of the following transformations.

a) $w = \dfrac{1}{z}$ **b)** $w = \dfrac{z - i}{z + i}$

c) $w = \sin z$ **d)** $w = z^2$

15. **a)** Show that if $P(\xi, \eta, \zeta)$ is a point on the stereographic sphere, then

$$\xi^2 + \eta^2 = \zeta(1 - \zeta).$$

b) If $P'(x, y)$ is the projection of P, then

$$z = x + iy = \frac{\xi + \eta i}{1 - \zeta}.$$

c) The points N, P, and P' are collinear; hence show that

$$\xi = tx, \qquad \eta = ty, \qquad \zeta - 1 = -t,$$

for some real constant t. Use the result in part (a) to find t and thus show that

$$\xi = \frac{x}{x^2 + y^2 + 1}, \qquad \eta = \frac{y}{x^2 + y^2 + 1}, \qquad \zeta = \frac{x^2 + y^2}{x^2 + y^2 + 1}.$$

16. Consider the mapping $w = \cos z$ having a fundamental strip $-\pi < x \leq \pi$, $-\infty < y < \infty$.

a) Map the half-line $x = \pi$, $-\infty < y \leq 0$.

b) Map the half-strip $0 < \text{Re } z < \pi$, $\text{Im } z < 0$.

c) By mapping the boundaries, map the rectangle $0 < x < \pi$, $-1 < y < 1$.

17. Consider the mapping $w = e^z$.

a) Where is the mapping conformal?

b) Map the fundamental strip $-\pi < y \leq \pi$, $-\infty < x < \infty$.

c) Map the line segment $x = a$, $-\pi < y \leq \pi$.

d) Map the semi-infinite strip $x \leq 0$, $0 \leq y \leq \pi$.

e) Map the rectangle $c_1 \leq x \leq c_2$, $0 \leq y \leq \pi$.

▶▶▶ 18. **a)** Show that the following problem of Example 10.4–1,

$$\text{P.D.E.: } U_{\xi\xi} + U_{\eta\eta} = 0, \quad \xi^2 + \eta^2 < 1;$$
$$\text{B.C.: } U = \xi + 1, \quad \xi^2 + \eta^2 = 1,$$

can be expressed in polar coordinates as

$$\text{P.D.E.: } \rho^2 U_{\rho\rho} + \rho U_\rho + U_{\phi\phi} = 0, \quad 0 < \rho < 1, \quad -\pi < \phi < \pi;$$
$$\text{B.C.: } U(1, \phi) = \cos \phi + 1, \quad -\pi < \phi < \pi.$$

b) Solve the above problem by separation of variables to obtain

$$U(\rho, \phi) = \rho \cos \phi + 1 = \xi + 1.$$

(*Note*: The function $\cos \phi + 1$ is an even function.)

19. Use the mapping $w = (1 + i)z$ to map the half-plane $y > 0$ by:

a) using rectangular coordinates;

b) using polar coordinates.

Sketch the resulting region.

20. Determine the bilinear transformation that maps the following points as shown:

$$z_1 = -1 \rightarrow w_1 = 1, \qquad z_2 = 0 \rightarrow w_2 = \infty, \qquad z_3 = 1 \rightarrow w_3 = i.$$

21. Consider the transformation $w = z^2$.

a) Map the first quadrant in the z-plane. (*Hint*: Use polar coordinates.)

b) Map the positive and negative real axes in the z-plane. How does the result show that the mapping is not one-to-one?

c) Map the circles $|z| = $ constant.

22. Show that the transformation of Example 10.4–2 maps the circle $|z| = 2$ into an ellipse. Draw a figure and locate corresponding points in the z- and w-planes.

10.5 THE COMPLEX INTEGRAL

Consider the definite integral of a function f of a complex variable z,

$$\int_\alpha^\beta f(z) \, dz,$$

where α and β are complex numbers. If we attach the usual meaning to this integral, that is, if z has an initial value $z = \alpha$ and a terminal value $z = \beta$, then this implies that z assumes values in the z-plane from the point α to the point β. Clearly, the definite integral above requires that some *path* between α and β be specified before it is truly a "definite" integral. This, in turn, implies that the definite integral of a complex function f is actually a *line integral*.

At this point you may find it worthwhile to review the material on line integrals given in Section 5.5. Especially helpful will be the concept of a conservative force field and the fact that certain line integrals evaluated around a closed path in a conservative field are zero. We will also use the term "smooth arc" as given in Definition 5.2–1. We define a **contour** as a continuous chain of a finite number of smooth arcs so that line integrals will also be called **contour integrals**.

EXAMPLE 10.5–1 Evaluate the contour integral

$$\oint_C \bar{z} \, dz$$

over the contour C consisting of the lower half of the closed unit semicircle centered at the origin.

Solution. The closed contour C described in the positive sense is shown in Fig. 10.5–1. We have $z = x + iy$, $\bar{z} = x - iy$, $dz = dx + i\,dy$, and, for the straight portion of the contour, $y = 0$, $dy = 0$. Hence for this portion,

$$\int_1^{-1} x \, dx = 0.$$

For the curved portion it is convenient to use polar coordinates so that $z = e^{i\phi}$, $dz = ie^{i\phi}\,d\phi$, $\bar{z} = e^{-i\phi}$, hence,

$$i \int_\pi^{2\pi} d\phi = \pi i.$$

Thus the integral around the closed contour is πi. ∎

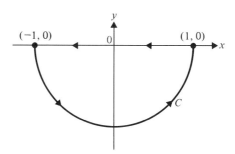

FIGURE 10.5–1

There is often an advantage in parameterizing the path of integration. When this is done, a contour C can be described as follows:

$$C: \quad x = x(t), \quad y = y(t), \quad a \le t \le b.$$

Then

$$w = u(x, y) + iv(x, y) = f(z)$$

and

$$\int_C f(z) \, dz = \int_a^b (u(x(t), y(t)) + iv(x(t), y(t))) \left(\frac{dx}{dt} + i \frac{dy}{dt} \right) dt$$

$$= \int_a^b \left(u(x(t), y(t)) \frac{dx}{dt} - v(x(t), y(t)) \frac{dy}{dt} \right) dt$$

$$+ i \int_a^b \left(v(x(t), y(t)) \frac{dx}{dt} + u(x(t), y(t)) \frac{dy}{dt} \right) dt$$

or, if we simplify the notation,

$$\int_C f(z) \, dz = \int_C (u \, dx - v \, dy) + i \int_C (v \, dx + u \, dy).$$

But both integrals on the right are now *real* integrals and, if we make C a *closed* contour and assume that u and v are continuously differentiable, then we can invoke Green's theorem in the plane (see Eq. 5.6–11) to write

$$\int_C (u \, dx - v \, dy) = -\iint_R \left(\frac{\partial v}{\partial x} + \frac{\partial u}{\partial y} \right) dx \, dy$$

and

$$\int_C (v \, dx + u \, dy) = \iint_R \left(\frac{\partial u}{\partial x} - \frac{\partial v}{\partial y} \right) dx \, dy,$$

where R is the closed region bounded by C, the latter being defined in a positive sense. If $f(z)$ is analytic in R, then the integrands in both double integrals are zero by virtue of the Cauchy–Riemann equations (Eqs. 10.3–2). Putting all these ideas together results in the following theorem.

Theorem 10.5–1 If f is analytic and f' is continuous at all points interior to and on a closed contour C, then

$$\oint_C f(z) \, dz = 0.$$

Theorem 10.5–1 is of interest since it implies that analytic functions have *unique integrals* as well as unique derivatives. More important, however, will be the extension of the theorem to nonanalytic functions when we discuss residues later in this section. Another consequence of the theorem is that the integral of an analytic function is an analytic function of its upper limit, provided that the path of integration lies in a simply connected domain throughout which the integrand is analytic.

The theorem was first proved by Cauchy and strengthened (by weakening the hypothesis) in 1883 by Edouard Goursat (1858–1936), a French mathematician. We present the strengthened version known today as Cauchy's theorem.*

Theorem 10.5–2 (*Cauchy's theorem*). If a function f is analytic at all points interior to and on a closed contour C, then

$$\oint_C f(z)\, dz = 0.$$

Cauchy's theorem has numerous applications and consequently merits being called *the fundamental theorem* in the theory of analytic functions. We remark that, although the theorem gives the result of zero for the integral around a closed contour, we define C in the positive sense; that is, as C is being traversed, the enclosed region is on the left. The reason for keeping to this convention is that later we will extend Cauchy's theorem to situations in which the result will *not* be zero.

EXAMPLE 10.5–2   Compute

$$\int_C \frac{dz}{z^2(z^2 + 9)},$$

where C is the boundary of the region $1 < |z| < 2$.

Solution. Figure 10.5–2 shows the contour C described in the positive sense. We show the contour beginning and ending at $z = 1$. The cut between $z = 1$ and $z = 2$ is purely artificial and could have been made to join the two circles at some other place to provide a simple closed curve C. The integrand of the integral to be computed is analytic everywhere except at $z = 0$ and $z = \pm 3i$. Since all three of these points are outside the closed region bounded by C, the integral in question is zero by Cauchy's theorem.   ■

The point z_0 is called a **singular point**, or **singularity**, of the function f if f is not analytic at z_0, but if every ε-neighborhood of z_0 contains points at which f is analytic. In particular, if f is analytic in every ε-neighborhood of z_0 except at z_0 itself, then z_0 is called an

---

* Also called the **Cauchy–Goursat theorem**.

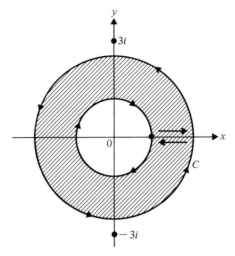

FIGURE 10.5–2

isolated singular point of f. The function

$$f(z) = \frac{1}{z^2(z^2 + 9)} \qquad (10.5–1)$$

has isolated singular points at $z = 0$ and $z = \pm 3i$.

If z_0 is an isolated singular point of f but

$$(z - z_0)^m f(z)$$

is analytic for some integer $m > 1$, then z_0 is called a **pole of order** m of f. If $m = 1$, the pole is called a **simple pole**. The function in Eq. (10.5–1) has a pole of order 2 at $z = 0$ and simple poles at $z = \pm 3i$.

It is apparent from Fig. 10.5–2 that the value of the integral does not depend on the particular circles chosen there. Any other closed path would have given the same result, provided only that the integrand remained analytic in the region enclosed and on its boundary.

Now suppose that an analytic function $f(z)$ has an isolated singular point at $z = z_0$. Then the quantity

$$\frac{1}{2\pi i} \int_C f(\zeta) \, d\zeta$$

is called the **residue** of $f(z)$ at z_0, provided that the path C enclosed no other singularity than z_0. We write

$$\text{Res}[f(z), z_0] = \frac{1}{2\pi i} \int_C f(\zeta) \, d\zeta. \qquad (10.5–2)$$

In order to understand the above definition of a residue more clearly we evaluate

$$\int_C (z - z_0)^m \, dz, \tag{10.5–3}$$

where C is a circle of radius ρ with center at z_0 described in the positive sense (Fig. 10.5–3). In terms of a parameter t the path C may be described by

$$z = z_0 + \rho(\cos t + i \sin t), \qquad 0 < t \leq 2\pi.$$

Hence the integral (10.5–3) becomes

$$\int_0^{2\pi} (\rho(\cos t + i \sin t))^m \rho(-\sin t + i \cos t) \, dt$$

$$= i\rho^{m+1} \int_0^{2\pi} (\cos (m + 1)t + i \sin (m + 1)t) \, dt$$

$$= 0$$

for all integral values of m except $m = -1$ (see Exercise 1). Thus

$$\int_C (z - z_0)^m \, dz = \begin{cases} 2\pi i, & \text{for } m = -1, \\ 0, & \text{for every other integral value of } m. \end{cases}$$

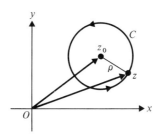

FIGURE 10.5–3 Path C for (10.5–3).

It now follows that if $f(z)$ is analytic in a region R except for a simple pole at the point z_0, then $f(z)$ can be expanded in a Taylor's series about the point z_0. The resulting series has the appearance

$$f(z) = \frac{a_{-1}}{z - z_0} + a_0 + a_1(z - z_0) + a_2(z - z_0)^2 + \cdots$$

and the integral around any simple closed path enclosing only the singularity z_0 shows that

$$a_{-1} = \frac{1}{2\pi i} \int_C f(\zeta) \, d\zeta = \text{Res}[f(z), z_0]. \tag{10.5–4}$$

Equation (10.5–4) suggests another way of finding the residue of a function $f(z)$ at z_0 where z_0 is a simple pole. We have

$$\lim_{z \to z_0} (z - z_0)f(z) = a_{-1} = \text{Res}[f(z), z_0]. \qquad \textbf{(10.5–5)}$$

This result is a special case of the following theorem, which we present without proof.

Theorem 10.5–3 Let $f(z)$ be analytic in a neighborhood of z_0 except at z_0 where it has a pole of order m. Then

$$\text{Res}[f(z), z_0] = \frac{1}{(m-1)!} \lim_{z \to z_0} \frac{d^{m-1}}{dz^{m-1}} ((z - z_0)^m f(z)). \qquad \textbf{(10.5–6)}$$

Residues are of great importance in evaluating contour integrals, as shown in the next theorem due to Cauchy.

Theorem 10.5–4 (*Residue theorem*). Let C be a simple closed contour and let $f(z)$ be analytic on C and in the region enclosed by C except at a finite number of singular points $z_1, z_2, \ldots, z_n$ in the region enclosed by C. Then

$$\int_C f(z)\, dz = 2\pi i \sum_{j=1}^{n} \text{Res}[f(z), z_j], \qquad \textbf{(10.5–7)}$$

where C is described in the positive sense.

EXAMPLE 10.5–3   Evaluate

$$\int_C \frac{3z + 1}{z^2 - z}\, dz,$$

where C is a circle of radius 2 centered at the origin and traversed in the positive sense (Fig. 10.5–4).

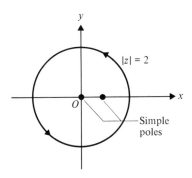

FIGURE 10.5–4

Solution. We have

$$\lim_{z \to 0} \frac{z(3z + 1)}{z(z - 1)} = -1$$

and

$$\lim_{z \to 1} \frac{(z - 1)(3z + 1)}{z(z - 1)} = 4;$$

hence the sum of the residues is 3 and, by Theorem 10.5–4, the value of the integral is $6\pi i$. ∎

EXAMPLE 10.5–4 Evaluate

$$\int_C z^2 \exp(2/z) \, dz,$$

where C is a circle of radius 2 centered at the origin.

Solution. We begin with the Taylor's series for $\exp z$ about the point $z = 0$:

$$\exp z = 1 + z + \frac{z^2}{2!} + \frac{z^3}{3!} + \frac{z^4}{4!} + \cdots,$$

$$\exp\left(\frac{2}{z}\right) = 1 + \frac{2}{z} + \frac{2^2}{2!z^2} + \frac{2^3}{3!z^3} + \frac{2^4}{4!z^4} + \cdots,$$

$$f(z) = z^2 \exp\left(\frac{2}{z}\right)$$

$$= z^2 + 2z + \frac{2^2}{2!} + \frac{2^3}{3!z} + \frac{2^4}{4!z^2} + \cdots. \qquad (10.5\text{–}8)$$

Thus $z = 0$ is the only singularity in the region enclosed by C. Hence the coefficient of z^{-1} in Eq. (10.5–8) is

$$a_{-1} = \text{Res}[f(z), 0] = \frac{2^3}{3!}$$

and the value of the given integral is $8\pi i/3$. ∎

A series such as in Eq. (10.5–8) which contains positive, negative, and zero integral powers of z is called a **Laurent* series**. It is a generalization of Taylor's series and it can be shown that the coefficients in the Laurent series for $f(z)$, that is,

$$f(z) = \sum_{n = -\infty}^{\infty} a_n (z - z_0)^n,$$

---

* After Pierre A. Laurent (1813–1854), a French military engineer, who published a paper on this topic in *Comptes Rendus*, **17** (1843), p. 348.

are given by

$$a_n = \frac{1}{2\pi i} \int_C \frac{f(z)}{(z - z_0)^{n+1}}\, dz \qquad (10.5\text{–}9)$$

and are unique. The portion of the series containing nonnegative powers of $(z - z_0)$ is called the *regular part*, whereas the portion containing negative powers of $(z - z_0)$ is called the *principal part* of the Laurent series. The series converges in an annular region, that is, the region between two circles (see Exercise 2).

One of the most important theorems in analytic function theory is Cauchy's integral formula. This theorem is called a *formula* because it shows that the value of an analytic function at an isolated singularity can be calculated by means of a contour integral. We give the theorem without proof.

Theorem 10.5–5 (*Cauchy's integral formula*). Let C be a simple, closed, positively oriented contour. Let $f(z)$ be analytic in the region enclosed by C and on C. Then

$$f(z_0) = \frac{1}{2\pi i} \int_C \frac{f(z)}{z - z_0}\, dz. \qquad (10.5\text{–}10)$$

If we differentiate Eq. (10.5–10) repeatedly with respect to z under the integral sign, it also follows that

$$f^{(n)}(z_0) = \frac{n!}{2\pi i} \int_C \frac{f(z)}{(z - z_0)^{n+1}}\, dz, \qquad n = 1, 2, 3, \ldots . \qquad (10.5\text{–}11)$$

EXAMPLE 10.5–5 Evaluate

$$\int_C \frac{5z^2 + 2z + 1}{(z - i)^3}\, dz,$$

where C is the circle $|z| = 2$ oriented in the clockwise sense.

Solution. We observe that $f(z) = 5z^2 + 2z + 1$ is an entire function and we can use Eq. (10.5–11) with $n = 2$ and $z_0 = i$. Since $f''(i) = 10$, the contour integral has the value $-10\pi i$, the negative sign coming from the negative orientation of the contour. ∎

We close this section with a theorem that places some rather severe restrictions on the behavior of analytic functions.

Theorem 10.5–6 (*Liouville's theorem*). The only bounded entire functions are the constant functions.

PROOF. Let $f(z)$ be a bounded entire function. Consequently, there exists a constant M such that

$$|f(z)| \leq M$$

for all values of z. Then,* by Eq. (10.5–11) with $n = 1$ and arbitrary z_0,

$$|f'(z_0)| = \left| \frac{1}{2\pi i} \int_C \frac{f(z)}{(z - z_0)^2} \, dz \right| \leq \frac{1}{2\pi} \int_C \left| \frac{f(z)}{(z - z_0)^2} \right| \, dz$$

$$= \frac{1}{2\pi} \cdot \frac{M}{r^2} \cdot 2\pi r = \frac{M}{r},$$

where $|z - z_0| = r$. Hence, by choosing r arbitrarily large, we can make $|f'(z_0)|$ arbitrarily small. In other words, $f'(z_0) = 0$ and, since z_0 was arbitrary, $f'(z) = 0$ for all z, that is, $f(z)$ is a constant. ∎

KEY WORDS AND PHRASES

contour
contour integral
singular point
pole of order m
simple pole

Cauchy's theorem
residue theorem
Laurent series
Cauchy's integral formula
Liouville's theorem

EXERCISES 10.5

▶ 1. a) Show that

$$\int_0^{2\pi} \cos (m + 1)t \, dt = \begin{cases} 2\pi, & \text{for } m = -1, \\ 0, & \text{for every other integral value of } m. \end{cases}$$

b) Show that

$$\int_0^{2\pi} \sin (m + 1)t \, dt = 0$$

for all integral values of m.

2. a) If the function

$$f(z) = \frac{z^2 + 1}{z(z^2 - 3z + 2)}$$

is to be expressed as a Laurent series using powers of $z + 1$, it is first necessary to write the function in terms of $z + 1$. Show that this results in

$$f(z) = \frac{(z + 1)^2 - 2(z + 1) + 2}{(z + 1 - 1)(z + 1 - 2)(z + 1 - 3)}.$$

* We are using a result from real analysis here, namely, if $f(x)$ is integrable on (a, b), then

$$\left| \int_a^b f(x) \, dx \right| \leq \int_a^b |f(x)| \, dx, \qquad a < b.$$

b) Put $\zeta = z + 1$ to obtain

$$\frac{\zeta^2 - 2\zeta + 2}{(\zeta - 1)(\zeta - 2)(\zeta - 3)} = \frac{1}{2(\zeta - 1)} - \frac{2}{\zeta - 2} + \frac{5}{2(\zeta - 3)}$$

upon making a partial-fraction decomposition.

c) Obtain the Laurent series

$$\frac{1}{2\zeta} \sum_{n=0}^{\infty} \frac{1}{\zeta^n} - \frac{2}{\zeta} \sum_{n=0}^{\infty} \left(\frac{2}{\zeta}\right)^n + \frac{5}{2\zeta} \sum_{n=0}^{\infty} \left(\frac{3}{\zeta}\right)^n,$$

valid for $|\zeta| > 3$.

d) Hence show that

$$\frac{z^2 + 1}{z(z^2 - 3z + 2)} = \frac{1}{2} \sum_{n=0}^{\infty} (1 - 2^{n+2} + 5(3)^n)(z + 1)^{-(n+1)},$$

valid for $|z + 1| > 3$.

3. Explain how the hypothesis of Theorem 10.5–2 has been strengthened as compared to that of Theorem 10.5–1.

4. Obtain Eq. (10.5–11) from Eq. (10.5–10).

▶▶ 5. Compute

$$\int_C f(z)\, dz$$

in each case if $f(z) = \bar{z}$ and C is as shown.

a) C: $z = e^{i\phi}$, $0 \le \phi \le \pi$; clockwise

b) C: $z = e^{i\phi}$, $0 \le \phi < 2\pi$; counterclockwise

c) C: square of side 2 with center at the origin and passing through $(1, 0)$; clockwise

6. Compute

$$\int_C f(z)\, dz$$

in each case if $f(z) = (z + 2)/z$ and C is as shown.

a) C: $x^2 + y^2 = 4$ from $(2, 0)$ to $(-2, 0)$; counterclockwise

b) C: $x^2 + y^2 = 4$ closed path; counterclockwise

c) C: $x^2 + y^2 = 4$ from $(2, 0)$ to $(-2, 0)$; clockwise

7. Explain why the functions of Exercises 5 and 6 do not seem to obey Theorem 10.5–1.

8. Evaluate

$$\int_C \frac{(9z^2 - iz + 4)}{z(z^2 + 1)}\, dz,$$

where C is the circle $|z| = 2$ in the positive sense. (*Hint*: Use partial-fraction decomposition.)

9. If $P_n(z)$ is any polynomial of degree n and C is any closed contour, explain why

$$\int_C P_n(z)\, dz = 0.$$

10. Let C be a closed contour described in the positive sense. Show that

$$\int_C \frac{z^3 + 2z}{(z - z_0)^3} \, dz = \begin{cases} 6\pi i z_0, & \text{if } z_0 \text{ is inside } C, \\ 0, & \text{if } z_0 \text{ is outside } C. \end{cases}$$

11. Evaluate

$$\int_C \frac{\sqrt{z}}{(z - 1)^3} \, dz,$$

where C is the closed contour $|z - 1| = \frac{1}{2}$ in the positive sense.

12. Evaluate

$$\int_C \frac{e^z + \sin z}{z} \, dz,$$

where C is the circle $|z - 2| = 3$ traversed in the positive sense.

13. Explain how it follows from Theorem 10.5–6 that the following functions are *not* bounded.

a) $P_n(z)$, polynomials of degree $n \geq 1$

b) e^z **c)** $\sin z$

d) $\sinh z$ **e)** $\cosh z$

14. By noting that the only pole of

$$\frac{1}{1 + z^2} = \frac{1}{(z + i)(z - i)}$$

in the *upper* half-plane is a simple pole at $z = i$, obtain the result

$$\int_{-\infty}^{\infty} \frac{dx}{1 + x^2} = \pi.$$

15. Using a method similar to the one suggested in Exercise 14, obtain the result

$$\int_{-\infty}^{\infty} \frac{dx}{1 + x^4} = \frac{\pi}{\sqrt{2}}.$$

16. Show that

$$\int_{-\infty}^{\infty} \frac{\cos \omega x}{a^2 + x^2} = \frac{\pi}{a} \exp(-\omega a), \qquad a > 0.$$

17. a) By long division obtain

$$\frac{1}{1 - z} = 1 + z + z^2 + z^3 + \cdots.$$

b) By long division obtain

$$\frac{1}{-z + 1} = -\left(\frac{1}{z} + \frac{1}{z^2} + \frac{1}{z^3} + \cdots \right).$$

c) Use the ratio test to show that the series in part (a) converges for $|z| < 1$ and the series in part (b) converges for $|z| > 1$.

18. **a)** Using partial-fraction decomposition, show that

$$\frac{1}{(1-z)(2-z)} = \frac{1}{z-1} + \frac{1}{z-2}.$$

b) By writing

$$-\frac{1}{z-1} = -\frac{1}{z\left(1 - \dfrac{1}{z}\right)},$$

obtain

$$-\frac{1}{z-1} = -\sum_{n=1}^{\infty} \frac{1}{z^n},$$

valid for $|z| > 1$.

c) Obtain by long division

$$\frac{1}{-2+z} = -\sum_{n=0}^{\infty} \frac{z^n}{2^{n+1}},$$

valid for $|z| < 2$.

d) Write the Laurent series for

$$\frac{1}{(1-z)(2-z)},$$

valid for $1 < |z| < 2$.

19. Obtain the Laurent series for

$$\frac{1}{(z-1)(z-3)},$$

valid for $1 < |z| < 3$.

▶▶▶ 20. The Legendre polynomials $P_n(z)$ are defined by Rodrigues' formula (see Eq. 9.4–18) as follows:

$$P_n(z) = \frac{1}{2^n n!} \frac{d^n (z^2 - 1)^n}{dz^n}, \qquad n = 0, 1, 2, \ldots .$$

By means of Eq. (10.5–11) obtain the following *integral representation*,

$$P_n(z) = \frac{1}{2\pi i} \int_C \frac{(\zeta^2 - 1)^n}{2^n (\zeta - z)^{n+1}} \, d\zeta,$$

where C is a simple, closed contour enclosing the point z.

21. The Bessel functions of the first kind of order n ($n \geq 0$) can be defined by the *generating function*,

$$\exp\left(z\left(t - \frac{1}{t}\right)\right) = \sum_{n=-\infty}^{\infty} J_n(z) t^n.$$

When the left member is expanded in powers of t and the coefficients of t^n are equated, the result is an expression for $J_n(z)$. Use Eq. (10.5–9) to obtain the *integral representation*,

$$J_n(z) = \frac{1}{\pi} \int_0^\pi \cos(n\theta - z \sin \theta) \, d\theta.$$

22. If $f(z)$ is an entire function, show that

$$\int_{z_1}^{z_2} f(z)\, dz$$

is independent of the contour joining z_1 and z_2.

10.6 APPLICATIONS

In this section we will present a few of the many applications of complex function theory. Since analytic function theory has a great variety of applications, the examples given here will by no means be exhaustive. Other examples can be found by consulting one of the books listed in the references at the end of this chapter.

Partial-fraction decomposition

Partial-fraction decomposition of rational functions, that is, functions of the form $P(z)/Q(z)$ where P and Q are polynomials, has a number of applications. In calculus the technique is studied as one of the methods of integration. In obtaining the inverse Laplace transform (see Section 3.3) the technique was useful. We have also used it in Section 10.5 in order to obtain the Laurent series expansion of a given function and to find residues.

For the present discussion we will assume that in $P(z)/Q(z)$ the degree of $P(z)$ is m whereas the degree of $Q(z)$ is n with $m < n$. If $m \geq n$, then the division can be performed and the remainder considered as a rational function whose numerator has degree less than the denominator. According to the fundamental theorem of algebra* every nonconstant polynomial $Q(z)$ can be factored into *linear* factors. First assume that these factors are all *distinct*. Then

$$f(z) = \frac{P(z)}{Q(z)} = \frac{A_1}{z - z_1} + \frac{A_2}{z - z_2} + \cdots + \frac{A_n}{z - z_n}$$

and we can form

$$(z - z_j)f(z) = \phi_j(z).$$

Hence

$$\phi_j(z_j) = A_j, \qquad j = 1, 2, \ldots, n,$$

that is,

$$A_j = \mathrm{Res}[f(z), z_j], \tag{10.6–1}$$

the residue of $f(z)$ at z_j (see Eq. 10.5–5).

---

* "A nonconstant polynomial $Q(z)$ with real or complex coefficients has at least one root in the complex plane." A proof of this theorem can be obtained by using Liouville's theorem (Theorem 10.5–6).

EXAMPLE 10.6–1 Decompose

$$f(z) = \frac{-7z - 1}{z^3 - 7z + 6}$$

into partial fractions.

Solution. Since $z = 1$ is an obvious root of $z^3 - 7z + 6 = 0$, $(z - 1)$ is one factor of the denominator $f(z)$. The other factors are readily found by synthetic division and by the quadratic formula. Thus

$$f(z) = \frac{-7z - 1}{z^3 - 7z + 6} = \frac{-7z - 1}{(z - 1)(z - 2)(z + 3)}$$

$$= \frac{A_1}{z - 1} + \frac{A_2}{z - 2} + \frac{A_3}{z + 3},$$

and, using Eq. (10.6–1), we have

$$A_1 = 2, \qquad A_2 = -3, \qquad A_3 = 1. \quad \blacksquare$$

Next we consider the case where $f(z) = P(z)/Q(z)$ has a pole of order n at z_1. Then

$$f(z) = \frac{P(z)}{Q(z)} = \frac{A_1}{z - z_1} + \frac{A_2}{(z - z_1)^2} + \cdots + \frac{A_n}{(z - z_1)^n}$$

so that, with

$$\phi(z) = (z - z_1)^n f(z)$$
$$= A_1(z - z_1)^{n-1} + A_2(z - z_1)^{n-2} + \cdots + A_n,$$

we have

$$A_n = \phi(z_1).$$

In general (Exercise 1),

$$A_j = \frac{\phi^{(n-j)}(z_1)}{(n - j)!} = \text{Res}[f(z), z_1],$$

which agrees with Eq. (10.5–6).

EXAMPLE 10.6–2 Decompose

$$f(z) = \frac{3z^2 + 8z + 6}{(z + 2)^3}$$

into partial fractions.

Solution. Here we have

$$\phi(z) = (z + 2)^3 f(z)$$
$$= 3z^2 + 8z + 6$$
$$= A_1(z + 2)^2 + A_2(z + 2) + A_3;$$

hence,
$$A_3 = \phi(-2) = 2.$$
Further,
$$\phi'(z) = 2A_1(z+2) + A_2$$
and
$$\phi''(z) = 2A_1,$$
so that
$$A_2 = \phi'(-2) = 6(-2) + 8 = -4$$
and
$$A_1 = \frac{\phi''(-2)}{2} = 3.$$
Thus
$$\frac{3z^2 + 8z + 6}{(z+2)^3} = \frac{3}{z+2} - \frac{4}{(z+2)^2} + \frac{2}{(z+2)^3}. \quad \blacksquare$$

Functions having both simple poles and poles of order $n > 1$ can be decomposed by using a combination of the methods illustrated in the last two examples. The exercises contain problems of this type.

Laplace transformation

Our study of the Laplace transformation in Chapter 3 was confined to the special case where s was *real*. We now extend that treatment in order to make the technique more flexible. At the same time, we show the relation between the Laplace transform and the Fourier transform. Recall that the latter was defined (see Eqs. 7.3–15) as

$$\bar{f}(\alpha) = \int_{-\infty}^{\infty} f(t)e^{i\alpha t}\, dt. \tag{10.6-2}$$

If we define $f(t)$ for all $t \geq 0$ and set $f(t) = 0$ for $t < 0$, then f is defined on the entire real line. Hence we can write Eq. (10.6–2) as

$$\bar{f}(\alpha) = \int_0^{\infty} f(t)e^{i\alpha t}\, dt,$$

and if we further replace α by $(i)(s)$, with $s > 0$, we obtain

$$F(s) = \int_0^{\infty} f(t)e^{-st}\, dt, \tag{10.6-3}$$

which is the definition of the Laplace transform (see Eq. 3.1–3). It can be shown that whenever*

$$|f(t)| \leq M \exp(at)$$

for positive numbers M and a, then the integral in Eq. (10.6–3) converges for any *complex* s such that $\text{Re}(s) > a$. Thus, in a general sense,

---

* In Definition 3.1–2 we said a function having this property was of **exponential order exp** (at).

the Laplace transform is a transformation from the *t*-domain to the *complex s*-domain. Accordingly, Laplace transform pairs such as the one in Section 3.1,

$$f(t) = \exp{(at)} \leftrightarrow F(s) = \frac{1}{s-a}, \qquad s > a, \qquad \textbf{(3.1–10)}$$

become

$$f(t) = \exp{(at)} \leftrightarrow F(s) = \frac{1}{s-a}, \qquad \mathrm{Re}(s) > \mathrm{Re}(a). \qquad \textbf{(10.6–4)}$$

It is the **Laplace inversion formula** that is of interest here, since we cannot hope to find every desired entry in a table of Laplace transforms. Knowing that the Fourier transform has an inversion formula and that the two transforms are related, we proceed as follows.

Recall that, in order for the Fourier transform of $f(t)$ to exist, $f(t)$ had to be absolutely integrable on the real line (see Theorem 7.3–1). Unfortunately, many functions of interest in applied mathematics are *not* absolutely integrable ($\sin \omega t$ is one simple example). On the other hand, if we define $f(t) = 0$ for $-\infty < t < 0$ and consider

$$\phi(t) = \exp{(-\gamma t)}f(t), \qquad \gamma > 0, \qquad \textbf{(10.6–5)}$$

then $\phi(t)$ *is* absolutely integrable on the real line (Exercise 2). Hence $\phi(t)$ has a Fourier integral representation given by (see Eq. 7.3–14)

$$\phi(t) = \frac{1}{2\pi} \int_{-\infty}^{\infty} \exp{(i\eta t)}\, d\eta \int_{0}^{\infty} \phi(\xi) \exp{(-i\xi\eta)}\, d\xi.$$

On substituting the value of $\phi(t)$ from Eq. (10.6–5), this becomes

$$f(t) = \frac{\exp{(\gamma t)}}{2\pi} \int_{-\infty}^{\infty} \exp{(i\eta t)}\, d\eta \int_{0}^{\infty} f(\xi) \exp{(-\xi(\gamma + i\eta))}\, d\xi.$$

Finally, making the substitutions

$$s = \gamma + i\eta, \qquad ds = i\, d\eta, \qquad F(s) = \int_{0}^{\infty} f(\xi) \exp{(-s\xi)}\, d\xi,$$

we obtain (Exercise 3)

$$f(t) = \frac{1}{2\pi i} \int_{\gamma - i\infty}^{\gamma + i\infty} F(s) \exp{(st)}\, ds, \qquad \textbf{(10.6–6)}$$

which is the **Laplace transform inversion formula**.

In Eq. (10.6–6) s is a complex variable and γ is a real positive number; hence the inversion integral is along a straight-line path in the complex plane. This path is also called the **Bromwich contour**.* Although the complex integral of Eq. (10.6–6) can be readily converted

---

* After T. J. I'A. Bromwich (1875–1929), an English mathematician.

into a real line integral, the evaluation of the latter is usually quite difficult. It is often much simpler to use contour integration and residue theory. We illustrate with an example.

EXAMPLE 10.6–3 Given that

$$F(s) = \frac{s}{s^2 + \omega^2},$$

find $f(t)$ by using the Laplace transform inversion formula (10.6–6).

Solution. We have

$$f(t) = \frac{1}{2\pi i} \int_{\gamma - i\infty}^{\gamma + i\infty} \frac{s \exp(st)}{s^2 + \omega^2} \, ds.$$

Consider the closed contour shown in Fig. 10.6–1. We have chosen γ as an arbitrary positive number and β as a number having sufficient magnitude so that a circle with center at the origin and passing through the points $\gamma \pm i\beta$ will enclose the simple poles $\pm i\omega$. Then we can write

$$f(t) = \lim_{\beta \to \infty} \frac{1}{2\pi i} \int_{C_\gamma + C_\beta} \frac{s \exp(st)}{s^2 + \omega^2} \, ds - \lim_{\beta \to \infty} \frac{1}{2\pi i} \int_{C_\beta} \frac{s \exp(st)}{s^2 + \omega^2} \, ds.$$

It can be shown that the second limit is zero in this case as well as for other functions $F(s)$ under certain conditions (see Exercise 26). The first integral is along a simple closed contour enclosing two simple poles, hence its value by Eq. (10.5–7) is

$$\frac{i\omega \exp(i\omega t)}{2i\omega} + \frac{-i\omega \exp(-i\omega t)}{-2i\omega} = \cos \omega t. \quad \blacksquare$$

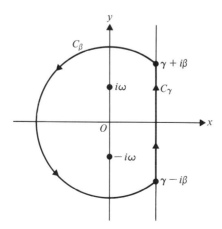

FIGURE 10.6–1 Bromwich contour.

In applications $F(s)$ often has the form

$$F(s) = \frac{p(s)}{Q(s)},$$

where $Q(s)$ is a polynomial of degree n less than the degree of $p(s)$. If $Q(s)$ can be factored into distinct linear factors $(s - s_j), j = 1, 2, \ldots, n$, then the Laplace transform inversion formula can be written as

$$f(t) = \mathscr{L}^{-1} \left\{ \frac{p(s)}{Q(s)} \right\} = \sum_{j=1}^{n} \frac{p(s_j) \exp (s_j t)}{Q'(s_j)}. \qquad (10.6-7)$$

This last is also known as the **Heaviside expansion formula**. Throughout the previous discussion we have assumed $p(s_j) \neq 0$.

Fluid mechanics

In fluid mechanics (aerodynamics and hydrodynamics) we make certain simplifying assumptions. We consider flows to be in a *plane*, that is, the fluid behaves in planes *parallel* to the xy-plane exactly as it does in the xy-plane. All flow barriers are also assumed to be in the xy-plane. We further assume that the fluid flow is independent of time, incompressible, and irrotational. Such a fluid is also called an **ideal fluid**.

Because the fluid is incompressible, its velocity vector $\mathbf{V}$ is the gradient of a scalar function ϕ, called the *velocity potential*. Moreover, ϕ satisfies Laplace's equation in any region devoid of sources and sinks. The foregoing can be seen by considering the velocity vector field $\mathbf{V}$ with components V_x and V_y. By Green's theorem in the plane (see Section 5.6),

$$\int_C (V_x \, dx + V_y \, dy) = \iint_R \left(\frac{\partial V_y}{\partial x} - \frac{\partial V_x}{\partial y} \right) dx \, dy,$$

where C is a simple closed curve enclosing a region R. If the fluid is irrotational ($\mathbf{V} \times \mathbf{V} = \mathbf{0}$), then the right side is zero (Exercise 4); hence the left side, which is the circulation around C, is also zero. This in turn implies the existence of a scalar function $\phi(x, y)$ such that

$$\frac{\partial \phi}{\partial x} = V_x \quad \text{and} \quad \frac{\partial \phi}{\partial y} = V_y.$$

Hence the vector $\mathbf{V}$ can be written as a complex variable,

$$\mathbf{V} = V_x + i V_y = \frac{\partial \phi}{\partial x} + i \frac{\partial \phi}{\partial y},$$

which shows that $\mathbf{V}$ is the gradient of ϕ.

Since the velocity vector $\mathbf{V}$ with components V_x and V_y is in the direction of the flow, the components of the vector normal to the flow are $-V_y$ and V_x (Exercise 5). The condition of incompressibility is that

$$\int_C (-V_y \, dx + V_x \, dy) = 0,$$

which expresses the *flux* through C. Again appealing to Green's theorem in the plane, we have

$$\int_C (-V_y \, dx + V_x \, dy) = \iint_R \left(\frac{\partial V_x}{\partial x} + \frac{\partial V_y}{\partial y} \right) dx \, dy.$$

Hence there exists a scalar function $\psi(x, y)$ such that

$$\frac{\partial \psi}{\partial x} = -V_y \quad \text{and} \quad \frac{\partial \psi}{\partial y} = V_x.$$

Moreover, $\phi(x, y)$ and $\psi(x, y)$ satisfy the Cauchy–Riemann equations and we can thus form the analytic function

$$w = f(z) = \phi(x, y) + i\psi(x, y),$$

which is called the **complex potential** of the flow. The curves $\phi(x, y) = $ constant are the **equipotentials** and the curves $\psi(x, y) = $ constant are the **streamlines**. The mapping $w = f(z)$ is conformal except at points where $f'(z) = 0$, and these points are called **stagnation points**.*

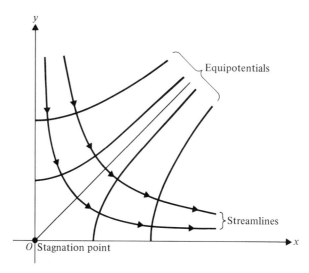

FIGURE 10.6–2 The flow $w = z^2$.

---

* Also called **critical points** (see Exercise 14 in Section 10.4).

EXAMPLE 10.6-4 Analyze the flow whose complex potential is $w = z^2$, $\mathrm{Re}(z) \geq 0$, $\mathrm{Im}(z) \geq 0$.

Solution. Here we have

$$w = z^2 = (x^2 - y^2) + i(2xy);$$

hence the equipotentials are the rectangular hyperbolas $x^2 - y^2 = c_1$ and the streamlines are the rectangular hyperbolas $xy = c_2$. The origin is a stagnation point and the fluid can be considered as flow around a corner with the x- and y-axes forming a barrier. Some of the equipotentials and streamlines are shown in Fig. 10.6-2. ∎

Boundary-value problems solved by mapping

We have already seen one application of conformal mapping in Example 10.4-1. Now we consider some more difficult problems in order to demonstrate the great advantages of mapping.

Mapping by means of analytic functions is especially important in problems involving the ubiquitous potential equation. The reason for this is contained in the following theorem, whose proof we omit.

Theorem 10.6-1 If f is an analytic function, then the change of variables

$$x + iy = f(u + iv)$$

transforms every harmonic function of x and y into a harmonic function of u and v.

Moreover, not only does mapping by means of an analytic function preserve the Laplacian, but it also preserves some of the common boundary conditions that occur. We illustrate with an example.

EXAMPLE 10.6-5 Find the bounded, steady-state temperatures in the semi-infinite plate $y > 0$ given the boundary conditions shown in Fig. 10.6-3.

Solution. This problem does not lend itself to the methods discussed in Chapter 8 because on a portion of the boundary the temperature is known, whereas on the remainder the normal derivative of the temperature is known. Boundary conditions of this kind are said to be of *mixed type** in contrast with Dirichlet and Neumann conditions.

---

* Conditions of mixed type are also called **Robin's conditions** (after Victor G. Robin (1855–1897), a French mathematician).

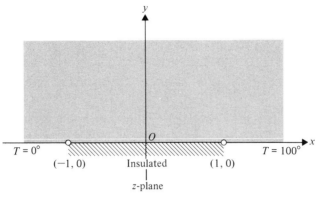

FIGURE 10.6–3

The given problem can be formulated mathematically as follows:

$$\text{P.D.E.:} \quad T_{xx} + T_{yy} = 0, \quad -\infty < x < \infty, \quad y > 0;$$
$$\text{B.C.:} \quad T(x, 0) = 0, \quad x < -1,$$
$$T_y(x, 0) = 0, \quad -1 < x < 1,$$
$$T(x, 0) = 100, \quad x > 1.$$

The mapping $z = \sin w$ or, equivalently, $w = \sin^{-1} z$ will transform the problem into one that is very simple to solve. This mapping was considered in detail in Example 10.4–3. We leave it as an exercise to show that the appropriate figure after the transformation is the one shown in Fig. 10.6–4 (Exercise 6).

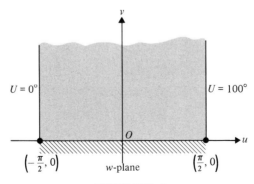

FIGURE 10.6–4

Sine U does not depend on v (why?), the transformed problem can be formulated as follows:

$$\frac{d^2U}{du^2} = 0, \qquad U\left(-\frac{\pi}{2}\right) = 0, \qquad U\left(\frac{\pi}{2}\right) = 100.$$

The solution is (Exercise 7)

$$U(u) = 50\left(1 + \frac{2}{\pi}u\right).$$

As so often happens, however, the ease with which the transformed problem was solved is offset somewhat by the difficulty encountered in going back to the original variables. We have

$$z = x + iy = \sin w = \sin (u + iv)$$
$$= \sin u \cosh v + i \cos u \sinh v;$$

hence

$$\frac{x}{\sin u} = \cosh v \quad \text{and} \quad \frac{y}{\cos u} = \sinh v.$$

From this we obtain

$$\frac{x^2}{\sin^2 u} - \frac{y^2}{\cos^2 u} = 1,$$

a hyperbola with center at the origin, vertices at $(\pm\sin u, 0)$, and foci at $(\pm 1, 0)$.* By definition, a point (x, y) on the hyperbola has the property that the difference in distances from (x, y) to the foci is $2 \sin u$ (the length of the transverse axis). In other words,

$$\sin u = \tfrac{1}{2}(\sqrt{(x + 1)^2 + y^2} - \sqrt{(x - 1)^2 + y^2}). \qquad \textbf{(10.6–8)}$$

Finally, the solution to the problem is

$$T(x, y) = 50\left(1 + \frac{2}{\pi}u\right), \qquad -\frac{\pi}{2} \le u \le \frac{\pi}{2},$$

with u defined in Eq. (10.6–8). We leave it for the exercises to verify this solution completely (Exercise 8). ∎

In the next example we present a problem dealing with electrostatic potential due to a conducting sheet.

EXAMPLE 10.6–6   In Fig. 10.6–5 is shown a semi-infinite, conducting sheet in the xz-plane. The strip between $x = -a$ and $x = a$ is insulated from the rest of the sheet so that the potential of this strip can be kept

---

* See George B. Thomas, Jr., and R. L. Finney, *Calculus and Analytic Geometry*, 5th ed., p. 418ff. (Reading, Mass.: Addison-Wesley, 1979).

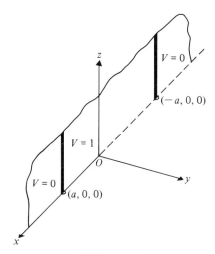

FIGURE 10.6–5

at 1 while the potential of the sheet on either side is kept at zero. Find the potential $V(x, y, z)$ at an arbitrary point (x, y, z) with $y > 0$, $z > 0$.

Solution. It is clear that the result will not depend on z (why?); hence the problem can be reduced to the following *two*-dimensional one.

$$\text{P.D.E.: } V_{xx} + V_{yy} = 0, \quad -\infty < x < \infty, \quad y > 0;$$
$$\text{B.C.: } V(x, 0) = 0, \quad |x| > a,$$
$$V(x, 0) = 1, \quad |x| < a,$$
$$|V(x, y)| < M, \quad -\infty < x < \infty, \quad y > 0.$$

The problem can be greatly simplified by making the transformation*

$$w = \text{Log} \left(\frac{z - a}{z + a} \right).$$

This transforms the xy-plane into the uv-plane and we now have the problem shown in Fig. 10.6–6 (Exercise 9). This is again a simple problem since the solution does not depend on u so that we have an ordinary differential equation. The solution is (Exercise 10) $U(u, v) = v/\pi$.

We now seek to determine v in terms of x and y. We have

$$\text{Log} (z - a) = \text{Log} \sqrt{(x - a)^2 + y^2} + i \arctan \frac{y}{x - a}$$

_____

* The choice of the proper transformation is important. For this purpose a table of transformations is most useful. Consult one of the references at the end of this chapter.

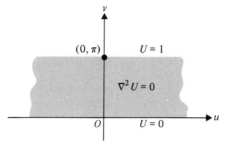

FIGURE 10.6-6

so that

$$v = \arctan \frac{y}{x - a} - \arctan \frac{y}{x + a}$$

$$= \theta_2 - \theta_1,$$

where θ_2 and θ_1 are shown in Fig. 10.6–7. In Exercise 11 you are asked to show that $\theta_2 - \theta_1 = \theta$ and to verify that

$$V(x, y) = \frac{1}{\pi} \arctan \left(\frac{2ay}{x^2 + y^2 - a^2} \right) \qquad \textbf{(10.6–9)}$$

with $0 \le \theta \le \pi$. This result shows that the potential at any point is a function only of the angle subtended by the strip of width $2a$. ■

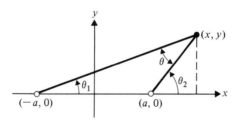

FIGURE 10.6-7

In this section we have attempted to give some idea of the many applications of complex analysis. The literature on this subject is fairly extensive. Annotated references at the end of this chapter should be consulted for further details.

KEY WORDS AND PHRASES

partial-fraction decomposition
Laplace transformation
Laplace transform inversion
    formula
Bromwich contour
Heaviside expansion formula

complex potential
equipotentials
streamlines
stagnation point
Robin's conditions

EXERCISES 10.6

▶ 1. Given that

$$\phi(z) = (z - z_1)^n f(z),$$

show that

$$\phi^{(n-j)}(z_1) = \text{Res}[f(z), z_1].$$

2. Show that if $f(t) = 0$ for $-\infty < t < 0$, then

$$\phi(t) = \exp(-\gamma t)f(t), \qquad \gamma > 0$$

is absolutely integrable on the real line.

3. Carry out the necessary details required to arrive at Eq. (10.6–6).

4. **a)** If the two-dimensional vector $\mathbf{V}$ has components V_x and V_y and the curl $\mathbf{V} \times \mathbf{V} = \mathbf{0}$, show that

$$\iint_R \left(\frac{\partial V_y}{\partial x} - \frac{\partial V_x}{\partial y} \right) dx\, dy = 0$$

over every region R.

b) If

$$\int_C (V_x\, dx + V_y\, dy) = 0$$

for every simple closed curve C, show that there exists a scalar function $\phi(x, y)$ such that

$$\frac{\partial \phi}{\partial x} = V_x \quad \text{and} \quad \frac{\partial \phi}{\partial y} = V_y.$$

5. Show that if a velocity vector $\mathbf{V}$ has components V_x and V_y, then a vector normal to $\mathbf{V}$ has components $-V_y$ and V_x.

6. In Example 10.6–5 use the mapping $w = \sin^{-1} z$ (that is, $z = \sin w$) to transform the given region and boundary conditions as shown. To what do the lines $x = $ constant and $y = $ constant correspond?

7. Solve the two-point boundary problem

$$\frac{d^2 U}{du^2} = 0, \qquad U\left(-\frac{\pi}{2}\right) = 0, \qquad U\left(\frac{\pi}{2}\right) = 100.$$

(Compare Example 10.6–5.)

8. **a)** Verify that the solution to Example 10.6–5 satisfies Laplace's equation and all boundary conditions.

b) Compute $T(0, 0)$, $T(-1, 0)$, and $T(1, 0)$. Are these results reasonable?

c) Compute $T(0, y)$ and explain the result.

9. In Example 10.6–6 verify that the transformation

$$w = \text{Log}\left(\frac{z - a}{z + a}\right)$$

transforms the given problem into the one in U as shown.

10. Solve the two-point boundary-value problem of Example 10.6–6,

$$\frac{d^2U}{du^2} = 0, \qquad U(0) = 0, \qquad U(\pi) = 1.$$

11. **a)** In Example 10.6–6 show that $\theta_2 - \theta_1 = \theta$, the angle subtended by the segment $-a \leq x \leq a$.

b) Obtain the relation

$$\tan \theta = \frac{2ay}{x^2 + y^2 - a^2}.$$

c) Verify that the solution (10.6–9) satisfies the potential equation and all boundary conditions. (*Hint*: Consider what happens to θ as $y \to 0$.)

▶▶ 12. **a)** Given

$$F(s) = \frac{s}{(s^2 + 1)^2},$$

decompose this function into partial fractions.

b) From the result of part (a) find the inverse Laplace transform of $F(s)$.

13. Decompose

$$f(z) = \frac{z^2 - 5}{z^3 - 6z^2 + 11z - 6}$$

into partial fractions.

14. Decompose

$$f(z) = \frac{z}{z^2 - 3z + 2}$$

into partial fractions.

15. Decompose each of the following into partial fractions.

a) $\dfrac{z^2 - 7z + 11}{(z - 2)(z^2 - 5z + 4)}$ **b)** $\dfrac{z^3 + 2}{z^3(z - 2)^2}$

c) $\dfrac{z^3 - 2}{z^2(z - 1)^2}$ **d)** $\dfrac{1}{(z^2 + 1)^3}$

e) $\dfrac{2z + 1}{z^2(z^2 + 1)}$ **f)** $\dfrac{1}{z^3(z^2 + 1)}$

16. Invert each of the following Laplace transforms without using partial-fraction decomposition.

a) $\dfrac{2s + 1}{s(s^2 + 1)}$ **b)** $\dfrac{s}{s^4 - 2s^2 + 1}$ **c)** $\dfrac{1}{s^4 - 1}$

17. Find the inverse Laplace transform in each case.

a) $F(s) = \dfrac{s}{(s+1)^2(s^2+3s-10)}$

b) $F(s) = \dfrac{s+1}{s(s+3)^2}$

18. Use the Laplace transform inversion formula (10.6–6) and the Heaviside expansion formula (10.6–7) to invert each of the following Laplace transforms.

a) $\dfrac{\omega}{s^2+\omega^2}$

b) $\dfrac{s-b}{(s-b)^2+\omega^2}$

c) $\dfrac{\omega}{(s-b)^2+\omega^2}$

d) $\dfrac{s}{s^2-a^2}$

e) $\dfrac{a}{s^2-a^2}$

19. Solve the problem of Example 10.6–3 by using a rectangular contour $\gamma - i\beta$ to $\gamma + i\beta$ to $-\gamma + ib$ to $-\gamma - ib$ to $\gamma - i\beta$. What assumptions must be made?

20. In Fig. 10.6–2 (Example 10.6–4), where are the equipotentials for which $c_1 > 0$, $c_1 = 0$, $c_1 < 0$?

21. Graph the streamlines and equipotentials given that the complex potential is $f(z) = z$.

22. a) Graph the streamlines and equipotentials given that the complex potential is $f(z) = \log |z|$.

b) What is the significance of the origin?

23. Analyze the flow whose potential function is given by

$$f(z) = \operatorname{Re}\left(\frac{1}{2}\left(z + \frac{1}{z}\right)\right).$$

(*Note*: A calculator may be helpful in plotting the curves. Use symmetry and consider only the region $x \geq 0$, $y \geq 0$.)

24. Find the bounded, steady-state temperatures $T(x, y)$ in the unbounded region $y > 0$ if $T(x, 0) = 0$ for $x > 0$ and $T(x, 0) = \pi$ for $x < 0$. (*Hint*: Use the mapping $z = e^w$; see Exercise 17 of Section 10.4.)

25. Solve the following Dirichlet problems.

a) $x > 0$, $y > 0$, $T(x, 0) = 0$, $T(0, y) = 100$
(*Hint*: Use the transformation $w = z^2$ followed by $\zeta = e^w$.)

b) $\rho < 1$, $0 < \phi < \pi$, $T_\rho(1, \phi) = 0$, $T(\rho, 0) = 0$, $T(\rho, \pi) = 100$
(*Hint*: Use the transformation $w = \log z$.)

c) $\rho < 1$ for $-\pi < \phi < \pi$, $T(1, \phi) = 1$ for $0 < \phi < \pi$, and $T(1, \phi) = 0$ for $-\pi < \phi < 0$ by using the transformation $w = \log \zeta$ after $w = (i - z)/(i + z)$.

▶▶▶ 26. In Example 10.6–3 if C_β is a circular arc, $F(s)$ is analytic except for isolated singular points in the half-plane $x < \gamma$, and

$$|F(s)| < \frac{M}{|s|^k}$$

for positive constants M and k, show that

$$\lim_{\beta \to \infty} \frac{1}{2\pi i} \int_{C_\beta} F(s) \exp(st) \, ds = 0.$$

27. Consider the exponential function

$$f(x) = A \exp\left(-\frac{x}{a}\right), \qquad x > 0,$$

where A and a are positive constants. Find the Fourier transform $\bar{f}(\alpha)$ of $f(x)$.

REFERENCES

Churchill, R. V., J. W. Brown, and R. F. Verhey, *Complex Variables and Applications*, 3d ed. New York: McGraw-Hill, 1974.
> *An excellent book at an intermediate level that includes many applications of residues and mapping. Also included is a useful table showing regions and their maps under various transformations.*

Churchill, R. V., *Operational Mathematics*, 3d ed. New York: McGraw-Hill, 1972.
> *Contains numerous applications of Fourier and Laplace transforms and mapping.*

Greenleaf, F. P., *Introduction to Complex Variables*. Philadelphia: W. B. Saunders, 1972.
> *Contains a chapter on harmonic functions and boundary value problems, one on physical applications of potential theory, and one on conformal mapping problems. A very complete treatment at an intermediate level.*

Henrici, P., *Applied and Computational Complex Analysis*, 2 vols. New York: John Wiley, 1977.
> *An advanced scholarly treatment of the subject.*

Holland, A. S., *Complex Function Theory*. New York: North Holland, 1980.
> *An intermediate-level text that contains many worked examples to illustrate different techniques, especially in the theory of residues where examples from quantum theory may be found.*

Marsden, J. E., *Basic Complex Analysis*. San Francisco: W. H. Freeman, 1973.
> *A clear exposition at an elementary level. Contains numerous applications of complex analysis and Laplace transformation to heat conduction, electrostatics, hydrodynamics, and theoretical physics. Includes an excellent bibliography.*

Saff, E. B., and A. D. Snider, *Fundamentals of Complex Analysis for Mathematics, Science, and Engineering*. Englewood Cliffs, N.J.: Prentice-Hall, 1976.
> *This is an easy-to-read elementary text. Applications are included in most of the chapters in separate sections. There is a chapter on Fourier series and integral transforms and two useful appendixes—one dealing with the equations of mathematical physics and one giving a brief table of conformal mappings. There are summaries and suggested readings after each chapter.*

Silverman, R. A., *Complex Analysis with Applications*. Englewood Cliffs, N.J.: Prentice-Hall, 1974.
> *A treatment at the intermediate level. The chapters on applications of residues and some physical applications are especially recommended.*

general references

Arfken, G., *Mathematical Methods for Physicists*, 2d ed. New York: Academic Press, 1973.

Hildebrand, F. B., *Advanced Calculus for Applications*, 2d ed. Englewood Cliffs, N.J.: Prentice-Hall, 1976.

Kaplan, W., *Advanced Mathematics for Engineers*. Reading, Mass.: Addison-Wesley, 1981.

Killingbeck, J., and G. H. A. Cole, *Mathematical Techniques and Physical Applications*. New York: Academic Press, 1971.

Lin, Chia Ch'iao, and L. A. Segel, *Mathematics Applied to Deterministic Problems in the Natural Sciences*. New York: Macmillan, 1974.

Meyer, R. M., *Essential Mathematics for Applied Fields*. New York: Springer-Verlag, 1979.

Riley, K. F., *Mathematical Methods for the Physical Sciences: An Informal Treatment for Students of Physics and Engineering*. New York: Cambridge University Press, 1974.

Rubinstein, M. F., *Patterns of Problem Solving*. Englewood Cliffs, N.J.: Prentice-Hall, 1975.

Rudin, W., *Principles of Mathematical Analysis*, 3d ed. New York: McGraw-Hill, 1976.

tables

TABLE 1 Laplace transforms

$F(s)$	$f(t)$
1. $\dfrac{a}{s}$	a
2. $\dfrac{1}{s^2}$	t
3. $\dfrac{2}{s^3}$	t^2
4. $\dfrac{n!}{s^{n+1}}$	t^n, n a positive integer
5. $\dfrac{\Gamma(\alpha + 1)}{s^{\alpha+1}}$	t^α, α a real number, $\alpha > -1$
6. $\dfrac{a}{s^2 + a^2}$	$\sin at$
7. $\dfrac{s}{s^2 + a^2}$	$\cos at$
8. $\dfrac{1}{s - a}$, $s > a$	$\exp(at)$
9. $\dfrac{1}{(s - a)^2}$	$t \exp(at)$
10. $\dfrac{2}{(s - a)^3}$	$t^2 \exp(at)$
11. $\dfrac{n!}{(s - a)^{n+1}}$	$t^n \exp(at)$, n a positive integer
12. $\dfrac{a}{s^2 - a^2}$	$\sinh at$
13. $\dfrac{s}{s^2 - a^2}$	$\cosh at$
14. $\dfrac{s^2 + 2a^2}{s(s^2 + 4a^2)}$	$\cos^2 at$
15. $\dfrac{s^2 - 2a^2}{s(s^2 - 4a^2)}$	$\cosh^2 at$
16. $\dfrac{2a^2}{s(s^2 + 4a^2)}$	$\sin^2 at$

TABLE 1 (*continued*)

$F(s)$	$f(t)$
17. $\dfrac{2a^2}{s(s^2 - 4a^2)}$	$\sinh^2 at$
18. $\dfrac{2a^2 s}{s^4 + 4a^4}$	$\sin at \sinh at$
19. $\dfrac{a(s^2 - 2a^2)}{s^4 + 4a^4}$	$\cos at \sinh at$
20. $\dfrac{a(s^2 + 2a^2)}{s^4 + 4a^4}$	$\sin at \cosh at$
21. $\dfrac{s^3}{s^4 + 4a^4}$	$\cos at \cosh at$
22. $\dfrac{2as}{(s^2 + a^2)^2}$	$t \sin at$
23. $\dfrac{s^2 - a^2}{(s^2 + a^2)^2}$	$t \cos at$
24. $\dfrac{2as}{(s^2 - a^2)^2}$	$t \sinh at$
25. $\dfrac{s^2 + a^2}{(s^2 - a^2)^2}$	$t \cosh at$
26. $\dfrac{a}{(s - b)^2 + a^2}$	$\exp(bt) \sin at$
27. $\dfrac{s - b}{(s - b)^2 + a^2}$	$\exp(bt) \cos at$
28. $\dfrac{1}{(s - a)(s - b)}$	$\dfrac{\exp(at) - \exp(bt)}{a - b}$
29. $\dfrac{s}{(s - a)(s - b)}$	$\dfrac{a \exp(at) - b \exp(bt)}{a - b}$
30. $\dfrac{1}{s(s - a)(s - b)}$	$\dfrac{1}{ab} + \dfrac{b \exp(at) - a \exp(bt)}{ab(a - b)}$
31. $\dfrac{1}{(s - a)(s - b)(s - c)}$	$\dfrac{(c - b)e^{at} + (a - c)e^{bt} + (b - a)e^{ct}}{(a - b)(a - c)(c - b)}$
32. $\dfrac{s}{(s - a)(s - b)(s - c)}$	$\dfrac{a(b - c)e^{at} + b(c - a)e^{bt} + c(a - b)e^{ct}}{(a - b)(b - c)(a - c)}$
33. $\dfrac{1}{s(s^2 + a^2)}$	$\dfrac{1}{a^2}(1 - \cos at)$
34. $\dfrac{1}{s(s^2 - a^2)}$	$\dfrac{1}{a^2}(\cosh at - 1)$
35. 1	$\delta(t)$
36. $\exp(-as)$	$\delta(t - a)$
37. $\dfrac{1}{s}\exp(-as)$	$u_a(t)$
38. $(s^2 + a^2)^{-1/2}$	$J_0(at)$
39. $\dfrac{1}{s}\log s$	$-\log t - C$

TABLE 1 (continued)

$F(s)$	$f(t)$
40. $\log \dfrac{s-a}{s}$	$\dfrac{1}{t}(1 - \exp(at))$
41. $\log \dfrac{s-a}{s-b}$	$\dfrac{1}{t}(\exp(bt) - \exp(at))$
42. $\log \dfrac{s+a}{s-a}$	$\dfrac{2}{t}\sinh at$
43. $\arctan\left(\dfrac{a}{s}\right)$	$\dfrac{1}{t}\sin at$
44. $\dfrac{\sqrt{\pi}}{2}\exp\left(\dfrac{s^2}{4}\right)\operatorname{erfc}\left(\dfrac{s}{2}\right)$	$\exp(-t^2)$
45. $\dfrac{1}{b}F\left(\dfrac{s}{b}\right),\quad b>0$	$f(bt)$
46. $F(s-b)$	$\exp(bt)f(t)$

$$\frac{1}{s-a}\ \frac{1}{s-b}$$

$$\frac{1}{A(s-b) + B(s-a)} = 1$$

$$B = \frac{1}{b-a}$$

$$A = B$$

$$s = a$$

$$A = \frac{1}{a-b}$$

$$\frac{1}{(a-b)(s-a)} \qquad \frac{1}{a-b}\left(e^{-at} + e^{-bt}\right)$$

TABLE 2 Exponential and hyperbolic functions

x	e^x	e^{-x}	sinh x	cosh x
0.00	1.0000	1.00000	0.0000	1.0000
0.05	1.0513	0.95123	0.0500	1.0013
0.10	1.1052	0.90484	0.1002	1.0050
0.15	1.1618	0.86071	0.1506	1.0113
0.20	1.2214	0.81873	0.2013	1.0201
0.25	1.2840	0.77880	0.2526	1.0314
0.30	1.3499	0.74082	0.3045	1.0453
0.35	1.4191	0.70469	0.3572	1.0619
0.40	1.4918	0.67032	0.4108	1.0811
0.45	1.5683	0.63763	0.4653	1.1030
0.50	1.6487	0.60653	0.5211	1.1276
0.55	1.7333	0.57695	0.5782	1.1551
0.60	1.8221	0.54881	0.6367	1.1855
0.65	1.9155	0.52205	0.6967	1.2188
0.70	2.0138	0.49659	0.7586	1.2552
0.75	2.1170	0.47237	0.8223	1.2947
0.80	2.2255	0.44933	0.8881	1.3374
0.85	2.3396	0.42741	0.9561	1.3835
0.90	2.4596	0.40657	1.0265	1.4331
0.95	2.5857	0.38674	1.0995	1.4862
1.00	2.7183	0.36788	1.1752	1.5431
1.05	2.8577	0.34994	1.2539	1.6038
1.10	3.0042	0.33287	1.3356	1.6685
1.15	3.1582	0.31664	1.4208	1.7374
1.20	3.3201	0.30119	1.5095	1.8107
1.25	3.4903	0.28650	1.6019	1.8884
1.30	3.6693	0.27253	1.6984	1.9709
1.35	3.8574	0.25924	1.7991	2.0583
1.40	4.0552	0.24660	1.9043	2.1509
1.45	4.2631	0.23457	2.0143	2.2488
1.50	4.4817	0.22313	2.1293	2.3524
1.55	4.7115	0.21225	2.2496	2.4619
1.60	4.9530	0.20190	2.3756	2.5775
1.65	5.2070	0.19205	2.5075	2.6995
1.70	5.4739	0.18268	2.6456	2.8283
1.75	5.7546	0.17377	2.7904	2.9642
1.80	6.0496	0.16530	2.9422	3.1075
1.85	6.3598	0.15724	3.1013	3.2585
1.90	6.6859	0.14957	3.2682	3.4177
1.95	7.0287	0.14227	3.4432	3.5855
2.00	7.3891	0.13534	3.6269	3.7622
2.05	7.7679	0.12873	3.8196	3.9483
2.10	8.1662	0.12246	4.0219	4.1443
2.15	8.5849	0.11648	4.2342	4.3507
2.20	9.0250	0.11080	4.4571	4.5679
2.25	9.4877	0.10540	4.6912	4.7966

TABLE 2 (*continued*)

x	e^x	e^{-x}	sinh x	cosh x
2.30	9.9742	0.10026	4.9370	5.0372
2.35	10.4856	0.09537	5.1951	5.2905
2.40	11.0232	0.09072	5.4662	5.5569
2.45	11.5883	0.08629	5.7510	5.8373
2.50	12.1825	0.08208	6.0502	6.1323
2.55	12.8071	0.07808	6.3645	6.4426
2.60	13.4637	0.07427	6.6947	6.7690
2.65	14.1540	0.07065	7.0417	7.1123
2.70	14.8797	0.06721	7.4063	7.4735
2.75	15.6426	0.06393	7.7894	7.8533
2.80	16.4446	0.06081	8.1919	8.2527
2.85	17.2878	0.05784	8.6150	8.6728
2.90	18.1741	0.05502	9.0596	9.1146
2.95	19.1060	0.05234	9.5268	9.5791
3.00	20.0855	0.04979	10.018	10.068
3.05	21.1153	0.04736	10.534	10.581
3.10	22.1980	0.04505	11.076	11.122
3.15	23.3361	0.04285	11.647	11.690
3.20	24.5325	0.04076	12.246	12.287
3.25	25.7903	0.03877	12.876	12.915
3.30	27.1126	0.03688	13.538	13.575
3.35	28.5027	0.03508	14.234	14.269
3.40	29.9641	0.03337	14.965	14.999
3.45	31.5004	0.03175	15.734	15.766
3.50	33.1155	0.03020	16.543	16.573
3.55	34.8133	0.02872	17.392	17.421
3.60	36.5982	0.02732	18.286	18.313
3.65	38.4747	0.02599	19.224	19.250
3.70	40.4473	0.02472	20.211	20.236
3.75	42.5211	0.02352	21.249	21.272
3.80	44.7012	0.02237	22.339	22.362
3.85	46.9931	0.02128	23.486	23.507
3.90	49.4024	0.02024	24.691	24.711
3.95	51.9354	0.01925	25.958	25.977
4.00	54.5982	0.01832	27.290	27.308
4.10	60.3403	0.01657	30.162	30.178
4.20	66.6863	0.01500	33.336	33.351
4.30	73.6998	0.01357	36.843	36.857
4.40	81.4509	0.01227	40.719	40.732
4.50	90.0171	0.01111	45.003	45.014
4.60	99.4843	0.01005	49.737	49.747
4.70	109.9472	0.00910	54.969	54.978
4.80	121.5104	0.00823	60.751	60.759
4.90	134.2898	0.00745	67.141	67.149
5.00	148.4132	0.00674	74.203	74.210

answers and hints
to selected exercises

CHAPTER 1

Section 1.1, p. 9

2. $\log C$ ranges over *all* real numbers as C varies over the positive real numbers.

3. Approximately 266 days; approximately 141 years.

5. **a)** $u(t) = 25° + 35° \exp(-0.0168t)$;   **c)** approximately $29.7°$

8. **a)** $y + t - \log\left|\dfrac{t+1}{y-1}\right| = C; y = 1$;   **b)** $(y+1)\exp(-y) = \dfrac{1}{x} + C$;

   c) $y = C \log x$;   **d)** $\arctan x - \sqrt{1+y^2} = C$

9. **a)** $y^2 + \cos\theta = 0$;   **b)** $\dfrac{1}{1+r} = \log|1-s| + \dfrac{1}{2}$;

   c) $5x^2 - 2y^2 = 2$;   **d)** $v^2 = 2g(t - t_0) + v_0^2$

11. **a)** $y = C \tan t$

12. **a)** $y = C \log x$;   **b)** $y = \log x^2$

13. $y = \dfrac{-2}{x^2 + C}; y = 0$

14. $\cos y = Cx \exp(-x^2/2)$

15. $u = \log \dfrac{2}{2C - e^{2t}}$

16. $\exp(y) = (e+1)\exp(t) - 1$

17. $x = \dfrac{C \exp(2t) - 1}{C \exp(2t) + 1}; x = 1$

18. $\cos u = -\sin t + C$

19. $x^2 + y^2 = 25$

20. $\arctan t + \arctan y + \dfrac{\pi}{4} = 0$

21. $\log s = 2r + \log \dfrac{C}{(r+1)^2}; s > 0, C > 0$

22. $u = u_0 \exp(-t^2)$

23. $2u + \sin 2u = v^2(2 + v^2) + C$

24. $x^2 + (y - C)^2 = 16$; circles of radius 4 centered at $(0, C)$

25. $y = Cy \exp (x^2/2) + 1$

26. Approximately 1 hr 41 min

29. Approximately 5.64 hr after it is hung out to dry

30. a) $A(t) = pe^{rt}$, where r is the annual rate of interest; **b)** \$3501.35

31. b) $y = Cx + \cos c$;
   c) the given y satisfies the differential equation in part (b) but it cannot be obtained from the general solution by assigning some constant value to c.

Section 1.2, p. 17

3. Not exact

4. Not exact

5. Not exact

6. Exact

7. Not exact

8. Not exact

9. $(1 + y)(1 + x) = C(1 - y)(1 - x)$; $y = \pm 1$

10. $u = \dfrac{1}{3} t^2 + \dfrac{C}{t}$

11. $y(y^2 - 3t) = C$

12. $(x^2 + 1)^2(2y^2 - 1) = C$

13. $y - \log (y + 1) + \frac{1}{2}x^2 = C$; $y = -1$

14. $r^2 - 6rs + 3s^2 = C$

15. $u^2 - 5uv + v^2 = C$

16. $2 \arctan t + \log (1 + y^2) = C$

17. $\sin x \cos y - \log |\sin x| = C$

18. $s(3r^2 + s^2) = C$

19. $r^2 + 2r(\sin \theta - \cos \theta) = C$

20. $2u \sin t - t^2 = C$

21. $x(4y - x^3) = C$

22. $r(r\theta - \tan \theta) = \dfrac{\pi}{4} - 1$

23. $r \cos \theta - \theta = 2$

24. $x(xy^2 - 4) + 4 = 0$

25. $(u^2 + v^2)^2 = 4v$

26. a) $xe^y + ye^x = C$; **b)** nonlinear because of the term e^y

27. $x^2(x^3 - 5y^3) = C$

28. The graph is an ellipse with center at $(3, 1)$.

Section 1.3, p. 23

1. $y = \dfrac{\exp{(x)}}{3} + C\exp{(-2x)}$

2. $y = \exp{(x^2/2)} - 1$

3. $y = \exp{(-x)} + C\exp{(-2x)}$

4. **a)** $y = \dfrac{x^2}{4} + \dfrac{1}{x^2}$

6. $\exp{(x^2)} = C(1 + 2y)$

7. $x = \frac{1}{2}(\sin y - \cos y) + \frac{1}{2}\exp{(\pi - y)}$

8. $y = (\frac{3}{2}x^2 + C)\csc x$

9. **a)** $y = x^{-2}\sin x$

10. $y = 2x^2 + Cx^{-2}$

11. $u = -\cos t + C\sin t$

12. $s = \dfrac{t - \cos t + C}{\sec t + \tan t}$

13. $x = \dfrac{2y^3 + 3y^2 + C}{6(y + 1)^2}$

14. $|1 - x|\exp{(y^2/2)} = C$

15. $\exp{(y)} = \exp{(x)} + C$

16. $y = Cx - \dfrac{1}{x}$

17. $y = \frac{1}{2}x^4 + 2x^2$

18. $y = \dfrac{1}{2x^2}((\log x)^2 + 6)$

19. $u = \sin\theta(2\theta\cot\theta + 1)$

20. $r\sin\theta = \dfrac{1}{2}\theta^2 + 1 - \dfrac{\pi^2}{32}$

21. $y = \frac{2}{3}\sin x(3 - \sin^2 x) + C\sec^3 x$

22. $x = \frac{1}{4} + C\exp{(-2y^2)}$

24. **a)** $y = -x^2\log|x|$;   **c)** all values except $x = 0$

Section 1.4, p. 29

3. $\dfrac{\sqrt{3}}{3}\log\left|\dfrac{X - Y\sqrt{3}}{X + Y\sqrt{3}}\right| - \dfrac{1}{2}\log\left|\dfrac{X^2 - 3Y^2}{Y^2}\right| = \log|CY|$

6. **a)** $x^2 + y^2 = Cx$;   **c)** begin with $\dfrac{\partial f}{\partial y} = \dfrac{2xy}{(x^2 + y^2)^2}$

7. **b)** $xy^2 - (y^2 - 2y + 2)\exp{(y)} = C$

8. **b)** $y = x^2\log|x| + Cx^2$

9. **b)** $x(x^2 + 3y^2) = C$

10. **b)** $y = x\tan{(x + C)}$

11. $\csc \dfrac{y}{x} - \cot \dfrac{y}{x} = Cx$

12. **a)** $2x^3 + 3xy^2 - 3y^3 = C$; **b)** $x = C \exp (2\sqrt{y/x})$;
    c) $\log Cx + \exp (-y/x) = 0$; **d)** $xy(y - x) = C$

13. **a)** Integrating factor: $(x^2 + y^2)^{-1}$; $2x + \log (x^2 + y^2) = C$, also $x = y = 0$;
    b) integrating factor: $(x^2 y^2)^{-1}$; $2Cxy - xy^3 + 2 = 0$, also $x = 0$ and $y = 0$;
    c) integrating factor: $(x^2 + y^2)^{-1}$; $2 \arctan \dfrac{x}{y} = y + C$, also $x = y = 0$

14. **a)** $x^2 - 2xy - 3y^2 = C$; **b)** $(y - 2x)^3 = C(y - x)^4$;
    c) $(x + 2y)(x - 2y - 4)^3 = C$; **d)** $\log |x - 3y + 8|^5 = x - 2y + C$

15. Let $u = x + y$: $y = C \exp (x) - x - 1$.

18. $y + \sqrt{y^2 - x^2} = Cx^2$ and $y = x$

Section 1.5, p. 45

1. 1.37×10^9 yr; 3×10^{-6} sec

2. Approximately 17,000 yr

3. **a)** Approximately 0.1155 hr^{-1}; approximately 9.5 hr

4. $u_0 = 7.5°C$ (Use the first set of data to express k in terms of u_0; then use the second set of data to obtain an equation that can be solved for u_0.)

5. Approximately 26.9 min; note that the ambient is now 20°C.

6. $u_0 = 30.5°C$

7. $T = T_0 \exp (0.4\theta)$; approximately 10,710 kg can be controlled, assuming a person uses a force equal to one-fourth of his or her weight

8. $x^2 - y^2 = C^2$, a family of hyperbolas

9. $k_1 = \dfrac{1}{t} \log \dfrac{a}{a - x}$; $x = a (1 - \exp (-k_1 t))$

10. $\dfrac{1}{k_1} \log 2$; approximately 7.6 min

11. $k_2 = \dfrac{1}{t(a - b)} \log \dfrac{b(a - x)}{a(b - x)}$; $x = a \dfrac{1 - \exp (k_2 t(a - b))}{1 - \dfrac{a}{b} \exp (k_2 t(a - b))}$

12. $k_2 a = \dfrac{1}{t} \log \dfrac{b}{b - x}$

13. $k_2 = \dfrac{1}{at} \dfrac{x}{a - x}$

14. $I(x) = I_0 \exp (-kx)$

15. 3.55 m

16. $i(t) = \dfrac{E}{R} (1 - \exp (-RT/L))$, using the initial condition $i(0) \doteq 0$

17. $i \to \dfrac{E}{R}$ as $t \to \infty$; $T_{1/2} = \dfrac{0.693L}{R}$

18. $i(t) = \dfrac{E_0}{R^2 + L^2\omega^2}\,(R\cos\omega t + L\omega\sin\omega t) + C\exp(-Rt/L)$

19. $i(0.1) \doteq 3.93$ amp

20. $i(0.1) \doteq -0.312$ amp

23. $u(t) = c\exp(a(1-k)t)$

24. k must satisfy $(1-kaT(1-k))\exp(a(1-k)T) = a$, which is obtained from $dp/dk = 0$

25. $(x-c)^2 - y^2 = k^2$

26. $\dfrac{di}{dt} + 2i = \sin t;\ i(t) = \dfrac{1}{5}(2\sin t - \cos t) + \dfrac{51}{5}\exp(-2t)$

27. a) Approximately 8.13 min; **b)** approximately 55.3°F

28. Letting $P = 1$, $t = 1$, and $r = 0.06$, we obtain the following equivalent annual simple interest rates.

n	1	2	4	8	16	32	64	128	256	512
%	6	6.09	6.136	6.160	6.172	6.178	6.181	6.182	6.183	6.183

29. 7.7%

30. Approximately 7.8%

31. 6.06%

32. The equation to be solved is $m\dfrac{dv}{dt} = mg - kv;\ v(t) = \dfrac{mg}{k}(1 - \exp(-kt/m))$; limiting velocity is mg/k.

33. The proportionality constant in the differential equation can be expressed as $(K\log_2 e)^{-1}$, which is positive if $K > 0$.

34. Recall the properties of logarithms:
$$\log y = (\log_2 y)(\log 2)\ \text{and}\ \log_2 y = \dfrac{1}{\log_y 2}.$$

35. $t = K$

36. The equation to be solved is $RI + \dfrac{1}{C}\displaystyle\int I\,dt = E,$

which, on differentiation, becomes $R\dfrac{dI}{dt} + \dfrac{I}{C} = 0$;

$I(t) = 0.01\exp(-100t);\ I(0.01) \doteq 0.0037$

37. $I(t) = -0.00239\exp(-100t) + 0.01239\cos(120\pi t) + 0.04671\sin(120\pi t);$
$I(0.01) \doteq -0.0384$

38. $\dfrac{dN}{dt} = kN(N_1 - N),\ N(0) = N_0,\ N_0 \le N \le N_1;$

$\dfrac{N_0}{N}\dfrac{N_1 - N}{N_1 - N_0} = \exp(-N_1 kt)$

40. a) $6\dfrac{dv}{dt} = 192 - \dfrac{1}{2}v;$ **b)** $v(t) = 384\left(1 - \exp\left(\dfrac{-t}{12}\right)\right)$

Section 1.6, p. 61

1. $y(4) \doteq -1.93$

2. $y(4) \doteq -1.93$

3. $y(4) \doteq -1.93$

5. At approximately $x = 3.6$ the three curves converge and become indistinguishable.

7. $y_{n+1} = y_n + h(y_n^2 - x_n) + \dfrac{h^2}{2}(2y_n^3 - 2x_n y_n - 1)$

12. $y(2) \doteq 2.24$

15. **a)** 1.7210;   **c)** 1.7987

17. $y_{n+1} = y_n + h(x_n + y_n) + \dfrac{h^2}{2}(1 + x_n + y_n) + \dfrac{h^3}{6}(1 + x_n + y_n) + \dfrac{h^4}{24}(1 + x_n + y_n)$

19. The solutions have the form of the normal probability distribution, that is, $y = c_1 \exp(-c_2 x^2)$, $c_2 > 0$.

21. The curves approach the line $y = 1$ since $y'(1) = 0$.

24. The curves approach $y = \pm 2$.

25. **a)** $y(0) \doteq 0.7$; $y(-1) \doteq 0$

Section 1.7, p. 67

1. **a)** $x \geq 0$ and $y > 0$ or $x \leq 0$ and $y < 0$;   **b)** $-2 < y < 2$;
   c) all y except $y = \pm 2$;   **d)** all values except $y = \pm x$

2. $y = 0$

3. **a)** $u = (\log t + \log 2)\left(\dfrac{1 + \sqrt{1 - t^2}}{t}\right)$;   **b)** $u = \dfrac{1}{t}\exp(t)$;

   c) $u = \dfrac{1}{1 - t^2}$ on $(-1, 1)$

4. $y = x^4/16$ and $y = 0$ are solutions

5. $y = x^2$ and $y = 0$ are solutions; no contradiction since $\partial f/\partial y$ does not exist at $y = 0$

11. **a)** $y = \dfrac{2}{2 - t^2}$, $y = \dfrac{2}{3 - t^2}$, $y = 0$, $y = 0$

15. Starting with $y_0 = 1$, we obtain $y_1 = 1 + x^2$, $y_2 = 1 + x^2 + x^4/2$,
    $y_3 = 1 + x^2 + x^4/2! + x^6/3!$.

17. The solution $y = F(x)$ is a continuous function of x and also a continuous function of x_0 and y_0.

CHAPTER 2

Section 2.1, p. 78

4. $W(\exp(rx), x\exp(rx)) = \exp(2rx) \neq 0$

6. The Wronskian is $\beta \exp(2\alpha x)$.

8. **a)** $y = c_1 \exp(x) + c_2 \exp(2x)$;   **b)** $y = (c_1 + c_2 x) \exp(3x)$;
    c) $y = (c_1 \cos 4x + c_2 \sin 4x) \exp(3x)$

9. $y = c_1 + c_2 \exp(3x)$

10. **a)** $y = (c_1 \cos x + c_2 \sin x) \exp(-x)$;

    b) $y = \left(c_1 \cos \dfrac{\sqrt{7}}{2} x + c_2 \sin \dfrac{\sqrt{7}}{2} x \right) \exp\left(-\dfrac{x}{2} \right)$;

    c) $y = 4 \sin\left(\dfrac{x}{4} \right) \exp\left(-\dfrac{x}{4} \right)$;   **d)** $x = -\dfrac{3}{2} \cos 2t + \sin 2t$

11. $y = 2 \exp(-x) - 2 \exp(-2x)$

12. $y = 2 \cos 3x + 3 \sin 3x$

13. $y = 3(1 - 4x) \exp(2x)$

14. **a)** $y = c_1 \cos 2t + c_2 \sin 2t$;   **b)** $u = (c_1 + c_2 v) \exp(-3v)$;
    c) $y = c_1 + c_2 \exp(-5r)$;   **d)** $y = c_1 \exp(-5x) + c_2 \exp(7x)$

15. **a)** $x = \dfrac{2\sqrt{13}}{13} \sin \dfrac{1}{2}\sqrt{13}t \, \exp\left(-\dfrac{t}{2} \right) \doteq 0.5547 \sin 1.8028t \, \exp\left(-\dfrac{t}{2} \right)$;

    b) $u = 5 \exp(-2t) - 4 \exp(-3t)$;

    c) $\theta = \dfrac{1}{\pi}(\pi^2 + \pi - 1 + t(1 - \pi^2)) \exp(\pi(1-t)) \doteq (3.8233 - 2.8233t) \exp(\pi(1-t))$;

    d) $y = \dfrac{1}{2}(2 + 9t) \exp\left(-\dfrac{5t}{2} \right)$

16. $y = -4.91t^2 + 5t + 80$

19. $y = \dfrac{(1 + e^{-2}) \exp(3x) - (1 + e^3) \exp(-2x)}{\exp(3) - \exp(-2)} \doteq 0.0569 \exp(3x) - 1.0569 \exp(-2x)$

Section 2.2, p. 84

1. The Wronskian has a value of 2.
3. **a)** $-2, -1, 1$;   **b)** $-3, 1, 1$;   **c)** $-2, -1, \pm i$;   **d)** $-1, -\frac{1}{2}, 2, 3$
4. The Wronskian is b^3.
5. $y = 7 - (5 + 9x) \exp(-3x)$
6. $y = c_1 \exp(-x) + c_2 \exp(x) + c_3 \exp(2x)$
7. $y = (c_1 + c_2 x) \exp(-2x) + (c_3 + c_4 x) \exp(2x)$
8. $y = c_1 \exp(-2x) + (c_2 + c_3 x) \exp(2x)$
9. $y = c_1 + c_2 \cos 3x + c_3 \sin 3x$
10. $y = c_1 + c_2 x + c_3 \cos 2x + c_4 \sin 2x$
11. $y = c_1 \exp(-2x) + c_2 \exp(x) + c_3 \exp(2x)$
12. $y = c_1 \exp(-2x) + c_2 \exp(2x) + c_3 \cos 3x + c_4 \sin 3x$
13. $y = c_1 \exp(-x) + c_2 + c_3 x + c_4 x^2$
14. $y = c_1 \exp(-3x) + c_2 \exp(-x) + c_3 \exp(x)$
15. $y = c_1 \exp(-x) + c_2 \exp(x) + c_3 \exp(2x)$
16. $y = -3 \exp(-x) + (3 \cos 3x + \sin 3x) \exp(2x)$

17. a) $y = c_1 \exp(-t) + (c_2 + c_3 t) \exp(t)$;
   b) $u = c_1 \exp(-5t) + c_2 \exp(-t) + c_3 \exp(t)$;
   c) $x = c_1 \exp(-t) + c_2 \exp(2t) + c_3 \exp(5t)$;
   d) $y = (c_1 + c_2 t + c_3 t^2 + c_4 t^3 + c_5 t^4) \exp(t)$

18. a) $y = (1 + t) \exp(t)$;
   b) $y = \frac{1}{9}(7 + \exp(3t) + \exp(-3t))$;
   c) $y = (1 + \sin t) \exp(t)$;
   d) $x = 2 \exp(2t) + (3t - 2) \exp(-t)$

19. $y = -2 + 6 \exp(-x) - \exp(-2x) + 2 \exp(2x)$

20. a) $y = c_1 + c_2 x + (c_3 + c_4 x) \exp(2x)$;
   b) $y = (c_1 + c_2 x) \cos \sqrt{3}x + (c_3 + c_4 x) \sin \sqrt{3}x$;
   c) $y = c_1 \exp(2x) + c_2 \exp(-2x) + c_3 \cos 2x + c_4 \sin 2x$;
   d) $y = (c_1 + c_2 x + c_3 x^2 + c_4 x^3) \exp(x)$

21. First verify that $y_1(x)$ satisfies the given equation. Then $(r - 1 - i)(r - 1 + i)$ must be a factor of the characteristic equation. Find the other factor to obtain $y = (c_1 \cos x + c_2 \sin x) \exp(x) + c_3 \exp(-3x) + c_4 \exp(2x)$.

22. Use synthetic division to find one factor (a factor of 6). Then use Exercise 3 to find a second factor.

$$y = c_1 \exp(-x) + c_2 \exp(3x) + (c_3 \cos x + c_4 \sin x) \exp(-x)$$

23. The characteristic equation is $r^4 + 1 = 0$ or $r^4 = -1$. To find the *four* fourth roots of -1 look ahead to Section 10.1 to obtain $(\sqrt{2}/2)(1 \pm i)$ and $-(\sqrt{2}/2)(1 \pm i)$.

24. $y = c_1 \exp(2x) + c_2 \exp(3x) + c_3 \exp(x/3)$

25. Use Exercise 3 to find the real root. Then let $\omega = a + bi$ and substitute this and ω^2 into the fourth-degree polynomial equation.

Section 2.3, p. 95

3. Find particular solutions for $y'' - 2y' - 3y = 2 \exp(x)$ and for $y'' - 2y' - 3y = 3 \exp(2x)$; then use Exercise 2.

5. a) $y_c(x) = (c_1 \exp(\sqrt{5}x) + c_2 \exp(-\sqrt{5}x)) \exp(-2x)$;
   b) $y_c(x) = c_1 \exp(-x) + c_2 \exp(3x)$;   **c)** $y_c(x) = (c_1 + c_2 x) \exp(x)$

6. $y = (c_1 + c_2 x + \frac{3}{2} x^2) \exp(x)$

9. Solve by Cramer's rule. Another method is to multiply Eq. (2.3–2) by $\sin x$ and Eq. (2.3–3) by $\cos x$ and add.

11. $y_p(x) = -2 \exp(x)$

12. $(Ax^2 + Bx + C) \sin 2x + (Dx^2 + Ex + F) \cos 2x$

13. $Ax \exp(x) + Bx \exp(-x)$

14. $(Ax^3 + Bx^2 + Cx) \sin 2x + (Dx^3 + Ex^2 + Fx) \cos 2x$

15. $(Ax^2 + Bx) \exp(x) + (Cx^2 + Dx) \exp(-x)$

16. $x(A \cos x + B \sin x) \exp(x) + x(C \cos x + D \sin x) \exp(-x)$

17. $y = c_1 \exp(2x) + c_2 \exp(-2x) + x \exp(2x)$

18. $N = c_1 \exp(2t) + c_2 t \exp(2t) + 2t^2 \exp(2t)$

19. Assume $y_p(x) = (Ax^3 + Bx^2) \exp(2x)$.

20. $y = c_1 \exp(-2x) + c_2 \exp(x) + c_3 \exp(2x) - \frac{1}{3}x \exp(x) + \frac{1}{4}x \exp(2x)$

21. $y = c_1 \exp(-2x) + c_2 \exp(-x) + c_3 \exp(x) + \frac{1}{4}(\cos 2x - \sin 2x)$

22. $y = (c_1 + c_2 x) \cos x + (c_3 + c_4 x) \sin x + x^2 - 2x - 4$

23. $y = (c_1 + c_2 x) \exp(-2x) + x^2 - 4x + \frac{7}{2}$

24. $y = c_1 \exp\left(-\dfrac{x}{2}\right) + c_2 \exp\left(\dfrac{3x}{2}\right) - \dfrac{1}{425}(19 \cos 2x + 8 \sin 2x)$;

$y_p(x) = -0.0447 \cos 2x - 0.0188 \sin 2x$

25. Find $y_1(x)$, the solution of $y'' + 2y' + y = \sin x$ and $y_2(x)$, the solution of $y'' + 2y' + y = \cos 2x$.

26. $y_p(x) = \frac{1}{10} \exp(3x)$

27. $y_p(x) = -\frac{3}{4}x \cos 2x$

29. $y = (c_1 \cos x + c_2 \sin x) \exp(2x) + \frac{1}{2}x \exp(2x) + (\frac{1}{34}x + \frac{4}{289}) \exp(-2x)$;
$y_p(x) = 0.5x \exp(2x) + (0.0294x + 0.0138) \exp(-2x)$

30. $y = (3 + 2x) \exp(x) + \frac{1}{6}x^3 \exp(x)$

31. a) $-\cos x (\log|\sec x + \tan x|)$;
b) $(\log|\cos x|) \cos x + x \sin x$;
c) $-(\log x + 1) \exp(x)$

32. a) $y = c_1 \exp(-x) + c_2 x \exp(-x) - \exp(-x)[\log(\exp(x) - 1) - x]$ (*Note:* Integration by parts may be helpful.);
b) $y = c_1 \exp(x) + c_2 \exp(2x) - x \exp(x) + \exp(x) \log(\exp(x) + 1) + \exp(2x) \log(1 + \exp(-x))$;
c) $y = c_1 \cos 2x + c_2 \sin 2x + \frac{1}{4} \sin 2x \log|\csc 2x - \cot 2x|$
(*Note:* Any $\cos 2x$ or $\sin 2x$ terms appearing in the particular solution may be combined with like terms in the complementary solution.);
d) $y = c_1 \cos 3x + c_2 \sin 3x - \frac{1}{3}x \cos 3x + (\frac{1}{9} \sin 3x)(\log(\sin 3x))$

33. $y = 2 \exp(x) + 3x \exp(x) - \exp(x) \log(1 - x)$

36. $y = c_1 x + \dfrac{c_2}{x} + \dfrac{1}{3}x^2 + \left(\dfrac{1 - x^2}{2x}\right) \log(x + 1) - \dfrac{1}{2}$

37. $y = c_1 \exp(-x) + c_2 \exp(x) - \frac{1}{2}(x \sin x + \cos x)$

Section 2.4, p. 102

1. $y_p(x) = (1 + 2x^{-1} + 2x^{-2}) \exp(-x)$

2. a) $y = c_1 x^{-2} + c_2 x + (1 + 2x^{-1} + 2x^{-2}) \exp(-x)$;

b) $y = \dfrac{c_1}{x} + \dfrac{c_2}{x} \log x + \dfrac{x^3}{16}$;

c) $y = c_1 \cos(\log x) + c_2 \sin(\log x) + \cos x$
$+ \cos(\log x) \displaystyle\int^x \sin t \cos(\log t)\, dt$
$+ \sin(\log x) \displaystyle\int^x \sin t \sin(\log t)\, dt$

3. $y = \dfrac{3}{x}(x^3 + 1)$

4. $y = c_1 x^{-5} + c_2 x$

5. $y = t^{-2}(c_1 \cos (\log t) + c_2 \sin (\log t))$

6. $y = c_1 x^{-1} + c_2 x^2 + c_3 x^4$

7. $u = c_1 r^{-1} + c_2 r^{-1} \log r$

8. $u = c_1 + c_2 \log t + \dfrac{c_3}{2} (\log t)^2$

9. $y = c_1 x^{-3} + c_2 x^3 + \frac{1}{4} x - \frac{1}{5} x^2$

10. $y = c_1 \cos (2 \log x) + c_2 \sin (2 \log x) + \frac{1}{4} \log x$
    (*Hint*: Make the substitution $u = \log x$.)

11. $y = \dfrac{x}{2} (3 \cos (\log x) + 5 \sin (\log x)) + \dfrac{5}{2} - 4x$

12. $y = c_1 x^{-4} + c_2 x + c_3 x^2$

13. $y = \frac{1}{6}(128 x^{-4} + 48x - 17x^2)$

14. **a)** $y = 1 + 2 \log x$;   **b)** $y = \frac{1}{3}(x^2 - x^{-1})$

16. $y = c_1 + c_2 x^{-1} + c_3 x^3 - \frac{1}{2} x^2$

17. $y = 2 + t + t \log t + \frac{3}{2} t(\log t)^2 - 5 \log t$

18. $y = c_1 x^{-1} + c_2 x^2 + c_3 x^4 - \frac{1}{6} x^2 \log x + \frac{1}{4}$

20. xy' becomes $cx(dy/d(cx)) = xy'$

Section 2.5, p. 115

1. $\theta = 0.1 \cos 3.5t$

2. $f \doteq 0.39$

3. $L \doteq 12.4$ cm

4. $i = c_1 \exp(-Rt/2L) + c_2 t \exp(-Rt/2L)$ for $R^2 C = 4L$;
   $i = c_1 \exp(r_1 t) + c_2 \exp(r_2 t)$, where

$$r_1 = \frac{-R + \sqrt{R^2 - \dfrac{4L}{C}}}{2L}, \qquad r_2 = \frac{-R - \sqrt{R^2 - \dfrac{4L}{C}}}{2L},$$

and $R^2 > 4L/C$;
$i = (c_1 \cos \omega t + c_2 \sin \omega t) \exp(-Rt/2L)$, where

$$\omega = \sqrt{\frac{4L - R^2 C}{4CL^2}}$$

and $R^2 < 4L/C$, in which case the solution is periodic.

5. The particular solution that must be added to the solutions in Exercise 4 is

$$\frac{E_0 \omega_0}{\left(\dfrac{1}{C} - \omega_0^2 L\right)^2 + \omega_0^2 R^2} \left(R\omega_0 \cos \omega_0 t + \left(\omega_0^2 L - \dfrac{1}{C}\right) \sin \omega_0 t\right),$$

provided $\omega_0 \neq \omega$.

6. $y = 3 \cos (0.7t)$

7. $T \doteq 1.43$ sec

8. $y = 2 \cos (0.4127t) \exp (-0.5654t)$

9. Approximately 18 sec^{-1}

10. 108 cm/sec

11. $y = \dfrac{bx^2}{24} (x^2 - 4Lx + 6L^2)$, where $b = 2.3/EI$; $y_{max} = bL^4/8$, the maximum deflection occurring when $x = L$

12. $y = \frac{2}{3}x^2(x^2 - 40x + 600)$

13. $t = \dfrac{1}{R} \sqrt{\dfrac{r_0}{2g}} \left(\sqrt{r_0 r - r^2} - \dfrac{r_0}{2} \arcsin \dfrac{2r - r_0}{r_0} \right) + C_1, \qquad 0 < r < r_0.$

(*Hint*: Multiply the numerator and denominator of the dr term by $\sqrt{r}$ after separating the variables.)

14. $C_1 = \dfrac{\pi r_0}{4R} \sqrt{\dfrac{r_0}{2g}}$

15. Approximately 208 sec

16. Approximately 264.8 ft/sec

17. $V = -gm/k$

18. $v = v_g \log \left(\dfrac{m_r + m_f - bt}{m_r + m_f} \right)$; if the restriction $t \le m_f/b$ is *not* met, then $m(t)$ becomes less than m_r, which is impossible.

19. a) $s = \dfrac{-v_g}{b} \left[(m_r + m_f - bt) \log \dfrac{m_r + m_f - bt}{m_r + m_f} + bt \right];$

   b) $\dfrac{-v_g}{b} \left(m_r \log \left(\dfrac{m_r}{m_r + m_f} \right) + m_f \right)$

20. $c_1 = c_3 \cos \delta, c_2 = -c_3 \sin \delta$

21. Transient is 0.19 amp; steady-state is -1.22 amp.

22. $i = (\sqrt{3}/3) \sin 5\sqrt{3}t \exp (-5t)$; $i_{max} \doteq 0.577$ amp

23. $y = 3 \cos \sqrt{\dfrac{k}{m}} t + 5 \sqrt{\dfrac{m}{k}} \sin \sqrt{\dfrac{k}{m}} t$

24. $y = (c_1 \cos \omega_0 t + c_2 \sin \omega_0 t) \exp (-ct/2m)$

$\qquad + \dfrac{3}{(k - 4m)^2 + 4c^2} ((k - 4m) \sin 2t - 6c \cos 2t),$

where $\omega_0 = \sqrt{4mk - c^2}/2m$ and $\omega_0 \ne 2$

25. $y = c_1 \cos \omega_0 t + c_2 \sin \omega_0 t + \dfrac{a}{m(\omega_0^2 - \omega^2)} \cos \omega t$, where $\omega_0 = \sqrt{k/m}$ and $\omega_0 \ne \omega$;

$\qquad y = c_1 \cos \omega t + c_2 \sin \omega t + \dfrac{a}{2m\omega} t \sin \omega t$, where $\omega = \sqrt{k/m}$

26. 124.83 cm

28. $y = \tan (C - x) - 1$

Section 2.6, p. 123

1. $\begin{cases} \dot{u}_1 = u_2, \\ \dot{u}_2 - (t + u_1)u_2 = t(2 - u_1^2); \end{cases}$
   $u_1(0) = -1, \, u_2(0) = 1$

2. $\begin{cases} u_1' = u_2, \\ EIu_2' = (1 + u_2^2)^{3/2} M(x); \end{cases}$
   $u_1(0) = u_2(0) = 0$

3. $\begin{cases} u_1' = u_2, \\ u_2' = u_3, \\ u_3' - 3u_3 - 6xu_2 - 12x^2u_1 = \cos x; \end{cases}$
   $u_1(0) = 1, \, u_2(0) = 2, \, u_3(0) = 3$

4. $y(0.3) = 1.2995, \, y'(0.3) = 0.9947$

6. $y(0.1) = 1.2, \, y(0.2) = 1.37$

11. $y(0.2) = 1.2015, \, y(0.4) = 1.4128, \, y(0.6) = 1.6468$

13. $y(0.5) = 1.500$

CHAPTER 3

Section 3.1, p. 137

1. **a)** Not defined;   **b)** 1;   **c)** -1;   **d)** -1;   **e)** -1;   **f)** 1/2;
   g) 1/2;   **h)** not defined

3. **a)** $2/s^3$;   **b)** $s/(s^2 + a^2)$

4. **a)** $s/(s^2 - a^2)$;   **b)** $a/(s^2 - a^2)$

6. $a/((s - b)^2 + a^2)$

8. **a)** $1, -1, 1, -1$;   **b)** $0, 2, 2, 0$;   **c)** $\pi, 0$, not defined, not defined

10. **a)** $\dfrac{6}{s^3} - \dfrac{2}{s^2} + \dfrac{5}{s}$;   **b)** $1/(s^2 + 1)$;   **c)** $1/(s - a)^2$

12. **a)** $\dfrac{\exp{(-s)}}{s^2}(1 + s)$;   **b)** $\exp{(-s\pi)}\left(\dfrac{1}{1 + s^2} + \dfrac{3}{s}\right) + \dfrac{1}{1 + s^2}$

13. **a)** $2t^2$;   **b)** $\frac{3}{2}\sin 2t$;   **c)** $2\cos 2t$;   **d)** $\sin t + 1$

15. **a)** $a_1 + a_2 t + \dfrac{a_3}{2!} t^2 + \cdots + \dfrac{a_n}{(n - 1)!} t^{n-1}$;

   b) $\cosh \sqrt{2}t - \sqrt{2} \sinh \sqrt{2}t$;

   c) $\cos \sqrt{3}t - \dfrac{2\sqrt{3}}{3} \sin \sqrt{3}t$

16. $\dfrac{1}{s}(1 - \exp{(-s)})$

18. $\mathscr{L}(\cos at) + i\mathscr{L}(\sin at) = \mathscr{L}(e^{iat}) = \dfrac{1}{s - ia} = \dfrac{s}{s^2 + a^2} + i\,\dfrac{a}{s^2 + a^2}$

19. **a)** $\dfrac{1}{2}\dfrac{s - a}{(s - a)^2 + a^2} + \dfrac{1}{2}\dfrac{s + a}{(s + a)^2 + a^2} = \dfrac{s^3}{s^4 + 4a^4}$;   **c)** $\dfrac{a(s^2 - 2a^2)}{s^4 + 4a^4}$

Section 3.2, p. 145

2. $\cos^2 at$ is continuous for all $t \geq 0$; $|\cos^2 at| = \cos^2 at \leq 1$; $\dfrac{d}{dt}(\cos^2 at) =$
$-a \sin 2at$, which is continuous for all $t \geq 0$.

7. a) $\mathscr{L}(f'(t)) = s\left(-\dfrac{1}{s}\right)(\exp(-2s) - \exp(-s)) - \exp(-s) + \exp(-2s) = 0$;

c) $\mathscr{L}(f'(t)) = \dfrac{1}{s}(\exp(-2s) - \exp(-s))$

8. a) $\dfrac{2a^2}{s^2 + 4a^2}$; **b)** $\dfrac{2}{(s-b)^3}$; **c)** $\dfrac{2as}{(s^2 + a^2)^2}$; **e)** $\dfrac{2as}{(s^2 - a^2)^2}$

9. a) $\log\sqrt{\dfrac{s^2 + a^2}{s^2}}$; **b)** $\log\left(\dfrac{s+1}{s}\right)$; **c)** $\log\dfrac{\sqrt{s^2 + 1}}{s - 1}$;

d) $\log\left(\dfrac{s^2 - a^2}{s^2}\right)$; **e)** $\log\left(\dfrac{s+b}{s+a}\right)$

10. a) $\dfrac{2s(s^2 - 3a^2)}{(s^2 + a^2)^3}$; **b)** $\dfrac{2a(3s^2 - a^2)}{(s^2 + a^2)^3}$; **c)** $\dfrac{2(s-1)}{((s-1)^2 + 1)^2}$

Section 3.3, p. 155

4. a) $\cos(\sqrt{5}t)$; **b)** $t\exp(-3t)$; **c)** $-\sqrt{2}\sinh(\sqrt{2}t)$;

d) $\dfrac{\sqrt{3}}{3}\sin(\sqrt{3}t)$;

e) $2\sin t - 3(1 - \cos t) + 5(t - \sin t) = 3(\cos t - \sin t - 1) + 5t$

5. b) $\frac{1}{2}(\cos 4t + 1) = \cos^2 2t$; **c)** $\frac{1}{5}(\exp(2t) - \exp(-3t))$

6. b) $\dfrac{\exp(-at)}{ab(b-a)}b(a-d) - \dfrac{\exp(-bt)}{ab(b-a)}a(d-b) + \dfrac{d}{ab}$;

d) $\dfrac{d-a}{(b-a)(c-a)}\exp(-at) + \dfrac{b-d}{(b-a)(c-b)}\exp(-bt) + \dfrac{c-d}{(c-a)(b-c)}\exp(-ct)$

8. a) $\dfrac{d-a}{a^2 + b^2}\dfrac{1}{s+a} + \dfrac{a-d}{a^2 + b^2}\dfrac{s}{s^2 + b^2} + \dfrac{b^2 + ad}{a^2 + b^2}\dfrac{1}{s^2 + b^2}$;

c) $\dfrac{1}{b^2 - a^2}\left(\dfrac{s}{s^2 + a^2} - \dfrac{s}{s^2 + b^2}\right)$

11. $(3\cos 3t + 2\sin 3t)\exp(-t)$

12. a) $(4\cos 2t + 5\sin 2t)\dfrac{\exp(t)}{2}$; **b)** $(8\cos 4t + 9\sin 4t)\dfrac{\exp(-3t)}{4}$;

c) $(\cos t + 2\sin t)\exp(-t)$

13. a) $y(t) = \exp(2t)$; **b)** $y(t) = y_0\cos\omega t + \dfrac{y_0'}{\omega}\sin\omega t$;

c) $y(t) = \frac{1}{6}(10\exp(t) - 3\exp(2t) - \exp(4t))$;

d) $y(t) = \frac{1}{8}(10t\exp(2t) - 3\exp(2t) + 2t^2 + 4t + 3)$;

e) $y(t) = \frac{1}{2}((3\cos t + 4\sin t)\exp(-t) + t - 1)$

14. a) $y(t) = t\exp(t)$; **b)** $y(t) = t^2/2$

18. a) $\int_0^t \dfrac{\sin x}{x}\,dx;$ **b)** $\dfrac{1}{t}(\exp(bt) - \exp(at));$ **c)** $\dfrac{2}{t}(1 - \cos at)$

22. a) The coefficients must be polynomials in t;
   b) the differential equation involving dY/ds can be solved for Y;
   c) the inverse transform of $Y(s)$ can be found.

Section 3.4, p. 162

4. b) $\dfrac{1}{a^2}(\exp(at) - at - 1);$ **d)** $\dfrac{1}{2b}(\sin bt - bt \cos bt);$ **e)** $t - \sin t$

5. a) $\dfrac{1}{a^2}(\exp(at) - at - 1);$ **b)** $\dfrac{1}{2a^3}(\sin at - at \cos at);$

   d) $\dfrac{1}{a^2}(at - \sin at);$ **e)** $\dfrac{1}{2}t^2 + \cos t - 1;$

   f) $\dfrac{\exp(-bt)}{2a^3}(\sin at - at \cos at)$

6. a) $\dfrac{3!}{s^4(s^2 + 1)};$ **b)** $\dfrac{5!}{s^6(s - 3)};$ **c)** $\dfrac{3!5!}{s^{10}}$

7. a) $f(t) = t^2 + \tfrac{1}{12}t^4;$ **c)** $x(t) = \exp(-t)(1 - t)^2$

8. $y(t) = \tfrac{1}{8}((3 - t^2)\sin t - 3t \cos t)$

9. $1 - \cos at$

14. $y(t) = (1 - t)\exp(-t)$

Section 3.5, p. 170

6. a) $\dfrac{1}{s};$ **b)** $(2s + 1)\dfrac{\exp(-2s)}{s^2}$ (*Hint*: Write t as $t - 2 + 2$.);

   c) $\dfrac{1}{s}\exp(-s(c + t_0))$ (*Note*: $u_c(t - t_0) = 0$ for $t - t_0 < c$, that is, $t < c + t_0$);

   d) $\dfrac{1}{s}(\exp(-2s) - \exp(-s))$

7. a) $1 - u_c(t);$ **b)** $u_a(t) - u_b(t);$

   c) $f(t) = \begin{cases} 0, & 0 \le t < 2a, \\ t - 2a, & 2a \le t < a + b, \\ 2a - t, & a + b \le t \le 2b, \\ 0, & 2b < t \end{cases}$

9. a) $\dfrac{-\exp(-\pi s)}{s^2 + 1}$ (*Note*: $\sin(t - \pi) = -\sin t$.)

10. a) $\exp(-2s)\left(\dfrac{6}{s^4} + \dfrac{6}{s^3} + \dfrac{4}{s^2} + \dfrac{8}{s}\right)$

   (*Note*: $t^3 - 3t^2 + 4t + 4 = (t - 2)^3 + 3(t - 2)^2 + 4(t - 2) + 8$);

   c) $\dfrac{\exp(-2s)}{s^2 + 1}$

12. a) $\dfrac{a \exp (-bs)}{s^2 - a^2}$;   **c)** $\dfrac{s \exp (-bs)}{s^2 + a^2}$

14. Put $f(t) = 1$ in Eq. (3.5–10).

15. a) $\delta(t) - 2a \exp (-at)$ (*Note*: Divide first.);   **b)** $\delta(t) - \sin t$;

   c) $\delta(t) - \dfrac{1}{3} \exp (-t) + \exp \left(\dfrac{t}{2}\right)\left(\dfrac{1}{3} \cos \dfrac{\sqrt{3}}{2} t - \dfrac{\sqrt{3}}{3} \sin \dfrac{\sqrt{3}}{2} t\right)$

16. a) $\dfrac{1}{s(1 + \exp (-as))}$;   **b)** $\dfrac{1}{s(1 + \exp (as))}$;

   c) $\dfrac{1 - \exp (-as)}{as^2(1 + \exp (-as))}$;   **d)** $\dfrac{a(1 + \exp (-\pi s/a))}{(s^2 + a^2)(1 - \exp (-\pi s/a))}$

17. a) $f(t) = \dfrac{1}{a}(a - t), 0 < t < a, f(t + a) = f(t)$;

   c) $f(t) = \dfrac{a}{t}, 0 < t < a, f(t + a) = f(t)$

Section 3.6, p. 179

1. $Y(s) = \dfrac{y(0)}{5}\left(\dfrac{1}{s + 4} + \dfrac{4}{s - 3}\right) + \dfrac{y'(0)}{5}\left(\dfrac{1}{s - 1} - \dfrac{1}{s + 4}\right)$

   $+ \dfrac{3}{25}\left(\dfrac{1}{s + 4} - \dfrac{1}{s - 1} + \dfrac{5}{(s - 1)^2}\right)$

8. $X(s) = \dfrac{A}{s} + \dfrac{B}{s^2} + \dfrac{C}{s^3} + \dfrac{D}{s - 1} + \dfrac{Es + F}{s^2 - 6s + 13}$, where $A = 0.03368, B = 0.20118,$

   $C = 0.76923, D = -0.25000, E = 0.21632, F = -3.24910$;

   $Y(s) = \dfrac{A'}{s} + \dfrac{B'}{s^2} + \dfrac{C'}{s^3} + \dfrac{D's + E'}{s^2 - 6s + 13}$, where $A' = -0.08375, B' = -0.28402,$

   $C' = -0.61538, D' = 1.08375, E' = -0.21849$

9. a) $y = \cos 2t + \frac{1}{3} \sin t(1 - u_\pi(t)) - \frac{1}{6} \sin 2t(1 + u_\pi(t))$;

   c) $y = \cos 2t + \frac{1}{3} \sin t(1 - u_\pi(t)) + \frac{1}{3} \sin 2t(1 - \frac{1}{2}u_\pi(t))$

10. a) $y = \frac{3}{5}x \exp (x) + \frac{1}{5} \exp (-4x)$;

   c) $y = \frac{1}{85}(12 \exp (-4x) + 68 \exp (x) + 5 \cos x - 20 \sin x)$

11. a) $y = t \exp (t)$;   **b)** $y = t^2 \exp (t)$ or $y = Ct^2 \exp (t)$;

   c) $t^2 - 1$

12. a) $y(t) = \mathscr{L}^{-1}\left(\dfrac{F(s)}{s + a}\right) + y_0 \exp (-at)$;

   b) provided the inverse transform exists

13. $t^2/2$; note that the transformed problem results in a first-order differential equation in $Y(s)$; the arbitrary constant in the solution of this must be zero to satisfy $\lim_{s \to \infty} Y(s) = 0$

14. $x = 2 - \exp (t), y = -2 + (4 - t) \exp (t), z = -2 + (5 - t) \exp (t)$

15. The transformed problem becomes $\dfrac{dY}{ds} + \dfrac{s}{s^2 + 1} Y = 0$, which is a first-order differential equation in which the variables can be separated.

16. $Y(s) = \dfrac{my_0 s + mv_0}{ms^2 + k}$

17. $Y(s) = \dfrac{mv_0}{ms^2 + cs + k}$

18. $x = \exp(t)$, $y = \exp(-t)$, $z = 2 \sinh t$

19. a) $1 + \exp(-2t)(2 \sin t)$

20. a) $u_0(t)(1 - \exp(-t)) - u_a(t)(1 - \exp(a - t))$

22. $\sin at$

25. $i_1 = -\frac{1}{2}(\exp(-t/3) + 5 \exp(-t)) + 3 \cos t + 4 \sin t$;
$i_2 = \frac{1}{2}(\exp(-t/3) - 5 \exp(-t)) + 2 \cos t + \sin t$

26. $i_1 = -\frac{1}{3}(5 \exp(-30t) + 15 \exp(-10t) - 20)$;
$i_2 = \frac{1}{3}(5 \exp(-30t) - 15 \exp(-10t) + 10)$

CHAPTER 4

Section 4.1, p. 192

5. a) $\begin{pmatrix} 4 & 2 \\ 2 & 8 \end{pmatrix}$, $\begin{pmatrix} 2 & -2 \\ -4 & 0 \end{pmatrix}$, $\begin{pmatrix} -2 & 2 \\ 4 & 0 \end{pmatrix}$;

c) $\begin{pmatrix} 9 & 6 \\ 7 & 20 \end{pmatrix}$, $(44, 64)$, $\begin{pmatrix} 34 & 52 \\ 94 & 124 \end{pmatrix}$, $\begin{pmatrix} 9 & 0 \\ -7 & 16 \end{pmatrix}$

6. b) $\begin{pmatrix} 3 & 5 & 10 \\ 5 & 9 & 2 \\ 2 & 1 & 2 \end{pmatrix}$, $\begin{pmatrix} -2 & 6 & -7 \\ 0 & 6 & 2 \\ 3 & 3 & -4 \end{pmatrix}$, $\begin{pmatrix} 19 & 0 & 17 \\ 6 & 5 & -4 \\ 15 & -4 & 19 \end{pmatrix}$;

d) $\begin{pmatrix} 32 & 12 & 6 \\ 12 & -4 & 12 \\ 6 & 12 & 5 \end{pmatrix}$, $\begin{pmatrix} 10 & 2 & 2 \\ 2 & 6 & 0 \\ 2 & 0 & 5 \end{pmatrix}$, $\begin{pmatrix} 0 & 0 & 0 \\ 0 & 0 & 0 \\ 0 & 0 & 0 \end{pmatrix}$

7. $\begin{pmatrix} 17 \\ 6 \end{pmatrix}$, $\begin{pmatrix} 3 \\ -14 \\ -1 \end{pmatrix}$, $\begin{pmatrix} 20 & 8 \\ 15 & 24 \end{pmatrix}$

9. a) $\begin{pmatrix} -1 & 2 & 1 \\ 0 & -2 & 0 \\ 1 & 2 & -1 \end{pmatrix}$, $\begin{pmatrix} -2 & -2 & 2 \\ 2 & 0 & -2 \\ -2 & 2 & 2 \end{pmatrix}$, $\begin{pmatrix} 4 & 4 & -4 \\ -4 & 0 & 4 \\ 4 & -4 & -4 \end{pmatrix}$, $16A^2$

b) O, the 3×3 zero matrix

10. If $CA = A$ and $AC = A$, then C is the identity.

11. Compute $B^2 - 7B + 6I$ and show that the result is the 2×2 zero matrix.

12. $\begin{pmatrix} a_{11} & a_{12} \\ 0 & 0 \end{pmatrix}$ with a_{11}, a_{12} arbitrary

13. $B = \begin{pmatrix} a_{11} & a_{12} \\ 0 & a_{11} + 2a_{12} \end{pmatrix}$

14. $A = \begin{pmatrix} a_{11} & a_{12} & a_{13} \\ 0 & 0 & 1 \end{pmatrix}$

16. a) $A^2 - AB + BA - B^2$;   **b)** if A and B commute

27. a) To show that AA^T is symmetric we must show that $AA^T = (AA^T)^T$;
   c) $(A - A^T)^T = A^T - A = -(A - A^T)$, showing that $A - A^T$ is skew-symmetric

Section 4.2, p. 207

2. 96

3. In part (a) multiply the second row by -2 and add the result to the first row; multiply the third row by -5 and add the result to the second row; multiply the third row by 7 and add the result to the first row. In part (c) multiply the second row by -4 and add the result to the first row.

4. To reduce to row echelon form: In part (d) multiply the second row by -1 and add the result to the third row; then multiply the third row by $-1/5$. In part (e) interchange the third and fourth rows. In part (f) interchange the first and second rows.

9. a) $\begin{pmatrix} a_{21} & a_{22} & a_{23} \\ a_{11} & a_{12} & a_{13} \\ a_{31} & a_{32} & a_{33} \end{pmatrix}$;   **d)** $\begin{pmatrix} a_{21} & a_{22} & a_{23} \\ a_{11} & a_{12} & a_{13} \\ a_{31} + ca_{21} & a_{32} + ca_{22} & a_{33} + ca_{23} \end{pmatrix}$

13. a) $\{(1, 2, 3)\}$;   **c)** $\{(2c, -5c, c)\}$

14. a) No solution; inconsistent system;   **b)** $\{(0, 0, 0)\}$

15. a) $\{(-2, 3)\}$;   **b)** $\{(0, 0, 0)\}$;   **c)** $\{(2, -1)\}$;   **d)** $\{(1, -3, -2)\}$

17. a) 12;   **b)** 1;   **c)** -42;   **d)** 27;   **e)** 8.3×10^{-2};
   f) 4.6×10^{-4};   **g)** -12;   **h)** 51

18. a) $\begin{pmatrix} 1 & 0 & 0 & 0 \\ 0 & 0 & 0 & 1 \\ 0 & 0 & 1 & 0 \\ 0 & 1 & 0 & 0 \end{pmatrix}$;   **b)** $\begin{pmatrix} 1 & 0 & 0 & 0 \\ 0 & 1 & 0 & 0 \\ 0 & 0 & c & 0 \\ 0 & 0 & 0 & 1 \end{pmatrix}$;

   c) $\begin{pmatrix} 1 & 0 & k & 0 \\ 0 & 1 & 0 & 0 \\ 0 & 0 & 1 & 0 \\ 0 & 0 & 0 & 1 \end{pmatrix}$;   **d)** $\begin{pmatrix} 1 & 0 & 0 & 0 \\ 0 & 1 & 0 & 0 \\ 0 & -2 & 1 & 0 \\ 0 & 3 & 0 & 1 \end{pmatrix}$

19. a) $\begin{pmatrix} 1 & 0 & 0 & 0 \\ 0 & 0 & 0 & 1 \\ 0 & 0 & 1 & 0 \\ 0 & 1 & 0 & 0 \end{pmatrix}$;   **b)** $\begin{pmatrix} 1 & 0 & 0 & 0 \\ 0 & 1 & 0 & 0 \\ 0 & 0 & 1/c & 0 \\ 0 & 0 & 0 & 1 \end{pmatrix}$;

   c) $\begin{pmatrix} 1 & 0 & -k & 0 \\ 0 & 1 & 0 & 0 \\ 0 & 0 & 1 & 0 \\ 0 & 0 & 0 & 1 \end{pmatrix}$;   **d)** $\begin{pmatrix} 1 & 0 & 0 & 0 \\ 0 & 1 & 0 & 0 \\ 0 & 2 & 1 & 0 \\ 0 & -3 & 0 & 1 \end{pmatrix}$

22. $\{(23/9, 31/18, -19/18)^T\}$ or approximately $\{(2.556, 1.722, -1.056)^T\}$

23. a) and **b)** are singular

25. a) $\{(-6.3333c, -4.8182c, -2.3939c, c)^T\};$
   c) $x_1 = -2x_2 - \frac{5}{2}x_4 + \frac{7}{2}, x_3 = \frac{1}{2}x_4 + \frac{1}{2}, x_2$ and x_4 arbitrary

26. c) $\{(1.45, -1.59, -0.27)^T\}$

27. a) $4b - a = 3;$ **b)** $4b - a \neq 3$

28. a) No solution;   **c)** $\{(8, 0, 20)\}$

29. $A = A^T \Rightarrow A^{-1} = (A^T)^{-1} = (A^{-1})^T$, hence A^{-1} is symmetric

30. $A^{-1}BA = A^{-1}AB = B \Rightarrow A^{-1}BAA^{-1} = BA^{-1}$, showing that $A^{-1}B = BA^{-1}$

Section 4.3, p. 222

1. The set is not closed under addition nor under scalar multiplication.

5. a) $c\mathbf{u} = \mathbf{0}$ only if $c = 0;$ **b)** $c\mathbf{0} = \mathbf{0}$ holds for nonzero c;
   c) $c_1\mathbf{0} + c_2\mathbf{u}_2 + c_3\mathbf{u}_3 + \cdots + c_n\mathbf{u}_n = \mathbf{0}$ holds for $c_1 \neq 0$ regardless of the values of the remaining c_i.

6. $c_1(1, 0, 0) + c_2(1, 1, 0) + c_3(1, 1, 1) = (0, 0, 0)$ leads to

$$\begin{cases} c_1 + c_2 + c_3 = 0, \\ \quad\;\; c_2 + c_3 = 0, \\ \quad\quad\;\;\; c_3 = 0, \end{cases}$$

which have the unique solution $c_1 = c_2 = c_3 = 0$.

7. a) $a(1, 0, 0) + b(1, 1, 0) + c(1, 1, 1) + d(-1, 2, 3) = (x_1, x_2, x_3)$ leads to

$$\begin{pmatrix} 1 & 1 & 1 & -1 & x_1 \\ 0 & 1 & 1 & 2 & x_2 \\ 0 & 0 & 1 & 3 & x_3 \end{pmatrix} \sim \begin{pmatrix} 1 & 0 & 0 & -3 & x_1 - x_2 \\ 0 & 1 & 0 & -1 & x_2 - x_3 \\ 0 & 0 & 1 & 3 & x_3 \end{pmatrix}$$

so that, taking $d = 0$, we have $a = x_1 - x_2$, $b = x_2 - x_3$, $c = x_3$, which shows that any vector in $\mathbb{R}^3$ can be uniquely expressed as a linear combination of $(1, 0, 0)$, $(1, 1, 0)$, and $(1, 1, 1)$.

8. b) $a(1, 0, 0) + b(1, 1, 1) = (x_1, x_2, x_3)$ cannot be solved uniquely for a and b unless $x_2 = x_3$

9. $L(u_1, u_2, u_3) + L(v_1, v_2, v_3) = L(u_1 + v_1, u_2 + v_2, 0)$ and also $L[(u_1, u_2, u_3) + (v_1, v_2, v_3)] = L(u_1 + v_1, u_2 + v_2, 0)$. Further, $L(cu_1, cu_2, cu_3) = L(cu_1, cu_2, 0) = cL(u_1, u_2, u_3)$.

10. Show that ker L is closed under addition and scalar multiplication.

12. a) Suppose $L(u_1, u_2) = (u_2, u_1, 1) = (v_1, v_2, v_3)$. Then $v_1 = u_2$, $v_2 = u_1$, $v_3 = 1$, showing that the mapping is single-valued.

   c) A point in the $u_1 u_2$-plane is projected into the plane $u_3 = 1$ along a line parallel to the u_3-axis, then reflected in the plane $u_1 = u_2$.

13. a) and **b)** span $\mathbb{R}^3$.

14. b) For example, $a(1, 1, 1) + b(1, 2, 3) + c(2, 2, 0) = (1, 0, 0)$ leads to $3\mathbf{v}_1 - \mathbf{v}_2 - \frac{1}{2}\mathbf{v}_3 = \mathbf{e}_1$.

18. The vectors span a subspace consisting of all vectors of the form $(x_1, x_2, 0)$, a subspace of dimension two.

19. **a)** and **c)**

20. $a(1, 0, 0, -1) + b(0, 1, 0, 1) + c(0, 0, 1, 1) = (x_1, x_2, x_3, x_4)$ leads to $a = x_1$, $b = x_2, c = x_3, x_4 = -x_1 + x_2 + x_3$.

21. **b)** The transformation is a reflection of a point in the xy-plane with respect to the y-axis.

22. **a)** The zero vector $(0, 0, 0)$ is a basis for the kernel; the set $\{(0, 0, 1),$ $(0, 1, 0), (1, 0, 0)\}$ forms a basis for the range.
  c) The vector $(1, -1, -1)$ is a basis for the kernel; the set $\{(1, 0, -3),$ $(0, 1, 1)\}$ forms a basis for the range.
  e) The transformation can be written

$$(x_1, x_2, x_3)\begin{pmatrix} 0 & 1 & 0 & 2 \\ 0 & 1 & 1 & 0 \\ 0 & 1 & -1 & 4 \end{pmatrix} = (0, x_1 + x_2 + x_3, x_2 - x_3, 2x_1 + 4x_3).$$

A basis for the kernel is the vector $(-2, 1, 1)$. A basis for the range is the set $\{(0, 1, 0, 2), (0, 0, 1, -2)\}$.

24. **a)** A plane determined by the three noncollinear points $(0, 0, 0), (-1, 2, 1)$, and $(2, 1, 3)$.

26. **a)** $0(1, 1, 0, 0) + 2(1, 0, 1, 0) - 3(0, 1, 1, 0) + 3(0, 1, 0, 1) = (2, 0, -1, 3)$

27. **b)**, **c)**, and **d)** are linearly dependent.

29. **a)** 3; **b)** 2

30. **a)**, **b)**, and **c)** are linearly independent.

31. **a)** Rotation through an angle $(3\pi/2) - 2\theta$, where $\theta = \arctan(x_2/x_1)$;
  c) reflection in the origin;
  e) projection on the x_1-axis along a line having a slope of -1

33. $\{(-1, 1, 0, 0), (-1, 0, 1, 0), (-1, 0, 0, 1)\}$

36. One pair is $(0, 1, 0, 0)$ and $(0, 0, 1, 0)$; another pair is $(1, 0, 0, 0)$ and $(0, 1, 0, 0)$.

37. **a)** $LM(x_1, x_2, x_3) = (x_1, x_2, -x_3)$,
  $ML(x_1, x_2, x_3) = (x_1, x_2, -2x_1 - 2x_2 - x_3)$;
  b) $LM(x_1, x_2, x_3) = (x_2, 0, x_2)$,
  $ML(x_1, x_2, x_3) = (2x_1 + 2x_2 + 2x_3, x_1 + x_3, -2x_1 - 2x_2 - 2x_3)$;
  neither pair commutes

39. $\{(1, 0, 0, 1), (0, 1, 0, -1), (0, 0, 1, 0)\}$

Section 4.4, p. 234

2. $\lambda_1 = 1, \lambda_2 = 1 + i, \lambda_3 = 1 - i$

4. A linear combination of $(1, 0, 1)^T$ and $(1, 1, 0)^T$ has the form

$$(x_3 + x_2, x_2, x_3)^T.$$

We need to show that vectors of this form are closed under addition and scalar multiplication. See Theorem 4.3–1.

5. Expand the determinant using the elements of the *third* row.

9. **a)** $\lambda_1 = -1, (1, -1)^T; \lambda_2 = 3, (1, 1)^T$;
  c) $\lambda_1 = 0, (2, -1)^T; \lambda_2 = 3, (1, 1)^T$

10. **a)** $\lambda_1 = -1, (6, 2, -7)^T$; $\lambda_2 = 1, (0, 1, -1)^T$; $\lambda_3 = 4, (3, 1, -1)^T$;
c) $\lambda_1 = 0, (1, 1, 1)^T$; $\lambda_2 = 1, (1, -1, 2)^T$; $\lambda_3 = 2, (2, 1, 2)^T$

11. **a)** $\lambda_1 = -1, (0, 1, -1)^T$; $\lambda_2 = 3, (1, 0, 0)^T$; $\lambda_3 = 7, (0, 1, 1)^T$;
c) $\lambda_1 = \lambda_2 = -2, (1, 0, -1)^T$ and $(1, -1, 0)^T$; $\lambda_3 = 4, (1, 1, 1)^T$

12. **a)** $\lambda_1 = 1, (4, 3, -1)^T$; $\lambda_2 = -i, (1 - i, 1, -1)^T$; $\lambda_3 = i, (1 + i, 1, -1)^T$;
c) $\lambda_1 = \lambda_2 = 0, (1, 0, -1)^T$ and $(1, -1, 0)^T$; $\lambda_3 = 3, (1, 1, 1)^T$

13. **a)** $\lambda_1 = \lambda_2 = 1, (1, 3, 0)^T$ and $(0, -2, 1)^T$; $\lambda_3 = 2, (2, 1, 2)^T$;
b) $\lambda_1 = \lambda_2 = -3, (-2, 1, 0)^T$ and $(3, 0, 1)^T$; $\lambda_3 = 5, (1, 2, -1)^T$;
d) $\lambda_1 = \lambda_2 = -2, (1, 0, 2)^T$ and $(0, 1, 2)^T$; $\lambda_3 = 7, (2, 2, -1)^T$;
f) $\lambda_1 = \lambda_2 = \lambda_3 = 1, (1, 0, 0)^T$;   **g)** $\lambda_1 = \lambda_2 = \lambda_3 = 1, (-3, 1, 1)^T$

14. **a)** $\lambda_1 = 0, (1, 0, 0)^T$; $\lambda_2 = 1, (0, 1, 0)^T$; $\lambda_3 = 2, (1, 0, 1)^T$;
b) $\lambda_1 = -\sqrt{2}, (1, 0, -1 - \sqrt{2})^T$; $\lambda_2 = \sqrt{2}, (1, 0, \sqrt{2} - 1)^T$; $\lambda_3 = 2, (0, 1, 0)^T$;
c) $\lambda_1 = 1, (3, -1, 3)^T$; $\lambda_2 = \lambda_3 = 2, (2, 1, 0)^T$ and $(2, 0, 1)^T$;
d) $\lambda_1 = 0, (3, -2, 1)^T$; $\lambda_2 = (5 + \sqrt{5})/2, (2, 2, \sqrt{5} - 1)^T$; $\lambda_3 = (5 - \sqrt{5})/2$,
$(2, 2, -\sqrt{5} - 1)^T$;
e) all are diagonalizable

15. **b)** $\lambda_1 = -1, (1, 0, 0, 0)^T$; $\lambda_2 = 0, (2, 1, 0, 0)^T$; $\lambda_3 = 2, (1, 3, -3, 0)^T$;
$\lambda_4 = 4, (7, 10, -15, 10)^T$

16. **a)** $P = \dfrac{\sqrt{2}}{2}\begin{pmatrix} 1 & -1 \\ 1 & 1 \end{pmatrix}$; diag $(4, 2)$;   **b)** $P = \dfrac{\sqrt{5}}{5}\begin{pmatrix} 2 & 1 \\ -1 & 2 \end{pmatrix}$; diag $(3, -2)$;

d) $P = \dfrac{\sqrt{2}}{2}\begin{pmatrix} 1 & 1 \\ 1 & -1 \end{pmatrix}$; diag $(0, 2)$

17. **b)** $P = \begin{pmatrix} \dfrac{-1}{\sqrt{6}} & \dfrac{1}{\sqrt{3}} & \dfrac{1}{\sqrt{2}} \\ \dfrac{1}{\sqrt{6}} & \dfrac{-1}{\sqrt{3}} & \dfrac{1}{\sqrt{2}} \\ \dfrac{2}{\sqrt{6}} & \dfrac{1}{\sqrt{3}} & 0 \end{pmatrix}$; diag $(3, 6, 9)$;

d) $P = \begin{pmatrix} \dfrac{2}{\sqrt{5}} & 0 & \dfrac{-1}{\sqrt{5}} \\ 0 & 1 & 0 \\ \dfrac{1}{\sqrt{5}} & 0 & \dfrac{2}{\sqrt{5}} \end{pmatrix}$; diag $(0, 2, 5)$

18. The characteristic equation is $\lambda^3 - 18\lambda^2 + 99\lambda - 162 = 0$, which can be factored as $(\lambda - 3)(\lambda - 6)(\lambda - 9) = 0$.

Section 4.5, p. 248

6. It is sufficient to show that the determinant of the matrix whose rows are the given vectors is different from zero.

15. $\lambda_1 = \dfrac{k}{2m}(-3 + \sqrt{13}), (2, 1 + \sqrt{13})^T$; $\lambda_2 = \dfrac{k}{2m}(-3 - \sqrt{13}), (2, 1 - \sqrt{13})^T$

16. **a)** $\mathbf{x} = c_1 \begin{pmatrix} 1 \\ -1 \end{pmatrix} \exp(t) + c_2 \begin{pmatrix} 2 \\ 1 \end{pmatrix} \exp(4t)$;

c) $\mathbf{x} = c_1 \begin{pmatrix} 1 \\ -1 \end{pmatrix} \exp(3t) + c_2 \begin{pmatrix} 1 \\ 2 \end{pmatrix} \exp(-3t);$

e) $\mathbf{x} = c_1 \begin{pmatrix} -4 \\ 5 \\ 7 \end{pmatrix} \exp(-2t) + c_2 \begin{pmatrix} -3 \\ 4 \\ 2 \end{pmatrix} \exp(-t) + c_3 \begin{pmatrix} 0 \\ 1 \\ -1 \end{pmatrix} \exp(2t)$

17. a) $\mathbf{x} = (\exp(t), 0)^T;$

c) $\mathbf{x} = 4 \begin{pmatrix} 3 \\ 1 \end{pmatrix} \exp(2t) - 8 \begin{pmatrix} 1 \\ -1 \end{pmatrix} \exp(-2t);$

e) $\mathbf{x} = \dfrac{2}{3} \begin{pmatrix} -1 \\ 1 \\ 1 \end{pmatrix} \exp(-2t) - \dfrac{13}{6} \begin{pmatrix} -1 \\ 4 \\ 1 \end{pmatrix} \exp(t) + \dfrac{3}{2} \begin{pmatrix} 1 \\ 2 \\ 1 \end{pmatrix} \exp(3t)$

18. a) $\mathbf{x} = c_1 \begin{pmatrix} 3 \\ 1 \end{pmatrix} \exp(2t) + c_2 \begin{pmatrix} -1 \\ 1 \end{pmatrix} \exp(-2t) - \dfrac{1}{5} \begin{pmatrix} 1 \\ 2 \end{pmatrix} \sin t + \dfrac{1}{5} \begin{pmatrix} 2 \\ -1 \end{pmatrix} \cos t;$

b) $\mathbf{x} = c_1 \begin{pmatrix} 1 \\ 1 \end{pmatrix} \exp(t) + c_2 \begin{pmatrix} 1 \\ 3 \end{pmatrix} \exp(-t) - \begin{pmatrix} 1 \\ 2 \end{pmatrix} + \begin{pmatrix} 0 \\ 1 \end{pmatrix} t$

21. b) $\mathbf{x} = c_1 \begin{pmatrix} 7 \\ 1 \end{pmatrix} \exp(5t) + c_2 \begin{pmatrix} 1 \\ 1 \end{pmatrix} \exp(-t);$

c) $\mathbf{x} = c_1 \begin{pmatrix} 1 \\ 1 \end{pmatrix} \exp(t) + c_2 \begin{pmatrix} 1 \\ 3 \end{pmatrix} \exp(-t)$

22. b) $\mathbf{x} = c_1 \begin{pmatrix} -1 \\ 4 \\ 1 \end{pmatrix} \exp(t) + c_2 \begin{pmatrix} 1 \\ -1 \\ -1 \end{pmatrix} \exp(-2t) + c_3 \begin{pmatrix} 1 \\ 2 \\ 1 \end{pmatrix} \exp(3t)$

23. a) $\mathbf{x} = c_1 \begin{pmatrix} 1 \\ -1 \end{pmatrix} \exp(t) + c_2 \left(\begin{pmatrix} 3 \\ -3 \end{pmatrix} t \exp(t) + \begin{pmatrix} 0 \\ -1 \end{pmatrix} \exp(t) \right);$

b) $\mathbf{x} = c_1 \left(\begin{pmatrix} 3 \\ 5 \end{pmatrix} \cos t - \begin{pmatrix} 1 \\ 0 \end{pmatrix} \sin t \right) + c_2 \left(\begin{pmatrix} 1 \\ 0 \end{pmatrix} \cos t + \begin{pmatrix} 3 \\ 5 \end{pmatrix} \sin t \right);$

c) $\mathbf{x} = c_1 \begin{pmatrix} 1 \\ -1 \end{pmatrix} \exp(2t) + c_2 \left(\begin{pmatrix} 1 \\ -1 \end{pmatrix} t \exp(2t) + \begin{pmatrix} 0 \\ -1 \end{pmatrix} \exp(2t) \right);$

d) $\mathbf{x} = c_1 \begin{pmatrix} 2 \\ -4 \end{pmatrix} + c_2 \left(\begin{pmatrix} 2 \\ -4 \end{pmatrix} t + \begin{pmatrix} 0 \\ 1 \end{pmatrix} \right)$

24. b) $\mathbf{x} = c_1 \begin{pmatrix} 0 \\ 1 \\ -1 \end{pmatrix} \exp(2t) + c_2 \left(\begin{pmatrix} 0 \\ 1 \\ -1 \end{pmatrix} t \exp(2t) + \begin{pmatrix} 1 \\ 0 \\ 1 \end{pmatrix} \exp(2t) \right)$

$+ c_3 \left(\begin{pmatrix} 0 \\ 1 \\ -1 \end{pmatrix} t^2 + \begin{pmatrix} 2 \\ 1 \\ 1 \end{pmatrix} t + \begin{pmatrix} 3 \\ 0 \\ 5 \end{pmatrix} \right) \exp(2t)$

26. a) $\mathbf{x} = c_1 \begin{pmatrix} 1 \\ 3 \end{pmatrix} t^2 + c_2 \begin{pmatrix} 1 \\ 1 \end{pmatrix} t^4$

27. a) $\mathbf{x} = c_1 \begin{pmatrix} \sin at \\ \cos at \end{pmatrix} + c_2 \begin{pmatrix} \cos at \\ -\sin at \end{pmatrix};$ **b)** $x_1^2 + x_2^2 = c_1^2 + c_2^2$

Section 4.6, p. 257

4. **a)** $x_1 = 1.200$, $x_2 = -0.400$, $x_3 = 0.200$;
  b) $x_1 = 1.125$, $x_2 = -0.344$, $x_3 = 0.125$;
  c) $x_1 = 1.189$, $x_2 = -0.395$, $x_3 = 0.198$

6. $x_1 = -0.591$, $x_2 = -1.340$, $x_3 = 4.500$, $x_4 = 3.477$

7. $x_1 = -1.500$, $x_2 = -3.625$, $x_3 = -2.875$

8. **a)** $x_1 = 1.00$, $x_2 = 1.09$, $x_3 = 0.94$; **b)** $x_1 = x_2 = x_3 = 1.00$

9. **a)** $x_1 = -1496.000$, $x_2 = 2.000$, $x_3 = 0.000$; **b)** $x_1 = x_2 = x_3 = 1.000$;
  c) The discrepancy is due to large round-off errors especially in the third row in part (a).

10. **b)** $x_1 = 0.880$, $x_2 = -2.35$, $x_3 = -2.66$;
  c) $x_1 = 0.9998$, $x_2 = 1.9995$, $x_3 = -1.002$

12. **a)** 55.385; **b)** $(0.464, 0.733, 0.323)^T$;
  c) $\begin{pmatrix} 0.190 & 0.260 & -0.107 \\ 0.009 & 0.184 & 0.149 \\ 0.173 & -0.110 & 0.144 \end{pmatrix}$

13. **a)** $\mathbf{x} = (2.556, 1.722, -1.056)^T$

14. $\mathbf{x} = (1.453, -1.589, -0.275)^T$

15. **a)** 7.004; **b)** $(1.000, 0.691, 1.812)^T$

16. **a)** $\lambda_1 = 4$, $(1, 0, 0)^T$; $\lambda_2 = 2$, $(-1/2, 1, 0)^T$; $\lambda_3 = -1$, $(1/15, -1/3, 1)^T$;
  b) 4.053, $(1.000, 0.053, -0.001)^T$

17. 3.491, $(1.000, 0.936, 0.777)^T$

19. $\lambda_1 = 8.387$, $(0.808, 0.772, 1)^T$; $\lambda_2 = 4.487$, $(0.217, 1, -0.947)^T$;
  $\lambda_3 = 2.126$, $(1, -0.567, -0.370)^T$

20. 19.29 is the dominant eigenvalue.

21. The coefficient matrices in Exercises 4 and 5 are positive definite.

25. The intermediate eigenvalue can be found by using the trace and the determinant as shown in Eq. 4.6–4.

26. **a)** $|H| \doteq 1.65 \times 10^{-7}$

Section 4.7, p. 270

9. All are vector spaces.

11. $\{(\cos x, 0), (0, \sin x)\}$

12. **a)** Two; **b)** two; **c)** two; **d)** four

13. **a)** $\text{adj } A = \begin{pmatrix} -26 & 7 & 16 \\ -2 & 7 & -2 \\ -4 & -7 & -4 \end{pmatrix}$, $|A| = -42$;

  b) $\text{adj } A = \begin{pmatrix} 6 & 21 & -9 & -36 \\ -1 & 1 & 6 & -3 \\ 10 & 17 & -6 & -51 \\ 4 & -4 & 3 & 12 \end{pmatrix}$, $|A| = 27$

14. a) $\operatorname{adj} A = \begin{pmatrix} 8 & -8 & -8 \\ 2 & -4 & 6 \\ -6 & 4 & -2 \end{pmatrix}$, $|A| = -16$

15. b) $\begin{pmatrix} 1 & 1 & 0 \\ -1 & 0 & 1 \\ 0 & -1 & 1 \end{pmatrix}$; **c)** $\dfrac{1}{2} \begin{pmatrix} 1 & -1 & 1 \\ 1 & 1 & -1 \\ 1 & 1 & 1 \end{pmatrix}$

26. c) The space spanned by the zero vector, that is, a zero-dimensional vector space.

27. a) $\{x^3, x^2, x, 1\}$; **b)** $\begin{pmatrix} 3 & 0 & 0 & 0 \\ 0 & 2 & 0 & 0 \\ 0 & 0 & 1 & 0 \end{pmatrix}$

CHAPTER 5

Section 5.1, p. 291

2. Suppose $\mathbf{v} = a\mathbf{i} + b\mathbf{j} = c\mathbf{i} + d\mathbf{j}$. Then $(a - c)\mathbf{i} + (b - d)\mathbf{j} = \mathbf{0}$ and it follows that $a = c$ and $b = d$.

7. $3\mathbf{i} - 3\mathbf{j} + 8\mathbf{k}$

9. Consider $\mathbf{a} = \mathbf{i} + \mathbf{j} + \mathbf{k}$, $\mathbf{b} = \mathbf{i}$, $\mathbf{c} = \mathbf{j}$. Then $(\mathbf{a} \times \mathbf{b}) \times \mathbf{c} = \mathbf{i}$, whereas $\mathbf{a} \times (\mathbf{b} \times \mathbf{c}) = \mathbf{i} - \mathbf{j}$.

11. c) 8

12. See Exercise 11(b).

13. Expand the determinant and show that the resulting terms agree with those obtained from $(\mathbf{u} \times \mathbf{v}) \cdot \mathbf{w}$.

14. a) $1/2$; **c)** 2

15. a) 0; **c)** $\sqrt{66} \doteq 8.124$

16. No; because the sum of the squares of the direction cosines would be greater than one.

17. a) $\cos \alpha \doteq 0.5774$, $\alpha \doteq 0.9553$

20. a) -2; **c)** $-(\mathbf{i} + 3\mathbf{j})$

22. a) $((x - 1)\mathbf{i} + (y + 2)\mathbf{j} + z\mathbf{k}) \cdot (-\mathbf{i} + 5\mathbf{j} - 5\mathbf{k}) = 0$;
 b) $y = t$, $z = s$, $x = 5t - 2s + 11$ (note that a space *curve* may be expressed in terms of a single parameter but a plane is a *surface* and thus requires two parameters.);
 c) $x - 5y + 2z = 11$

23. b) $x = t$, $z = s$, $y = 2t + s - 5$; **c)** $2x - y + z = 5$

24. b) $\dfrac{x - 1}{3} = \dfrac{y - 2}{0} = \dfrac{z + 1}{-2}$ (note that $(y - 2)/0$ is to be interpreted as $y = 2$.);
 c) $x = 1 + 3t$, $y = 2$, $z = -1 - 2t$

27. $\mathbf{0}$; the vectors are perpendicular

28. The vector itself

30. $2x + 2y + z = 5$

31. $\cos \alpha = 3/\sqrt{13} \doteq 0.8321$, $\cos \beta = 0$, $\cos \gamma = -2/\sqrt{13} \doteq -0.5547$

32. -82

33. $\pm(\sqrt{29}/29)(2\mathbf{i} + 3\mathbf{j} + 4\mathbf{k})$

34. No

35. **c)** $10/\sqrt{29} \doteq 1.857$

Section 5.2, p. 305

6. See Fig. 5.2–2.

7. Use Eq. (5.2–10).

8. What kind of vector is **B** for a plane curve? Then use Eq. (5.2–12).

9. Similar to Exercise 8 using Eq. (5.2–9).

11. a) Let $\mathbf{r}_0 = x_0\mathbf{i} + y_0\mathbf{j} + z_0\mathbf{k}$ be the point of tangency. Then the position vector $\mathbf{r} = x\mathbf{i} + y\mathbf{j} + z\mathbf{k}$ of an arbitrary point on the tangent vector is given by $\mathbf{r} = \mathbf{r}_0 + t\mathbf{T}$, where t is a scalar parameter.

 c) If $\mathbf{r} = x\mathbf{i} + y\mathbf{j} + z\mathbf{k}$ is the position vector of an arbitrary point in the osculating plane, then $(\mathbf{r} - \mathbf{r}_0) \cdot \mathbf{B} = 0$, with $\mathbf{r}_0$ as in part (a).

12. $v = \sqrt{6}$, $v^2/\rho = \frac{36}{7}\sqrt{14} \doteq 19.243$, $dv/dt = 3\sqrt{6} \doteq 7.348$

13. $\mathbf{T} = \frac{1}{13}(3\mathbf{i} - 4\mathbf{j} + 12\mathbf{k})$

14. $\mathbf{T} = \dfrac{1}{\sqrt{5}}(\sin t\mathbf{i} + \cos t\mathbf{j} + 2\mathbf{k})$

15. $a_n = t$, $a_t = \sqrt{5}$

16. $\kappa = \dfrac{1}{5t}$

19. $\kappa = \tau = \dfrac{1}{3(1 + t^2)^2}$

20. $\tau = \dfrac{1}{9t^4 + t^2 + 1}$

21. a) -5;   **c)** $3x^2$

23. $\mathbf{r} = (2 + s)\mathbf{i} + 12(s + 1)\mathbf{j} + 4(2 + 3s)\mathbf{k}$, where s is a scalar parameter

28. $\rho^2 + s^2 = 16a^2$, where s is measured from the top of the arch, that is, $s = 0$ at $t = \pi$.

29. $d\mathbf{T}/ds = \kappa\mathbf{N} = \kappa(\mathbf{B} \times \mathbf{T}) = (\tau\mathbf{T} + \kappa\mathbf{B}) \times \mathbf{T} = \mathbf{D} \times \mathbf{T}$

Section 5.3, p. 316

2. The maximum value is $\sqrt{62} \doteq 7.874$; the minimum value is the negative of this.

6. The gradient is $2x\mathbf{i} - \mathbf{j}$, which is a vector making an angle with the x-axis whose tangent is $-1/2x$; the derivative is $2x$ which is the negative reciprocal of the slope of the gradient.

7. a) Concentric circles centered at the origin;

 c) straight lines of slope -1;

 e) rectangular hyperbolas (the x- and y-axes are asymptotes)

8. **a)** Concentric spheres centered at the origin;
   c) parallel planes with normals having direction numbers (a, b, c)

9. Tangent plane: $3x + 5y + 4z = 18$;

   normal line: $\dfrac{x-3}{3} = \dfrac{y-5}{5} = \dfrac{z+4}{4}$

10. Normal line: $\dfrac{x-3}{3} = \dfrac{y-4}{4} = \dfrac{z+5}{5}$

11. **a)** $4/7 \doteq 0.5714$;   **b)** $9/1183 \doteq 0.0076$;   **c)** $2/3 \doteq 0.6667$

12. $\cos \theta \doteq 0.8124$, $\theta \doteq 0.6878$

13. **a)** $\dfrac{-3}{196\sqrt{14}}(\mathbf{i} + 2\mathbf{j} + 3\mathbf{k}) \doteq -0.0041(\mathbf{i} + 2\mathbf{j} + 3\mathbf{k})$;

   c) $\left(\dfrac{-\sqrt{14}}{14}, \dfrac{-\sqrt{14}}{7}, \dfrac{-3\sqrt{14}}{14}\right) \doteq (-0.2673, -0.5345, -0.8018)$

14. $\mathbf{r}/|\mathbf{r}|$

15. $\cos y + y(2z - x)$

18. **b)** $y \cos (xy) - z \sin (yz) + 2x \cos (2xz)$;   **d)** $6xyz$

20. $2(xy + yz + xz)$

22. **a)** $(1 - z^2 - y^2) \cos (yz) \exp (x)$;   **c)** $4x$

24. They are all irrotational.

Section 5.4, p. 326

2. $\mathbf{i} = \cos \phi \mathbf{e}_\rho - \sin \phi \mathbf{e}_\phi$, $\mathbf{j} = \cos \phi \mathbf{e}_\phi + \sin \phi \mathbf{e}_\rho$

11. $\mathbf{i} = \sin \theta \cos \phi \mathbf{e}_r + \cos \theta \cos \phi \mathbf{e}_\theta - \sin \phi \mathbf{e}_\phi$,
    $\mathbf{j} = \sin \theta \sin \phi \mathbf{e}_r + \cos \theta \sin \phi \mathbf{e}_\theta + \cos \phi \mathbf{e}_\phi$, $\mathbf{k} = \cos \theta \mathbf{e}_r - \sin \theta \mathbf{e}_\theta$

16. $\rho^2 = z(4 - z)$; $r = 4 \cos \theta$

17. $x^2 + y^2 = z^2$; $\tan^2 \theta = 1$ or $\theta = \pi/4, 3\pi/4$

19. $h_u = h_v = a\sqrt{\sinh^2 u + \sin^2 v}$

20. $h_u = h_v = a\sqrt{\sinh^2 u + \sin^2 v}$, $h_\theta = a \sinh u \sin v$

21. $dV = uv(u^2 + v^2) \, du \, dv \, d\theta$

23. **a)** The vectors $\mathbf{w}_i$ are linearly independent since $\mathbf{w}_1 \times \mathbf{w}_2 \cdot \mathbf{w}_3 = 1 \neq 0$;
    b) $\mathbf{v}_1 = (1, 3, -1)$, $\mathbf{v}_2 = (-3, 2, 3)$, $\mathbf{v}_3 = (1, 0, 1)$;

    c) $\mathbf{u}_1 = \dfrac{\sqrt{11}}{11}(1, 3, -1)$, $\mathbf{u}_2 = \dfrac{\sqrt{22}}{22}(-3, 2, 3)$, $\mathbf{u}_3 = \dfrac{\sqrt{2}}{2}(1, 0, 1)$;

    d) $\mathbf{w}_2 = \dfrac{3\sqrt{11}}{11}\mathbf{u}_1 + \dfrac{\sqrt{22}}{11}\mathbf{u}_2$

25. $\rho = \sqrt{x^2 + y^2}$, $\phi = \arctan (y/x)$, $z = z$

Section 5.5, p. 345

4. Compute $\mathbf{V} \times \mathbf{F}$ to obtain $\mathbf{0}$.

5. Zero

9. $\dfrac{\pi a^2}{4}(b^2 - 2)$

14. **b)** Zero;   **d)** $16\sqrt{2}/3 \doteq 7.5425$

15. **a)** $\sin 1 \doteq 0.8415$;   **c)** $\frac{1}{2} + \sin 1 - \frac{1}{2}\cos 1 \doteq 1.0713$

16. **a)** $f(x, y, z) = \exp(xyz) + xz + C$;   **b)** not conservative;
    c) $f(x, y, z) = -\arctan(y/z),\ z > 0$;   **d)** not conservative

17. **a)** 2π;   **b)** 0;   **c)** 2π;
    d) no, because the work for a particular path may be zero even though the field is not irrotational. The field in part (b), however, *is* irrotational.

18. **b)** 0;   **c)** 0;   **d)** -3

19. $\sqrt{3}\pi \doteq 5.4414$

20. 8π

21. -8π

22. **a)** 24π;   **b)** 135π;   **c)** $4\pi a^3$

23. The volume integral to be evaluated is

$$\int_0^2 \int_0^3 \int_0^{4-x^2} (x + 2y)\, dz\, dy\, dx$$

or an equivalent of this.

24. The volume integral to be evaluated is

$$\int_0^2 \int_0^{2-x} \int_0^{4-2x-2y} (2x - 4)\, dz\, dy\, dx$$

or an equivalent of this.

26. **b)** $f(x, y, z) = x^2 yz + C$

Section 5.6, p. 360

5. Take $\mathbf{F} = az\mathbf{k}$ and use the divergence theorem.

9. **a)** Since $\mathbf{r} = x\mathbf{i} + y\mathbf{j} + z\mathbf{k}$, $\nabla \cdot \mathbf{r} = 3$.

10. **a)** 0;   **b)** 24;   **c)** 0

11. $1/8$

12. Use the divergence theorem and subtract the surface integral over the missing face, that is, subtract $1/12$ from the result in Exercise 11.

15. -8π

16. **a)** 0;   **c)** 0

17. **a)** 4π

19. 20π

20. -4π

23. πab

24. $3\pi a^2$

25. 12π

28. $M(x, y)$ and $N(x, y)$ are not continuous at the origin.

Section 5.7, p. 374

8. ρ is in g/cm^3, k is in cal/cm sec °C, and h^2 is in cm^2/sec

13. See Theorem 5.3–3(g).

14. First, obtain

$$\nabla f(r) = f'(r)\left(\frac{\partial r}{\partial x}\mathbf{i} + \frac{\partial r}{\partial y}\mathbf{j} + \frac{\partial r}{\partial z}\mathbf{k}\right) = \frac{f'(r)\mathbf{r}}{r};$$

then find the divergence of

$$\frac{f'(r)(x\mathbf{i} + y\mathbf{j} + z\mathbf{k})}{\sqrt{x^2 + y^2 + z^2}}.$$

17. Use the definition of $\nabla \times$; then add and subtract the terms

$$\frac{\partial^2 A_1}{\partial x^2}\mathbf{i}, \qquad \frac{\partial^2 A_2}{\partial y^2}\mathbf{j}, \qquad \frac{\partial^2 A_3}{\partial z^2}\mathbf{k},$$

where $\mathbf{A} = A_1\mathbf{i} + A_2\mathbf{j} + A_3\mathbf{k}$.

19. See Theorem 5.3–3(e).

22. a) Due west;   **b)** due south

CHAPTER 6

Section 6.1, p. 385

10. a) $F\left(z, \dfrac{x+z}{y+z}\right) = 0;$ **b)** $F\left(\dfrac{y}{x}, x^2 + y^2 + z^2\right) = 0;$

c) $F(2y - x^2, z \exp(-x)) = 0;$ **d)** $F(2x - y^2, z \exp(-y)) = 0;$

f) $F\left(\dfrac{xy}{z}, \dfrac{x-y}{z}\right) = 0;$ **g)** $F(x^2 + y^2 - z^2, xy + z) = 0;$ **h)** $z = f\left(\dfrac{y}{x}\right);$

i) $F\left(\dfrac{y}{x}, \dfrac{z}{x}\right) = 0$

11. a) $z = f(x)\exp(-y^2);$ **b)** $z = f(y)\exp(x^2 y);$
   c) $z = -x\cos(y/x) + f(x);$ **d)** $z = y^3/3 + x^2 y + f(x);$
   e) $z = x^3/3 + f(2x + y)$

12. $z = (x + y)^2$

13. a) $F(x - z, y - z^2/2) = 0;$ **b)** $z = 1 + \sqrt{1 - 2(x - y)}$

14. a) $F(xy, x^2 + y^2 + z^2) = 0$

16. a) $2z_x + z_y = 0;$ **b)** $yz_y - z = 0;$ **c)** $zz_x + yz_y = x;$
   d) $x^2 z_x + y^2 z_y = xy$

19. a) $z = f(x^3 y^2)$

22. $z = f(bx - ay)$

23. a) $x = ky$, planes containing the z-axis

Section 6.2, p. 394

5. **b)** Elliptic when $x > 0$; hyperbolic when $x < 0$; parabolic when $x = 0$;
   d) elliptic outside the unit circle with center at the origin; hyperbolic inside this unit circle; parabolic on the unit circle.

11. **b), d)**, and **e)** are not linear

17. **a)** Parabolic; **b)** hyperbolic; **c)** elliptic;
    d) parabolic if $x = 0$ or $y = 0$, otherwise hyperbolic

Section 6.3, p. 403

7. $u(x, y) = \dfrac{3 \sin x \sinh (b - y)}{\sinh b}$

8. **a)** 3; **b)** 0; **c)** 0; **d)** 0.9721

12. $u_n(x, y) = \dfrac{B_n}{\cosh (n\pi^2/b)} \sin \dfrac{n\pi}{b} y \sinh \dfrac{n\pi}{b} (x - \pi)$ with $u_n(0, y) = g_1(y)$

Section 6.4, p. 411

4. $y(x, t) = \frac{1}{2}(\sin (x + at) + \sin (x - at)) = \sin x \cos at$

5. $y(x, t) = \dfrac{1}{a} \cos x \sin at$

13. $y(x, t) = 3 \sin \dfrac{2\pi}{L} x \cos \dfrac{2\pi at}{L}$ (Note that $\exp (i\omega t) = \cos \omega t + i \sin \omega t$ and the condition $y_t(x, 0) = 0$ requires that only the cosine term be retained.)

14. $y(x, t) = \dfrac{2L}{3\pi a} \sin \dfrac{3\pi x}{L} \sin \dfrac{3\pi at}{L}$

17. The boundary-value problem becomes (see Exercise 16) $a^2 y''(x) - g = 0$, $y(0) = y(L) = 0$, whose solution is $4g(x - L/2)^2 = 8a^2 y + gL^2$, which is a parabola. Differentiating, we find $dy/dx = 0$ when $x = L/2$ and the maximum displacement is $-gL^2/8a^2$.

CHAPTER 7

Section 7.1, p. 428

9. If $f(x)$ is odd, then $f(x) = -f(-x)$. If $g(x)$ is even, then $g(x) = g(-x)$. Hence $f(x)g(x) = -f(-x)g(-x)$, showing that the product $f(x)g(x)$ is odd.

11. **a)** 4π; **c)** 2/3; **e)** 4

12. **a)** $u(x, y) = \dfrac{4}{\pi} \sum_{m=1}^{\infty} \dfrac{\sinh (2m - 1)(b - y) \sin (2m - 1)x}{(2m - 1) \sinh (2m - 1)b}$, using the fact that $\cos n\pi = (-1)^n$.

13. $u(x, y) = 2 \sum_{n=1}^{\infty} \dfrac{(-1)^{n+1} \sinh n(b - y) \sin nx}{n \sinh (nb)}$

14. 0.648

15. **b)** 10 terms yield 1.2087 compared with 1.2337

17. $f(x) = f(x + 1) = (x + 1) - (x + 1)^2 = -x^2 - x = f(-x)$

Section 7.2, p. 436

3. The set is orthonormal on $[-1, 1]$ with weight function one.

4. $f(x)$ is defined as $-x$ for $-\pi < x < 0$, which is the same as $-f(x)$.

6. $f(x)$ is defined as x for $-\pi < x < 0$, which is the same as $-(-x) = -f(-x)$.

9. **a)** $\dfrac{4}{\pi} \sum\limits_{n=1}^{\infty} \dfrac{\sin \dfrac{n\pi x}{2}}{n}$

10. **b)** 3

11. **a)** $\dfrac{2}{\pi} \sum\limits_{n=1}^{\infty} \dfrac{1}{n}\left((-1)^{n+1} - 2\dfrac{1 + (-1)^{n+1}}{n^2\pi^2}\right) \sin n\pi x$;

   b) $\dfrac{1}{3} + \dfrac{4}{\pi^2} \sum\limits_{n=1}^{\infty} \dfrac{(-1)^n}{n^2} \cos n\pi x$

12. **b)** $\dfrac{1}{2}(e^2 - 1) + 4 \sum\limits_{n=1}^{\infty} \dfrac{(-1)^n(e^2 - 1)}{4 + n^2\pi^2} \cos \dfrac{n\pi x}{2}$

13. **a)** $\sin \pi x$;   **b)** $\dfrac{2}{\pi} + \dfrac{2}{\pi} \sum\limits_{n=2}^{\infty} \dfrac{\cos n\pi x}{1 - n^2}$

14. **a)** $\dfrac{8}{\pi} \sum\limits_{n=1}^{\infty} \dfrac{n \sin 2nx}{4n^2 - 1}$;   **b)** $\dfrac{2}{\pi} + \dfrac{4}{\pi} \sum\limits_{n=1}^{\infty} \dfrac{(-1)^n \cos 2nx}{1 - 4n^2}$

18. **a)** $\dfrac{1}{2} - \dfrac{4}{\pi^2} \sum\limits_{n=1}^{\infty} \dfrac{\cos (2n - 1)\pi x}{(2n - 1)^2}$;   **b)** put $x = 1$ in the result for part (a).

19. $\dfrac{4}{\pi} \sum\limits_{n=1}^{\infty} \dfrac{\sin (2n - 1)x}{2n - 1}$

20. **a)** Put $x = \pi/2$ in the result for Exercise 19;
   b) put $x = \pi/4$ in the result for Exercise 19.

21. **a)** One approach is to define

$$f(s) = \begin{cases} 1 - s, & 0 < s < 1, \\ s - 1, & 1 < s < 2. \end{cases}$$

The odd periodic extension of this function is found to have the Fourier series representation,

$$f(s) \sim \frac{4}{\pi}\left(1 - \frac{2}{\pi}\right) \sin \frac{\pi s}{2} + \frac{4}{3\pi}\left(1 + \frac{2}{3\pi}\right) \sin \frac{3\pi s}{2}$$
$$+ \frac{4}{5\pi}\left(1 - \frac{2}{5\pi}\right) \sin \frac{5\pi s}{2} + \cdots.$$

Replacing s by $x - 1$ in the above will represent the function specified. A sketch of $f(s)$ will be helpful. Observe that the Fourier series in x is actually a *cosine* series in this case.

22. $c_n = 0$ if n is even (including $n = 0$) and $c_n = 2/(in\pi)$ if n is odd;

$$f(x) \sim \frac{2}{i\pi} \sum_{n=-\infty}^{\infty} \frac{1}{2n-1} \exp\left(i(2n-1)\pi x/2\right).$$

Note that this can be written as

$$f(x) \sim \frac{2}{i\pi} \sum_{n=-\infty}^{\infty} \frac{1}{2n-1} \left(\cos \frac{(2n-1)\pi}{2} x + i \sin \frac{(2n-1)\pi}{2} x\right).$$

Now observe that the cosine terms vanish while the sine terms double so that we have

$$f(x) \sim \frac{4}{\pi} \sum_{n=1}^{\infty} \frac{1}{2n-1} \sin \frac{(2n-1)\pi}{2} x,$$

which agrees with the result obtained by using Eq. (7.2–6).

23. Make an even periodic extension of the function $f(x) = 1$, $-1 < x < 1$. Then $c_0 = 1$ and all other $c_n = 0$.

28. Compute $\sin \dfrac{n\pi x}{L} \sin \dfrac{m\pi x}{L}$, integrate, and note that the result is zero unless $m = n$.

31. b) $\dfrac{4h}{\pi} \sum_{n=1}^{\infty} \dfrac{\sin(2n-1)\pi x}{2n-1}$

Section 7.3, p. 446

5. Only the function in part (a) is absolutely integrable.

11. $\dfrac{2}{\pi} \displaystyle\int_0^\infty \dfrac{\cos \alpha x}{1 + \alpha^2} d\alpha$ is valid for $-\infty < x < \infty$

18. $\bar{f}(\alpha) = a \dfrac{k \sin^2(a\alpha/2)}{(a\alpha/2)^2}$

Section 7.4, p. 456

2. $u(x, t) = \dfrac{2}{\pi} \displaystyle\int_0^\infty \dfrac{\alpha \exp(-\alpha^2 kt) \sin \alpha x \, d\alpha}{\alpha^2 + 1}$

3. $u(x, t) = \dfrac{2}{\pi} \displaystyle\int_0^\infty f(s) \cos \alpha s \, ds \int_0^\infty \exp(-\alpha^2 kt) \cos \alpha x \, d\alpha$

7. $u(x, t) = \dfrac{2u_0}{\pi} \displaystyle\int_0^\infty \left(1 - \exp(-\alpha^2 kt)\right) \dfrac{\sin \alpha x}{\alpha} d\alpha$

9. $u(x, t) = \dfrac{2u_0}{\pi} \displaystyle\int_0^\infty (1 - \cos \alpha L) \exp(-\alpha^2 kt) \dfrac{\sin \alpha x}{\alpha} d\alpha$

10. a) $\bar{f}(\alpha) = \dfrac{2 \cos(\alpha\pi/2)}{1 - \alpha^2}$

12. $u(x, t) = u_0 \left(\mathrm{erf}\left(\dfrac{x}{\sqrt{4kt}}\right) - \dfrac{1}{2}\mathrm{erf}\left(\dfrac{x+L}{\sqrt{4kt}}\right) - \dfrac{1}{2}\mathrm{erf}\left(\dfrac{x-L}{\sqrt{4kt}}\right)\right)$

13. $\bar{f}_s(\omega_0) = N/\omega_0$ which represents the maximum value of the transformed finite wave.

17. $\mathscr{F}(xy) = -i \dfrac{d\bar{y}(\alpha)}{d\alpha}$

CHAPTER 8

Section 8.1, p. 467

2. Use the identity $\sinh (A - B) = \sinh A \cosh B - \cosh A \sinh B$.

3. $u(x, y, z) = \displaystyle\sum_{n=1}^{\infty} \sum_{m=1}^{\infty} B_{mn} \sinh \omega_{mn}(\pi - z) \sin (my) \sin (nx)$, where

$$B_{mn} = \frac{4}{\pi^2 \sinh (\pi \omega_{mn})} \int_0^\pi \int_0^\pi f(s, t) \sin (ns) \sin (mt) \, ds \, dt \text{ and } \omega_{mn} = \sqrt{n^2 + m^2}.$$

4. $B_{mn} = \dfrac{4ab}{\pi^2 mn \sinh (c\omega_{mn})}$

5. Use the same identity as in Exercise 2.

6. $V(x, y) = \dfrac{2}{\pi} \displaystyle\int_0^\infty \dfrac{\alpha}{\alpha^2 + 1} \dfrac{\sinh \alpha(b - y)}{\sinh (\alpha b)} \sin (\alpha x) \, d\alpha$, using the result

$\displaystyle\int_0^\infty \exp (-ax) \sin bx \, dx = \dfrac{a}{a^2 + b^2}$, if $a > 0$, found in integral tables.

7. $V(x, y) = \dfrac{2}{\pi} \displaystyle\int_0^\infty \dfrac{\sin \alpha \pi}{1 - \alpha^2} \dfrac{\sinh \alpha(b - y)}{\sinh (\alpha b)} \sin \alpha x \, d\alpha$

14. $u(x, y) = \dfrac{2a}{\pi} \displaystyle\int_0^\infty \dfrac{\sinh (\alpha(1 - x)) \cos (\alpha y)}{(a^2 + \alpha^2) \sinh \alpha} \, d\alpha$

15. $u(x, y) = \dfrac{2}{\pi} \displaystyle\sum \dfrac{\cos nx \sinh n(\pi - y)}{1 - n^2 \sinh n\pi}$

21. If $\phi_x(x, y) = -\phi_y(x, y)$

Section 8.2, p. 477

4. $B_{mn} = \dfrac{64a^2 b^2}{\pi^6 (2n - 1)^3 (2m - 1)^3}$, $n = 1, 2, \ldots, m = 1, 2, \ldots$

5. $u(x, y, t) = \dfrac{4\pi^2 k}{ab} \sin \left(\dfrac{\pi y}{b}\right) \sin \left(\dfrac{\pi x}{a}\right) \cos \left(\dfrac{c\pi}{ab} \sqrt{a^2 + b^2} t\right)$;

$f = \dfrac{c}{2ab} \sqrt{a^2 + b^2} \text{ sec}^{-1}$

16. $u(x, y, t) = \displaystyle\sum_{m=1}^{\infty} \sum_{n=1}^{\infty} A_{mn} \sin \left(\dfrac{m\pi y}{b}\right) \sin \left(\dfrac{n\pi a}{x}\right) \sin (c\omega_{mn} t)$, where

$A_{mn} = \dfrac{4}{abc\omega_{mn}} \displaystyle\int_0^a \int_0^b g(x, y) \sin \left(\dfrac{m\pi x}{a}\right) \sin \left(\dfrac{n\pi y}{b}\right) dy \, dx$

19. $b_{2n-1} = 4k/\pi(2n-1)$, $n = 1, 2, \ldots$

20. $u(x, t) = \dfrac{8h}{\pi^2}\left(\sin\left(\dfrac{\pi x}{L}\right)\cos\left(\dfrac{c\pi t}{L}\right)\right) - \dfrac{1}{9}\sin(3\pi x/L)\cos(3c\pi t/L) + -\cdots$

Section 8.3, p. 484

4. The steady-state solution is zero.

11. $u(x, t) = \displaystyle\sum_{n=1} b_{2n-1}\exp\left(\dfrac{-\pi^2(2n-1)^2}{4L^2}t\right)\sin\dfrac{(2n-1)\pi}{2L}x$, where

$b_{2n-1} = \dfrac{2}{\pi(2n-1)}\displaystyle\int_0^L f(s)\sin\dfrac{(2n-1)\pi}{2L}s\,ds$, $n = 1, 2, 3, \ldots$

12. $b_1 \doteq 0.250$

13. $u(x, t) = \dfrac{1}{2}a_0 + \displaystyle\sum_{n=1} a_n\exp\left(\dfrac{-n^2\pi^2 t}{L^2}\right)\cos\dfrac{n\pi x}{L}$, where

$a_n = \dfrac{2}{L}\displaystyle\int_0^L f(s)\cos\dfrac{n\pi s}{L}\,ds$, $n = 0, 1, 2, \ldots$

14. $u(x, t) = \dfrac{Cx(L-x)}{2} + \displaystyle\sum_{n=1} b_n\exp(-n^2 t)\sin\dfrac{n\pi x}{L}$, where

$b_n = \dfrac{2}{L}\displaystyle\int_0^L\left(\dfrac{Cs(s-L)}{2} + f(s)\right)\sin\dfrac{n\pi s}{L}\,ds$

15. $u(x, t) = \dfrac{4C}{L}\displaystyle\sum_{n=1}\dfrac{1 - \exp(-(2n-1)^2 t)}{(2n-1)^3}\sin(2n-1)\dfrac{\pi x}{L}$

17. $u(x, t) = \displaystyle\sum_{n=1} c_n\exp\left(\dfrac{-(2n-1)^2 t}{4}\right)\sin\left(\dfrac{(2n-1)\pi x}{2}\right)$. In order to find c_n, however, we must have the condition that the functions $\sin((2n-1)\pi x/2)$ are orthogonal on the interval $0 < x < 1$. Verify that this is so. Then find

$$c_n = \dfrac{(-1)^{n+1}8u_0}{(2n-1)^2\pi^2}.$$

19. $u(x, y, t) = \displaystyle\sum_{m=1}\sum_{n=1} B_{mn}\exp\left(-k\pi^2 t\left(\dfrac{m^2}{a^2} + \dfrac{n^2}{b^2}\right)\right)\sin\left(\dfrac{m\pi x}{a}\right)\sin\left(\dfrac{n\pi y}{b}\right)$,

where $B_{mn} = \dfrac{4}{ab}\displaystyle\int_0^b\sin\left(\dfrac{n\pi v}{c}\right)dv\int_0^a f(s, v)\sin\left(\dfrac{m\pi s}{a}\right)ds$

20. $c_n = \dfrac{16a}{\pi^3(2n-1)^3}((-1)^{n+1}\pi(2n-1) - 2)$

Section 8.4, p. 494

2. $v(x, y) = \dfrac{2}{\pi}\displaystyle\int_0^\infty\dfrac{\sinh(\alpha x)\cos(\alpha y)}{(1+\alpha^2)\cosh(\alpha c)}\,d\alpha$

3. Use the method of undetermined coefficients to obtain a particular solution.

7. Use the method of undetermined coefficients as in Exercise 2. Note that the boundary conditions are such that the complementary solution vanishes.

13. $V(x, y) = \dfrac{2}{\pi} \displaystyle\int_0^\infty \dfrac{\sinh{(\alpha x)} \cos{(\alpha y)}}{\alpha \cosh \alpha} \int_0^\infty f(s) \cos \alpha s \, ds \, d\alpha$

14. $V(x, y) = \dfrac{2}{\pi} \displaystyle\int_0^\infty \dfrac{\sin \alpha \cos{(\alpha x)} \cosh{(\alpha y)}}{\alpha \cosh \alpha} \, d\alpha$

15. $u(x, t) = 1 + \dfrac{2}{\pi} \displaystyle\sum_{n=1}^\infty \dfrac{(-1)^n}{n} (\sin{(n\pi x)} \sin n\pi(1 - x)) \exp{(-n^2\pi^2 t)}$

Section 8.5, p. 503

6. $\lambda_n = \left(\dfrac{2n - 1}{2}\right)^2, n = 1, 2, \ldots ; y_n(x) = \cos\left(\dfrac{2n - 1}{2}\right)x$

7. $\lambda_n = n^2, \quad n = 0, 1, 2, \ldots ; \quad y_n(x) = \cos nx;$ also $\lambda_n = (2n + 1)^2/4, \quad n = 0, 1, 2, \ldots ; y_n(x) = \sin{(2n + 1)x/2}$

9. $\lambda_n = n^2 - 1, n = 1, 2, 3, \ldots ; y_n(x) = \sin nx$

10. $\lambda_n = -n^2\pi^2, n = 1, 2, 3, \ldots ; y_n(x) = \exp{(-x)} \sin n\pi x$

11. $\lambda_n = -n^2, n = 1, 2, 3, \ldots ; y_n(x) = (n \cos nx + \sin nx) \exp{(-x)};$ also $\lambda = 1,$ $y(x) = 1$

18. **b)** $\exp{(-x)}$

19. **b)** $\exp{(-x^2)}$

20. **b)** $(1 - x^2)^{-1/2}$

CHAPTER 9

Section 9.1, p. 511

2. Divide by ρ, then make the substitution $v = dR/d\rho$.

5. One interpretation is to find the steady-state temperatures in an infinitely long pipe having inner radius b and outer radius c. The outside of the pipe is kept at temperature zero, while the inside surface has a temperature given by $f(\phi)$.

6. The pipe is infinitely long, that is, $-\infty < z < \infty$, so that there are no boundaries in the z-direction. Moreover, the boundary values of the inside and outside surfaces are independent of z.

8. Note that the differential equation is a Cauchy–Euler equation (see Section 2.4).

12. $u(\rho, \phi) = \dfrac{4u_0}{\pi} \displaystyle\sum_{n=1}^\infty \left(\dfrac{\rho^{2n-1} - \rho^{-(2n-1)}}{c^{2n-1} - c^{-(2n-1)}}\right) \dfrac{\sin{(2n - 1)\phi}}{2n - 1}$

14. $u(\rho, \phi) = \dfrac{\log \rho}{2 \log c} a_0 + \displaystyle\sum_{n=1}^\infty a_n \dfrac{\rho^{2n} - \rho^{-2n}}{c^{2n} - c^{-2n}} \cos{(2n\phi)},$ where

$a_0 = \dfrac{4}{\pi} \displaystyle\int_0^{\pi/2} f(s) \, ds$ and $a_n = \dfrac{4}{\pi} \displaystyle\int_0^{\pi/2} f(s) \cos 2ns \, ds, n = 1, 2, 3, \ldots$

18. a) $z(\rho) = z_0 \dfrac{\log \rho}{\log \rho_0}$, $1 \le \rho \le \rho_0$;

b) The solution represents the static transverse displacements of a homogeneous membrane in the shape of a washer. The membrane is fastened to a frame along the circles $\rho = 1$ and $\rho = \rho_0$. The inner frame is held fixed while the outer frame is given a displacement z_0.

Section 9.2, p. 527

1. Use the criteria for convergence of an alternating series to establish convergence when $x = 2$.

2. b) $y_2(x) = a_1 \displaystyle\sum_{n=0}^{\infty} \dfrac{2 \cdot 5 \cdot 8 \cdots (3n - 1)}{(3n + 1)!} x^{3n+1}$ (*Note*: Here and in part (a) use 1 for the numerator when $n = 0$.)

7. a) $(2^n - 1)/2^n$

9. a) This is a geometric series with first term $1/10$ and common ratio $1/10$;
b) $1/9$

10. a) 1; **b)** ∞; **c)** 1; **d)** e; **e)** 3; **f)** ∞

11. a) $x = 0$ is a regular singular point;
d) $x = 1$ is a regular singular point; $x = 0$ is an irregular singular point.

12. a) $y = a_0 \cos x + a_1 \sin x$; **b)** $y = a_0 \cosh x + a_1 \sinh x$;
c) $y = (a_0 + 2) \exp(x) - x^2 - 2x - 2$; **d)** $y = a_0 \exp(x^2/2)$;
e) $y = a_0(1 + x \arctan x) + a_1 x$

13. a) $y = a_0 J_0(x) = a_0 \left(1 - \dfrac{x^2}{2^2} + \dfrac{x^4}{2^2 \cdot 4^2} - \dfrac{x^6}{2^2 \cdot 4^2 \cdot 6^2} + - \cdots \right)$;

b) $y = a_0 \cos \sqrt{x} + a_1 \sin \sqrt{x}$; **c)** $y = a_0 x + a_1 x^{-2}$

15. $y = a_0(1 + \frac{1}{2}(x - 1)^2 + \frac{1}{6}(x - 1)^3 + \frac{1}{6}(x - 1)^4 + \cdots)$
$\qquad + a_1((x - 1) + \frac{1}{2}(x - 1)^2 + \frac{1}{2}(x - 1)^3 + \frac{1}{4}(x - 1)^4 + \cdots)$

Section 9.3, p. 539

6. a) $y = c_1 J_0(x) + c_2 Y_0(x)$; **b)** $y = c_1 J_0(\sqrt{x}) + c_2 Y_0(\sqrt{x})$;
c) $y = c_1 J_0(e^x) + c_2 Y_0(e^x)$

7. a) Make the substitution $u = \lambda_j s$, $du = \lambda_j \, ds$ and the integral becomes

$$\frac{1}{\lambda_j} \int_0^{\lambda_j} J_1(u) \, du = \frac{1}{\lambda_j} (-J_0(u)) \Big|_0^{\lambda_j} = \frac{1}{\lambda_j};$$

c) evaluate $\int_0^b J_1(\lambda_j s) \, ds$ as in part (a), then let $b \to \infty$.

8. a) Use integration by parts with $u = J_0(s)$, $dv = J_1(s) \, ds$.

9. a) $\dfrac{2}{c} \displaystyle\sum_{j=1}^{\infty} \dfrac{J_0(\lambda_j x)}{\lambda_j J_1(\lambda_j c)}$; **b)** $\dfrac{2}{c} \displaystyle\sum_{j=1}^{\infty} \dfrac{(\lambda_j c)^2 - 4}{\lambda_j^3 J_1(\lambda_j c)} J_0(\lambda_j x)$

10. a) $\dfrac{dy}{dx} = \dfrac{1}{az} \dfrac{d^2 z}{dx^2} - \dfrac{1}{az^2} \left(\dfrac{dz}{dx}\right)^2$;

b) the transformed equation is $r^2 \dfrac{d^2 Z}{dr^2} + r \dfrac{dZ}{dr} + \left(\lambda^2 r^2 - \left(n + \dfrac{1}{2}\right)^2\right) Z = 0$.

12. a) $\dfrac{1}{2}\displaystyle\sum_{j=1}^{\infty} \dfrac{J_0(\lambda_j) - J_0(2\lambda_j)}{\lambda_j (J_2(2\lambda_j))^2}\, J_1(\lambda_j x);$

b) $\frac{1}{2}$, the average value of the jump discontinuity at $x = 1$

14. $x^n J_n(x)$

15. b) $-x^{-n} J_n(x)$

Section 9.4, p. 559

3. b) From Eq. (9.4–14) it follows that $P_n(-x) = (-1)^n P_n(x)$, $n = 0, 1, 2, \ldots$;

d) $P'_{2n}(x)$ contains terms in $x^{2n-2k-1}$ so the only way to obtain nonzero terms when $x = 0$ is to have $2n - 2k - 1 = 0$. But this is impossible since it implies that $2(n - k) = 1$. Hence $P'_{2n}(0) = 0$.

e) From Eq. (9.4–14),

$$P_{2n}(x) = \frac{1}{2^{2n}} \sum_{k=0}^{N} \frac{(-1)^k (4n - 2k)!}{(2n - 2k)!(2n - k)!}\, x^{2n-2k}.$$

When $x = 0$, the only nonzero term in this sum occurs when $k = n$ and then $P_{2n}(0)$ has the required value.

7. Show that P_0 and P_1, P_1 and P_2, and P_0 and P_2 are orthogonal pairs.

12. a) $a P_1(x) + b P_0(x);$ **b)** $\dfrac{2}{3} a P_2(x) + b P_1(x) + \left(c + \dfrac{a}{3}\right) P_0(x);$

c) $\dfrac{2}{5} a P_3(x) + \dfrac{2}{3} b P_2(x) + \left(c + \dfrac{3a}{5}\right) P_1(x) + \left(d + \dfrac{b}{3}\right) P_0(x)$

15. b) $A_{2n+1} = (4n + 3)\displaystyle\int_0^1 x P_{2n+1}(x)\, dx, \quad n = 0, 1, 2, \ldots$

$\qquad\qquad\;\; = (4n + 3)\displaystyle\int_0^1 P_1(x) P_{2n+1}(x)\, dx = 0$ for all n except $n = 0$.

18. Use Rodrigues' formula.

Section 9.5, p. 570

11. $u(r, \theta) = \dfrac{r}{b} \cos \theta$

12. a) $u(\rho, z) = 28.68 J_0(2.405\rho) \cosh (2.405z)$
$\qquad\qquad - 0.85 J_0(5.520\rho) \cosh (5.520z)$
$\qquad\qquad + 0.03 J_0(8.654\rho) \cosh (8.654z) + \cdots ;$

b) $u(0, 0) \doteq 27.86$

13. $u(\rho, z) = \dfrac{200}{c} \displaystyle\sum_{j=1}^{\infty} \dfrac{J_0(\lambda_j \rho) \sinh (\lambda_j z)}{\lambda_j \sinh (\lambda_j b) J_1(\lambda_j c)}$

15. $z(\rho, t) = \dfrac{2}{c} \displaystyle\sum_{j=1}^{\infty} \dfrac{J_0(\lambda_j \rho) \cos (\lambda_j at)}{\lambda_j J_1(\lambda_j c)}$

18. $u(r, \theta) = \displaystyle\sum_{m=0}^{\infty} (4m+3)\left(\dfrac{r}{b}\right)^{2m+1} P_{2m+1}(\cos \theta) \int_0^{\pi/2} f(\cos \theta) P_{2m+1}(\cos \theta) \sin \theta\, d\theta$

20. $u(r, \theta) = 100$; the result should be obvious, but work through the appropriate steps of Example 9.5–5.

21. $u(r, \theta) = -\dfrac{3E}{K+2} r \cos \theta, \ r < b;$

$U(r, \theta) = -Er \cos \theta + Eb^3 \left(\dfrac{K-1}{K+2}\right) r^{-2} \cos \theta, \ r > b$

22. $u(r, \theta) = -Er \cos \theta + E\dfrac{b^3}{r^2} \cos \theta$

23. $u(r) = \dfrac{1}{r(b-a)} (u_1 a(b-r) + u_2 b(r-a))$

Section 9.6, p. 575

3. $u(5, 5) = 1.786, \ u(10, 5) = 7.143, \ u(15, 5) = 26.786$

4. 21

5. $u(10/3, 10/3) = 0.69, \ u(20/3, 10/3) = 2.08, \ u(10, 10/3) = 5.56,$
$u(40/3, 10/3) = 14.58, \ u(50/3, 10/3) = 38.19;$
$u(x, 20/3) = u(x, 10/3)$ by symmetry. Note that the resulting *five* equations can be solved readily by elimination.

CHAPTER 10

Section 10.1, p. 585

4. The modulus is changed; the argument remains the same.

8. a) $13 + 7i;$ **c)** $\frac{1}{5}(1 - 3i);$ **d)** $22 - 4i$

9. b) $\sqrt{2} \operatorname{cis} (-\pi/4);$ **d)** $\operatorname{cis} 3\pi/2;$ **f)** $2 \operatorname{cis} (-\pi/3)$

10. a) $2, \pi/3;$ **c)** $5\sqrt{5}, \arctan(-11/2)$ or $11.180, -1.391;$
d) $5, 0;$ **f)** $8, 3\pi/2$

11. a) $\pm\sqrt{2}(-1 + i)/2;$
b) $y = \exp(\sqrt{2}x/2)(c_1 \cos(\sqrt{2}x/2) + c_2 \sin(\sqrt{2}x/2))$
$+ \exp(-\sqrt{2}x/2)(c_3 \cos(\sqrt{2}x/2) + c_4 \sin(\sqrt{2}x/2))$

12. b) The half-plane to the right of the line $x = -1;$
d) the region between the circles $x^2 + y^2 = 4$ and $x^2 + y^2 = 25$

13. a) No side of a triangle can have a length greater than the sum of the lengths of the other two sides;
b) no side of a triangle can be less in length than the difference of the lengths of the other two sides.

14. $(z^2 + \sqrt{2}z + 1)(z^2 - \sqrt{2}z + 1)$

15. $2 \sin \dfrac{\phi}{2} \operatorname{cis} \left(\dfrac{\pi - \phi}{2}\right)$

16. a) $2 \operatorname{cis} \dfrac{(2k+1)\pi}{6}, \ k = 0, 1, 2, 3, 4, 5;$

c) $2^{1/6} \operatorname{cis} (8k + 3) \dfrac{\pi}{12}, \ k = 0, 1, 2$

17. a) $\pm 2 + i$;   **b)** $1 + i, 1 + 2i$

19. a) $z^2 - 4z + 5 = 0$

20. a) $2(1 + \sqrt{3}i)$;   **b)** $-3(1 - i)$;   **c)** -3;   **d)** $1 + i$

28. $\lambda_1 = 1, (1, 0, 0)^T$; $\lambda_2 = 2i, (0, 1, 2i)^T$; $\lambda_3 = -2i, (0, 1, -2i)^T$

Section 10.2, p. 593

3. Use Eq. (10.1–9) to express $z^{1/2}$.

5. a) Use the trigonometric identities

$$\cos(A + B) = \cos A \cos B - \sin A \sin B,$$
$$\sin(A + B) = \sin A \cos B + \sin B \cos A.$$

  c) Show that $\exp(z + 2\pi i) = \exp(z)$, using Eq. (10.2–1).
  d) If $e^z = 1$, then $|e^z| = e^x = 1$ and $x = 0$; hence $e^z = e^{iy} = \cos y + i \sin y = 1$ from which we deduce that $\cos y = 1$, $\sin y = 0$, and $y = 2n\pi$. Conversely, if $z = 2n\pi i$, then $e^z = 1$.
  e) Using part (b), we have $\exp(z_1) = \exp(z_2)$ if and only if $\exp(z_1 - z_2) = \exp(0) = 1$ and this last holds by part (d) whenever $z_1 - z_2 = 2n\pi i$.

9. Similar to Example 10.2–2.

12. a) 0.0001;   **b)** $23.1361 + 0.4443i$

15. a) $\log 2 \doteq 0.693 + 2n\pi i$, where n is an integer, $\text{Log } 2 \doteq 0.693$;
  b) $\log(i - 1) \doteq 0.346 + (3\pi/4 + 2n\pi)i$, $\text{Log }(i - 1) \doteq 0.346 + 2.356i$;
  c) $\log(-1) = (2n + 1)\pi i$, $\text{Log }(-1)$ is undefined.

17. b) $(x^3 - 3xy^2) + i(3x^2y - y^3)$;   **d)** $((x - 1)^2 + y^2)^{-1/2}(x^2 - y^2 - 3 + 2xyi)$

21. a) $i \sinh e$;   **d)** $(e^3\sqrt{2}/2)(1 - i) \doteq 14.203(1 - i)$

22. $\exp(i \text{ Log } 2) \exp(-\pi - 2n\pi)$, $n = 0, \pm 1, \pm 2, \ldots$

25. $(a^z)^w = (\exp(z \log a))^w = \exp(w \log(\exp(z \log a)))$
$\qquad = \exp(w(z \log a + 2n\pi i)) = \exp(wz \log a) \exp(2n\pi i)w$
$\qquad = a^{wz} \exp(2n\pi i)w \neq a^{wz}$

28. $\dfrac{1}{s^2 + 1} = -\dfrac{1}{2i}\dfrac{1}{s + i} + \dfrac{1}{2i}\dfrac{1}{s - i}$, $f(t) = \dfrac{i}{2}\exp(-it) + \dfrac{1}{2i}\exp(it)$ or

$f(t) = \dfrac{i}{2}(\cos t - i \sin t) + \dfrac{1}{2i}(\cos t + i \sin t) = \sin t$

Section 10.3, p. 602

1. All z such that $z = 1 + iy$ or $z = 2 + iy$. The set is not closed since it has no boundary in the y-direction.

3. a) $\sin z = \sin(x + iy) = \sin x \cosh y + i \cos x \sinh y$, which is analytic for all x and y.

5. b) The set of points in the interior of a circle of radius ρ with center at z_0;
  d) the left half of the z-plane including the y-axis.

6. a) This is a closed set consisting of all points on and in the interior of the unit circle with center at $z = i$. The boundary points are the points on the circle, that is, all points (x, y) satisfying $x^2 + (y - 1)^2 = 1$.

11. **a)** Log $z = $ Log $|z| + i$ Arg z from Eq. (10.2–8), so Log z is not continuous at $z = 0$. Let x_0 be an arbitrary point on the negative real axis. Then, since $-\pi < $ Arg $z \le \pi$, the function Arg $z \to \pi$ as $z \to x_0$ from the upper half-plane and Arg $z \to -\pi$ as $z \to x_0$ from the lower half-plane. Consequently, Log z is not continuous at $z = x_0$, hence not analytic.

13. **a)** $3x^2y - y^3 - x + C$;   **c)** arctan $(y/x) + C$

17. **a)** $z^3 - i(z - C)$;   **c)** Log $z + \alpha$, where α is a complex constant.

Section 10.4, p. 611

5. **a)** $y = mx$ is mapped into $v = -mu$;
    d) There are two points of intersection, $(0, 0)$ and (a, a). The origin maps into the point at infinity, and the point (a, a) maps into $(1/2a, -1/2a)$.

6. **a)** The circle $|z| = 1$ maps into $|u| \ge 1$;
    c) Limiting the mapping to $|z| > 1$ produces a one-to-one mapping, which is the desired situation in applications.

7. The rectangular region is mapped into $\dfrac{u^2}{\cosh^2 c} + \dfrac{v^2}{\sinh^2 c} = 1, v \ge 0$.

13. **a)** ± 1;   **c)** the point at infinity

15. **a)** The equation of the sphere is $\xi^2 + \eta^2 + (\zeta - \tfrac{1}{2})^2 = \tfrac{1}{4}$.

16. **a)** $-\infty < u \le -1, v = 0$;   **b)** Im $w > 0$;

    c) the interior of the ellipse $\dfrac{u^2}{\cosh^2 1} + \dfrac{v^2}{\sinh^2 1} = 1$, excluding the real segments, $-\cosh 1 \le x \le -1$ and $1 \le x \le \cosh 1$.

17. **a)** Everywhere;   **c)** the unit circle with center at the origin;
    e) the semicircular ring centered at the origin above the v-axis and bounded by circles of radius c_1 and c_2.

22. $|z| = 2$ maps into $\frac{16}{25}u^2 + \frac{16}{9}v^2 = 1$.

Section 10.5, p. 622

2. **d)** Apply the ratio test and divide numerator and denominator by 3^n before taking the limit.

5. **a)** $-\pi i$;   **c)** zero

6. **a)** $-4 + 2\pi i$;   **b)** $4\pi i$

7. The function $f(z) = \bar{z}$ is nowhere differentiable, hence not analytic; the function $(z + 2)/z$ is not analytic at the origin.

8. $18\pi i$

11. $-i\pi/12$

12. $2\pi i$

Section 10.6, p. 638

5. The dot product is zero.

12. **a)** $F(s) = \dfrac{i}{4} \dfrac{1}{(s + i)^2} - \dfrac{i}{4} \dfrac{1}{(s - i)^2}$;

b) $\dfrac{i}{4}(te^{-it} - te^{it}) = \dfrac{t}{2}\left(\dfrac{e^{it} - e^{-it}}{2i}\right) = \dfrac{t}{2}\sinh(it) = \dfrac{t}{2}\sin t$

13. $\dfrac{-2}{z-1} + \dfrac{1}{z-2} + \dfrac{2}{z-3}$

14. $\dfrac{2}{z-2} - \dfrac{1}{z-1}$

15. a) $-\dfrac{1}{2}\dfrac{1}{z-2} + \dfrac{5}{3}\dfrac{1}{z-1} - \dfrac{1}{6}\dfrac{1}{z-4}$;

b) $\dfrac{3}{8}\dfrac{1}{z} + \dfrac{1}{2}\dfrac{1}{z^2} + \dfrac{1}{2}\dfrac{1}{z^3} - \dfrac{3}{8}\dfrac{1}{z-2} + \dfrac{5}{4}\dfrac{1}{(z-2)^2}$;

c) $\dfrac{-4}{z} - \dfrac{2}{z^2} + \dfrac{5}{z-1} - \dfrac{1}{(z-1)^2}$

16. a) $1 + 2\sin t - \cos t$; **c)** $\tfrac{1}{2}(\sinh t - \sin t)$

17. a) $\dfrac{e^{-t}}{12}\left(t - \dfrac{11}{12}\right) + \dfrac{5}{112}e^{-5t} + \dfrac{2}{63}e^{2t}$; **b)** $\dfrac{1}{9}(6te^{-3t} - e^{-3t} + 1)$

18. a) $\sin \omega t$; **c)** $\exp(bt)\sin \omega t$; **e)** $\sinh at$

20. $c_1 = 0$ corresponds to $y = x$, $c_1 > 0$ corresponds to equipotentials below $y = x$, and $c_1 < 0$ to those above $y = x$.

21. $y = $ constant are the streamlines, $x = $ constant are the equipotentials.

22. a) Rays emanating from the origin are the streamlines, circles centered at the origin are the equipotentials;
b) the origin is a sink.

24. $T(x, y) = \arctan(y/x)$

25. a) $T(x, y) = \dfrac{100}{\pi}\arctan\left(\dfrac{2xy}{x^2 - y^2}\right)$; **b)** $T(\rho, \phi) = \dfrac{100}{\pi}\phi$;

c) $T(x, y) = 1 - \dfrac{1}{\pi}\arctan\left(\dfrac{1 - x^2 - y^2}{2y}\right)$

27. $\bar{f}(\alpha) = \dfrac{aA}{1 + ia\alpha}$

index

Abel, Niels Henrik (1802–1829), 82
Absolute value of complex number, 582
Absolutely integrable, 440
Absorption of light, 41
  coefficient, 42
Abstract vector space, 262
Acceleration
  centripetal, 372
  Coriolis, 372
  normal component of, 299
  tangential component of, 299
  vector, 298
Adams, John C. (1819–1892), 54n
Adams–Bashforth method, 55
Adams–Moulton method, 55
Addition
  of complex numbers, 579
  of matrices, 185
  of vectors, 278
Additive identity, 186
  inverse, 186, 580
Adjoint
  of a differential operator, 504
  of a matrix, 266
Aerodynamics, 397, 631
Airy, Sir George (1801–1892), 528n
Airy function, 529
Airy's equation, 528
Algebra, fundamental theorem of, 626n
Algebraic equations, systems of, 251
Alternating series, 515
Ambient temperature, 4
Analytic function, 518, 598
Angular velocity vector, 289
Annihilator property, 188
Anticommutative property, 283

Approximately equal to, 4n
Arbitrary constant of integration, 3, 7, 70
Arbitrary function, 380
Arc length, 298
  in cylindrical coordinates, 321
  smooth, 297, 614
  in spherical coordinates, 327
Arden, B. W., 257n
Area, 29, 284, 358, 362
  element, projection of, 341
  surface, 340
Argand, Jean R. (1768–1822), 578n
Argand diagram, 578
Argument of complex number, 582
Assumptions, simplifying, 78, 104, 107, 405, 547
Astill, K. N., 257
Atled, *see* Del operator
Atomic physics, 411
Augmented matrix, 200
Auxiliary equation, 100
Average value of a function, 423
Axes
  principal, 233
  rotation of, 129
  translation of, 128

Back substitution, 202
Bacterial growth, 33
Bandwidth, spectral, 252
Bashforth, Francis (1819–1912), 55n
Basis
  change of, 267
  natural, 215
  of a vector space, 214
Beam, bending of, 109
Bellman, R., 50

687

Bernoulli, Jakob (James)
  (1654–1705), 110n
Bernoulli–Euler law, 110
Bessel, Friedrich Wilhelm
  (1784–1846), 522
Bessel equation, modified, 542
Bessel function(s), 180
  of the first kind, 532, 530, 625
  integral representation of, 625
  modified, 542
  orthogonality of, 533
  of the second kind, 530, 538
  zeros of, 567
Bessel's differential equation, 522
Bicontinuous mapping, 610
Biharmonic equation, 396n
Bilinear transformation, 605
Binormal, 300
Biokinetics, 35
Biot, M. A., 110n
Block multiplication, 190
Bôcher, Maxime (1867–1918), 424n
Borel, Émile (1871–1956), 159n
Borel's theorem, 159. *See also*
  Convolution.
Bouguer, Pierre (1698–1758), 41
Bound vector, 277. *See also* Position
  vector.
Boundary of a set, 596
Boundary condition, 389
  nonhomogeneous, 482
  periodic, 504, 512
Boundary point, 596
Boundary-value problem, 389
  two-point, 80, 111, 400, 465, 476,
  639
Bounded function, 330n, 508
Box product, 285
Boyce, W. E., 14n
Braun, M., 65
Bromwich, T. J. I'A. (1875–1929),
  629n
Bromwich contour, 629
Brown, J. W., 551n

Capping surface, 355
Capstan, 36
Carbon-12, 32
Carbon-14, 32
Carrying capacity of environment, 43

Carslaw, H. S., 414
Cauchy, Augustin–Louis
  (1789–1857), 98n, 496, 598n
Cauchy–Euler equation, 98, 509, 519,
  546
Cauchy–Goursat theorem, 616
Cauchy–Riemann equations, 598,
  615, 632
Cauchy–Schwarz inequality, 294
Cauchy's integral formula, 621
Cauchy's theorem, 616
Cayley, Arthur (1821–1895), 184, 227
Center of curvature, 299
Central axis theorem, 290
Central field, 363
Central force, 364
Centrifugal force, 364
  potential of, 365
Centripetal acceleration, 372
Chandraratna, P. A. N., 455n
Change of basis, 267
Change of variable, 474
Characteristic curve, 381
Characteristic equation, 72, 81, 227,
  393, 583
Characteristic frequency, 477
Characteristic value, *see* Eigenvalue
  vector, 226
Chebyshev's differential equation,
  505
Chemical kinetics, 39
Chemical reaction
  first-order, 39
  second-order, 40
Chemotherapy, 161
Chernoff, P. R., 387
Chu, W. K., 455n
Churchill, R. V., 551n
Circle, osculating, 299
Circle of curvature, 299
Circuit, transformer-coupled, 182
Circular symmetry, 569
Circulation, 335, 631
Cis, 581
Clairaut, Alexis C. (1713–1765), 12
Clairaut's equation, 12
Closed form of power series, 518
Closed set, 596
Codomain, 220
Coefficient of sliding friction, 37

Coefficients, undetermined, 87, 109, 242, 492
Cofactor, 266
Column vector, 184
Columns of a matrix, 184
Compartmental analysis, 34
Compatible orientations, 354
Common ratio, geometric series, 515
Complementary error function, 490
Complementary function, 86
Complementary solution, 86, 245, 492
Complete orthonormal set, 551
Complete set, 427, 502
   of modified Bessel functions, 535
Completing the square, 173
Complex conjugate, 583
Complex Fourier coefficients, 434
Complex number(s)
   absolute value of, 582
   addition of, 579
   argument of, 582
   division of, 581, 584
   equality of, 578
   imaginary part of, 584
   multiplication of, 581
   as ordered pairs, 578
   polar form of, 581
   real part of, 584
   subtraction of, 580
   as vectors, 578
Complex period, 591
Complex plane, extended, 610
Complex potential, 632
Complex variables, references, 641
Compressible fluid, 349
Compressibility of a fluid, 312
Computer, 51, 83, 123, 251
Condition
   boundary, 389
   initial, 3, 153, 389
   Lipschitz, 66
   nonhomogeneous boundary, 482
   periodic boundary, 504
   Robin's, 633n
Conditions, Dirichlet, 421, 439
Conductivity, thermal, 373, 479
Cone, 316
Confidence interval, 33
Conformable matrices, 187n

Conformal mapping, 606
Conjugate
   complex, 583
   harmonic, 602
Conjugation, properties of, 584
Connected set, 600
Conservation of energy, 337
Conservation of mass, 349
Conservative field, 315, 335, 337, 364
Consistent system, 197
Constant
   Euler's, 539
   gravitational, 365
   of integration, 3, 7, 70
   Lipschitz, 67
   radioactive decay, 3, 32
   separation, 460, 530, 545, 564
   spring, 107
Continuity
   equation of, 349
   of solution, 66
Continuous at point, 597
Continuous on a set, 597
Continuous scalar field, 303
Continuously compounded interest, 12, 48
Contour, 614
   Bromwich, 629
Contour integral, 614
Contraction, 606
Convergence
   interval of, 516
   in the mean, 428
   pointwise, 424
   radius of, 515
Convergent method, 122
Convergent series, 515
Convolution, 158
Cooling, law of, 4, 35, 465, 480
Coordinate system
   dextral, 275
   spherical, 322, 543
Coordinate transformation, 128, 129
Coordinates
   curvilinear, 324
   cylindrical, 318
   elliptic, 327
   parabolic cylindrical, 327
   paraboloidal, 329
   polar, 506

prolate spheroidal, 328
spherical, 322, 543
Coplanar vectors, 285, 292
Coriolis, Gaspard G. de (1792–1843),
370n
Coriolis acceleration, 372
Corson, D. R., 370n
Cosh x, 80
table of, 646
Cosine series, 466, 481
Cosine transform, 448, 488
Cosmic radiation, 33
Coulomb, Charles A. de (1736–1806),
366
Coulomb, 366
Coulomb constant, 366
Coulomb's law, 366
Couple, 290
Cramer, Gabriel (1704–1752), 197n
Cramer's rule, 197, 267
Criminology, 49
Critical damping, 119
Critical point, 632n
Critical points of a transformation,
612
Cross product, 282
properties of, 283
triple scalar, 285
triple vector, 284
Cycloid, 362
Cylindrical coordinates, 318
Cylindrical harmonics, 531
Cubic, twisted, 306
Curl
coordinate-free definition, 354
in cylindrical coordinates, 322
properties of, 315
in rectangular coordinates, 314
in spherical coordinates, 324
Current, electric
steady-state, 117
transient, 117
Currents in Pacific Ocean, 304
Curvature, 299
center of, 299
circle of, 299
radius of, 299
Curve(s)
characteristic, 381
deflection, 110

normal to, 299
one-parameter family of, 7, 38
oriented, 297
orthogonal families of, 601
space, 297
twisted, 301
Curvilinear coordinate systems, 324

D'Alembert, Jean-Le-Rond
(1717–1783), 409n, 414
D'Alembert's solution, 409
Damper, viscous, 108
Danese, A. E., 168n, 502n
Darboux, Jean Gaston (1842–1917),
307n
Darboux vector, 307
Dashpot, 108
Dating, radiocarbon, 32
Davis, H. T., 162n
de Moivre, Abraham (1667–1754),
582n
de Moivre's formula, 582
Decay
exponential, 4
radioactive, 32
Decomposition
of nitrogen pentoxide, 39
into partial fractions, 25, 150, 623,
626
of radioactive substance, 1, 32
Definite integral, 3, 330, 613
Deflection curve, 110
Del operator
in cylindrical coordinates, 320
in rectangular coordinates, 308
in spherical coordinates, 324
Delta
function, 168, 453
Kronecker, 420
Dependence, linear, 214
Derivative
directional, 309
normal, 373, 390, 633
partial, 308
of a vector function, 295
Derrick, W. R., 52n, 67
Determinant, 198
Dettman, J. W., 215
Dextral coordinate system, 275
Diagonal, principal, 189

Diagonal matrix, 189, 230
Diagonalizable matrix, 230
Diagonalization, 237
Diagram, Argand, 578
Dielectric sphere, 571
Difference quotient, 573
Differentiable function, 598
Differential, exact, 13, 335
Differential equation(s)
  Bessel's, 522
  Cauchy–Euler, 98, 509, 519, 546
  Chebyshev's, 505
  Clairaut's, 12
  exact, 13
  Fourier's, 541
  Helmholtz, 541
  Hermite's, 505
  homogeneous, 72
  Lagrange's, 379
  Laguerre's, 505
  Legendre's, 524, 547
  Legendre's associated, 547
  linear, 8
  nonlinear, 8, 50
  order of, 7
  ordinary, 1
  partial, 1, 13, 335
  Pfaffian, 380
  references, 125
  Riccati, 50, 114, 541
  separable, 3
  systems of, 177, 237
Differential geometry, 301
Differential operator, 501
Differentiation, 273
Diffusion equation, 373, 464
Diffusion of neutrons, 484
Diffusivity, thermal, 373, 465, 479
Dimension, 215
DiPrima, R. C., 14n
Dirac delta function, 168, 453
Dirac, Paul A. M. (1902–   ), 168n
Direction angles, 279
Direction cosines, 279
Direction numbers, 280
Directional derivative, 309
Dirichlet, Peter G. L. (1805–1859),
  390n
Dirichlet conditions, 421, 439
Dirichlet problem, 390, 464, 572

Discontinuity, jump, 127
Discrete scalar field, 303
Displacement(s)
  simultaneous, 254
  static, 413, 514
  successive, 254
Distance formula, 277
Distribution theory, 168
Divergence
  coordinate-free definition, 352
  in cylindrical coordinates, 322
  of the gradient, *see* Laplacian
  properties of, 312
  in rectangular coordinates, 312,
    349
  in spherical coordinates, 324
Divergence theorem, 350
Divergent series, 514
Division
  of complex numbers, 581, 584
  synthetic, 83
Domain, 220, 600
  of definition, 588
  of a function, 588
Dominant eigenvalue, 256
Dot product, 232n, 281
  properties of, 281
Double Fourier series, 461
Double infinite series, 461
Double subscripts, 185
Dummy index of summation, 517
Dummy variable of integration, 21
Dynamics, fluid, 314, 609

Echocardiogram, 455
Economics, 44
Ecosystem, 33
Eigenfunctions, 401, 460, 466
  normalized, 502
  orthogonality of, 498
  of Sturm–Liouville problem, 497
Eigenspace, 229
Eigenvalue(s), 226, 401, 460
  of Sturm–Liouville problem, 497
Eigenvector, 226
Elastic curve, 110
  limit, 107
Elasticity, 109
  laws of, 472
Electric potential, 368, 396

Electromagnetic induction, law of, 370

Electromagnetic vector wave function, 375

Electrostatic potential, 635

Electrostatics, 366, 396

Element of volume, 321, 327, 344, 349

Elementary function, 22n, 590

Elementary matrix, 204

Elementary operation, 590

Elementary row operation, 200

Elements of a matrix, 184

Elliptic coordinates, 327

Elliptic equation, 388, 575

Empty set, 197

Energy, conservation of, 337

Entire function, 598

Environment, carrying capacity of, 43

Epsilon-neighborhood, 595

Equality

  of complex numbers, 578

  of matrices, 185

  of vectors, 277

Equation(s)

  auxiliary, 100

  Bessel, modified, 542

  Cauchy–Euler, 98, 509, 519, 546

  Cauchy–Riemann, 598, 615, 632

  characteristic, 72, 81, 227, 393, 583

  of continuity, 349

  difference, 577n

  diffusion, 373, 464

  elliptic, 388, 575

  Fourier's, 541

  Helmholtz, 541

  hyperbolic, 388, 391, 393, 407, 575

  indicial, 521

  integral, 161

  integrodifferential, 106, 161

  intrinsic, 307

  Lagrange's, 379

  Laguerre's, 505

  Laplace's, 313, 368, 389, 631

  Legendre's, 524, 547

  of line, 287

  logistic, 43

  Maxwell's, 313, 370

  parabolic, 387, 391, 479, 575

  parametric, 287

  of plane, 287

  Poisson's, 368, 397

  Schrödinger wave, 411

  Volterra integral, 161

  wave, 391, 470

Equilibrium, static, 291

Equipotential surfaces, 310, 384

Equipotentials, 39, 601, 632

Equivalence relation, 210

Equivalence row, 200

Erf x, 490

Erfc x, 490

Error

  in Euler method, 52

  local, 54

  roundoff, 54

Escape velocity, 113

Esser, M., 91n

Euler, Leonhard (1707–1783), 51n, 414

Euler method, 51

  error in, 52

  improved, 52

Euler–Fourier coefficients, 418, 434, 511

Euler's constant, 539

Euler's formula, 77, 97, 433

Even extension, 554

Even function, 416

Even periodic extension, 431, 466

Exact differential, 13, 335

Exact differential equation, 13

Excitation function, 153

Existence

  of Laplace transform, 133

  of solution, 64

Exp u, 20

Exp x, table of, 646

Exp $(-x)$, table of, 646

Explicit solution, 8

Explosion, population, 49

Exponential decay, 4

Exponential function, 590

Exponential order, 133, 628n

Extended complex plane, 610

Extension

  even, 554

  even periodic, 431, 466

  odd, 554

odd periodic, 432, 475
periodic, 430, 466

Factor
  integrating, 19, 28, 536, 556
  of polynomial equation, 83
Factorial function, 140
Factors, scale, 321, 323
Faddeeva, V. N., 254n
Faltung, 162
Faraday, Michael (1791–1867), 369n
Faraday's law of induction, 369
Fast Fourier transform (FFT), 446
Fermi, Enrico (1901–1954), 484n
Fermi age, 484
Field
  central, 363
  conservative, 315, 335, 337, 364
  continuous scalar, 303
  discrete scalar, 303
  electromagnetic wave, 375
  irrotational, 315
  scalar, 303
  solenoidal, 313, 369
  source, 307
  steady, 305, 375
  time-independent, 305
  vector, 304
  velocity, 631
Finite-difference method, 573
Finney, R. L., 330, 635n
First-order numerical method, 54
Fission, nuclear, 49
Fixed points of a transformation, 612
Fixed vector, 277
Flow, transonic, 388
Fluid
  compressibility of, 312
  compressible, 349
  ideal, 631
  incompressible, 312, 631
  irrotational, 631
Fluid dynamics, 314, 609
Fluid mechanics, 631
Flux, 352, 479, 488, 632
  magnetic, 369
  SI units of, 480
Food chain, 32
Force
  central, 364

centrifugal, 364
frictional, 37
gravitational, 365
impulsive, 167
inverse square, 313
lines of, 384, 601
normal, 37
tangential, 105
Forced vibrations, 109
Forces, resolution of, 36
Forcing function, 109, 153, 181
Formula
  Cauchy's integral, 621
  De Moivre's, 582
  distance, 277
  Euler's, 77, 97, 433
  Heaviside expansion, 631
  recursion, 517, 523
  Rodrigues', 552, 625
Foucault, Jean L. (1819–1868), 372n
Foucault pendulum, 372
Fourier, Jean B. J., (1768–1830),
      414n
Fourier coefficients, 418, 434, 511
Fourier integral
  complex form, 442
  representation, 441
Fourier series
  complex form, 434
  cosine, 431, 466, 481
  differentiation of, 435
  double, 461
  exponential form, 433
  integration of, 436
  representation, 418, 509, 511
  sine, 432, 462
Fourier transform
  cosine, 448, 488
  fast (FFT), 446
  inverse of, 443
  pair, 443
  sine, 448, 463, 489
Fourier–Bessel series, 535, 564
Fourier's equation, 541
Fourth-order Runge–Kutta method,
      57
Fox, A. H., 122n
Fraleigh, J. B., 366n
Free fall, 113
Free space, 366n

Free vector, 277
Free vibrations, 108
Frenet, Jean-Frédéric (1816–1900), 301n
Frenet–Serret formulas, 301
Frequency characteristic, 477
Frequency of vibrating membrane, 472
Friction, sliding, 37
Frictional force, 37
Frobenius, Georg (1849–1917), 520
Frobenius, method of, 520
Fuchs, Lazarus (1833–1902), 520n
Function(s)
  Airy, 529
  analytic, 518, 598
  arbitrary, 318
  average value of, 423
  Bessel, modified, 542
  bounded, 330n
  complementary, 86
  complementary error, 490
  differentiable, 598
  elementary, 590
  entire, 598
  error, 457, 490
  even, 416
  excitation, 153
  exponential, 590
  of exponential order, 133
  forcing, 109, 153, 181
  gamma, 140
  generalized factorial, 140
  harmonic, 396, 464, 487, 531, 600, 604
  homogeneous, 26
  hyperbolic, 80
  impulse, 167
  integer-valued, 33
  Legendre, 547, 557
  linearly independent, 73
  as a mapping, 588
  normalized, 420
  null, 149
  odd, 416
  orthogonal, 419
  orthonormal, 420
  period of, 420
  periodic, 169, 420, 508
  piecewise continuous, 126
  piecewise smooth, 420
  pseudoimpulse, 167
  range of, 588
  of a real variable, 588
  scalar, 295
  scalar point, 303
  spectrum of, 452
  step, 439
  strength of, 167
  system, 153
  unit step, 164
  vector, 295
  wave, 411
  weight, 419, 534
Functions of a complex variable, 588
  exponential, 590
  hyperbolic, 591
  logarithmic, 592
  multiple-valued, 590
  trigonometric, 591
Fundamental period, 429
Fundamental planes, 300
Fundamental set of solutions, 239
Fundamental strip, 609
Fundamental theorem of algebra, 626n

Gamma function, 140
Gaussian elimination, 199
Gaussian reduction, 202
Gauss, Karl F. (1777–1855), 199n
Gauss's law, 367
Gauss's theorem, 350
Gauss–Jordan reduction, 203
Gauss–Seidel method, 254
General solution
  first-order partial differential equation, 380
  ordinary differential equation, 7
  second-order partial differential equation, 388
  systems of ordinary differential equations, 239
Generalized factorial function, 140
Generalized product, 158
Geometric series, 515
Geometry, differential, 301
Gerald, C. F., 55, 251n, 254, 575
Gibbs, Josiah W., (1839–1903), 424n
Gibbs phenomenon, 424

Golomb, M., 28n
Goursat, Edouard (1858–1936), 616
Gradient
    coordinate-free definition, 354
    in cylindrical coordinates, 320
    properties of, 312
    in rectangular coordinates, 309, 631
    in spherical coordinates, 324
Gradient field, *see* Conservative field
Gram, Jorgen P. (1850–1916), 326n
Gram–Schmidt orthogonalization, 326
Grassmann, Hermann G. (1809–1877), 283
Grattan-Guinness, I., 414
Gravitation
    law of, 111
    universal law of, 365
Gravitational constant, 365
Gravitational force, 365
Gravitational potential, 366, 397
Gray, A., 535n
Gray, J. D., 598n
Green, George (1793–1841), 356n
Green's theorem in the plane, 356, 615, 631, 632
Grid, 573
Grossman, S. I., 52n, 67
Growth
    bacterial, 33
    population, 43, 161
Growth rate, 43
Gyroscope, 372n

Hamilton, Sir William Rowan (1805–1865), 227, 275, 308
Hamming, Richard W. (1915–   ), 122n
Hamming's method, 122
Harmonic conjugate, 602
Harmonic function, 396, 454n, 464, 487, 531, 600, 604
Harmonic motion, 108
Harmonic oscillator, 375
Harmonics
    cylindrical, 531
    spherical, 546
Head of a vector, 275
Heat equation, 373, 464

Heaviside, Oliver (1850–1925), 130
Heaviside function, *see* Unit step function
Heaviside's expansion formula, 631
Heisenberg, Werner (Carl) (1901–1976), 453n
Heisenberg uncertainty principle, 453
Helix, right circular, 301, 339
Helmholtz, Hermann von (1821–1894), 541n
Helmholtz equation, 541
Henrici, P., 60
Henry, Joseph (1797–1878), 369n
Hermite, Charles (1822–1901), 437n
Hermite's differential equation, 505
Hermitian orthogonality, 437
Heun, Karl L. W. (1859–1929), 52n
Heun's method, 52
Hilbert, David (1862–1943), 261n
Hilbert matrix, 261
Holomorphic function, 518, 598
Homogeneous differential equation, 72
Homogeneous function, 26
Homogeneous system, 205, 243
Hooke, Robert (1635–1703), 106n
Hooke's law, 106
Hurricane, 372
Hydraulics, 41
Hydrodynamics, 41, 349, 397, 631
Hyperbolic cosine, table of, 646
Hyperbolic equation, 388, 391, 393, 407, 575
Hyperbolic functions, 80
Hyperbolic sine, table of, 646
Hyperboloid, 316
Hypocycloid, four-cusped, 359

Ideal fluid, 631
Idempotent matrix, 236
Identity, additive, 186
Identity matrix, 189
Identity relation, 151
Ill-conditioned system, 252
Imaginary part of complex number, 584
Implicit solution, 8
Improved Euler method, 52

Impulse function, 157
Impulse response, 181
Impulsive force, 167
Incompressible fluid, 312, 631
Inconsistent system, 196
Independence, linear, 214
Independent of path, 16, 335
Index of summation, 517
Indicial equation, 521
Induction
  law of, 369
  magnetic, 368
Inequality, Cauchy–Schwarz, 294
Inertial frame, 371n
Infinite series, 514
  double, 461
Initial condition, 3, 153, 389
Initial point, 275
Initial velocity, 71
Initial-value problem, 3
Inner product, 232n. *See also* Dot
  product.
Insulation, 465
Integer-valued function, 33
Integral
  definite, 3, 330, 613
  formula, Cauchy's, 621
  Fourier, 441
  particular, 87, 492
  Riemann, 330
  volume, 344
Integral equation, 161
Integral surface, 381
Integral test, 515
Integral transform, 129
  reference, 458
Integrating factor, 19, 28, 536, 556
Integrodifferential equation, 106, 161
Interest, continuously compounded,
  12, 48
Interior point, 596
Interval of convergence, 516
Intrinsic equations of a curve, 307
Invariant point, 612
Invariant vector, 227
Inverse
  additive, 186, 580
  of a matrix, 203
Inverse Fourier transform, 443
Inverse Laplace transform, 148

Inversion, 604
Iron metabolism, 34
Irregular singular point, 520
Irrotational field, 315
Irrotational fluid, 631
Isobar, 39
Isocline, 60
Isogonal mapping, 610
Isolated singular point, 617
Isomorphic spaces, 263
Isothermal, 573
Isothermal surface, 310
Isotropic, thermally, 372

Jacobi, Carl G. J. (1804–1851), 253n
Jacobi iteration method, 253
James, R. C., 14n, 350, 395n
Jones, R. B. A., 123
Jordan, Camille (1838–1922), 203n
Jump discontinuity, 127

Katz, V. J., 354n
Kepler, Johannes (1571–1630), 366n
Kepler's laws, 366
Kernel, 218
  of integral transform, 129
Kirchhoff, Gustave R. (1824–1887), 42n
Kirchhoff's second law, 42, 106
Kinetics
  chemical, 39
  tracer, 34
Kreider, D. L., 14
Kronecker, Leopold (1823–1891),
  420n
Kronecker delta, 420
Kutta, Martin W. (1867–1944), 57n

Lagrange, Joseph Louis (1736–1813),
  75, 379
Lagrange's equation, 379
Laguerre's equation, 505
Lambert, Johann H. (1728–1777), 41
Lambert's law, 41
Lancaster, P., 255n
Laplace, Pierre Simon de
  (1749–1827), 313n, 389n
Laplace inversion formula, 629
Laplace transform, 130, 490, 628
  existence of, 133
  inverse of, 148

pair, 131
references, 183
table of, 137, 643
Laplace's equation, 313, 368, 389, 631
  in cylindrical coordinates, 529
  finite-difference approximation, 574
Laplacian, 313
  in cylindrical coordinates, 510, 542
  in polar coordinates, 507
  in rectangular coordinates, 507
  in spherical coordinates, 544
Laurent, Pierre A. (1813–1854), 620n
Laurent series, 620
  principal part, 621
  regular part, 621
Law(s)
  of conservation of energy, 337
  of conservation of mass, 349
  of cooling, 4, 35, 465, 480
  Coulomb's, 366
  of elasticity, 472
  of electromagnetic induction, 370
  Faraday's, 369
  Gauss's, 367
  Hooke's, 106
  Kepler's, 366
  Kirchhoff's, 42, 106
  Lambert's, 41
  of Newtonian mechanics, 366
  Newton's first, 114
  Newton's gravitation, 111
  Newton's second, 71, 105, 108, 112, 247, 406
  Newton's third, 365
  Newton's universal, 365
  of nullity, 221
  of thermodynamics, 35
  Toricelli's, 41
Lead-204, 32
Left-hand limit, 127
Legendre, Adrien Marie (1752–1833), 524n
Legendre function, 547
  of the second kind, 557
  singularities of, 558
Legendre polynomials, 526, 547, 548, 625
  definition of, 549

integral representation, 625
  orthogonality relation, 550
  properties of, 549
Legendre polynomials of even degree, 526, 549
  orthogonality relations, 551
Legendre polynomials of odd degree, 526
  orthogonality relations, 559
Legendre series, 551, 563
Legendre's associated differential equation, 547
Legendre's differential equation, 524, 547
Leibniz's rule, 24
Length of a vector, 276
Lerch, M. (1860–1922), 149
Lerch's theorem, 149
Level surface, 310
Libby, W. F., 32
Liebmann, Karl O. H. (1874–1939), 573n
Liebmann's method, 573
Limit
  left-hand, 127
  in the mean, 428
  point of a sequence, 515n
  right-hand, 127
Limits, properties of, 597
Line, equations of
  in parametric form, 287
  in symmetric form, 287
  in vector form, 286
Line integral, 332, 614
Linear algebra, references, 273
Linear combination, 73, 213
Linear differential equation, 8
Linear fractional transformation, 605
Linear independence, 214
Linear operator, 136, 504
Linear oscillator, 505
Linear transformation, 136, 186, 216
Linear velocity vector, 289
Linearly dependent functions, 73
Linearly dependent vectors, 214
Linearly independent functions, 73
Lines of force, 384, 601
Lipschitz, Rudolf O. S. (1832–1903), 66n
Lipschitz condition, 66

Lipschitz constant, 67
Liouville, Joseph (1809–1882), 496
Liouville's theorem, 621, 626n
Local error, 54
Log defined, 2
Logistic equation, 43
Longitudinal waves, 472
Lorrain, P., 370n
Lower triangular matrix, 191

Magnetic flux, 369
Magnetic induction vector, 368
Magnetic potential, 397
Magnification, 605
Malthus, Thomas R. (1766–1834), 34
Mapping, 216, 588
    bicontinuous, 610
    conformal, 606
    isogonal, 610
Marion, J. B., 366n
Markov, A. A. (1856–1922), 194n
Markov matrix, 194
Mass
    conservation of, 349
    variable, 114
Mass–spring system, 247
Mathews, G. B., 535n
Matrices
    addition of, 185
    conformable, 187n
    equality of, 185
    partitioning of, 190
Matrix
    adjoint, 266
    augmented, 200
    columns of, 184
    diagonal, 189, 230
    diagonalizable, 230
    elementary, 204
    elements of, 184
    Hilbert, 261
    idempotent, 236
    identity, 189
    inverse of, 203
    lower triangular, 191
    Markov, 194
    nilpotent, 195
    nonsingular, 204
    order of, 265
    orthogonal, 233

positive definite, 254
    rank of, 265
    rows of, 184
    scalar, 189
    singular, 204
    size of, 184
    sparse, 191
    square, 184
    stochastic, 194
    symmetric, 189, 231
    trace of, 255
    triangular, 191
    zero, 186
Maxwell, James Clerk (1831–1879),
    313n, 349
Maxwell's equations, 313, 370
Mechanical advantage, 36
Mechanics, 289
    fluid, 609
    Newtonian, 366
Membrane, vibrating, 470
Metabolism, iron, 34
Meteorology, 39
Method of Frobenius, 520
Method of undetermined coefficients,
    87, 242, 492
Method of variation of parameters,
    75, 92, 94, 556
Milne, William E. (1890–1971), 120n
Milne's method, 120
Möbius, August F. (1790–1868), 351n
Möbius strip, 351n
Möbius transformation, 605
Modified Bessel equation, 542
Modulus, 582
Moment, vector, 289
Monte Carlo method, 573
Morris, S. A., 598n
Motion
    critically damped, 119
    overdamped, 118
    underdamped, 119
Moulton, Forest R. (1872–1952),
    55n
Moving trihedral, 300
Multiplication
    block, 190
    of complex numbers, 581
    row-by-column, 187
Multiplicative unit, 172

Nabla, *see* Del operator
Natural basis, 215
Natural equations of a curve, 307
Neighborhood of a point, 595
Network analysis, 181
Network parallel, 182
Neumann, Carl G. (1832–1925), 390n
Neumann function, 530, 538
Neumann problem, 390, 464
Neutron transport, 457
Neutrons, diffusion of, 484
Newton, Sir Isaac (1642–1727), 4n
Newtonian mechanics, laws of, 366
Newton's first law, 114
Newton's law of cooling, 4, 35, 465, 480
Newton's law of gravitation, 111
Newton's second law, 71, 105, 108, 112, 247, 406
Newton's third law, 365
Newton's universal law of gravitation, 365
Nilpotent matrix, 195
Nitrogen pentoxide, 39
Nitrogen tetroxide, 39
Nonhomogeneous boundary condition, 482
Nonlinear differential equation, 8, 50
Nonsingular matrix, 204
Norm, 537, 552
Normal
   to a curve, 299
   to a plane, 287
   principal, 299
   to a surface, 310
Normal component of acceleration, 299
Normal derivative, 373, 390, 633
Normal force, 37
Normal plane, 300
Normalized eigenfunctions, 502
Normalized functions, 420
Nuclear fission, 49
Nuclear physics, 457
Nuclear reactor, 161, 484n
Null function, 149
Null space, 220
Nullity, 220
   law of, 221
Number, complex, 578

Numerical method
   Adams–Bashforth, 55
   Adams–Moulton, 55
   Cramer's rule, 197, 267
   Euler, 51
   finite-difference, 573
   first-order, 54
   Gaussian elimination, 199
   Gauss–Jordan reduction, 203
   Gauss–Seidel, 254
   Hamming's, 122
   Heun's, 52
   improved Euler, 52
   Jacobi iteration, 253
   Milne's, 120
   Monte Carlo, 573
   order of, 53
   Picard's, 68
   power, 256
   predictor-corrector, 54
   Runge–Kutta, 57
   simultaneous displacements, 254
   successive displacements, 254

Ocean currents, 304
Odd extension, 554
Odd function, 416
Odd periodic extension, 432, 475
Olsson, I. U., 33n
One-parameter family
   of curves, 7, 38
   of surfaces, 384
One-to-one transformation, 218
Open set, 596
Operation
   elementary, 590
   elementary row, 200
Operator
   del, 308
   differential, 501
   linear, 136, 504
   self-adjoint, 504
Order
   of a differential equation, 7
   of a numerical method, 53
   reduction of, 76, 97, 509, 556
   of a square matrix, 265
Ordinary differential equation, 1
Ordinary point, 520
Orientable surface, 351

Orientations, compatible, 354
Oriented curve, 297
Orthogonal families of curves, 601
Orthogonal functions, 419
Orthogonal matrix, 233
Orthogonal surfaces, 384
Orthogonal trajectories, 38
Orthogonal vectors, 232
Orthogonality
  of Bessel functions, 533
  of eigenfunctions, 498
  Hermitian, 437
  of Legendre polynomials, 550
  relations, 418
Orthogonalization, Gram–Schmidt,
  326
Orthonormal functions, 420
Orthonormal set, 554
  of vectors, 324
Oscillator, linear, 505
Osculating circle, 299
Osculating plane, 300
Outer product, *see* Cross product
Overdamped motion, 118
Overshoot, 424

Parabolic cylindrical coordinates, 327
Parabolic equation, 388, 391, 479,
  575
Paraboloid, 311, 347, 361
Paraboloidal coordinates, 328
Parallel network, 182
Parameter(s), 287
  variation of, 75, 92, 94, 556
Parametric equations, 287
Partial derivative, 308
Partial differential equation, 13, 335,
  379
Partial differential vector operator,
  308
Partial fraction decomposition, 25,
  150, 623, 626
Partial sums, 424, 427, 514
Particular integral, 87, 492
Particular solution, 7, 87, 243, 380,
  388
Partitioning of matrices, 190
Pascal's triangle, 85
Path in space, 297
Patzert, W. C., 304

Pendulum
  Foucault, 372
  simple, 104
Period
  complex, 591
  of a function, 420
  fundamental, 429
Periodic boundary conditions, 504,
  512
Periodic extension, 430, 466
Periodic function, 169, 420, 508
Pfaff, Johann F. (1765–1825), 380n
Pfaffian differential equation, 380
Pharmacokinetics, 35
Phase, 582
Photosynthesis, 32
Physics
  atomic, 411
  nuclear, 457
Picard, Charles Émile (1856–1941),
  68n
Picard's method, 68
Piecewise continuous function,
  126
Piecewise smooth function, 420
Piecewise smooth surface, 351
Pivoting, 252
Plane
  complex extended, 610
  equation of, 287
  fundamental, 300
  normal, 300
  normal to, 287
  osculating, 300
  rectifying, 300
Plucked string, 478
Plywood, manufacture of, 486
Point
  boundary, 596
  at infinity, 610
  initial, 275
  interior, 596
  irregular singular, 520
  isolated singular, 617
  limit, 515n
  ordinary, 520
  regular singular, 520
  singular, 325, 616
  terminal, 275
Pointwise convergence, 424

Poisson, Siméon de (1781–1840), 368n
Poisson's equation, 368, 397
Polar coordinates, 506
Polar form of complex number, 581
Pole
  of order m, 617
  simple, 617
Polonium-212, 32
Polynomial, Legendre, 526, 549
Polynomial equation
  factor of, 83
  root of, 83
Population
  dynamics, 44
  explosion, 49
  growth, 43, 161
Position vector
  in cylindrical coordinates, 318
  in rectangular coordinates, 297
  in spherical coordinates, 323
Positive definite matrix, 254
Postmultiply, 191
Potential
  of a centrifugal force, 365
  complex, 632
  electric, 368, 396
  electrostatic, 635
  gravitational, 366, 397
  magnetic, 397
  of a simple harmonic oscillator, 375
  vector, 375
  velocity, 631
Potential equation, *see* Laplace's equation
Potential function, 396, 454n, 464, 487, 531, 600, 604
Power method, 256
Power series, 515
  closed form, 518
  differentiation of, 516
  interval of convergence, 516
  radius of convergence, 515
Poynting, John H. (1852–1914), 370n
Poynting vector, 370
Predictor-corrector method, 54
Premultiply, 191
Principal axes, 233
Principal diagonal, 189
Principal normal, 299

Principal value of log z, 592
Principle of superposition, 88, 97, 398, 409, 467
Probability theory, 194, 572
Problem
  boundary-value, 389
  Dirichlet, 390, 464, 572
  initial-value, 3
  Neumann, 390, 464
  Sturm–Liouville, 496
Product
  box, 285
  dot, 232n, 281
  generalized, 158
  scalar, 232
  scalar triple, 285
Projection, 217, 282
  of area element, 341
  stereographic, 610
Prolate spheroidal coordinates, 328
Propagation, wave, 370, 410
Pseudofunction, 168
Pseudoimpulse function, 167
Pythagorean triple, 273

Quadratic form, 231
Quantum mechanics, 168n, 184, 453, 505
Quasilinear partial differential equation, 379

Rabinowitz, P., 57n, 122, 251n, 254, 446
Radiation, cosmic, 33
Radioactive decay, 4, 32
Radiocarbon C-14, 32
Radiocarbon dating, 32
Radium-88, 3
Radius
  of convergence, 515
  of curvature, 299
  of torsion, 301
Raeside, D. E., 455n
Ralston, A., 57n, 122, 251n, 254, 446
Random walk, 572
Range, 220
  of a function, 588
Rank of a matrix, 265
Ratio test, 515
Reaction, chemical, 39

Reactor, nuclear, 161, 484n
Real part of complex number, 584
Reciprocal set of vectors, 329
Rectifying plane, 300
Recurrence relation, 525
Recursion formula, 517, 523
Recursive relation, 53
Reduced row echelon form, 203
Reduction of order, 76, 97, 509, 556
Reference, integral transforms, 458
References
    complex variables, 641
    differential equations, 125
    general, 642
    Laplace transform, 183
    linear algebra, 273
    vector calculus, 376
Reflexive property, 210
Regular function, 518, 598
Regular singular point, 520, 522, 524
Regular Sturm–Liouville problem,
    497
Relation
    equivalence, 210
    recurrence, 523
    recursive, 53
Relaxation method, 573
Renfrew, C., 33n
Residue, 617, 626
Residue theorem, 619
Resolution of forces, 36
Response, impulse, 181
Response transform, 153
Resultant, 282, 291, 579
Retrothrust, 114
Riccati, Jacopo (1676–1754), 50n
Riccati equation, 50, 114, 541
Riemann, Georg F. B. (1826–1866),
    330n, 598n
Riemann integral, 330
Riemann sphere, 610n
Right circular helix, 301, 339
Right-hand coordinate system, 275
Right-hand limit, 127
Right-hand rule, 289
Robin, Victor G. (1855–1897), 633n
Robin's conditions, 633n
Rocket, 114
Rodrigues, Olinde (1794–1851), 552n
Rodrigues' formula, 552, 625

Root of a polynomial equation, 83
Rotation, 226, 605
    of axes, 129
Roundoff error, 54
Row echelon form, 202
    reduced, 203
Row equivalence, 200
Row vector, 184
Row-by-column multiplication, 187
Rows of a matrix, 184
Rudin, W., 330
Rule
    Leibniz's, 24
    right-hand, 289
Runge, Carl (1856–1927), 57n
Runge–Kutta methods, 57

Saff, E. B., 598, 606
Sagan, H., 428n
Satellite, 112, 375
    photograph, 372
Scalar field, 303
Scalar function, 295
Scalar matrix, 189
Scalar point function, 303
Scalar product, 232
Scalar triple product, 285
Scalars, 184
Scale change, 134
Scale factor
    cylindrical coordinates, 321
    spherical coordinates, 323
Scaling, 252
Schmidt, Erhard (1876–1959), 326n
Schrödinger, Erwin (1887–1961), 411n
Schrödinger wave equation, 411
Schwarz, Hermann A. (1843–1921),
    294n
Second law of thermodynamics, 35
Second-order numerical method, 54
Sectionally continuous, 126
Seidel, P. L. (1821–1896), 254n
Self-adjoint operator, 504
Separable differential equation, 3
Separation constant, 460, 530, 545,
    564, 566, 567
Separation of variables, 2, 399, 545,
    562, 566
Sequence, 514
    limit point of, 515n

of partial sums, 514
Series
  alternating, 515
  of constants, 514
  cosine, 431, 466, 481
  Fourier, 418, 433, 461
  Fourier–Bessel, 535, 564
  geometric, 515
  Laurent, 620
  Legendre, 551, 563
  power, 515
  sine, 432, 462
Series circuit, 42
Serret, Joseph A. (1819–1885),
    301n
Set
  boundary of, 596
  closed, 596
  complete orthonormal, 551
  connected, 600
  divergent, 514
  empty, 197
  fundamental, 239
  open, 596
  orthonormal, 554
  reciprocal, 329
  solution, 195
  spanning, 213
Shanks, M. E., 28n
Shelupsky, D., 425n
Shifting theorem, 135
SI units, 365
Sifting property, 168
Similarity, 230
Simple harmonic motion, 108
Simple harmonic oscillator, 375
Simple pendulum, 104
Simple pole, 617
Simple surface, 351
Simplifying assumptions, 78, 104,
    107, 405, 547
Simultaneous displacements, 254
Singular matrix, 204
Singular point, 329, 616
  irregular, 520
  isolated, 617
  regular, 520, 522, 524
Singular solution, 617
Singular Sturm–Liouville problem,
    503, 534

Singularity, *see* Singular point
Sinh x, 80
  table of, 646
Sink, 349n
Size of a matrix, 184
Skew-symmetric, 194
Skydiver, 49
Skydiving, 113
Smooth arc, 297, 614
Smooth surface, 339
Sneddon, I. N., 379n, 380, 439
Snider, A. D., 598, 606
Sokolnikoff, I. S., 110n
Solenoidal field, 313, 369
Solution
  complementary, 86, 245, 492
  continuity of, 66
  explicit, 8
  general, 7
  implicit, 8
  particular, 7, 87, 243, 380, 388
  set, 195
  singular, 7
  steady-state, 464, 483
  transient, 483
  trivial, 205, 227, 400, 497
  updated, 463, 497
Source, 349n
Source field, 307
Space(s)
  isomorphic, 263
  null, 220
  vector, 211
Space curve, 297
Spanning set, 213
Sparse matrix, 191
Specific heat, 479
Spectral bandwidth, 453
Spectrum of a function, 452
Speed, 298
Sphere, 320
  dielectric, 571
  Riemann, 610n
Spherical coordinate system, 322
Spherical coordinates, 543
Spherical harmonics, 546
Spherical symmetry, 569
Spinning top, 289
Spivak, M., 351n
Spring constant, 107

Square, completing the, 173
Square matrix, 184
Stable method, 122
Stagnation point, 632n
Staib, J. H., 82
Starting values, 55
Static displacement, 413, 514
Static equilibrium, 291
Statics, 290
Statistics, 236
Steady field, 305, 375
Steady-state current, 117
Steady-state solution, 464, 483
Steady-state temperature, 465, 488
Steinberg, D. I., 252n, 254, 255n
Step function, 439
Stereographic projection, 610
Stochastic matrix, 194
Stoker, J. J., 50
Stokes, George G. (1819–1903), 354
Stokes's theorem, 354
Streamlines, 39, 632
Stretching, 606
String, plucked, 478
Strip, fundamental, 609
Strontium-90, 3
Sturm, Jacques C. F. (1803–1855), 496
Sturm–Liouville problem, 496
    eigenfunctions of, 497
    eigenvalues of, 497
    regular, 497
    singular, 503, 534
Sturm–Liouville system, 496
Submatrix, 190
Subscripts, double, 185
Subspace, 212
Substitution, back, 202
Subtraction, complex numbers, 580
Successive displacements, 254
Suess, H. E., 33n
Summation, index of, 517
Sums, partial, 424, 427, 514
Superposition, 88, 97, 398, 409, 467
Surface(s)
    equipotential, 310, 384
    isothermal, 310
    level, 310
    normal to, 310
    one-parameter family of, 384
    orientable, 351

orthogonal, 384
piecewise smooth, 351
simple, 351
smooth, 339
Surface area, 340
Surface integral, 340
Sylvester, James J. (1814–1897),
    221n
Sylvester's law of nullity, 221
Symmetric matrix, 189, 231
Symmetric property, 210
Symmetry, circular, 569
Synthetic division, 83
System(s)
    consistent, 196
    function, 153
    homogeneous, 205, 243
    ill-conditioned, 252
    inconsistent, 196
    of linear algebraic equations, 251
    of ordinary differential equations,
        177, 237
    Sturm–Liouville, 496
Système International d'Unites (SI),
    365

Table
    of hyperbolic functions, 646
    of Laplace transforms, 137, 643
Tail of a vector, 275
Tait, J. H., 457n
Tangent vector, 298
Tangential component, 299
Tangential force, 105
Taylor's series method, 56
Temperature
    ambient, 4
    steady-state, 465, 488
Terminal point, 275
Test
    integral, 515
    ratio, 515
Theorem
    Cauchy's, 616
    central axis, 290
    divergence, 350
    fundamental, 626n
    Gauss's, 350
    Green's, 356, 615, 631
    Lerch's, 149

Liouville's, 621, 626n
  residue, 619, 626
  shifting, 135
  Stokes's, 354
Thermal conductivity, 373, 479
Thermal diffusivity, 373, 465, 479
Thermally isotropic, 372
Thermodynamics, 161
  second law, 35
Thomas, G. B., 330, 635n
Three-fold root, 81
Time-independent field, 305
Torricelli, Evangelista (1608–1647),
  41n
Torricelli's law, 41
Torsion, 301
Trace of a matrix, 255
Tracer kinetics, 34
Trajectories, orthogonal, 38
Transform
  cosine, 448, 488
  fast Fourier, 446
  Fourier, 442, 628
  Laplace, 130
  response, 153
  sine, 448, 463, 489
Transformation
  bilinear, 605
  coordinate, 128, 129
  fixed points of, 612
  linear, 136, 186, 216
  Möbius, 605
  one-to-one, 218
Transformer-coupled circuit, 182
Transient current, 117
Transient solution, 483
Transitive property, 210
Translation, 605
  of axes, 128
Transonic flow, 388
Transport, neutron, 457
Transpose, 189
Transposition, 188
Transverse waves, 472
Triangle, Pascal's, 85
Triangular matrix, 191
Tricomi, Francesco G. (1897–   ),
  388n
Tricomi equation, 388
Trihedral, moving, 300

Triple, Pythagorean, 273
Triple scalar product, 285
Triple vector product, 284
Trivial solution, 205, 227, 400, 497
Twisted cubic, 306
Twisted curve, 301
Two-point boundary-value problem,
  80, 111, 400, 465, 476, 639

Uncertainty principle, 453
Underdamped motion, 119
Undershoot, 424
Undetermined coefficients, 87, 109,
  242, 492
Uniqueness, 64
Unit step function, 164
Unit tangent vector, 298
Unit vectors
  in cylindrical coordinates, 319
  in rectangular coordinates, 275
  in spherical coordinates, 323
Unstable difference equation, 577n
Updated solution, 463, 497
Updating a solution, 400, 569
Upper triangular matrix, 191

Value, principal of log z, 592
van Kármán, T., 110n
Variable(s)
  change of, 474
  of integration, 21
  separation of, 2, 399, 545, 562
Variable mass, 114
Variation of parameters, 75, 92, 94, 556
Vector(s), 211, 275
  acceleration, 298
  addition, 278
  column, 184
  complex numbers as, 578
  coplanar, 285
  Darboux, 307
  electromagnetic wave, 375
  free, 277
  head of, 275
  invariant, 227
  length of, 276
  linear velocity, 289
  linearly dependent, 214
  magnetic induction, 368
  orthogonal, 232

orthonormal, 324
Poynting, 370
reciprocal set of, 329
row, 184
space, 211
space, abstract, 262
tail of, 275
unit, 275, 319, 323
vorticity, 314n
zero, 277
Vector calculus, references, 376
Vector field, 304
Vector function, 295
Vector moment, 289
Vector point function, 303
Vector potential, 375
Vector product, *see* Cross product
Vector tangent, 298
Velocity
  angular, 288
  of escape, 113
  initial, 71
  linear, 289
  potential, 631
  vector, 298
  vector field, 631
Vena contracta, 41
Verhulst, Pierre-François (1804–1849), 44
Vibrating membrane, 470
  frequency of, 472
Vibrations
  forced, 109
  free, 108
Viscous damper, 108
Volterra, Vito (1860–1940), 161

Volterra integral equation, 161
Volume, 285
Volume element
  in cylindrical coordinates, 321
  in rectangular coordinates, 344, 349
  in spherical coordinates, 327
Volume integral, 344
Von Mises, Richard E. (1883–1953), 256n
Vorticity vector, 314n

Wave equation
  one-dimensional, 391
  two-dimensional, 470
Wave function, 411
Wave propagation, 370, 410
Waves
  longitudinal, 472
  transverse, 472
Weber's Bessel function, *see* Bessel function(s)
Weight function, 419, 534
Widder, D. V., 105n
Wiley, C. R., 134n
Wilhelmy, Ludwig F. (1812–1864), 39n
Winkel, B. J., 49
Wolf, K. B., 426, 453n
Work, 288, 334
Wrench, 291
Wronski, Hoene (1778–1853), 74n
Wronskian, 74, 239

Zero matrix, 186
Zero vector, 277
Zeros of $J_0(x)$ and $J_1(x)$, 532